1st International Conference on New Material and Chemical Industry (NMCI2016)

IOP Conference Series: Materials Science and Engineering
Volume 167

Sanya, China
19 – 21 November 2016

Editor:

Min Li

ISBN: 978-1-5108-3671-6
ISSN: 1757-8981

Printed from e-media with permission by:

Curran Associates, Inc.
57 Morehouse Lane
Red Hook, NY 12571

Some format issues inherent in the e-media version may also appear in this print version.

Printed by Curran Associates, Inc. (2017)

For permission requests, please contact the Institute of Physics
at the address below.

Institute of Physics
Dirac House, Temple Back
Bristol BS1 6BE UK

Phone: 44 1 17 929 7481
Fax: 44 1 17 920 0979

techtracking@iop.org

Additional copies of this publication are available from:

Curran Associates, Inc.
57 Morehouse Lane
Red Hook, NY 12571 USA
Phone: 845-758-0400
Fax: 845-758-2633
Email: curran@proceedings.com
Web: www.proceedings.com

Table of contents

Volume 167

1st International Conference on New Material and Chemical Industry (NMCI2016)
19–21 November 2016, Sanya, China

Accepted papers received: 16 January 2017
Published online: 13 February 2017

Preface

Biological Materials and Ecological Materials

Energy Materials

Nano Materials and Technology

Non-Ferrous Alloys, Metals and Rare Element

Materials and Chemical Industry

IOP Conf. Series: Materials Science and Engineering **167** (2017) 011001 doi:10.1088/1757-899X/167/1/011001

Preface

Min Li

Editor, Proceedings of the First International Conference on New Material and Chemical Industry

School of Materials Science and Engineering, International Institute for Urban Systems Engineering, Southeast University, China

E-mail: limin.li@163.com

The First International Conference on New Material and Chemical Industry (NMCI2016) was successfully held in Sanya, China during November 19-21, 2016. The conference was an annual forum for researchers and application developers in the area of material and chemical industry.

This conference proceeding included 76 accepted articles selected from 197 submissions.

We would like to express our gratitude to the reviewers of these manuscripts, who provided constructive criticism and stimulated comments and suggestions to the authors. We are extremely grateful as organizers, technical program committee and editors and extend our most sincere thanks to all the authors for their excellent contribution and work. Our sincere gratitude also goes to the IOP Publishing editors and managers for their helpful cooperation during the preparation of the proceeding.

On behalf of The Organizing Committees of NMCI2016

IOP Conf. Series: Materials Science and Engineering **167** (2017) 011001 doi:10.1088/1757-899X/167/1/011001

NMCI2016-photos from the conference

Peer review statement

All papers published in this volume of *IOP Conference Series: Materials Science and Engineering* have been peer reviewed through processes administered by the proceedings Editors. Reviews were conducted by expert referees to the professional and scientific standards expected of a proceedings journal published by IOP Publishing.

IOP Conf. Series: Materials Science and Engineering **167** (2017) 012001 doi:10.1088/1757-899X/167/1/012001

Pluronic F127/chitosan blend microspheres for mucoadhesive drug delivery

W Z Gu and X F Hu[1]

College of Life Science, Zhejiang Chinese Medical University, Hangzhou 310053, China

E-mail: xiaofenhu@sina.cn

Abstract. Pluronic F127/chitosan blend microspheres were prepared via emulsification and cross-linking process using glutaraldehyde as a cross-linker. Compared with chitosan microspheres fabricated under the same experimental conditions, blend microspheres exhibited better physical stability and higher swelling capacity. Puerarin, a traditional Chinese medicine, was incorporated into microparticlesas the model drug. The in vitro release of puerarin from blend microspheres was reduced because of the improved compatibility of the drug with the matrices. According to the results from in vitro adhesion experiments, mucoadhesive behavior of blend microspheres on a mucosa-like surface was similar to that of chitosan microspheres, despite their good ability of anti-protein absorption in solution.

1. Introduction

In recent years, microspheres based on chitosan have been extensively investigated as carriers formucosaldrug delivery because of the unique physicochemical properties and the safety of chitosan [1]. Aiming to improve and optimize the characteristics of chitosan microspheres, several researches have been carried out on microspheres based on chitosan derivatives [2, 3]. In our previously published studies, PEGylated chitosan derivatives were synthesized and used to preparemicrospheres for the transmucosal delivery of puerarin [4]. The PEGylation of chitosan conferred valuable benefits to the propertiesof the microspheres even at a low degree of PEG substitution. Compared with the development of novel chitosan derivatives, polymer blending is a much more convenient approach for the desirable new properties of chitosan. Blends of chitosan with numerous natural or synthetic polymers, such as silk fibroin [5], polyurethane [6], polyethylene glycol [7] and polyvinylpyrrolidone [8], have been fabricated in the form of microspheres in order to enhance the physicochemical, mechanical, and functional properties for biomedical applications.

Yoo and colleagues used Pluronic F127 (F127), a hydrophilic ABA-type triblock copolymer consisting of polyoxyethylene units (A) and polyoxypropyleneunits (B), to prepared F127/chitosan blend microspheres via an ionic gelation process with tripolyphosphate for vaccine delivery [9]. Compared with chitosan microspheres, the blend microspheres had improved stability and immune activity, and also exhibited potential protection against infection. It has been reported that chitosan-based materials prepared using tripolyphosphate as the cross-linker have poor mechanical strength [10]. F127/chitosan blend microspheres fabricated by the emulsion-crosslinking method employing glutaraldehyde as a cross-linker were studied [11]. The published work investigated the influence of formulations on the incorporation and release behavior of 5-fluorouracil in the blend microspheres;

however, investigations focused on the mucoadhesiveness of F127/chitosan blend microspheres have yet to be reported.

Since the mucoadhesive behavior of the microspheres formulations is extremely important for a potential mucosal delivery system, the interactions between the F127/chitosan blend microspheres and mucin in vitro was investigated in this study. Puerarin, an effective traditional Chinese medicine treatment of cerebrovascular and cardiovascular diseases, was chosen as the model drug. Researches on swelling, dissolution, drug loading and release behavior of the blend microspheres were also carried out.

2. Materials and methods

2.1. Materials

Pluronic F127(F127) was purchased from Aldrich. Chitosan (200-400 mPa·s, >90%deacetylation) was obtained from Aladdin Reagent Co.. Puerarin (>98%) was provided by Nanjing Zelang Medical Technology Co.. Glutaraldehyde (50% in H_2O) were obtained from Sinopharm Chemical Reagent Co.. Mucin (from porcine stomach) was purchased from Amresco. All other reagents used were of analytical grade and used as received.

2.2. Preparation and characterization of microspheres

Microspheres were produced by emulsification and cross-linking process. Firstly, F127 and 50 mg of chitosan were dissolved in 3% (v/v) acetic acid solution to obtain a miscible polymer solution (aqueous phase). This solution (3.3 mL) was added slowly into liquid paraffin (100 mL, oil phase) containing 2% (v/v) span-80 under constant stirring at 1000 rpm for 10 min using a homogenizer (RW20 Digital Mechanical Overhead Stirrer, IKA, Germany). At the end of sufficient emulsification, the W/O emulsion was solidified by slowly dropping glutaraldehyde in 1,4-dioxane solution (5%, v/v, 1.25 mL) at 450 rpm for 1.5 h. The hardened microspheres were collected by filtration, separately washed with petroleum ether and isopropyl alcohol, and then dried in a vacuum oven overnight. Chitosan microparticles were prepared as a comparison.

The surface morphology of microspheres was examined under a scanning electron microscope (SEM, S-3000N, Hitachi, Japan). Microspheres were placed on a copper stub, coated with gold and observed.

The particle size analysis was performed on samples of the microspheres redispersed in distilled water (~1.0%, wt%) using the WQL particle size analyzer (LKY-2, SPSIC, China).

2.3. Swelling and dissolution experiments

The swelling properties of the microspheres were evaluated measuring the equilibrium water uptake of samples incubated in deionized water at 37 °C. At predetermined times (6 h, 12 h and 24 h), samples were taken out, and the excess surface adhered water was removed carefully. The swollen microspheres were weighed (W_s), and then oven dried at 60 °C until there was no change in the weight of the dried samples. The weight of dried microspheres was recorded (W_d). Each experiment was repeated six times andthe average valuewas used to calculate the swelling capacity (%) by the following equation.

$$\text{Swelling capacity (\%)} = \frac{W_s - W_d}{W_d} \times 100\% \qquad (1)$$

The percentage dissolution of microparticles was calculated from the following equation. Each dissolution value was averaged from the six measurements.

$$\text{Dissolution (\%)} = \frac{W_0 - W_d}{W_0} \times 100\% \qquad (2)$$

Where W_0 is the initial weight of microspheres and W_d is the remaining weight of dried samples.

2.4. Evaluation of puerarin-loading capacity of microspheres

Puerarin was dissolved directly into the polymer solution to form a drug-containingaqueous phase. Drug-loaded microspheres were prepared via the same emulsification and cross-linking process mentioned above. Ten milligrams of puerarin-loaded microspheres was dissolved in hydrochloric acid solution before the quantitative determination of the drug-loading capacity of samples using Agilent 1200 series HPLC (Chemstation System, DAD detector, Eclipse XDB-C18 column (4.6 mm × 150 mm, 5 μm)) as reported[12]. All measurements were performed in triplicate and averaged.

Drug-loading capacity (LC) was calculated as the following equation.

$$\text{LC}\ (\%) = \frac{\text{incorporated puerarin weight}}{\text{microspheres weight}} \times 100\% \tag{3}$$

2.5. X-ray diffraction analysis

X-ray diffraction patterns of pure puerarin, puerarin-loaded chitosan microspheres and puerarin-loaded F127/chitosan blend microspheres were obtained using an X-ray diffractometer (SHIMADZU XRD-7000, Japan) with Cu-Kα radiation source operating at 40 kV and 30 mA. Diffractograms were performed from the initial angle $2\theta = 10°$ to the final angle $2\theta = 80°$ with a step width of 0.02°.

2.6. In vitro release studies

Release studies of puerarinfrom microspheres were performed under sink conditions. Puerarin-loaded samples (10 mg) were suspended in 5 mL of pure water at 37 °C with horizontal shaking. At predetermined time points, samples were centrifuged at 4000 rpm for 4 min. Then, 1 mL of supernatant was withdrawn and immediately replaced with equal volumes of fresh water. The collected solutions were subjected to further HPLC analysis using the same conditions reported above. All experiments were carried out in triplicate.

2.7. Adsorption of mucin on microspheres

Mucin stock solutions with a concentration of 1 mg/mL in pure water were prepared. Ten milligrams of microspheres were dispersed in the above mucin aqueous solution (6 mL), and shaken at 37 °C for 2 h. Then, the dispersion was centrifuged at 4000 rpm for 2 min, and the supernatant was removed for the measurement of the free mucin content. The protein estimation was done by Periodic acid/Schiff colorimetric method[13]. All measurements were performed in triplicate and averaged. The adsorption of mucin on microspheres was assessed as follows.

$$\text{Adsorption ratio}\ (\%) = \frac{\text{amount of total mucin-amount of free mucin}}{\text{amount of total mucin}} \times 100\% \tag{4}$$

2.8. In vitro mucoadhesion studies

Mucoadhesion testing was conducted usingmucin saturated filter paper as the in vitro experimental model according to the previously described method[14]. Firstly, the filter paper (2.5 cm x 2.5 cm) was wetted completely with mucin solution (2% w/v in pure water) and fixed inside a tube under controlled conditions of humidity (75.8 ± 1.6%) and temperature (28 ± 1°C). Puerarin-loaded microspheres (10 mg) were spread out onto themucin saturated filter paper, and then a stream of air (6.2 ± 1 m/s) was blown over the filter paper with samples face downward for 15 s. Microparticles sticking to the surface were recovered by washing the filter paper fully with 3% (v/v) hydrochloric acid solution. The amount of drug in the remaining samples was quantified via HPLC as mentioned above. All measurements were performed in triplicate and averaged. The in vitro mucoadhesion behavior of the microspheres was expressed as the following equation.

$$\text{Mucoadhesivity}\ (\%) = \frac{\text{the amount of puerarin in the remaining microspheres}}{\text{the total amount of puerarin}} \tag{5}$$

3. Results

3.1. Morphology observation
Mixture of F127 and chitosan (1/5, 2/5 and 3/5, w/w) in acetic acid solution was used as the water phase for the preparation of blend microspheres (CS-10F, CS-20F and CS-30F). Blank chitosan microspheres (CS) were also prepared as a comparison. Both blend microspheres and CS had a spherical shape and a smooth surface (figure 1). Decreased degree of particle aggregation was observed in the SEM images of blend microspheres. The mean diameters of the CS, CS-10F, CS-20F and CS-30F were 3.42±0.21， 3.50±0.27， 3.69±0.18 and 3.20±0.10 μm, respectively.

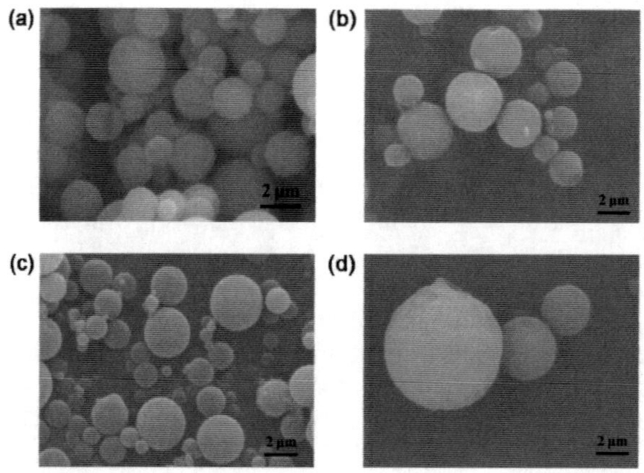

Figure 1. SEM images of (a) CS and (b) CS-10F, (c) CS-20F and (d) CS-30F.

3.2. Swelling behavior and dissolution assessment
All of the investigated samples exhibited considerable swelling properties, and a significant amount of water was absorbed within the first 6 h (figure 2). After immersion for 6 h, the swelling capacities of the CS, CS-10F, CS-20F and CS-30F were 172.22 ± 12.23, 245.45 ± 14.10, 249.65 ± 18.88 and 347.9 ± 18.32 %, respectively. After 24 h, the water uptake of microspheres approached saturation.

The dissolution values of the CS in water (37 °C) for 24 h was 11.7 ± 6.7 %. The dissolutions of composite microspheres CS-10F, CS-20F and CS-30F were 21.6 ± 3.6, 20.0 ± 2.8 and 22.2 ± 3.1 %, respectively.

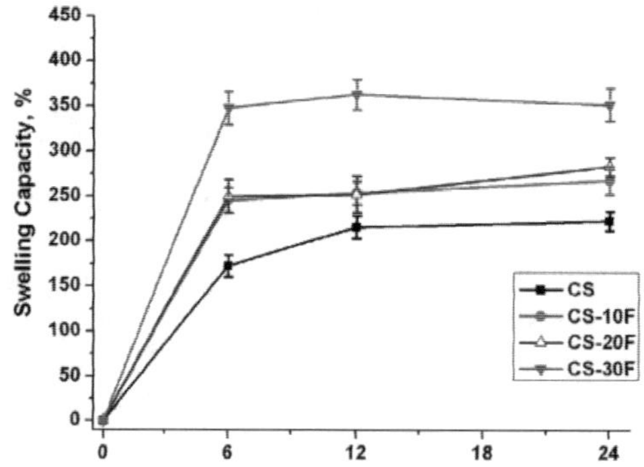

Figure 2. Swelling capacities of CS, CS-10F, CS-20F and CS-30F with respect to time.

3.3. Drug-loading capacity determination

The LC of the blank microspheres CS was 3.48±0.08%, which was 2.26±0.03%, 0.28±0.03% and 0.26 ±0.01% for the blend microspheres CS-10F, CS-20F and CS-30F.

3.4. X-ray diffraction analysis

X-ray diffraction patterns for puerarin alone, puerarin-loaded CS and puerarin-loaded CS-10F were shown in figure 3. The spectrum obtained from puerarin alone gave quite sharp and distinct diffraction peaks with a low and flat background as expected from a crystalline sample. The microspheres containing puerarin exhibited notable broadening and overlap of the diffraction peaks, and the sharp peaks corresponding to the drug in the crystalline state weakened and even disappeared. The peak intensity of CS-10F in the range of 12° - 30° was obviously weaker in comparison to that of CS.

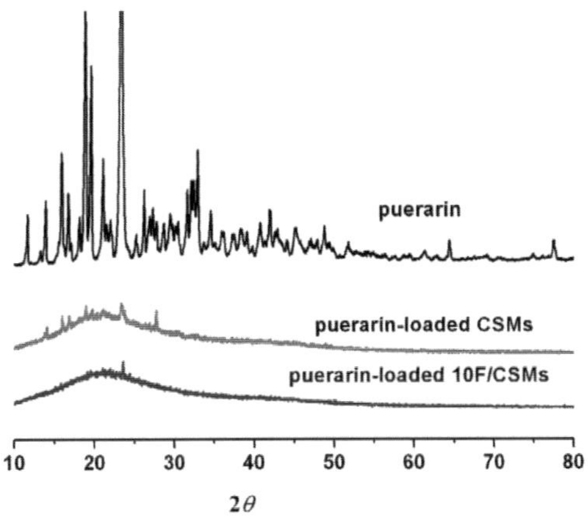

Figure 3. X-ray diffraction patterns of puerarin and puerarin-loaded microspheres.

3.5. In vitro release

In vitro release ratio studies from CS indicated severe initial burst releases, 75% within the first 15 min, and 81% at 2 h. The release rates of puerarin from CS-10F were obviously delayed, 53% at 15 min; the initial release was later followed by a more controlled release in the selected experimental time window, releasing accumulative 70% of drug from the microspheres (figure 4).

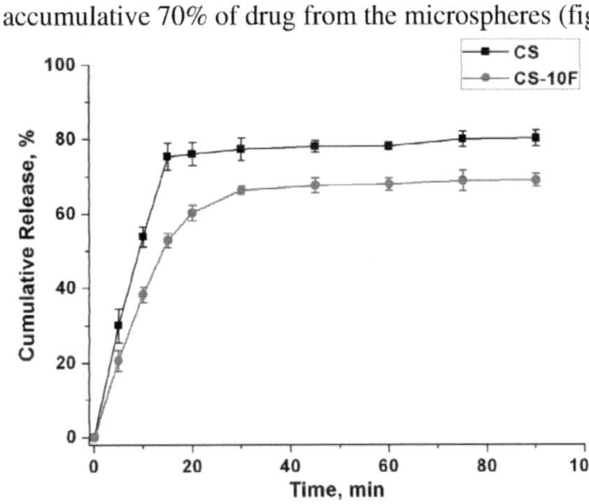

Figure 4. Release profiles of puerarin from CS and CS-10F.

3.6. In vitro mucoadhesivebehavior assessment

The amount of mucin adsorbed onto the surface of microspheres was determined from the change in the free concentration of mucin in the reaction solution according to equation (4). The average adsorption ratio of mucin on CS and CS-10F was 67.52±0.49% and 40.71±0.63%, respectively.

The amount of puerarin-loaded microspheres adhering to the mucin saturated surface after the application of a stream of air was further determined according to equation (5) via HPLC measurement of the concentration of the drug in the particles adhered on the surface. The average adhesion ratio of the CS and CS-10F on the mucosa-like surface was 73.97±1.09% and 72.10±1.25%, respectively.

4. Discussion

F127/chitosan blend microspheres were prepared by the emulsification and cross-linking method. The preparation parameters containing the stirring rate, cross-linking degree and the chitosan solution concentration, which generally affected the physicochemical and morphological properties of chitosan microspheres, remained constant based on our previous work. The influence of F127 content in polymer blends on microsphere morphology, drug loading and release profile characteristics, and mucoadhessiveness was investigated.

The presence of hydrophilic PEO chains of F127 hindered the self-aggregation of chitosan microspheres as expected, but increasing the concentration of F127 in the polymer blends solution did not affect the particle size. The particle size of microspheres was mainly determined by the droplets formed during emulsification which were later solidified in the presence of glutaraldehyde. It could be noted that the viscosity of polymer solution increased with increasing the amount of polymer blends, and highly viscous polymer solution was apt to form big droplets during emulsification [7]. Rokhade et al. reported that formulations containing higher amounts of F127 prepared under constant stirring at 400 rpm speed had bigger particle size [11]. In this work, the emulsion was violently stirred at a high speed of 1000 rpm to produce numerous stable and small droplets (about 20 times smaller than previously reported); thus the increase of the viscosity of polymer blends solution had no obviously effect on the particle size.

Results from swelling experiments demonstrated that blend microspheres had higher swelling rates than the blank CS, which were quite different from the reported ones [11]. On the basis of previously published researches on composite particles [15], it could be speculated that the formation of relatively loose structure of blend microspheres due to the entanglement of soft hydrophilic PEO chains of F127 with chitosan molecules resulted in an increase in the swelling capacity of microspheres. Herein, the maximum equilibrium water uptake value was obtained in the case of CS-30F.

F127 has been widely used as a pharmaceutical excipient for its surfactant and stabilizing properties. Because of the amphiphilic nature, F127 could self-assemble at high concentration, and exhibit its capability to increase the solubility of drugs. Puerarin-loaded microspheres formulated with a drug/chitosan mass ratio of 3/5 were preparedaccording to our previous research [4]. Blending of F127 with chitosan led to a significant decrease in the LC compared with the blank CS, especially in the case of CS-20F and CS-30F. The leak of puerarin from droplets during emulsification process attributing to the surfactant property of F127 perhaps was the main reason for the low LC of blend microspheres. CS-10F with acceptable drug loading was chosen tofurther investigate the properties of blend microspheres.

Drug release from the microspheres generally follows several types of mechanisms, including release from the surface of microspheres, diffusion through the swollen microspheres, and release due to the dissolution of microspheres. Hydration and dissolution behavior of the microspheres played an important role in their drug release profiles [7]. As mentioned previously, the presence of F127 promoted the water uptake of the microspheres. Chemical cross-linking with glutaraldehyde provided higher stability for chitosan microspheres, meanwhile, F127 diffusion from the polymer matrix occurred after immersion of blend microspheres in water at 37 °C. Both formulation CS and CS-10F showed an observed initial burst release which would be helpful in achieving the therapeutic plasma concentration of the drug in a short time. Nevertheless, a significant slowdown of the drug release rate was observed in case of CS-10F. The results from X-ray analysis revealed a molecular rearrangement of puerarin in the polymer matrix and an amorphous state. Polymer blends increased the degree of miscibility between the drug and the polymer matrix. It could be hypothesized that a certain amount of puerarin was entrapped within the hydrophobic nanodomainsin the blend microspheres due to the self-assembly of F127 at high concentration. This was well in agreement with the results from the reported researches [11].

Because of the strong electrostatic interaction, mucin could be adsorbed onto the surface of the chitosan microspheres [16]. Extensive researches have shown that PEG-based polymers exhibit excellent protein resistant properties because of the PEG chain mobility and steric stabilization force [17]. In this study, the interaction between blend microspheres and mucin in the aqueous solution was decreased due to the presence of PEG chains on the surface of microspheres, which resulted in a dramatic decrease in the adsorption ratio of mucin on CS-10F as compared with CS. The mucoadhesive properties of chitosan microspheres were thought to be the results of the combined efforts of electrostatic interaction between chitosan and mucin, and increased viscosity of mucus via a dehydration process [18, 19].There was no obvious difference in the mucoadhesive properties of formulations CS and CS-10F, although the electrostatic interaction between CS-10F with mucin was significantly weaker. Combined with the results from the experiments of water uptake, it could be concluded that the improved dehydration of mucus played a considerable important role in the adherence of CS-10F onto the mucin saturated surface.

5. Conclusions

In this study, Pluronic F127/chitosan were prepared via an emulsification and cross-linking process. Blend microspheres exhibited more physical stability and drug-polymer compatibility than chitosan microspheres. Moreover, the excellent mucoadhesive properties of microspheres remained as F127/chitosan polymer blends instead of blank chitosan was employed in formulations. Therefore, blending of F127 with chitosan permit animproved sustained drug release from CS and these blend microspheres seem to be an acceptable mucosal drug delivery system.

Acknowledgements

This work was supported by the National Natural Science Foundation of China (NO. 51303163) and the Zhejiang Provincial Natural Science Foundation of China (NO. LQ12H28002).

References

[1] Islam MA, Firdous J, Choi YJ, Yun CH and Cho CS 2012 Design and application of chitosan microspheres as oral and nasal vaccine carriers: an updated review *International Journal of Nanomedicine* **7** 6077-93

[2] Casettari L and Illum L 2014 Chitosan in nasal delivery systems for therapeutic drugs *Journal of ControlledRelease* **190** 189-200

[3] Rekha MR and Sharma CP 2011 Glutamine-chitosan microparticles as oral insulin delivery matrix: in vitro characterization *Journal of Applied Polymer Science* **122** 2374-82

[4] Hu XF, Zhang YY, Zhou HF and Wan HT 2015 PEGylated chitosan microspheres as mucoadhesive drug-deliverycarriers for puerarin *Journal of Applied Polymer Science* **132** 42623

[5] Cheerarot O and Baimark Y 2015 Biodegradable silk fibroin/chitosan blend microparticles prepared by emulsification-diffusion method *e-Polymers* **15** 67-74

[6] Sullad AG, Manjeshwar LS and Aminabhavi TM 2015 Blend microspheres of chitosan and polyurethanefor controlled release of water-solubleantihypertensive drugs*PolymerBulletin* **72** 265-80

[7] Gunbas ID, Sezer UA,İz SG, GürhanİD and Hasirci N 2012 Semi-IPN chitosan/PEG microspheres and films for biomedical applications: characterization and sustained release

optimization*Industrial & Engineering Chemistry Research***51** 11946-54

[8] Ozerkan T, Sezer UA, Gurhan İD, İz SG and N Hasirci 2013 Semi-IPNchitosan/polyvinylpyrrolidone microspheres and films:sustained release and property optimisation *Journal of Microencapsulation***30** 762-70

[9] Kang ML, Jiang HL, Kang SG, Guo DD, Lee DY, Cho CS and Yoo HS 2007 Pluronic® F127 enhances the effect as an adjuvant ofchitosan microspheres in the intranasal delivery of Bordetellabronchiseptica antigens containing dermonecrotoxin *Vaccine***25** 4602-10

[10] Sun Y, Gu L, Gao Y and Gao F 2010 Preparation and characterization of 5-fluorouracil loaded chitosanmicrospheres by a two-step solidification method*Chemical & Pharmaceutical Bulletin***58** 891-5

[11] Rokhade AP, Shelke NB, Patil SA and Aminabhavi TM 2007 Novel hydrogelmicrospheres of chitosan andpluronic F-127 for controlled release of 5-fluorouracil *Journal of Microencapsulation***24** 274-88

[12] Yan ZH, Hu XF, Zhang YY, Yang JH and Wan HT 2013 Determination of loading efficiency of puerarin inchitosanmicrospheres by HPLC*Chinese Journal of Pharmaceutical Analysis***33** 599-602

[13] He P, Davis SS and Illum L 1998 In vitro evaluation of the mucoadhesive properties of chitosan microspheres*International Journal of Pharmaceutics***166** 75-88

[14] Gavini E, Rassu G, Sanna V, Cossu M and Giunchedi P 2005 Mucoadhesive microspheres for nasal administration of an antiemetic drug, metoclopramide: in-vitro/ex-vivo studies *Journal of Pharmacy and Pharmacology***57** 287-94

[15] Xu YM and Du Y 2003 Effect of molecular structure of chitosan on protein deliveryproperties of chitosan nanoparticles*International Journal of Pharmaceutics***250** 215-26

[16] Soane RJ, Hinchcliffe M, Davis SS and Illum L 2001 Clearance characteristics of chitosan based formulations in the sheep nasal cavity*International Journal of Pharmaceutics***217** 183-91

[17] Banerjee I, Pangule RC andKane RS 2011 Antifouling coatings: recent developments in the design of surfaces that prevent fouling by proteins, bacteria, and marine organisms *AdvancedMaterials***23** 690-718

[18] Sonia TA, RekhaMR and Sharma CP 2011 Bioadhesive hydrophobic chitosan microparticles for oral delivery of insulin: in vitro characterization and in vivo uptake studies *Journal of Applied Polymer Science***119** 2902-10

[19] Rekha MR and Sharma CP 2015 Simultaneous effect of thiolation and carboxylation of chitosan particles towards mucoadhesive oral insulin delivery applications: an in vitro and in vivo evaluation*Journal of Biomedical Nanotechnology***11** 165-76

The research of far infrared flame retardant polyester staple fiber

Qingshan Li[1,5], Kaijun Zhang[1], Jinqong Luo[1], Ji'an Li[2] Jian Jiang[2], Qianqian Liang[3], Yongxia Jin[1] and Bing Liu[4]

[1]State Key Laboratory of Metastable Materials Science and Technology, Yanshan University, Qinhuangdao, 066004, China P.R;
[2]Jiangsu zhangjiagang anshun technology development co., LTD, Jiangsu, China, 215000;
[3]Sichuan east of material science and technology group co., LTD In mianyang, China, 6210004
[4]Chengdu rainbow (group) co., LTD Chengdu, China, 610000

E-mail: liqs126@126.com

Abstract: Far infrared flame retardant slices was prepared, fiber with far infrared flame retardant composite function was also prepared by the method of melt spinning. Scanning electron microscopy (SEM) was used to observe the fibrous microscopic structure. In the SEM images, functional ultrafine powder particle size and distribution in the fiber were visible. The results show that the functional ultrafine powder is evenly distributed on the fibrous surface, which is closely combined with fiber, and the far infrared emissivity is F, which is more than (8 to 14 microns) 0.88. Far infrared flame retardant polyester fiber has not only good flame retardant, but also environmental health effect: releasing negative ions and launch far-infrared, which shows wide application prospect. The fiber was processed into far-infrared flame retardant electric blanket, whose functional indicators and flame retardant properties are not reduced.

1. Introduction

Electric blanket is a contact heating appliance. Electric blankets use special insulative cable-type electric heating elements, which were snake-like woven into or seam into the blanket. The heat is issued when the electric heating elements was powered on. Electric blanket was mainly used to improve the bed temperature to achieve the purpose of keeping warm for sleeping person. In addition, electric blanket can also be used for bedding dehumidification. Electric blanket has many advantages, such as less power consumption, temperature adjustable, easy to use, and so on. Therefore, electric blanke is popular with consumers, especially in cold northern China and southern humid areas.

However, electric blanket also has some disadvantages: drying, radiation and fire safe hidden troubles. The use of textiles is one of the important factors that threaten the fire. It is an important measure to reduce the risk of fire by adopting flame retardant method to deal with textiles. Combining the flame retardant requirements of textile products, the development and production of flame retardant, low smoke and excellent flame retardant products can prevent the threat of textiles fire fundamentally and provide safety and security for people's lives and property [1-3].

In this experiment, flame retardant polyester staple fiber with far-infrared function was prepared. Far infrared flame retardant slices was prepared, fiber with far infrared flame retardant composite function was also prepared by the method of melt spinning. Far infrared emissivity and flame retardance at room temperature were measured.

Electric blanket with far infrared flame retardant fibers has a unique far-infrared function, which can effectively improve the metabolism of human body, promote blood circulation, anti-inflammatory analgesic and eliminate fatigue. Moreover, the natural far infrared emission function of the far infrared flame retardant fibers has a higher infrared absorption and thermal storage capacity, which can not only save energy, but also greatly reduce the side effects of radiation. And its good flame retardant properties can effectively prevent the occurrence of fire accidents, and because of the addition of powder formed irregular fiber surface, making the electric blanket has a good breathable properties, which make the electric blanket can maintain and regulate the body's normal water to avoid throat thirst feeling [4-6].

2. Experiment

2.1. Experimental materials and equipments

The main experimental materials used in the experiment are shown in Table 1.

Table 1. Experimental Materials

Name	Factory
Superfine functional powder	Beijing General Research Institute of Mining and Metallurgy
Far infrared flame retardant polyester chips	Sichuan East wood Technology Group Co., Ltd
Additives	Market procurement
Cationic modified polyester chips	Zhejiang Chemical Fiber Joint Group

The main experimental equipments used in the experiment are shown in Table 2.

Table 2. Experimental equipments

Name	Factory
Vacuum drum dryer	Jiangyin, Jiangsu Chemical Machinery Co
Spinning machine	Jiangsu new pure Chemical Fiber Equipment Co., Ltd

2.2. The preparation of far infrared flame retardant polyester chips

The terephthalic acid($C_8H_6O_4$), ethylene glycol($(CH_2OH)_2$), flame retardants, far-infrared functional powder, catalyst and other additives were added into the esterification kettle according to a certain proportion. The esterification temperature was controlled between 245-258℃. When the water content reaching the predetermined value, and the top temperature of the column lowering than 100℃, is the end point of esterification. After esterification, the esterification liquid was introduced into polycondensation kettle, the condensation temperature was controlled between 275-285℃, vacuum <50Pa. Stop stirring when the motor stirring power meetting the technical requirements, and after cooling, pelletizing and drying, the far infrared flame retardant polyester masterbatch was prepared.

2.3. The preparation of far infrared flame retardant polyester staple fiber

Far infrared flame retardant polyester chips were dried in the temperature of 150℃-170℃ for 8~10h, and then well mixed according to a certain proportion of the mixing machine before melt spinning.

The spinning process conditions (1.67dtex × 38mm):

(a) The set temperature of screw extruder: I ~ V were 260 ℃, 270℃, 278℃, 275℃, 275℃ respectively;

(b) melt pipe: 275℃;
(c) pre-filter: 270℃;
(d) spinning box: 275℃;
(e) Metering pump speed: 35 rpm; Specifications: 12 cc;
(f) pump supply: 487.2 g / min;
(g) spinning speed: 900 m / min;
(h) Amount of oil: 0.7%.

The drafting process is set as follows :
(a) Draft speed: 120 m / min;
(b) draft multiple: 3.8 times;
(c) Drafting temperature: oil and water bath 72 ℃; steam 105 ℃;
(d) oven heat setting 115 ~ 120 ℃;
(e) number of crimp: 8-9 pieces / 25mm;
(f) heat setting temperature: 120 ℃.

3. Results and discussion

3.1. The test results of far infrared flame retardant polyester staple fiber

3.1.1. Ash measurement. By measuring the ash, on the one hand, the content of inorganic powder in the fiber and the maintain situation of the active ingredient can be determined; on the other hand, the process of granulation and spinning filtration performance can be determined. There are three groups of parallel experiments according to the national standard GB / T9345-88, the results were in Table 3:

Table 3. Ash measurement results of far infrared anion flame retardant composite functional fiber

Number	Initial weight (g)	After burning constant weight (g)	Ash (%)
1	10.3465	0.3325	3.2136
2	10.7149	0.3444	3.2142
3	10.3100	0.3304	3.2046

The average ash content of the far infrared flame retardant polyester staple fiber is 3.21%. By calculating the content of each substance in the fiber, the results show that the mineral powders were whole into the fiber, and the filter performance is good among the whole process in this study, and the active ingredient was well maintained.

3.1.2. Measurement of flame retardant properties. The flame retardant properties of the far infrared flame retardant polyester staple was measured by the authority, and the test results: limit oxygen index ≥ 30%, flame retardant UL94 V-2, with a very high limit oxygen index, low smoke, difficult to lead retardant, flame retardant effect of permanent, non-toxic, non-corrosive, non-polluting, is environmentally friendly halogen-free flame retardant products.

3.1.3. Far infrared emission function. The far infrared emission rate of the far infrared flame retardant polyester staple was measured by the authority, and the test results: far infrared emission rate is F(8-14μm)> 0.88, and the match rate with human skin's wavelength is high, so it can be well absorbed by the body.

3.2. Analysis of far infrared flame retardant polyester staple fiber by SEM

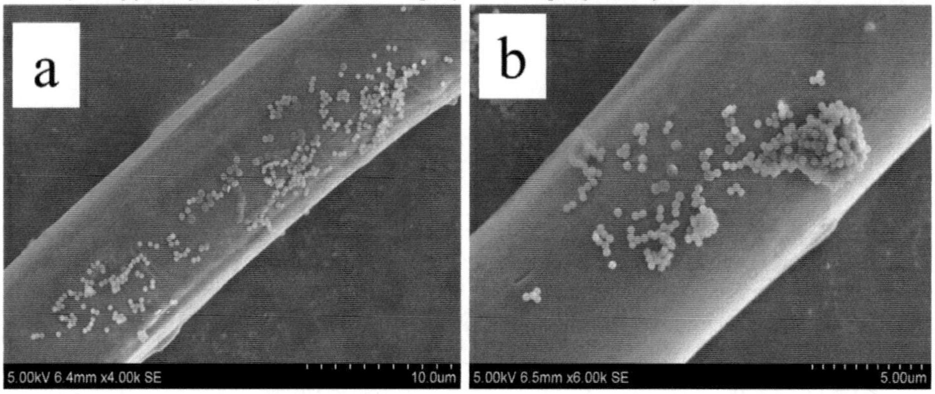

Figure 1. SEM images of far infrared flame retardant polyester staple

Figure 1 shows different magnification scanning electron microscopy photos of far infrared flame retardant PET staple, it can be observed that the distribution of far-infrared powder in the jurisdiction of the shape. From a, b figures, far infrared powders and fiber can be seen closely. The surface of fiber has many folds, which increase the surface area and improve the distribution area of far infrared powder materials and nano-additives.

4. Conclusion

The prepared far infrared flame retardant polyester staple fiber by this project has far infrared emissivity F (8-14 μm) > 0.88, limiting oxygen index ⩾30% and flame retardant UL94 V-2. The flame retardant can withstand high-temperature melt spinning and has good compatibility with the host polymer. Therefore, the post-spinning treatment can not be affected and the physical and mechanical properties of the fiber has no big impact. The far infrared flame retardant polyester staple fiber has good flame-retardant performance durability. From the SEM figures, it can be seen that the ultra-fine far infrared particles were more evenly distributed in the fibrous surface and has a close combination with the fiber. The filtration and spinning performance of the far infrared modified polyester masterbatch is good. The far infrared flame retardant polyester staple fiber not only has far infrared effect and flame retardant properties, but also has a higher breaking strength and better functional fibrous performance, which make it has widely promotion and application value. The fiber was processed into far-infrared flame retardant electric blanket by the Chengdu Rainbow (Group) Co., Ltd., whose functional indicators and flame retardant properties are not reduced.

Acknowledgements

The authors are grateful for State Key Laboratory of Metastable Materials Science and Technology of Yanshang University, and the support of Jiangsu zhangjiagang anshun technology development co., LTD, Sichuan east of material science and technology group co., LTD In mianyang, China, and chengdu rainbow (group) co., LTD Chengdu, China.

Reference

[1] Ren W J 2014 Development of Functional Fibers at Home and Abroad. *Science and technology.* **18** 128.
[2] Li J H, Li X X and Zhang Y 2014 Fire retardant STC. *Printing and Dyeing.* **21** 35-37.
[3] Li R 2014 Development trend of flame retardants. *Printing and dyeing.* **23** 53-54.
[4] Zheng F J 2012 Study on detection method of flame-retardant viscose fiber oxygen index. *Artificial fibers.* **02** 21-22.

[5] Wang C Y, Li X, Xie T T, et al 2011 Simultaneous Determination of Six Kinds of Organophosphorus Flame Retardants in Textiles by Microwave-Assisted Extraction-Gas Chromatography-Tandem Mass Spectrometry. *Chromatography* .**08** 731-736.

[6] Zhang H K 2009 A Review of Flame Retardancy of Textiles. *Printing and Dyeing auxiliaries.* **02** 7-15.

Effects of magnesium potassium phosphate cements mixed with silica fume on the solidification and reduction of municipal sludge

Yanchang Chen, Lixin Wang, Peng Song and Qi Wang[1]

Material Science and Engineering, University of Jinan, Jinan, China, 250022

E-mail: mse_wangq@ujn.edu.cn; wangqi_515@sina.com

Abstract. In order to investigate Magnesium-Potassium Phosphate Cement (MKPC) mixed with silica fume on the solidification and reduction of high moisture content (94%) of municipal sludge. The moisture content of the solidified sludge, volume shrinkage, compressive strength and other properties were studied, and the mechanism was discussed based on phase and morphology analysis. The results indicates that the solidified sludge (30% MKPC) with 30% silica fume can obtain with the early compressive strength of 430 KPa, which reached the requirements of the landfill (≥ 350 KPa). The moisture content of solidified sludge decrease to 30% and the volume shrinkage of solidified sludge can reduce to 68% or less at 7d, which played a key role for sludge reduction. MKPC and silica fume form skeleton structure that can promote the evaporation of moisture in the sludge.

1. Introduction

Solidification/stabilization (S/S) [1-2] technology is a common and effective method for treating a variety of wastes. However, it is difficult to dewater due to the nature of the sludge [3]. The moisture of sludge should be below 60% when being disposed of in a landfill. Traditional dewatering methods for sludge include adding cement, silica fume, lime and other inorganic support materials [4]. Because of the physical and chemical properties of the sludge with high moisture content and hydrophilic [5], it is easy to bring serious questions, such as the sludge clumps, uneven mixing and volume increase.

Magnesium potassium phosphate cement (MKPC) is a new type of environmentally friendly cementitious materials formed by an acid-base reaction between magnesia and potassium dihydrogen phosphate in the presence of water according to the following reaction [6]:

$$MgO + KH_2PO_4 + H_2O \rightarrow MgKPO_4 \cdot 6H_2O \quad (1)$$

MKPC has many excellent properties, such as good volume stability, higher propensity for binding with waste inclusion and higher early age compressive strength [7]. Traditional magnesia-phosphate cement is produced through chemical reactions between dead burnt MgO and potassium phosphate [8]. The paper proposed to use light-burned MgO to replace dead burnt MgO, which could overcome the high cost disadvantage. The residual material of MKPC not only will not cause pollution to the soil, but also can provide some nutrition, so it is a kind of green environmental cementitious material [9]. In recent years, MKPC has been gradually applied to hazardous waste stabilization [10-11]. Various mineral admixtures such as ground granulated blast furnace slag, fly ash and coal bottom ash [6] are usually added into MKPC because of their low cost and effective performance. Silica fume can improve sludge flow and improve the performance of sludge dewatering.

Therefore, this study aims to investigate the effects of MKPC and silica fume on the solidification and reduction (S/R) of municipal sludge with high moisture content (94%). MKPC mixed with silica fume as a mineral admixture was used to solidify/reduce sludge. The performance of solidified sludge was compared based on the moisture content, compressive strength and the volume shrinkage. The aims were to explore an effective chemical dewatering and S/R technique for the dewatered sludge treatment.

2. Materials and methods

2.1. Raw materials

The urban sludge was obtained from the sedimentation tank of sewage treatment plant in the University of Jinan. The specific surface area and the particle size distribution of sludge were measured by LS13320 laser particle size analyzer produced in Beckman Coulter (American). The chemical composition were estimated by X-ray Fluorescence. Its physicochemical properties and chemical compositions of sludge are depicted Table 1 and Table 2.

Table 1. Physical properties of municipal sludge

Property	Moisture content (wt. %)	pH	Organic matter content (wt. %)	Average particle size (μm)	Specific surface area (cm^2/mL)
Value	94.15	7.05	26.77	69.90	3185

Table 2. Chemical properties of municipal sludge

Chemical composition (wt. %) (dewatered sludge)	SiO_2	Al_2O_3	CaO	Fe_2O_3	MgO	K_2O	P_2O_5	SO_3	K_2O
	22.83	19.25	10.9	5.75	1.63	0.76	4.07	4.48	1.13
	TiO_2	Cl	BaO	ZnO	MnO	SrO	ZrO_2	Other	Total
	0.95	0.51	0.08	0.08	0.06	0.04	0.02	0.69	73.23

Light-burned magnesium oxide (MgO) with specific surface of 15834 cm^2/mL and over 95% MgO content was obtained by Minmetals Industrial Co., Ltd, in Liaoning province. Its activity is greater than 68, the particle size ranges from 100 to 200 mesh. The potassium phosphate (KH_2PO_4) used was ananalytic grade chemical. Silica fume with particle size of 10.48 μm. The chemical compositions of MgO and silica fume are given in Table 3.

Table 2. Chemical composition of MgO and Silica fume (wt. %)

	SiO_2	Fe_2O_3	SO_3	CaO	Al_2O_3	MgO	K_2O	P_2O_5	Other
MgO	1.33	0.59	0.28	1.37	0.15	95.49	0.20	0.46	0.13
Silica fume	97.16	0.03	1.04	0.18	0.09	0.46	0.73	0.13	0.17

2.2. Preparation of Solidified sludge

Solidified effect was characterized by compressive strength. The effect of reduction were characterized by moisture content and volume shrinkage. The MKPC were made based on previous studies [12], the basic components of MKPC cement (mass ratio) were P/M = 1:4. The ratio of silica fume was calculated based on a weight percentage of MKPC. MKPC content accounted for the sludge of 20-60 wt. % (There is no need to add additional water). The proportions of silica fume accounted for the MKPC of 10-50 wt. %. Specimens were cast 20 × 20 × 20 mm moulds, vibrated to remove air bubbles, and cured at natural conditions (15-25℃, 50-70% RH) to 3d, 7d and 28d.

2.3. Experimental Methods

2.3.1. Compressive strength Compressive strength of the specimens were tested using the universal testing machine with a maximum load of 5kN. Six samples were tested for each mixture at each designated age under compression and the average value was taken as the representative compressive strength of the mixture.

2.3.2. Moisture content and volume shrinkage According to soil test procedures, access to different proportions of sludge moisture of experimental methods were referring to GB/T50123-1999 [13]. The volume shrinkage of the solidified sludge also were referring to GB/T50123-1999 [13]. The solidified time of the samples was 1weeks.

2.3.3. Phase and morphology analysis XRD datas were obtained by D8 ADVANCE X-ray diffractometer using strictly monochromatic CuKα radiation produced in Bruker (Germany) for analysing the phase composition of the samples and the data for each sample was collected from 5° to 70° (2θ). The SEM pictures were acquired to observe the microtopography of solidified sludge by using Quanta FEG 250 field emission scanning electron microscope produced in FEI (American) whose resolution was 1.04 nm.

3. Results and discussion

3.1. Solidification of municipal sludge

Compressive strength is an important factor to decide whether solidified sludge were put into a landfill or used as a construction material. In the United States, the strength of waste landfill treated by the S/S method must be higher than 0.35 MPa [14]. The Waste water Technology Center of Canada has adopted a higher value of 3.45 MPa [15].

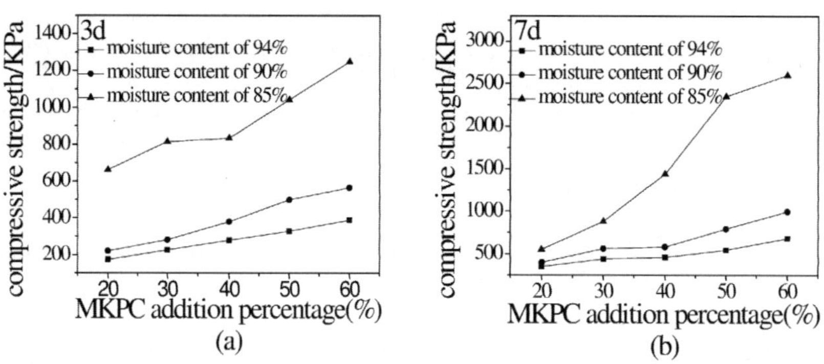

Figure 1. Effects of MKPC on the compressive strengths of samples with different moisture contents. (a) 3d; (b) 7d

Figure 1 shows the effects of MKPC content on the compressive strengths of sample with different moisture content of sludge at 3d and 7d. The compressive strength of the solidified sludge increased with the increased content of MKPC at 3d. But the rate of increase (slope) is closely related to the initial moisture content of the sludge. However, the solidified effect is limited, there are still a large number of pores in the sample, which caused an unsatisfactory curing strength.

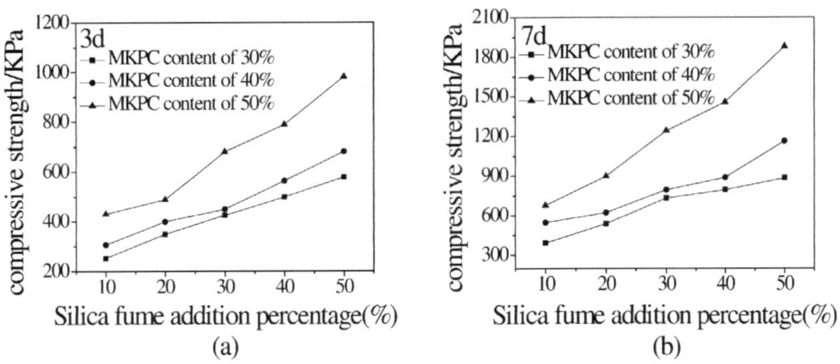

Figure 2. Effects of silica fume on the compressive strengths of samples with 30%, 40%, 50% MKPC. (a) 3d; (b) 7d

Figure 2 shows the effects of the silica fume on the compressive strengths at 3d and 7d. The sludge of 94% moisture content were used in the samples. The compressive strengths of the solidified sludge showed a trend of increase with the increase silica fume contents. Figure 1-2 show that the compressive strengths of solidified sludge (MKPC with silica fume) were higher than that of solidified sludge without silica fume. For example, the compressive strengths of samples (30% MKPC) with 30% silica fume reached 423 KPa at 3d, showing an increase of 196 KPa compared with 30% MKPC in Figure 1(a).

Through the above analysis, compressive strength is related to the sludge moisture content, cement content and the content of silica fume. Analysis showed solidified matrix became more rigid framework by adding silica fume and MKPC. When the silica fume reached to 30%, the strengths of all samples could reach 350 KPa at 3d, indicating that they were suitable for use as waste landfill materials [16].

3.2. Moisture reduction of solidified sludge

According to the compressive strength analysis, the specimens with only 30% MKPC content were sufficient to achieve the target compressive strength of 0.35MPa required for landfill disposal. Thus, only MKPC in 30% and Silica fume were selected to study the moisture content of solidified sludge. The effect of MKPC (with 20%, 40%, 50% and 60%) with silica fume on the moisture content of solidified sludge will be carried out in soon.

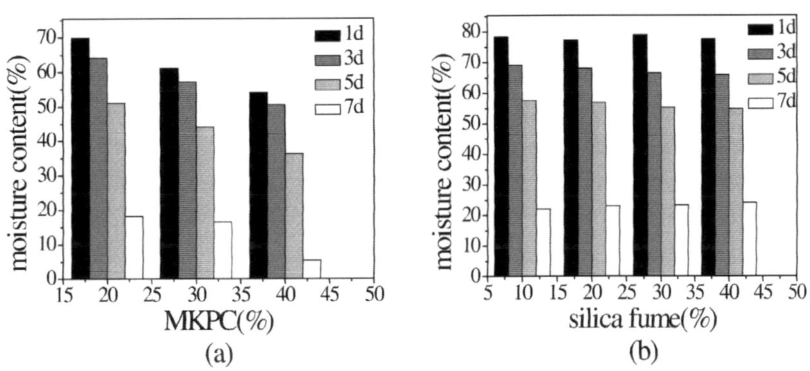

Figure 3. Effects of MKPC and silica fume on the moisture content
(a) MKPC; (b) 30%MKPC+silica fume

The effect of MKPC and combination of MKPC with silica fume on the moisture content of the solidified sludge are shown in Figure 3. The results show that MKPC can reduce the moisture content of the sludge. All the moisture content of sample reduce to 30% or less at 7d. The moisture content of the solidified sludge decrease as the MKPC ratio increasing. Figure 3 (b) shows the same ratio sludge, the effect of silica fume content on the moisture content of the solidified sludge is not obvious.

Because of the big specific surface area, silica fume can absorb some water. Thus, the moisture content of samples with silica fume higher the samples without silica fume. It can be observed that moisture content of solidified sludge decrease to 30% in a short time, which played an important role in the follow-up treatment of sludge.

3.3. The volume shrinkage of solidified sludge

The volume shrinkage of solidified sludge is also an important indicator. If the volume is too large, it will significantly increase the costs of after-treatment. As shown in Figure 4, the volume shrinkage rate of different proportions of solidified sludge show a trend of increase as the curing age increasing. Before 3d, the volume shrinkage of different samples are similar in Figure 4 (a). The volume shrinkage rate began to stabilize at 6d. The volume shrinkage rate of the sample decreased from 37% to 63% as MKPC content increasing from 20 to 40 wt.% at 7d in Figure 4 (a). When the MKPC ratio was 30%, the volume shrinkage rate of solidified sludge with 30% silica fume could reach 37% at 7d in Figure 4 (b). The longer the curing period of solidified sludge, the smaller the volume change.

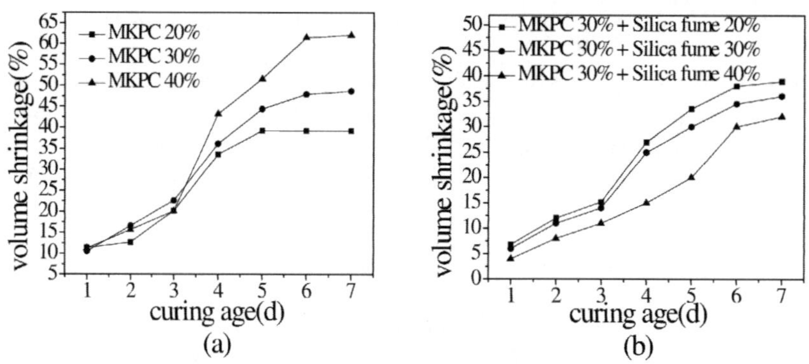

Figure 4. The volume shrinkage rate of samples as a function of curing time.
(a) MKPC; (b) 30% MKPC + Silica fume

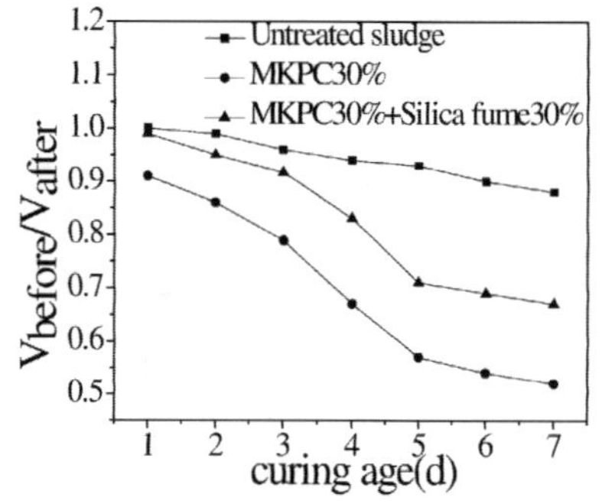

Figure 5. The volume ratio of solidified sludge before and after curing

The rate of natural sedimentation for sludge is very slow, as show in Figure 5, MKPC and silica fume change the water evaporation characteristics of the sludge. After adding MKPC, the sludge volume began to decline. The volume ratio of solidified sludge (with 30% MKPC) before and after curing reduce to 0.53 at 7d; the volume ratio of solidified sludge (with 30% silica fume) before and after curing can reduce to about 0.68 at 7d. Due to the rapid discharge of water resulting in the volume shrinkage of solidified sludge. Thus, the composite curing agent played a key role for sludge reduction.

3.4. Mineral phase analysis

The XRD test results of MKPC mixed with 30% silica fume and 30% MKPC at 7 d are shown in Figure 6. MKPC had an intense diffraction peak, suggesting a high degree of crystallization. Additionally, the results indicated that its main composition was unreacted MgO, as well as the hydration products $MgKPO_4 \cdot 6H_2O$. However, the addition of silica fume makes MKP and MgO diffraction peak intensities weaker than that of group with 30% MKPC.

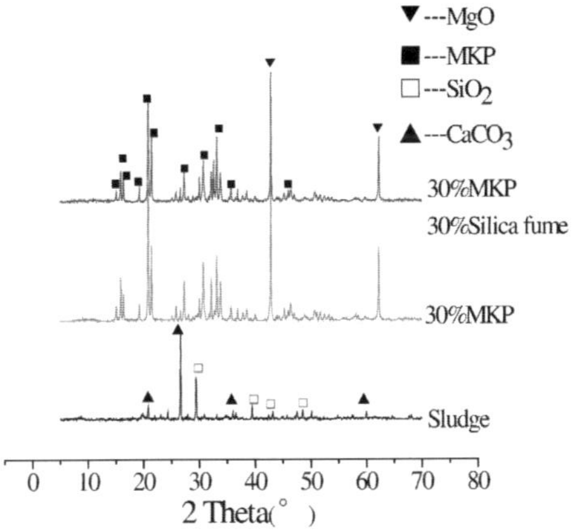

Figure 6. XRD patterns of the MKPC pastes and MKPC with silica fume (7 d).

3.5. Micrograph of solidified sludge

Based on the solution-diffusion mechanism [16], sludge solidification process is divided into two stages. The first stage of MKPC hydration reaction, absorb large amounts of free water, produce amorphous colloid and deposit on the surface of clay particles. The second stage is unstable hydration products gradually transform into crystals, intersperse with each other and fills in the gap between the clay particles, thereby improving the compactness of sludge.

Figure 7. SEM of various proportions of solidified sludge (7d).
(a) 30% MPC; (b) 30% MKPC+30% silica fume

Figure 7 shows that hydration of MKPC generate rod-shaped products. The surface and internal of MKPC filled with unreacted MgO particles. The unreacted MgO can fill the pores of MKPC to improve the pore structure and compactness [17]. The hydration products crystallize and generate the skeleton structure with the hydration of MKPC and silica fume. However, a lot of water of solidified sludge were volatilized, resulting in more pores. The image of sample (Figure 7a) shows that the pores formed are approximately 2-5 μm long. The pores provide a path for the water discharge. The micrograph of MKPC with silica fume show in Figure 8 (b), the pores are relatively less.

The effect of silica fume on MKPC can be analyzed from physical and chemical phases [18]. Chemical effect is embodied by the hydrated silica fume and the reaction product can enhance the bonding of solid phase and improve the density of MKPC. Physical effect is embodied in the filling effect. Because of the small fineness, silica fume can fill the pores of MKPC.

4. Conclusions

This paper presents the effect of MKPC with silica fume on solidification of sludge with high moisture content. Based on the experimental results, the following conclusions are drawn:

- Municipal sludge can be solidified and reduced in a matrix of MKPC.
- Compressive strength is related to the sludge moisture content and cement content. The compressive strengths of samples (30% MKPC) with 30% silica fume reached 430 KPa at 3d, which meet the requirements of the landfill (350 KPa).
- It can be noted that the moisture content of solidified sludge can be reduced from 94% to 30% or less at 7d. The volume shrinkage of solidified sludge with MKPC can reduce to 68% at 7d. The volume shrinkage of solidified sludge with 30% MKPC and silica fume can reduce to 40% or less at 7d.
- It is observed that MKPC and silica fume formed skeleton structure, which can promote the evaporation of moisture in the sludge.

Acknowledgements

This work is supported by the Science and Technology Development Project of Shandong Province (No. 2013GGB01156).

References

[1] Malviya R and Chaudhary R 2006 *J. Hazard. Mater* **137** 267
[2] Barth E, Percin P, Arozarena M 1990 Noyes Data 80
[3] Thapa K, Qi Y, Clayton S and Hoadley A 2009 *Water Research* **43** 623
[4] Asavapisit S and Chotklang D 2004 *Cem. Concr. Res* **34** 349
[5] Stevens J, Northcott G, Stern G, Tomy G and Jones K 2003 Environ. Sci. Technol 37 462
[6] Gardner L, Bernal S, Walling S,Corkhill C, Provis J and Hyatt N 2015 Cem. Concr. Res 74 78
[7] Abdelrazig B, Sharp J and El-Jazairi B 1989 *Cem. Concr. Res* **19** 247
[8] Yang J M, Qian C X 2010 *Journal of Wuhan University of Technology-Mater. Sci* **25** 613
[9] Yang J M, Luo L M Qian C X and Xun Y 2008 *New building materials* **9** 7 (in chinese)
[10] Yin J H 2001*Geotechnical Testing Journal* **24** 299
[11] Yang J, Shin J, Lee C, Heo C, Jeno M and Kang K 2013 *J. Radioanal. Nucl.Chem* **295** 211
[12] Shi C, Yang J, Yang N and Chang Y 2014 *Cem. Concr. Compos* **53** 83
[13] GB/T 50123 Stand for soil test method 1999
[14] USEPA, U.S. EPA, Prohibition on the Placement of Bulk Liquid Hazardous Waste in Landfills-Statutory Interpretive Guidance (OSWER) Policy Directive No.9487.00-2A, EPA/530-SW-016 (1986)
[15] Stegemann J and Cote P. 1996 *Sci. Total Environ* **178** 103
[16] Zheng D, Ji T,Wang C, Sun C, Li X and Hossin K 2016 *Constr. Build. Mater* **106** 415
[17] Xu B, Ma H, Li Z 2015 *Cem. Concr. Res* **68** 1
[18] Lin W, Sun W and Li Z 2010 *J. Build. Mater* **13** 716

Preparation of the sludge activated carbon with domestic sludge mixed agricultural straw

Laifu Wang, Yan Wang and Jingyan Lian[1]

Tianjin University of Technology, Tianjin 300384, China.

E-mail: swzy@tjut.edu.cn

Abstract. Urban sewage sludge with complicated composition produce largely each year, pollution problem and resource utilization has increasingly become the focus of attention. Sewage sludge is utilized to prepare adsorbent that is a new type method. Agricultural stalks was added to material (urban sewage sludge) and activator ($ZnCl_2$), calcined under the condition of no inert gas, and obtained domestic sludge activated carbon. The properties were measured by iodine adsorption value and BET, discussed influence factors of sludge activated carbon preparation, including activator concentration, solid-liquid ratio, calcific temperature and calcific time. The best process condition of orthogonal experiment had explored that activated time is 10 minutes, calcific temperature is 350℃, the activator concentration $ZnCl_2$ is 3 mol/L and the mixing ratio of raw materials and activator is approximately 1:5. The iodine adsorption value and the optimal BET of as-obtained domestic sludge activated carbon is 445.06 mg/g, 525.31m^2/g, respectively.

1. Introduction

Urban sludge is composed of organic and inorganic substance that contains a variety of fungi and protozoa constitute, its content was 60%~70%. The activated carbon can be made of residual activated sludge through high temperature carbonization and activation, for instance, Rozada [1] Jevaseelan [2] prepared sludge activated carbon with $ZnCl_2$, the results showed that the $ZnCl_2$ activation effect is superior to other activator. As-obtained sludge adsorbent with $ZnCl_2$ as activating agent, research think $ZnCl_2$ chemical activation is very outstanding that the effect is very good as activator [3]. Tay [4] on the premise of $ZnCl_2$ as activator, to explore the domestic sludge optimum process condition of preparation of activated sludge carbon. The results show that the activated sludge carbon specific surface area of 867.61 m^2/g, when concentration of $ZnCl_2$ is 5 mol/L, the activation temperature is 500℃, the activation time is 2 h. And so Ren Ai-ling [5] utilize the traditional activated carbon preparation technology, with $ZnCl_2$ as activator, under optimal technological condition that 40% of $ZnCl_2$ solution, the activation temperature is 600℃, the activation time 20 minutes, solid-liquid ratio of 1:2~1:3, as-obtained product iodine adsorption value is 514-542 mg/g.

In this study, the activator is zinc chloride solution, the preparation of the sludge adsorbent under the condition of without inert gas protection and use of urban sludge and agricultural straw to calcine. We could determine the product quality by iodine adsorption value, however, optimal reaction condition was necessary to be determinated by orthogonal experiment. The influence factors of sludge activated carbon preparation, including activator concentration, solid-liquid ratio, calcific temperature and calcific time were investigated systematically.

2. Experiment section

2.1. Experimental materials and Instrument

Urban sludge taken sample from the xianyang road wastewater treatment plant in tianjin, the experiment used instruments including: disintegrator, electric heating air-blowing drier, muffle furnace, desktop thermostatic oscillator, molecular sieve. The reagents used in the experiments including: $ZnCl_2$(AR), HCl(LR), $Na_2S_2O_3$(AR), iodine standard solution(AR).

2.2. The preparation of sludge activated carbon

The experimental materials adopt domestic sludge moisture content of 89.2%, the reagents used in the test, respectively, 0.5 mol/L hydrochloric acid standard solution configuration, 0.5 mol/L sodium thiosulfate solution configuration, 1 g/L starch indicator configuration, the configuration of a solution of different concentrations of $ZnCl_2$, iodine standard solution configuration.

2.3. The preparation of activated carbon from digested sewage sludge test

In a typical procedure, the crushed reed straw was put into the solution containing domestic sludge with 3 mol/L concentration of zinc chloride solution impregnated and activated the mixture. The sample was reacted for a day and filtered, and then calcined at 400℃ for 15 min in a muffle furnace. After it was cooled down to room temperature, the sample was removed from the autoclave, rinsed with dilute hydrochloric acid, and dried at 60℃ in an oven. Through the single factor experiment to determine the single factor effect on the properties of adsorbent, this experiment by orthogonal experiments to determine the experimental parameters: the optimum process conditions of activator concentration, solid-liquid ratio, calcific temperature and calcific time.

3. Results and discussion

3.1. No agricultural straw doping to prepare sludge adsorbent

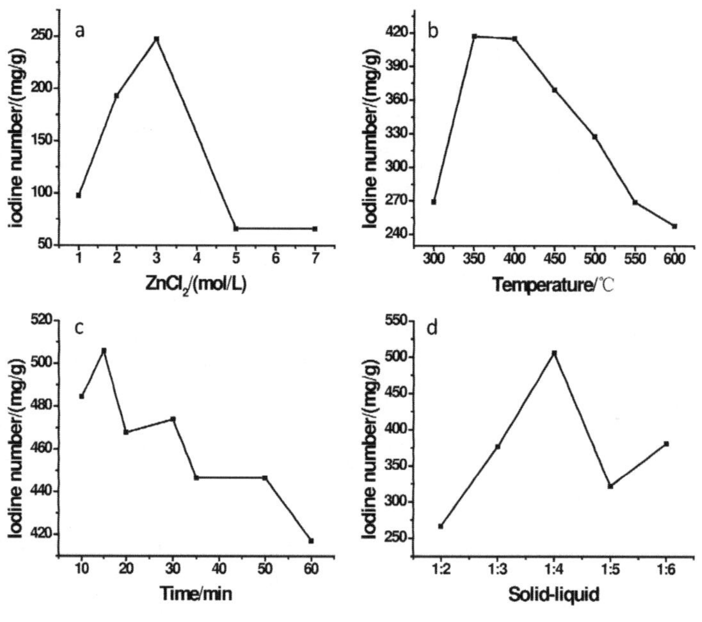

Figure 1. The influence curve of different factors on the iodine adsorption value: (a) $ZnCl_2$ concentration, (b) temperature, (c) time and (d) solid-liquid ratio

The single factor experiment curve graph of sludge adsorbent without agricultural straw doping obviously reveals different factors impact on the iodine adsorption value, as shown as figure 1. The optional experimental conditions of sludge adsorbent without agricultural straw doping can be clear shown that activator concentration of $ZnCl_2$ is 3 mol/L, calcific temperature is 350℃, calcific time is 15 minutes and the mixing ratio of raw materials and activator is 1:4.

3.2. Doping of agricultural straw to prepare sludge adsorbent

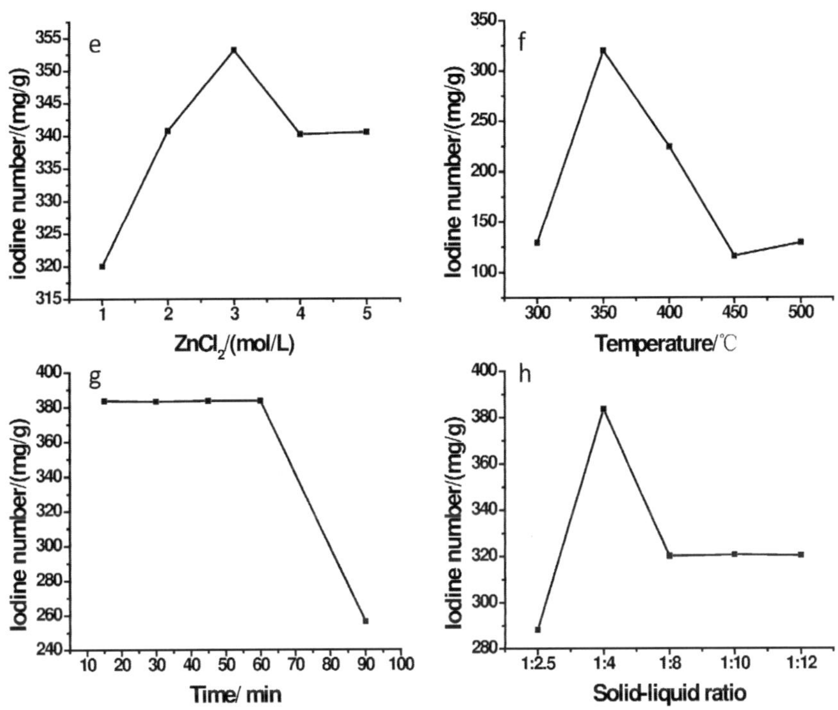

Figure 2. The influence curve of different factors on the iodine adsorption value: (e) $ZnCl_2$ concentration, (f) temperature, (g) time and (h) solid-liquid ratio

The single factor experiment curve graph of sludge adsorbent with agricultural straw doping obviously reveals different factors impact on the iodine adsorption value, as shown as figure 2. The optional experimental conditions of sludge adsorbent without agricultural straw doping can be clear shown that activator concentration of $ZnCl_2$ is 3 mol/L, calcific temperature is 350℃, calcific time is 10 minutes and the mixing ratio of raw materials and activator is 1:4. Compared with no agricultural straw doping, the optional experimental conditions of sludge adsorbent are similar, the difference is that calcific time becomes shorter. This may increase the porosity of the sludge, tend to maximize the specific surface area.

3.3. Orthogonal experimental design and analysis

The orthogonal experiment of preparation of sludge adsorbent with domestic sludge and reed straw, as summarized in the table 1 and table 2. Data analysis and experimental results show the important sequence of factors are calcification temperature, calcification time, solid-liquid ratio, activator concentration, respectively. The optional experimental conditions of sludge adsorbent can be determined by orthogonal test that activated time of 15 minutes, calcific temperature is 350℃, activator concentration of $ZnCl_2$ is 2 mol/L, the mixing ratio of reed straw and domestic sludge is 1:5. The highest adsorption capacity of the best product in standard iodine is 380.90 mg/g.

Table 1. Factors and levels of orthogonal experiment

Level	A==Calcific time/min	B==Calcific temperature/℃	C==Activator concentration/(mol/L)	D==Solid-liquid ratio
1	10	300	2	1:3
2	15	350	3	1:4
3	20	400	4	1:5

Tabel 2. Results of orthogonal experiment

Test number	Factor				Iodine adsorption capacity (mg/g)
	A	B	C	D	
1	1	1	1	1	317.56
2	1	2	2	2	351.79
3	1	3	3	3	378.86
4	2	1	2	3	351.78
5	2	2	3	1	379.99
6	2	3	1	2	349.68
7	3	1	3	2	285.35
8	3	2	1	3	380.9
9	3	3	2	1	288.79
K_1	1048.21	954.69	1048.14	986.34	
K_2	1081.45	1112.68	992.36	986.82	
K_3	955.04	1017.33	1044.2	1111.54	
k_1	349.4	318.23	349.38	328.78	
k_2	360.48	370.89	330.79	328.94	
k_3	318.35	339.11	348.07	370.51	
Range	42.13	52.66	18.59	41.73	
Important order	B>A>D>C				
Optimal levels	A_2	B_2	C_1	D_3	
Optimal combination	$A_2B_2C_1D_3$				

The orthogonal experiment of preparation of sludge adsorbent with domestic sludge and peanut shells, as summarized in the table 3 and table 4. Data analysis and experimental results show the important sequence of factors are calcific time, solid-liquid ratio, calcific temperature and activator concentration. The optional experimental conditions of sludge adsorbent can be determined by orthogonal test that calcific time of 10 minutes, calcific temperature is 350℃, activator concentration $ZnCl_2$ is 3 mol/L, the mixing ratio of activator and raw materials is 1:5. The highest adsorption capacity of the best product in standard iodine is 445.06 mg/g.

Table 3. Factors and levels of orthogonal experiment

Level	A==Calcific temperature/℃	B==Solid-liquid ratio	C==Activator concentration/(mol/L)	A==Calcific time/min
1	300	1:3	2	10
2	350	1:4	3	15
3	400	1:5	4	20

Table 4. Results of doping reed lever of orthogonal experiment

Test number	Factor				Iodine adsorption capacity (mg/g)
	A	B	C	D	
1	1	1	1	1	349.82
2	1	2	2	2	318.01
3	1	3	3	3	349.61
4	2	1	2	3	381.58
5	2	2	3	1	381.81
6	2	3	1	2	380.82
7	3	1	3	2	318.01
8	3	2	1	3	317.44
9	3	3	2	1	445.06
K_1	1017.44	1049.41	1048.08	1176.69	
K_2	1144.21	1017.26	1144.65	1016.85	
K_3	1080.51	1175.49	1049.43	1048.63	
k_1	339.15	349.8	349.36	392.23	
k_2	381.4	339.09	381.55	338.95	
k_3	360.17	391.83	349.81	349.54	
Range	42.25	52.74	32.19	53.28	
Important order	D>B>A>C				
Optimal levels	A_2	B_3	C_2	D_1	
Optimal combination	A_2 B_3 C_2 D_1				

3.4. The test result of SEM

Figure 3. SEM images of sludge adsorbent

The SEM images of sludge adsorbent (figure 3) shows similar spherical particles, particle size of nanoscale materials with the size of 200 nm. It is noticed that the major pore size distribution of sludge adsorbent relatively disperse, the formation of this humble pore structure may be the evaporation of moisture and organic small molecules escape in the process of thermal decomposition. The mixture of urban sludge and agricultural straw under the catalysis of activator $ZnCl_2$ at high temperature calcination, restrained effectively organic carbon decomposition and escape, in favour of adequate mixture and carbonization of organic at high temperature.

3.5. The test result of BET

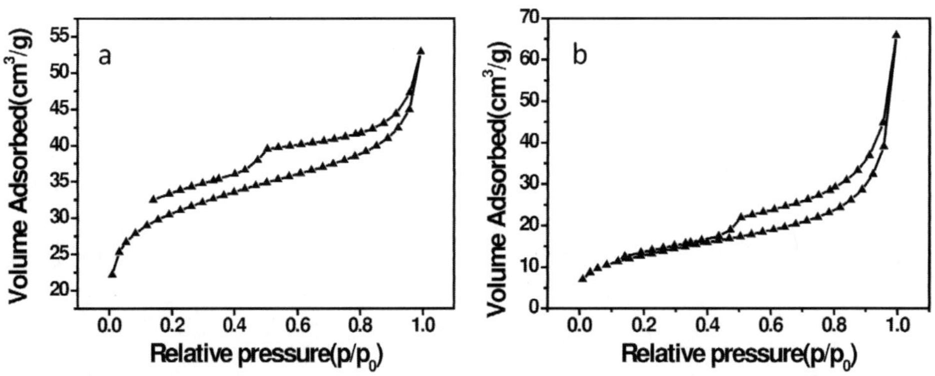

Figure 4 N_2 stripping - adsorption isotherm of the sludge adsorbent
(a) the urban sludge mixed reed straw and (b) the urban sludge mixed peanut shells

Specific surface area of the adsorbent (BET) is one of the important factors affecting its properties, so we test the sludge adsorbent mass N_2 stripping - adsorption isotherm. As shown in figure 4, compared to the urban sludge mixed reed straw, coincidence degree of the urban sludge mixed peanut shells is better, this isotherm type is similar to mesoporous material type. The specific surface area of sludge adsorbent materials are 428.15 m^2/g and 525.31 m^2/g, respectively. High specific surface area and mesoporous structure can improve the contact area of tested material and sludge adsorbent, therefore, we need to prepare as-obtained sludge adsorbent with good adsorption properties.

4. Conclusion

This experiment provides a new way for utilization of the domestic sludge resource and agricultural straw resource. This novel of this experiment is that preparation of calcific sludge adsorbent under no inert gas protection made a systemic analysis of affecting factors of the experiment. Considered fully the effect of single factor on preparation of sludge adsorbent, the single factor tests are individually established. Then optimum condition of sludge adsorbent with doping agricultural straw is that $ZnCl_2$ is 3 mol/L, calcific temperature is 350℃, calcific time is 10 minutes and the mixing ratio of raw materials and activator is 1:4. And optimum condition of sludge adsorbent without doping agricultural straw is that activator concentration $ZnCl_2$ is 3 mol/L, calcific temperature is 350℃, calcific time is 15 minutes and the mixing ratio of raw materials and activator is 1:4.

The above-mentioned orthogonal experiments were adopted to determine the optimum process conditions of sludge activated carbon prepared by domestic sludge mixed reed straw or peanut shell. Analysing orthogonal experiment data, optimal combination and data of the factors are determined. The SEM images of sludge adsorbent with optimal combination show that sludge adsorbent particle size is nanoscale and the material aperture is mesopore, the formation of material tends to maximize specific surface area and pore size. And two groups BET tests indicate that BET of product with domestic sludge mixed reed straw is 428.15 m^2/g, BET of product with domestic sludge mixed peanut shell is 525.31 m^2/g. Thus, the sample with domestic sludge mixed peanut shell is the best sludge adsorbent.

References

[1] Rozada F, Otero M and Moran A 2005 *J. Hazard. Mater.* Activated carbons from sewage sludge and discarded tyres: production and optimization **124** 181
[2] Jeyaseelan S and Gao Q L 1996 *Wat. Sci. Tech.* Development of Adsorbent/Catalyst from Municipal Wastewater sludge **34** 499
[3] Zan W B, 2005 *Chang sha:Hu nan university* Academic research on adsorbent derived from sewage sludge based on chemical activation method and its application **57** 58
[4] Tay J H, Jeyaseelan S and Graham N S 2001 *Chemosphere* Optimising the preparation of activated carbon from digested sewage sludge and coconut husk **44** 45
[5] Ren A L, Wang Q S and He J 2004 *Environ. Sci.* Study on activated carbon produced with sewage sludge **25** 48

IOP Conf. Series: Materials Science and Engineering 167 (2017) 012005 doi:10.1088/1757-899X/167/1/012005

Separation and recovery of heavy metals from waste water using synergistic solvent extraction

Yan Li[1], Limei Yang, Zheng Xu and Qi Sun

National Engineering Laboratory of Biohydrometallurgy, General Research Institute for Non-Ferrous Metals (GRINM), China

E-mail: Ly_wolfish@163.com

Abstract. Heavy metal wastewater pollution is one of the three major water pollutions in the world. The zinc hydrometallurgy smelting process usually discharge large quantities of heavy metal wastewater into the environment. In this paper, a synergistic solvent extraction process has been developed to recover copper, nickel, zinc and cadmium respectively from calcium and magnesium.The synergistic organic system contained 0.50 M Versatic 10 and 0.5 M Mextral 984H in DT100. Adjusting pH to 2.0 at 40 °C, the copper will be extracted preferentially with the extraction rate more than 99%. Continuing to adjust pH to 4.2 at 40 °C, the nickel will be extracted secondly with an extraction rate more than 98%; the zinc and cadmium in raffinate could be extracted separately while pH is about 6.5.

1. Introduction

With rapid development of the mining and smelting industry in China, wastewater generated by this process was largely emitted into the soil and water bodies. This waste water contains heavy metals such as Zn (II), Cu (II), Ni (II), Hg (II), Cd (II), Co (II) and As (II). It not only affects the quality of the atmosphere, but also threatens the health of human beings by way of the food chain[1,2]. Heavy metal wastewater pollution had been one of the three major water pollutions in the world, and the zinc hydrometallurgy smelting process usually discharges large quantities of wastewater into the environment[3]. Currently, the most common heavy metal wastewater treatment techniques are chemical reaction methods[4,5]. This usually engenders hydroxide precipitation, sulfide precipitation and heavy metal chelating precipitation. So far, the most widely used chemical precipitation technique is hydroxide precipitation[6], because it can remove any heavy metals other than mercury, and the process is usually simple and low cost. However, hydroxide precipitation generates large volumes of relatively low density sludge and both the metals and acid resource cannot be efficiently recovered for recycle[7]. Finding a way to clean the heavy metals from wastewater and recover resourceful metals is the most significant task.

Recently, people began to apply solvent extraction technology into wastewater treatment, as it can provide the selectivity necessary to create valuable product streams suitable for recycle or reuse. Compared to the solvent extraction system, the synergistic solvent extraction system can simplify the extraction process, and recover three or more metal ions respectively at the same time in the complex solution system[8].

In this paper, the synergistic solvent extraction system was developed using Versatic 10, Mextral 984H and DT 100 for the effective and efficient recover copper, nickel, zinc and cadmium respectively from the main impurities, magnesium, calcium. All three reagents are commercially available.

2. Materials and methods

2.1. Feed Solutions

The composition of a synthetic aqueous feed solution is shown in Table 1. It was prepared by dissolving analytical grade metal sulphate salts in de-ionized water including $CuSO_4 \cdot 5H_2O$, $NiSO_4 \cdot 6H_2O$, $ZnSO_4 \cdot 7H_2O$, $3CdSO_4 \cdot 8H_2O$, $MgSO_4 \cdot 7H_2O$, and $CaSO_4 \cdot 2H_2O$. Its profile is similar to a certain zinc hydrometallurgical wastewater.

Table 1. The composition of the feed solution

Concentrations (mg/L)						
Cu	Ni	Zn	Cd	Ca	Mg	pH
1803.6	507.5	5288.5	736.7	481.4	11084.6	3.3

2.2. Organic Solutions

The synergistic organic system consisted of 0.50 M Versatic 10 and 0.5 M Mextral 984H in DT100. Versatic10 was supplied by Shell Chemicals, Mextral 984H and Mextral DT100 were supplied by Kopperchem of China. Versatic10 was 2-methyl-2-ethylheptanoic acid. Mextral 984H was a mixture of 2-hydroxy-5-nonyl-acetophenone oxime and 5-nonyl-salicylaldoxime in a high flash point hydrocarbon diluent. Mextral DT100 (containing >99% aliphatics and <0.5% aromatics) was used as the diluent for all organic systems tested.

2.3. Experimental Procedure

All batch experiments were carried out in glass bottles. The pH of the solution, adjusted to the desired value by adding drops of 5 M NaOH, was measured by an on-line pH monitor. The two phases were mixed for enough time to ensure that equilibrium was reached. After phase separation, the metal concentrations in the raffinate were detected by inductively coupled plasmaatomic emission spectroscopy (ICP-AES), and the metal concentrations in the organic phase were calculated by mass balance.

3. Results and discussion

3.1. Extraction pH Isotherms of Versatic10

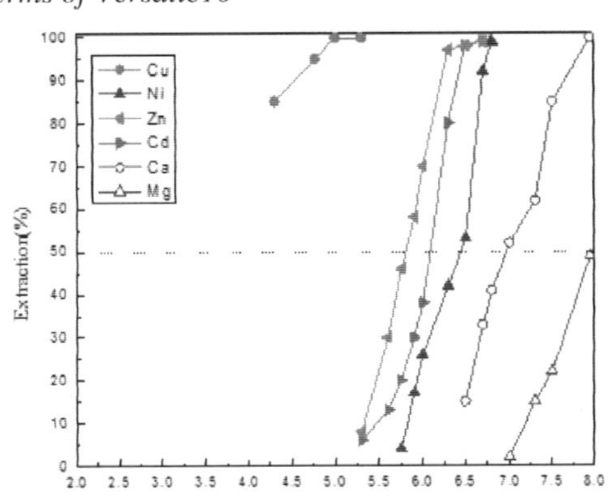

Figure 1. Extraction pH isotherms of metals with 0.5 M Versatic 10 in Mextral DT100 and the synthetic solution at an A/O ratio of 1:1 and 40C.

As it is shown in Figure 1, in low pH values, Cu was well extracted; in a pH profile of <8.0. The metal extraction order is Cu>Zn>Cd>Ni>Ca>Mg, and it is shown a better extraction for a single metal irons. But the heavy metal ions were extracted within the range of high pH values, where metal hydrolysis occured. The separation factors between the desirable metals with calcite are small which could introduce the generation of gypsum in aqueous solution if operated under this condition. By comparing the half extraction(pH_{50}), pH_{50}^{Zn}=5.8, pH_{50}^{Cd}=6.1, pH_{50}^{Ni}=6.45, pH_{50}^{Ca}=6.9, the pH values are so close that it would be difficult to recover the valuable metals from others with only a few steps, and using the cascade process to extract these metals would be costly.

3.2. Extraction pH Isotherms of Mextral 984H

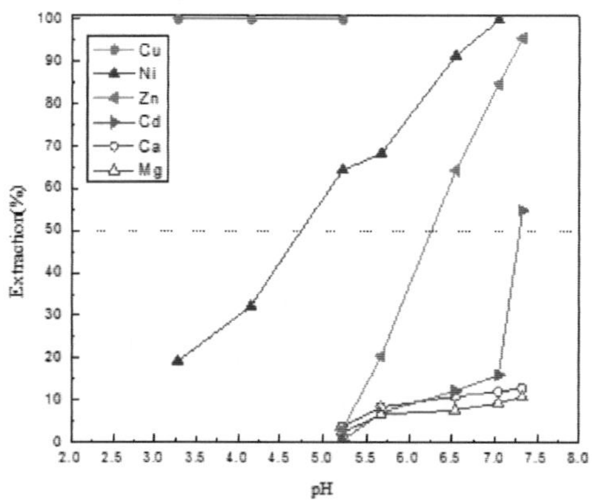

Figure 2. Extraction pH isotherms of metals with 0.5 M Mextral 984H in Mextral DT100 and the synthetic solution at an A/O ratio of 1:1 and 40C.

Figure 2 shows that in a pH profile of <7.5, the metal extraction order is Cu>Ni>Zn>Cd>Ca> Mg. The extraction of Ca and Mg were lower than 15%. Cu was well extracted at low pH values; the others with higher extraction only occurred at the high pH values. It was also difficult to recover copper, nickel, zinc, and cadmium by using Mextral 984H.

3.3. Extraction pH Isotherms of the Synergistic system

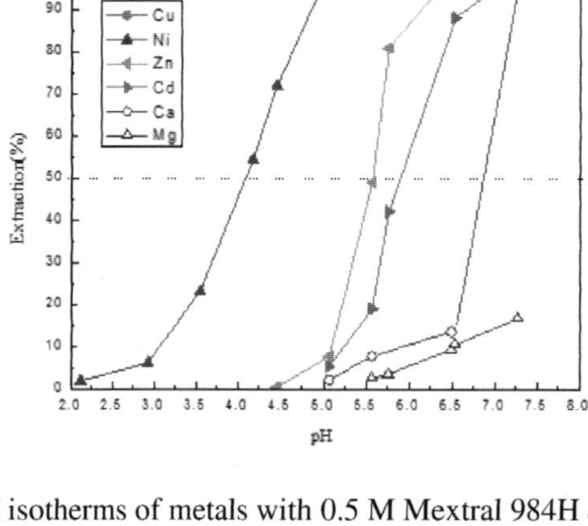

Figure 3. Extraction pH isotherms of metals with 0.5 M Mextral 984H & 0.5 M Versatic 10 in Mextral DT100 and the synthetic solution at an A/O ratio of 1:1 and 40C.

Figure 3 shows that adding Mextral 984H to Versatic 10 resulted in very large synergistic shifts with heavy metal ions. As a general rule, it is unusual for an extractant to cause a ΔpH_{50} greater than 1. The synergistic shifts for copper were greater than 2. However, the nickel ΔpH_{50} shifted scale was small, only 0.5 pH units, the ΔpH_{99} increased more than 2 pH units. The pH isotherms of zinc also shifted, but not very significantly, towards lower pH. The pH isotherms of cadmium also shifted 1.5 pH units to lower pH. Magnesium and calcium shifted either but the change scale is small. According to figure 3, the technique process could be divided into 3 stages. In the 1st, the copper in feed solution could be extracted totally by controlling the pH at about 2.0, while other ions were hardly extracted. In the 2nd, nickel is extracted at a pH of about 5.0, and finally zinc and cadmium at a pH of about 6.5.

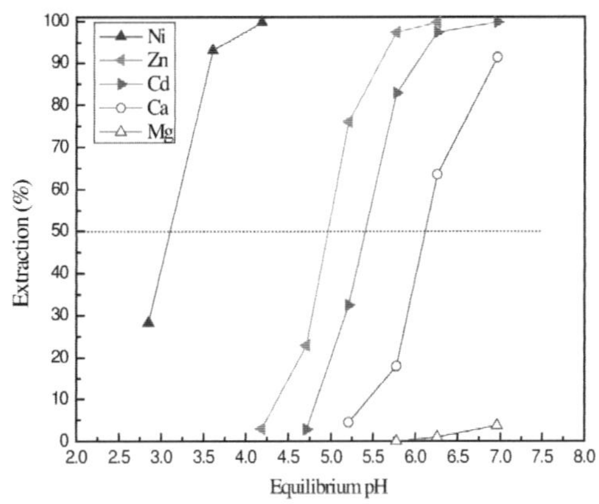

Figure 4. Extraction pH isotherms of metals without copper.

As is shown in figure 4, after extracting copper, the extraction pH of isotherms of other metals shifted. The pH isotherm of nickel shifted 1.0 pH units to a lower pH. The pH isotherm of zinc and cadmium also shifted a little to a lower pH. According to figure 4, after extracting copper, the nickel could be extracted totally by controlling the pH at about 4.0, while other ions are hardly extracted. The zinc and cadmium could be extracted at a pH of about 6.25.

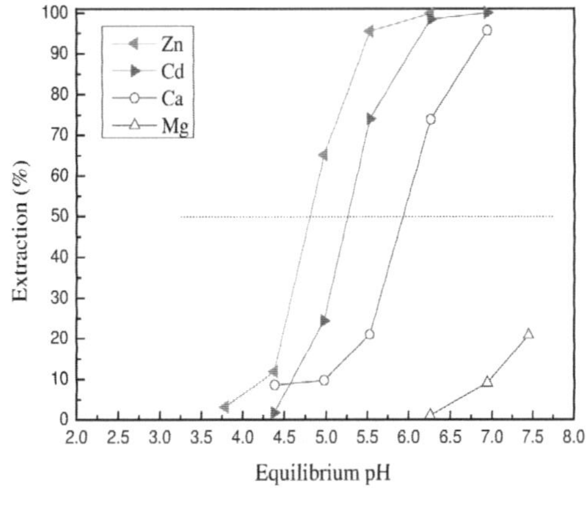

Figure 5. Extraction pH isotherms of metals without copper and nickel.

Figure 5 shows that after extracting copper and nickel, the extraction pH of isotherms of zinc and cadmium did not shifted again. They would be extracted at a pH of about 6.0. However this synergistic system cannot recover zinc and cadmium directly, and calcium will be significantly extracted under these conditions. Many other extraction systems which had been used in industry, such as D2EHPA, can extract zinc and cadmium separately from calcium and magnesium[9].

4. Conclusions

A single solvent extraction system such as Versatic10 or Mextral 984H, cannot recycle heavy metals separately in complex waste water. The synergistic organic system consisting of 0.50 M Versatic10 and 0.5 M Mextral 984H in DT100 can deal with a certain waste water which contains Zn (II), Cu (II), Ni (II), Ca(II), Mg(II) efficiently, and as extraction proceeds, some improved change of extraction pH isotherms of metal ions will occur. At an A/O ratio of 1:1 and 40℃, the copper could be extracted while pH is about 2.0, nickel will be preferentially extracted while pH is about 4.2, zinc and cadmium could be recovered while pH is about 6.5.

References

[1] Cheng S P. Heavy metal pollution in China: Origin, pattern and control[J]. *Environmental Science and Pollution Research*, 2003, **10**(3): 192-198.

[2] Guo J F, Lu Y J. Progress of heavy metal wastewater treatment technology[J]. *Nonferrous Metals Processing*, 2006, **35**(4): 48-50.

[3] Wang T, Li X G, Du Q Y. Research progresses in treatment technologies for heavy metal ion containing wastewater[J]. *Environmental Protection of Chemical Industry*, 2008, **28**(4): 323-325.

[4] Li Y, Yang G Y. Present status of treatment and utilization of heavy metal wastewater[J]. *Environmental Protection for Electric Power*, 2009, **25**(4): 50-51.

[5] Ma Q, Zhang X L. Advances in new technology for heavy metal wastewater treatment at home and abroad[J]. *Chinese Journal of Environmental Engineering*, 2007, **1**(7): 10.

[6] Chai L Y, Li Q Z, Li M; Liang Y J, Peng B, Min X B. Research advances on reduction and treatment technology of pollutants discharged from zinc smelting[J]. *Nonferrous Metals Science and Engineering*, 2013, **4**(4): 1-10.

[7] Fu F, Wang Q. Removal of heavy metal ions from wastewaters: A review[J]. *Journal of Environmental Management*. 2011, **92**(3): 407-418.

[8] Zhang M J. The Extraction of Metal and Other Useful Components in the Solution[M]. *Metallurgical Industry Press*, 1995.

[9] Ma R J. Principle of Hydrometallurgy[M]. *Metallurgical Industry Press*, 2007.

Preparation and characterization of bio-composite PEEK/nHA

Y S Jin[1], C C Bian[1,2,3], Z Q Zhang[1], Y Zhao[1] and L Yang[1]

[1]Department of Materials Science and Engineering, Beijing Institute of Petrochemical Technology, Beijing 102617, China
[2]College of Materials Science and Engineering, Beijing University of Chemical Technology, Beijing 100029, China

E-mail: 1135308203@qq.com

Abstract. PEEK/nHA composite material, with excellent mechanical property as polyetheretherketone (PEEK) and biological activity as hydroxyapatite (HA), has attracted wide attention of medical experts and materials science experts. The addition of hydroxyapatite was the decisive factor for biological activity in PEEK/nHA composite. In this paper, acicular nanohydroxyapatite was prepared by chemical precipitation method with $Ca(NO_3)_2$, $(NH_4)_2HPO_4$ as raw material; PEEK/nHA composite was prepared by solution blending and vacuum sintering method. The composite was characterized with FT-IR, XRD, DSC, TG and mechanical property test. Results showed that the composite has good thermal stability and compressive property when the mass ratio of PEEK to nHA is 10:3; and high nHA content can improve the biological activity of the composite, which can meet the basic requirements for bone tissue engineering scaffold.

1. Introduction

Polyetheretherketone (PEEK) is a kind of semi-crystalline thermoplastic material with high performance which has not only good processing performance and excellent high temperature resistance, but also many other excellent performance such as good abrasion performance [1-2], chemical resistance, fire resistance, hydrolysis resistance, radiation resistance [3]. For its biocompatibility and characteristics similar to human bone strength and modulus, artificial bone made of PEEK has not only advantages as light-weight, non-toxic, corrosion resistance and so on, but also can be integrated with organism. So using PEEK to fabricate human skeleton instead of metal is one of the most important applications in medical field [3-6]. But the lack of biological activity in PEEK limits its use. Hydroxyapatite (HA), a kind of bioactive material with excellent biological compatibility, which can form strong bond with bone tissue and has osteoconductivity and osteoinductivity, is widely used in clinic [7-8]. But pure HA with low bending strength and large brittleness can't meet the requirements for clinical applications [9]. To overcome these shortcomings, HA and PEEK were compounded, which can not only increase the toughness of composite, but also endow the composite with biological activity, while increase the rigidity of PEEK. Thus a composite can not only play a supporting role but also has certain biological function in vivo was obtained.

Traditional PEEK/nHA composites were prepared by in-situ polymerization and melt-blending mostly. For the different shapes of human bones and complex surfaces composing the bones, the precise tailor can't be achieved according to traditional processing method. Therefore, in this paper, nanohydroxyapatite with biological activity was prepared, and then the nanohydroxyapatite slurry was blended with PEEK. The composite powder was put into mold for extrusion forming after drying. The formed solid was sintered in a Muffle furnace. After all that, composites PEEK/nHA with good dispersion and thermal stability were prepared.

2. Experiment

2.1. Reagents

Calcium nitrate, purity quotient≥99%, Tianjin Fuchen Chemical Reagents Factory; ammonium phosphate, purity quotient=98%, J&K Scientific (Beijing); anhydrous methanol, analytial reagent, Beijing Chemical Works; ammonium hydroxide, analytial reagent, Beijing Chemical Works; PEEK, made by our own; acetone, analytial reagent, Beijing Chemical Works; ethanol, analytial reagent, Beijing Chemical Works.

2.2. Preparation of PEEK/nHA Composite

2.2.1. Preparation of nano HA. 0.6mol/L calcium nitrate aqueous solution and 0.36mol/L ammonium phosphate aqueous solution were prepared. 50mL calcium nitrate aqueous solution and 50mL ammonium phosphate aqueous solution were respectively added into different constant pressure funnels. Flask with three necks was put in thermostatic waterbath at a certain temperature. Calcium nitrate aqueous solution and ammonium phosphate aqueous solution were dropped slowly into the flask with three necks at the same speed while stirring. In the process of reaction, add ammonium hydroxide to maintain the pH value of the reaction system at about 11, and continue to react for 6 hours after two kinds of solution finished dropping. Keep the system static for 24 hours at room temperature. Wash it with de-ionized water repeatedly until the liquid poured out is neutral. White powder of nanohydroxyapatite was obtained after vacuum filtration and drying.

2.2.2. Preparation of PEEK/nHA. 4 conical flasks (250mL) with PEEK powder and HA in certain quality as 4 group tests, appropriate amount of ethanol as dispersant, magnetic stir at room temperature for 4 hours hermetically, and homogeneous dispersion liquid was obtained. Then vacuum filter it and dry the solid, PEEK/nHA composite powder with mass ratios of PEEK to nHA as 10:1, 10:2, 10:3 and 10:4 in the 4 groups respectively were prepared. Then the 4 groups of PEEK/nHA composite powder were milled and pressed into blocks in mold to form certain shapes. Finally, the blocks were put in vacuum annealing furnace for vacuum sintering with a sintering temperature at 350℃. In addition, a group of pure PEEK was prepared and sintered in the above manner as contrast.

2.3. Performance and Characterization

Nicolet-380 Fourier transform infrared spectrometer (FT-IR) from Thermo Electron Corporation was used to determine the molecular structure of nanohydroxyapatite and its composite; D8 KP2025-0NS 16-1-Z X-ray diffractometer from Bruker AXS Diffraktometer were used to determine the crystalline structure (scanning range as 10-60°, scanning speed as 4 deg/min); TGA/DSC1/SF/1100 thermogravimetric analyzer from METTLER TOLEDO Instrument Co., Ltd. was used to determine the thermal stability (heat from room temperature to 800℃ at a rate of 10℃/min); differential scanning calorimeter DSC131 from SETARAM Instrumentation was used to determine the melting point and crystallization temperature (nitrogen flow rate at 50mL/min, keep in 35℃ for 35s, then heat up to 400℃ at a rate of 10℃/min and keep for 20s, then cool down to 30℃ at rate of 10℃/min); microcomputer control electronic universal testing machine RG-300 from Shenzhen Reger Instrument Co., Ltd. was used to determine the mechanical property (according to GB/1041-2008 standard, test temperature at 23℃, humidity as 20%, at a test speed as 1mm/min).

3. Results and Discussion

3.1. Tests and characterization of nHA

3.1.1. Characterization of nHA. Figure 1 is the IR spectrum of HA sample we prepared. In the figure, the absorption peak at 3434.40cm^{-1} is corresponding to the contracting vibration peak of O-H in H_2O; the absorption peaks at 962.20cm^{-1} and 1032.20cm^{-1} are corresponding to the key band of PO_4^{3}-; the absorption peak at 873.07cm^{-1} is corresponding to the key band of PO_4^{2-}; the absorption peaks at 563.30 cm^{-1} and 602.83cm^{-1} belong to the bending vibration absorption peak of PO_4^{3-}. From the analysis above, it is confirmed that the prepared sample do be hydroxyapatite.

In figure 2, HA-40, HA-60, HA-70 are the XRD diffraction patterns of nanohydroxyapatite obtained in 40℃, 60℃ and 70℃ respectively. The sharp peaks at 25.8°, 31.5°, 32.8°, 34.1°, 39.5°, 46.6°, 49.3° as 2θ are the characteristic diffraction peaks of HA. As seen in table 1, the grain size of HA gradually increased with the increase of reaction temperature.

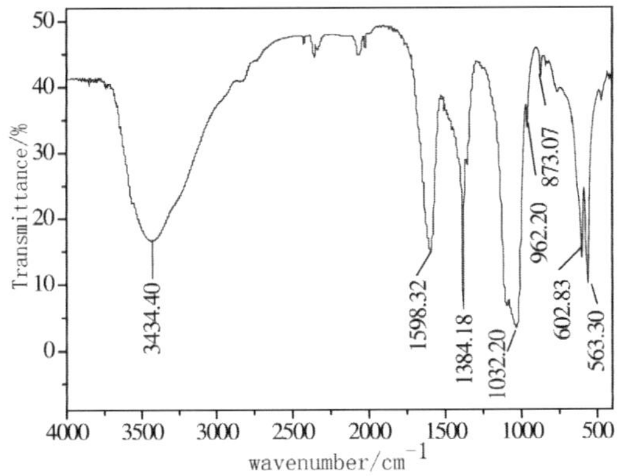

Figure 1. IR spectrum of nHA.

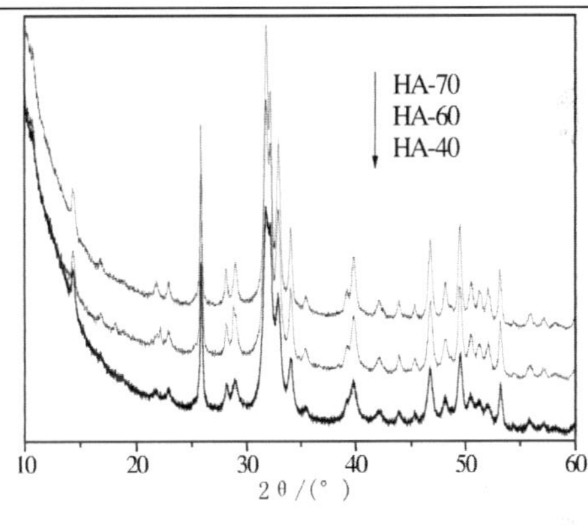

Figure 2. XRD diffraction of HA.

Table 1. Grain size of HA (2θ=31.5°).

Sample	Grain size (nm)
HA-40	10.6
HA-60	15.9
HA-70	21.6

3.1.2. SEM. As seen in figure 3, the nanohydroxyapatite we prepared is acicular crystal, and its acicular section diameter is less than 100nm, which can be proved that the sample we prepared is nanohydroxyapatite.

Figure 3. SEM of HA.

3.2. Tests and characterization of PEEK /nHA composite

3.2.1. Characterization of PEEK /nHA. Figure 4 is the IR spectrum of PEEK/nHA. In it, the stretching vibration peak for C=O is at 1649.80 cm^{-1}; the aromatic ring framework vibration peak is at 1597.62 cm^{-1} and 1490.41 cm^{-1}; asymmetric stretching vibration peak for R-O-R is at 1222.91cm^{-1}; symmetric stretching vibration peak for R-CO-R is at 927.07cm^{-1}; bending vibration absorption peaks for C-H out of the benzene ring plane is at 836.92cm^{-1} and 767.34cm^{-1}, meanwhile, at 836.92cm^{-1} is the characteristic peak of aromatic ring para-position substitution; from the analysis above, it can be confirmed that there is PEEK in the sample. Absorption peaks at 964.50cm^{-1} and 1035.84cm^{-1} are corresponding to the key band of PO_4^{3-} ; absorption peaks at 564.740 cm^{-1} and 603.75cm^{-1} belong to the bending vibration peaks of PO_4^{3-}. From the analysis above, it can be confirmed that there is HA in the sample. No appearance of obvious shifting or new peaks, which indicates that the original HA and PEEK matrix is physically combined, which is propitious to maintain the original biological characteristics of HA and PEEK.

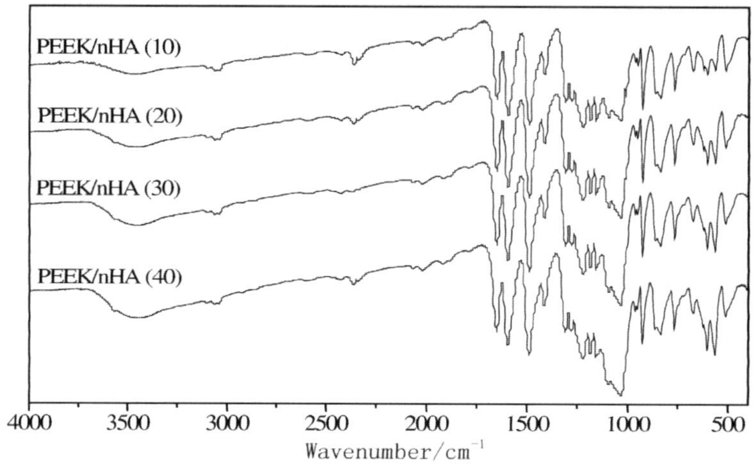

Figure 4. IR spectra of PEEK/nHA.

Figure 5 is the XRD diffraction of PEEK/nHA. In it, the peaks at 18.9°, 20.9°, 22.9°, 28.9° are the characteristic diffraction peaks of PEEK; the peaks at 25.8°, 31.5°, 32.9°, 34.4°, 46.9° are the characteristic diffraction peaks of HA. No new diffraction peaks appearance in the pattern further illustrated that the HA and PEEK matrix were physically combined.

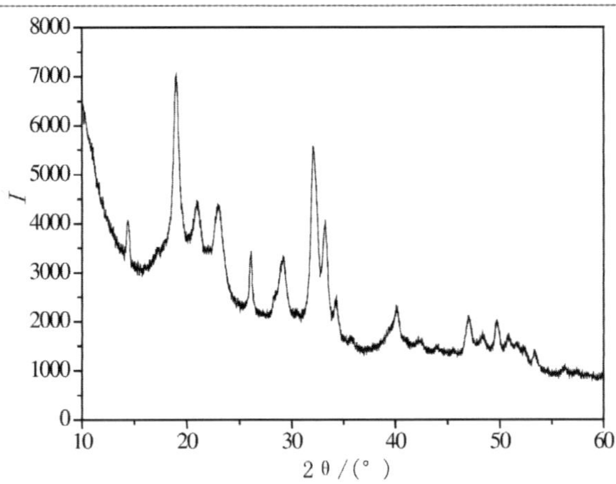

Figure 5. XRD diffraction of PEEK/nHA.

3.2.2. Thermal property of PEEK/nHA. Good thermal stability is required when biological composite material PEEK/nHA as bone tissue substitute material. In long-term organism environment and molding process it can keep its heat-resisting, thermostability and property stable. TG and DSC test results of PEEK/nHA composites are shown in table 2. It can be seen that the decomposition temperature and the maximum weight loss rate of the composites gradually increase with the increase of HA content, but the decomposition temperature decreased slightly when the addition of hydroxyapatite exceeds 30%, which showed that the thermal stability of the composite is better than that of PEEK. Melting point of PEEK matrix in PEEK/nHA composite is lower than pure PEEK, while the crystallization temperature didn't change basically. Synthesizing the TG and DSC analysis, it can be proved that the PEEK /nHA composite material has good thermal stability.

Table 2. Thermal property of PEEK /nHA.

Sample number	m_{nHA}/m_{PEEK}	T_m[a]/℃	T_c[b]/℃	T_{max}[c]/℃	T_5[d]/℃	T_{10}[d]/℃	T_{20}[d]/℃
PEEK/nHA(0)	0	341.9	281.2	583.1	572.1	579.1	587.2
PEEK/nHA(10)	10	337.7	283.3	592.9	570.9	581.9	592.4
PEEK/nHA(20)	20	337.7	281.3	595.1	572.2	583.0	593.2
PEEK/nHA(30)	30	337.6	279.9	594.1	575.1	585.3	595.8
PEEK/nHA(40)	40	339.1	279.2	593.4	569.6	582.9	594.9

[a] Melting point.
[b] Crystallization temperature.
[c] Temperature in maximum weight loss rate.
[d] Temperature when weight loss rate on TG curve was 5%, 10% and 20%, respectively.

3.2.3. Mechanical properties. As scaffold for sclerous tissues, material should maintain its shape within a period of time in order to ensure that it can withstand proper external forces in the process of new bone formation, so mechanical property of material is one of the important factors. The compressive strength values of PEEK/nHA composites with different nHA contents are listed in table 3. It can be seen from table 3, compressive strength of material decreased with the increase of hydroxyapatite addition but decreased slightly. When mass fraction of HA is 40 %, the compressive strength of PEEK/nHA is more than 235MPa, which indicated that PEEK/nHA has good compression performance and can meet the requirements of mechanical strength for bone tissue engineering.

Table 3. Mechanical property of PEEK/nHA.

Sample	Compressive strength(MPa)
PEEK/nHA(30)	236.8
PEEK/nHA(40)	235.9

4. Conclusions

Acicular nanohydroxyapatite with particle size under 100nm was prepared by chemical precipitation method; PEEK/nHA composite material was prepared by solution blending and vacuum sintering method. The composite material has good thermal stability and compressive property, which is potential to be used as scaffold material in bone repair and bone tissue engineering.

References

[1] Pei X Q, Bennewitz R, Busse M and Schlarb A K 2013 Effects of single asperity geometry on friction and wear of PEEK *Wear* **304** 1-2 109-117

[2] Pan Y S and Xiong D S 2006 Research status on tribological performance of PEEK composite *Engineering Plastics Application* **34** 2 68-70

[3] Bai S and Zhou J 2004 Application status of PEEK resin *Chemical Technology Market* **08** 21-23.

[4] Shuang Y 2013 Market analysis of PEEK *Chemical Industry* **31** 12 35-40

[5] Kurtz S M and Devine J N 2007 PEEK biomaterials in trauma, orthopedic, and spinal implants *Biomaterials* **28** 32 4845-4869

[6] Toth J M, Wang M, Estes B T, Scifert J L, Seim III H B and Turner A S 2006 Polyetheretherketone as a biomaterial for spinal applications *Biomaterials* **27** 324-334

[7] Roeder R K, Converse G L, Kane R J and Yue W M 2008 Hydroxyapatite reinforced polymer biocomposites for synthetic bone substitutes *JOM* **60** 3 38-45

[8] Andriano K P, Daniels A U and Heller J 1992 Biocompatibility and mechanical properties of a totally absorbable composite material for orthopedic fixation devices *J.Biomed.Mater.Res.*B *Appl. Biomater* **3** 3 197-206

[9] Villacampa A I and Garcí a-Ruiz J M 2000 Synthesis of a new hydroxyapatite-silica composite material *Journal of Crystal Growth* **211** 4 111-115

Preparation and characterization of composite material $Na_2SO_4 \cdot 10H_2O$ - $KAl(SO_4)_2 \cdot 12H_2O$ for thermal storage

M Zheng[1], J W Luo, Y H Zhang and P Chen

School of Chemical Engineering, Northwest University, Xi'an, 710069, China

E-mail: mszheng2@yahoo.com

Abstract. In the present paper, the preparation and characterization of composite material $Na_2SO_4 \cdot 10H_2O$ - $KAl(SO_4)_2 \cdot 12H_2O$ for thermal storage are reported. The co - melting method is employed to perform the preparation. The thermal properties and the step cooling behaviors of the composites are studied. The results show that: the phenomenon of phase separation disappears in the composite system with the mass ratio of 1 : 9 to 3 : 7 ($Na_2SO_4 \cdot 10H_2O$: $KAl(SO_4)_2 \cdot 12H_2O$), and the optimized mass ratio is about 2 : 8, which results in the lowest phase change temperature, the longest heat releasing platform of 41 min and a higher latent heat of 819.93J/g at 72.7 °C; the enthalpy remains higher value after 50 thermal cycles. Thus, this composite material has a potential as the phase change thermal storage material for low temperature application.

1. Introduction

Phase change thermal storage material has the advantage of higher latent heat with endothermic and exothermic processes at almost constant temperatures [1-2]. Therefore, it plays a significant role in some industrial fields, including aerospace, energy-efficient buildings, solar energy utilization, etc.

Inorganic hydrated salts have been the effective phase change thermal storage materials owing to their higher latent heat of fusion and melting point ranging from a few degrees to more than a hundred degree, which includes metallic halide salts of alkali and alkaline earth metals, sulphates, phosphates, vinegar salt and other salt hydrate. However, the problems in these materials are phase separation and supercooling in sole component inorganic hydrated salt once melting. Many scholars are doing research to improve their property in order to solve these problems [3-5]. The effective approaches include doping, thickening agent adding and microencapsulated.

The supercooling phenomena of $KAl(SO_4)_2 \cdot 12H_2O$ was studied by Song through cold fingering and adding nucleating agent. The results indicated that the supercooling in $KAl(SO_4)_2 \cdot 12H_2O$ can disappear as the nucleating agent $MgCl_2 \cdot 6H_2O$ being added about 2%, in addition to the unchangeable of the phase transition temperature of $KAl(SO_4)_2 \cdot 12H_2O$ [6]. Kuznik prepared composite materials by combing zeolite and dehydrated $MgSO_4$, the actual thermal storage density of this composite reaches to 45% of the theoretical value, and its property remains unchanged after more than three cycles [7].

Wang et al prepared composite hydrated salts by $KAl(SO_4)_2 \cdot 12H_2O$ and $MgSO_4 \cdot 7H_2O$, which has stable performance and no phase separation phenomenon [8]. Liang et al studied the melting behaviour of the binary hydrated salts system for cold storage by DSC (Differential Scanning Calorimeter) test, they showed that a higher phase change latent heat with a primary endothermic peak could be obtained through proper combination of multiple salts [9].

Wang et al studied the thermal properties of $KAl(SO_4)_2\cdot 12H_2O$ experimentally [10], the results indicate that the endothermic enthalpy is 449 kJ/mol in the temperature range of 60～150℃, and the averaged exothermic enthalpy is 341.7kJ/mol in the temperature range of 25～60℃. The applicable efficiency of energy is 76.1% within 3 times cyclic operation. The mixture of graphite and $KAl(SO_4)_2\cdot 12H_2O$ could obtain a better the heat conductivity [11].

In the present article, thermal storage composite material of $Na_2SO_4\cdot 10H_2O$ and $KAl(SO_4)_2\cdot 12H_2O$ is prepared and characterized. The co-melting is employed to perform the preparation. The characterization includes DTA (Differential Thermal Analysis), step cooling and heating – cooling cyclic tests. It aims to provide a technical support to the study of phase change thermal storage material at low temperature.

2. Fundamental properties of the initial materials

2.1. Initial materials

Experimental material. The main chemicals employed are: $Na_2SO_4\cdot 10H_2O$, Tianjin Kemiao Chemical Reagent Co. Ltd., Tianjin, purity of 99%; KAl $(SO_4)_2\cdot 12H_2O$, Zhengzhou Paiey Chemical Reagent, Zhengzhou, purity of 99.5%.

Laboratory instrument. The experimental Instruments are: JJ124BC electronic balance (Max = 120g, Min = 2mg), DF-101S constant temperature heater with magnetic stirrer, a thermometer (apuhua TM-902C, −50 ℃ ~ 1300 ℃, accuracy 0.1 ℃), HCT-1 differential scanning calorimetric balance, DZF-6030 vacuum oven.

2.2. Basic properties of $Na_2SO_4\cdot 10H_2O$ and KAl $(SO_4)_2\cdot 12H_2O$

The experimental procedure is as follows,

(1) Pour 10g of $Na_2SO_4\cdot 10H_2O$ and KAl $(SO_4)_2\cdot 12H_2O$ into the mill ground milling to fine powder, respectively, and then pour each fine powder into individual test tube;

(2) Place each test tube in the thermostat heating magnetic stirrer with a constant temperature heater, and keep half an hour after the material fully melting;

(3) Remove the test tube from the thermostat heating magnetic stirrer and stand at room temperature, record its temperature data once every 10s, drawing step cooling curve after the temperature tests;

(4) Take about 10mg samples of $Na_2SO_4\cdot 10H_2O$ and KAl $(SO_4)_2\cdot 12H_2O$ to conduct DTA tests, respectively, the temperature ranges from room temperature to 150 ℃, and the heating rate is 1 ℃ / min.

The experimental step cooling curve and DTA curve for $Na_2SO_4\cdot 10H_2O$ are shown in figure 1 and figure 2, respectively.

Figure 1 presents that the temperature is decreasing continuously during step cooling process of $Na_2SO_4\cdot 10H_2O$. Figure 2 shows that the endothermic peak during heating process appears at 37.8 ℃, with a total phase change latent heat of 870.65J/g, which exhibits a high latent heat of phase change material.

Figure 3 is the step cooling curve of $KAl(SO_4)_2\cdot 12H_2O$, it shows that the molten of $KAl(SO_4)_2\cdot 12H_2O$ begins to crystallize till cooling to 63.0℃, then there is a temperature rising to 73.4 ℃ due to the releasing of latent heat, which implies a supercooling of 10.4℃ (= 73.4 ℃ − 63.0 ℃). Figure 4 shows the DTA curve of $KAl(SO_4)_2\cdot 12H_2O$, it shows a large latent heat of 882.58J/g with the phase transition temperature of 75.7 ℃.

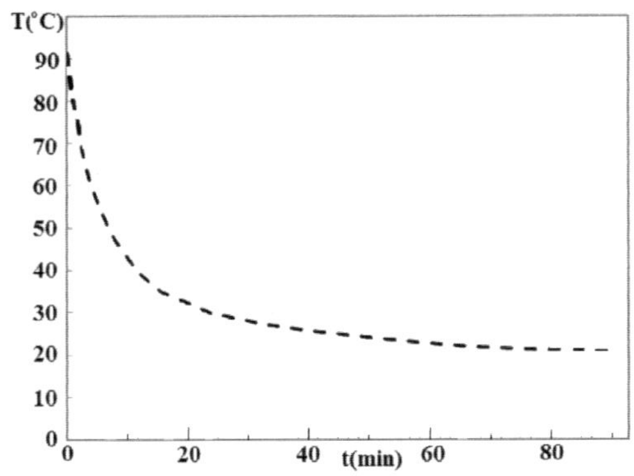

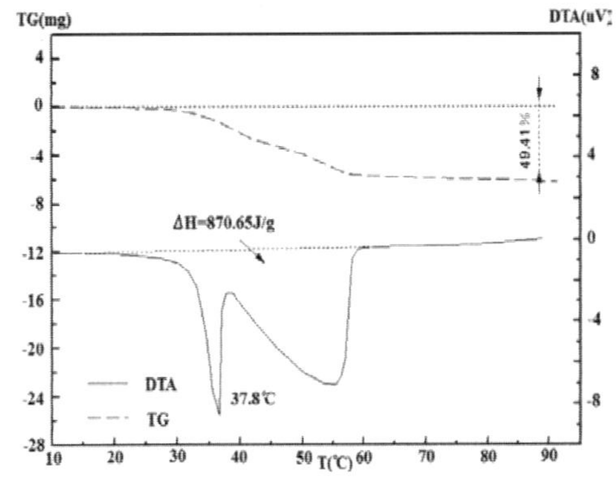

Figure 1. Step cooling curve of $Na_2SO_4 \cdot 10H_2O$ **Figure 2.** DTA curve of $Na_2SO_4 \cdot 10H_2O$

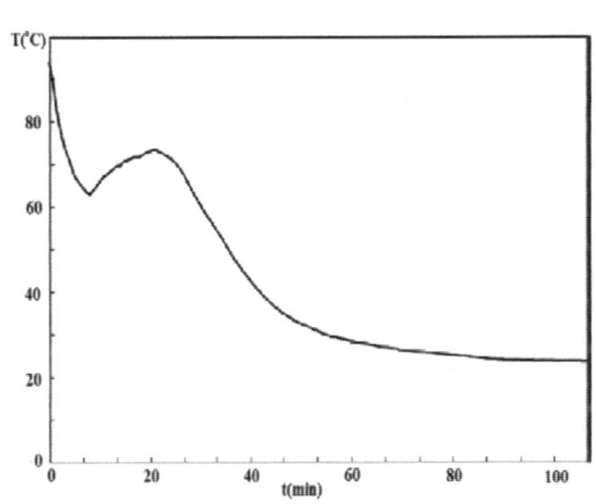

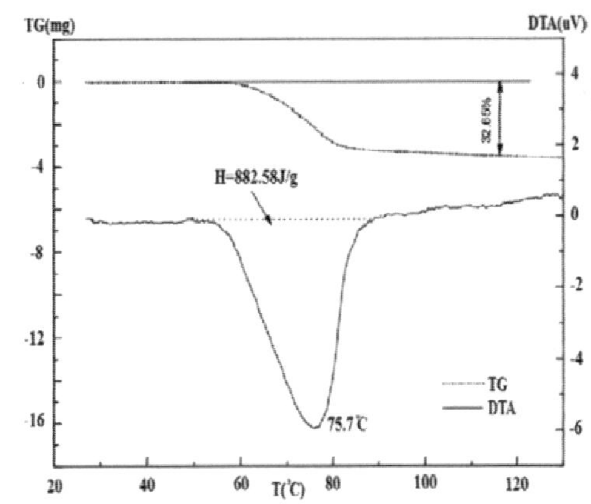

Figure 3. Step cooling curve of $KAl(SO_4)_2 \cdot 12H_2O$ **Figure 4.** DTA curve of $KAl(SO_4)_2 \cdot 12H_2O$

3. Preparation and property test of the composite materials

3.1. Preparation of phase change thermal storage composite material.

The preparation process of composite (or mixture) material is as follows,

(1) Mix and pour the milled $Na_2SO_4 \cdot 10H_2O$ and $KAl(SO_4)_2 \cdot 12H_2O$ powders into a test tube, the mass ratio of the salts is shown in table 1. The total amount is 10g, the test tube is a $\phi 20 \times 50$ml plastic tube;

(2) Place the test tube in the thermostat heating magnetic stirrer with a constant temperature heater, and keep half an hour after the material fully melting;

(3) Remove the test tube from the thermostat heating magnetic stirrer and stand at room temperature, record its temperature data once every 10s, drawing step cooling curve after the temperature tests.

Table 1. Mass ratio of $Na_2SO_4 \cdot 10H_2O$ to $KAl(SO_4)_2 \cdot 12H_2O$

No.	S1	S10	S2	S21	S3	S4	S5	S6	S7	S8	S9
$m_1:m_2$	1:9	1.5:8.5	2:8	2.5:7.5	3:7	4:6	5:5	6:4	7:3	8:2	9:1

The heating-cooling test indicates that samples S4-S9 exhibit phase separate, i.e., there is liquid in upper part and solid in the lower part of the test tubes. However, there is no phase separate in S1, S10, S2, S21 and S3.

3.2. Step cooling curve analysis of composite materials.
Figure 5 is the step - cooling curves of the samples with no phase separate. The tested supercooling temperature T_{sc}, the phase transition temperature T_c and the duration of heat releasing platform T_l for the phase separate free samples are listed in table 2.

Table 2. Variations of T_{sc}, T_c and T_l v.s. mass ratio

No / Contents	S1	S10	S2	S11	S3
T_{sc} /℃	1.8	0	0	4	0
Tc/℃	59.9	64.0	37.6	47.8	43.9
T_l/min	8.0	10.0	41.0	17.0	9.0

As can be seen from table 2, the phase transition temperature T_c behaves a "wavy" shape with the increase of content. The lowest phase transition temperature 37.6 ℃ appears at the mass ratio of 2:8. The samples appear obvious exothermic platform with different lengths during cooling process, the length of exothermic platforms is more than 8 min, and the longest exothermic platforms of S2 lasts 41 min.

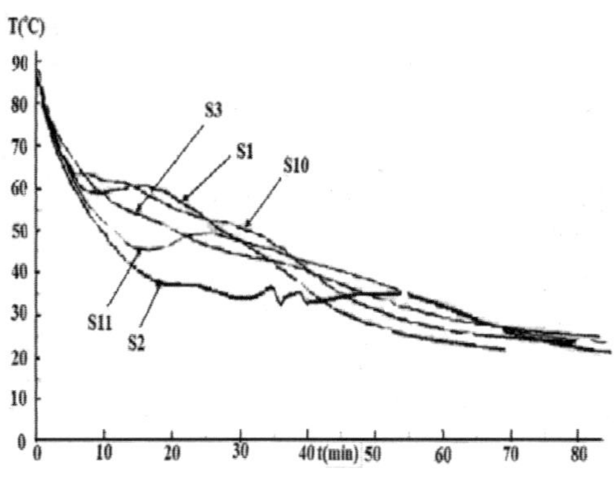

Figure 5. Step cooling curve of some composites

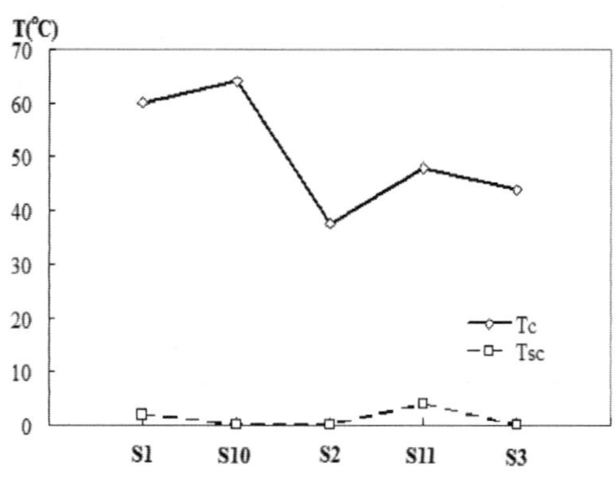

Figure 6. Tc and supercooling of the composites

Figure 6 shows the variation of phase transition temperature Tc and supercooling degree Tsc with respect to the mass ratio of the composite. As can be seen from figure 6, the degree of supercooling of S2 is comparatively smaller. Meanwhile, the thermal releasing platform of S2 is the longest, which lasts about 41 min with the phase transition temperature of 37.6 ℃.

Through above analysis, it can be seen that all the composites behave significant phase transition phenomenon. The composite of the mass ratio of 2:8 behaves a stable phase transition temperature, longer exothermic platform, no phase separation and supercooling during step cooling.

3.3. DTA Analysis of $Na_2SO_4 \cdot 10H_2O$ and $KAl(SO_4)_2 \cdot 12H_2O$ Composite
Take about 10mg composite samples of $Na_2SO_4 \cdot 10H_2O$ and $KAl(SO_4)_2 \cdot 12H_2O$ to conduct their DTA test, the temperature ranges from room temperature to 150 °C with heating rate of 1 °C/min.

3.4. DTA analysis of the composites.

Table 3 lists the test data of phase change temperature and latent heat of the composite materials with respect to content. Figure 7 shows the variation of the latent heat with respect to the mass ratio. As can be seen from table 3 and figure 7 that the biggest phase change latent heat change exhibits at the mass ratio of 1:9 (i.e. S1), its phase change latent heat is 968.26J/g; the lowest phase transition temperature Tcc is at the mass ratio of 2:8 (i.e., S2), 72.7 ℃, which is between those of pure Na_2SO_4•$10H_2O$ (37.8 ℃) and KAl $(SO_4)_2$•$12H_2O$ (75.7 ℃).

Table 3. DTA of phase change thermal storage composite materials

Sample No.	mass ratio	latent heat (J/g)	Tcc (℃)
S_1	1:9	968.26	79.7
S_{10}	1.5:8.5	912.22	79.4
S_2	2:8	819.93	72.7
S_{11}	2.5:7.5	830.96	77.6
S_3	3:7	955.86	77.9

Comprehensively, the composite S2 with the mass ratio of 2:8 behaves excellent property, of which the phase change latent heat is 819.93J/g, and the phase transition temperature is 72.7 ℃. The DTA curve of the composite S2 with mass ratio of 2:8 is shown in figure 8.

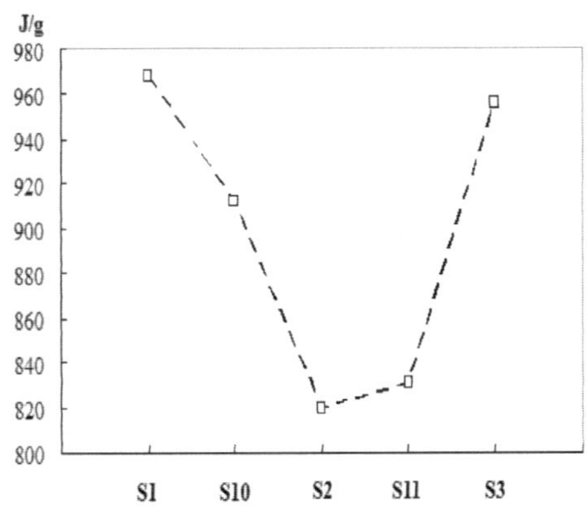

Figure 7. Thermal characteristics of the composites

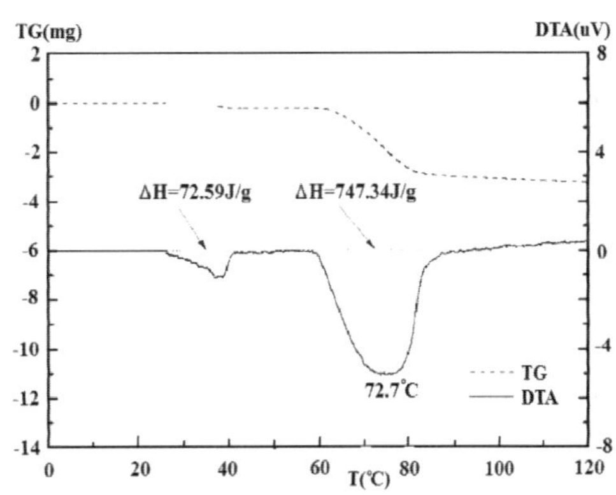

Figure 8. DTA curve of composite S2

3.5. Cyclic Performance Test of Composite S2

The cyclic performance test of composite S2 is conducted. The procedure of the cyclic performance test is as follows,

Take 10g of the sample material into a test tube, and heat it in a constant-temperature bath at 95 ℃ till melting completely, and then remove the test tube from the bath and cool it in air to room temperature, this is one heating – cooling cycle. A total of 50 cycles was conducted.

DTA test was carried out for the sample of about 10mg after every 10 cycles. The heating process of the DTA test is from room temperature to 150 ℃ with heating rate 1 ℃ / min.

IOP Conf. Series: Materials Science and Engineering **167** (2017) 012007 doi:10.1088/1757-899X/167/1/012007

3.6. Results.

From the DTA test, it can be seen that the latent heat of composite S2 decreases slightly with the increase of cyclic number, while the phase transition temperature maintains in the range of 70~80 ℃.

4. Concluding Remarks

Through this study, following conclusion could be obtained:

(1) The step cooling curve analysis shows that the composite sample with the mass ratio of 2 : 8 behaves longer step cooling platform of 41 min and free of supercooling, as well as lower the phase transition temperature of 37.6 ℃ during step cooling process.

(2) DTA analysis shows that the composite sample with the mass ratio of 2 : 8 has latent heat of 819.93J/g and transition temperature of 72.7 ℃ during heating.

(3) The heating – cooling cycle performance testing indicates that the composite sample with mass ratio of 2:8 maintains phase transition temperature within 70 ~ 80℃, stable performance and no stratification process within 50 cycles.

References

[1] Wang J and Cao X 2008 *Guangdong Chemistry* **36** 75
[2] Hasnain S M 1998 *Energy Conversion and Management* **39** 1127
[3] Fan S, Liang D and Yang X 2004 *Energy Storage Materials & Technology* (Beijing: Chemical Industry Press) p 100-151
[4] Ling S, Gao X and Yin H 2007 *Guangdong Chemistry* **34** 48
[5] Lu J 2005 *Liaoning Chemical Engineering* **34** 304
[6] Song J, Zeng L, Shui A, Liu Y and Wang H 2007 *Journal of Synthetic Crystals* **36** 358
[7] Hongois S, Kuznik F, Stevens P and Roux J J 2011 *Solar Energy Materials and Solar Cells* **95** 1831
[8] Wang Z, Tian H, Wang K, Zhao J and Xiong W 2013 *Journal of Synthetic Crystals* **42** 491
[9] Liang D, Fan S, Guo K and Wang R 2001 *Journal of Instrumental Analysis* **20** 15
[10] Wang T, Liu L, Song Y, Su Y and Liu P 2001 *Journal of Yunnan Normal University* **21** 65
[11] Song J, Zeng L, Chen B, Liu Y, Wand H, Liu P and Cheng X 2009 *Functional Materials* **38** 1307

Fabrication of micro-nano composite textured surface for slurry sawn mc-Si wafers cell

Y C Niu[1,2,4], Z liu[2], X K Ren[3], X J Liu[2], H T Liu[3] and Y S Jiang[3]

[1]Co-innovation Center for Green Building, Shandong Jianzhu University, Jinan 250101, China
[2]School of Materials Science and Engineering, Shandong Jianzhu University, Jinan 250101, China
[3]Shandong Linuo Solar Power Holdings Co., Ltd, Shandong, Jinan 250103, China

E-mail: niuyuchao@sdjzu.edu.cn

Abstract. In order to enhance the PV efficiency of the cell made from slurry sawn (SS) mc-Si wafers, using a Ag-assisted electroless etching ($AgNO_3$+HF+H_2O_2) combined with an auxiliary etching (HF+HNO_3) the RENA textured SS mc-Si wafers (called as RENA wafers) were further textured (nano pores were formed on the original micro pits) to change into micro-nano composite textured wafers (called as MN-RENA wafers). The solar cells made from the MN-RENA wafers had a better PV efficiency than that of RENA wafers. This is mainly attributed to the higher light-trapping of the micro-nano composite texture. The nano size texture enhanced the light-trap of wafer surface and, at the same time, the micro size texture maintained the light-trap uniformity of different gains of RENA wafer. However, there still exist a potential for optimization, such as, the SiNx passviation coating should be improved to be deposited more uniformly in order to passivate the bottom of pits better and to reduce the reflectance of the obtuse tips of pits.

1. Introduction

RENA wafers have been widely used to fabricate PV cells, but they have a higher surface light reflectance, which results in a lower PV efficiency. Therefore, the black Si cell has been developed using metal-assisted electroless texturing and has been got a much attention [1]. Recently, diamond wire sawn (DWS) is expected to be the mainstream wafering technology in Si-based photovoltaic industry. The advantages of this technique over the conventional slurry sawn (SS) include higher productivity, higher precision enabling production of thinner wafers, less wire consumption, lower environmental impact, easier recovery of silicon kerf loss and less surface mechanical damages [2-5]. However, DWS Wafers differ from SS ones in saw damage, making standard or traditional HF+HNO_3 etching mixture less suitable to achieve the desirable surface textures in the wafers [6-8]. The optimization of etching mixture formula and etching parameters (e.g. etching time, temperature) based on the traditional HF+HNO_3 etching mixtures system are also not suitable for a successful texture of DWS wafers [6]. Therefore, the DWS wafer texturing has been implemented successfully using Ag-assisted electroless etching recently [9,10]. In the work [9], $AgNO_3$+HF+H_2O_2 etching was applied to ensure forming pores on the smooth surface of a DWS wafer to form nano-porous Si layer, the auxiliary etching with HF+HNO_3 solution was needed to make this porous Si layer develop a macro-porous Si layer with sub-micron pits in order to avoid heavily doped defects on the nano-porous Si layer during the afterwards phosphorus doping process [11,12]. Unfortunately, among the grains with different crystal orientation there existed the differences in the axial direction and the depth of nano-pores during $AgNO_3$+HF+ H_2O_2 etching. These differences resulted in the final different texture structures after HF+HNO_3 etching, which limits the efficiency of Si cells [13]. In this paper, SS mc-Si wafers were textured using RENA texturing at first, then, the RENA textured wafer was further textured (nano pores formed on the micro pits of RENA wafer using Ag-assisted electroless etching ($AgNO_3$+HF+H_2O_2) and, at last, the nano pores were enlarged into submicron pits using an auxiliary etching (HF+HNO_3). As a result, there formed a micro-nano composite textured surface. This composite textured surface eliminate both the higher surface light reflectance of RENA textured surface and the different textures between different grains on a flat surface etched using Ag-assisted electroless etching combined with (HF+HNO_3) an auxiliary etching, so that the cells made from the wafers with the composite textured surface had a higher efficiency. Also, the problems resulted from submicron pits on the composite textured surface were investigated in detail.

2. Experimental

SS mc-Si wafers were supplied by Linuo Solar Power Holdings Co., Ltd and textured using traditional HF+HNO_3 etching (called as RENA texturing) on the standard texturing line in this company.

Ag-assisted electroless etching ($AgNO_3$+HF+H_2O_2) combined with an auxiliary etching (HF+HNO_3) was used for RENA wafers. Ag-assisted electroless etching ($AgNO_3$+HF+H_2O_2) was divided into two sub-steps: HF+$AgNO_3$ etching and HF+H_2O_2 etching. The whole texturing process consisted of seven steps. From Step 1 to Step 6, RENA wafers were immersed successively in the aqueous solution specified as flowing and rinsed in deionized water (DIW) at room temperature (RT) for 20 s after each step.

(1) Several concentrations (0.1, 0.2, 0.5, 1 and 2mM/L) $AgNO_3$: 1.15M/L HF, RT for different times, to deposite Ag catalysts.
(2) 4M/L HF: 0.7M/L H_2O_2, RT for different times, to form nano-pores.
(3) 1.3M/L HF: 4.9M/L HNO_3, RT for different times, to enlarge nano-pores into sub-micron pits and remove Ag catalysts [9].
(4) 6.95% (w/w) KOH, RT for 10 s, to remove Si stain [9].
(5) 0.2mM/L H_2O_2: 0.2mM/L HCl, RT for 5 min, to remove metal ions.
(6) 4.9M/L HF, RT for 1min, to remove SiO_2 layer.
(7) Hot air drying, 135 °C for 2 min.

The above MN-RENA wafers were made into solar cells on a standard solar cell manufacturing line.

The morphologies of the wafers were observed using a laser scanning confocal microscope (LSCM, OLS4000) and a S4800 scanning electron microscope (SEM). The reflectivity of the textured wafers and the PV properties of solar cells were characterized respectively using a reflectometer (R9000-2DMA) and a photovoltaic cell testing apparatus (Berger PSS10).

3. Results and discussion

Figure 1(a) shows the surface structure of as cut SS mc-Si wafer. The sizes of structures on the Si surface range from 1 to 3μm. The structure is called mechanical damages resulted from the wafer cutting and must be removed before texturing. After RENA texturing, the surface of RENA wafer contains wormlike pores with width ranging from 2 to 4μm and random length, as shown in Figure 1(b).

The RENA wafer was deposited with Ag particles and etched in HF+H_2O_2 etching solution, as described in experimental section. Its surface structure is shown in Figure 2. It can be seen that the nano pores formed on the RENA wafer. The average diameter of the nanopores on the surface of RENA wafer is about 10nm [14]. Figure 3 shows the spectra of light reflectivity of the as etched wafers. For the sake of comparison, the spectra of light reflectivity of the RENA wafer are also shown in both Figure 3(a) and (b).

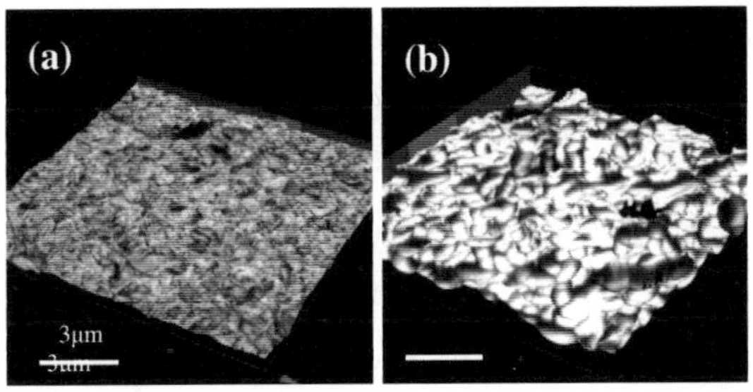

Figure 1. Surface micrographs of the as cut SS wafer (a) and RENA wafer (b)

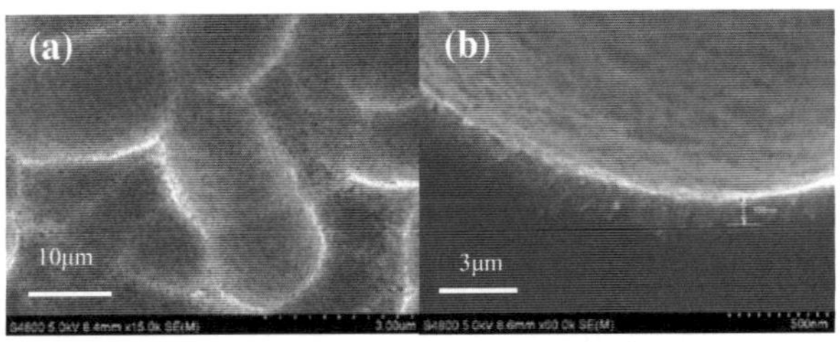

Figure 2. Micrographs of micro-nano size structured surface of wafer
(a) plan view (b) section view

In Figure 3(a), with the increase of $AgNO_3$ concentration from 0.1mM/L to 0.5mM/L, the average reflectivity of the silicon wafer surface decreased from 7.7% to 4.4%. When the $AgNO_3$ concentration increased from 0.5mM/L to 2mM/L, the average reflectivity of the silicon wafer increased slightly from 4.4% to 4.9%. In Figure3(b), the $AgNO_3$ concentration of 0.1mM/L was selected and the deposition time of Ag was from 60s to 180s. Under such etched condition, the RENA wafers were deposited with Ag particles and etched in $HF+H_2O_2$ etching solution, as described in experimental section. With the deposition time of Ag increased from 60s to 120s, the surface reflectivity of the wafers decreased from 7.7% to 3.9%. While with the deposition time of Ag continued increase to 180s, the surface reflectivity increased from 3.9% to 4.1%.

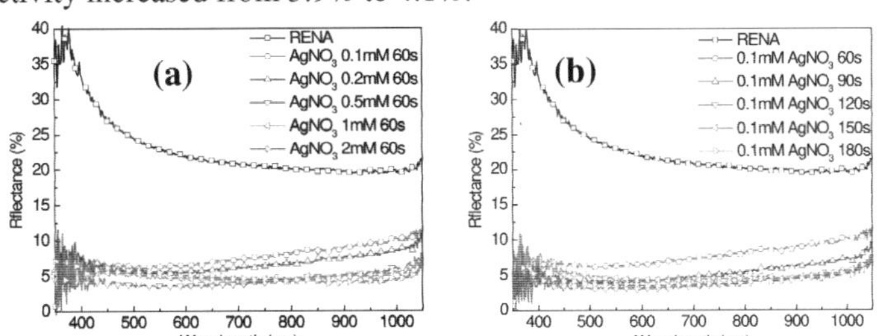

Figure 3. Spectra of light reflectivity of the as etched wafers
(a) the wafer etched at different concentration of $AgNO_3$ for the same etching time
(b) the wafer etched at the same concentration of $AgNO_3$ for different etching time

Since Ag deposition and $HF+H_2O_2$ etching under Ag catalytic effect were carried out separately, the distribution, shape and size of Ag partials, and the depth of the pores etched under Ag particles could be controlled separately. Furthermore, since the distribution, shapes and size of pore depend on those of Ag particles, the controlling of Ag deposition is a key issue. Whether with increasing the $AgNO_3$ concentration under the same deposition time or with increasing the deposition time under the same $AgNO_3$ concentration, the amount of Ag particles increased. Therefore, the number of the nano pores formed under Ag particles increased and the reflectivity of wafer surface decreased. When the amount of deposited Ag reached a certain value, the Ag particles accumulated to form dendrites, which hindered the contact between the etching solution and the silicon substrate, and then hindered the Ag catalytic etching in next step. Therefore, when the $AgNO_3$ concentration was over 0.5mM/L under the deposition time of 60s or the Ag deposition time was over 120s under the $AgNO_3$ concentration of 0.1mM/L, the reflectivity of the wafer surface increased, as shown in Figure 3a and b respectively. Therefore, for the lower or higher concentration of $AgNO_3$ solution, the appropriate amount of Ag deposition of can be obtained by adjusting the Ag deposition time.

On the micro-nano size structured surface described above, there existed nano pores, which could be heavily doped and became defects during the afterwards phosphorus doping process[11,12]. Thus, the auxiliary etching with HF-HNO_3 solution was needed to make this porous Si layer develop into a macroporous Si layer with sub-micron pits, the surface with which is called a sub-micro size surface [13], so that the Step 3 and 4 of texturing process as described in experimental section were carried out. Meanwhile, Ag particles in the nano pores were dug out [9].

Figure 4 shows the micrographs of sub-micro size surface of wafer. It can be seen that, there formed sub-micron pits on the surfaces of the micro size pits. The formation process of the sub-micron pits is the combination of pores or pits step by step [13]. The aim of sub-micron pits formation on the surfaces of the micro size pits was to enhance the effect of light-trapping and reduce the color difference between different Si grains [15].

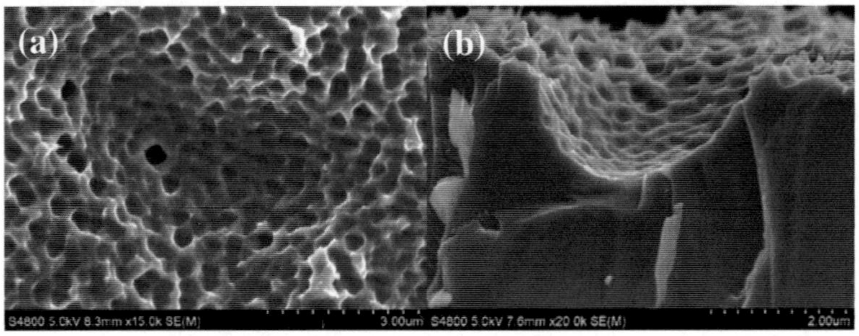

Figure 4. Micrographs of sub-micro size surface of wafer
(a) plan view (b) section view

The spectra of light reflectivity of the wafers with the sub-micro size surface after different HF-HNO_3 etching times are shown in figure 5. The as textured wafer after etching Step 2 described in experimental section has a lowest reflectance, as shown by Curve A. With increasing the HF-HNO_3 etching time, the reflectance of the samples increases. The curve of the reflectance rises up from Curve A to Curve E. Curve F represents the reflectance of the RENA wafer. The increase of reflectance of the sample is resulted from the enlargement of diameter of pores or pits on the surface of wafer during the HF-HNO_3 etching and the longer the etching time, the bigger the diameter is [13].

In the preparation process of Si solar cells, the SiNx coatings on the surface of the wafer using PECVD have two main functions: (1) as an antireflection, further reduce the surface reflectance according to the principle of optical interference; (2) as the surface passivation, reducing the surface recombination rate to improve the effective minority carrier lifetime.

The passivated surface is shown in Figure 6. It can be seen from the figure that after coated, the pits on the surface of the wafer changed into smaller ones, and that there is an agglomeration phenomenon of SiNx on the peaks between pits. During the PECVD of the coating, the active groups involved in the reaction diffused into the pits. The active groups were more difficult to diffuse into the bottom of pits than other positions. On the contrary, near the pits opening the reaction gases were easy to exchange, so that the film formed easily. As a result, as the reaction was carried on the coating near the pits opening became thicker and thicker, so that the agglomeration of SiNx formed. Meanwhile, the agglomeration hindered the active groups to enter into the pits and the passivation at the bottom of pits became worse.

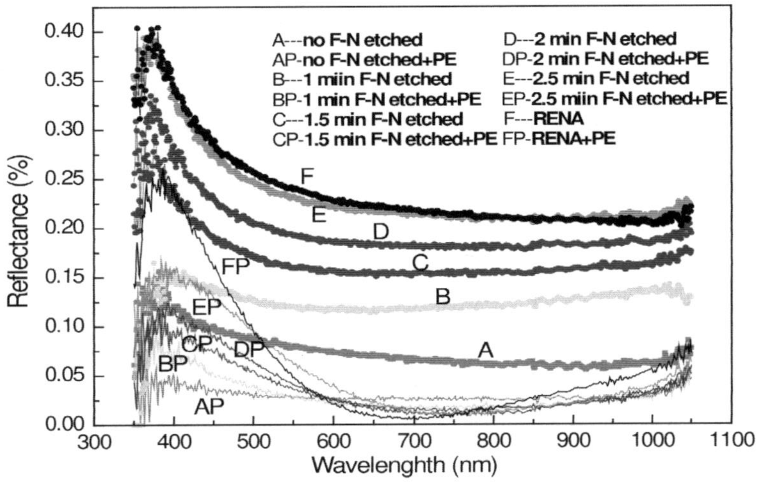

Figure 5. The spectra of light reflectivity of the wafers with the micro-nano size texture after different HF-HNO_3 etching times

Notes: 1) no F-N means no HF-HNO_3 etched: the as textured wafer after etching Step 2

2) F-N means HF-HNO_3 etched: the as textured wafer described in Note 1 were further etched in 1.3M/L HF: 4.9M/L HNO_3, RT for different times.

3) +PE means samples were deposited with SiN_x passive coating using PECVD.

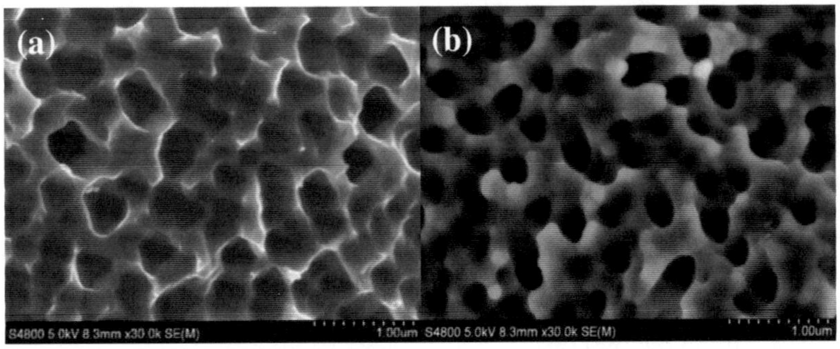

Figure 6. Comparison between as sub-micro size textured surface (a) and SiNx coated sub-micro size textured surface (b)

Through the experiment and optimization, the main process of RENA wafers cells and its parameters are listed as follows: 0.1mM/L $AgNO_3$-1.15M/L HF solution for 60s, 0.7M/L H_2O_2-4M/L HF solution for 120s, 1.3M/L HF-4.9M/L HNO_3 solution for 60s, 1% NaOH solution for 10s etching, the square resistance of the wafer about 90Ω/□ after P diffusion, thickness of SiNx passivation coating about 90nm and the rest of the process of SS mc-Si wafers cells is as the same as that used in traditional SS mc-Si wafers cell production line. The average value of PV properties of solar cells made from RENA wafers, which acted as a reference group, and micro-nano textured wafers are listed in Table 1.

Table 1. PV properties of solar cells made from RENA wafer and micro-nano textured wafers

	Count	*Uoc*/V	*Isc*/A	*Rs*/Ω	*Rsh*/Ω	*Ff*/%	*Ncell*/%	*Irev*/1
Reference Group	385	0.6287	8.847	0.0020	283.4	78.76	18.00	0.088
Experimental group	375	0.6291	9.031	0.0022	233.1	78.64	18.36	0.091

As can be seen from the table, *Uoc* of the experimental group and the reference group was almost the same, but *Isc* of experimental group increases obviously compared with the reference group. This is mainly due to the lower reflectivity of silicon wafer surface of the experimental group. As a result, the average *Ncell* of the experimental group was 18.36%, 0.36% higher than that of the reference group.

4. Summary

The RENA wafers have been textured successfully using a Ag-assisted electroless etching ($AgNO_3$+HF+ H_2O_2) combined with an auxiliary etching (HF+HNO_3). The solar cells made from the MN-RENA wafers had a better PV efficiency than that of RENA wafers. This is mainly attributed to the higher light-trapping of the micron-nano composite texture. The nano size texture enhanced the light-trap of wafer surface and, at the same time, the micro size texture maintained the light-trap uniformity of different gains of RENA wafer. However, there still exist a potential for optimization, such as, the SiNx passviation coating should be improved to be deposited more uniformly in order to passivate the bottom of pits better and to reduce the reflectance of the obtuse tips of pits.

Acknwledgments

This work was supported by Shandong government financial supporting (L37002013098), Jinan government financial supporting (JK20130307 and 301305033). The authors are grateful to Prof. Xiao-Lin Wang from University of Wollongong for his valuable discussions and assistance.

References

[1] Huang Z P, Geyer N, Werner P, Boor J D and Gösele U 2011 *Adv. Mater.* **23** 285
[2] Watanabe N, Kondo Y, Ide D, Matsuki T, Takato H and Sakata I 2010 *Prog. Photovolt: Res. Appl.* **18** 485
[3] Chen W H, Liu X M, Li M, Yin Q C and Zhou L2014 *Mater. Sci. Semicond. Process.* **27** 220
[4] Teomete E 2013*Arab J. Sci. Eng.* **38** 1209
[5] Meng H C and Zhou L 2014 *Silicon* **6** 129
[6] Meinel B, Koschwitz T and Acker J 2012 *Energy Procedia* **27** 330
[7] Lippold M, Buchholz F, Gondek Ch, Honeit F, Wefringhau E and Kroke E 2014 *Sol. Energy Mater.Sol. Cells* **127** 104
[8] Meinel B, Koschwitz T, Blocks C and Acker J 2014 *Mater. Sci. Semicond. Process.* **26** 93
[9] Kumagai A 2015 *Sol. Energy Mater. Sol. Cells* **133** 216
[10] Cao F, Chen K X, Zhang J J, Ye X Y, Li J J, Zou S and Su X D 2015 *Sol. Energy Mater. Sol. Cells* **141** 132
[11] Shirasawa K, Fukui K, Okada K, Inomata Y, Takahashi H, Fukawa Y and Fujii S 1997 *Fourteenth European Photovoltaic Solar Energy Conferences Barcelona (Spain)*, p384
[12] Oh J, Yuan H C and Branz H M 2012 *Nat. nanotechnol.* **7** 743
[13] Niu Y C, Liu H T, Liu X J, Jiang Y S, Ren X K, Cai P and Zhai T G 2016 *Materials Science in Semiconductor Processing* **56** 119
[14] Liu X J, Niu Y C, Zhai T G, Ma Y Y, Zhen Y T, Ma X Y and Gao Y 2016 *Appl. Phys. A* **122** 127:1
[15] Han C A，Zou S，Li J J，Cao F and Wang X S 2013 *Acta energlae solaris sinica* **34** 2164

Experimental study on heat transfer performance of aluminium foam parallel-flow condenser in air conditioner

X Zhang[1], Z M Wan[2,3,4] , H W Chang[2] and Y D Wang[3]

[1]China Ship Design and Research Center, Wuhan 430064, China
[2]School of Energy and Power Engineering, Huazhong University of Science and Technology, Wuhan 430074, China
[3]Guangdong Chigo Air Conditioning Co. Ltd. Foshan, China

E-mail: zhongminwan@hotmail.com

Abstract: Open cell aluminium foam was used in parallel-flow condenser in air conditioner, and two condensers with different pore density were fabricated. The experimental study was conducted on the heat transfer performance and temperature distribution. The experimental results show that both of the heat transfer load and air pressure drop increase with the increase of pore density, air velocity is 2.5m/s, the heat transfer capacities of the condenser with 10PPI and 8PPI are 4.786kw and 3.344kW respectively. Along the flow direction of refrigerant, the outlet temperatures of refrigerant drop with the rise of air velocity when the inlet temperature is constant.The outlet temperature of the refrigerant decreases with the increase of pore density.

1. Introduction

It is well known that Open-cell foam aluminum have many interesting physical properties, such as low weight high specific surface area combined with relatively high thermal conductivity. Consequently, they have become a focus in industrial applications[1]. With the development of compact heat exchanger, the air flow in the metal foams could form turbulence, thus it could enhance the heat transfer performance and be used in heat exchanger more and more.

Tadrist Let.al[2] investigated the flow and heat exchange characteristics in aluminum foam, the best structural parameter of aluminum foam was investigated based on the heat exchanger performance and pressure drop. NohJ S et.al[3] investigated the aluminum foam heat exchanger. It was found that the aluminum foam could enhance the performance significantly. T'JoenC et.al[4] investigated the heat exchanger design with metal foams through wind tunnel testing experimentally. The results indicated that a good metallic bonding between the foam and the tubes can be achieved. Comparing with helically finned tubes ,metal foam covered tubes with small tube spacing, small foam heights and high specific surface area potentially offer more benefits at higher air velocities (>4 m/s) Han X et.al[5] et.al used aluminum foam for the dehumidifying heat exchangers, both of the dry condition and wet condition were experimental investigated. The results showed that the drainage from metal foam is as good as or better than drainage from the louver-fin exchangers, (the heat transfer efficiency enhanced with the PPI). RibeiroG B et.al[6] experimentally investigated three different copper metal foam structures with distinct pore densities (10 and 20 PPI) and porosities (0.893 and 0.947). The results showed air-side pressure drop increases with the increase/decrease? of pore density and decreases with porosity, and the air-side thermal conductance increases with pore density and decreases with porosity. Ranut P et.al[7] adopt high resolution μ-CT-based CFD simulation, performed on three different open-cell aluminum foams samples. The results demonstrate demonstrated that open-cell aluminum foams are effective means for enhancing heat transfer. Moreover, the procedure proved that μ-CT is a valid tool for capturing the peculiar details of the foam structure, thus to overcome the limits associated to the use of economical, but simplified, geometric models. Hua H et.al [8] analyzed the effect of tube diameter on pressure drop characteristics of refrigerant–oil mixture flow boiling in metal-foam filled tubes. Experiments on metal-foam filled tubes with an inner diameter of 7.9 mm were performed. The research results show showed that, the pressure drop increases with increasing PPI, and the impact of PPI becomes more insignificant as the tube diameter decreases. A new pressure drop correlation was developed, and it agrees well with the experimental data for different diameter tubes.

The aluminum foam is used for fin in air conditioner parallel-flow heat transfer. Two condensers with different porosity were fabricated. The heat transfer performance was investigated in different air velocity and refrigerant flow.

2. Test Sample

The model of parallel-flow condenser is shown in Figure 1, it consists of header, louver fin and multi-channel flat tubes. The refrigerant flows from header to the flat tubes while the air flow is passing the louver fin. The refrigerant flow direction is perpendicular to the direction of air flow. The overall size of the condensers is 608mm*390mm*16mm. The specific parameters are as shown in Table 1.

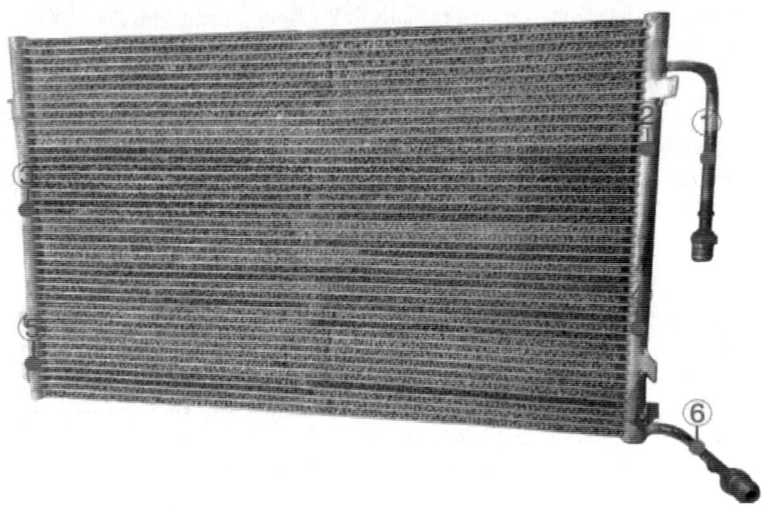

Figure 1. Structure of the aluminium foam parallel-flow condenser

Table 1 Structure parameter of the aluminium foam parallel-flow condenser

Parameter	Value	Parameter	Value
Micro-channel width CHw (mm)	0.94	Flat tube height Th (mm)	2.0
Micro-channel height CHh (mm)	0.6	Flat tube Width Tw (mm)	16
Micro-channel number	10		

The tube number from first pass to forth pass is t 14、12、8 and 6. The specific parameters are shown in Table 2.

Table 2 Structure parameter of the aluminium foam parallel-flow condenser

Parameter	Value	Parameter	Value
Pores Per Inch	10PPI	Pores Per Inch	8PPI
Prosity	0.92	Prosity	0.935
Foam height	8	Foam height	8
Foam width	16	Foam width	16

3. Experiment apparatus

The experiment is conducted in the high accurate laboratory, and the schematic is shown in Figure 2. The fan and compressor are controlled by PLC, data collector is DA100 produced by Heng He corporation, the test error of temperature sensor is less than ±0.1℃.The temperature sensors are set in the inlet and outlet of the condenser, the enthalpy difference is obtained, the heat transfer is calculated.

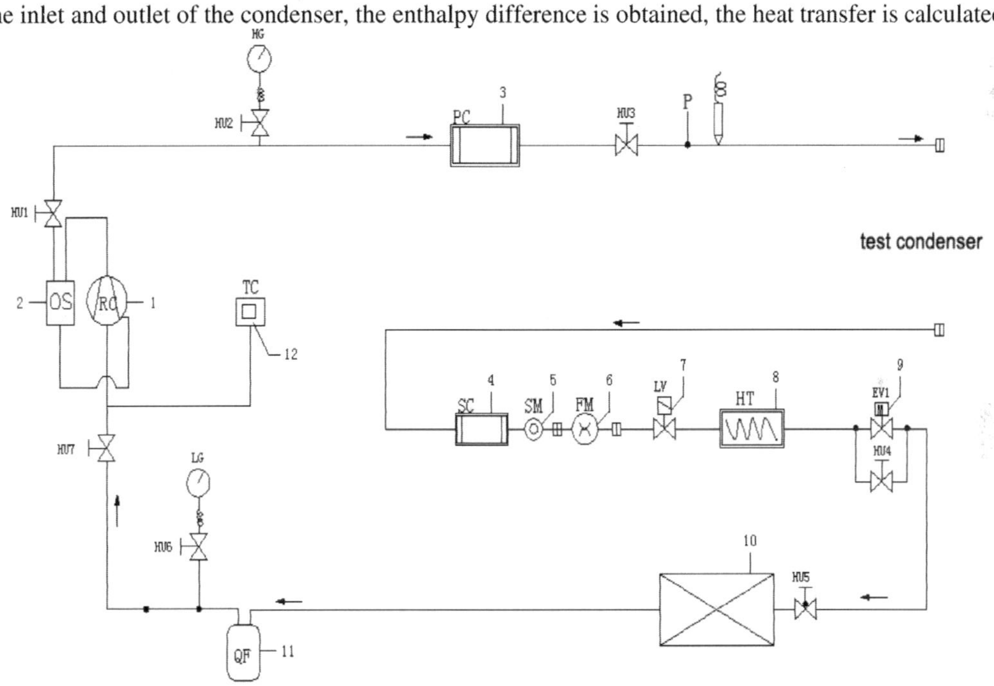

Figure 2. Schematic view of the experimental system

1-compressor,2-oil separator, 3-precooler,4-subcooler,5-sight mirror,6-flow meter,7-solenoid valve,8-heater,9-Electronic expansion valve,10-auxillary evaporator,11-gas separator,12-temperature controller

4. Experiment result discussions

4.1. Effect of face air velocity on heat transfer performance

Two condensers with different porosity are used to conduct experiment. Experimental condition: the temperature of refrigerant at condenser inlet is 65℃, the condensate temperature is 50℃, the air dry and wet temperature at inlet is 35℃, 24℃ respectively. The air velocity is 1.0m/s, 1.5m/s, 2.0m/s, 2.5m/s respectively.

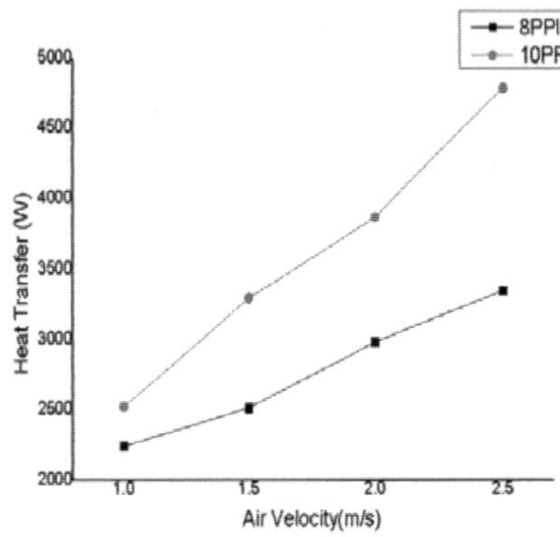

Figure 3. Heat transfer capacity against the air velocity

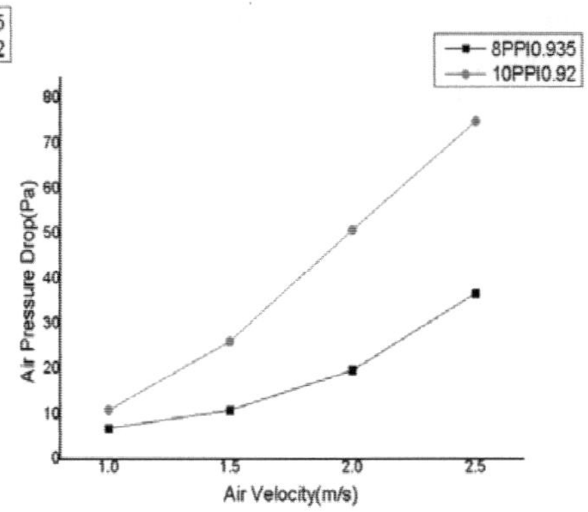

Figure 4. Pressure drop against the air velocity

The experimental results at different air velocity are shown in Figure 3 and Figure 4.The heat transfer capacity increases with the increase of air velocity. When air velocity is 2.5m/s, the heat transfer quality of 10PPI is 4.786kW while the 8PPI is 3.344kW， the difference is 43.1%. Because of the pore diameter is small, air in the aluminum foam with 10PPI could form turbulent three dimension flow, thus the heat transfer capacity increases rapidly with increase of the air velocity. When the pore diameter is bigger, the capacity of heat transmission is not enough. Thus the heat transfer capacity increases. As is shown in Figure 4, the pressure drop increases rapidly. Thus the air velocity should not be too high to avoid the increase of the energy consumption.

4.2. Effect of flow rate of refrigerant on heat transfer performance

Two condensers with different porosity are used to conduct experiment. Experiment condition: the temperature of refrigerant at condenser inlet is 65℃, the condensing temperature is 50℃, the inlet air dry and wet temperature is 35℃, 24℃ respectively. The flow rates of refrigerant are 70Kg/h, 80Kg/h, 90Kg/h, and 100Kg/h, 110Kg /h， 120Kg /h, respectively.

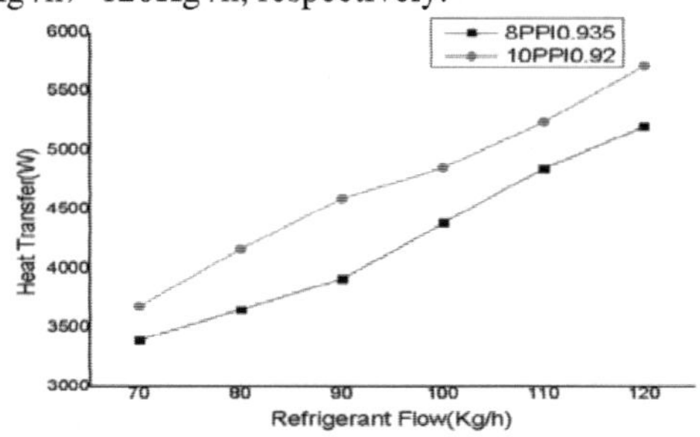

Figure 5. Heat transfer capacity against the refrigerant flow

The experimental results of six kinds of flow rate of refrigerant are shown in Figure 5.The heat transfer capacity increases with the raise of refrigerant flow rate. The increasing trend is almost the same, when flow rate of refrigerant is 120Kg/h, the heat transfer capacity of 10PPI is 5.714kW while the 8PPI is 5.203kW, and the former is 9.8% higher than the latter.

4.3. Temperature distribution in different face air velocity

The heat exchanger has 4 pass, the thermocouples are set at condenser inlet, first pass inlet, second pass inlet, third pass inlet, forth pass inlet and the condenser outlet, as is shown in Figure 1, the condenser temperature distributions at different air velocity are experimental investigated.

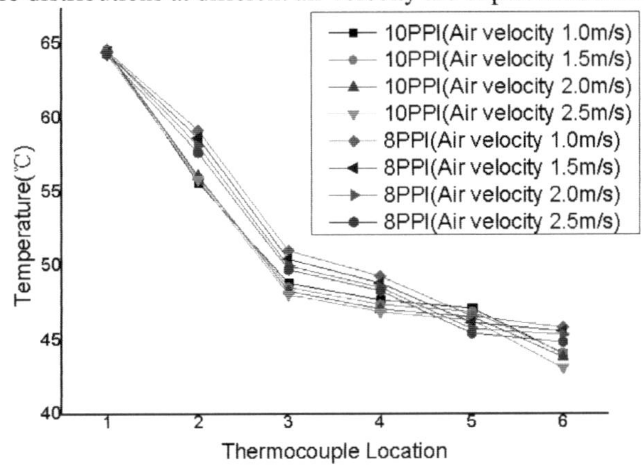

Figure 6. Temperature distribution of the aluminium foam parallel-flow condenser

The experimental results of temperature distribution are shown in Figure 6.The temperature decreases with the increase of air velocity in refrigerant flow direction. The temperature between the first pass inlet and the second pass inlet decreases the fastest. It is found the capacity of heat transmission in the first pass is larger than other three pass. The outlet temperature of 10PPI condenser is lower than that of 8PPI, that is to say heat transfer of 10PPI condenser is better than that of 8PPI.

The temperature reaches 50 ℃ in the third pass, it shows the refrigerant begin to subcooling in the third pass. When the air velocity is 2.5m/s, the dropping speed of temperature in 10PPI condenser toward condensing temperature is the fastest while the 8PPI one is the slowest.

5. Conclusions

Open cell aluminum foam is used in parallel-flow condenser in air conditioner, two condensers with different pore density are fabricated, and the experimental study is conducted on the heat transfer performance and temperature distribution. The experimental results show that both of the heat transfer capacity and air pressure drop increase with the increase of pore density. At the air velocity is 2.5m/s, the heat transfer capacities of the condenser with 10PPI and 8PPI are 4.786kw and 3.344kW respectively. The heat transfer capacity increases with the raise of refrigerant flow rate. When flow rate of refrigerant is 120Kg/h, the heat transfer quality of 10PPI is 5.714kW while the 8PPI is 5.203kW, and the former is 9.8% higher than the latter. In the flow direction of the refrigerant, the outlet temperature of refrigerant decreases with the raise of air velocity when the inlet temperature is constant. The outlet temperature of the refrigerant decreases with the increase of pore density.

Acknowledgements

The current work is supported by the National Natural Science Foundation of China (Nos. 51376058 and 51676067), Guangdong Province Strategic emerging industry government-corporation-bank combination Fund (2012720090) and Scientific Research Fund of Hunan Provincial Education Department (15C0616).

Refrences

[1] Aderhold J, Davydov V Yu, Fedler F, Klausing H, Mistele D, Rotter T, Semchinova O, Stemmer J and Graul J 2001 *J. Cryst. Growth* **222** 701
[2] Tadrist L,et al 2004 About the use of fibrous materials in compact heat exchangers, *Exp Therm*

IOP Conf. Series: Materials Science and Engineering **167** (2017) 012009 doi:10.1088/1757-899X/167/1/012009

Fluid Sci, **28**193–99

[3] Noh J S, LeeK B and LeeC G 2006 Pressure loss and forced convectiveheat transfer in an annulus filled with aluminum foam, *Int Commun Heat Mass* **33** 434–44

[4] T'JoenC,JaegerPD,Huisseune H, Herzeele S V, VorstcN and PaepeMD 2010 Thermo-hydraulic study of a single row heat exchanger consisting of metal foam covered round tubes*Int. J. Heat Mass Tran* **54** 3262-74

[5] Han X, et al 2012 Air-side performance of open-cell metal foams for use in dehumidifying heat exchangers*Int. Refrigeration and Air Conditioning Conf. at Purdue*, **7** 16-19

[6] RibeiroGB, Barbosa JRand PrataAT 2012 Performance of microchannel condensers with metal foams on the air-side: Application in small-scale refrigeration systems*Appl Therm Eng* **36**152–60

[7] RanutaP, Nobile E and Mancini L 2014 High resolution microtomography-based CFD simulation of flow and heat transfer in aluminum metalfoams*Appl Therm Eng* **69** 230–40

[8] HuH and Zhu Y 2014 Effect of tube diameter on pressure drop characteristics of refrigerant–oil mixture flow boiling inside metal-foam filled tube*Appl Therm Eng* **62** 433–51

IOP Conf. Series: Materials Science and Engineering **167** (2017) 012010 doi:10.1088/1757-899X/167/1/012010

Understanding electronic and optical properties of strontium titanate at both PBE and HSE06 levels

Jianhui Yang, Qiang Fan [1]

School of Physics and Electronic Engineering, Leshan Normal University, Leshan, 614004 P. R. China

E-mail: yjh20021220@foxmail.com

Abstract. The structural parameters, electronic structure and optical properties of strontium titanate have been investigated by the first-principles. Exchange–correlation effects are treated by the generalized gradient approximation(GGA) employing Perdew–Burke–Ernzerhof (PBE) functional and hybrid density functional Heyd-Scuseria-Ernzerhof(HSE06). The direct-band gaps are equal to 2.06 and 3.73 eV under PBE and HSE06 level, respectively. The indirect-band gaps are equal to 1.67 and 3.33 eV under PBE and HSE06 level, respectively. The optical properties including complex dielectric function, absorption coefficient, refractivity and reflectivity index have been calculated. Meanwhile, the origin of the spectral peaks on the basis of the electronic band structures has been interpreted.

1. Introduction

Strontium titanate ($SrTiO_3$ or STO for short) has cubic structure and space group $Pm\bar{3}m$ stable from 23 to 1100°C with wide band gap (~3.3 eV) and large dielectric constant (300) [1-2]. As a chemistry and thermally stable ceramic oxide, which is widely studied due to the important applications in ferroelectricity[3], thermoelectricity[4-9], photoelectricity[2], superconductivity[10], semiconductivity [11-12], photoconductivity [13-14], microelectronics [15] and so on. Recently, cubic STO are widely studied as a promising optical material, such as photoluminescence emission and rear-earth doped cubic STO phosphors, which significantly attracted material scientists due to the enhanced luminescence properties [16-19]. However, so far, most of the works have been addressed from experiment; we undertake a theoretical study to learn more about it in this paper. We investigate the electronic and optical properties of the cubic structures of the perovskite STO. First, we calculate the electronic structures, because the optical properties depend on the electronic transitions between occupied bands and empty bands. Second, we calculate the optical properties, including complex dielectric function, absorption coefficient, refractive index and reflectivity.

The reliable optical property is decided by accurate electronic structure. Band gap is very challenging for first principle theory, since most methods, including local density approximation (LDA) and generalized gradient approximation (GGA), the common exchange–correlation functional in the density functional theory (DFT), sometimes underestimate band gap by as large as 3 eV compared with the experimental results [20-22]. Hybrid functional, which mixes Hartree-Fock exchange functional with LDA or GGA exchange-correlation functional, can provide decent prediction for band gap [23-24]. In this paper, we are presenting a detailed investigate for STO on the electronic structure and optical properties using the Heyd-Scuseria-Ernzerhof (HSE06) hybrid functional based on DFT rather than the scissors operator for standard DFT functional, to obtain accurate and reliable optical properties for STO system. To compare the effect of electronic and optical properties by hybrid function, the GGA of DFT employing Perdew–Burke–Ernzerhof (PBE) functional form has been chose as well.

2. Method of calculation

All calculations in our work were performed based on the DFT within the projector augmented wave (PAW) method as implemented in the Vienna Ab-initio Simulation Package (VASP) code [25-26]. The structure optimization was performed using the GGA of DFT employing PBE functional form [27]. The electronic structures and optical properties were calculated on the optimized structures using the hybrid density functional HSE06, which replaces the slowly decaying long-ranged part of the Fock exchange, by the corresponding density functional counterpart. The resulting expression for the exchange-correlation energy is given by:

$$E_{XC}^{HSE} = aE_X^{HF,SR}(\omega) + (1-a)E_X^{PBE,SR}(\omega) + E_X^{PBE,LR}(\omega) + E_C^{PBE} \quad (1)$$

where, the screening parameter ω defines the range-separation of the electron–electron interaction into a short-(SR) and long- ranged (LR) part, is assigned a value of 0.207$Å^{-1}$ [28]. The short-range interactions become negligible when the distance exceeds 2/ω. Hartree-Fock mixing constant α is set to 0.28.

The valence electron for the Sr, Ti, O atoms were chosen as $4s^24p^65s^2$, $3p^63d^24s^2$, and $2s^22p^4$, respectively. Monkhorst-Pack grids of 11×11×11 was used for Brillouin zone sampling. The electron wave function was expanded in plane waves to a cutoff energy of 450 eV with total energy converging to lower than 1×10^{-5} eV. The lattice constants and all atomic positions for supercell were fully relaxed until the maximal force on each atom was less than 0.01 eV/Å. The higher cutoff energies and more k-points have been chosen and the results barely changed, which is in accordance with other theoretical calculations [15-29-30].

3. Results and discussion

The optimized lattice constants are a = b = c = 3.945 Å for STO, which are in good agreement with the experimental and theoretical data [29, 31, 32]. The result shows that the calculation parameters are reasonable. As well known, electronic transport and optical properties of semiconductors are dominated by the electronic structures. Thus, it is necessary to understand the nature of the electronic structures before explaining and optimizing the properties of the systems. In the band structure calculation, a path along higher symmetry points Γ(0, 0, 0)-X(0, 0.5, 0)-M(0.5, 0.5, 0)-Γ(0, 0, 0)-R(0.5, 0.5, 0.5)-X(0, 0.5, 0) has been chosen in the first Brillouin zone. The calculated band structure along higher symmetry direction in the Brillouin zone of pure STO using PBE and HSE06 are shown in figure 1.

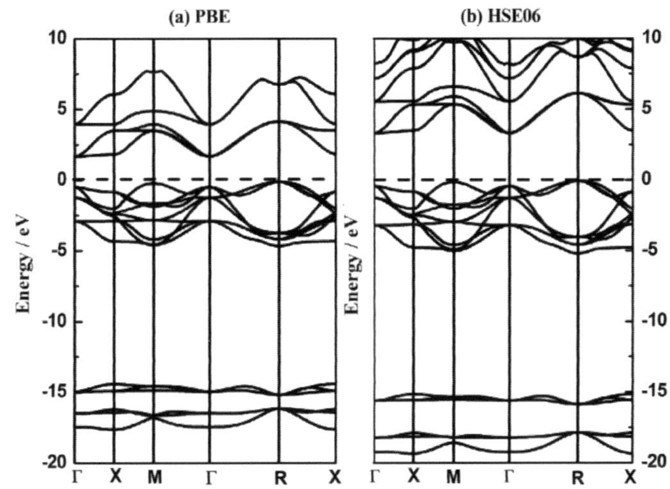

Figure 1. Band structure of pure STO (The Fermi level is set at 0 eV).

As shown in figure.1, for pure STO, the calculated band structure shows that the indirect (Γ→R) band gap energies are 1.67 and 3.33 eV for PBE and HSE06, respectively. The direct (Γ→Γ) band gap energies are 2.06 and 3.73 eV for PBE and HSE06, respectively. Compared with the energy band of PBE, the conduction band minimum (CBM) of HSE06 move into high energy range. Meanwhile both the indirect and direct band gap energies of PBE are smaller than that of HSE06. The results of HSE06 are in excellent agreement with experimental results [1- 2], which means HSE06 hybrid density functional can truly predict the electronic structure for pure STO.

More detail property about the electronic structure can be observed from the partial and total DOS. The partial DOS (PDOS) from PBE and HSE06 functional are shown in figure 2(a) and figure 2(b). In figure 2(c) the total DOS (TDOS) is plotted as well.

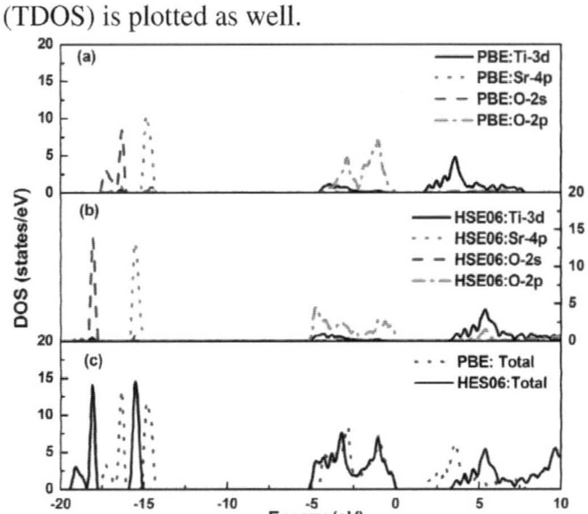

Figure 2. Partial (PDOS), local (LDOS), and total (TDOS) densities of states of cubic STO.

From the TDOS curves figure 2, it is obvious that the valence band maximum (VBM) and conduction band minimum (CBM) are contributed mainly by O-2p states and Ti-3d states, respectively.

Comparing with PBE level results, the conduction band of HSE06 moves into higher energy range and the valence band moves into lower energy range, which makes the energy band of HSE06 larger than that of PBE. Valence band (VB) of HSE06 can be divided into three parts: firstly, from -19.61 to

-18.04 eV is mostly formed by O-2s states; secondly, range from -16.08 to -15.23 eV, resulting from Sr-4p; thirdly, between -5.07 to 0 eV is formed by O-2p states with slight admixture of Ti-3d. As regards for the conduction band (CB), can be assigned to Ti-3d states with some admixture of O-2p states.

It is known that the optical properties can be calculated by the complex dielectric function $\varepsilon(\omega)=\varepsilon_1(\omega)+i\varepsilon_2(\omega)$, where $\varepsilon_1(\omega)$ and $\varepsilon_2(\omega)$ are the real and imaginary parts of the dielectric function, respectively. The imaginary part $\varepsilon_2(\omega)$ is a critical characteristic of the optical properties for all the optical materials which is related to energy loss of electric field in solid. In general, the electronic structure is directly associated with imaginary part $\varepsilon_2(\omega)$ and it can be calculated by adding all possible occupied states to the unoccupied states of transitions. The imaginary part of the dielectric function $\varepsilon_2(\omega)$ of the pure STO at both PBE and HSE06 levels, are shown in figure 3 with energy up to 50 eV.

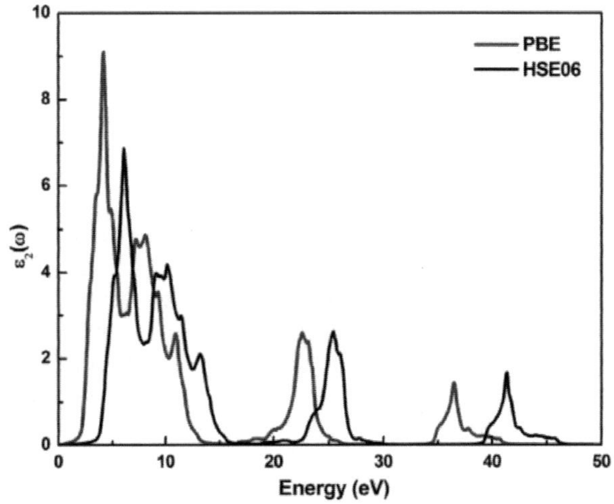

Figure 3. The imaginary part of dielectric function for cubic STO.

It can be seen clearly from figure 3, For pure STO, there are three main peaks. The energy threshold for PBE level is 2.09eV, according with the direct band gap energy. The first peak of PBE functional is at 4.03 eV, which mainly be caused by optical transitions between O-2p states in the highest valence band and Ti-3d states in the lowest conduction band. The second peak at 22.50 eV corresponds to transfers from the Sr-4p valence bands to the Ti-3d conduction bands. The peak at 36.50 eV corresponds to the deep electron excitations from Sr-4p, Ti-3d to the conduction bands.

Compared with the imaginary part of dielectric function of PBE, the results of HSE06 have same tendency, also have three mainly peaks. The most remarkable difference for imaginary part of dielectric function of PBE and HSE06 is the curve moving into higher energy and the energy threshold for HSE06 level is 3.79eV. There are three main peaks at 6.00, 25.41 and 41.45 eV which have a good agreement with the previous report [33] due to the conduction band of Ti-3d moving into higher energy.

The real part $\varepsilon_1(\omega)$ can be derived from the imaginary part using the Kramers–Kronig transformation [34] and the optical properties such as refractivity index ($n(\omega)$), reflectivity($R(\omega)$) index, and absorption coefficient $\alpha(\omega)$ can be derived from $\varepsilon_1(\omega)$ and $\varepsilon_2(\omega)$. In this paper, the absorption coefficient $\alpha(\omega)$ calculated using the relation as follows.

$$\alpha(\omega)=\sqrt{2}\omega\left[\sqrt{\varepsilon_1^2(\omega)+\varepsilon_2^2(\omega)}-\varepsilon_1(\omega)\right]^{1/2} \tag{2}$$

$$n(\omega)=\left(\frac{\sqrt{\varepsilon_r^2(\omega)+\varepsilon_i^2(\omega)}+\varepsilon_r(\omega)}{2}\right)^{1/2} \quad (3)$$

$$R=\left|\frac{\sqrt{\varepsilon(\omega)-1}}{\sqrt{\varepsilon(\omega)+1}}\right|^2 \quad (4)$$

For the promising applications in optoelectronic devices, the optical absorption should be taken into account, the absorption coefficients, refractivity, reflectivity index of pure STO under PBE and HSE06 levels are plotted in figure 4.

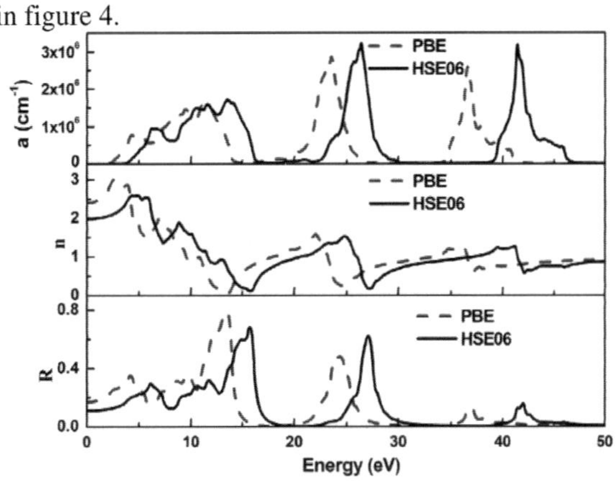

Figure 4. The absorption coefficients refractivity, reflectivity index for pure STO.

As can be seen from figure 4, the curves of absorption coefficients refractivity and reflectivity index are similar between PBE and HSE06 levels. The absorption, refractivity and reflectivity peaks moves into higher energy range as the tendency of dielectric function. The absorption thresholds for PBE and HSE06 levels of pure STO are at 2.09 and 3.79eV, which is determined by the transition between VBM and CBM as same with dielectric function. The absorption coefficient for pure STO is low in the visible region.

4. Conclusion

In this work, the electronic structure and optical properties of pure STO have been systematically investigated using first-principles calculations based on DFT. The calculated band structure shows that the indirect (Γ→R) and direct (Γ→Γ) band gap energies are 3.33 eV and 3.73 eV, respectively for Heyd-Scuseria-Ernzerhof (HSE06) hybrid functional, which are in excellent agreement with experimental results. The results show that HSE06 corrects the shortcomings of standard Perdew–Burke–Ernzerhof (PBE) approximation for band structure and truly predict the electronic structure for pure STO. We conclude that the accurate and systematic description of the electronic, optical and mechanical properties of such type of materials can be achieved with the use of HSE06. We hope that our work will provide a comprehensive understanding of perovskite STO, and provide assistance and guide to design and development of such type of materials, which are very important for the industrial and scientific point of view.

Acknowledgments
We thank the financial supports from the Scientific Research Foundation of the Education Department of Sichuan Province, China (No. 15ZA0273), Science and Technology Bureau of Leshan city, China (No.15GZD111,15GZD113), the Open Research Fund of Computational Physics Key Laboratory of

Sichuan Province, Yibin University(No.JSWL2015KF01) and Leshan Normal University Research Program, China (No. Z1273).

References

[1] van Benthem K, Elsässer C and French R H 2001 *J. Appl. Phys.* **90** 6156
[2] Xu K, Yao M, Chen J, Zou P, Peng Y, Li F and Yao X 2015 *J. Alloy. Compd* **653** 7
[3] Weng S et al. 2014 *Europhysics Letters* **107** 36006
[4] Yaremchenko A A, Populoh S, Patrício S G, Macías J, Thiel P, Fagg D P, Weidenkaff A, Frade J R and Kovalevsky A V 2015 *Chem. Mater.* **27** 4995
[5] Kovalevsky A V, Populoh S, Patrício S G, Thiel P, Ferro M C, Fagg D P, Frade J R and Weidenkaff A 2015 *The Journal of Physical Chemistry C* **119** 4466
[6] Ozdogan K, Upadhyay Kahaly M, Sarath Kumar S R, Alshareef H N and Schwingenschlögl U 2012 *J. Appl. Phys.* **111** 54313
[7] Sahli B, Bouafia H, Abidri B, Abdellaoui A, Hiadsi S, Akriche A, Benkhettou N and Rached D 2015 *J. Alloy. Compd* **635** 163
[8] Kovalevsky A V, Yaremchenko A A, Populoh S, Weidenkaff A and Frade J R 2013 *J. Appl. Phys.* **113** 53704
[9] Zhang R, Hu X, Guo P and Wang C 2012 *Physica B: Condensed Matter* **407** 1114
[10] Olaya D, Pan F, Rogers C T and Price J C 2004 *Appl. Phys. Lett.* **84** 4020
[11] Wang C, Qiu H, Inoue T and Yao Q 2014 *Int. J. Hydrogen Energ.* **39** 12507
[12] Bellingeri E, Pellegrino L, Marré D, Pallecchi I and Siri A S 2003 *J. Appl. Phys.* **94** 5976
[13] Zhang H, Yan L and Habermeier H 2013 *Journal of Physics: Condensed Matter* **25** 35802
[14] Liu P, Nisar J, Pathak B and Ahuja R 2012 *Int. J. Hydrogen Energ.* **37** 11611
[15] Samantaray C B, Sim H and Hwang H 2005 *Microelectron. J.* **36** 725
[16] Modak B and Ghosh S K 2015 *The Journal of Physical Chemistry C* **119** 7215
[17] Potůček Z, Skvortsov A P, Poletaev N K, Bryknar Z, Jastrabik L, Dejneka A and Trepakov V A 2014 *J. Lumin.* **154** 437
[18] Choi M et al. 2014 *J. Appl. Phys.* **116** 43705
[19] Marí B, Singh K C, Cembrero-Coca P, Singh I, Singh D and Chand S 2013 *Displays* **34** 346
[20] Liu Y, Hou Q Y, Xu H P, Li L M and Zhang Y 2012 *Physica B* **407** 2359
[21] Ruan X X, Zhang F C and Zhang W H 2014 *J. Theor. Comput. Chem.* **13** 1450070
[22] Zhang W B et al. 2013 *J. Theor. Comput. Chem.* **12** 1350065
[23] Wan R, Zhang H and Peng J 2011 *Chem. Phys. Lett.* **513** 17
[24] Carvalho A, Alkauskas A, Pasquarello A, Tagantsev A K and Setter N 2009 *Physica B* **404** 4797
[25] Kresse G and Furthmuller J 1996 *Phys. Rev. B Condens Matter* **54** 11169
[26] Kresse G 1999 *Phys. Rev. B* **59** 1758
[27] Perdew J P, Burke K and Ernzerhof M 1997 *Phys. Rev. Lett.* **78** 1396
[28] Krukau A V, Vydrov O A, Izmaylov A F and Scuseria G E 2006 *J. Chem. Phys.* **125** 224106
[29] Wang C, Qiu H, Inoue T and Yao Q 2014 *Solid State Commun.* **181** 5
[30] Li N and Yao K L 2012 *AIP Adv.* **2** 32135
[31] Schmidbauer M, Kwasniewski A and Schwarzkopf J 2012 *Acta Crystallographica Section B* **68** 8
[32] Guo Y, Qiu X, Dong H and Zhou X 2015 *Phys. Chem. Chem. Phys.* **17** 21611
[33] Khaber L, Beniaiche A and Hachemi A 2014 *Solid State Commun.* **189** 32
[34] Gajdoš M, Hummer K, Kresse G, Furthmüller J and Bechstedt F 2006 *Phys. Rev. B* **73** 45112

Investigation of the abnormal Zn diffusion phenomenon in III-V compound semiconductors induced by the surface self-diffusion of matrix atoms

LiangliangTang[1,3], Chang Xu[1]and Zhuming Liu[2]

[1]College of Energy and Electricity, Hohai University, Nanjing 210098, China
[2]Institute of Electrical Engineering, Chinese Academy of Sciences, Beijing 100190, China
Email: tangll@hhu.edu.cn

Abstract. Zn diffusion in III-V compound semiconductorsare commonly processed under group V-atoms rich conditions because thevapor pressure of group V-atoms is relativelyhigh. In this paper, we found that group V-atoms in the diffusion sources would not change the shaped of Zn profiles, while the Zn diffusion would change dramatically undergroup III-atoms rich conditions. The Zn diffusions were investigated in typical III-V semiconductors: GaAs, GaSb and InAs. We found that under group V-atoms rich or pure Zn conditions, the double-hump Zn profiles would be formed in all materials except InAs. While under group III-atoms rich conditions, single-hump Zn profiles would be formed in all materials. Detailed diffusion models were established to explain the Zn diffusionprocess;the surface self-diffusion of matrix atoms is the origin of the abnormal Zn diffusion phenomenon.

1. Introduction

Zn diffusion into III-V compound semiconductor is an important process for fabrication of junction field-effect transistors[1],thermophotovoltaic (TPV)cells[2-5] and infrared detectors[6]. The shape ofZn profiles has a direct impact on device performance. However, under constant Zn pressure, Zn profiles in III-V semiconductorsdon't follow the complementary error function[7]; it is difficult to predict the precise diffusion profiles. During Zndiffusion process, the self-diffusion of matrix atoms occurred at the same time, thus the self-diffusion should be taken into account during Zn diffusion.

Take typicalIII-V semiconductor - GaAs for example, the vapor pressure of As is higher than the Ga, thus the As atomsare commonly added into the diffusion source. Under As-rich conditions, the double-hump Zn profileis obtained in GaAs[1]. In this paper, we found that if Ga atoms were added into the diffusion source, the first hump inZn profile was "removed". Based on the above phenomenon, the investigations were extended to the materials of GaSb and InAs. The surface self-diffusion of matrix atoms should be responsible for the abnormal Zn diffusion phenomenon, details about our experiments and analyses are presented below.

2. Experimental procedure

The n-type GaSb and InAs wafers were selected for the experiments. There are several internal defects in n-GaAs (Si doped)which would have effect on Zn diffusion [10], thus the semi-insulating GaAs with littledefects was selected. The specificparameters of wafers and the diffusion sources are listed in Table 1; the purity of all the diffusion sources is 99.9999%.

Table1. Parameters of wafers and diffusion sources used for experiments.

Wafer	Doping and orientation	Diffusion sources		
		Pure Zn	III-atoms rich	V-atoms rich
GaAs	Semi-insulating <100> -oriented	Zn	Zn-Ga (3wt%Zn)	Zn-As (Zn_3Sb_2)
GaSb	Te-doped (n-type) <100> -oriented	Zn	Zn -Ga (3wt%Zn)	Zn -Sb (Zn_3Sb_2)
InAs	S-doped (n-type) <100> -oriented	Zn	Zn-In (3wt%Zn)	n/a

Zn diffusion experiments were performed in vacuum sealed ampoules. The ampoules were placed in a small-sized OTF-1200X tube furnace. After the diffusion, the ampoules were cooled in N_2. Zn profileswere measured using secondary-ion-mass spectrometry.

3. Experimental results and analysis

3.1. Zn diffusion in GaAs

Figure 1a shows the Zn profiles using Zn, Zn-Ga and Zn-As alloy sources.Clear disparities can be found. Under pure Zn condition, the Zn profile has double humps; while under Ga-rich condition, the Zn profile has only a single hump, the firsthump has been "removed".We found that $D \propto C$(D:Zn diffusion coefficient, C:Zn diffusion coefficient)is fit for the main region of the single-hump profile and the second hump of double-hump profile; while $D \propto C^2$is fit for the first hump of the double-hump profile.

The Zn-As alloyis commonly used for the diffusion source due to the high pressure of As[11], while our experimental results showed that the As-rich condition was similar to that of pure Zn. The As atoms in the diffusion sources can only shorten the first hump of the double-hump profile.

Figure 1b shows the PL peaks at different etching depth. The main region of single-hump profile (A1) showed the similar PL peaks with the second hump of double-hump profile (B2). The PL peaks at 0.75eV demonstrate interstitial Zn toms; the PL peaks at 1.5eV denotein-band transitions[12, 13]. Compared to the above two regions, another PL peak at 1.32eV (B1) exists in the first hump of double-hump profile, which represents the Zn-V_{As} complex.

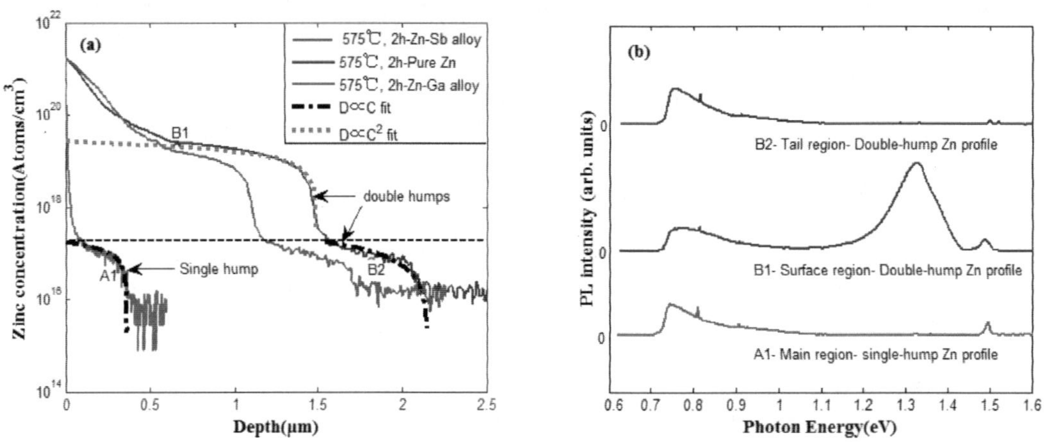

Figure1.(a) Zn profiles in GaAs with different sources and simulations.**(b)**PL spectra at 15K at the point of A1, B1 and B2.

Considering the diffusion and PL results, we may conclude that Zn diffusion mechanism is determined by diffusion sources. Under Ga-rich conditions, Ga vacancies in the surface of wafers were occupied by Ga atoms. Therefore, Zn atoms diffuse by kicking out Ga atoms, as shown in Figure 2.

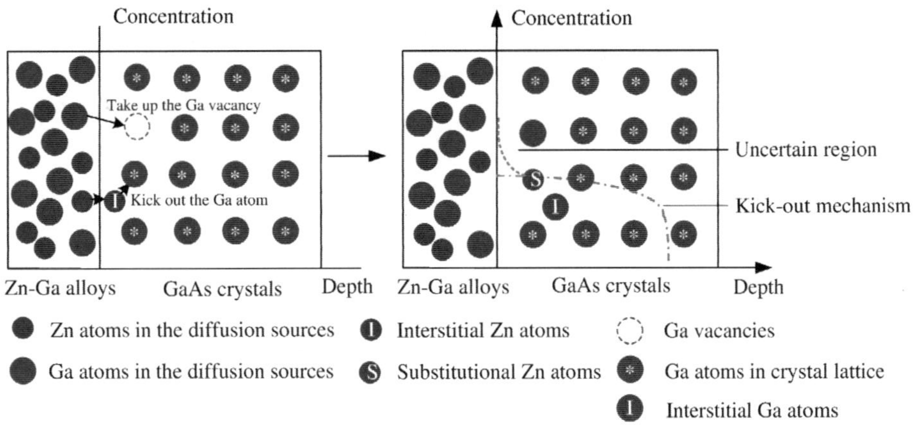

Figure 2. Schematic diagram of the Zn diffusion under Ga-rich conditions

The above diffusion can be described as an interstitial Zn atom (Zn_i^+) kicking out a Ga atom (Ga_{Ga}^0) to the interstitial state (I_{Ga}^+):

$$Zn_i^+ + Ga_{Ga}^0 \rightarrow Zn_{Ga}^-(Zn_s^-) + I_{Ga}^+ + h^+ \quad (1)$$

The function of $D \propto C$ can be got from Equation 1.The simulated $D \propto C$ profile coincides well with the main regions of the single-hump Zn profile, as shown in Figure 1a.

Under pure Zn conditions, Ga vacancies are generatedowing to the self-diffusionfrom the GaAs wafers. Zn atoms would diffuse by taking up the vacancies firstly;therefore the dissociative diffusion would operate in the first hump. This diffusion can be described as an interstitial Zn (Zn_i^+) taking up a Ga vacancy (V_{Ga}^0) and turningto anacceptor (Zn_{Ga}^-):

$$Zn_i^+ + V_{Ga}^0 \rightarrow Zn_{Ga}^-(Zn_s^-) + 2h^+ \quad (2)$$

The function of $D \propto C^2$ can be got from Equation 2, whichcoincides with the first hump of the double-hump profiles.

The Ga vacancies were generated by self-diffusion; its concentration would decrease to aintrinsic equilibrium value at the frontier of diffusion profile. The short ofGa vacancies would prevent further Zn diffusion. Therefore, theZn atoms would be compelled to kick the Ga atoms. The overall diffusionprocess under pure Zn condition is shown in Figure3.

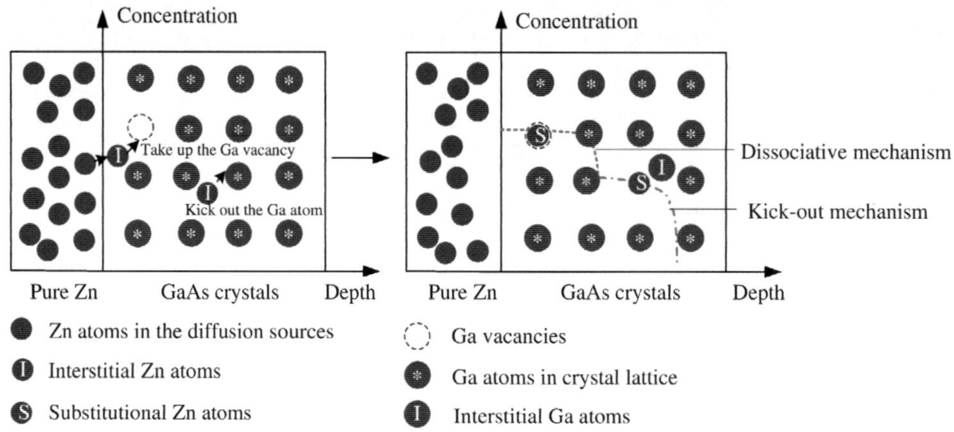

Figure3. Schematic diagram of the Zn diffusion under pure Zn conditions

In summary, the shape of Zn profile inGaAs is depended by the diffusion source. WhenZn-Ga alloy sources are used, the self-diffusion of Ga atoms ishold back, the kick-out mechanism operates owing to short of Ga vacancies, a single-hump profile will be generated. When pure Zn or Zn-As alloy sources are used, Ga vacancies will generateowing to the self-diffusion. Therefore, the dissociative mechanism operates until the concentration of the Ga vacancies reduced to anintrinsicequilibrium value, after which the kick-out mechanism operated. The different relation between D and C for the above two mechanisms causes the distortion of the Zn profile, thus the double-hump profile will be formed.

3.2. Zn diffusion in GaSb

Zn diffusion in GaSb has been investigated intensively due to theimportant application in GaSbTPV cells[14-16]. In our previous study, the similar Zn diffusion phenomenon in GaSb have been found as that in GaAs[17]. Recently, several auxiliary experiments have been done to prove the impact of the Ga-self diffusion from matrix atoms. In the following, we will show three groups of diffusion results processed at 500, 550and 610℃.

Figure 4a shows the Zn profiles after diffusion at 500℃ for 2h using different sources. The Zn-Sb sourcecan only shortens the first hump. While the Zn-Ga alloy sources can "remove" the first hump dramatically;this result will play a direct role for fabricating ofGaSb cells. The first hump should be precisely etched to increase the quantum efficiency of GaSb cells[3, 15], this etching process could be omitted based on the above results. Our PL analyses (Figure 4b) showed that a PL peak located at 0.78eV, which demonstrated the Ga excess existed in the single-hump profile and the second hump of double-hump profile, while this PL peaks didn't exist in the first hump of the double-hump profile. This demonstrates that the Ga atoms have been escaped by self-diffusion under pure Zn or Sb-rich conditions.

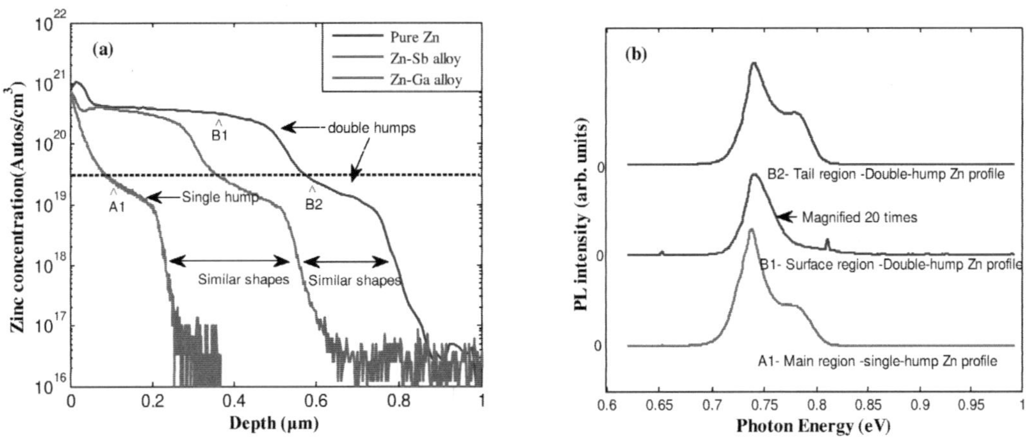

Figure 4.(a) Zn profiles in n-GaSb under different sources at 500℃ for 2h.**(b)** PL spectra at 15K at the point of A1, B1 and B2.

Figure 5a shows the diffusion results at 550℃. The GaSb wafer showsa mirrored surface under Ga-rich conditions, and a single-hump profile isobtained. The GaSbwafer shows the blue surface under the pure Zn or Sb-rich conditions. Figure 6 shows the surface morphology and elemental analysis for the blue surface; there weren't other elements on the surface except the Ga and Sb atoms. This may be a"color center" due to component deviation. This "blue-color center" is similar to a"yellow-color center"in ZnOwhen it is heated in vacuum, which causes the deviation of Zn and O atoms.

Figure 5b shows the diffusion results at 610℃.The wafer after Ga-rich diffusion conditions also showsa mirrored surface and has a single-humpprofile inside. While under pure Zn or Sb-rich conditions, the wafers are dissolved due to the escaping of large amounts of Ga atoms.

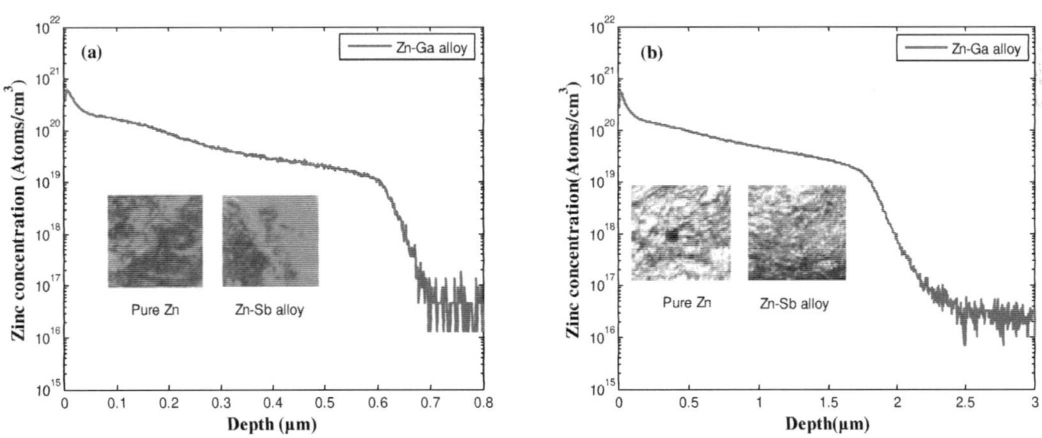

Figure 5. Zn diffusion results in n-GaSb under different sources.**(a)**T= 550℃, 2h; **(b)** T=610℃, 2h.

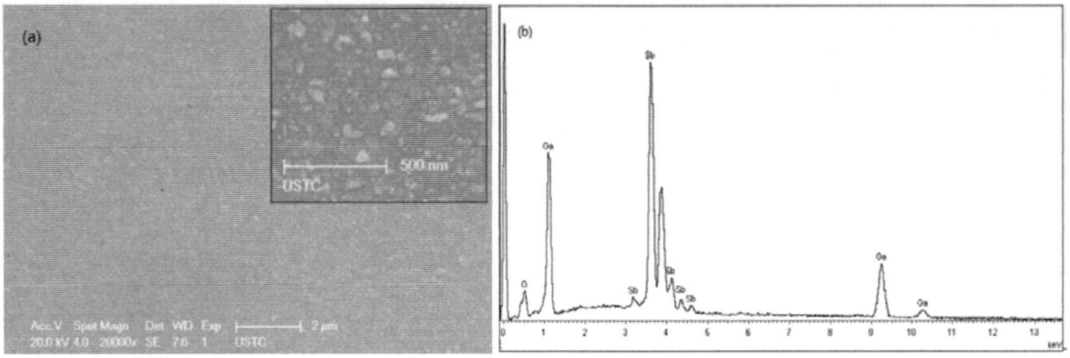

Figure 6. (**a**) Surface morphology of the GaSb wafer with the blue surface.(**b**)Elemental analysis.

3.3. Zn diffusion in InAs

The bandgap of InAs is about 0.35eV, the InAsTPV cells fabricated by the Zn diffusion method absorb the photons in the range of 1~3.5 μm [5, 18], thus they have import applications in low temperature TPV systems.

Figure 7a shows the Zn profiles in n-InAs under Ga-rich conditions. Over 450~500℃, all InAs wafers show mirrored surfaces and have single-hump profiles inside. The simulated $D \propto C$ profiles fit well for the profiles. Under pure Zn conditions, the InAs wafer shows the yellow surface above the temperature of 450℃, which demonstrates that the component deviationoccurs and produces the "color center". The temperature is decreased to 440℃ for comparison, as shown in Figure 7b. Single-hump Zn profiles are obtained under both conditions; this demonstrates that the In atoms can't escape from InAs substrates easily.

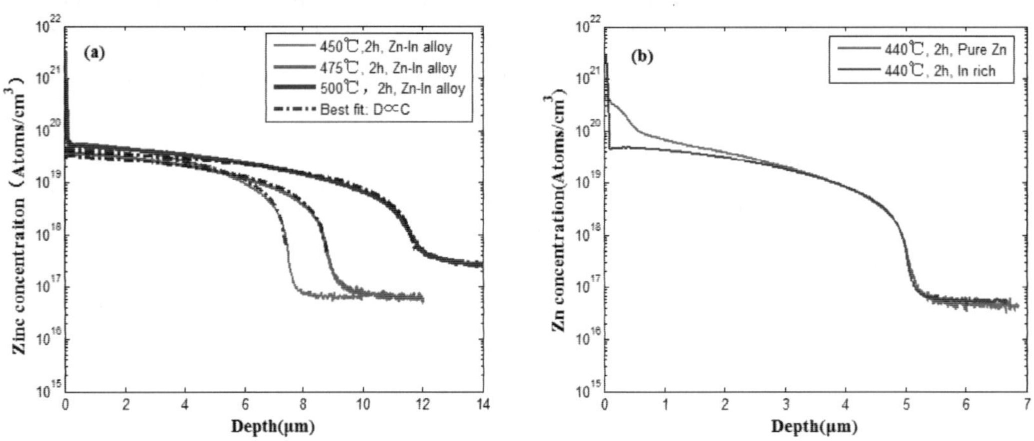

Figure 7. The Zn profiles in InAs under Ga-rich (**a**) and pure Zn conditions(**b**)

4. Conclusions

Different diffusion sources are used to study the diffusion mechanisms of Zn diffusion in III-V compound semiconductors. The III-atoms arecrucial to obtain the single-hump Zn profiles, the diffusion results under V-atoms richconditionsaresimilar to that of pure Zn conditions. Under pure Zn or V-atoms rich conditions, the double-hump profiles will be obtained in GaAs and GaSb. The existing of group III-atoms in diffusion sources would "cut off" the first hump, thus single-hump Zn profileswould be obtained. The In atoms are stable in InAs substrates, thus the single-hump Zn profiles would be obtained under pure Zn or In rich conditions.

Acknowledgment
This work was funded by the Natural Science Foundation of Jiangsu Province (No. BK20150805) and the National Natural Science Foundation of China (No. 51506045).

References

[1] Reynolds S, Vook D W and Gibbons J F 1988 *J. Appl. Phys.***63** 1052
[2] Sulima OV, Bett AW, Mauk MG, Ber BY, Dutta PS 2003 Fifth Conference on Thermophotovoltaic Generation of Electricity p 402
[3] Sulima O V and Bett A W 2001*Sol. Energy Mat. Sol. Cells***66**533
[4] Khvostikov V, Sorokina S, Khvostikova O, Timoshina NK, Potapovich N, Ber BY 2013 Semiconductors**47**307
[5] Khvostikov V, Khostikov O, Oliva E, Rumyantsev V, Shvarts M, Tabarov 2002 Photovoltaic Specialists Conferencep943
[6] Kunitsyna EV, Andreev I, Sherstnev V, L'vova T, Mikhailova M, Yakovlev YP 2010 *Opt. Mater.***32**1573
[7] Tuck B 1988 Atomic Diffusion in *III-V Semiconductors*HilgerBristol
[8] Kahen K B 1989 *Appl. Phys. Lett.***55** 2117
[9] Kahen KB, Spence JP, Rajeswaran G 1991 *Appl. Phys. Lett.***70**2464
[10] Domke C, Ebert P, Heinrich M, Urban K 1996 *Phy. Rev. B.***54**10288
[11] Ky NH, Pavesi L, Araujo D, Ganiere J, Reinhart F 1991*J. Appl. Phys.***69**7585
[12] Hudait M, Modak P, Rao K, Krupanidhi S 1998*Mat. Sci. Eng. B***57**:62
[13] Nathan MI, Burns G, Blum SE, Marinace JC 1963*Phy. Rev***132**:1482.
[14] Sundaram VS, Gruenbaum PE 1993J *J. Appl. Phys.***73**3787
[15] Rajagopalan G, Reddy N S, Ehsani H, Bhat I B, Dutta P S, Gutmann R J, Nichols G and Sulima O V 2003 *J. Electron. Mater.***32**1317
[16] Bett AW, Keser S, Sulima OV 1997 .**18**19
[17] Ye H, Tang L, Li K 2013*Semicond. Sci. Tech***28**015001
[18] Andreev V, Khvostikov V, Khvostikova O, Kaluzhniy N, Oliva E, Rumyantsev V 2003 Proceedings of 3rd World Conference onIEEEp 15

Dual phase $Li_4Ti_5O_{12}$-TiO_2 hierarchical hollow microspheres as anode materials for high rate lithium-ion batteries

Kunxu Zhu and Guoxin Hu[1]

School of Mechanical & Power Engineering, Shanghai Jiao Tong University, Shanghai 200240, P.R. China

E-mail: hugx@sjtu.edu.cn

Abstract.Dual phase $Li_4Ti_5O_{12}$-TiO_2 hierarchical hollow microspheres composed of nanosheets are successfully fabricated by the calcination of hydrothermal product obtained from lithium peroxotitanate complex solution. Low-cost industrial H_2TiO_3 particles are chosen as titanium sources, which is significant for the inexpensive and large-scale production of $Li_4Ti_5O_{12}$-TiO_2 composite material. The $Li_4Ti_5O_{12}$-TiO_2 electrode yields excellent rate capability (151, 139 and 134 mA h g^{-1} at 10, 20 and 25 C, respectively) and good cycling stability (96% capacity retention after 500 cycles at 10 C). The mesoporous hierarchical morphology and high grain boundary density are likely the contributing factors to the excellent electrochemical performance of $Li_4Ti_5O_{12}$-TiO_2 composite.

1. Introduction

Rechargeable lithium ion batteries (LIBs) have been used as the energy sources for most portable electronic devices due to their large energy density and long cycle life, whereas their practical application in electric vehicles (EVs) and stationary energy storage systems highly depends on the electrochemical performance of electrode materials, such as large capacity, high power and energy density, and good cycling stability [1, 2]. Graphite-based materials are currently the dominant anode materials for commercial LIBs, however, poor rate performance resulted from their low Li^+ diffusion coefficient, serious safety issues induced by lithium dendrite formation when the Li-insertion potential approaches almost 0 V vs. Li/Li^+, and poor cycle stability caused by volume change (ca. 9-13%) during lithiation/delithiation processes limits their applications for the next generation high power and energy LIBs [3-5].

Among different mushrooming advanced materials that substitute the traditional carbonaceous anode materials, titanium oxide based materials (e.g., $Li_4Ti_5O_{12}$ and TiO_2) have been proven to be the potential alternative anode materials for LIBs because they are inherently safe and compatible with the electrolyte [6, 7]. Spinel $Li_4Ti_5O_{12}$ (LTO) features a stable operating voltage plateau of about 1.55 V vs. Li/Li^+, which is high enough to avoid most electrolyte reduction and thus enhances the safety performance. Also, as a "zero-strain" material, LTO has excellent structural stability with negligible volume change during Li^+ insertion/extraction process, which results in the good reversibility and cycling stability. So LTO was the first commercialized non-carbonaceous anode material for LIBs in EVs and advanced energy storage devices by companies such as Toshiba [8, 9] and Altairnano [10]. Nevertheless, the drawbacks of LTO should not be ignored, i.e., the intrinsic low electronic conductivity (ca. 10^{-13} S cm^{-1}) and poor lithium ion diffusion coefficient (10^{-9}-10^{-12} cm^2 s^{-1}) restrict the rate capability of LTO.

So far, many efforts have been made to improve the rate performance of LTO, such as surface conductive coating [11-13], doping with heteroatoms [14, 15], and preparing nanostructured LTO materials [16-18]. Among these approaches, preparing LTO nanomaterials is an attractive and efficient way, because reducing LTO particle size to nanometer scale significantly shortens the electrons/lithium ions transport length and increases the contact area between active material and electrolyte. On the other hand, interfacial design has been demonstrated another way to improve the lithium ion transport properties and electronic conductivity of the electrode material [19, 20]. When the spacing of interfaces is reduced down to the nanoscale regime, the grain boundaries would act as channels to transport electrons and lithium ions in the particles. And the grain boundaries with large interfacial areas can store a large number of lithium ions, increasing the specific capacities of the materials [21]. Therefore, LTO materials featuring a high boundary density with large interfacial areas are expected to improve their rate capability. To increase grain boundary density, it is reasonable to choose dual- or multi-phase compounds rather than single phase compounds. This is consistent with several recent reports that dual phase $Li_4Ti_5O_{12}$-TiO_2 materials showed excellent electrochemical performance as anode materials for LIBs [22-24].

Herein, our purpose was to combine the advantages of both nanostructure and dual phase to improve the electrochemical performance of LTO. Instead of using organometallic titanium compounds such as tetrabutyl titanate or titanium isopropoxide as Ti sources, we chose low-cost and abundant industrial H_2TiO_3 particles to supply Ti element, which occupies more than 52.2% mass percentage in $Li_4Ti_5O_{12}$-TiO_2 composite. This is significant for the inexpensive and large-scale production of $Li_4Ti_5O_{12}$-TiO_2 nanomaterials. By dissolving H_2TiO_3 particles in LiOH solution with the help of H_2O_2, peroxotitanate complex solution was obtained. Dual phase $Li_4Ti_5O_{12}$-TiO_2 hierarchical hollow spheres composed of nanosheets were finally prepared by calcining the hydrothermal product from the peroxotitanate solution. The $Li_4Ti_5O_{12}$-TiO_2 composite electrode showed excellent electrochemical performance.

2. Experimental

2.1 Materials synthesis

The $Li_4Ti_5O_{12}$-TiO_2 materials were synthesized by a hydrothermal reaction of peroxotitanate complex precursor solution followed by a calcination treatment. The precursor solution was obtained by the same way reported in our previous work [25, 26]. Typically, 1.0 g of $LiOH\cdot H_2O$ was dissolved in 20 ml deionized water. 1.3 g of commercial H_2TiO_3 powder (ca. 1 μm in diameter) and 8 ml of H_2O_2 (30%) were then added into the solution. After magnetic stirring for 1.5 h at 33 °C, a pale yellow solution was obtained and then transferred into a 50 ml Teflon-lined stainless steel autoclave, which was subsequently heated at 110 °C for 6 h. After cooling down naturally, the white precipitate was filtered and washed by deionized water several times, and then dried at 80 °C for 3 h. Finally, the dried powders were calcined at 550 °C for 4 h to obtain the $Li_4Ti_5O_{12}$-TiO_2 materials.

2.2 Materials characterization

The phase compositions of the samples were identified by X-ray diffraction (XRD, D8 Advance, Bruker) with Cu K_a (λ = 0.15418 nm) radiation. The morphologies were characterized using field emission scanning electron microscope (FE-SEM, JEOL JSM-7800F Prime) and transmission electron microscope (TEM, JEOL JEM-2100, 200Kv). The nitrogen adsorption/desorption isotherms were obtained by a Micromeritics TriStar II 3020 apparatus.

2.3 Electrochemical measurement

The working electrodes were prepared by mixing the active materials, carbon black, and polyvinylidene difluoride (PVDF) at a weight ratio of 80: 10: 10 in *N*-methylpyrrolidone (NMP). The obtained slurry was coated onto a Cu foil and vacuum dried at 80 °C for 12 h, and then cut into disks. The electrochemical measurements were carried out using CR2032 coin-type cells which were assembled in an Ar-filled glove box (H_2O, O_2< 1 ppm), with lithium metal as the counter electrode, Celgard 2400 as the separator, and 1M $LiPF_6$ in a mixture of ethylene carbonate (EC)/dimethyl carbonate(DMC) (1/1 in volume) as the electrolyte. Galvanostatic charge/discharge cycles of the cells were tested between 1.0 and 3.0 V vs. Li/Li^+ on a Land CT2001A battery tester. Cyclic voltammetry (CV) measurement was performed on a CHI 660D electrochemical workstation.

3. Results and Discussion

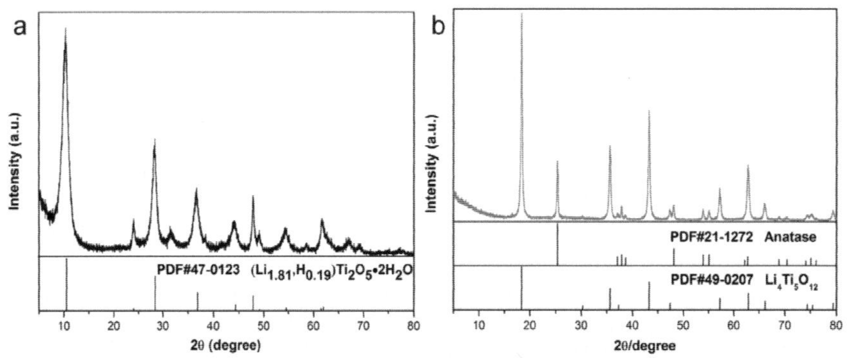

Figure 1. XRD patterns of (a) hydrothermal product obtained from peroxotitanate complex solution and (b) as-prepared $Li_4Ti_5O_{12}$-TiO_2 composite material.

H_2TiO_3 particles with poor crystallinity were easily dissolved in the LiOH solution in the presence of H_2O_2 to form peroxotitanate complex solution. Figure 1a shows the XRD pattern of hydrothermal product obtained from the peroxotitanate precursor solution, which is ascribed to layered lithium titanate $Li_{1.81}H_{0.19}Ti_2O_5{\cdot}2H_2O$ (JCPDS NO. 47-0123) [22]. After calcined at 550 °C in the air, $Li_4Ti_5O_{12}$ with a cubic spinel structure (JCPDS NO. 49-0207) as well as anatase TiO_2 (JCPDS NO. 21-1272) were detected in the product, as presented in figure 1b. The XRD analysis confirms that both spinel $Li_4Ti_5O_{12}$ and anatase TiO_2 exist in the calcined product, forming a dual-phase $Li_4Ti_5O_{12}$-TiO_2 composite material. The formation of anatase TiO_2 may be caused by the crystallization of residual amorphous titania in the hydrothermal product. Moreover, the weight percent of the spinel $Li_4Ti_5O_{12}$ is caculated about 84.9% according to the XRD pattern.

Figure 2. SEM images of (a, b) $Li_{1.81}H_{0.19}Ti_2O_5{\cdot}2H_2O$ hierarchical hollow microspheres and (c, d) $Li_4Ti_5O_{12}$-TiO_2 hierarchical hollow microspheres.

The morphologies of the $Li_{1.81}H_{0.19}Ti_2O_5{\cdot}2H_2O$ and $Li_4Ti_5O_{12}$-TiO_2 products were revealed by SEM and TEM observations. SEM images in figure 2a and b show that the $Li_{1.81}H_{0.19}Ti_2O_5{\cdot}2H_2O$ lithium titanate exhibits a hierarchical sphere-shaped morphology assembled by primary interleaving nanosheets. The diameters of the microspheres are around 1.2 μm and the thickness of nanosheets ranges from 10 to 20 nm. Actually, these spheres are hollow inside, as is shown in figure 3a. The assembly of the nanosheets results in the formation of the large interior void space and the well-defined shell. Although calcined at 550 °C for 4 hours, the $Li_4Ti_5O_{12}$-TiO_2 sample still retains the initial microstructures of the hydrothermal product, as presented in figure 2c. And the inside hollow structure of the $Li_4Ti_5O_{12}$-TiO_2 sample is confirmed by a cross section of a damaged microsphere (figure 2d).

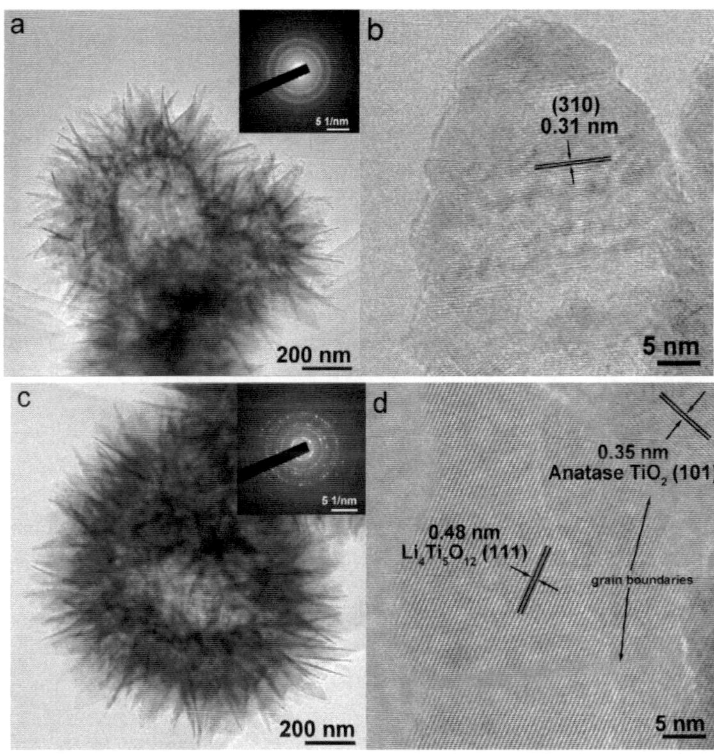

Figure 3. TEM images of (a, b) $Li_{1.81}H_{0.19}Ti_2O_5{\cdot}2H_2O$ hierarchical hollow microspheres and (c, d) $Li_4Ti_5O_{12}$-TiO_2 hierarchical hollow microspheres.

Figure 3 shows further structural features of the hydrothermal and calcined products. A typical single nanosheet of the $Li_{1.81}H_{0.19}Ti_2O_5{\cdot}2H_2O$ microsphere (figure 3a) has a length of approximately 200 nm and width of 70-110 nm. A top view HRTEM image of a nanosheet in figure 3b exhibits a lattice spacing of 0.31 nm, which is assigned to (310) plane of $Li_{1.81}H_{0.19}Ti_2O_5{\cdot}2H_2O$, consistent with the XRD result. For $Li_4Ti_5O_{12}$-TiO_2 material, negligible change to the hierarchical hollow sphere-shaped structure indicates a good structural stability of the spheres (figure 3c). It should be mentioned that the primary nanosheets can be destroyed into particle aggregates when calcined at higher temperature (700 °C). A top view TEM image shown in figure 3d reveals the crystal structure of the nanosheets. Two sets of lattice fringe spacing of 0.48 and 0.35 nm are observed, which are ascribed to the (111) plane of $Li_4Ti_5O_{12}$ and (101) plane of anatase TiO_2, respectively, confirming the coexisting phases of $Li_4Ti_5O_{12}$ and TiO_2 in the sample. Moreover, some grain boundaries are observed among the crystalline domains in the nanosheets, which are considered beneficial for the improvement of lithium ions and electrons transport.

Figure 4 shows the N_2 adsorption/desorption isotherms of the $Li_4Ti_5O_{12}$-TiO_2 hierarchical hollow microspheres. A typical Ⅳ curves with an H1 hysteresis loop is observed for this sample, indicating that the $Li_4Ti_5O_{12}$-TiO_2 material has open mesoporous structures. The inset in figure 4 is the pore size distribution curve calculated by Barrett-Joyner-Halenda (BJH) method. The dual phase $Li_4Ti_5O_{12}$-TiO_2 microspheres exhibit a pore size distribution with an average pore diameter at ca. 20 nm. The $Li_4Ti_5O_{12}$-TiO_2 microspheres have a Brunauer-Emmett-Teller (BET) specific surface area of 51 m^2 g^{-1} and a pore volume of 0.26 m^3 g^{-1}. Hereby, depending on the demonstrated structural features, the $Li_4Ti_5O_{12}$-TiO_2 hierarchical hollow microspheres are expected to make the contact between active material and electrolyte more efficiently and shorten the electrons/lithium ions transport length, thus enhancing the electrochemical performance.

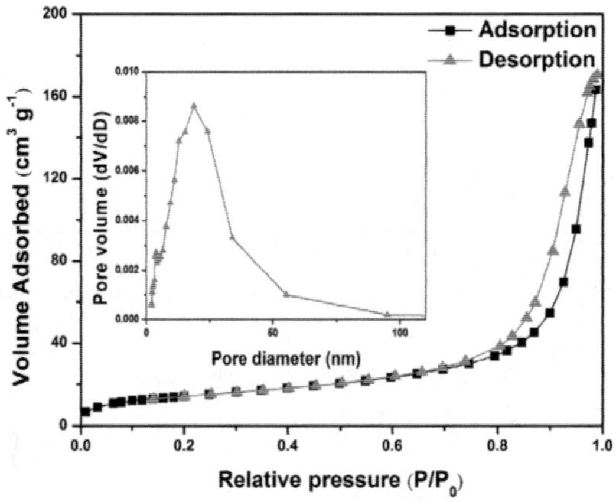

Figure 4. Nitrogen adsorption-desorption isotherms of the $Li_4Ti_5O_{12}$-TiO_2 microspheres. The inset is the pore size distribution calculated by BJH method.

The electrochemical performance of the dual phase $Li_4Ti_5O_{12}$-TiO_2 materials was evaluated by coin-type cells, using metallic lithium foil as both counter and reference electrodes. Figure 5a shows the first two CV cycles conducted between 1.0 and 3.0 V at a scan rate of 0.2 mV s^{-1}. In the first cycle, two pairs of redox peaks located at 1.435/1.697 V and 1.668/2.054 V (vs. Li/Li^+) are associated with Li^+ insertion/extraction within the spinel $Li_4Ti_5O_{12}$ and anatase TiO_2, respectively. In the second cycle, the curve shape and redox peak potentials are substantially unchanged, and the redox peaks have good symmetry, suggesting excellent reversibility and high coulombic efficiency of the $Li_4Ti_5O_{12}$-TiO_2 electrode [14, 17, 22].

Figure 5b presents the galvanostatic discharge-charge curves of the $Li_4Ti_5O_{12}$-TiO_2 electrode at C-rates of 0.5-25 C (1C = 175 mA g^{-1}). At relatively low rates (0.5, 1, 2 C), besides a voltage platform around 1.55 V (vs. Li/Li^+) related to Li^+ insertion/extraction in spinel $Li_4Ti_5O_{12}$, a voltage plateau at around 1.7 and 2.0 V is observed, which is the characteristic of Li^+ ion insertion/extraction of anatase TiO_2. The $Li_4Ti_5O_{12}$-TiO_2 materials can deliver a discharge capacity of 181, 174, 165, 157, 151, 139, 134 mA h g^{-1} when cycled at rate of 0.5, 1, 2, 5, 10, 20, and 25 C, respectively. According to the rate performance in figure 5c, when the rate is put back to 0.5 C, the discharge capacity of the $Li_4Ti_5O_{12}$-TiO_2 electrode returns back to 180 mA h g^{-1}, which further confirms the excellent reversibility of the $Li_4Ti_5O_{12}$-TiO_2 electrode. The capacities of $Li_4Ti_5O_{12}$-TiO_2 are much higher than the compared commercial $Li_4Ti_5O_{12}$ sample, especially at high rates, which only delivers discharge capacity of 165, 140, 111, 82, 50, 32 and 25 mA h g^{-1} at 0.5, 1, 2, 5, 10, 20 and 25 C, respectively. In addition, the $Li_4Ti_5O_{12}$-TiO_2 electrode exhibits an outstanding cycling stability over long-term cycles at high rate of 10 C. As shown in figure 5d, the $Li_4Ti_5O_{12}$-TiO_2 electrode gives a discharge capacity of 145 mA h g^{-1} after 500 cycles, which is 96% of the initial 151 mA h g^{-1}. The periodic fluctuation of the charge/discharge capacity is caused by the temperature difference between day and night. And the coulombic efficiency reaches to 100% after the first two cycles, confirming the great cycling stability of the material.

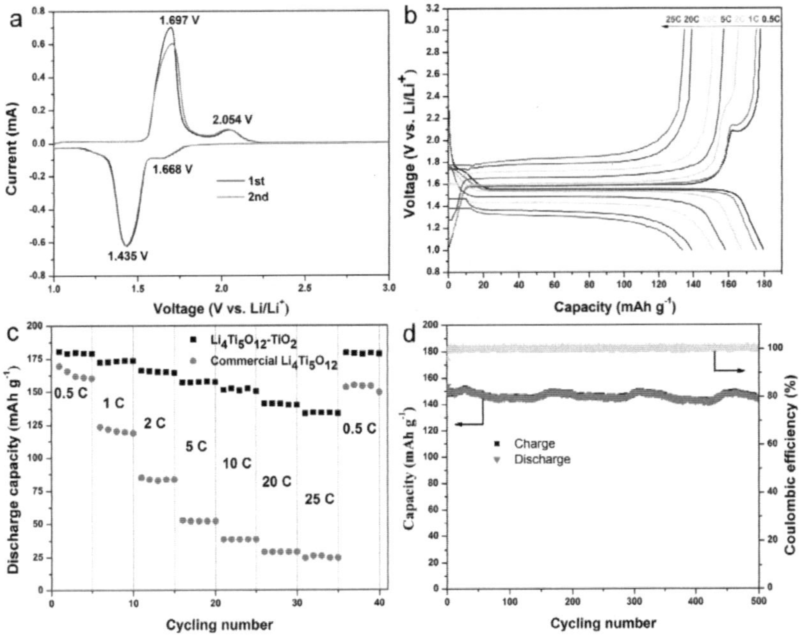

Figure 5. (a) The initial two CV curves of the dual phase $Li_4Ti_5O_{12}$-TiO_2 electrode obtained at a scan rate of 0.2 mV s^{-1}; (b) the galvanostatic discharge-charge profiles of the $Li_4Ti_5O_{12}$-TiO_2 electrode at different C-rates from 0.5 to 25 C; (c) Rate performances of the $Li_4Ti_5O_{12}$-TiO_2 electrode and commercial $Li_4Ti_5O_{12}$ electrode at rates of 0.5-25 C; (d) cycling performance of $Li_4Ti_5O_{12}$-TiO_2 electrode and the corresponding coulombic efficiency at a rate of 10 C; All the electrochemical measurements were conducted within a voltage window of 1.0-3.0 V (vs. Li/Li^+).

4. Conclusion

In conclusion, we have successfully prepared dual phase $Li_4Ti_5O_{12}$-TiO_2 hierarchical hollow microspheres assembled by nanosheets through hydrothermal treatment of peroxotitanate complex solution and a followed calcination. The dual phase composition of the $Li_4Ti_5O_{12}$-TiO_2 material results in the formation of some grain boundaries among the crystalline domains in the nanosheets. Combining with the unique mesoporous hierarchical nanostructure, the $Li_4Ti_5O_{12}$-TiO_2 electrode exhibits excellent rate capability (134 mA h g^{-1} at 25 C) and cycling stability (145 mA h g^{-1} retained after 500 cycles at 10 C). Furthermore, instead of organometallic titanium compounds, industrial low-cost H_2TiO_3 particles were chosen as the Ti sources, which is significant for the inexpensive and large-scale production of the $Li_4Ti_5O_{12}$-TiO_2 materials.

Acknowledgement

This work was supported by the National Natural Science Foundation of China (No. 51476100, 51676128 and 51506124). The authors thank the Instrumental Analysis Center of Shanghai Jiao Tong University (SJTU) for SEM measurements.

References

[1] Tang Y, Zhang Y, Li W, Ma B and Chen X 2015 *Chem. Soc. Rev.* **44** 5926-40
[2] Choi J W and Aurbach D 2016 *Nat. Rev. Mater.* **1** 16013
[3] Shen L, Uchaker E, Zhang X and Cao G 2012 *Adv. Mater.* **24** 6502-6
[4] Xu H, Hu X, Sun Y, Luo W, Chen C, Liu Y and Huang Y 2014 *Nano Energy* **10** 163-71
[5] Shi Y, Gao J, Abruña H D, Liu H, Li H, Wang J and Wu Y 2014 *Nano Energy* **8** 297-304
[6] Zhu G N, Wang Y G and Xia Y Y 2012 *Energy Environ. Sci.* **5** 6652-67
[7] Yi T F, Yang S Y and Xie Y 2015 *J. Mater. Chem.* **3** 5750-77
[8] Takami N, Kosugi S and Honda K 2008 *Toshiba Rev.* **63** 54-7
[9] Toshiba, Toshiba's SCiB™ Rechargeable Battery, *http://www.scib.jp/en/index.htm.*
[10] Manev V, Spitler T, Stewart M and Shelburne J 2010 Negative electrode for lithium ion battery. In: Altairnano, editor.: *Google Patents.*
[11] Li B, Han C, He Y B, Yang C, Du H, Yang Q H and Kang F 2012 *Energy Environ. Sci.* **5** 9595-602
[12] Ma Y, Ding B, Ji G and Lee J Y 2013 *ACS Nano* **7** 10870-8
[13] Sha Y, Xu X, Li L, Cai R and Shao Z 2016 J. *Power Sources* **314** 18-27
[14] Chen C, Huang Y, An C, Zhang H, Wang Y, Jiao L and Yuan H 2015 *ChemSusChem* **8** 114-22
[15] Kim J G, Shi D, Park M S, Jeong G, Heo Y U, Seo M, Kim Y J, Kim J H and Dou S X 2013 *Nano Res.* **6** 365-72
[16] Chen J, Yang L, Fang S and Tang Y 2010 *Electrochim. Acta* **55** 6596-600
[17] Cheng J, Che R, Liang C, Liu J, Wang M and Xu J 2014 *Nano Res.* **7** 1043-53
[18] Haridas A K, Sharma C S and Rao T N 2015 *Small* **11** 290-4
[19] Liao J-Y, Chabot V, Gu M, Wang C, Xiao X and Chen Z 2014 *Nano Energy* **9** 383-91
[20] Rahman M M, Wang J-Z, Hassan M F, Wexler D and Liu H K 2011 *Adv. Energy Mater.* **1** 212-20
[21] Jiang Y M, Wang K X, Zhang H J, Wang J F and Chen J S 2013 *Sci. Rep.* **3** 3490-4.
[22] Jiang Y M, Wang K X, Wu X Y, Zhang H J, Bartlett B M and Chen J S 2014 *ACS Appl. Mater. Interfaces* **6** 19791-6
[23] Yi T F, Fang Z K, Xie Y, Zhu Y R and Yang S Y 2014 *ACS Appl. Mater. Interfaces* **6** 20205-13
[24] Xu C, Xue L, Zhang W, Fan X, Yan Y, Li Q, Huang Y and Zhang W 2014 *Electrochim. Acta* **147** 506-12
[25] Zhu K X and Hu G X 2014 *J. Supercrit. Fluids* **94** 165-73
[26] Zhu K X, Gao H Y, Hu G X, Liu M J and Wang H C 2017 *J. Power Sources* **340** 263-72

IOP Conf. Series: Materials Science and Engineering **167** (2017) 012013 doi:10.1088/1757-899X/167/1/012013

Quantum chemical calculations of anion complex $[B_{12}H_x(NF_2)_{12-x}]^{2-}$, x = 9 – 12

E A Koblova[1,3], **V I Saldin**[1], **A Yu Ustinov**[1,2]

[1] Institute of Chemistry, Russian Academy of Sciences, Far Eastern Branch, 159 Prosp. 100-letiya Vladivostoka, Vladivostok, 690022, Russian Federation
[2] Far Eastern Federal University, 8 Sukhanova St., Vladivostok, 690950, Russian Federation

E-mail: les@ich.dvo.ru

Abstract. The geometric, energetic, spectral and electronic properties of various isomers of $B_{12}H_x(NF_2)_{12-x}^{2-}$ anion complex with x = 9 – 12 have been studied using Density Functional Theory (B3LYP/6-311++G**). It was shown that the most stable isomers are characterized by the preference to form the most symmetric structures with uniformly distributed charge densities. However, when replacing a hydrogen atom with difluoramino group, an inductive effect occurs. NF_2 group pulls a part of electron density that leads to the polarization of the boron core. Blue shifts in the IR spectrum compared to the vibrations of the free radical NF_2 ranging from 5 to 69 cm^{-1} for the most stable isomers with the minimum total energy are characteristic and points to the stability of $B_{12}H_x(NF_2)_{12-x}^{2-}$ anions. The obtained results broaden the idea of aromaticity of $B_{12}H_{12}^{2-}$ anion and will be useful in synthesis of new $B_{12}H_{12}^{2-}$ derivatives.

1. Introduction

There are currently synthesized and investigated hundreds of compounds of dodecahydro-*closo*-dodecaborate anions ($B_{12}H_{12}^{2-}$, figure 1) with simple (almost all metals of the Periodic Table), onium (alkyl-ammonium, -sulfonium, -phosphonium), and complex (transition metals with various organic ligands) cations [1-4]. The elemental composition of $B_{12}H_{12}^{2-}$ holds some promise for the preparation of compounds that can be used as energy-intensive components of energy-saturated materials for various purposes. However, because of the superior stability of the $B_{12}H_{12}^{2-}$ anion, the vast majority of its compounds are difficult to ignite and burn poorly because of the formation of boron-oxycompounds on the surface of the burning particle melt. This fact does not allow the realization of the high energy intensity of these compounds. One of the ways to overcome this drawback is to incorporate the anions in mixed formulations with ultrafine polytetrafluoroethylene (UPTFE) [5]. Replacing the oxygen-containing oxidant (fully or partially) to UPTFE improves combustion of the borohydride fragment because the burning surface is constantly renewed by release of volatile boron trifluoride BF_3 or boron oxofluoride $(BOF)_3$. In pure oxygen systems, boron release in the form of a boron oxide melt creates a protective film on the burning surface.

One of the most important features of the $B_{12}H_{12}^{2-}$ anion is its aromaticity, i.e. the ability to replace the terminal hydrogen atoms with other atoms or groups of atoms without destroying the core. Therefore, another possible method of improving the flammability of the boron anion may be the introduction of fluorinated groups to the exo-environment of $B_{12}H_{12}^{2-}$.

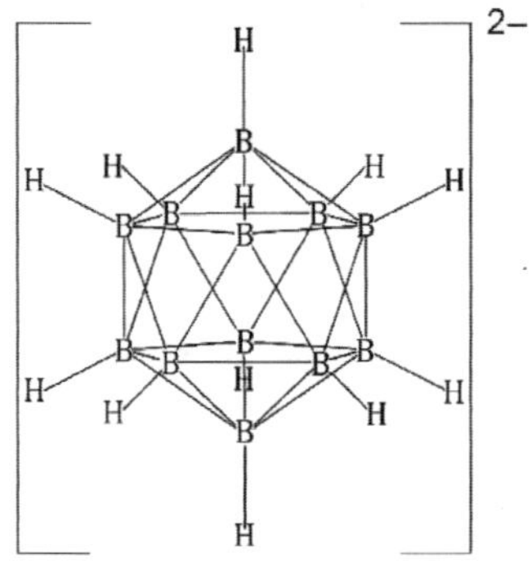

Figure 1. Structure of the $B_{12}H_{12}^{2-}$ anion.

Of particular fundamental and practical interest is the investigation of $B_{12}H_{12}^{2-}$ derivatives – structures with the hydrogen atoms replaced with other atoms or groups [1-9]. Previous quantum chemical calculations showed that the replacement of some hydrogen atoms in $B_{12}H_{12}^{2-}$ anion structure by fluoromethyl groups leads to the existence of stable anions [10]. Known is a number of nitrogen-substituted derivatives of $B_{12}H_{12}^{2-}$, for example, $B_{12}H_{11}NO^{2-}$, $B_{12}H_{11}NH_3^-$, $B_{12}H_{10}(NH_3)_2$, etc. [11]. Of particular theoretical interest is the substitution of oxygen and hydrogen atoms associated with a nitrogen atom to fluorine atoms. Nitrogen-fluorinated derivatives of $B_{12}H_{12}^{2-}$ anion can be practically valuable as a power-consuming fuel, the combustion of which will lead to the release of boron in the form of volatile fluorides. In this case the amount of combustion gas can be increased by 25% in comparison with carbon-fluorinated derivatives since nitrogen will be evolved in a gas form.

The aim of the present work is to characterize the stability and some physico-chemical properties of the anion complex $B_{12}H_x(NF_2)_{12-x}^{2-}$ (x = 9 – 12) with the use of quantum-chemical calculations.

2. Computational Details

All the calculations were carried out within the GAMESS-US package [12] using the hybrid functional B3LYP [13, 14] and the split-valence basis set 6-311++G** with diffuse and polarization functions. The choice of this computational method is based on the available literature data and our personal results. On the figure 2 we illustrated some of the results obtained with the use of various methods and split-valence basis sets. From these data it is well seen that even with the parallel computing, the computation time significantly increases with the addition of diffuse and polarization functions. But as far as we deal with the dodecahydro-*closo*-dodecaborate anion $B_{12}H_{12}^{2-}$ and its derivatives, the addition of diffuse functions is necessary for the correct description of the ionic nature, and the addition of polarization functions is important for the correct modeling of chemical bonds. Also, the calculated frequencies of the free radical NF_2 are in a better agreement with the known experimental value of ~1075 cm^{-1} [15] when using DFT method of calculations. Furthermore, B3LYP/6-311++G** shows a good correlation between accuracy and computational time.

The initial geometry of the $B_{12}H_{12}^{2-}$ structure was formed with I_h symmetry. Further optimization of $B_{12}H_{12}^{2-}$ and $B_{12}H_x(NF_2)_{12-x}^{2-}$ (x = 9 – 12) was executed without symmetry constraints. Anion structures with substituted difluoramino groups $B_{12}H_x(NF_2)_{12-x}^{2-}$, (x = 9 – 12) were formed by replacing of one to three hydrogen atoms with the appropriate number of NF_2 groups. To find the most stable structures, we compared total energies of all possible isomers of the given size. Harmonic vibrational frequency analysis was used to confirm the true local minima of the potential energy surface. If no imaginary frequencies were found, the structure was considered to be a global minimum. All energies were corrected by their zero-point energies (ZPE), and scaled by a factor of 0.9877 [16]. Bond orders between atoms and valence electron configurations were computed using Natural Bond Orbital (NBO) analysis [17].

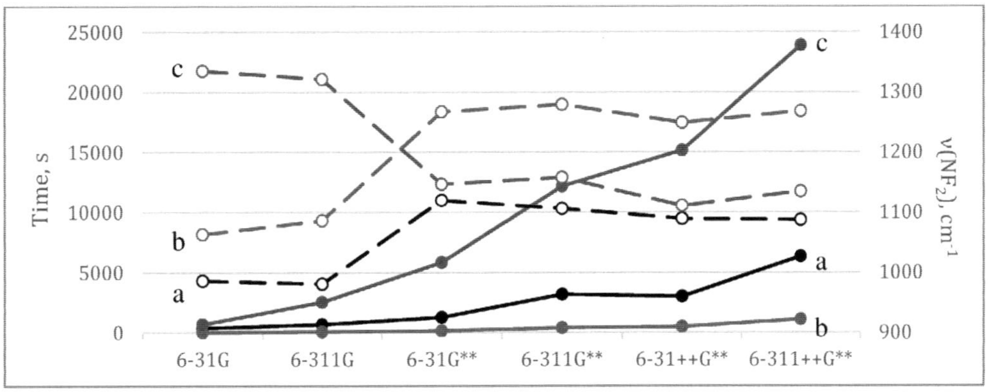

Figure 2. Time (s) needed for the optimization and IR frequency calculations of the $B_{12}H_{12}^{2-}$ anion structure (solid lines with filled circles); values of the stretching vibrations (cm^{-1}) of the free radical NF_2 (dashed lines with opened circles). The results obtained using DFT (a), HF (b) and MP2 (c) methods.

3. Results and Discussion

3.1. X = 12.

Before studying the anion structures with substituted difluoramino groups, it was necessary to analyse the electronic and geometric structure of the initial $B_{12}H_{12}^{2-}$ anion. To be sure that the singlet is the most energetically stable structure, we also computed the $B_{12}H_{12}^{2-}$ cluster in the triplet state. The results of the calculations confirmed the 0.2 eV preference in energy of the singlet over the triplet, so the singlet anion structures will be considered in further discussion.

In the initial $B_{12}H_{12}^{2-}$ cluster (figure 1) boron atoms are bonded to each other with the bond length of 1.787 or 1.789 Å, wherein the B-H bond length is equal to 1.203 Å. The formed B-B-B angles are equal to 60.0° and 108.0°, and B-B-H angles are equal to 121.7°. The total energy of the $B_{12}H_{12}^{2-}$ anion is −305.6596 au. Analysis of the frequencies in the IR-spectrum of the icosahedral $B_{12}H_{12}^{2-}$ anion reveals that the oscillation near 2490 cm^{-1} is characteristic of the nearly isolated B-H stretch. These given geometric and spectroscopic parameters are in a good agreement with the known experimental and theoretical results [3, 4, 6, 8, 11, 18, 19].

The calculated Mulliken charges of the boron and hydrogen atoms are listed in table 1. As seen, in the initial $B_{12}H_{12}^{2-}$ anion the charge is uniformly distributed in the core: boron atoms are positively charged, whereas negative charge is distributed over the hydrogen atoms. According to the NBO analysis, the valence electron configuration of B and H is: $2s^{0.65}2p^{2.50}3p^{0.02}$ and $1s^{0.99}$, respectively.

3.2. X = 11.

The $B_{12}H_{11}NF_2^{2-}$ anion structure, shown in figure 3a have been found by replacing of one hydrogen atom of the optimized $B_{12}H_{12}^{2-}$ structure by difluoramino group. The anion structure with one NF_2 group is seen to have decreased B-B and B-H bond distances near the B atom related to the NF_2 group, when compared to the initial $B_{12}H_{12}^{2-}$ anion and are in the range of 1.776 – 1.788 and 1.194 – 1.196 Å, respectively. The bond distances between boron atoms that are under the influence of fluorine atoms of NF_2 group change most of all. The optimized B-N and N-F bond distances are 1.558 Å and 1.450, 1.451 Å, respectively (table 2).

In any case, when replacing a hydrogen atom with an electron-withdrawing difluoramino group, an inductive effect occurs due to the transfer of electron density from the boron core. As a result, the charge on the boron atom related to the substituted NF_2 group (B7, figure 3a) and on the neighboring to the B7 boron atoms (connected by a common edge, i.e. B3, B4, B8, B11 and B6) increases. At the

IOP Conf. Series: Materials Science and Engineering **167** (2017) 012013 doi:10.1088/1757-899X/167/1/012013

Table 1. Calculated Mulliken charges of the most stable structures of $B_{12}H_x(NF_2)_{12-x}^{2-}$ (x = 9 – 12).

Atom number	Mulliken charge	Atom number	Mulliken charge	Atom number	Mulliken charge	Atom number	Mulliken charge
x = 12		x = 11		x = 10		x = 9	
B1	0.149	B1	0.538	B1	0.383	B1	0.521
B2	0.109	B2	0.060	B2	0.369	B2	0.356
B3	0.118	B3	0.124	B3	-0.083	B3	-0.424
B4	0.132	B4	0.133	B4	0.143	B4	0.514
B5	0.091	B5	-0.098	B5	0.123	B5	0.083
B6	0.086	B6	0.185	B6	-0.014	B6	0.467
B7	0.079	B7	0.231	B7	0.358	B7	0.241
B8	0.105	B8	0.227	B8	0.319	B8	0.224
B9	0.081	B9	-0.077	B9	-0.108	B9	-0.330
B10	0.076	B10	-0.031	B10	0.011	B10	0.038
B11	0.096	B11	0.479	B11	0.102	B11	0.186
B12	0.041	B12	-0.062	B12	0.056	B12	0.262
H13	-0.261	H13	-0.256	H13	-0.236	H13	-0.405
H14	-0.265	H14	-0.440	H14	-0.274	H14	-0.268
H15	-0.265	H15	-0.266	H15	-0.192	H15	-0.109
H16	-0.264	H16	-0.249	H16	-0.237	H16	-0.393
H17	-0.261	H17	-0.282	H17	-0.276	H17	-0.237
H18	-0.266	H18	-0.296	H18	-0.189	H18	-0.064
H19	-0.265	H19	-0.262	H19	-0.258	H19	-0.394
H20	-0.264	H20	-0.258	H20	-0.254	H20	-0.227
H21	-0.267	H21	-0.232	H21	-0.227	H21	-0.258
H22	-0.266	H22	-0.309	H22	-0.226	N22	-0.640
H23	-0.259	H23	-0.272	N23	-0.667	N23	-0.669
H24	-0.261	N24	-0.576	N24	-0.670	N24	-0.644
		F25	-0.008	F25	0.012	F25	0.022
		F26	-0.004	F26	0.012	F26	0.028
				F27	0.012	F27	0.034
				F28	0.013	F28	0.034
						F29	0.022
						F30	0.028

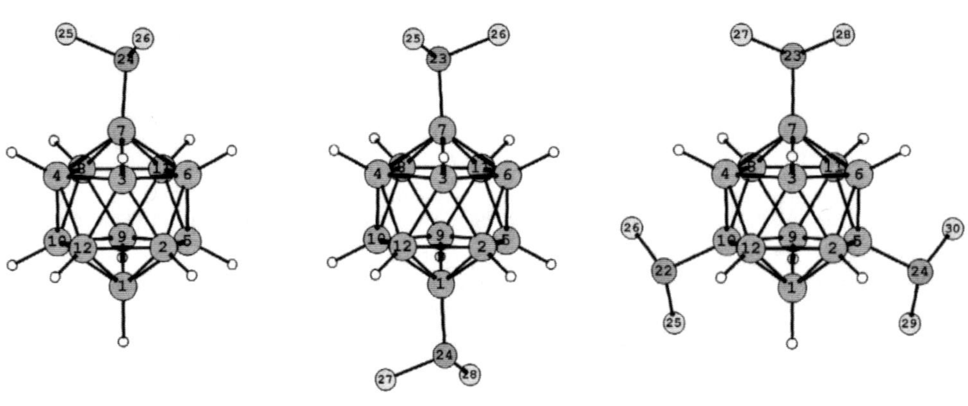

(a) (b) (c)

Figure 3. The most stable structures of $B_{12}H_x(NF_2)_{12-x}^{2-}$: (a) x = 11, (b) x = 10, (c) x = 9.

same time, the charge on the distant boron atoms (not connected by an edge, i.e. B2, B12, B10, B9 and B5) decreases (table 1). The total Mulliken charge on the boron atoms of the upper layer increases to 1.148, instead of 0.537 *e* for the initial $B_{12}H_{12}^{2-}$ anion, and the boron atoms of the bottom layer become charged negatively with the total Mulliken charge equal to −0.208 instead of 0.398 *e* (table 1). Herewith, the largest increase in the charge on the diagonally arranged boron atoms B1 and B11 is observed. These changes ultimately lead to the polarization of the boron core. As a result, boron atoms of the bottom layer have an excess negative charge while the closest to the substituted boron atoms of the upper layer have an excess positive charge (in comparison with the initial $B_{12}H_{12}^{2-}$). The electron configuration of $B_{12}H_{11}NF_2^{2-}$ according to the results of NBO analysis can be described as following: B $2s^{0.66}2p^{2.49}3p^{0.02}$, B* $2s^{0.58}2p^{2.22}3s^{0.01}3p^{0.03}$, H $1s^{0.98}$, N $2s^{1.53}2p^{3.31}3p^{0.02}3d^{0.01}$, F $2s^{1.90}2p^{5.40}$, where B* indicates the boron atom related to the NF_2 group (B7).

Table 2. Calculated bond lengths (Å) of $B_{12}H_x(NF_2)_{12-x}^{2-}$ structures, x = 9 – 11.

Label	Atom number	*r*(B-B)	*r*(B-H)	*r*(B-N)	*r*(N-F)
1	7	1.776-1.802	1.194-1.200	1.558	1.450, 1.451
2-1	7, 1	1.775-1.803	1.192-1.195	1.553, 1.553	1.447, 1.449
2-2	7, 2	1.772-1.800	1.191-1.198	1.553, 1.560	1.446-1.449
2-3	7, 8	1.770-1.801	1.189-1.198	1.555, 1.551	1.443-1.447
3-1	8, 2, 7	1.769-1.803	1.186-1.193	1.551, 1.547, 1.551	1.436-1.447
3-2	8, 7, 4	1.769-1.819	1.186-1.196	1.554, 1.544, 1.549	1.432-1.443
3-3	8, 12, 3	1.772-1.814	1.184-1.196	1.559, 1.549, 1.549	1.439-1.446
3-4	10, 4, 3	1.769-1.800	1.184-1.196	1.554, 1.548, 1.549	1.438-1.446
3-5	10, 4, 11	1.774-1.803	1.186-1.196	1.547, 1.547, 1.558	1.443-1.448
3-6	10, 7, 5	1.772-1.808	1.188-1.193	1.553, 1.548, 1.553	1.442-1.445

3.3. X = 10.

To find the most stable anion structures with two or three substituted difluoramino groups of $B_{12}H_x(NF_2)_{12-x}^{2-}$, various isomers fully optimized without symmetry constraint were obtained. Some of the results are summarized in tables 2, 3 and the energetically most favourable structures are shown in figure 3. By analogy with the disubstituted benzenes, the following prefixes indicating the positions of the substituent can be used: ortho-, meta-, para-. Thus, the position of NF_2 groups substituted on the adjacent boron atoms (connected by a common edge, e.g. B7 and B8) will be called ortho-, the position with NF_2 groups substituted on the boron atoms separated by one or two boron atoms of the same plane (not connected by an edge, e.g. B7 and B2) – meta-, and the position with NF_2 groups that are substituted at a maximum distance from each other (e.g. B7 and B1) – para-position.

The calculated features of the possible isomers of $B_{12}H_x(NF_2)_{12-x}^{2-}$ with x = 9, 10 show a large influence of the position of NF_2 groups. As seen from the figure 3, where the most stable isomers of $B_{12}H_x(NF_2)_{12-x}^{2-}$ with x = 9 – 11 are shown, para-position with NF_2 groups substituted on hydrogen atoms bonded to B7 and B1, is the most favourable one (table 3). The energy preference of this position over ortho- and meta-positions is 0.27 and 0.16 eV, respectively. A broader range of B-B and B-H bond lengths in the $B_{12}H_{10}(NF_2)_2^{2-}$ anion is observed when compared to the structure with one NF_2 group (x = 11). Thus, substitution of hydrogen atoms by NF_2 groups leads to the increase of the charge on the boron atoms conjugated with nitrogen atoms of difluoramino groups (table 1) and polarization of the boron atoms of the core. For the most stable isomers 1 and 2-1 the total charges on NF_2 groups are equal to −0.588 and −0.644 *e* (the average value), respectively. For other possible isomers of $B_{12}H_x(NF_2)_{12-x}^{2-}$ with x = 10 the total charges on NF_2 groups are equal −0.642 *e* for meta- and −0.575 *e* for ortho-positions. As seen from the table 3, for the most stable isomer 2-1 the dipole moment is equal to 0.00 D while for others (less symmetric structures) the dipole moment is much higher. The difference in values of the dipole moments is the greater, the greater the difference in energies among the energetically most favorable isomer with the energy taken as a zero, and other isomers.

Table 3. Calculated relative energies (ΔE), dipole moments (d), IR frequencies ($\nu(NF_2)$) and frequency shifts ($\Delta\nu(NF_2)$) of $B_{12}H_x(NF_2)_{12-x}^{2-}$ structures, x = 9 – 11.

Label	Atom number	ΔE, eV	d, D	$\nu(NF_2)_i$, cm^{-1}	$\Delta\nu(NF_2)_i$, cm^{-1}
1	7	0.00	4.42	1092	5
2-1	7, 1	0.00	0.00	1092, 1120	5, 33
2-2	7, 2	0.16	2.78	1076, 1090	-11, 3
2-3	7, 8	0.27	4.66	1076, 1102	-11, 15
3-1	8, 2, 7	0.23	1.73	1092, 1103, 1129	5, 16, 42
3-2	8, 7, 4	0.65	4.24	1085, 1095, 1118	-2, 8, 31
3-3	8, 12, 3	0.38	2.40	1088, 1099, 1121	1, 12, 34
3-4	10, 4, 3	0.32	3.21	1099, 1120, 1130	12, 33, 43
3-5	10, 4, 11	0.41	2.32	1081, 1083, 1098	-6, -4, 11
3-6	10, 7, 5	0.00	1.01	1121, 1156, 1156	34, 69, 69

3.4. X = 9.

As the number of NF_2 groups increases the substituted groups prefer to occupy positions such that the electron density in the core is most uniformly distributed. This corresponds to the symmetrical arrangement of NF_2 groups on the vertices of the core. Among all the isomers that meet this condition, 3-6 isomer with three substituted NF_2 groups bonded to B10, B7 and B5 shown on the figure 3c has the lowest total energy.

Behaviour similar to the above-mentioned features was observed for this $B_{12}H_{10}(NF_2)_3^{2-}$ structure. Boron atoms related to the NF_2 group, and especially NF_2 groups itself, pull a part of electron density from the boron core. As a result, an excess positive charge appears around the boron atom related to the NF_2 group. The diagonally arranged boron atoms B3 and B9 have a negative charge that, presumably, partly compensate the large positive charge. The total charges on the NF_2 groups decrease with the increasing number of NF_2 groups and are equal to −0.588 *e* for the anion structure with x = 11 and −0.594 *e* for x = 9. The electron configuration of the most stable 3-6 isomer according to the results of NBO analysis can be described as following: B $2s^{0.67}2p^{2.48}3p^{0.02}$, B* $2s^{0.57}2p^{2.21}3p^{0.04}$, B** $2s^{0.57}2p^{2.21}3s^{0.01}3p^{0.03}$, H $1s^{0.95}$, N $2s^{1.52}2p^{3.32}3p^{0.02}3d^{0.01}$, F $2s^{1.90}2p^{5.39}$, where B* indicates the B5, B10 and B** – B7. In comparison with the electron configuration of the most stable structure of $B_{12}H_x(NF_2)_{12-x}^{2-}$ with the larger x, there is a tendency of increasing electrons on 2s orbitals and decreasing of electrons on 2p orbitals of B atoms, wherein hydrogen atoms donate electrons, that reflects in the decreasing of electrons on its 1s orbitals.

Changes in the electron structure of $B_{12}H_x(NF_2)_{12-x}^{2-}$ reflect in the changes of its geometry. Thus, substitution of one or more hydrogen atoms with the appropriate number of NF_2 groups leads to the weakening of the bonds in the core, and the formation of strong ionic bond B-NF_2. This is indicated by the calculated bond distances and bond orders. With the increasing number of NF_2 groups bond length between the boron atoms of the core is increased, the length of the B-H, B-N and N-F decreases; and, on the contrary, the order of the B-B bond decreases, B-N and B-F – increases. These changes indicate to the better binding of NF_2 groups with the core for $B_{12}H_x(NF_2)_{12-x}^{2-}$ with x = 9 in comparison with the anion structures with x = 10 and 11.

In the infrared spectra of $B_{12}H_x(NF_2)_{12-x}^{2-}$ with x = 9 – 11 the most intense bands appear in the 1076 – 1156 cm^{-1} region and are assigned to the vibrations of the difluoramino group. The calculated shifts compared to the vibrations of the free radical NF_2 are accompanied with the structural changes of $B_{12}H_x(NF_2)_{12-x}^{2-}$. Blue shifts in the IR spectrum compared to the vibrations of the free radical NF_2

are characteristic for the most stable isomers with the minimum total energy, namely, 1, 2-1 and 3-6 isomer, and points to a stronger bonding between NF_2 and the core. For x = 9 four isomers with blue shifts are observed but as it is seen from the table 3, the most stable isomer 3-6 is characterized by the greater values of $\nu(NF_2)$. The smaller values of frequency shifts and red shifts are typical for the less stable structures.

4. Conclusions

Within the present study, we examined the geometric, energetic, spectral and electronic features of the dodecahydro-closo-dodecaborate anion $B_{12}H_{12}^{2-}$ and its difluoramino-substituted derivatives. The results of quantum chemical calculations of the most stable isomers of $B_{12}H_x(NF_2)_{12-x}^{2-}$ (x = 9 – 12) at the B3LYP/6-311++G** level of theory showed that as the number of NF_2 groups increases, they preferentially occupy symmetrical positions relatively to the core, such that the charge density in the complex achieves the most uniform distribution. The details of such distribution and the stability of the anion complex $B_{12}H_x(NF_2)_{12-x}^{2-}$ (x = 9 – 12) have been discussed. We hope that these results will aid in the investigation of $B_{12}H_{12}^{2-}$ derivatives and will broaden the idea of one of the most important features of $B_{12}H_{12}^{2-}$ anion – its aromaticity.

Acknowledgments

This work was supported by "Dalniy Vostok", the Grant project № 0265-2015-0011.

References

[1] Ellis I A, Gaines D F and Schaeffer R 1963 *J. Amer. Chem. Soc.* **85** 3885
[2] Miller H C, Miller N E and Muetterties E L 1964 *Inorg. Chem.* **3** 1456-1463
[3] Mikhailov B M 1967 *Chimiya borovodorodov* (Moscow: Nauka) pp 419-424
[4] Kuznetsov N T 1988 *Chimiya poliedricheskih borovodorodnyh soedinenyi* (Sbornik trudov IOIC) ed A Yu Civadze (Moscow: Nauka) pp 78-97
[5] Saldin V I and Sukhovey V V 2013 *Sposob polucheniya borftorsoderzhashchej ehnergoemkoj kompozicii* Russia Patent No. 2479560
[6] Kuznetsov N T, Ionov S P and Solntsev K A 2009 *Razvitie koncepcii aromatichnosti: Poliedricheskie structure* (Moscow: Nauka) chapter 3 pp 264-395
[7] Kaczmarczyk A and Kolskiand G B 1965 *Inorg. Chem.* **4** 665-671
[8] Hoffmann R and Lipscomb W N 1962 *J. Chem. Phys.* **37** 520-523
[9] Allis D G and Hudson B S 2006 *J. Phys. Chem.* A **110** 3744-9
[10] Koblova E A, Saldin V I and Ustinov A Yu 2016 *AIP Conf. Proc.* **1790** 020014
[11] Kuznetsov N T, Ionov S P and Solntsev K A 2009 *Razvitie koncepcii aromatichnosti: Poliedricheskie structure* (Moscow: Nauka) chapter 3 pp 296-306
[12] Schmidt M W *et al.* 1993 *J. Comput. Chem.* **14** 1347-1363
[13] Becke A D 1993 *J. Chem. Phys.* **98** 5648-5652
[14] Lee C, Yang W and Parr R G 1988 *Phys. Rev.* B **37** 785-789
[15] Harmony M D, Myers R J, Schoen L J, Lide D R and Mann D E 1961 *J. Chem. Phys.* **35** 1129-1130
[16] Andersson M P and Uvdal P 2005 *J. Phys. Chem.* A **109** 2937-2941
[17] Glendening E D, Badenhoop J K, Reed A E, Carpenter J E, Bohmann J A, Morales C M, Landis C R and Weinhold F 2013 *NBO Version 6.0*
[18] Grundkotter-Stock B, Bezugly V, Kunstmann J, Cuniberti G, Frauenheim T and Niehaus T A 2012 *J. Chem. Theory Comput.* **8** 1153-1163
[19] Muetterties E L, Merrifield R E, Miller H C, Knoth W H and Downing J R 1962 *J. Am. Chem. Soc.* **84** 2506-2508

Preparation and characterization of Ni-doped carbon aerogel for supercapacitor

Shasha Wang[1,2,4], Meifang Yan[1,2], Haihua Liu[1,3], Yuelong Xu[1,2], Lihui Zhang[1,2] and Zhenfa Liu[1,2]

[1]Institute of Energy Resources, Hebei Academy of Science, Shijiazhuang Hebei 050081, China
[2]Hebei Engineering Research Center for Water Saving in Industry, Shijiazhuang Hebei 050081, China
[3]College of Chemistry & Material Science, Hebei Normal University, Shijiazhuang Hebei 050024, China

E-mail: 517263436@qq.com

Abstract. Ni-doped carbon aerogel was prepared by impregnation methods, physical structure, and electrochemical properties were investigated. Electrochemical properties of prepared Ni-doped carbon aerogel and carbon aerogel electrodes were measured by galvanostatic charge/discharge measurements. The results show Ni-doped carbon aerogels maintain the elementary structure of carbon aerogel, but they exhibited higher specific capacitance than carbon aerogel.

1. Introduction

Carbon aerogel can be obtained by sol–gel polycondensation of certain organic monomers such as resorcinol, formaldehyde, and polymerization. Because of the interesting properties such as low density, high surface area, continuous open porosity and electrical conductivity, carbon aerogel has been widely employed as an electrode material of electrochemical capacitors. Supercapacitors are electrochemical energy storage devices that have attracted considerable attention due to their excellent charge and discharge properties and long cycle life [1-2]. Double-layer capacitance and pseudocapacitance are utilized for energy storage. The double-layer capacitance derives from the accumulation of ions on the double-layer at the electrode/electrolyte interface and is dependent on the surface area and the accessibility of the electrolyte to the electrode porosity. Pseudocapacitance derives from eversible faradaic reactions at the electrode/electrolyte interface thanks to the presence of surface functionalities and dopants in the carbon electrodes [3].

In this work, carbon aerogel was prepared by a sol-gel polymerization. To combine the excellent electrochemical performance of carbon aerogel with a pseudo-capacitive property of metal oxide, nickel oxide was doped on carbon aerogel by an incipient wetness impregnation. To evaluate the effect of nickel oxide on the electrochemical properties, the performance of carbon aerogel and Ni-doped carbon aerogel for pseudo-capacitive supercapacitor electrode were investigated.

2. Experimental

2.1. Preparation of Ni-doped carbon aerogel
Carbon aerogel was prepared by a sol-gel formaldehyde, polymerization, and resorcinol according to the method in previous work[4]. To obtain Ni-doped carbon aerogel, after pyrolysis, the PRF (polymerization-resorcinol-formaldehyde) hydrogel was soaked in 0.05 M solution of nickel nitrate hexahydrate for three days (the saline solution was changed after every 24 h). Solvent exchange for acetone was carried out via changing the solution after every 24 h for 3 times. Drying was carried out at ambient pressure. The Ni-doped carbon aerogel was obtained at 900 °C for three h under a flowing high purity nitrogen atmosphere (100 mL/min).

2.2. Fabrication of carbon aerogel electrode
The prepared Ni-doped carbon aerogels and carbon aerogels were casted respectively using polytetrafluoroethylene (PTFE) as a binder with a weight ratio of 90:10. The resultant was rolled to be 8-10 μm thickness. The electrode material was pressed onto foamed nickel, which was used as a working electrode.

2.3. Characterizations
BET surface area and N_2 adsorption–desorption isotherm were measured with an ASAP 2420 (Micromeritics, America) instrument. The samples were degassed at 350 °C overnight before the adsorption measurements. Pore size distribution was determined by the BJH method applied to the desorption branch of the nitrogen isotherm. XRD patterns on aerogel and carbon aerogel were examined by Ultima IV X-ray diffractometer (Rigaku, Japan) using Cu Kα radiation.

2.4. Measurement of electrochemical property for carbon aerogel
Electrochemical property of Ni-doped carbon aerogel electrode was measured with a conventional three-electrode cell system in 6 M KOH electrolyte. A platinum plate and saturated calomel electrode were used as a counter electrode and a reference electrode, respectively. Galvanostatic charge/discharge measurements were carried out at a constant current of 1 A/g within voltage range of 0–0.8 V.

3. Results and discussion

3.1. Physical characteristics
The pore structure of the carbon aerogels was investigated by low-temperature nitrogen adsorption measurements. The N_2 adsorption/desorption isotherms and the pore size distribution (PSD) curves are showed in Figure 1(a) and Figure 1(b), respectively. The porosity data deduced from the isotherms are listed in Table 1.

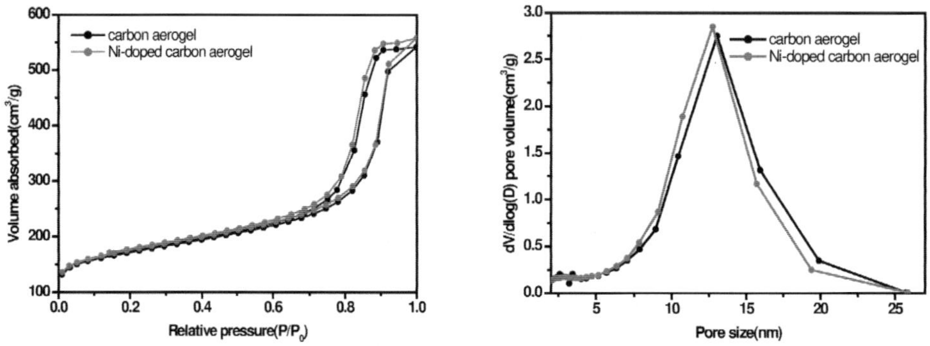

(a) (b)

Figure 1. (a) Low-temperature nitrogen isotherms and (b) pore size distribution of carbon aerogels.

Table 1. Textual properties of carbon aerogel and Ni-doped carbon aerogel.

	BET surface area (m^2/g)	Average pore diameter (nm)	Pore volume (cm^3/g)
Carbon aerogel	658	6.11	0.89
Ni-doped carbon aerogel	601	6.02	0.83

Figure 1(a) compares the N2 adsorption–desorption isotherms and pore size distributions of carbon aerogel and Ni-doped carbon aerogel. Both samples exhibited type IV isotherms with type H2 hysteresis loops. This indicates that porous structure of carbon aerogel was still maintained even after Ni doping. Table 1 shows the textual properties of carbon aerogel and Ni-doped carbon aerogel. Although surface area, average pore diameter, and pore volume of carbon aerogel slightly decreased after the impregnation of Ni(NO3)2, Ni-doped carbon aerogel still retained high surface area (601 m2/g), fine average pore diameter (6.02 nm), and large pore volume (0.83 cm3/g). From the results in Figure 1(a) and Figure 1(b), it can be indicated that network structures of carbon aerogel are maintained even after the metal doping.

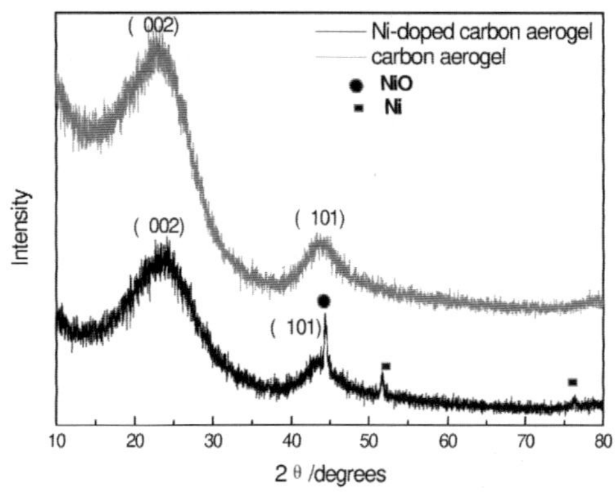

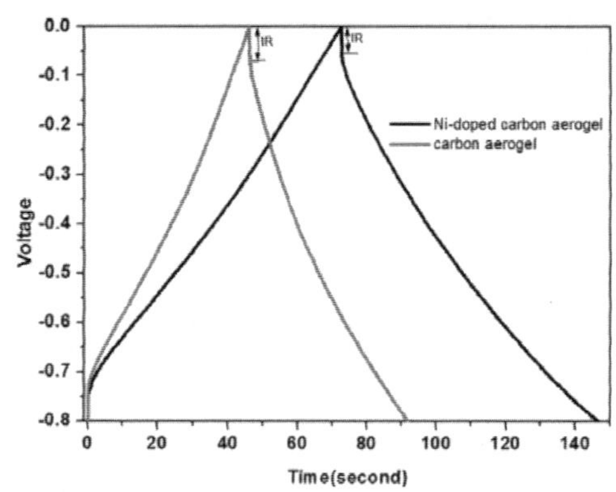

Figure 2. XRD patterns of carbon aerogels **Figure 3.** Charge/discharge profiles of carbon aerogels

XRD patterns of carbon aerogels are characterized, as shown in Figure. 2. Successful impregnation of metal species on carbon aerogel was confirmed by XRD analysis in Figure. 2. Carbon aerogel showed two broad peaks at 23.5° (0 0 2) and 43.8° (1 0 1), which were attributed to the reflections of graphite carbon, respectively [5].A mixture of metal and metal oxides was observed in the XRD patterns of Ni-doped carbon aerogels [6]. Considering that metal Ni and metal oxides NiO are closely related to the faradaic redox reactions, it is expected that Ni-doped carbon aerogel would show an enhanced capacitance than carbon aerogel.

3.2. Electrochemical properties of carbon aerogels

Table 2. Specific capacitance of carbon aerogel electrodes

	carbon aerogel	Ni-doped carbon aerogel
Specific capacitance (F/g)	57.5	92.6

Figure 3 shows the galvanostatic charge/discharge profiles (current density = 1 A/g) of carbon aerogel and Ni-doped carbon aerogel electrodes, and the calculated specific capacitances are summarized in Table 2. Ni-doped carbon aerogel showed higher specific capacitance (92.6 F/g) than carbon aerogel (57.5 F/g). Moreover, inner resistance (IR) drop intensity of Ni-doped carbon aerogels was negligible. This indicates that capacitive property of Ni-doped carbon aerogels was excellent. On the other hand, IR of carbon aerogel was clearly observed. This potential drop derives from the resistance of electrolyte and the IR of ion diffusion in carbon micropores [7-9]. Thus, the specific capacitance of carbon aerogel could be improved by adding pseudocapacitive material.

4. Conclusions

Ni-doped carbon aerogel was prepared by wet impregnation methods to compare the physical structure and electrochemical properties with carbon aerogel. Electrochemical properties of prepared Ni-doped carbon aerogel and carbon aerogel electrodes were investigated by galvanostatic charge/discharge measurements. Ni-doped carbon aerogels exhibited higher specific capacitance than carbon aerogel. The improved performance of Ni-doped carbon aerogels was attributed to pseudocapacitive characteristics of nickel and nickel oxide.

Acknowledgements

This research was financially supported by Science-technology Support Plan Projects of Hebei Province (16214402D) and (15291407D).

References

[1] B.E. Conway, Electrochemical Supercapacitors. Scientific Fundamentals and Technologic- al Applications, Kluwer Academic/Plenum, New York, 1999.
[2] L.L. Zhang, X.S. Zhao, Chem. Soc. Rev. **38** (2009) 2520.
[3] F. Beguin, E. Frackowiak, Carbon Materials for Electrochemical Energy Storage Systems, CRC Press, *Taylor & Francis, Boca Raton*, **2010**.
[4] M.F. Yan, L.H. Zhang, R. He, Z.F. Liu, *J Porous Mater* **22** (2015) 699.
[5] L.H. Ai, J. Jiang, *Powder Technol.* **195** (2009) 11.
[6] L.C. Cotet, M. Gich, A. Roig, I.C. Popescu, V. Cosoveanu, E. Molins, V. Danciu. *Journal of Non-Crystalline Solids* **352** (2006) 2772.
[7] X. Liu, T. Osaka, J. Electrochem. Soc. **144** (1997) 3066.
[8] M. Endo, T. Maeda, T. Takeda, Y.J. Kim, K. Koshiba, H. Hara, M.S. Dresselhaus, J.Electrochem. Soc. **148** (2001) 910.
[9] Y.R. Nian, H. Teng, J. Electrochem. Soc. **149** (2002) 1008.

The wettability between etching solutions and the surface of multicrystalline silicon wafer during metal-assisted chemical etching process

Y C Niu[1,2,4], Z Liu[2], X J Liu[2], Y Gao[2], W L Lin[2], H T Liu[3], Y S Jiang[3] and X K Ren[3]

[1]Co-innovation Center for Green Building, Shandong Jianzhu University, Jinan 250101, China
[2]School of Materials Science and Engineering, Shandong Jianzhu University, Jinan 250101, China
[3]Shandong Linuo Solar Power Holdings Co., Ltd, Shandong, Jinan 250103, China

E-mail: niuyuchao@sdjzu.edu.cn

Abstract. In order to investigate the wettability of multicrystalline silicon (mc-Si) with the etching solutions during metal-assisted chemical etching process, different surface structures were fabricated on the p-type multi-wire slurry sawn mc-Si wafers, such as as-cut wafers, polished wafers, and wafers etched in different solutions. The contact angles of different etching solutions on the surfaces of the wafers were measured. It was noted that all contact angles of etching solutions were smaller than the corresponding ones of deionized water, but the contact angles of different etching solutions were quite different. Among the contact angles of the etching solutions of $AgNO_3$-HF, H_2O_2-HF, TMAH and HNO_3-HF, the contact angle of TMAH solution was much larger than the others and that of HNO_3-HF solution was much smaller. It is suggested that the larger contact angle may lead to an unevenly etching of silicon wafer due to the long retention of big bubbles on the wafers in the etching reaction, which should be paid attention to and overcome.

1. Introduction

Since silicon surface with a nano texture (also called a black Si) have a very strong light-trapping and can improve the power conversion efficiency of solar cells by reducing light reflection [1,2], the surface nano texture fabrications are becoming a focus of attention. To save the manufacturing costs, nano texture can be fabricated using wet etching texturing method, which is also compatibility with mass production and absence of a residual mechanically damaged layer after etching [3]. Wet etching texturing includes alkaline etching, acid etching and metal-assisted chemical etching (MACE) [1,4,5]. Wet etching usually contains several different chemical solution treatments and between each treatment the deionizedwater (DI water) cleaning must be used. In addition, almost all of the wet etching will be accompanied by the generation of gases on the Si surface. Then, the gases will grow up to bubbles and adsorb on the Si surface as the reaction carrying on [6, 7]. Whether the bubble releasing fast or not from the Si surface depends on the wettability between etching solution or cleaning solution and Si surface [8, 9], so that the wettability is a very important issue and the studying it is indispensable on whether for the etching or cleaning. In reports [10-15], the contact angles of pure water on the different surfaces of Si were studied. These surfaces included the polished surface, the nicro-structure surfaces with different roughness and the nano-structure surfaces. In our previous research work [16], the wettability between deionized water and different surface of Si wafer has been investigated systemically. Also, the contact angles of chemical solutions on the different surfaces of Si were studied in some researches [17,18]. These solutions involved pure and surfactant-added KOH etchants, NH_4OH, TMAH, tetrabutylammonium hydroxide (TBAH), and a mixture of isopropyl alcohol (IPA) and TMAH. Unfortunately, up to now, there has been rare report about a systemic research on the wettability between etching solutions and Si wafers during black silicon fabrication using MACE, during which the solutions of $AgNO_3$-HF, H_2O_2-HF, TMAH and HNO_3-HF solutions were often used [19,20]. In this paper, the contact angles of these etching solutions on different surfaces of multicrystalline silicon wafers were systemically measured and the wetting characteristics of these solutions on the silicon surface were investigated in detail.

2. Experimental

The traditional p-type multi-wire slurry sawn mc-Si wafers (180-μm-thick, 156×156 mm2) were supplied by Linuo Solar Power Holdings Co., Ltd. The wafers were divided into four groups (a, b, c and d). The four groups of samples were treated separately to form different surface structures, which are shown as follows. In order to emphasize application significance, the etching solutions were made according to chemical compositions used in the production process of the black silicon.

a: As-cut wafer sample with mechanical damages layer resulted from the wafer cutting.

b: Polished in a mixed acid solution (V/V: 70%HNO_3: 49%HF= 6:1) for 3 min at room temperature.

c: Etched (textured) in a mixed solution (V/V: 70%HNO_3: 49%HF: DI-water= 4:1:2) for 3 min at 8°C to obtain the micro-structured surface. The etched samples are called RENA textured wafer hereafter.

d: Micro-nano etched (textured) using MACE. Sample of Group c was chosen as a substrate and textured using MACE. Silver nanoparticles were electrolessly deposited on the silicon wafers in a solution of 8mM/L $AgNO_3$ and 4 vol.% HF for 15s at room temperature. Nanoporous silicon was formed in a solution (V/V: 30%H_2O_2: 49%HF: DI-water= 2:2:7) for 60 s at room temperature. In this work, the surface of Sample d is called a micro-nano textured surface hereafter.

During surface treatment of the samples, the samples were rinsed two times in deionized water (DIW) at room temperature after the polishing or etching.

Before the contact angle measurements, all samples were immersed in 5 vol.% HF for 60s at room temperature to remove the SiO_2 layer formed in air and rinsed two times in deionized water (DIW) at room temperature. The SL1000 automatic contact angle meter was used to test the contact angle of all samples; the surface morphologies of samples were analyzed by means of OLS4000 confocal lasers scanning microscope (CLSM) and S4800 scanning electron microscope (SEM).

3. Results and discussion

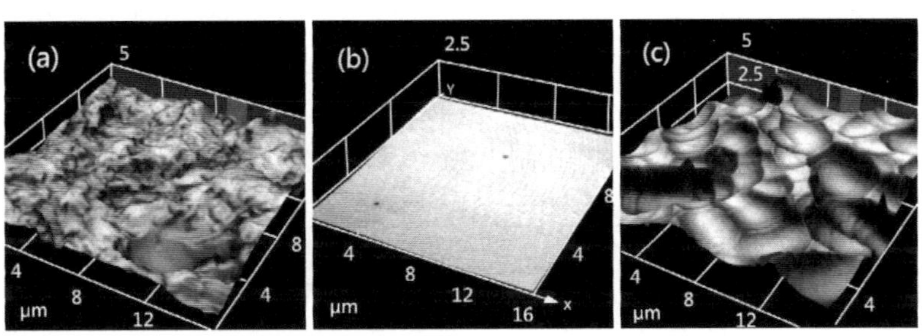

Figure 1. Micrographs of wafer surfaces
a as cut wafer b polished wafer c RENA textured wafer

As the studied in our previous research work [16], the surface structure of the silicon wafer has a strong influence on the contact angle. Therefore, the surface morphology of the as cut, RENA textured and the polished samples of the slurry sawn mc-Si wafers are shown in Figure 1. The polished one has a very smooth surface, the as cut and RENA ones present the micron size structure and the fluctuation of the latter is more intense.

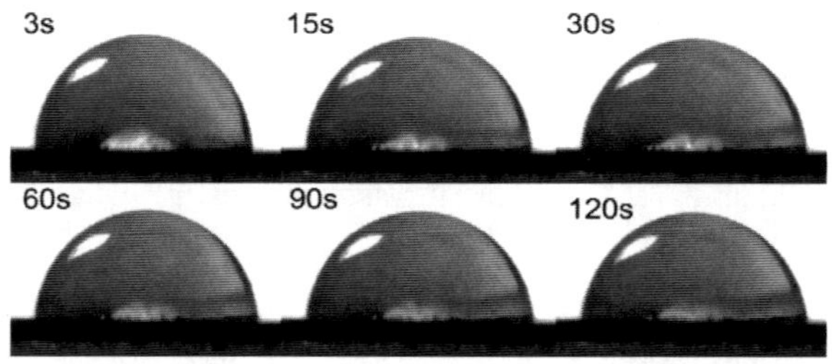

Figure 2. The shape change of 8mM/L $AgNO_3$ and 4 vol.% HF solution droplet on the RENA wafer with the time variation

Figure 2 shows the shape change of the $AgNO_3$-HF solution droplet on the surface of RENA textured wafer with time. With the increase of time, the shape of the droplet does not change obviously and no evident bubble appears in the droplet. This is mainly because the concentrations of $AgNO_3$ and HF are very low, the precipitation reaction of Ag is weak and the gas generated in the reaction is less. The optical photographs of the shape change of the droplet on the as cut and the polished samples is similar to that on the RENA textured wafer, but the contact angles of the $AgNO_3$-HF solution droplet on the surfaces of these three samples are obvious different as shown in Figure 3. It can be seen that three contact angles of $AgNO_3$-HF solution on three wafers are lower respectively than corresponding contact angles of water. The ordering of the solution contact angles is in agreement with that of the water contact angle [16], *i.e.*, the contact angle on textured wafer > the contact angle on as cut wafer > the contact angle on polished wafer. With the increase of time, three solution contact angles decrease slightly.

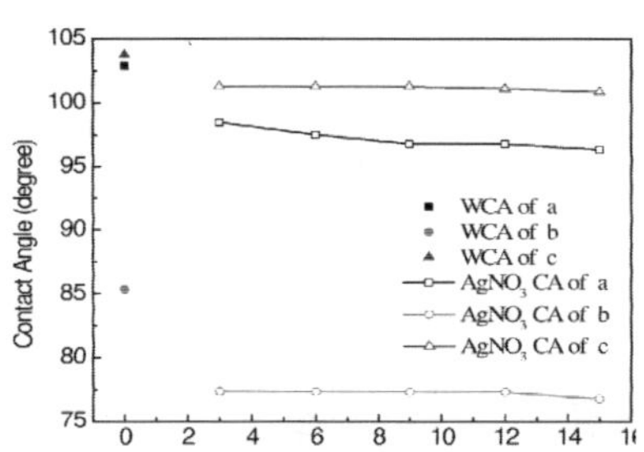

Figure 3. The water contact angles and change of 8mM/L $AgNO_3$-4 vol.% HF solution contact angle with time variation
Note: WCA means contact angle of water on the wafer; AgNO3 CA means contact angle of AgNO3-HF solution on the wafer
a as cut wafer b polished wafer c textured wafer

The appearance of wafers after deposition of Ag is shown in Figure 4. Because the surface of the as cut sample or the RENA sample with a micron size structure is in a big fluctuation, the light trapping effect is relative stronger, so that it has a darker color (shown in Figure 4b) compared with the polished sample (shown in Figure 4a). Due to the deposition amount of Ag is very little, Ag particles can not change the surface fluctuation of the three samples except for the appearance color. Figure 5 shows the water contact angle and the change of the solution contact angles with time variation. The solution contact angle of solution is much smaller than that of water on the corresponding wafer, and the contact angle decreases with the increase of time. Figure 6 shows the pictures of H_2O_2-HF solution droplet on the surface of a polished silicon wafer, on which the Ag particles had been deposited. As shown in the pictures, there exist many bubbles in the droplet, which are generated during etching reaction between H_2O_2-HF solution and Si wafer under the catalysis of Ag. The bubbles grow to about 0.2mm, rise and break away from the droplet at last.

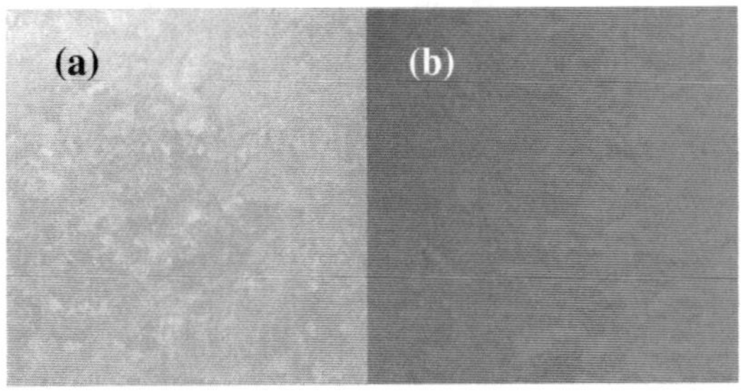

Figure 4. Appearance of mc-Si wafers after the deposition of Ag
a polished wafer b as cut wafer or RENA textured wafer

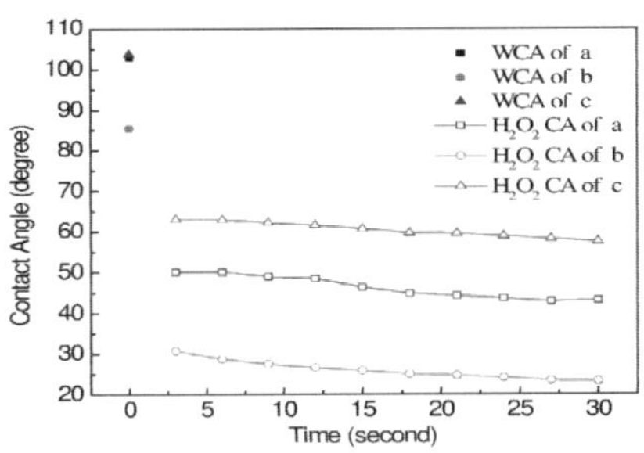

Figure 5. The water contact angles and changes of H_2O_2-HF (volume ratio: 30%H_2O_2: 49%HF: DI-water= 2:2:7) solution contact angle with time variation
a as cut wafer; b polished wafer;
c RENA textured wafer

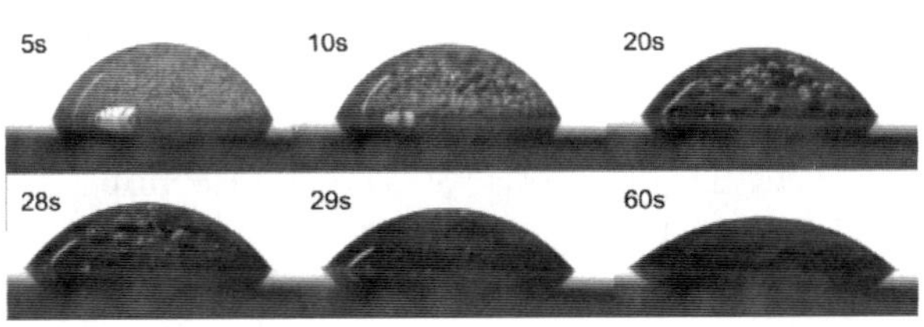

Figure 6. The shape change of H_2O_2-HF solution droplet with the time variation

Textured structure of the sample of Group d that described in experimental section is shown in Figure 7. The diameter of nano pores on the surface are about 10nm [16], so that their light trapping is very strong and the color of the wafer is black. As can be seen from the Figure 8, the initial contact angle of TMAH solution on the silicon wafer is very large, about 140 degree and as the reaction time increases to 60s the contact angle decreases to 113 degree. In contrast, the initial contact angle of the HNO_3-HF mixed acid solution on the silicon wafer is small, about 49 degree, and it decreases slightly with the time increases.

Figure 7. The surface morphology of nano-textured wafer (Group d)

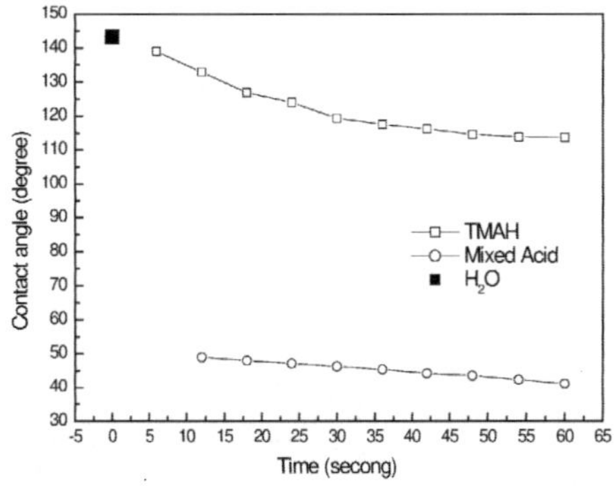

Figure 8. The changes of contact angle of 1%TMAH solution and 4.8M/L HNO_3-1.3M/L HF mixed acid solution on the nano textured surfaces (Sample d) with time variation

Figure 9 shows the pictures of TMAH solution droplet on the surface of sample of Group d. There exist some big bubbles in the droplet. Different from the bubbles in H_2O_2-HF liquid droplet, the size of bubbles in TMAH solution droplet is in a larger range from about 0.2mm to about 3mm. The formation of the big bubbles is attributed to the big contact angle (about 120-140 degree) between the TMAH solution and the wafer. Because of the agitation of the reaction gas, some of the little bubbles are separated from the wafer surface before they grow up, so that the size of bubbles is scattered. Contrarily, as shown in Figure 10, the contact angle of the HNO_3-HF mixed acid solution on the wafer is only about 50 degree (shown in Figure 9) and the bubbles are separated easily from the wafer surface, so that the bubbles in the mixed acid solution droplet are small and in uniform size.

Figure 9. The shape change of 1%TMAH solution drop on Sample d with the time variation

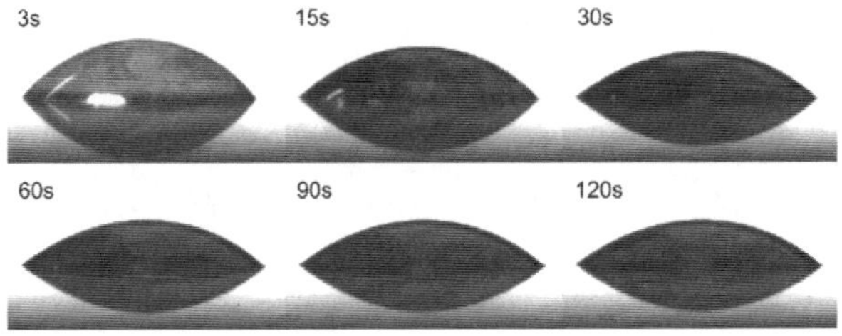

Figure 10. The shape change of 4.8M/L HNO_3-1.3M/L HF mixed acid solution drop on Sample d with the time variation

The measurement condition of the contact angle for each solution mentioned above were tried to keep similar to those in the actual production process, including wafer surface state and the composition of the solution, in order to correspond to the actual situation in production. The initial contact angles of the different solutions are shown in Figure 11.

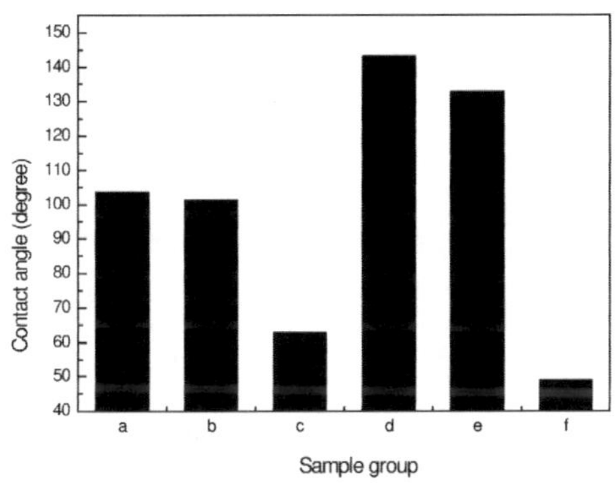

Figure 11. The average contact angles for six cases
a RENA textured sample contacting with water;
b RENA textured sample contacting with AgNO3-HF solution;
c Ag deposited sample contacting with H202-HF solution;
d Nano size textured sample contacting with water;
e Nano size textured sample contacting with TMAH solution;
f Nano size textured sample contacting with HNO3-HF solution

Due to the surface structure of the sample has a great influence on the contact angle, the contact angle of pure water on the nano size textured surface is obviously much larger than that on the micron size textured surface [16]. Compared with the contact angle of pure water, different etching solutions appear different contact angles. In case of the micro size textured surface (Case a-c), the contact angle of H_20_2-HF solution is much smaller than that of water, but the contact angle of $AgNO_3$-HF solution is close to it since the concentration of $AgNO_3$-HF in solution is very low. In case of the nano size textured surface (Cases d-f), the contact angle of HNO_3-HF solution is the smallest among the three cases. Similar to the situation of $AgNO_3$-HF solution, 1%TMAH solution is also a low concentration solution and its contact angle is close to the contact angle of water.

Generally, after inorganic substances are added into water, surface tension (γ_{LG}) will increase to some extent. The more substances added, the greater the surface tension. A large surface tension often leads to a larger contact angle [21]. However, the experiments in this paper shows opposite results, as shown in Figure 11. Combined with Figure 12 and formula 1[22], it can be analyzed as follows. After acid or alkali or salt is added into water, assuming that γ_{SL} does not change, then the increase of γ_{LG}

will results in the decrease of $\cos\theta_C$ or the increase of θ_C due to γ_{SG} is not affected by the addition of solutes. However, the experimental results in this paper show that instead of increasing, θ_C decreases, which indicates γ_{SL} reduced. In the actual case of Si etching in the solution, the reaction takes place between the etching solutions and Si wafers, which suggests that there are affinity between Si atoms and the related atoms in the solutions, and hence γ_{SL} decreases. Thus, the balance of the three forces shown in Figure 12 is broken and θ_C must decrease to build up a new balance. Therefore, no matter whether the solution concentration is high or low, the contact angles of $AgNO_3$-HF, H_2O_2-HF, TMAH and HNO_3-HF solutions are less than that of water due to affinity between Si atoms and the atoms in solutions. Since both H_2O_2-HF and HNO_3-HF solutions are higher concentration solution and there exists strong affinity, their contact angles on the Si wafers are much smaller.

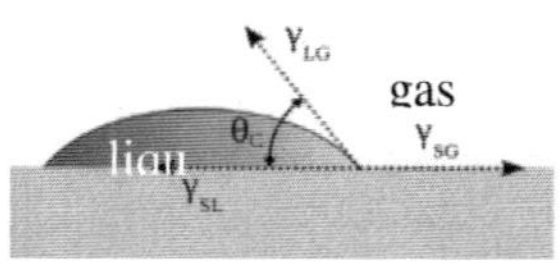

Figure 12. Schematic of a liquid drop showing the quantities in Young's equation.

$$\gamma_{SG} = \gamma_{SL} + \gamma_{LG} \cdot \cos\theta_C \quad (1)$$

Where, γ_{SG} γ_{SL} and γ_{LG} are three interface tensions (solid–gas, solid–liquid and liquid–gas) respectively, and θ_C is the included angle between γ_{SL} and γ_{LG}.

According to the degree of the initial contact angle size, the order of the solutions, from big to small, is TMAH>$AgNO_3$-HF>H_2O_2-HF>HNO_3-HF as shown in Figure 11. As a result, according to the size of bubble generated in droplet, the order of the solutions, from big to small, also is TMAH >$AgNO_3$-HF>H_2O_2-HF>HNO_3-HF (shown in the related figures above), The bigger contact angle or the bigger bubble means that the bubbles are more difficult to leave from the surface of wafer and detain longer on the surface. Meanwhile, the bigger bubbles cover larger area of the surface of wafer, so that the big detained bubbles hinder the etching of wafer under the bubbles. From this point of view, the etching of Si wafer by TMAH solution may be more likely to cause uneven etching [23].

It should be pointed that, from Figure 3, 5 and 8, with increasing time the contact angles of etching solutions decrease at different speeds. The reason for the decrease of contact angle may come from two aspects. Firstly, because Si is very easy to be oxidized the surface might be contaminated in the atmosphere before dropping. When the surface is dropped on a droplet, the contamination is gradually etched and the solid-liquid contact becomes better, so that the affinity is gradually strengthened. This takes place at the initial stage. Secondly, since every solution mentioned above can etch Si wafer, with increasing residence time of the etching solution droplet on the Si wafer the surface structure of the wafer changes gradually. For the $AgNO_3$-HF solution droplet, the surface structure of the wafer under the droplet changed very little, but Ag deposited on the surface improved the wettability of the surface. With increasing deposition time the amount of deposited Ag increased and the contact angle of the droplet decreased. For H_2O_2-HF solution droplet, the surface structure of the wafer tends to form nano size texture because nano pores were dug under Ag particles [24] and the specific surface area of the wafer increased. Since Si wafer was in a hydrophilic state in case of contacting with H_2O_2-HF solution, the increase of specific surface area of the wafer might improve its hydrophilic property, and the larger the surface area, the smaller the contact angle [16]. Therefore, with the increasing the residence time of the droplet, the specific surface area increases, and the contact angle decreases continuously. It should be pointed out that, during the production process, the surface texture after the etching of H_2O_2-HF solution must be adjusted from nano size texture to a submicron one using the etching of TMAH or HNO_3-HF in order to reduce surface defects and heavy doping [1, 24]. Therefore, the contact angle measurements of TMAH or HNO_3-HF solution need a nano size textured surface. Thus, the surface used in the measurement was an already prepared nano size texture surface and there existed gas in the pores [25]. With increasing the residence time of TMAH or HNO_3-HF solution droplet, the gas trapped in nano pores would run out gradually together with the reaction gas. As a result, the wafer surface area contacting with the solution are increased and the contact angles of the etching solutions reduced.

4. Conclusions

During metal-assisted chemical etching process of multicrystalline silicon wafer, several different etching solutions and cleaning deionized water (DI water) must be used. In this paper, the contact angles of different etching solutions on the surfaces of the wafers were systemically measured. It was noted that all contact angles of etching solutions were smaller than the corresponding ones of deionized water, but the contact angles of different etching solutions were quite different. Among the contact angles of the etching solutions of $AgNO_3$-HF, H_2O_2-HF, TMAH and HNO_3-HF, the contact angle of TMAH solution was much larger than the others and that of HNO_3-HF solution was much smaller. It suggests that the larger contact angle may make the silicon wafer etched unevenly due to the long retention of big bubbles on the surface in the etching reaction and it should be paid attention to and overcome.

Acknowledgments

This work was supported by Shandong government financial supporting (L37002013098), Jinan government financial supporting (JK20130307 and 301305033). The authors are grateful to Prof. Xiao-Lin Wang from University of Wollongong for his valuable discussions and assistance.

References

[1] Oh J, Yuan H C and Branz H M 2012 *Nature Nanotechnology* **7** 743
[2] Savin H, Repo P, Gastrow G V, Ortega P, Calle E, Garín M and Alcubilla R 2015 *Nature Nanotechnology* **10** 624
[3] Geng X W, Qi Z, Li M C, Duan B K, Zhao L C and Bohn P W 2012 *Solar Energy Materials and Solar Cells* **103** 98
[4] Sanjay K. Srivastava, Dinesh K, Vandana, Mukul S, Ravi K and Singh P K 2012 *Solar Energy Materials and Solar Cells* **100** 33
[5] Zhao J, Wang A, Green M A and Ferrazza F 1998 *Applied Physics Letters* **73** 1991
[6] Li X, Bohn P W. 2000 Metal-assisted chemical etching in HF/H_2O_2 produces porous silicon *Applied Physics Letters* **77** 2572
[7] Chartier C, Bastide S and Lévy-Clément C. 2008 *Electrochimical Acta* **53** 5509
[8] Brussieux C, Viers P, Roustan H and Rakib M 2011 *Electrochimical Acta* **56** 7194
[9] Yoshiteru M 2001 *ISIJ International* **41** S56
[10] Nishioka K, Sueto T, Yoshino K and Saito N 2011 *Japanese Journal of Applied Physics* **50** 128003
[11] Togonal A S, He L M, Roca P, Cabarrocas and Rusl 2014 *Langmuir* **30** 10290
[12] Chen C Y and Chen C Y 2013 *Nanotechnology* **24** 265101
[13] Kim B S, Shin S W, Shin S J, Kim K M and Cho H H 2011 *Nanoscale Research Letters* **6** 333
[14] Heib F, Hempelmann R, Munief W M, Ingebrandt S, Fug F, Possart W, Groß K, and Schmitt M 2015 *Applied Surface Science* **342** 11
[15] Wang W, Li D, Tian M, Lee Y C and Yang R G 2012 *Applied Surface Science* **258** 8649
[16] Liu X J, Niu Y C, Zhai T G, Ma Y Y, Zhen Y T, Ma X Y and Gao Y 2016 *Appl. Phys. A* **122** 127:1
[17] Yang C R, Chen P Y, Chiou Y C and Lee R T 2005 *Sensors and Actuators A* **119** 263
[18] Ko K, Song M G, Jeon H, Han J, Yoon B U, Koh Y, Ahn C and Kim T 2016 *Microelectronic Engineering* **149** 85

[19] Chen K, Liu Y Y, Wang X S, Zhang L J and Su X D 2015 *Solar Energy Materials & Solar Cells* **133** 148
[20] Niu Y C, Liu H T, Liu X J, Jiang Y S, Ren X K, Cai P and Zhai T G 2016 *Materials Science in Semiconductor Processing* **56** 119
[21] Adamson A W 1976 *Physical Chemistry of Surfaces* (New York: John Wiley & Sons)
[22] Young T 1805 *Phil. Trans. R. Soc. Lond* **95** 65
[23] Zubei I, Kramk, Wska M and Rola K P 2012 *Sensors and Actuators A: Physical* **178** 126
[24] Huang Z P, Geyer N, Werner P, Boor J de and U. Gösele 2011 *Adv. Mater.* **23** 285
[25] Cassie A B D and Baxter S 1944 *Trans Faraday Soc* **40** 546

Preparation techniques of the submicron lithium titanate materials by electro-spinning

Yudong Wang[1], Kai Yang[2,3], Fei Gao[2], Hao Liu[2] and Mingjie Zhang[2]

[1]University of Science and Technology Beijing, 30 Xueyuan Road, Haidian District, Beijing, PR China, 100083
[2]China Electric Power Research Institute, Haidian District, Beijing, PR China, 100192

E-mail: yangkai@epri.sgcc.com.cn

Abstract. Combining sol-gel process and electro-spinning, the submicron lithium titanate materials are prepared with lithium acetate and titanium tetraisopropanolate respectively as the lithium and titanium sources, and polyvinylpyrrolidone (PVP) as the template. It's found by scanning electron microscope(SEM)that, the prepared lithium titanate materials are characterized by the fiber diameter 150~200nm, a large number of irregular indentations in the surface, and the larger specific surface area than that before calcination. The lithium titanate cell receives charge-discharge test and cyclic voltammetry. The capacity of the submicron lithium titanate materials is up to 160mAh·g-1 at the rate of 0.1C, and it's revealed by cyclic voltammetry that the cell in the charge or discharge process undergoes a single redox reaction, but having good reversibility.

1. Introduction

The recently developed lithium titanate anode material has attracted widespread attention at home and abroad with its great strengths, such as long cycle life, structural stability among others.[1] There have been many lithium titanate preparation techniques over the world, mainly including solid phase method, sol-gel process, hydrothermal synthesis, templating, molten salt method, electro-spinning among others.[2], having great impacts on the materials properties to a varying degree.

Electro-spinning is an up-and-coming star among the preparation techniques of the fibers of different diameters. In the electric field formed by a high-voltage power supply, the precursor solution driven by the electric force is drawn into filaments, and such filaments eventually fall to a grounded collector. The fibers of different diameters can be obtained by adjusting different spinning parameters. Compared with solid phase method, electro-spinning has the submicron lithium titanate fibers of a smaller diameter prepared without agglomeration but having a larger surface area by which the specific capacity can be improved with a full contact with the electrolyte and the complete lithium intercalation/de-intercalation. In contrast with other techniques of preparing nano-materials, the water absorption of the sub-micron materials prepared by electro-spinning is significantly reduced, thereby effectively controlling the side reaction of any gas generated by the decomposition of water in the lithium titanate cell. [3,4]

2. Experiment scheme

2.1. Preparation of the lithium titanate fibers

The present experiment combines sol-gel process and electro-spinning, with lithium acetate and titanium Isopropoxide respectively as the lithium and the titanium source, the molar ratio of lithium to titanium of 0.82, polyvinyl pyrrolidone (PVP), a high molecular polymer, as the template for reaction and absolute ethyl alcohol as the solvent. Spinning parameter settings: 20kV, and 15cm(spinning distance). The preliminary experimental results show, a sufficient electric force is assured by 20kV, and combining with the given electro-spinning distance, the precursor of lithium titanate is stretched into submicron lithium titanate precursor filaments.

The prepared precursor filaments are placed into a muffle furnace, with the calcination conditions as follows: the temperature rises from room temperature to 600℃ at the rate of 2 ℃·min^{-1}, and to 750℃ at the same rate after constant temperature for 2h, and after constant temperature for 5h, lithium titanate filaments are finally obtained.

2.2. Morphological characterization

The filament morphology is observed by SEM(Hitachi X−650) before and after calcination.

2.3. Electrochemical test

The prepared lithium titanate, acetylene black and PVDF are evenly mixed by the mass ratio of 8:1:1, and uniformly coated on the copper foil, and the button cell is fabricated with a lithium metal sheet as the counter electrode. The cell is tested by cyclic voltammetry under the following conditions: under the open-circuit potential, the voltage range is 1-2.5V[5], and the scanning is 0.5mV·s^{-1}, with a total of 5 cycles. Then, at room temperature, the cell undergoes the charge-discharge test, with the aim of characterizing the electrochemical performance of lithium titanate fibers.

3. Results and discussions

3.1. Morphological characterization

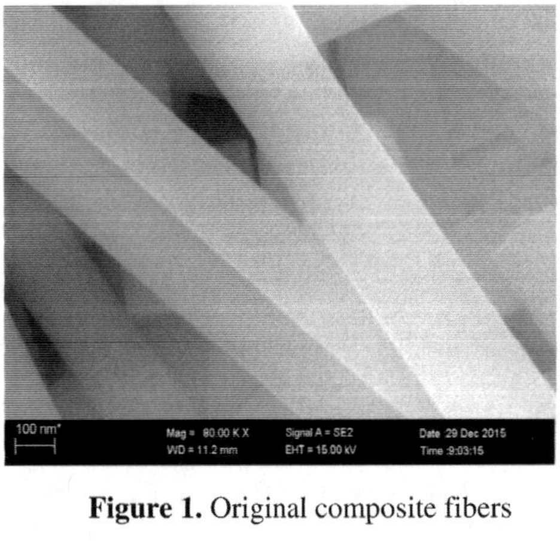

Figure 1. Original composite fibers

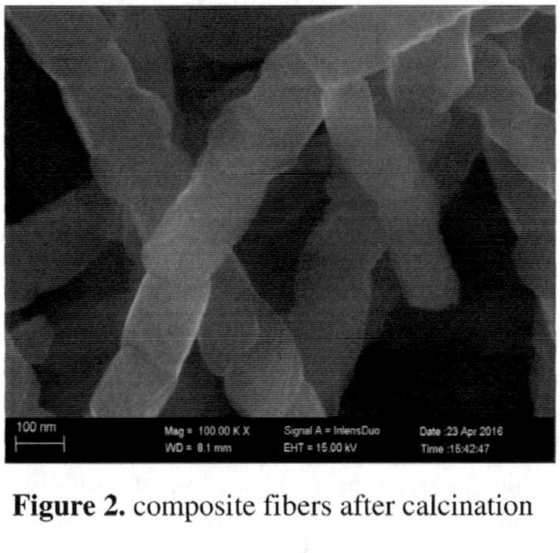

Figure 2. composite fibers after calcination

Figure 1 & 2 are respectively the SEM image of the lithium titanate precursor fibers and that of the lithium titanate fibers obtained by calcination, and it's found that, the diameter of the lithium titanate fibers prepared by electro-spinning is 150~200nm. The lithium titanate precursor fibers before calcination is smooth silk-like under the electrostatic interaction and the bonding effect of the polymer, without agglomeration; however, after calcination, because lithium titanate is produced in the reaction between the lithium and titanium sources at high temperature and the polymer at a high temperature generate gas while being calcined, thus forming surface irregularities which enlarge the specific surface area of the submicron lithium titanate materials. In contrast with nanometer materials, the submicron lithium titanate reduces the water absorption, thereby cut back the side reaction on the decomposition of water in the cell.

3.2. Electrochemical Test

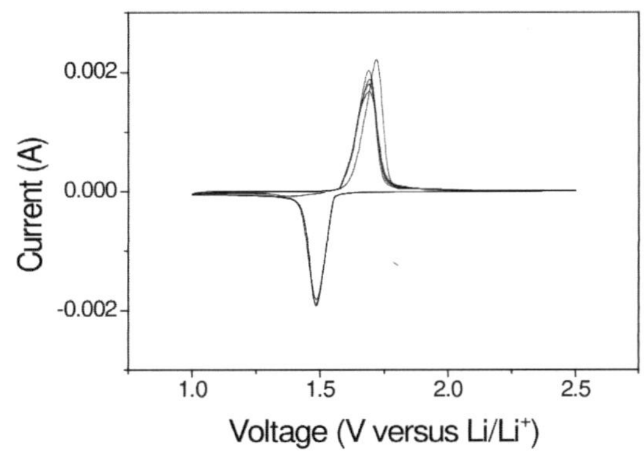

Figure 3. cyclic voltammetry curves of the button cell of submicron lithium titanate

Figure 3 shows the cyclic voltammetry curves of the button cell of submicron lithium titanate, with a singlet oxidation peak and a singlet reduction peak, indicating there is no side reaction but intercalation/de-intercalation in the charge-discharge process; and, both the oxidation peak and the reduction peak are symmetric in the peak pattern, and the peak current ratio is approximately 1, which indicates the redox reaction is reversible.

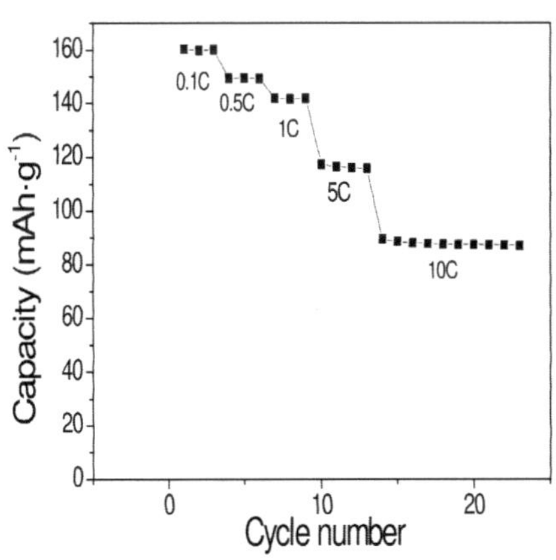

Figure 4. Cycle-Capacity Graph the button cell of submicron lithium titanate

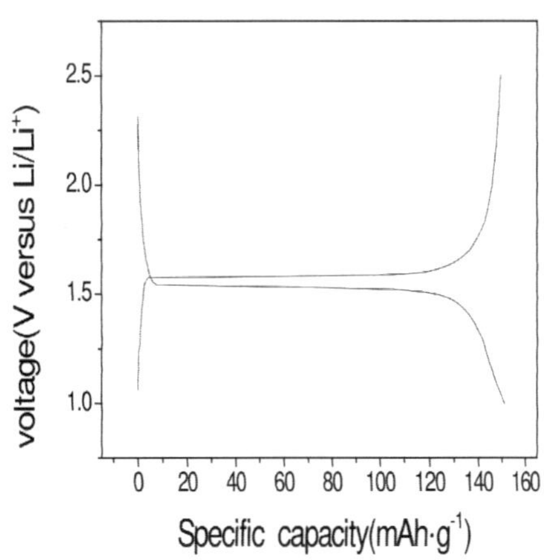

Figure 5. Capacity-Voltage Graph the button cell of submicron lithium titanate titanate

Figure 4 is the cycle-capacity graph of a button cell of lithium titanate at 0.1, 0.5, 1 and 5 and 10C after being assembled, and it's observed thatthe specific capacity of the submicron lithium titanate fibers at low rate is about 160mAh·g^{-1}. As the rate increases, the intercalation/de-intercalation time is shortened, and part of the sites for lithium ions lose their intercalation / de-intercalation abilities due to interfacial reaction rate is less than lithium ion transport rate, leading to a decline of the capacity. Figure 5 is the capacity-voltage graph of the button cell at 0.5C, the capacity retention rate in the charge-discharge process is relatively high, and the voltage difference between the charge platform and the discharge platform is smaller, indicating less polarization in the process.

4. Conclusions

In this paper, the submicron lithium titanate fibers were prepared by combining electro-spinning with certain calcination condition. Seen from the SEM images, the fibers with the diameter in the sub-micron level and free of aggregation, have a larger surface area. The specific capacity of the prepared button cell at low rate can reach 160mAh·g^{-1}. It's revealed by cyclic voltammetry that, the lithium titanate fibers in the intercalation/de-intercalation does not have any side reaction, but having high redox reversibility.

Acknowledgment

Thanks for being funded by the Technology Research Program of the State Grid about Preparing Submicron Lithium Titanate-based Materials for Energy Storage Batteries (DG71-15-042) .

References:

[1] Zhong H, XU H. *The Chinese Journal of Nonferrous Metals*, 2004, **14(2)**: 157−161.

[2] Bing Y, Dianbo R, Guansheng F, Zhaorong C, Chengyang W. Advance in Lithium Titanate Anode Materials [J]. *Materials Review*, 2015, **S1**: 157-162.

[3] BELHAROUAK I, KOENIG JR G M, TAN T S, et al. Performance Degradation and Gassing of Li4Ti5O12/LiMn2O4 Lithium-Ion Cells [J].*J Electrochem Soc*, 2012, **159(8)**: A1165–A1170.

[4] DING Z J, ZHAO L, SUO L M, et al. Understanding the Effects of Carbon and Nitrogen-doped Carbon Coating on the Electrochemical Performance of Li4Ti5O12 in Lithium Ion Batteries: A Combined Experimental and Theoretical Study [J]. *Phys Chem Chem Phys*, 2011, **13**: 15127–15133.

[5] Xu R, Li J R, Tan A, Tang Z L, Zhang Z T. Novel lithium titanate hydrate nanotubes with outstanding rate capabilities and long cycle life[J]. *Journal of Power Sources*, 2011,**196**: 2283–2288.

XPS and NMR analysis on $12CaO\cdot7Al_2O_3$

R K Pan[1,2], S Feng[2] and H Z Tao[2,3]

[1]School of Materials Science and Engineering, Hubei University, Wuhan 430062, China
[2]State Key Laboratory of Silicate Materials for Architectures, Wuhan University of Technology, Wuhan 430070, China

E-mail: thz@whut.edu.cn

Abstract. $12CaO\cdot7Al_2O_3$ (C12A7) glass was prepared by the melt quenching method. The glass transition temperature (T_g) and the crystallization temperature (T_c) of C12A7 glass are about 1050 K and 1194 K, respectively, measured by the differential scanning calorimetry (DSC). The structure of C12A7 glass was investigated by X-ray photoelectron spectroscopy (XPS) as well as magic angle spinning-nuclear magnetic resonance spectroscopy (MAS-NMR). Analysis shows that Al coordination number is about four in C12A7 glass, in which AlO_4 tetrahedrons and bridging oxygens (BO) constitute the glass network. Ca^{2+} produces a few of non-bridging oxygens (NBO), which become neighbours of Al.

1. Introduction

Calcium aluminates $(CaO)_x(Al_2O_3)_{1-x}$ are main component materials in the Earth's mantle and aluminous cement, which have been focused by many researchers [1,2]. Calcium aluminates are very fragile glass former and do not contain typical network-forming cations. The structure of liquid $(CaO)_x(Al_2O_3)_{1-x}$ with x=0.33, 0.5, 0.75 was investigated in a laser heated aerodynamically levitation furnace by neutron and high energy x-ray diffractions [3]. Reports show that the coordination number (CN) of Al and Ca is around 4.1 and 5-6, respectively, and the bond length of Al-O and Ca-O is 1.765 Å and 2.38 Å, respectively [4]. This implies that Al is in tetrahedral coordination and Ca is in distorted octahedral sites.

C12A7 also can be used as transparent electride materials because there are 12 sub-nanometer-sized cages in every unit cell, which can accommodate electrons and get high electronic conductivity. H. Hosono prepared room-tenperature-stable electrodes ($C12A7:e^-$) from C12A7 glass, which exhibits unique electronic properties such as high electronic conductivity [5]. C12A7 glass was annealed in reduced atmosphere to be crystallized. The temperature is important during the annealing process. In this paper, C12A7 glass was prepared. The glass transition temperature (T_g) and crystallization temperature (T_c) were determined by DSC test. The structure of C12A7 was discussed by XPS and NMR tests.

2. Experimental

2.1. Glass preparation

C12A7 glass samples were prepared by the conventional melt quenching method. The raw materials $CaCO_3$ and Al_2O_3 (99.9%) were thoroughly mixed and then melted in a corundum crucible. The melting process was done at the temperature 1900 K for 10 h. The melt was quenched between two iron plates to obtain amorphous C12A7 samples.

2.2. DSC and XRD tests

T_g and T_c were determined using a standard calorimetric method. The calorimetric measurement was performed using a differential scanning calorimetric (DSC) instrument (STA 449 F1, Netzsch). The glass samples were cut to dimensions of approximately 4×4×0.8 mm^3 and the bottom surface was optically polished. The measurements were conducted in a purged argon atmosphere (40 mL/min). The upscan rate was 20 K/min. The crystalline structure of C12A7 samples after DSC test was analyzed by an X-ray diffractometer (Bruker, D8 Discover) with Cu K_α radiation (λ = 0.15405 nm).

2.3. XPS test

XPS measurement was carried out on an X-ray photoelectron spectrometer (Thermo Scientific, VG Multilab 2000) with Al Kα radiation ($h\nu$=1486.6 eV) under UHV (1.33x10^{-8} Pa). The energy resolution was set to 0.5 eV for wide scans and 0.05 eV for narrow scans, respectively. The analyzer pass energy was set to 20 eV for narrow scans. All peaks are calibrated by the carbon deposit C 1s binding energy at 284.8 eV.

2.4. NMR test

^{27}Al MAS-NMR experiment of C12A7 glass was carried out using a 9.8 Tesla spectrometer (Bruker BioSpin GmbH) corresponding to a ^{27}Al resonance frequency of 104.26 MHz. The spinning speed was set at 12 kHz for the MAS rotor (4 mm ZrO_2). The recycle time is 1 s and pulse width is 0.9700 μs.

3. Results and discussion

Figure 1 shows the DSC up-scanning curve of the C12A7 glass sample from 330K up to 1370 K at 20 K/min. During heating, C12A7 glass experienced an obvious glass transition range from 1000 to 1100 K, and began to crystallize after 1120 K. The glass transition temperature is 1050 K. There are two sharp and intense crystallization peaks at 1194K and 1280 K, respectively. The temperature gap between the offset of glass transition and the onset of crystallization is very narrow, which indicates that C12A7 glass might have relatively poor glass forming ability.

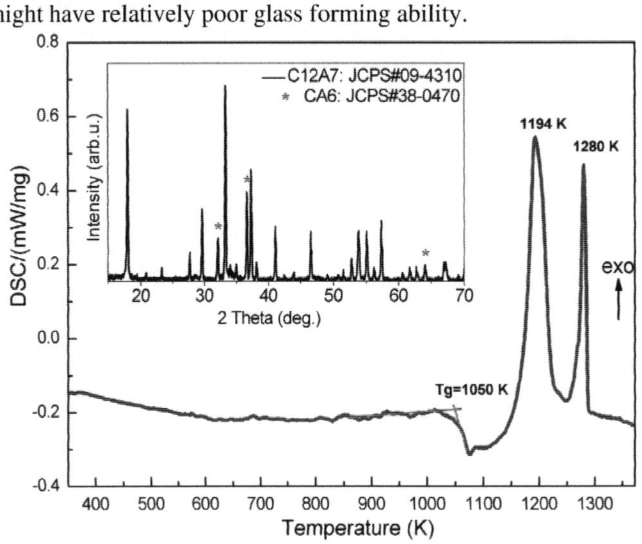

Figure 1. DSC curve of C12A7 glass sample from 350 to 1380 K up-scanned at 20 K/min. The inset is XRD pattern of C12A7 glass after DSC test.

To identify the peaks of crystallization, XRD measurement was done for the glass sample after DSC annealing. The inset in Figure 1 shows that there are two crystalline phases. The main phase is C12A7 and the other is CA6. So it can deduce that the first crystallization peak at 1194K is related to C12A7. The second peak at 1280 K is related to CA6. Reports give that the exothermic peak near 1090 K is due to the crystallization of CaO [6], the crystallization of γ -Al_2O_3 and formation of α-Al_2O_3 occur at about 1220 K [7]. The transformation of non-equilibrium $CaO{\cdot}Al_2O_3$ to equilibrium C12A7 has up to the temperature 1320 K [8]. But these exothermic peaks with higher temperatures in above reports were got from annealing dried gels using sol-gel method, not glass samples. It seems the crystallization temperatures are much lower using crystallizing glass method than using sol-gel method.

Figure 2 shows the XPS survey spectrum of C12A7 glass sample, in which Ca, Al, and O can be detected. The stoichiometry verified well as the nominated C12A7, as calculated from Ca 2p, Al 2p and O 1s peak areas amounts.

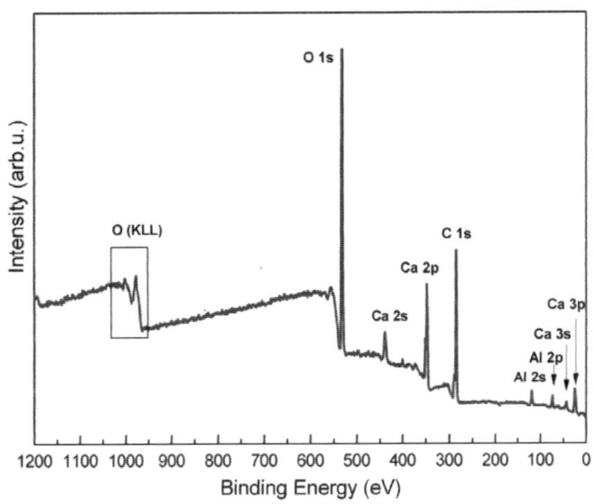

Figure 2. XPS survey spectrum of C12A7 glass.

Figure 3 shows the XPS spectra of O 1s, Ca 2p, Al 2s and Al 2p, respectively. The O 1s spectrum in Figure 3(a) can be clearly deconvoluted into two distinct contributions, which peaks are at around 531 eV and 531.7 eV, respectively. The high energy component is attributed to bridging oxygen (BO) atoms, while the low energy component is assigned to non-bridging oxygen (NBO) atoms [9]. The ratio of NBO/BO is about 1:4 by calculating the areas of deconvoluted O 1s XPS spectrum.

Ca 2p narrow XPS spectrum in Figure 3(b) has two peaks at around 350.7 eV and 347 eV, which are related to Ca $2p_{1/2}$ and Ca $2p_{3/2}$, respectively. Due to spin orbit splitting, Ca 2p XPS spectrum is clearly distinguishable two peaks. The narrow and sharp peaks indicate that Ca connected with O and without any hydrated specimen [10, 11]. There is a report shows that Ca 2p XPS peaks shift to lower binding energy in $Pr_{1-x}Ca_xMnO_3$ crystalline [12]. Comparing with that of glass sample, there are less O atoms around Ca in $Pr_{1-x}Ca_xMnO_3$ crystalline, which caused the peak position shift.

The Al 2s and 2p XPS spectra in Figure 3(c) and (d) have peak positions at around 118.8 eV and 74 eV, respectively, which are consistent with the reports [13, 14]. That indicates that the valence state of Al in C12A7 is as same as that of Al_2O_3. So we can conclude that only Ca-O bonds and Al-O bonds exist in C12A7 glass sample.

Figure 4 shows the NMR spectrum of C12A7 glass, in which a sharp and narrow peak (69.9 ppm) indicates that Al in C12A7 glass is in four-fold (tetrahedral) coordination connecting to oxygen. The asymmetric central peak is due to the effects of the second order quradrupolar interaction, which can not be averaged in the solid-state MAS-NMR. Using the method mentioned by P F McMillan et al.[15], the isotropic chemical shift can be got: δ_{iso}=81.4 ppm, which is nearly equal to the report in Ref.[15]. P F McMillan pointed out there was no evidence for Al in any other coordination state in the glasses with $CaO:Al_2O_3 \geqslant 1$, independent of the quench rate. So C12A7 glass can be described in terms of polymerization state of the aluminate framework with Ca^{2+} ions stuffing sites within the framework structure. It mainly contains some six-membered rings of AlO_4 tetrahedra, in which each AlO_4 group has two terminals or non-bridging. Al-O bonds and two bridging oxygens share in intertetrahedral Al-O-Al linkages to form Q^2 units.

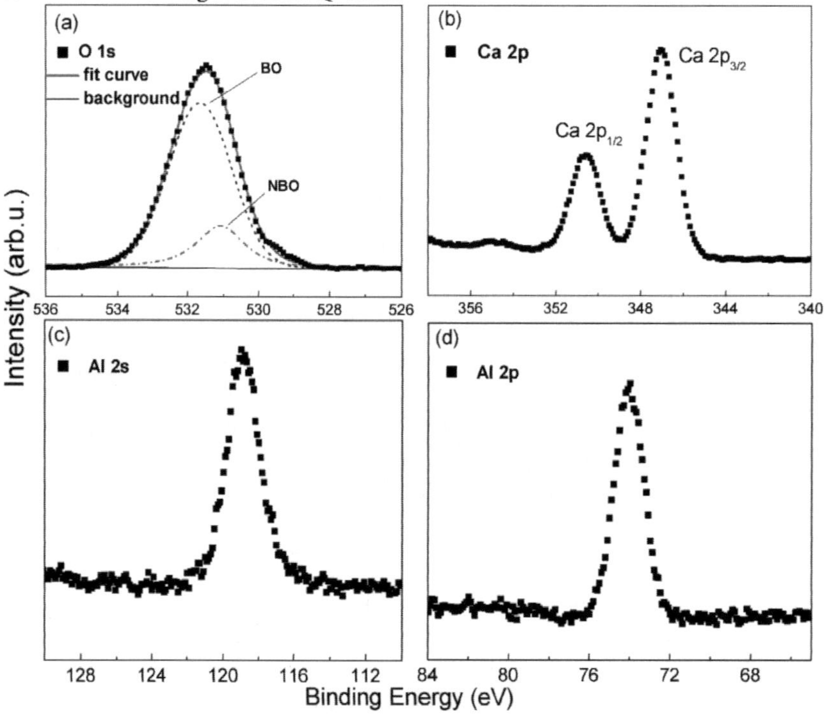

Figure 3. XPS spectra of O 1s (a), Ca 2p (b), Al 2s (c) and Al 2p (d), respectively.

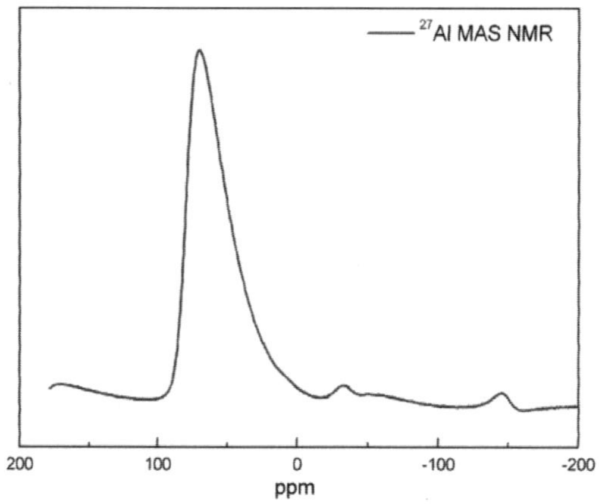

Figure 4. ^{27}Al MAS-NMR spectrum for C12A7 glass tested at room temperature.

4. Conclusions

In this work, C12A7 glass was prepared. DSC and XRD tests indicate that the glass transition temperature and the crystallization temperature of C12A7 are about 1050 K and 1194 K, respectively. The ability of glass formation of C12A7 is poor. Analysis on XPS and NMR indicates that Al coordination number is about four in C12A7 glass. And AlO_4 tetrahedron and bridging oxygens (BO) constitute the glass network. Ca^{2+} produces a few of non-bridging oxygens (NBO), which become the neighbours of Al.

Acknowledgments

We are grateful to the National Natural Science Foundation of China (no. 51272072) and the funding of State Key Laboratory of Silicate Materials for Architectures (SYSJJ2016-14).

References

[1] Jerebtsov D A and Mikhailov G G 2001 *Ceram. Inter.* 27 25-28
[2] Drewitt J W E, Hennet L, Zeidler A, Jahn S, Salmon P S, Neuville D R and Fischer H E 2012 *Rev. Lett.* 109 235501
[3] Drewitt J W E, Jahn S, Cristiglio V, Bytchkov A, Leydier M, Brassamin S, Fischer H E and Hennet L 2012 *J. Phys.: Condens. Matter* 24 099501
[4] Neuville D R, Henderson SG, Cormier L and Massiot D 2010 *Am. Miner.* 95 1580–1589.
[5] Hosono H, Kim S W, Miyakawa M, Matsuishi S and Kamiya T 2008 *J Non-Cryst. Solids* 354 2772-2776
[6] Meng Q, Lin J, Fu L, Zhang H, Wang S and Zhou Y 2001 J. Mater. Chem. **11** 3382-6
[7] Assem E E 2005 J. Phys. D: Appl. Phys. **38** 942-5
[8] Sung Y M and Dunn S A 1996 J. Mater. Sci. **31** 4741–4
[9] Chowdari B V R, Rao G V S and Lee G Y H 2000 *Solid State Ionics* 136–137 1067–1075
[10] Voigts F, Bebensee F, Dahle S, Volgmann K and Friedrichs W M 2009 *Surface Science* 603 40–49
[11] Comans R N J, Eighmy T T and Shaw E L1996 *Surf. Sci. Spectra* 4 150-156
[12] Dalai M K, Kundu R, Pal P, Bhanja M, Sekhar B R and Martin C 2011 *J Alloys Compd.* **509** 7674–7676
[13] Benito N, Díaz D, Vergara L, Galindo R E, Sánchez O and Palacio C 2011 *Surf. Coat. Tech.* 206 1484–1489
[14] Brand J V D, Sloof W G, Terryn H and Wit J H W 2004 *Surf. Interface Anal.* **36** 81–88
[15] McMillan P F, Petuskey W T, Coté B, Massiot D, Landron C and Coutures J P 1996 *J Non-Cryst. Solids* 195 261-271

Effects of hot-rolling reduction on microstructure, texture and magnetic properties of high silicon steel produced by strip casting

D Y Hou, H J Xu, H T Jiao, C W Zhao, W Xiong, J P Yang, W Z Qiu and Y B Xu[1]

State Key Laboratory of Rolling and Automation, Northeastern University, Shenyang 110819, China

E-mail: yunbo_xu@126.com

Abstract. Non-oriented Fe-7.1wt.% Si as-cast strips were produced by twin-roll strip casting process. Then the as-cast strips were hot rolled with different reductions, followed by warm rolling and final annealing. The microstructure, texture evolution and magnetic properties were investigated in detail. The texture of hot rolled sheets with 40% reduction showed strongest {001}<110> texture, whereas the dominated texture was turned into {110}<001> and {110}<112>as the reduction was increased to 56% and 68%. After warm rolling and annealing, the average grain size was decreased firstly and then increased with an increase in hot rolling reduction. In the case of 40% hot rolling reduction, the recrystallization texture was dominated by strong γ (<111>//ND) texture. With an increase in hot rolling reduction, the γ texture was gradually weakened while α (<110>//RD) texture was enhanced. In addition, relatively stronger {100} texture was presented in the sheet of 68% hot rolling reduction. The highest B_{50} value attained was 1.66 T and the lowest $P_{10/400}$ was 24.26 W/kg at a reduction of 56%.

1. Introduction

High silicon electrical steels with high resistivity and almost zero magnetostriction are excellent soft magnetic materials for high-frequency motors, transformers, and magnetic shields [Refs]. However, it is extremely hard to produce high silicon steel by conventional rolling process because of the decreasing workability with the increase of silicon content. In order to overcome the poor workability, several methods have been proposed, such as rapid quenching [1], rapid solidification [2], spray forming [3], chemical vapor deposition (CVD) [4] and strip casting [5]. The only successful commercial manufacturing process is CVD technology developed by NKK Corp. In addition, the twin-roll strip casting (TRSC) technology is another promising alternate, which produces thin sheet directly from molten steel.

Extensive researches on the strip casting high silicon steel had been carried out [6-11]. Li [6] et al. investigated the microstructure and texture evolution along the whole processing routes. The beneficial recrystallization textures were formed in the products, such as {001}<210>, {001}<010>, {115}<5 10 1> and {410} <001>. Liu [7-8] et al. reported the properties of strip casting high silicon steels with different forming routes, and the microstructure, texture, formability and magnetic properties of the as-cast strips were studied in detail in the research. In addition, the properties of the alloy doped with cerium, the effects of warm-rolling temperature, and the tensile properties of the sheets rolled at elevated temperatures were also investigated [9-11]. Although TRSC has mostly solved the problem of workability, the brittleness is still an obstacle for industrialization of high silicon steels. Hot rolling process is an important procedure to reduce the thickness of sheet. While, different processes may have different effects on the microstructure and texture. Thus, the further study on the hot rolling process in high silicon manufacture is necessary.

In the present work, high silicon as-cast strips were produced by twin-roll strip casting and followed by different hot rolling processes. Then the hot-rolled sheets were warm-rolled and annealed. The effects of rolling reduction on the evolution of microstructure, texture and magnetic properties were investigated.

2. Experimental procedures

The chemical composition of the alloyed strip was examined with Optima 8300DV inductively coupled plasma-atomic emission spectrometer (Table 1.). The as-cast strips with 140 mm width and 2.5 mm thickness were fabricated by a vertical type twin-roll strip caster. The as-cast strips were hot rolled to 1.5 mm, 1.1 mm and 0.8 mm at 1050°C respectively. Then the sheets were warm-rolled to 0.5 mm at 650 °C. After that, all the rolled sheets were finally annealed at 1050°C for 5 min.

Table 1. Chemical composition of the as-cast strip (mass fraction, %)

C	Si	Al	Mn	S
0.005	7.09	0.11	0.23	0.016

The microstructures of the as-cast, rolled and annealed sheets were characterized by metallographic investigations. Textures were quantitatively examined with $Co_{K\alpha 1}$ radiation in the Bruker D8 Discover X-ray diffraction (XRD). Electron backscattered diffraction (EBSD) analysis was partly performed to investigate the microstructure and micro-texture of hot-rolled samples on the FEI Quanta 600 scanning electron microscopy (SEM). Results of the texture measurements were presented in the form of Orientation Distribution Function (ODF). The magnetic inductions at 800 Am^{-1} (B_8), 5000 Am^{-1} (B_{50}) and the iron losses at 50 Hz and 1.5 T ($P_{15/50}$), 400 Hz and 1 T ($P_{10/400}$), were respectively measured using a single sheet tester in both rolling and transverse directions of the annealed specimens with 100 mm length and 30 mm width.

3. Results and discussion

3.1. Microstructure and texture of the as-cast strip

Figure 1 shows the microstructure and texture of the as-cast strip. The microstructure of as-cast strip was consisted of some fine columnar grains at the subsurface layer and coarser equiaxed grains at the center layer (Figure 1a). The average grain size was 134 μm. The texture of as-cast strip was characterized by pronounced λ-fiber (<100>//ND) texture and weak α-fiber texture, as show in Figure 1b and c.

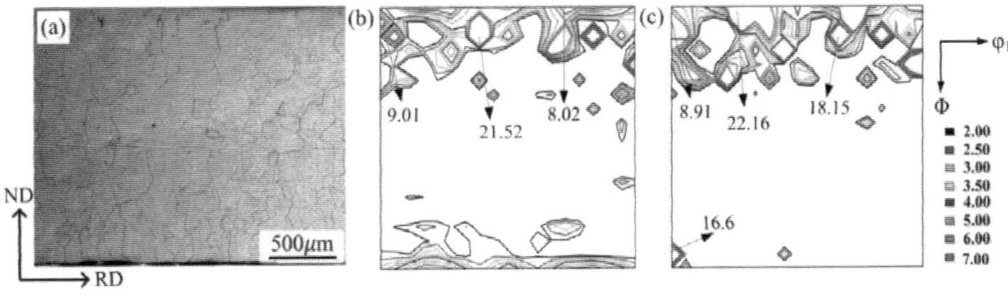

Figure 1. Microstructure (a) and ODF ((b) φ_2=0°, (c) φ_2=45°) of the as-cast strip.

3.2. Microstructure and texture of the hot rolled sheets

Figure 2 presents the microstructure of the hot rolled sheets. With the increase of the hot-rolling reduction, the average grain size increased gradually. The large grains in sample of 56% and 68% reduction were induced by the heat treatment during rolling passes. Figure 3 shows the texture of the hot rolled sheets. The hot rolling reduction had a significant influence on the texture evolution. The texture of the sheets with 40 % reduction was similar to the initial as-cast strip, which were characterized by strong {001}<110> and weak {001}<100> texture, as shown in Figure 3a. When the hot rolling reduction increased from 56% to 68%, λ-fibre texture almost disappeared, replaced by strong Goss ({110} <001>) and Brass ({110}<112>) texture respectively. In addition, the intensity of α-fiber texture increased firstly and then decreased slightly, as shown in Figure 3b and c.

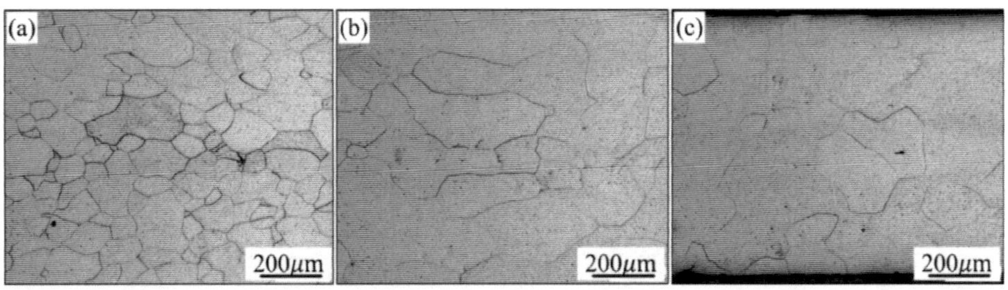

Figure 2. Microstructure of hot rolled sheets with reductions of: (a) 40%; (b) 56%; (c) 68%.

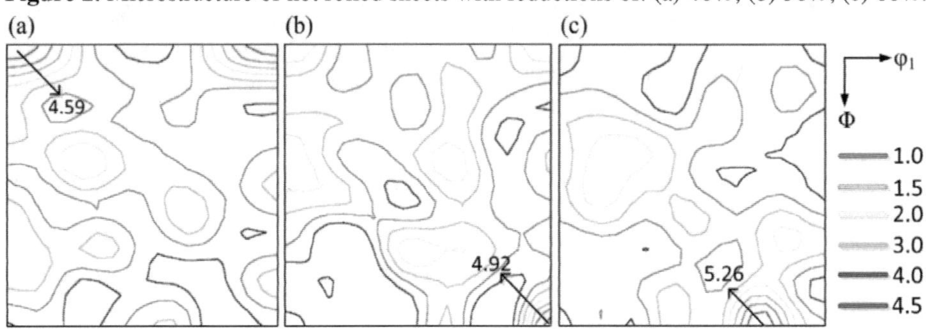

Figure 3. ODF (φ_2=45°) of hot rolled sheets with reductions of: (a) 40%; (b) 56%; (c) 68%.

3.3. Microstructure and texture of the warm-rolled sheets

Figure 4 shows the microstructure of the warm-rolled sheets. It was clearly found that the microstructure evolution highly depended on the rolling reduction. The shape of grains was the obvious elongated grains along the rolling direction. Some grains went through the whole visual field. As the hot rolling reduction of increased, the elongation of grains was more evident, and the average grain size was increased from 97.3 μm to 178.4 μm. Evident in-grain shear bands were not observed because the warm-rolling was carried out by multiple passes with small reduction and heat preservation for 4 min between passes. Figure 5 shows the texture of the warm-rolled sheets. The warm-rolled texture was characterized by strong γ-fiber (<111>//ND) texture, λ-fiber and weak α-fiber component. As the increase in hot rolling reduction, the intensity of γ-fiber texture was increased firstly and then decreased. While, the {100} texture was weakened gradually.

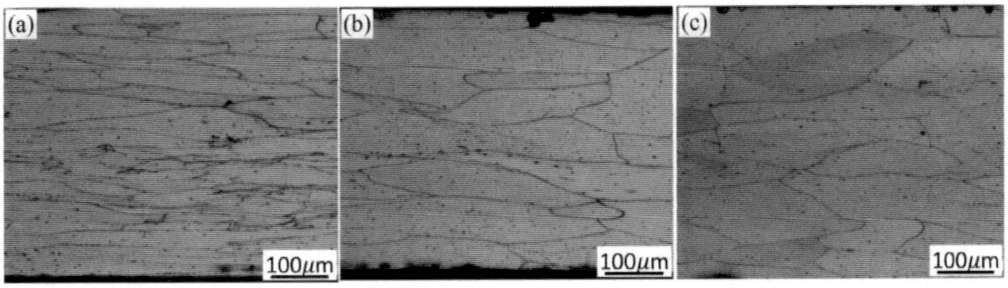

Figure 4. Microstructure of the warm-rolled sheets with hot rolling reduction of: (a) 40%; (b) 56%; (c) 68%.

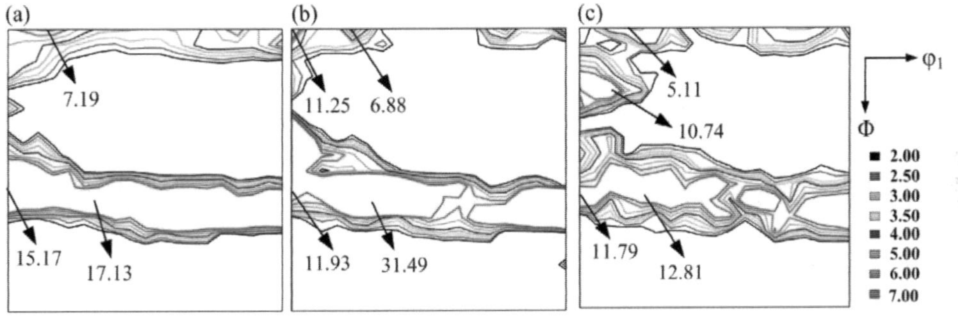

Figure 5. ODF (φ_2=45°) of the warm-rolled sheets with hot rolling reduction of: (a) 40%; (b) 56%; (c) 68%.

3.4. Microstructure and texture of the annealed sheets

Figure 6 shows the microstructure of the annealed sheets. It was found that the hot rolling reduction had an important role on the evolution of the final recrystallization microstructure. With the increase of hot rolling reduction, the average grain size of the annealed sheet was decreased at first and then increased. In the condition of 56% hot rolling reduction, the microstructural heterogeneity was the most obvious, as shown in Figure 6b. Figure 7 shows the texture of the annealed sheets. In the case of light hot rolling reduction (<56%), the texture was characterized by strong γ-fiber texture and weak α-fiber texture. The γ-fiber texture was reduced with the increase of rolling reduction. When the reduction reached 68%, the γ-fiber texture was weakest. It is interesting to note that strong γ-fiber texture was presented in the sample with 56% hot rolling reduction. As the reduction reached 68%, the final texture was more random, as shown in Figure 7c.

Figure 6. Microstructure of the annealed sheets with hot rolling reduction of: (a) 40%; (b) 56%; (c) 68%.

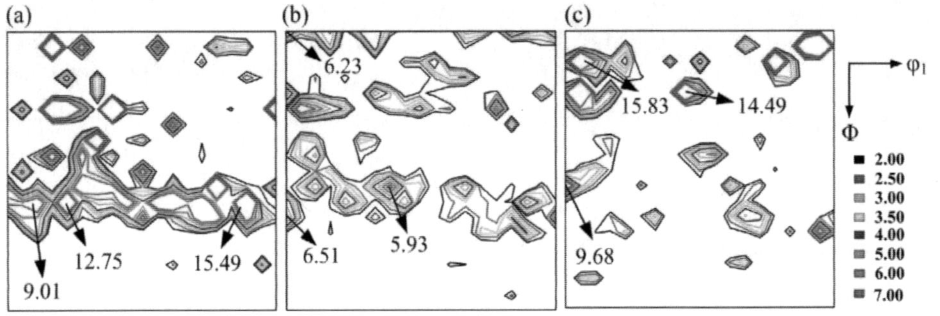

Figure 7. ODF (φ_2=45°) of the annealed sheets with hot rolling reduction of: (a) 40%; (b) 56%; (c) 68%.

3.5. Magnetic properties of the annealed sheets

Figure 8 show the magnetic properties of the final annealed sheets. It was found that the magnetic induction B_{50} increased at first and then decreased. As the reduction reached 56%, the magnetic induction B_{50} was 1.66 T. The magnetic induction B_8 without significant change decreased from 1.33 T to 1.31 T, as shown in the Figure 8a. The iron loss $P_{15/50}$ was reduced with increase in hot rolling reduction. As the reduction increased, however, the iron loss $P_{10/400}$ was decreased at first and then increased.

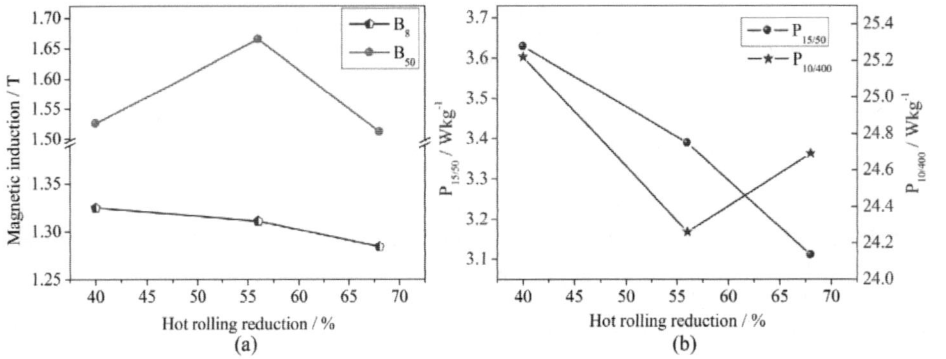

Figure 8. Magnetic properties of the annealed sheets: (a) magnetic induction; (b) iron loss.

4. Conclusions

It was found that the evolution of microstructure, texture and the magnetic properties highly depended on the hot rolling reduction. The initial as-cast strip was characterized by strong λ-fiber texture. The microstructure of hot rolled sheets was inhomogeneous. The λ-fiber texture decreased and the α-fiber, Goss texture increased with increasing hot rolling reduction. The warm-rolled texture was characterized by strong γ-fiber and weaker λ-fiber texture. After annealing, the texture of sample with 40% hot rolling reduction was dominated by strong γ-fiber, which was weakened with the increase in reduction. In addition, strong λ-fiber texture was developed in the sample with 56% hot rolling reduction. The magnetic induction B_{50} increased at first and then decreased, while the iron loss decreased at first and then increased. The highest B_{50} value attained was 1.66 T and the lowest $P_{10/400}$ was 24.26 W/kg at a reduction of 56%.

Acknowledgments
This work was financially supported by the National Natural Science Foundation of China (Nos.51674080, 51404155 and U1260204), the Program for New Century Excellent Talents in University (NCET-13-0111) and the Program for Liaoning Excellent Talents in University (LR2014007)。

References
[1] Roy R K, Panda A K, Ghosh M, Mitra A and Ghosh R N 2009 *J. Magn. Magn. Mater* **321** 2865-70
[2] Viala B, Degauque J, Baricco M, Ferrara E, Pasquale M and Fiorillo F 1996 *J. Magn. Magn. Mater.* **160** 315-7
[3] Kasama A H, Bolfarini C, Kiminami C S and Botta Filho W J 2007 *Mater. Sci. Eng. A* **449** 375-7
[4] Okada K, Yamaji T and Kasai K 1996 *ISIJ Int.* **36** 706-13
[5] Zapuskalov N 1999 *ISIJ Int.* **39** 463-70
[6] Li H Z, Liu H T, Liu Z Y, Lu H H, Song H Y and Wang G D 2014 *Mater. Charact.* **88** 1-6
[7] Liu H T, Li H Z, Li H L, Gao F, Liu G H, Luo Z H, Zhang F Q, Chen S L, Cao G M, Liu Z Y and Wang G D 2015 *J. Magn. Magn. Mater.* **391** 65-74
[8] Liu H T, Liu Z Y, Qiu Y Q, Sun Y and Wang G D 2012 J. *Mater Process.* Tec. **212** 1941-5
[9] Li H Z, Liu H T, Liu Z Y and Wang G D 2015 *Mater. Charact.* **103** 101-6
[10] Li H Z, Liu H T, Liu Y, Liu Z Y, Cao G M, Luo Z H, Zhang F Q, Chen S L, Lyu L and Wang G D 2014 *J. Magn. Magn. Mater.* **370** 6-12
[11] Li H Z, Liu Z Y 2015 *Mater. Sci. Eng. A* **639** 412-6

IOP Conf. Series: Materials Science and Engineering **167** (2017) 012019 doi:10.1088/1757-899X/167/1/012019

Preparation and characterization of high-quality perovskite $CH_3NH_3PbX_3$ (I, Br) single crystal

Yangyang Zhou, Chunmei Li, Yu Wang, Xiang Du, Pengyi Liu, Weiguang Xie[1]

Siyuan Laboratory, Guangzhou Key Laboratory of Vacuum Coating Technologies and New Energy Materials, Department of Physics, Jinan University, Guangzhou, Guangdong, 510632, P. R. China

E-mail: wgxie@email.jnu.edu.cn

Abstract. In this paper, inverse temperature crystallization and top-seed solution growth method are used to grow high-quality perovskite $CH_3NH_3PbX_3$ (I, Br) single crystal. The maximum crystal diameter is 6 mm for $CH_3NH_3PbI_3$, and 3 mm for $CH_3NH_3PbBr_3$. The results of XRD show that the crystals are tetragonal structure for $CH_3NH_3PbI_3$, cubic crystal orientation for $CH_3NH_3PbBr_3$. What is more, crystals exhibit excellent stability in air and can be maintained for eight months. Furthermore, the crystal growth processes are described in detail. The results demonstrate that $CH_3NH_3PbBr_3$ single crystal growth is more sensitive to the environment.

1. Introduction

As the world energy crisis and environmental pollution, it is increasingly urgent to find a clean, abundant energy. Scientists are increasingly concerned about the use of solar energy because of its clean and renewable. Solar cells as a new type of photoelectric conversion devices can effectively solve the energy crisis. In recent years, development of perovskite $CH_3NH_3PbX_3$ solar cell is rapid and the solar-to-energy conversion efficiency of solar cells based on perovskite active layer has exceeded 22% [1-5]. Although perovskite solar cells have a breakthrough in efficiency, there are still many problems to be solved in practical terms. On the one hand, the size of the solar cell is only a few centimeters. There is a big technical problem for the large-size solar cells. On the other hand, perovskite is less stable in water and oxygen environments [6]. In order to optimize the performance of solar cells, the study of physical and chemical properties, the optical properties and stability of perovskite active layer are necessary[7-8]. The crystal material is very important for us to understand the nature of this material. So it is necessary to study perovskite crystals.

At present, many methods have been used to grow perovskite crystals [9] including adding PbI_2 crystals to the solution of CH_3NH_3I [10], anti-solvent vapor-assisted crystallization (AVC) [11], inverse temperature crystallization (ITC) [12-13], top-seed solution growth (TSSG) [14-15] and so on. High- quality, large-size crystals can be grown by these methods, but they only focus on the study of its photoelectric properties. It paid little attention to its stability and specific growth processes. In this study, high-quality perovskite crystals were grown by ITC and TSSG methods, and their growth processes and stability were studied.

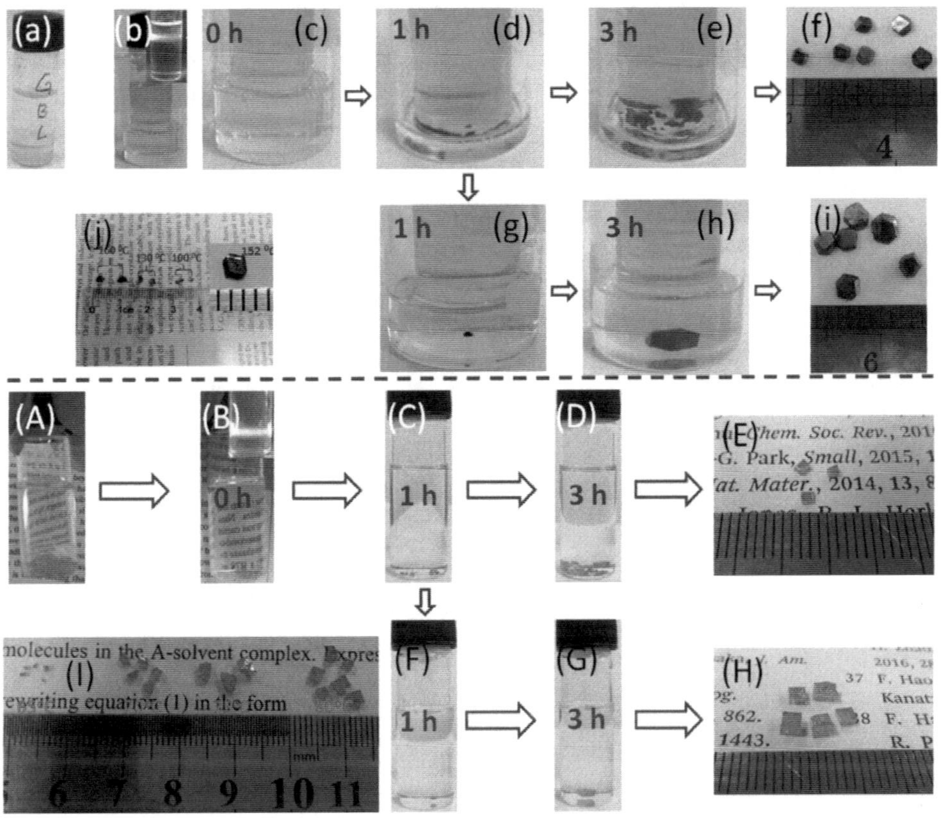

Figure 1. (a) & (A) is not completely dissolved perovskite $CH_3NH_3PbX_3$ (I, Br) precursor solution. (b) & (B) $CH_3NH_3PbX_3$ (I, Br) precursor solution, inset images show that solutions have Tyndall effect. (c)-(f) & (C)-(E) $CH_3NH_3PbX_3$ (I, Br) crystals growth processes directly at 1 hour and 3 hours. (g)-(i) & (F)-(H) $CH_3NH_3PbX_3$ (I, Br) seed growth processes at 1 hour and 3 hours. (j) & (I) The grown $CH_3NH_3PbX_3$ (I, Br) crystals at different temperatures.

2. Crystal growth

For perovskite $CH_3NH_3PbI_3$ crystals, 0.3975g CH_3NH_3I and 1.1525g PbI_2 were thoroughly mixed in a glass vial, then added 5ml ϒ-GBL in Figure 1 (a). At this time, CH_3NH_3I and PbI_2 were not completely dissolved in the glass vital. It needs to be heated in oil bath at 60 °C to dissolve completely in Figure 1 (b). The completely dissolved perovskite $CH_3NH_3PbI_3$ solution shows a bright yellow color in Figure 1 (c). In addition, the solution also exhibits a Tyndall effect in Figure 1 (b) inset. A bright pathway indicates the presence of colloidal particles in solution. And this is beneficial to the crystal growth. At last, the perovskite solution was heated in oil bath at 152 °C in Figure 1 (b)-(e). A regular, the diameter of 3 mm size of the crystals can be grown in 3 hours in Figure 1 (f). In order to obtain larger crystals, the seed is removed from Figure 1 (d) and placed in a fresh perovskite solution to continue growing in Figure 1 (g). Larger crystals were obtained in 3 hours in Figure 1 (h). By repeating this step, a crystal of desired size can be obtained in Figure 1 (i). However, in the process of growing crystals, the temperature needs to be constantly adjusted in Figure 1 (j). When the temperature is below 150 °C, the growth of the crystal is small and irregular. When the temperature is higher than 160 °C, the crystal is large, but irregular. And the temperature is too high, the cap is easy to loose so as to affect the crystallization. Therefore, 152 °C is appropriate to grow large and regular crystals.

IOP Conf. Series: Materials Science and Engineering **167** (2017) 012019 doi:10.1088/1757-899X/167/1/012019

For perovskite $CH_3NH_3PbBr_3$ crystals, 0.183g $PbBr_2$ and 0.056gCH_3NH_3Br were mixed and 5ml DMF was added in Figure 1 (A). The solution is transparent, and also has a Tyndall effect after complete dissolution in Figure 1 (B). Unlike the growth processes of $CH_3NH_3PbI_3$ single crystal, the processes of dissolution and growth of the crystal processes need to be carried out in a constant temperature oven, because the $CH_3NH_3PbBr_3$ single crystal growth processes are more sensitive to the external environment. Figures 1 (C)-(D), (F)-(G) are crystal growth processes. The orange crystals were grown out 1 hour later. And 3 hours later, the volume of crystals reached the maximum. When the temperature is 80 °C, the crystal is large and regular. It is worth noting that the glass vital can not be shaken during seed crystals growth processes in Figures 1 (F)-(G).

In order to obtain high-quality crystals, the concentration of perovskite solution, the heating temperature and the external environment must be taken into account.

3. Results & discussion

Fresh crystal powders X-ray diffraction (XRD) spectra result shows that $CH_3NH_3PbI_3$ crystals are tetragonal structure in Figure 2 (a). The XRD peaks are strong and the half-peak width is narrow. All of these indicate that the perovskite $CH_3NH_3PbI_3$ crystal has good crystallinity. UV-visible reflectance spectra was used to calculate the band gap of samples. According to the Kubelka-munk equation [16], $F(R) = (1-R)^2/(2R)$, where R represents reflectance (%). Figure 2 (c) demonstrates the band gap of $CH_3NH_3PbI_3$ single crystal, Eg=1.50eV, by calculating. In addition to band gap, the photoluminescence spectra of $CH_3NH_3PbI_3$ single crystals is also measured in Figure 2 (c) red line. The result shows that the position of photoluminescence peak of $CH_3NH_3PbI_3$ single crystal is located at 820 nm when excitation wavelength is 545 nm.

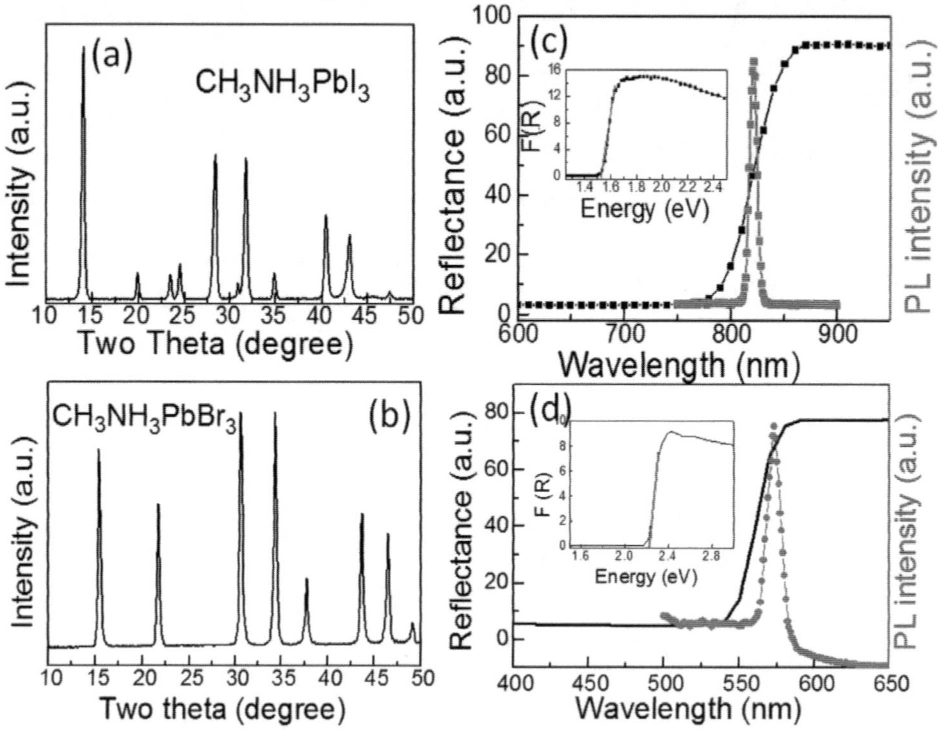

Figure 2. (a) & (b) XRD spectra of perovskite $CH_3NH_3PbX_3$ (I, Br) crystals. Photoluminescence spectra and reflectance spectra of perovskite crystals, $CH_3NH_3PbI_3$ (c), $CH_3NH_3PbBr_3$ (d).

In the same way, XRD result shows that perovskite $CH_3NH_3PbBr_3$ belongs to the cubic structure in Figure 2 (b). Its bandgap is 2.21eV and the position of photoluminescence peak of samples is located at 572 nm in Figure 2 (d). This is consistent with the values of report by using inverse temperature crystallization (ITC) [13].

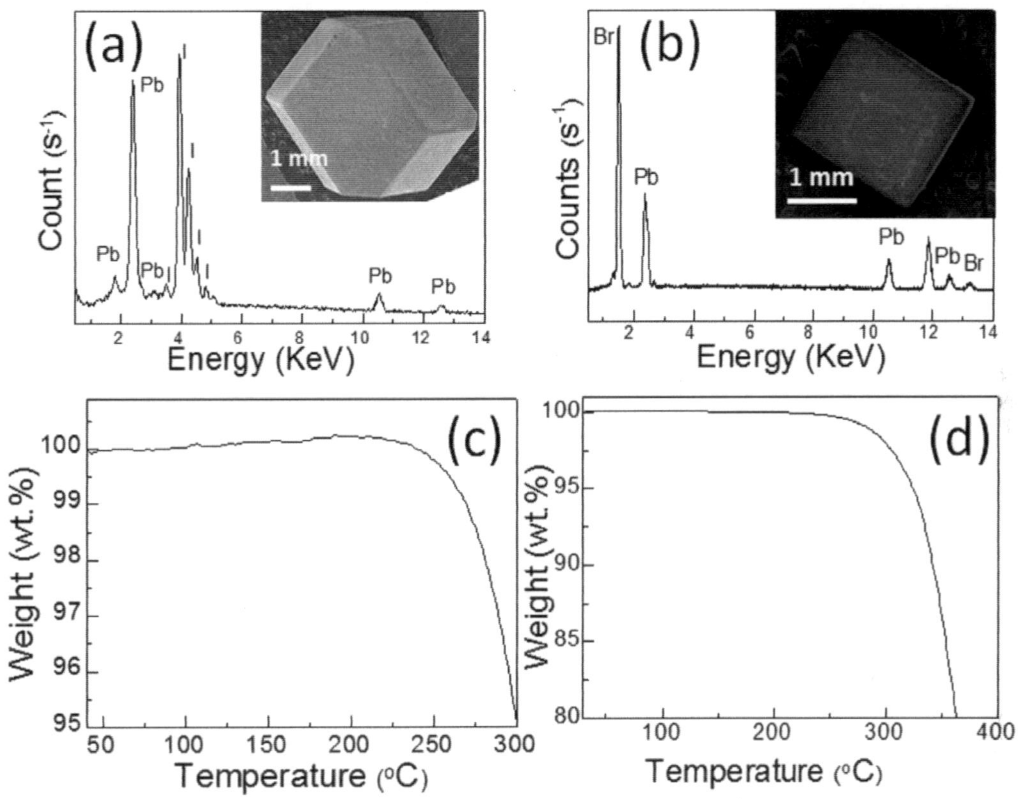

Figure 3. (a) & (b) EDS spectra of $CH_3NH_3PbX_3$ (I, Br).The inset is SEM image of $CH_3NH_3PbX_3$ (I, Br). (c) & (d) Thermogravimetric curve for $CH_3NH_3PbX_3$ (I, Br) respectively.

Figure 3 (a), (b) are the SEM images and EDS analysis of perovskite single crystal. The EDS results indicate that the atomic ratio of I and Pb is about 3.01,Br and Pb is 2.94, complying with its chemical formula. SEM images show that no grain boundaries are found on the surface in $CH_3NH_3PbX_3$ (I, Br). In order to study the thermal stability of the samples, the thermogravimetric (TG) was used. In Figure 3 (c),(d) show that $CH_3NH_3PbX_3$ (I, Br) can be stabilized to 240 °C. The perovskite $CH_3NH_3PbX_3$ (I, Br) single crystals exhibit great stability because a lot of $CH_3NH_3PbI_3$ material decompose at about 150 °C [14].

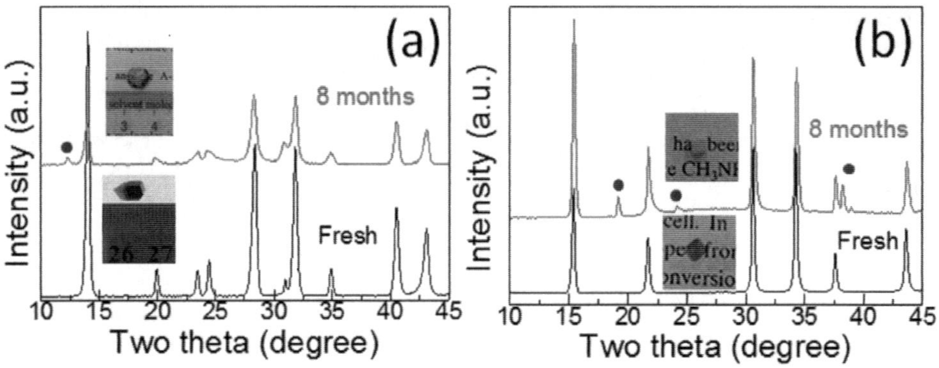

Figure 4. (a) & (b) XRD spectra of the as-prepared $CH_3NH_3PbX_3$ (I, Br) and placing eight months in the air. The insets are photos of samples in as-prepared and eight months.

In order to study the stability of crystal in air, the perovskite $CH_3NH_3PbX_3$ (I, Br) crystals were placed in air (with temperature at 25 °C and humidity of 55% to 65%) for eight months. Fresh perovskite $CH_3NH_3PbI_3$ are regular and appear black. After eight months later, the surface of samples was rough and had a large amount of yellow powders in Figure 4 (a) inset. XRD results demonstrate that the diffraction peak intensity decreases, the full width at half maximum becomes wide and PbI_2 peak appears (at blue circle) when the sample is left for eight months in Figure 4 (a). For $CH_3NH_3PbBr_3$, the fresh sample was orange-red, and after eight months it turned to yellow. There are also $PbBr_2$ peaks (at blue circle) of X-ray diffraction spectra. We speculate that the change of crystal surface morphology and color is due to CH_3NH_3X (I, Br) escaping. In the air, as many other $CH_3NH_3PbX_3$ (I, Br) materials beginning to decay a few days later, perovskite single crystals show a better stability.

4. Conclusion

In conclusion, according to inverse temperature crystallization and top-seed solution growth methods, high-quality perovskite $CH_3NH_3PbX_3$ (I, Br) crystal are grown. The crystal growth processes are described in detail. Powder XRD and UV-visible reflectance spectrum confirmed that the crystals were $CH_3NH_3PbX_3$ (I, Br) single crystals, and their band gaps are 1.50 eV, 2.21 eV respectively. The thermal behaviors of the crystals show that the material are stable up to 240 °C and can be stored for eight months in air.

Materials and characterization

CH_3NH_3I (>99.8%) and CH_3NH_3Br (>99.8%) are purchased from Xi'an Polymer Light Technology Corp. PbI_2 (99.99%) and $PbBr_2$ (99.99%) are purchased from Aldrich. ϒ-GBL (99%) and DMF (99%) are purchased from Aladdin.

The Model of X-Ray Diffractometer (Cu Kα radiation (15.46 nm), with voltage of 40kV and current of 40 mA) is Rigaku-MiniFlex600. UV-visible reflectance spectra was measured by Enlitech's QE-R system. Shimadzu RF-5301PC was used to measure the photoluminescence spectra. Scanning electron microscope (SEM) is used to characterize the morphology and EDS of CH3NH3PbX3. Thermogravimetric (TG) was carried out on Netzsch TG209 F1 at constant N2 flow of 50ml min^{-1} with heating and cooling rates of 5 °C min^{-1}.

Author Contributions

Yangyang Zhou and Chunmei Li contributed equally to this work.

Acknowledgements

This work was financially supported by the National Natural Science Foundation of China (Grants Nos. 11574119, 61674070), the Guangdong Natural Science Foundation (Grants Nos. 2014A030313381),

References

[1] Kojima A, Teshima K, Shirai Y, et al. Organometal halide perovskites as visible-light sensitizers for photovoltaic cells[J]. *Journal of the American Chemical Society*, 2009, **131**(17): 6050-6051.

[2] Lee M M, Teuscher J, Miyasaka T, et al. Efficient hybrid solar cells based on meso-superstructured organometal halide perovskites[J]. Science, 2012, **338**(6107): 643-647.

[3] Liu M, Johnston M B, Snaith H J. Efficient planar heterojunction perovskite solar cells by vapour deposition[J]. Nature, 2013, **501**(7467): 395-398.

[4] Burschka J, Pellet N, Moon S J, et al. Sequential deposition as a route to high-performance perovskite-sensitized solar cells[J]. Nature, 2013, **499**(7458): 316-319.

[5] National Renewable Energy Laboratory. Best Research-Cell Efficiencies: http://www.nrel.gov/ncpv/images/efficiency_chart.jpg

[6] Zhao J, Cai B, Luo Z, et al. Investigation of the Hydrolysis of Perovskite Organometallic Halide CH3NH3PbI3 in Humidity Environment[J]. Scientific reports, 2016, 6.

[7] Snaith H J, Abate A, Ball J M, et al. Anomalous hysteresis in perovskite solar cells[J]. *The journal of physical chemistry letters*, 2014, **5**(9): 1511-1515.

[8] Dualeh A, Tétreault N, Moehl T, et al. Effect of Annealing Temperature on Film Morphology of Organic - Inorganic Hybrid Pervoskite Solid - State Solar Cells[J]. *Advanced Functional Materials*, 2014, **24**(21): 3250-3258.

[9] Dang Y, Ju D, Wang L, et al. Recent progress in the synthesis of hybrid halide perovskite single crystals[J]. CrystEngComm, 2016.

[10] Yang S, Zheng Y C, Hou Y, et al. Formation mechanism of freestanding CH3NH3PbI3 functional crystals: in situ transformation vs dissolution–crystallization[J]. *Chemistry of Materials*, 2014, **26**(23): 6705-6710.

[11] Shi D, Adinolfi V, Comin R, et al. Low trap-state density and long carrier diffusion in organolead trihalide perovskite single crystals[J]. Science, 2015, **347**(6221): 519-522.

[12] Kadro J M, Nonomura K, Gachet D, et al. Facile route to freestanding CH3NH3PbI3 crystals using inverse solubility[J]. Scientific reports, 2015, 5.

[13] Saidaminov M I, Abdelhady A L, Murali B, et al. High-quality bulk hybrid perovskite single crystals within minutes by inverse temperature crystallization[J]. Nature communications, 2015, 6.

[14] Liu Y, Yang Z, Cui D, et al. Two - Inch - Sized Perovskite CH3NH3PbX3 (X= Cl, Br, I) Crystals: Growth and Characterization[J]. *Advanced Materials*, 2015, **27**(35): 5176-5183.

[15] Dong Q, Fang Y, Shao Y, et al. Electron-hole diffusion lengths> 175 μm in solution-grown CH3NH3PbI3 single crystals[J]. Science, 2015, **347**(6225): 967-970.

[16] J. Torrent and V. Barron, Encyclopedia of Surface and Colloid Science, Marcel Dekker, Inc., New York, 2002

IOP Conf. Series: Materials Science and Engineering **167** (2017) 012020 doi:10.1088/1757-899X/167/1/012020

Synthesis of the few layered two-dimensional molybdenum oxide atomic crystal

Yu Wang, Ruihui He, Mingze Su, Weiguang Xie[1]

Siyuan Laboratory, Guangzhou Key Laboratory of Vacuum Coating Technologies and New Energy Materials, Department of Physics, Jinan University, Guangzhou, Guangdong, 510632, P. R. China

E-mail: wgxie@email.jnu.edu.cn

Abstract. We study the effect of substrate type, source temperature, and substrate position on the morphology of Alfa molybdenum oxide (α-MoO_3) in physical vapour deposition. We found that the orientation and thickness of the MoO_3 is strongly affected by the evaporated flow and temperature. Lying down, few layered (FL) MoO_3 can be grown directly on technical substrate such as SiO_2 with typical length scale of tens of micrometres. Our studies provide a simple, catalyst-free and low cost synthesis way to prepared large size FL MoO_3 for further optical and electrical investigation such as photodetector, field effect transistors, and electrochromism.

1. Introduction

Researchers have paid much attention to two-dimensional metal oxide nanomaterials recently because of its unique optical and electric properties [1]. Among various kinds of mental oxide, molybdenum oxide (MoO_3) is a multifunctional materials that has potential application in photovoltaic devices[2], electrochromic window[3], gas sensors[4] and so on. There are three main crystalline polymorphs: orthorhombic molybdenum oxide (α-MoO_3), monoclinic molybdenum oxide (β-MoO_3), and hexagonal molybdenum oxide (h-MoO_3). α-MoO_3 is a thermodynamically stable phase with layered structures, which is made from MoO_6 octahedra (Figure 1(a)). MoO_6 octahedra forms corner-sharing chains along the [100] direction and edge-sharing chains along the [001] direction. The layers are stacked by weak van der Waals forces along the [010] direction with a layer thickness of 0.7 nm.

It's been found that the Coulomb scattering can be screened effectively due to its high dielectric constant. As a result, MoO_3 nanosheet has carrier mobility of more than 1000 cm^2/Vs, which is attractive in field effect transistors[5]. The layered MoO_3 has been fabricated by a variety of synthesis methods, including thermal evaporation[6], electrodeposition[7], spray pyrolysis[8], sol-gel[9], sputtering[10], and mechanical exfoliation[11]. Recently, Di Wang prepared α-MoO_3 sheet with thickness of single-unit-cell and length scale of tens of micrometers on mica by physical vapor deposition[12]. However, direct synthesis of few layered (FL) MoO_3 on a technical substrate, for example, SiO_2 is still lacking. In this study, a simple, high-efficiency, catalyst-free synthesis technique for the preparation of FL two-dimensional MoO_3 was developed.

IOP Conf. Series: Materials Science and Engineering **167** (2017) 012020 doi:10.1088/1757-899X/167/1/012020

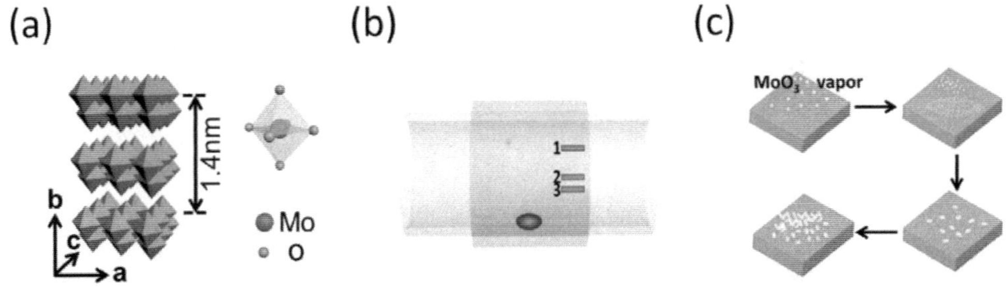

Figure 1 (a) The schematic structure of α-MoO_3. The thickness of monolayer is 0.7 nm and a unit cell content two layers. (b) Schematic setup of tube furnace for physical vapour deposition; (c) The growth process of MoO_3.

2. Experimental and Results

The MoO_3 sheets were synthesized by thermal physical vapor deposition method under air condition. The substrates of SiO_2 were ultrasonic cleaned in acetone, alcohol, and distilled water successively. Each cleaning step lasted for 10 minutes. The substrates were then removed from the distilled water and were dried with nitrogen gas. 0.1 g of MoO_3 powder was placed in an Al_2O_3 crucible at the center of a quartz tube (Figure 1(b)). The substrates were placed at the low temperature region. The distance between the substrate and powder is 12-16 cm. The temperature of the source was increased to 780 ℃ from room temperature in 70 minutes, maintaining for 120 minutes, and then cooled down to room temperature naturally. The crystallization process of MoO_3 is realized by gas-solid mechanism. After the powder of MoO_3 is sublimated as gaseous species in the high temperature region, it will diffuse to the low temperature region, and then be recrystallized on the substrate[13].

There are different morphologies on diverse substrates, for example, silicon, silicon oxide, textured silicon (T-Si) substrate, and Mo plate at same temperature in Figure 2. The α-MoO_3 plates are densely packed on Si, SiO_2, and T-Si. On the Si substrate in the Figure 2(a), the α-MoO_3 nanosheets are rectangular with thickness of ~200 nm, width of ~1 μm, and length of ~12 μm. Some small nanosheets were also observed; maybe due to short growth time, or the finite space for crystallization. On SiO_2 surface (Figure 2(b)), the nanosheets with thickness of ~150 nm, width of ~2 μm, and length of ~5 μm were observed. There are many nanosheets with thickness of ~500 nm, width of ~3 μm, and length of ~10 μm on the T-Si (Figure 2(c)). There are obvious differences between MoO_3 on the Mo plate (Figure 2(d)) and MoO_3 nanosheets on Si, SiO_2, and T-Si. It grows longer with typical length of 30 μm, and thicker and smaller in width. The morphology difference may be due to the difference in surface roughness[14]. The Mo surface is rougher than the Si and SiO_2 in submicrometer scale. The deposited MoO_x species on the surface is not easy to diffuse on the Mo surface, which leads to slower growth in width and thickness, but faster growth along the length.

Figure 2. SEM images of MoO_3 on the different substrates at 780 °C. (a) silicon wafer; (b) silicon wafer with 300 nm thick silicon oxide; (c) textured silicon wafer; (d) Mo plate.

In order to confirm the crystal structure of α-MoO_3 nanosheets, the Raman of MoO_3 nanosheets (a confocal Raman microscope by Renishaw inVia Reflex with a 532 nm laser as excitation) was conducted, the X-ray diffraction (XRD) spectroscopy in a Rigaku-MiniFlex600 using the Cu κα radiation (1.546Å), with 40 kV and current of 40 mA was carried out, and high-resolution transmission electron microscopy (HRTEM) and selected area electron diffraction (SAED) experiments were conducted on JEOL 2100F operated at 200 kV (HRTEM). Figure 3(a) presents a typical Raman spectrum of MoO_3 nanosheets. The peaks at 339 and 666 cm^{-1} are corresponding to Mo-O_3 stretching mode, which is along [001]. The peak at 819 cm^{-1} is corresponding to Mo-O_2 stretching mode. The peak at 995 cm^{-1} is corresponding to Mo=O stretching mode of the terminal oxygen. The XRD pattern (Figure 3(b)) shows the strongest diffraction peaks for MoO_3, for example, 12.5°, 25.4°, 38.7°, which correspond to (020), (040), (060) planes, respectively. In the Figure 3(c), the lattice spacing is 0.391 nm and 0.374 nm, which is in accord with the orthorhombic structure of MoO_3 phase, and with the lattice constants of a=3.96 Å, b=13.86 Å, c=3.7 Å (JGPDS 05-0508). The

SAED diffraction pattern indicates that the growth direction along (001) and (100) planes, which is normal to (010) plane.

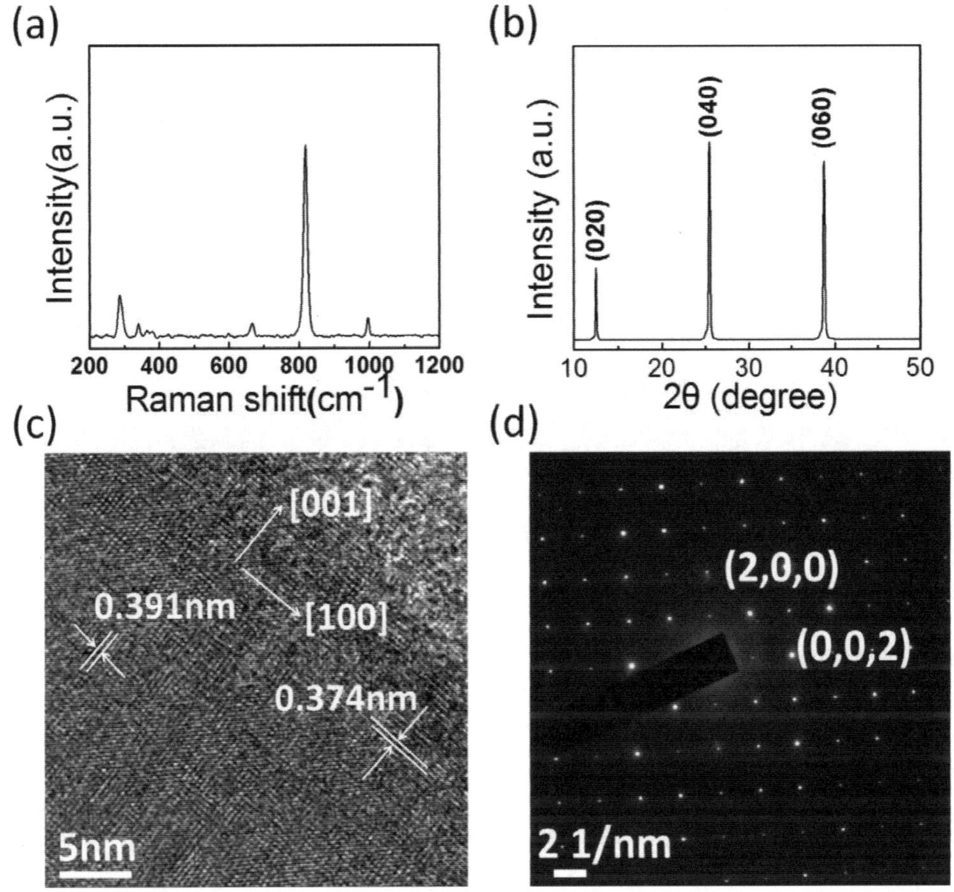

Figure 3. (a) Raman spectrum of α-MoO_3. (b) XRD pattern of the α-MoO_3. (c) HRTEM image of a single nanosheet. (d) The corresponding SAED pattern.

The morphology of MoO_3 depends on the source temperature. The nanosheets are sparsely dispersed on the SiO_2 (Figure 4(a)) at the source temperature of 580 ℃. This may be due to the very low evaporated flow of MoO_3. As the source temperature is raised to 680 ℃, the nanosheets are more densely packed (Figure 4(b)). Typical length of the nanosheet increases form ~1 μm to 2 μm, and the width and the thickness increase at the same time. When the source temperature is 780 ℃, a dense thick film composed of long MoO_3 nanoplates formed.

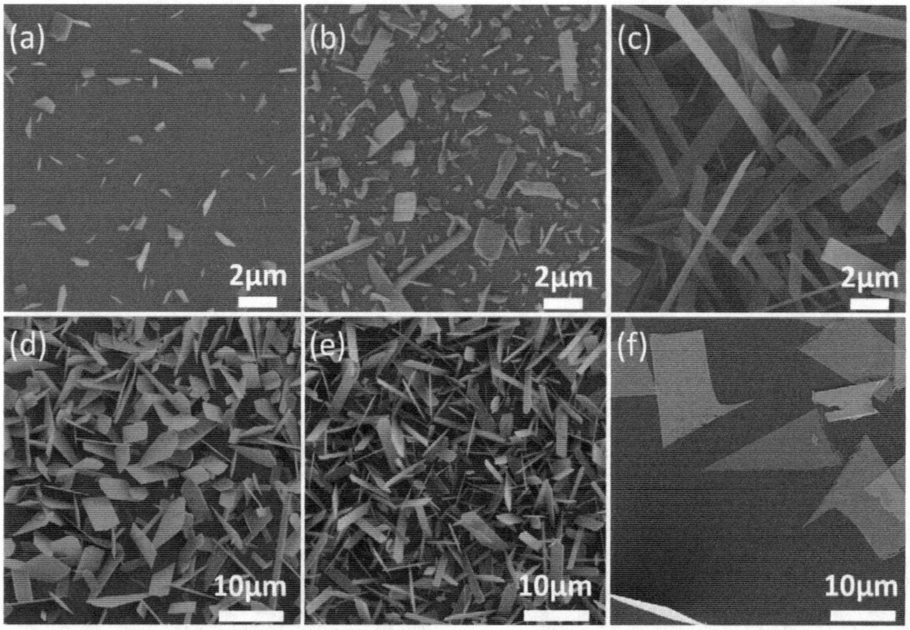

Figure 4 Effects of the source temperature and the substrate position on the morphology of the MoO_3 nanosheets. SEM images of the MoO_3 nanosheets at different source temperature of (a) 580℃, (b) 680 ℃, and (c) 780℃. (d), (e) and (f) are all 12.5 cm away from the source, but 0.8 cm, 2.5 cm, and 3.0 cm away from the top of the tube wall, the positions of the three substrates correspond to the positions 1, 2, and 3 in the Figure 1(c).

There are different morphologies at the different deposition position at same source temperature (Figure 4(d), (e), and (f)). Near the top of the tube wall, the nanosheets are randomly orientated, with width of ~3 µm. The width gets smaller as the position gets away from the top. But at 3.0 cm away from the top, we found that most of nanosheets get parallel to the substrate. Nanosheets are sparsely distributed. The length and width are normally bigger than 10 µm. This may offer opportunity to make device easily. By observing the morphology along the tube axis, we found that the lying down MoO_3 nanosheets have grown at the position close to the source (Figure 5(a)). Normally, the evaporated MoO_3 gas will accumulate at the top of the wall and then diffuse from the center to both sides. The MoO_3 nanosheets are denser near the top of the wall, and decrease as the position gets lower. On the other hand, when the position gets near the source, the temperature gets higher. Therefore, the morphology of the as-grown MoO_3 is controlled by gas flow and the temperature.

To determine the thickness of the nanosheet, atomic force microscope was carried out. The thickness of the lying down nanosheets is ranging from several unit cells to ~ 100 nm. Figure 5(b) shows that the thickness of nanosheet is 5.42 nm. A layer thickness is 0.7 nm, so it is about 8 layers. We also found that the FL MoO_3 can also be grown on Si as well. Our results show that FL MoO_3 nanosheet with tens of micrometres in length can be prepared by physical vapour deposition on technical substrate.

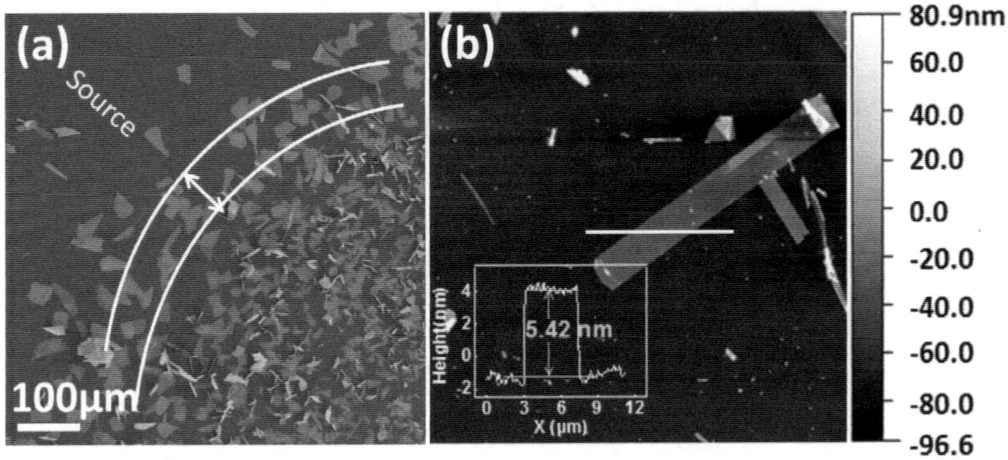

Figure 5. (a) SEM image of MoO_3 nanosheets distribution. (b) AFM image of a FL MoO_3 nanosheet, the thickness is 5.42 nm.

3. Conclusion

In conclusion, we have grown MoO_3 nanosheets on different substrate by physical vapor deposition. The morphology of MoO_3 nanoplates is controlled by the substrate type, source temperature, and substrate position. When the temperature of the source is 780 ℃, FL MoO_3 atomic crystal can be fabricated on SiO_2 and Si substrate at about 12.5 cm from the source and 3 cm away from the top wall of the tube. Our findings provide a low cost and feasible way to fabricate FL MoO_3 on technical substrate, and offer the possibility to investigate the photoelectric properties of two dimensional MoO_3.

Acknowledgements

This work was financially supported by the National Natural Science Foundation of China (Grants Nos. 11574119, 61674070), the Guangdong Natural Science Foundation (Grants Nos. 2014A030313381),

References

[1] Xiang D, Han C, Zhang J and Chen W 2014 *Sci. Rep.* **4** 4891
[2] Liang Z, Su M, Zhou Y, Gong L, Zhao C, Chen K, Xie F, Zhang W, Chen J, Liu P and Xie W 2015 *Phys. Chem. Chem. Phys.* **17** 27409-13
[3] Buyan M, Brühwiler P A, Azens A, Gustavsson G, Karmhag R and Granqvist C G 2006 *Int. J. Indu. Ergon.* **36** 11-16
[4] Okumu J 2004 *J. Appl. Phys.* **95** 7632
[5] Balendhran S, Deng J, Ou J Z, Walia S, Scott J, Tang J, Wang K L, Field M R, Russo S, Zhuiykov S, Strano M S, Medhekar N, Sriram S, Bhaskaran M and Kalantar-zadeh K 2013 *Adv. Mater.* **25** 109-14
[6] Xie W, Su M, Zheng Z, Wang Y, Gong L, Xie F, Zhang W, Luo Z, Luo J, Liu P, Xu N, Deng S, Chen H and Chen J 2016 *ACS Nano* **10** 1662-70
[7] Patil R S, Uplane M D and Patil P S 2006 *Appl. Surf. Sci.* **252** 8050-56
[8] Bouzidia A, Benramdanea N, Tabet-Derraza H, Mathieub C, Khelifab B and Desfeuxc R 2003 *Mater. Sci. Eng. B* **78** 5-8
[9] Prasad A K, Kubinski D J and Gouma P I 2003 *Sens. Actuators, B* **93** 25-30
[10] Navas I, Vinodkumarl R, Lethy K J, Detty A P, Ganesan V, Sathe V and Mahadevan Pillai V P 2009 *J. Phys. D: Appl. Phys.* **42** 175305

[11] Kalantar-zadeh K, Tang J, Wang M, Wang K L, Shailos A, Galatsis K, Kojima R, Strong V, Lech A, Wlodarski W and Kaner R B 2010 *Nanoscale* **2** 429-33
[12] Wang D, Li J-N, Zhou Y, Xu D-H, Xiong X, Peng R-W and Wang M 2016 *Appl. Phys. Lett.* **108** 053107
[13] Shi Y, Li H and Li L J 2015 *Chem. Soc. Rev.* **44** 2744-56
[14] Cai L, Rao P M and Zheng X 2011 *Nano Lett.* **11** 872-7

IOP Conf. Series: Materials Science and Engineering **167** (2017) 012021 doi:10.1088/1757-899X/167/1/012021

Dependence of seed layer thickness on sensitivity of nano-ZnO cholesterol biosensor

Yang-Ming Lu[1,3],Po-Chin Wang[1],Jian-Fu Tang[2], Sheng-Yuan Chu[2]

[1]Department of Electrical Engineering, National University of Tainan,Tainan,Taiwan
[2]Department of Electrical Engineering, National Cheng Kung University, Tainan,Taiwan.

E-mail: ymlu@mail.nutn.edu.tw

Abstract. The anemone-like ZnO nanostructures have been synthesized by hydrothermal method and were further adsorbed immobilized cholesterol oxidase (ChOx) as a nano-biosensor.In this study,the sensitivity of biosensor were improved by varying the thickness of the ZnO seed layer.The SEM analysis showed changes in thickness of seed layer will not affect the morphologies of anemone-like ZnOnanostructures.The X-ray Diffraction patterns showed that the (002) plane of anemone-like ZnO grown on various thickness of the seed layer was more prouded than other crystal plane.Abioelectrode (ChOx/ZnO/ITO/glass) grown on the 30nm of ZnO seed layer with high sensitivity of 57.533μAmM-1cm-2 (1.488 μA (mg/dl) -1cm-2), a wide sensitive range from 25 to 500 mg/dl. It is concluded that the thinner sputtered ZnO seed layer for growing anemone-like ZnO nanostructure can effectively improve the sensitivity of the ZnO biosensor.

1. Introduction

According to the World Health Organization [1], heart and cerebrovascular diseases accounted for 20% of the world's cause of death. Many clinical diseases have a strong correlation with cholesterol concentration in the blood, such as heart disease, coronary artery disease, atherosclerosis, hypertension, thrombosis, etc. [2-3].Excess cholesterol is a major cause of heart disease and cerebrovascular disease, and therefore it is important to monitor cholesterol levels in the blood. Development of fast and accurate cholesterol biosensor is extremely important to find those diseases. Due to their exceptional optical and electrical properties, the materials with nanostructures are receiving great attention for enzyme immobilization to improve stability and sensitivity of biosensors[4-5].

Enzymes directly immobilized on the electrode surface is easy to bedeteriorated[6].In order to improve this situation, using appropriate intermediate to modify the electrode,such as polymer, metal oxide or organic-inorganic composite film is necessary. These modifications can be used to secure the enzyme, keep the essence of the enzyme and increase the electron transfer.Therefore selection a suitable intermediate is very important for biosensor. Zinc oxide (ZnO) is an excellent intermediate in the enzyme electrode due toits unique characters [7-11]：(1) It is a semiconductor material with a wide bandgap of 3.37eV. (2) ZnO has a high excitation binding energy of 60mV. (3) The Zn elementis easy to be obtained and cheaper. (4) Zinc oxides can be grown using various deposition methodsat a low temperatures. (6)It has excellent biological safety and biocompatibility. (7) Zinc oxide will be positively charged in a neutral solution due to its high isoelectric point (IEP ~ 9.5).It can let protein or enzyme adsorbed by electrostatic attraction force.

There have been a number of reports on the fabrication of cholesterol biosensor using ZnO thin films and nanostructures [12-14], but the sensitivity is too low or the detection range is too narrow. In this study, thesensitivity of ZnO biosensor was improved by varying the thickness of the ZnOseed layer.The anemone-like ZnO nanostructures had been grown on the seed layer by hydrothermal method. The cholesterol oxidase (ChOx) was immobilized on the anemone-like ZnO nanostructures as a cholesterol nano-biosensor.

2. Experimental

ZnO seeds layer was pre-sputtered on the surface of ITO (4.4Ω/□) glasses to grow a high orientation anemone-like ZnO nanostructures. Different thickness of seed layers with 30nm,50nm and 100nm were prepared by changing sputtering time. In hydrothermal growth, all deionized water was pre-heated to a temperature of 90 °C before the hydrothermal reaction executed. The 0.05 M zinc nitrate solution made in 45 ml deionized water was mixed with 0.05M hexamethylenetetramine (HMTA) solution made in 45 ml deionized water in serum bottles. Then the electrodes were dipped in solution at 90 °C for 9 hrs. After synthesis was completed, the samples were rinsed with deionized water and dried. After drying the samples, 10μl of ChOx solution (1.0 mg/mL) prepared in phosphate buffer saline(PBS)(50mM,pH ~7.4,0.9% NaCl) was spread over the surfaces of anemone-like ZnO structures and dried at room temperature for about an hour. The sample was then kept overnight for enzyme immobilization at a temperature of 4 °C by physical adsorption technique and subsequently washed with buffer solution to remove the unbound enzyme molecules. After hydrothermal growth, all the sample were stored at 4 °C.Electrochemical analysis is carried out on an Jiehan 5000 potentiostat/galvanostat using a three-electrode configuration with Ag/AgCl electrode as a reference electrode in PBS solution(50 mM, pH ~7.4,0.9% NaCl) containing 5 mM $[FeCN_6]^{3-/4-}$. The ZnO sample with nanostructures was used as the working electrode and platinum foil as a counter electrode. The cholesterol solutions have been prepared from PBS solution(50 mM, pH ~7.4,0.9% NaCl) containing 1% of triton-X100 and stored at 4 °C.

3. Results and discussion

3.1. Materials analysis

Fig.1 shows the results of grown ZnOnanostructures by a modified hydrothermal method in this study. The SEM images show the morphologies of these ZnO nanostructures which look like anemones appearances.Fig.1(a) shows that each anemone-like structure has three tofive elongated tentacles.There are lots of spacesamong the tentacles which provide plenty of surfaces for enzyme adsorption. These anemone-like structures basically can be divided into two parts,the stubbybasement structures and elongated tentacles structures, such as shown in Fig.1(b)(d)(f). TheseZnO nanostructures were densely and well arranged on the surfaces of ITO substrates.Thepre-sputtered ZnO seeds layerson ITO top surfaces help to overcome the activation energy for heterogeneous nucleation of ZnO nanostructures on the ITO surfaces.

Different thicknesses of ZnO seed layer were prepared by changing the sputtering time.TheZnO seed layers of 30nm,50nm and 100nmwere prepared for following hydrothermal growth. The thickness of the seed layer did not have obvious effect on the morphologies of ZnO nanostructures. The morphologies are all similar as shown in Fig.1 except with different lengths.All the appearanceslook like anemones. The variations of ZnOaspect ratio(length/diameter) of anemone tentacles depend on seed layer thickness. The aspect ratio decreases with decreasing the thickness of seed layer as seen in Fig.1(b),(d),(f).A lower aspect ratio leads to lower sensitivity when the ZnO nanostructures were assembled as cholesterol biosensors according to the literature [15].

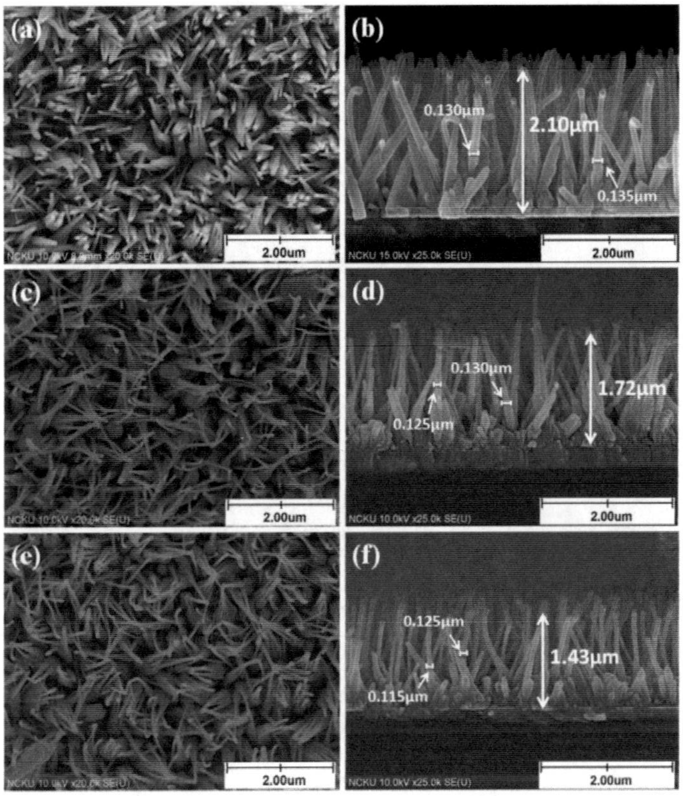

Figure 1.HR-SEM images of anemone-like ZnO nanostructures grown on 100nmZnO seedlayer ,50nmZnO seed layer and 30nm ZnO seedslayer respectively. (a),(b)with 100nm ZnO seed layer ,(c),(d)with 50nm ZnO seed layer and (e),(f)with 30nm ZnO seed layer.

Fig.2 was the X-ray diffraction patterns of anemone-like ZnO nanostructures grown on 100nm ZnO seed layer,50nmZnO seed layer and 30nm ZnO seed layer. The results show that thecrystallinities of anemone-likeZnO were good because the diffraction peaks are sharp and strong. A prouded(002) peakswere observed in the XRD patterns imply the ZnO nanostructures have a strong growth directions along the c axis direction. The SEM and XRD results were consistent as presented in Fig.1 and Fig.2. In Fig.2, there are some other small peaks(denoted by triangles)observed in the Figures which are the diffraction peaks of ITO substrates.

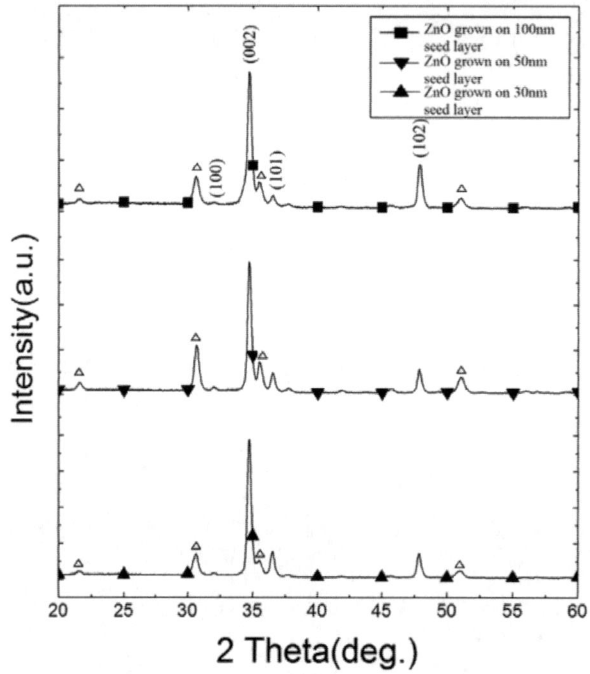

Figure1. Shows X-ray diffraction patterns of anemone-like ZnO nanostructuresgrown on 100nm ZnO seed layer,50nmZnO seed layer and 30nm ZnO seed layer respectively

3.2. Electochemical analysis

The anemone-like ZnO nanostructures were assembled as cholesterol biosensor. The sensitivity and linear sensing range of the biosensor were used to calculate the value of the Michaelis-Menten constant (Km). The electrochemical measurement was employed in a PBS solution(50 mM, pH ~7.4,0.9% NaCl) containing 5 mM $[FeCN_6]^{3-/4-}$,the scan range of cyclic voltammetry from -0.2V to 0.6V and with sweep rate of 50 mV/s. Fig.3 show thesethree biosensors had good current responses. Thepotential difference between reduction and oxidation peaksinversely proportional to the number of electrons transported between anode and cathodein the solution[16].The thinner the thickness of seed layer, the less electrical resistance it is which means more electronic transportation in the solution. The oxidation current increases with decreasing the thickness of the seed layer due to less electrical resistance with thinner seed layer as seen in Fig.3 from approximate 1x10-4 to around 3x10-4amper. The anemone-like ZnO nanostructures grown on 30nmZnO seed layer has the maximum amount of oxidation current will lead to a higher sensitivity as shown in Fig.4. The sensitivity is defined as the peak oxidation current change per unit cholesterol concentration in the linear range as shown in the following equation:

$$\text{Sensitivity}=\Delta I_{pa}/(\Delta C \times A) \quad (1)$$

Where ΔIpais the variation of oxidation current peak.△C is theanalyte concentration change.A is the working area of electrode. The sensitivity has increased significantly by reducing the thickness of theZnO seed layer from 100nm to 30nm. The best sensitivity obtained in this study is 57.5 μA/mM/cm2 as the ZnO nanostructure prepared on the 30nm seed layer.

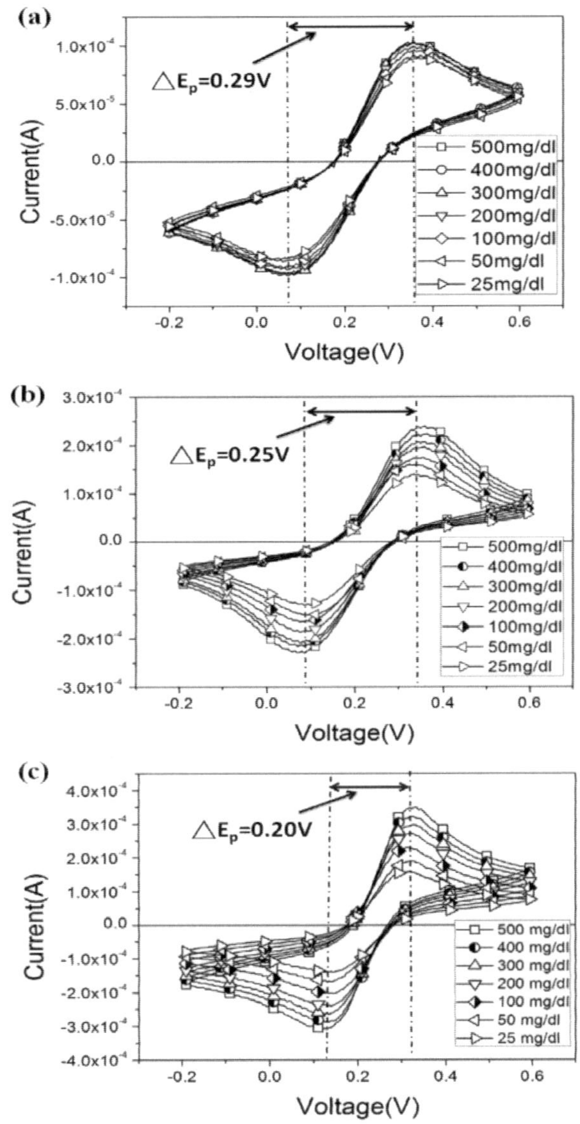

Figure3. Cyclic voltammetry of the ChOx/ZnO/ITO/glass bioelectrode grown on 100nmZnOseed layer,50nmZnOseed layer and 30nmZnOseed layermeasured at different cholesterol concentration.(a)with100nm ZnO seeds layer ,(b)with 50nm ZnO seeds layer and (c)with 30nm ZnO seeds layer

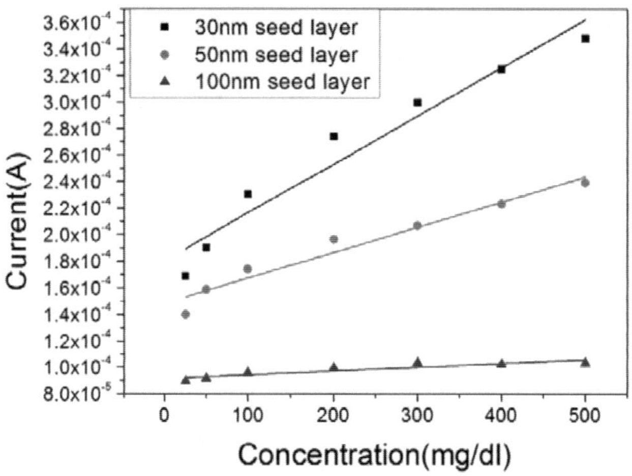

Figure 4. Variation of sensitivity of anemone-like ZnO biosensors prepared on different thickness of seed layers

4. Conclusion

In this study,the thinner sputtered ZnO seed layer for growing anemone-like ZnO nanostructure can effectively improve the sensitivity of the ZnO biosensor. The best sensitivity can be obtained in this study is 57.5 μA/mM/cm2 as the ZnO nanostructure grown on the 30nm seed layer. Thelinear sensitive range is from 25 to 500 mg/dl.

Acknowledgement

The authors are grateful to National Science Coucil(NSC) in Taiwan for its financial supporting this research under projects-MOST104-2221-E-024-011，MOST105-2221-E-024-001.

References

[1] World Health Organization. http://www.who.int/mediacentre/factsheets/fs310/en/index.html
[2] Aghaei A., Hosseini M.R.M. and Najafi M. 2010 *Electrochim. Acta.***55** 1503-1508.
[3] Shih W. C., Yang M. C. and M. S. Lin 2009 *Biosens. Bioelectron* **24** 1679-1684.
[4] Singh S.P., Arya S.K., Pandey P., Malhotra B.D. and Saha S. 2007 *Appl. Phys. Lett.* 91063901.
[5] Pandey P., Singh S. P., Arya S. K., Gupta V., Datta M., Singh S.and Malhotra B. D. 2007 *Langmuir* **23** 3333-3337.
[6] Wang G. , Thai N.M. and S.T. Yau 2006 *Electrochem. Commun.* **8** 987-992.
[7] Jeong W. J. 2005 PVSEC 11 International Photovoltaic Science and Engineering Conference No.11 65 37-42.
[8] Li Y. J., Kwon Y. W., Jones M., Heo Y. W., Zhou J., Luo S. C., Holloway P. H., Douglas E., Norton D. P., Park Z. and S. Li 2005 *Semicond. Sci. Technol.* **20** 720-725.
[9] Umar A., Rahman M.M., Vaseem M. and Y.B. Hahn 2009 *Electrochem. Commun* **11** 118-121.
[10] Khan R., Kaushik A., Solanki P. R. and Ansari A. A. 2008 *Anal. Chim. Acta* **66** 207-213.
[11] Israr M.Q., Sadaf J.R., Nur O., Willander M., Salman S. and Danielsson B. 2011 *Appl. Phys. Lett.* 98253705.
[12] Solanki P.R., Kaushik A., Ansari A. A.,and Malhotra B. D. 2009 *Appl. Phys. Lett.* **94** 143901 .
[13] Zhang C., Wang G., Liu M., Feng Y., Zhang Z.and Fang B. 2010 *Electrochim. Acta* **55** 2835–2840.
[14] Umar A., Rahman M. M., Hajry A. A.and Hahn Y. B. 2009 *Talanta* **78** 284–289.

[15] Ahmad R., Tripathy N. and Y.B. Hahn ,2012 *Sen. Actuators B* **169** 382– 386.

IOP Conf. Series: Materials Science and Engineering **167** (2017) 012022 doi:10.1088/1757-899X/167/1/012022

Bandgap control for the lattice-matched InGaAs/InAlAs/InP single quantum wells

Y Wang[1, 2, 3], X Z Sheng[1], Q L Guo[2] and B L Liang[2]

[1] School of Science, Beijing Jiaotong University, Beijing 100044,P. R. China
[2] College of Physics Science & Technology, Hebei University, Baoding 071002, P.R. China

E-mail: hbuwangying@126.com

Abstract. $In_{0.53}Ga_{0.47}As/In_{0.52}Al_{0.48}As$ single quantum well (SQW) grown by molecular-beam epitaxy on the InP (001) substrate is investigated by using photoluminescence (PL) spectroscopy and time-resolved photoluminescence (TRPL). At the low temperature of 8 K and excitation intensity of 1 W/cm^2, the emitting wavelength of QWs shifts from 1387 nm to 1537 nm when the well width increases from 5 nm to 50 nm, approaching the emitting wavelength of bulk $In_{0.52}Al_{0.48}As$. The emitting wavelength of QWs can be flexibly tailored by varying the thickness of the InGaAs layer.

1. Introduction

The $In_{0.53}Ga_{0.47}As/In_{0.52}Al_{0.48}As$ heterostructures lattice-matched to InP (001) substrate have been studied extensively due to their unique properties.[1-3] The InGaAs/InAlAs QW heterostructures, including single-quantum well, multiple-quantum wells, and superlattices, have special advantages for applications in long-wavelength optical communications as well as high-speed electronic and photonic devices.[4] First, InGaAs has a high electron mobility and a high peak velocity which make the material very attractive for high-speed devices.[5] Second, The InGaAs/InAlAs quantum well (QW) structure are very suitable for optical communication applications because they allow the emission wavelength to be shifted from the 1.65 μm (bulk InGaAs lattice-matched to InP) to the 1.3-1.55 μm region by varying well thicknesses. [6-8] Third, InGaAs/InAlAs system which has a large conduction-band offset enables a strong electron confinement, and subsequently a high-temperature stability and a high-speed modulation capabilities, making the system very attractive and suitable for quantum cascade lasers, inter-subband detectors, and devices based on nonlinear optical properties. [9, 10]

In this experiment, we have fabricated lattice-matched $In_{0.53}Ga_{0.47}As/In_{0.52}Al_{0.48}As/InP$ quantum well with different well width of 5 nm, 15 nm, 25 nm and 50 nm by molecular beam epitaxy (MBE) to tailor the bandgap of the heterostructures. The photoluminescence (PL) and time-resolved photoluminescence (TRPL) were performed to investigate the effect of well width on PL emission.

2. Experiments

The QW samples were grown by a solid-source molecular beam epitaxy on semi-insulating InP (001) substrate at a growth temperature of 515 °C. The samples consist of a lattice-matched $In_{0.53}Ga_{0.47}As$ single quantum well (SQW) with nominal well width of 5 nm, 15 nm, 25 nm and 50 nm, respectively. The InGaAs QW is sandwiched between InAlAs barrier with a 60 nm InAlAs layer on the top and an 80 nm layer underneath, as shown in Figure 1. In order to protect the top InAlAs layer from oxidation effect, a 3 nm lattice-matched InGaAs layer was growth on the top for all the QW samples. For reference, one InGaAs layer only sample and one InAlAs layer only sample were also grown to calibrate the composition and to measure the bandgap of bulk InAlAs and InGaAs. After growth, all samples show a mirror-like surface. We did not observe haziness and cross-hatching under a Normaski microscope.

Then the QW samples were characterized by photoluminescence (PL) and time-resolved photoluminescence (TRPL). For the PL measurements, the samples were mounted in a closed-cycle cryostat and excited by a 532 nm laser. The laser beam is focused to a spot of ~20 μm in diameter. The PL signal was detected by a liquid nitrogen cooled CCD detector array attached to a 50 cm Acton spectrometer. The TRPL was measured by a PicoHarp 300 time-correlated-single-photon-counting (TCSPC) system with a NKT super-continuum pulse laser as the excitation.

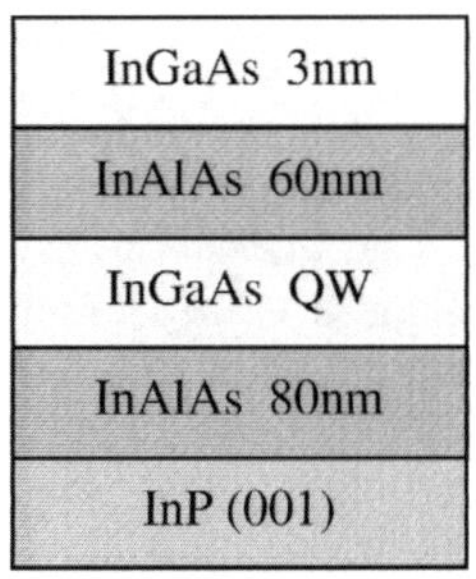

Figure 1. Schematic of QW structures

3. Results and discussions

3.1. Low temperature photoluminescence

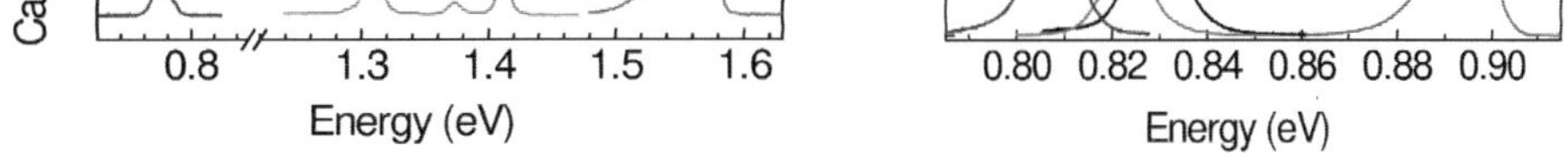

Figure 2. (a) and (b) are PL spectra for InGaAs bulk, InAlAs bulk and all the QW samples obtained at 8K with an excitation intensity of 1 W/cm^2

Figure 2 shows PL spectrum for InGaAs bulk, InAlAs bulk and all the QW samples measured at 8 K with a laser intensity of 1 W/cm^2. From Figure 2(a), the bandgap of InGaAs material is estimated to be 0.772 eV and the bandgap of InAlAs material is 1.562 eV. It means that the total band offset between InGaAs and InAlAs is 0.79 eV. Figure 2(a) also shows that the PL linewidth (Full Width at the Half Maximum, FWHM) of two reference samples are 6.3 meV and 24.3 meV, respectively. Meanwhile their PL spectra show symmetric Gaussian profile. [11] Under the same excitation intensity, the emission of InAlAs bulk material is stronger than that of the InGaAs bulk. It is likely that the radiative recombination centers in the interface between the InAlAs layer and InP (001) substrate have better confinement.

Figure 2(b) also shows that the emission peaks are quite different for the QW of different width. The PL peaks at 0.894 eV, 0.829 eV, 0.824 eV and 0.807 eV, are attributed to the e_1-hh_1 excitonic transition between the first electron level and the first heavy hole level in the 5 nm, 15 nm, 25 nm and 50 nm InGaAs/InAlAs QW, respectively. It can be observed that the QW PL peaks gradually approach to that of the InGaAs bulk when the QW width increases from 5 nm to 50 nm, but they stand far away from PL peak of the InAlAs bulk. It means that the carriers corresponding to the e_1-hh_1 excitonic transition in InGaAs/InAlAs QWs have very strong confinement while we flexibly tailor the thickness of QWs.

The PL intensity of the four QW samples have the same order of magnitude, but FWHM of the emission peak increases with decreasing the well width, being 6.60 meV, 11.10 meV, 8.44 meV and 13.69 meV for 50 nm, 25 nm, 15 nm and 5 nm QWs, respectively. It can be explained that the QW samples have better performance of interface roughness. The FWHM of the PL peak is related to the fluctuations in the confinement potential of InGaAs/InAlAs interface. Among the samples, the FWHM of 25 nm QW shows a slight deviation, maybe due to the fluctuation of the growth condition.

In order to test the quality of the QWs, we have measured the excitation-intensity dependent PL. Figure 3 shows the FWHM as a function of the laser excitation intensity. It can be seen that the FWHM of 15 nm, 25 nm and 50 nm QW have almost no change when the excitation intensity is less than 100 W/cm^2. But the FWHM of 5 nm QW exhibit a decrease. It means that the emission of 5 nm QW is strongly affected by the fluctuations of InGaAs/InAlAs interface. For the 5 nm QW, more excitons are trapped at the interface, resulting in a broad spectrum even under the low excitation intensity at low temperature.

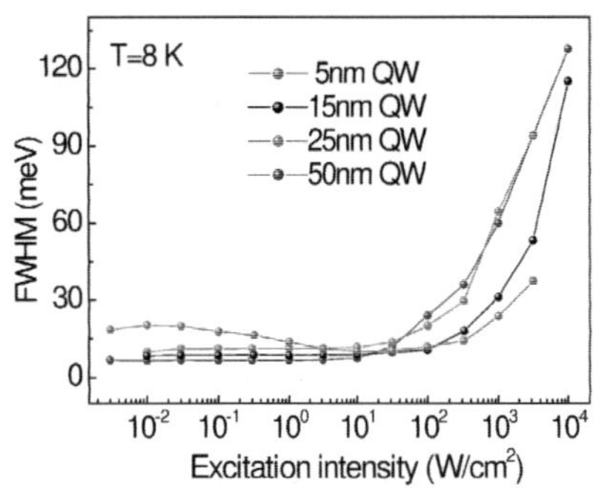

Figure 3. The FWHM of PL spectra as a function of excitation intensity

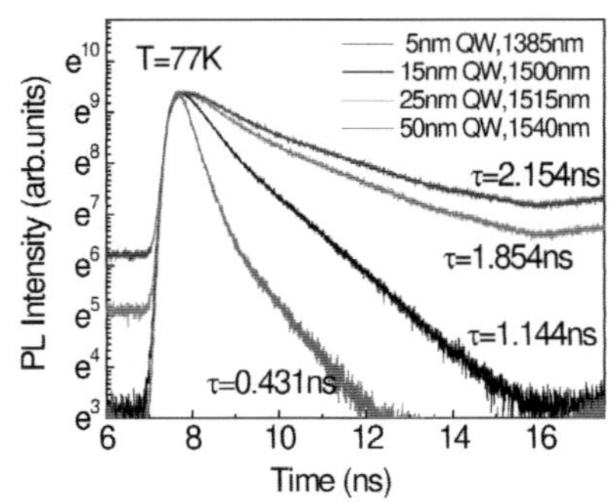

Figure 4. TRPL spectra of all QW samples

3.2. *Time-resolved photoluminescece*

To investigate the QW emission decay time, TRPL measurement was carried out at 77 K using a TCSPC module with a NKT super-continue pulse laser as the excitation. The PL decay curve detected at the peak wavelength of 1385 nm, 1500 nm, 1515 nm, and 1540 nm, respectively. As shown in Figure 4, the rise time of the transient photoluminescence almost coincides for all samples. Therefore, the carrier capture and relaxation in the QW samples are independent of the well width. Through single exponential fitting, the lifetime is determined to be 0.431 ns, 1.144 ns, 1.854 ns and 2.154 ns for the 5 nm, 15 nm, 25 nm and 50 nm QWs, respectively. The PL decay time increases monotonically with increasing emission wavelength, which corresponds to the well width. This result can be explained that the exciton wave function in a thinner QW layer has a better overlap due to the carrier localization. As a result, the oscillator strength increases, leading to a decrease of exciton lifetime.

4. Conclusion

In conclusion, we have fabricated the lattice-matched InGaAs/InAlAs QW samples on InP (001) substrate. The photoluminescence and time-resolved photoluminescence are investigated for all samples. The FWHM of PL spectrum indicates that all samples have good crystal quality, and the emission from thicker QW is less affected by the fluctuations at the InGaAs/InAlAs interface. The PL decay time of InGaAs QW increases monotonically with increasing emission wavelength, indicating that a thinner well has a better exciton oscillator strength.

Acknowledgement:
This research is supported by the "Hebei Province 100-Talents Program" (Grant # E2013100013), the Natural Science Foundation of Hebei Province (Grant # A2012201013), and the Natural Science Foundation of China (Grant #61575016).

References

[1] Poças L C, Duarte J L, Lopes E M, Dias I F L, Laureto E, César D F, and Harmand J C. 2006 *J. Appl. Phys.* **100** 053519
[2] Mourokh L G, Malikova L, Pollak F H, Shi B Q, and Nguyen C, 2001 *J. Appl. Phys.* **89** 2500
[3] Galiev G B, Vasil'evskii I S, Klimov E A, Klochkov A N, Lavruhin D V, Pushkarev S S, and Maltsev P. P. 2015 *Semiconductors* **49** (9), 1207–1217
[4] Kim D H and Alamo J A del, 2010 *IEEE Electron Dev. Lett.* **31** 806
[5] Monéger S, Baltagi Y, Benyattou T, Tabata A, Ragot B, Guillot G, Georgakilas A, Zekentes K, and Halkias G. 1993 *J. Appl. Phys.***74**(2) 1437
[6] Shen W Z, Shen S C, Chang Y, Tang W G, Chen J X, and Li A Z 1996 *J. Appl. Phys.* **80** 5348
[7] Ng G I, Hong W P, Pavlidis D, Tutt M, and Bhattacharya P K, 1988 *IEEE Electron Dev. Lett.* **EDL-9** 439
[8] Wakita K, Kawamura Y, Yoshikuri Y, and Asaki H, 1985 *Electron. Lett.* **21** 574
[9] Galiev G B, Vasil'evskii I S, Klimov E A, Klochkov A N, Lavruhin D V, Pushkarev S S., and Maltsev P. P. 2015 *Semiconductors* **49** (2), 234–241
[10] Kim D H and Alamo J A del, 2010 *IEEE Electron Dev. Lett.* **57** 1504
[11] Ouadjaout D and Marfaing Y, 1992 *Phys. Rev. B* **46** 7908

IOP Conf. Series: Materials Science and Engineering **167** (2017) 012023 doi:10.1088/1757-899X/167/1/012023

Effects of flexoelectricity and surface elasticity on piezoelectric potential in a bent ZnO nanowire

Shuangzhe Zhang, Haiyan Yao[1], Wenliang Fan, Yu Hao, Xudong Wu and Dongyuan Hou

College of Ordos, Inner Mongolia University, Ordos, 017000，China

E-mail: ndyaohaiyan@163.com

Abstract. In this work, a rapid model is established to study the effects of flexoelectricity and surface elasticity on the piezoelectric potential of a bent ZnO nanowire. Based on the piezoelectric theory and core-surface model, the distribution of piezoelectric potential of the ZnO nanowire is investigated. The analytical solution shows that the flexoelectricity and surface elasticity both significantly influence the piezoelectric potential. However, the effect of flexoelectricity is longitudinal dependent, which vanishes on the top side of nanowire, but only left surface elasticity effect on the potential. Simulation results show that the maximum value of potential on the top side of nanowire is about ± 220.5mV, of which result is lower compared to other theoretical models, but it should be more reasonable.

1. Introduction

One-dimensional zinc oxide (ZnO) nanowire is an important semiconducting piezoelectric material that has been widely used as a biomedical [1], environment monitoring, and even personal electronic nanodevices and nanosystem. In recent years, ZnO nanowires are widely used as nanoscale power generating systems. Many theoretical papers are reported to analyze the electrostatic potential of ZnO nanowires. Gao and Wang [2] established a classical piezoelelctric perturbation theory, which provides a way to estimate the potential generating in ZnO nanowire. Other groups [3, 4] also reported some similar studies to quantify the piezoelectric potential on a bent ZnO nanowire. It is well known that, with decreasing size of nanowire, strong size dependencies are displayed. Besides, flexoeletricity also becomes prominent at nanoscale, Liu et al. [5] took flexoelectric effect into account, analyzed the piezoelectric potential generated in a ZnO nanowire. Yan et al. [6, 7] utilized an extended linear theory of piezoelectricity to discuss the influence of flexoelectric effect on the electroelastic responses, and the bending and vibration behavior of piezoelectric nanobeams.

In this paper, with the consideration of the flexoelectricity and surface elasticity effect, a simple and comprehensive theoretical framework for bending ZnO nanowires will be established.

2. Theoretical Model And Derivation

2.1. Problem identification and flexoelectric equation

Based on the piezoelectricity of ZnO nanowires, Wang et al. [8] proposed a nanogenerator which can convert the mechanical energy into electrical energy by bending the ZnO nanowires. Figure 1 shows the configuration of a nanogenerator using an atomic force microscope (AFM) tip scanning over a ZnO nanowire. A lateral force f_y from the AFM tip applied to the top side of nanowire, result in a mechanical deflection and piezoelectric polarization. Flexoelectric polarization shows apparently in the nanoscaled material, and it should not be ignored. Liu et al [5] have set a model which describes the flexoelectricity affects electrostatic potential in a bent piezoelectric nanowire. However, the derivation process is complex, thus, it will be useful to simplify such a derivation process to establish a rapid model. In this research, for granting a most close realistic conclusion, the size dependent effect has been taken into consideration.

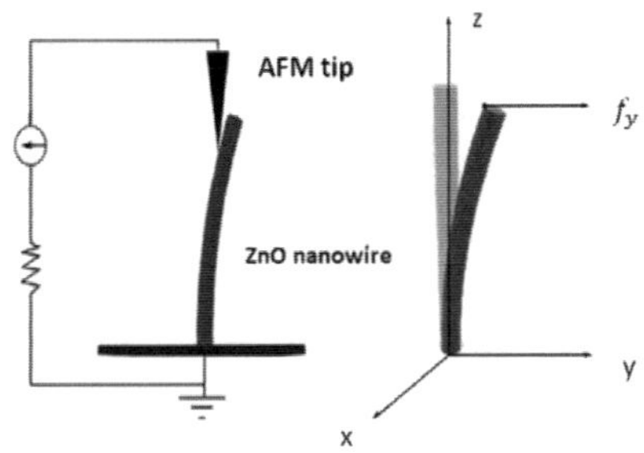

Figure 1. Sketch of nanogenerator based on ZnO nanowire.

Flexoelectricity describes the generation of electric polarization in dielectrics by the application of the strain gradient. For a wurtzite ZnO crystal, the constitutive equation for flexoelectricity from reference [5] takes the form

$$P_i = e_{ijk}\varepsilon_{jk} + \mu_{ijkl}\varepsilon_{jk,l} \quad (1)$$

where P_i (i=1, 2, 3) are components of the piezoelectric polarization vector $\mathbf{P}$, e_{ijk} is the piezoelectric stress constant , ε is strain tensor, and μ is fourth-order flexoeletric tensor. For simplicity, we give the constitutive relationship between the piezoelectric polarization vector **P** and stress tensor σ_{jk} as follow

$$P_i = d_{ijk}\sigma_{jk} + f_{ijkl}\sigma_{jk,l} \quad (2)$$

σ_{jk} (j, k=1, 2, 3) are the components of stress tensor. For a wurtzite crystal, the piezoelectric deformation constant matrix is

$$d_{ijk} = \begin{bmatrix} 0 & 0 & 0 & 0 & d_{15} & 0 \\ 0 & 0 & 0 & d_{15} & 0 & 0 \\ d_{31} & d_{31} & d_{33} & 0 & 0 & 0 \end{bmatrix} \quad (3)$$

f_{ijkl} is a so-called fourth-order flexoeletric tensor related to stresses. The transformed flexoelectric coefficients f_{ijkl} with the relation to μ_{imnl} can be written as follow:

$$f_{ijkl} = \mu_{imnl} c^{-1}_{mnjk} \quad (4)$$

Here, c^{-1}_{mnjk} is the transposition of linear elastic constant c_{mnjk} . The number and types of all possible non-zero flexoelectric coefficients μ_{imnl} was analyzed by Quang and He [9]. For cubic crystal, the non-zreo flexoelectric coefficients are expresses as follow [10]:

$$\mu_{1111} = \mu_{2222} = \mu_{3333} = \mu_{11}$$

$$\mu_{1133} = \mu_{2233} = \mu_{1122} = \mu_{2121} = \mu_{3232} = \mu_{3131} = \mu_{111}$$

$$\mu_{1221} = \mu_{1331} = \mu_{2112} = \mu_{2332} = \mu_{3223} = \mu_{3113} = \mu_{14} \quad (5)$$

It is well known that the polarization charges induced by piezoelectric effect and flexoelectric effect are bound charges rather than free charges. By Gauss's law, it gives

$$\nabla \cdot \mathbf{D} = \rho_s = 0 \tag{6}$$

Where ρ_s is the cylinder surface free charge density in nanowire, and **D** is the electric displacement, which is defined as

$$\mathbf{D} = -\varepsilon \nabla \varphi + \mathbf{P} \tag{7}$$

Here ε is the dielectric constant, and φ is the electric potential.

2.2. Model derivation

The ZnO nanowire is a hexagonal prism in geometry. To simplify the process of analytical solution, we assume the nanowire has a cylindrical shape with a uniform cross section of diameter $2a$ and length l. By Saint-Venant's principle of pure bending in electricity [11], the stress induced in the nanowire is

$$\sigma = \begin{pmatrix} 0 \\ 0 \\ -\frac{f_y}{I_x} y(l-z) \\ \frac{f_y}{I_x}\frac{3+2\nu}{8(1+\nu)}\left[a^2 - y^2 - \frac{1-2\nu}{3+2\nu}x^2\right] \\ -\frac{f_y}{I_x}\frac{1+2\nu}{4(1+\nu)}xy \\ 0 \end{pmatrix} \tag{8}$$

where $I_x = \frac{\pi a^4}{4}$ is the cross-sectional inertial moment of the nanowire.

Hence, non-zero stress gradients are derived as

$$\sigma_{33,2} = -\frac{f_y}{I_x}(l-z)\ ;\ \sigma_{33,3} = \frac{f_y}{I_x}y\ ;$$
$$\sigma_{23,1} = \frac{f_y}{I_x}\frac{2\nu-1}{4(1+\nu)}x\ ;\ \sigma_{23,2} = -\frac{f_y}{I_x}\frac{3+2\nu}{4(1+\nu)}y\ ;$$
$$\sigma_{13,1} = -\frac{f_y}{I_x}\frac{1+2\nu}{4(1+\nu)}y\ ;\ \sigma_{13,2} = -\frac{f_y}{I_x}\frac{1+2\nu}{4(1+\nu)}x. \tag{9}$$

It is known that the material elastic constants can be obtained by an isotropic elastic modulus E and Poisson radio ν, hence, the non-zero flexoelectric coefficients f_{ijkl} can be derived from the equation (4) as

$$\begin{cases} f_{2332} = \frac{1}{E}\left[(1-\nu)\mu_{14} - \nu\mu_{11}\right] = f_{23} \\ f_{3232} = f_{3131} = \frac{2}{E}\mu_{111}(1+\nu) = f_{31} \\ f_{3333} = \frac{1}{E}\left[\mu_{11} - 2\mu_{14}\nu\right] = f_{33} \end{cases} \tag{10}$$

By solving the simultaneous equations (2), (3), (8), (9), and (10), we have

$$\boldsymbol{P}=\frac{f_y}{I_x}\begin{pmatrix} -d_{15}\dfrac{(1+2\nu)}{4\ (1+\nu)}xy \\ d_{15}\dfrac{(3+2\nu)(a^2-y^2-\dfrac{1-2\nu}{3+2\nu}x^2)}{8(1+\nu)}-f_{23}(l\text{-}z) \\ -d_{33}y(l-z)+\left[f_{33}-2f_{31}\right]y \end{pmatrix}. \tag{11}$$

Then we can easily get the piezoelectric charge density

$$\rho_v=-\nabla\cdot\boldsymbol{P}=-\frac{f_y}{I_x}(d_{33}-d_{15})y\approx Ay\ . \tag{12}$$

Where $A=-\frac{f_y}{I_x}(d_{33}-d_{15})$. This equation shows the polarization charges distribution is only dependent on y. Combining the equations (6) and (7), the relationship between piezoelectric potential φ and piezoelectric charge density ρ_v is derived as

$$\nabla^2\varphi=-\frac{\rho_v}{\varepsilon}\ . \tag{13}$$

In cylindrical coordinates, the non-homogeneous differential equation (13) becomes

$$\frac{1}{r}\frac{\partial}{\partial r}(r\frac{\partial\varphi}{\partial r})+\frac{1}{r^2}\frac{\partial^2\varphi}{\partial\theta^2}+\frac{\partial^2\varphi}{\partial z^2}=-\frac{\rho_v}{\varepsilon}=-\frac{Ar}{\varepsilon}\sin\theta\ . \tag{14}$$

Then, we follow the solving steps of paper [4]. Finally, we can get the analytical solution of the piezoelectric potential distribution of the ZnO nanowire

$$\begin{cases} \varphi_{in}=\dfrac{f_y}{\pi}\left[\dfrac{d_{15}-d_{33}}{2\varepsilon}(\dfrac{3\varepsilon+\varepsilon_0}{\varepsilon_0+\varepsilon}\dfrac{1}{a^2}-\dfrac{r^2}{a^4})-\dfrac{4}{a^4}\dfrac{f_{23}}{\varepsilon_0+\varepsilon}(l\text{-}z)\right]r\sin\theta & (r\le a) \\ \varphi_{out}=\dfrac{f_y}{\pi}\left[\dfrac{d_{15}-d_{33}}{\varepsilon_0+\varepsilon}-\dfrac{4}{a^2}\dfrac{f_{23}}{\varepsilon_0+\varepsilon}(l\text{-}z)\right]\dfrac{1}{r}\sin\theta & (r>a) \end{cases} \tag{15}$$

This equation is completely equal to the solution of paper [5].

As we known, when the characteristic sizes of materials shrink to nanometer, surface effects often plays an important role in their material properties due to the increasing ratio of surface area to volume. For the effect of surface elasticity, we adopt the core-surface model, which essentially assumes that a nanowire consists of a core with elastic modulus E_0 and a surface (zero thickness) with so-called surface elastic modulus E_s (the unit is N/m). We can use a concept of effective modulus E^* to describe the elasticity of nanowire with surface effect. Under bending, the expression of effective elasticity E^* with the bulk elasticity E_0 and surface elasticity E_s is written as follows [12]:

$$E^*=E_0+\frac{4E_s}{a}=E_0(1+\frac{4\gamma}{a})\ . \tag{16}$$

Here, γ is a dimensionless elastic modulus ratio $\gamma=E_s/E_0$.

If the influence of surface elasticity is taken into account in our model, the solutions of potential (15) should be modified. Firstly, surface elasticity will affect the flexoelectric coefficients f_{ijkl}, thus,

$$f_{23}^{*}=\frac{1}{E^{*}}[(1-\nu)\mu_{14}-\nu\mu_{11}]=\frac{a}{E_0(a+4\gamma)}[(1-\nu)\mu_{14}-\nu\mu_{11}]. \quad (17)$$

Secondly, surface elasticity will affect the piezoelectric coefficients d_{ijk}, the piezoelectric coefficients d_{ijk} with surface effect can be represented by $d_{ijk}^{*}=\frac{E_0}{E^{*}}d_{ijk}^{0}$, then,

$$\begin{cases} d_{15}^{*}=\frac{E_0}{E^{*}}d_{15}^{0}=\frac{a}{a+4\gamma}d_{15}^{0} \\ d_{33}^{*}=\frac{E_0}{E^{*}}d_{33}^{0}=\frac{a}{a+4\gamma}d_{33}^{0} \end{cases}. \quad (18)$$

Therefore, if the effect of surface elasticity is taken into account, the solutions of piezoelectric potential are modified into

$$\begin{cases} \varphi_{\text{in}}=\frac{f_y}{\pi}\frac{a}{a+4\gamma}\left[\frac{d_{15}^{0}-d_{33}^{0}}{2\varepsilon}\left(\frac{3\varepsilon+\varepsilon_0}{\varepsilon_0+\varepsilon}\frac{1}{a^2}-\frac{r^2}{a^4}\right)-\frac{4}{a^4}\frac{\mathrm{f}_{23}^{0}}{\varepsilon_0+\varepsilon}(l\text{-z})\right]\mathrm{r}\sin\theta & (r\le a) \\ \varphi_{\text{out}}=\frac{f_y}{\pi}\frac{a}{a+4\gamma}\left[\frac{d_{15}^{0}-d_{33}^{0}}{\varepsilon_0+\varepsilon}-\frac{4}{a^2}\frac{\mathrm{f}_{23}^{0}}{\varepsilon_0+\varepsilon}(l\text{-z})\right]\frac{1}{r}\sin\theta & (r>a) \end{cases} \quad (19)$$

If the effect of surface elasticity is neglected, surface elasticity $E_s=0$, namely, the dimensionless elastic modulus ratio $\gamma=0$, the analytical solution can be reduced to the equation (15), which is the results without surface elasticity effect, but only with the effect of flexoelectricity.

3. Numerical Results And Discussion

We established a rapid model which simplifies the derivation process, and also incorporating the effects of flexoelctricity and surface elasticity on the electric potential of ZnO nanowire. For a bulk ZnO, the Young's modulus E_0= 129GPa, Poisson's ratio ν= 0.349 [2, 4], relative dielectric constant $\varepsilon/\varepsilon_0$= 7.7 [13], and the surface elastic modulus E_s takes the value of 235N/m [12]. The piezoelectric constants $d_{15}^{0}=-8.3$ pC/N and $d_{33}^{0}=12.3$ pC/N [4]. And the strain-related flexoelectric coefficient μ_{14} $=\mu_{11}=10^{-9}$ C/m [5], by the values of μ_{11} and μ_{14}, we can get the flexoelectric coefficient f_{23}^{0}. Figure 2 gives the potential distribution along the y-axis direction at z = 300 nm for a nanowire with a = 25nm and l = 600nm bent by a force of 80 nN under different effect conditions. It is shown that flexoelectric effect strengthens the value of electric potential, and this curve with flexoelectric effect is consistent with the previous work [5]. However, the effect of surface elasticity weakens the value of electric potential. It is found from the curve that the value of potential decreases when the effects of flexoelectricity and surface elasticity are both taken into consideration. The curve with no surface elasticity and flexoelectricity comes to the classic result of piezoelectric polarization [2, 4].

Moreover, we simulate the side and top cross-sectional potential distribution at an applied force of 80nN on the nanowire, of which the effects of flexoelectricity and surface elasticity are both taken into account, shown in figure 3(a) and 3(b). It is clearly found that the electric potential due to flexoelectricity is dependent on z. More closely to the top side, more weak effect of flexoelectricity. According to the equation (19), the effect from flexoelectricity is vanished on the top side. Therefore, on the top of nanowire, the main effect is only from surface elasticity. The maximum output of potential on the top side of nanowire is about ± 220.5mV, but at the bottom side is ± 312mV. In Gao and Wang's model [2] the calculated potential value is ± 284mV, and Shao et al. obtained the voltage theoretically is ± 271mV [4]. Though our theoretical result is decreased but still has a wide gap from the experimental value. However, our theoretical result should be more reasonable, for the reason that, firstly, actually, in the process of experimental operation, the primary captured voltage comes from the top side or its vicinity of the ZnO nanowire when the AFM tip scanning over the nanowire, that is to say, the influence of flexoelectricity is almost zero or very weak. Secondly, most researches have proved that surface effects become the important effect for the nanoscale elements [14-16]; especially the effect of surface elasticity cannot be ignored. Thus, it is possible that surface elasticity becomes one of key factors influencing the piezoelectric potential.

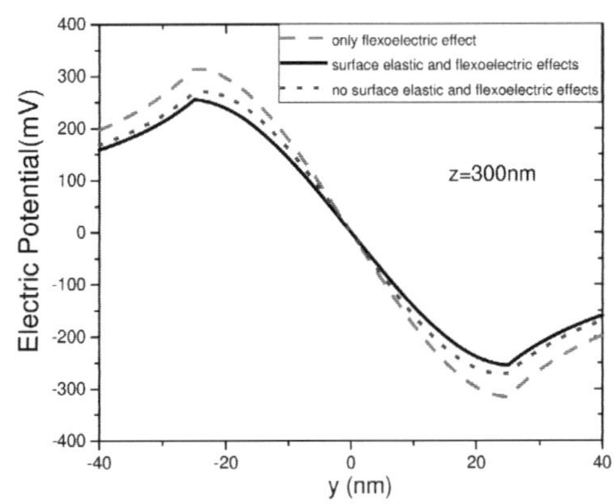

Figure 2. The electric potential distribution on the y-axis with different conditions for a nanowire of a =25nm at z =300nm.

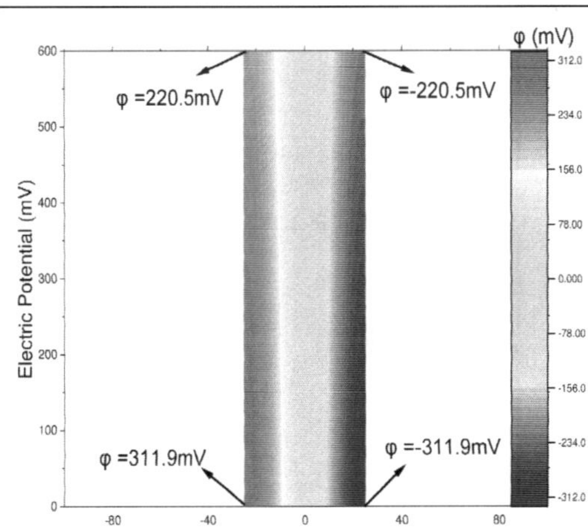

Figure 3(a). Side cross-sectional output of the piezoelectric potential for the nanowire with a =25nm and l =600nm by a lateral force of 80 nN,. due to both flecxoelectric and surface elastic effects.

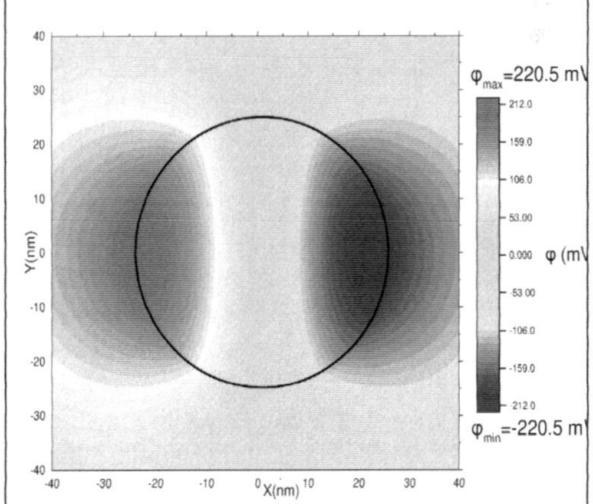

3(b). Top cross-sectional output of the piezoelectric potential of the nanowire, due to both flecxoelectric and surface elastic effects.

4. Conclusions

In summary, we discuss a rapid theoretical model to describe the effects of flexoelectricity and surface elasticity on the piezoelectric potential generated in a bent ZnO nanowire. An analytical solution of potential distribution is deduced, and the results indicate that the effect of flexoelectricity is longitudinal dependent, which vanishes on the top side of nanowire, but surface elasticity demonstrates significant effect on the captured potential. Therefore, the primary contribution to the piezoelectric potential of ZnO nanowire comes from surface elasticity. Simulation result show that the maximum output of potential on the top side of nanowire is about ± 220.5mV, which is more reasonable than other theoretical models. Though our result still has a wide gap between the experimental data, there must be other influencing factors for further study in the future.

Acknowledgements

This paper is supported by the Inner Mongolia National Science Foundation (Grant No. 2014BS0102), Scientific Research Program for Higher Schools of Inner Mongolia (Grant No. NJZY16371) and College Student Innovative Training Program of Inner Mongolia University (Grant No. 201519004).

References

[1] Zhu G, Wang A C, Liu Y, Zhou Y and Wang Z L 2012 Functional Electrical Stimulation by Nanogenerator with 58 V Output Voltage *Nano Lett.*, **12** (6), 3086–3090

[2] Gao Y F and Wang Z L 2007 Electrostatic potential in a bent piezoelectricity nanowire. The fundamental theory of nanogenerator and nanopiezotrotronics *Nano Lett.* **7** 2499-505

[3] Tong H, Wang B L and Ou-Yang Z C 2008 Electric potential generated in ZnO nanowire due to piezoelectric effect *Thin Soild Films* **516** 2708

[4] Shao Z Z, Wen L Y, Wu D M, Wang X F, Zhang X A, and Chang S L 2010 A Continuum Model of Piezoelectric Potential Generated in a Bent ZnO Nanorod *J. Phy. D: Appl. Phys.* **43** 245403

[5] Liu C C, Hu S L and Shen S P 2012 Effect of flexoelectricity on electrostatic potential in a bent piezoelectric nanowire *Smart Mater. Struct.* **21** 115024(9pp)

[6] Yan Z and Jiang L Y 2013 Flexoelectric on the Electroelastic Responses of Bending Piezoelectric Nanobeams *J. Appl. Phys.* **113** 19410

[7] Yan Z and Jiang L Y 2013 Size-dependent Bending and Vibration Behavior of Piezoelectric Nanobeams due to Flexoelectricity *J. Phy. D:Appl. Phys.* **46** 355502 (7pp)

[8] Wang Z L and Song J H 2006 Piezoelectric nanogenerators based on zinc oxide nanowire arrays *Science* **312** 242–6

[9] Quang H L and He Q C 2011 The number and types of all possible rotational symmetries for flexoelectric tensors *Proc. R. Soc. A* **467** 2369-86

[10] Shu L L, Wei X Y, Pang T, Yao X and Wang C L 2011 Symmetry of flexoelectric coefficients in crystalline medium *Journal of Applied Physics* **110** 10 104-106

[11] Green A E and Zerna W 2002 theoretical elasticity (New York: Dover)

[12] Yao H Y, Yun G H, Bai N and Li J G 2012 Effect of Surface Elasticity on the Pizoeelectric Potential of a Bant ZnO Nanowire *Jpn. J. Appl. Phys.* **51** 075001

[13] Ashkenov N, Mbenkum B N and Bundesmann C 2011 Infrared Dielectic Functions and Phonon Modes of high-quality ZnO Film *J. Appl. Phys.* **93** 126-33

[14] Chen C Q, Shi Y, Zhu J and Yan Y J 2006 Size Dependence of Young's Modulus in ZnO Nanowires *Phys. Rev. Lett.* **96** 075505

[15] Asthana A, Momeni K, Prasad A, Yap Y K and Yassar R S 2011 In situ observation of size-scale effects on the mechanical properties of ZnO nanowires *Nanotechnology* **22** 265712

[16] Yao H Y, Yun G H, Bai N and Li J G 2012 Surface elasticity effect on the size-dependent elastic property of nanowires *J. Appl. Phys.* **111** 083506

Photonic crystal hydrogel sensor for detection of nerve agent

Jiayu Xu[1,3], Chunxiao Yan[1], Chao Liu[1], Chaohua Zhou[1], Xiaochun Hu[1], Fenglian Qi[2]

[1]Department of Chemical and Biological Defence, Institute of NBC Defence, Beijing, 102205, China
[2]School of Chemical Engineering and Environment, Beijing Institute of Technology

E-mail: fhxujiayu@163.com

Abstract. Nowadays the photonic crystal hydrogel materials have shown great promise in the detection of different chemical analytes, including creatinine, glucose, metal ions and so on. In this paper, we developed a novel three-dimensional photonic crystal hydrogel, which was hydrolyzed by sodium hydroxide (NaOH) and immobilized with butyrylcholinesterase (BuChE) by 1-(3-Dimethylaminopropyl)-3-ethylcarbodiimide hydrochloride(EDC). They are demonstrated to be excellent in response to sarin and a limit of detection(LOD) of 1×10^{-9} mg mL^{-1} was achieved.

1. Introduction

There is currently an urgent need for efficient, rapid detection of nerve agents. Nerve agents are a class of highly toxic organophosphates which disrupt nerve conduction by blocking acetylcholin esterase (AChE), an enzyme that normally destroys acetylcholine, a neurotransmitter. Among various sensor technologies, photonic crystal hydrogel is a novel technology which now attracts more and more attentions due to its unique ability for the detection [1-3].The term “photonic crystals(PhC)” relates to a periodical arrangement of a regularly shaped material consisting of nanoparticles [4]. A much more rugged polymerized crystalline colloidal array (PCCA) was prepared by embedding the crystalline colloidal array (CCA) within a hydrogel matrix. These three-dimensionally ordered CCAs can intensely Bragg diffract light in the ultraviolet, visible, and infrared regions of the spectrum. The diffracted wavelength of the photonic crystal hydrogel shifts as the hydrogel volume changes due to the alterations in the spacing between lattice planes [5-7].

Liu et al. developed a molecularly imprinted photonic crystal for the detection of methyl phosphonic acid (MPA), which was released from the hydrolysis of nerve agents, with the LODs of 3.5×10^{-6}, 2.5×10^{-5}, 7.5×10^{-5} and 7.5×10^{-5} mol L^{-1} for Sarin, Soman, VX and R-VX respectively [8]. Walker and Asher attached AChE to photonic crystal, and AChE binds organophosphorus compounds irreversibly, creating an anionic phosphonyl species. These AChE-PhC acted as dosimeters for parathion concentrations as low as 4.26×10^{-15} mol L^{-1} [9]. Although AChE has good recognition ability to organophosphates, exploration of other cheaper and more efficient enzymes is still necessary.

In this study, we developed a method of synthesizing monodisperse polystyrene colloids of diameters between 100~440 nm. These monodisperse colloidal particles readily self-assembled into robust three-dimensionally ordered CCAs. We then polymerized a hydrogel around the CCA to form a PCCA. The PhC is immobilized with BuChE from duck blood, and was characterized optical fiber spectrometer. The LOD of Sarin agent was also calculated. The simple method for the detection of Sarin agent represented its potential in response to other nerve agents.

2. Experimental Section

2.1. Preparation of photonic crystal hydrogel

Figure 1 depicts the preparation of the photonic crystal hydrogel. Monodisperse polystyrene spheres were synthesized by emulsion polymerization with sodium-dodecyl sulphate (SDS, Beijing Chemical Works) as emulsifier. After 8 h of reaction, all the suspension were centrifuged for purification. The particle sizes were measured using quasi-elastic light scattering (QELS; Malvern Nano Series) and the morphology was measured by scanning electron microscope (SEM; HITACHI Quanta 250 FEG). Acrylamide (Amresco), N,N′-methylenebisacrylamide (Sigma-Aldrich), a colloid suspension (polystyrene spheres) in nanopure water, and 10% diethoxyacetophenone (DEAP, Alfa Aesar) in DMSO (Sigma-Aldrich) were mixed and injected between two quartz disks. The colloidal particles self-assembled into a CCA, leading to a liquid film that diffracts light. The film was exposed to 365-nm UV light (Ultra-Violet Products Ltd) for 2 h. A polyacrylamide hydrogel network formed around the CCA, resulting in a PCCA. The diffraction of the PCCA was monitored using a fiber-optic diode spectrometer with a tungsten halogen light source (Avaspec-2048TEC, Avantes) using a reflectance probe.

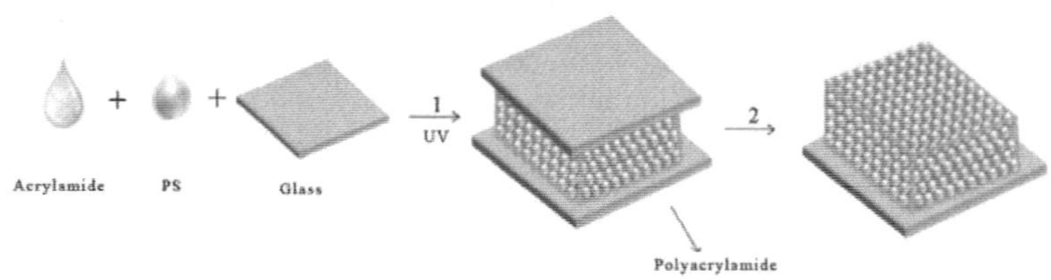

Figure 1. Preparation of photonic crystal hydrogel.

2.2. Functionalization of PCCA

The PCCA was then hydrolyzed in a solution of NaOH (Beijing Chemical Works) containing 10% v/v N,N,N',N'-tetramethylethylenediamine (TEMED, Sigma-Aldrich) at 25 °C for different times. The hydrolyzed PCCA was washed for 2 h with 150 mM NaCl (Beijing Chemical Works) [10]. The surface profile of hydrolyzed PCCA was measured using scanning electron microscope (SEM; HITACHI Quanta 250 FEG). The diffraction of the PCCA was monitored using a fiber-optic diode spectrometer.

To immobilize BuChE onto the hydrolyzed PCCA, 2000 units BuChE was dissolved into Tris buffer (0.15 mol L^{-1}, pH=7.4), and the hydrolyzed PCCA was incubated in it for 48 h before immersed into 1-(3-Dimethylaminopropyl)-3-ethylcarbodiimide hydrochloride(EDC•HCl) solution for 2 h. After that, the PCCA was re-incubated into the enzyme solution for another 2 h, and finally was washed in Tris solution for 48 h to clean up the unreacted BuChE. The functionalized BuChE-PCCA was cut into small pieces before usage.

2.3. Measurement of Sarin agents

The diffraction of BuChE-PCCA was recorded using Avaspec-2048TEC optical fiber spectrometer (Avantes, Netherlands). The BuChE-PCCA was incubated into nanopure water for 15 min, and the diffraction spectrum was recorded as a reference. To detect Sarin agent, the concentration of Sarin solution was adjusted from 1×10^{-9} mg mL^{-1} to 1×10^{-2} mg mL^{-1} using nanopure water, and its pH was adjusted to NaOH solutions, the BuChE-PCCA was immersed into the Sarin solution for 10 min, and the diffraction spectrum was recorded.

3. Results and discussion

3.1. Characterization of PCCA

Monodisperse, highly charged polystyrene colloidal particles with diameters between 100–400 nm were synthesized that listed in Table 1. Because of electrostatic interactions, all of these colloids readily formed CCAs, which exhibited sharp diffraction peaks in the UV/Vis region. It is known that the diffraction of PCCA depends on the self-assembly procedure. With the increasing diameter of colloidal particles, the diffraction of PCCA red shifted (Fig. 2), which agreed with the Bragg's law.

Table 1. Characterization of polystyrene nanospheres with different dosage of emulsifier by QELS

number	1	2	3	4	5	6	7
SDS/g	0.100	0.125	0.150	0.175	0.200	0.225	0.250
Diameter/nm	441.2	404.3	355.7	289.6	223.5	180.0	101.0
PDI	0.007	0.060	0.094	0.069	0.079	0.007	0.035

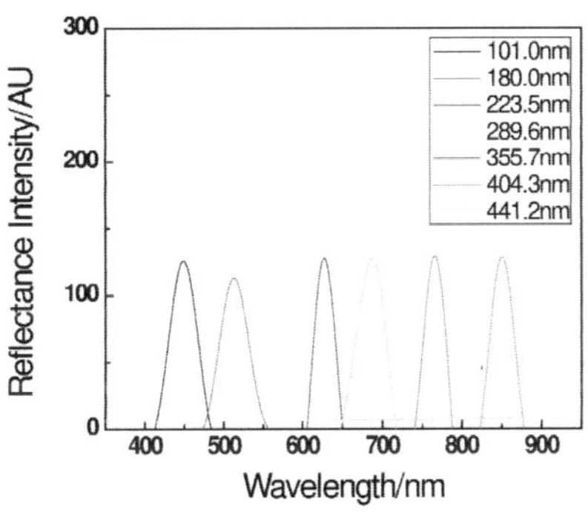

Figure 2. The diffraction spectra of PCCA in related to the colloidal diameter.

3.2. Characterization of Functionalized PCCA

For immobilizing BuChE onto the PCCA, the amide group of the photonic crystal hydrogel needed to be converted into carboxylic group via hydrolysis with NaOH. Figure 3a showed SEM images of functionalized PCCA. The polystyrene spheres in the hydrogel appeared neat and showed high ordering with unclose packed structure. Via monitoring the diffraction spectrum of functionalized PCCA, we found that the longer the hydrolysis time, the more amide groups were converted to carboxylic groups, and the hydrogel became more hydrophilic and red shifted the diffraction (Fig. 3b). We found that modification could make PCCA volume increased. Too short modification time may make the modification inadequately. Too long modification time can give rise to the deterioration in PCCA mechanical strength, because the PCCA volume increased too much. For chemical sensing, the PCCA is required to have proper mechanical strength, so 40 min was optimized as the appropriate modification time.

BuChE is a substitute of AChE, which can serve as a stoichiometric bioscavenger of organophosphate AChE inhibitors in the human body to detoxify organophosphorus molecules. BuChE obtained from duck blood is cheaper than AChE. The unimmobilized PCCA diffracted the light at 632 nm, when BuChE was immobilized on the photonic crystal hydrogel via condensation with EDC, the diffraction of BuChE-PCCA red shifted (Fig.4). With the immobilization of BuChE, the hydrogel became more hydrophilic and the free energy of mixing increased, thus the diffraction red shifted.

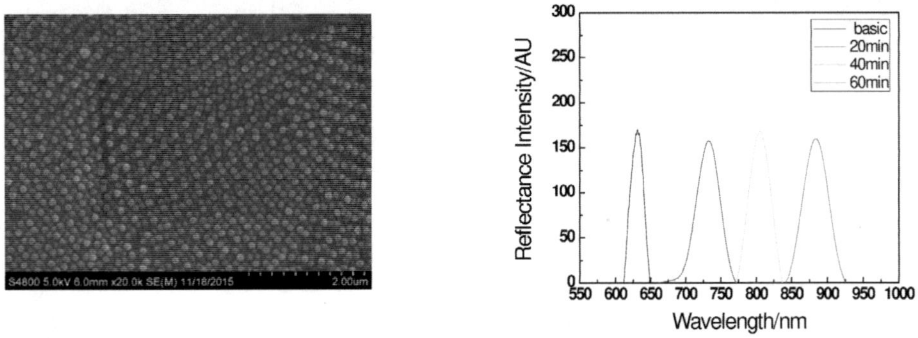

Figure 3.(a) SEM of the functionalized PCCA; **(b)**The diffraction spectrum of PhC in related to the hydrolysis time.

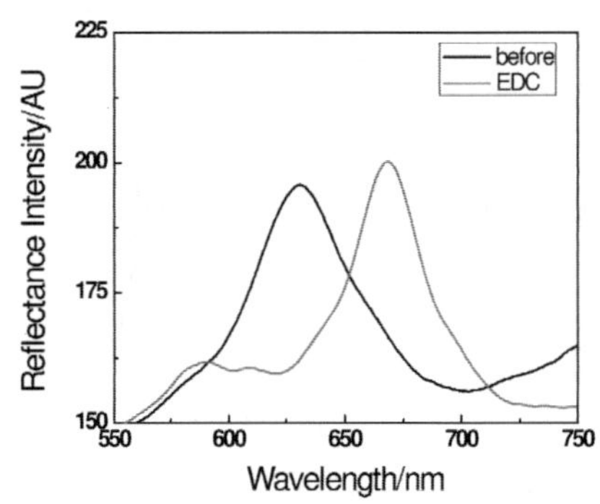

Figure 4. Diffraction of the PCCA before and after immobilization with BuChE.

3.3. Sensor Properties of BuChE-PCCA

Figure 5 showed the diffraction response of unimmobilized PCCA to sarin. Figure 6 showed the diffraction response of BuChE-PCCA to sarin. The results showed that as Sarin concentration increased from 1×10^{-9} mg mL^{-1} to 1×10^{-2} mg mL^{-1}, the diffraction of the unimmobilized PCCA blue shifted. However, the diffractions of the BuChE-PCCA which response to Sarin concentration from 1×10^{-9} mg mL^{-1} to 1×10^{-7} mg mL^{-1} red shifted compared with the diffractions of BuChE-PCCA in nanopure water, and blue shifted for Sarin concentration from 1×10^{-6} mg mL^{-1} to 1×10^{-2} mg mL^{-1}. A LOD of 1×10^{-9} mg mL^{-1} was achieved. A BuChE-free PCCA was exposed to Sarin solutions in the same way as a control experiment and this indicated that the resulting diffraction shift was not triggered by the amide or carboxyl groups on the polymer backbone.

The blue shift of diffraction in response to Sarin agent is different from that of Walker and Asher's report [4].The molecular recognition agent for the sensor was the BuChE, which bound sarin irreversibly, creating an anionic phosphonyl species. This charged species creates a Donnan potential, which swells the hydrogel network, which increases the embedded particle array lattice spacing and causes a red-shift in the wavelength of light diffracted [9].But as the concentration of sarin increased, the amount of red-shift decreased continuously. While in this case, the response of the PhC was due to the HF formed when Sarin inhibited the BuChE [11]. The formation of HF made the solution acidic, and thus caused the decreasing of pH value and the increasing of ionic strength, both of which were able to make the decrease of red-shift.

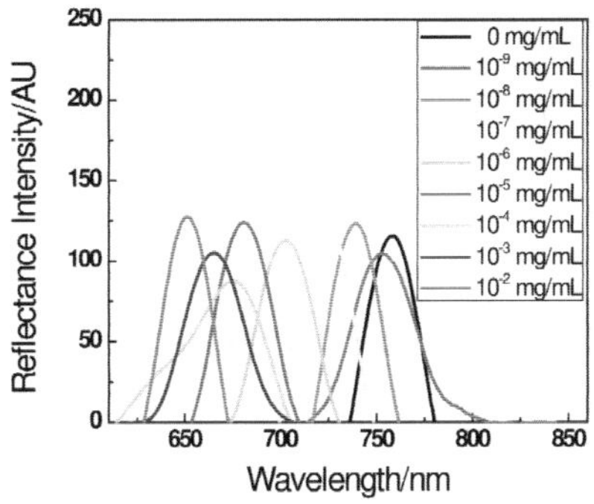

Figure 5. Diffraction spectra of the BuChE-free PCCA at various concentration of Sarin solutions.

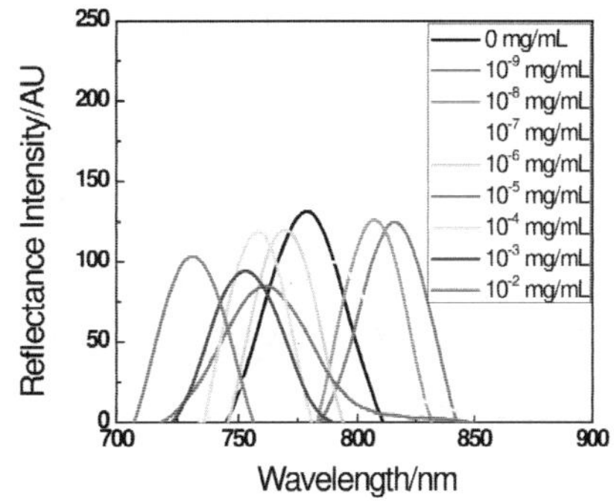

Figure 6. Diffraction spectra of the BuChE- PCCA at various concentration of Sarin solutions.

4. Conclusions.

We developed a novel material to determine various concentrations of sarin, which were immobilized with BuChE by EDC onto functionalize PCCA. After chemical modification of hydrogel backbone and immobilization, the diffraction of the functionalized photonic crystal hydrogel red shifted compared with the diffractions of BuChE-PCCA in nanopure water in response to Sarin concentration from 1×10^{-9} mg mL^{-1} to 1×10^{-7} mg mL^{-1} and blue shifted for Sarin concentration from 1×10^{-6} mg mL^{-1} to 1×10^{-2} mg mL^{-1}. This detection limit was 1×10^{-9} mg mL^{-1} lower than current enzyme detection methods for organophosphorus agents. We were developing this sensor for use in detecting other organophosphorus agents in this field.

References

[1] Moshe M B, Alexeev V L and Asher S A, Fast responsive crystalline colloidal array photonic crystal glucose sensors, *Anal. Chem.* **78** (2006) 5149–5157

[2] Reese C E and Asher S A. Photonic Crystal Optrode Sensor for Detection of Pb^{2+} in High Ionic Strength Environments. *Analytical Chemistry*, **75**(2003) 3915-3918

[3] Holtz J H and Asher S A. Polymerized Colloidal Crystal Hydrogel Films as Intelligent Chemical Sensing Materials. *Nature*, **389**(1997) 829-832

[4] Fenzl C, Wilhelm S, Hirsch T and Wolfbeis O S, Optical Sensing of the Ionic Strength Using Photonic Crystals in a Hydrogel Matrix, *Applied Materials & Interfaces*, (2013) 173–178

[5] Kamenjicki M, Kesavamoorthy R and Asher S A, Photonic Crystal Devices, *Ionics.* (2004), 233–236

[6] Ge J P and Yin Y D, Responsive photonic crystals, *Angew. Chem. Int .Ed.* **50** (2011) 1492-1522

[7] Kamenjicki M and Asher S A. Epoxide Functionalized Polymerized Crystalline Colloidal Arrays. *Sensors and Actuators B*, **106**(2005) 373-377

[8] Liu F, Huang S Y, Xue F, Wang Y F, Detection of organophosphorus compounds using a molecularly imprinted photonic crystal, *Biosens. Bioelectron.* **32** (2012) 273–277

[9] Walker J P and Asher S A, Acetylcholinesterase-Based Organophosphate, *Analytical Chemistry*, (2005) 1596–1600

[10] Walker J P, Kimble K W and Asher S A, Photonic Crystal Sensor for Organophosphate Nerve Agents Utilizing the Organophosphorus Hydrolase Enzyme, *Anal Bioanal Chem*, (2007) 2115–2124

[11] Hu X B, Li G T, Li M H, Huang J, Li Y, Gao Y B and Zhang Y H, Ultrasensitive specific stimulant assay based on molecularly imprinted photonic hydrogels, *Adv. Funct. Mater*, **18** (2008) 575–583

Ferrum nano particles and multiwall carbon nano tubes based electrode as FIA detector for determination of amino acids in hypothalamus microdialysis fluids

L Sun, J Wang[1], Y T Wang, L Yu, H Peng and J Z Zhu

School of Chemistry and Chemical Engineering, East China Normal University, 3663 North Zhongshan Road, Shanghai 200062, China

E-mail: jwang@nbic.ecnu.edu.cn

Abstract. An amperometric electrode based on multiwall carbon nanotubes (MWCNTs) and Fe nanoparticles (NPs) has been successfully fabricated. Combined with Flow Injection Analysis (FIA) and chromatography separation column, the electrode exhibits linear response in the concentration range of 0.1 -12 μM and the sensitivity of 30.0 nA μM^{-1} for most of amino acids. The determination of 17 amino acids in the hypothalamus microdialysis fluids of guinea pigs, illustrates that the electrode is a powerful tool to investigate physiology and pathology mechanisms

1. Introduction

Amino acids (AAs) are essential building blocks of biological molecules and play key roles in many neurochemical response mechanisms such as memory, appetite control, and pain transmission. The accurate determination of amino acids can improve our understanding of its biological role and facilitate the treatment of diseases caused by its disruption. Because most of amino acids lack natural chromophore or fluorophore for direct photometric and fluorometric detection, derivatization procedures are necessary for the sensitive and selective determination. However, derivatization procedures generally suffer from the instability of the derivatization product and time-consuming. The presence of relatively low concentrations of amino acids in microdialysis samples requires a sensitive analytical method for their determination. Microdialysis is a powerful technique which permits determination of the extracellular concentration of amino acid neurotransmitters during various pharmacological or physiological manipulations. The amperometric electrode detector combined with FIA possesses many advantages such as high sensitivity, rapid response, and simplicity in operating procedure, and much attention has been paid to the development of electrochemical detection of amino acids. However, many amino acids are nonelectroactive at the conventional carbon electrodes. The noble metal electrodes exhibit a greater degree of activity toward amino acids oxidation, but they are subject to rapid surface fouling [1]. The alternative method is using non-noble transition metal, such as nickel and copper [2], chemically modified electrodes that can mediate fast electron transfer between amino acids and the substrate electrode and can reduce surface fouling. These modifiers include copper nanoparticles and zinc oxide nanorod array [3], copper–manganese alloy [4], $Cu_2V_2O_7$, nickel–copper alloy [5], nickel hydroxide nanoflakes [6], copper microparticle [7], Cu_2O [8], cobalt and cadmium doped nickel hydroxide [9], and nickel–curcumin complex [10]. Furthermore, carbon nanotubes (CNTs) are of good electrocatalytic properties as electrode materials. Iron is also a transitional element, should possess similar properties with nickel and copper, but rarely reported. Fe nanoparticles (NPs) are complex centre with AAs like Cu NPs, while MWCNTs can disperse iron nanoparticles to avoid reunion and fleetly transfer electrons from AAs to the electrode. The bulk sensing layer is convenient for repeatable renew of electrode. Thus, Fe NPs and MWCNTs were employed to fabricate a novel AAs sensor. The electrochemical behaviour of Fe NPs/MWCNTs electrode and the oxidation of AAs' complex with Fe NPs were investigated by electrochemistry methods. Moreover, the electrode combined with FIA method and separation technology was proposed for the determination of AAs with fast response, good stability, and sensitivity. The highlights of this work are that the electrode proposed used an electrolyte with near neutral pH value (8.0) and applied low work potential (0.6 V) as well as determined successfully 17 amino acids in the hypothalamus micro-dialysis fluid of guinea pigs.

2. Experimental

2.1. Reagents and animals

MWCNTs (GMWCNT2, 10–20 nm diameter, 10–20 μm length, purity > 99.9%) were purchased from Shanghai Jikart Chemical engineering Sci & Tech Co. Ltd. (Shanghai, China) and Nano-Fe (50 nm average diameter, purity 99.99%) was purchased from Beijing Deke Daojin Science & Technology Co. Ltd (Beijing, China). They were used without further purification. Adult guinea pigs weighing 400–500 g purchased from the Experimental Animal Centre of Medical School, Shanghai Jiao Tong University. All amino acids were obtained from Sigma Co. (USA). Other reagents were of at least analytical-reagent grade. All solutions were prepared using doubly distilled water. Standard stock solution of amino acids (20 mM) were prepared in a mixture of water–methanol (1:1) and stored at 4°C for 1 month. Working standard solutions were prepared daily by diluting the stock solutions with PBS buffer (pH 8.0).

2.2. Measurement

Electrochemical experiments were performed on a CHI 832A workstation (CH Instruments, Shanghai, China) at room temperature. A conventional three electrode system was employed, including a homemade Fe–MWCNTs electrode (3.0-mm-diameter) as a working electrode (WE), an Ag/AgCl reference electrode (RE), and a platinum wire counter electrode (CE). All of the potentials quoted in this work were referred to SCE. The FIA system was consisted of a Cole–Parmer microprocessor pump drive, a Rehodyne model 7125-sample injection valve (20 mL loop) with interconnecting Teflon tube and a Zensor SF-100 thin layer detecting electrochemical system (The dimension of the electrode was specifically designed for it). Chromatographic separations were performed using a silica based HPLC column (Prevail organic acid 5 u, Alltech, 100×4.6 mm) with a mobile phase of 0.1 M, pH 8.0 PBS.

Acquity UPLC system (Waters Co.) with a Series 200 IC pump and a diode array detector 235C (Perkin Elmer, Norwalk, CT) was also applied to determine amino acids. The column used was Shim-pack C_{18}, 5 μm, 150 mm×4.6 mm (Shimadzu, Japan). The micro-dialysis fluids were derived with 6-aminoquinolyl-N-hydroxysuccinimidylcarbamate (AQC) reagent and the derivatives were separated by a two gradient elution and detected by fluorescence detector. The wavelength used for the measurement of the derivatives was 286 nm. The HPLC flow rate was 1.5 ml/min. All the measurements were performed at room temperature. Comparison between results of two analysis methods was realized.

The experiment system was shown in Figure 1. It consists of three parts: A) microdialysis, B) AAs separation column, and C) Flow Injection Analysis (FIA) with Fe NPs/MWCNTs electrode as a detector. Fe nanoparticles and MWCNTs with weight ratios ($W_{Fe}/ W_{Fe+CNTs}$) of 0.30, 0.25, 0.20, 0.15 and 0.10 were mixed and grinded in a mortar to obtain a homogeneous mixture. The Fe-MWCNTs electrode was prepared by mixing the mixture and silicone oil in a ratio of 3:2 (w/w) in a mortar and grind until a uniform paste was obtained. A portion of the resulting paste was packed firmly into the

IOP Conf. Series: Materials Science and Engineering **167** (2017) 012025 doi:10.1088/1757-899X/167/1/012025

cavity of a polytetrafluoroethylene (PTFE) tube (2 mm depth, 3.0 and 1.0 mm diameter for electrochemical and application investigation, respectively). The electric contact was established via a copper wire with the aid of Wood's alloy as shown in the inset of Figure 1C. The electrode surface was gently smoothed by rubbing on a piece of weighing paper just prior to using. This procedure was also used to regenerate the surface of the electrode proposed. SEM image of the surface of spherical Fe/MWNTs composite structure bulk was shown in Figure 2.

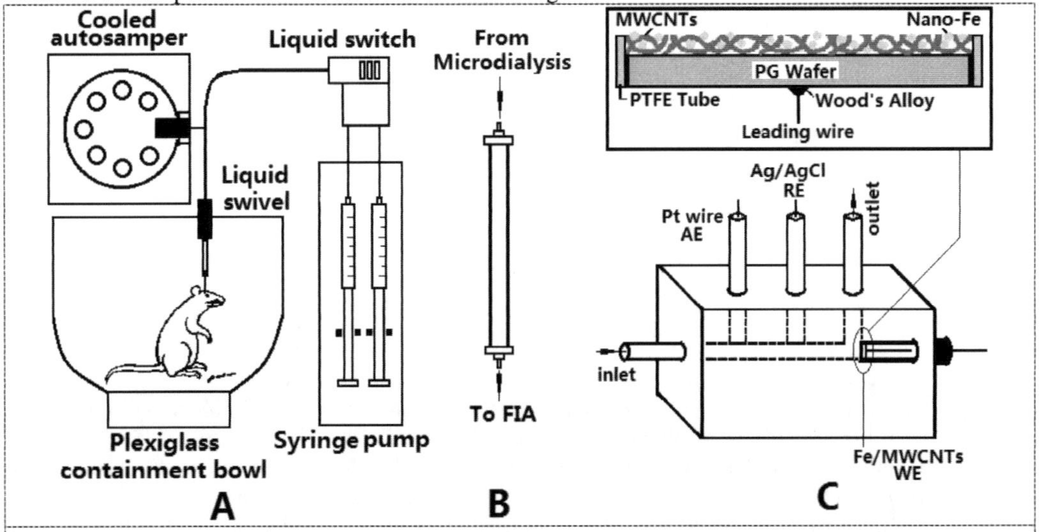

Figure 1. Schematic diagram of the electrochemical measurement system with Fe NPs/MWCNTs electrode for determining amino acids in the hypothalamus microdialysis fluids of guinea pigs. It included A) microdialysis, B) AAs separation column, and C) Flow Injection Analysis (FIA) with Fe NPs/MWCNTs electrode as a detector (a larger view is above).

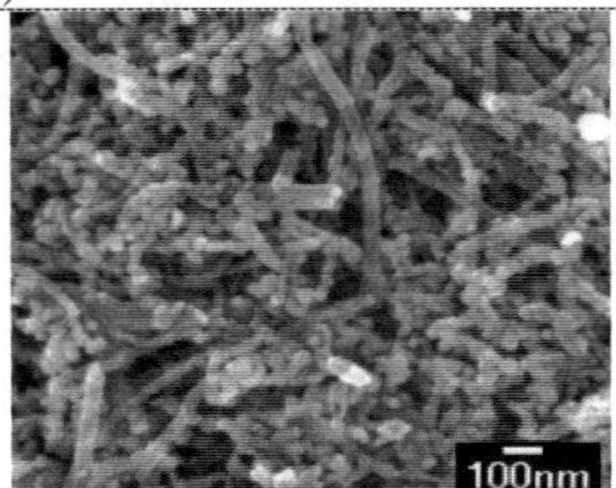

Figure 2. SEM image of the surface of the electrode proposed. It is evident that the MWCNTs are very long and present as a highly entangled network structure and the images reveal that the Fe NPs were distributed randomly and uniformly on the surface of CNTs. The MWCNTs in the modified electrode play an important rule to disperse Fe NPs and enhance conductivity of the electrode.

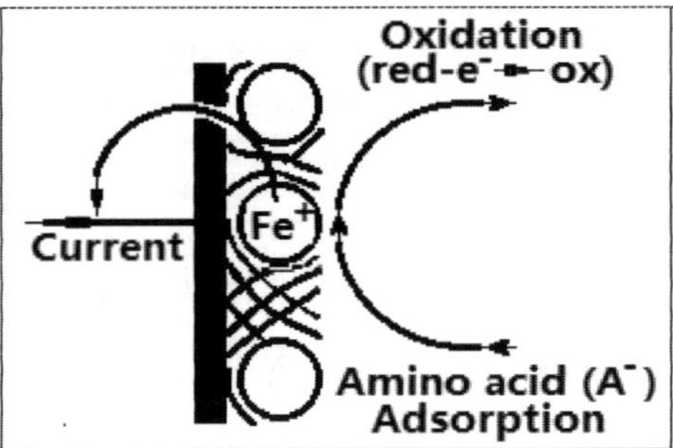

Figure 3. Schematic diagram of electrochemical reaction of the Fe NPs/MWCNTs electrode. The negative charged AAs- molecular (generator) were adsorbed on Fe NPs (acceptor) with positive charge due to electrostatic interaction. An electron can fleetly be transferred from AAs to the electrode and formed electrochemical current.

2.3. Sample treatment

Guinea pigs were narcotized by 3 g/dl pentobarbital of intraperitoneal injection and located on the brain locator. The dialysis micro-needle was imported into a predetermined depth of hypothalamus and fixed by denture acrylic. A Ringer's perfusion fluid was perfused via the needle at 20 μl/min. Different period of dialysates were collected, respectively. All procedures of animal experiments complied with the regulations of the ''protection of animals used for experimental and other scientific purposes'' from the relevant directives of Good Laboratory Practices (GLP). The study protocols were approved by the Ethical Committee at our institutes.

2.4. Sensing mechanism of Fe-MWCNTs electrode

Isoelectric points of most AAs are smaller than 8, so they show negative charged in PBS of pH 8.0. Positive charged Fe NPs at the surface of the electrode interact with AAs through the electrostatic interaction. Thus Fe NPs at the surface of the electrode become the complex center. The MWCNTs in the modified electrode play an important rule to disperse Fe NPs and enhance conductivity of the electrode. It can fleetly transfer electron from AAs to the electrode. Sensing mechanism of Fe-MWCNTs electrode was shown in Figure 3.

3. Results and Discussion

3.1. Characterization of Fe-MWCNTs electrode

The dependence of response current on the ratio of W_{Fe}/W_{Fe+CNT} was tested by using linear sweep voltammetry, as shown in Figure 4. The response current increased with increasing of Fe nanoparticles proportions until the ratio reached 25%, because of the increase of the complexation center. When the ratio was 30%, the response current became smaller (second-highest in Fig.4). Therefore, 25% of ratio was used in the following experiments.

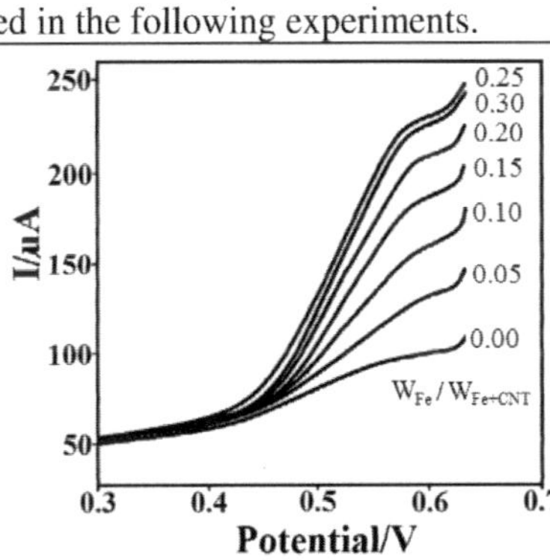

Figure 4. Linear sweep voltammetry curves of 2 mM glycine at the electrode with different ration of W_{Fe}/W_{Fe+CNT} in pH 8.0 PBS. Scan rate: 100 mVs^{-1}. The response current increased with increasing of Fe nanoparticles proportions until the ratio reached 25%. When the ratio was 30%, the response current became smaller. Therefore, 25% of ratio was used in the following experiments.

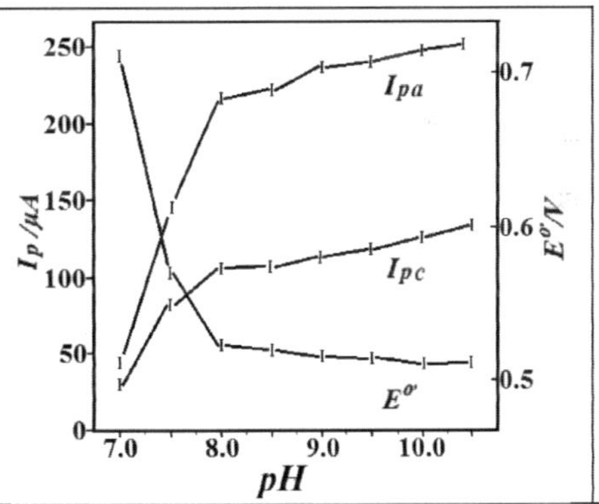

Figure 5. The peak current I_p and peak formal potential E^0 of Fe-MWCNT electrode were depended on the pH value of the PBS. With the increase of pH value of the solution, both anodic peak current, I_{pa} and cathode peak current, I_{pc} were enhanced. The peak formal potential E^0 of the redox couple was negatively shifted. Scan rate: 100 mV/s.

3.2. Effect of pH on the Sensor Response

To investigate the influence of pH value on the voltammetric behaviour of glycine on Fe–MWCNTs electrode, the electrode was examined in a pH range from 7.0 to 10.5, and the results were shown in

Figure 5. As shown, with the increase of pH value of the solution, both anodic peak current I_{pa} and cathode peak current I_{pc} were enhanced and the formal potential E^0 of the redox couple negatively shifted, suggesting that OH^- participated in the redox process of the modified electrode.

3.3. Cyclic Voltamograms

The cyclic voltammograms of glycine on Fe-MWCNTs electrode was shown in Figure 6. The oxidation and reduction peaks of glycine appear at 0.6 and 0.08 V, respectively. The peak currents has a linear relationship with the concentration of glycine in a range from 0.1 to 20 mM.

3.4. Chronoamperogram and Calibration curve

Figure 7 represents the chronoamperogram of the electrode obtained in 0.1 M PBS solution (pH8.0) at 0.6 V by successively adding aliquots of glycine. The sensor proposed displayed a fast amperometric response time of less than 2.5 s. The calibration curve is also shown as the inset (linear response range: 0.1 - 20 mM, sensitivity: 1.1 μA mM^{-1}, RSD: 3.5%).

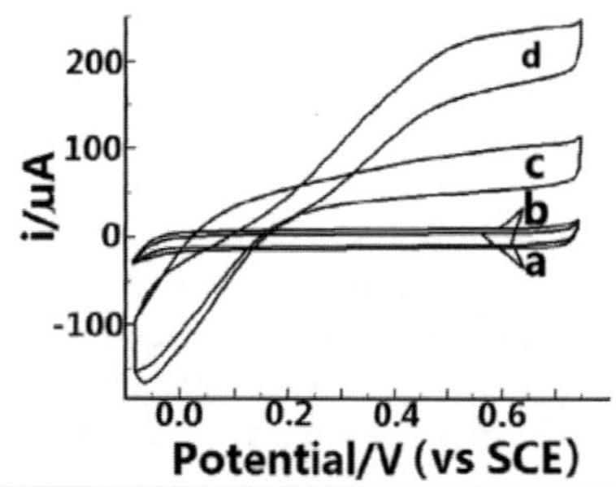

Figure 6. Cyclic voltammograms of the electrodes proposed without Fe NPs and with 25% Fe NPs in the absence (a and b) and presence of 2 mM glycine (c and d), respectively in 0.1 M PBS solution（pH 8.0）at a scan rate of 100 mV/s. The anodic peak height for the electrode with W_{Fe}/W_{Fe+CNT} = 25% is 2.3 times one for the electrode without FE NPs (W_{Fe}/W_{Fe+CNT} = 0).

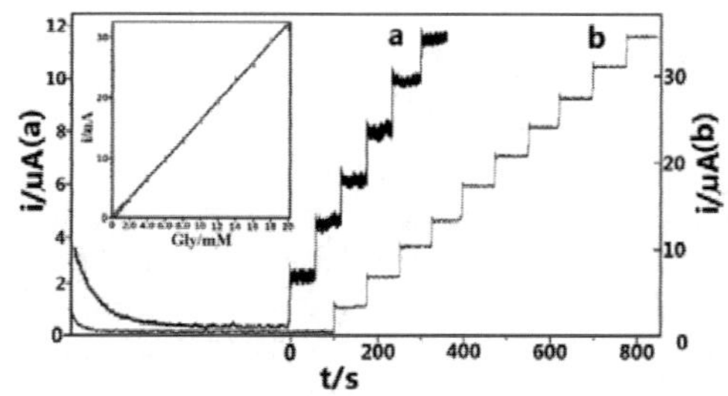

Figure 7. Amperometric responses of the electrode proposed in 0.1 M PBS solution (pH8.0) by successive addition of 0.2 (a) and 2.0 (b) mM glycine. Applied potential: 0.6 V. Inset: calibration curve. The response currents has a linear relationship with the concentration of glycine in a range from 0.2 to 20.0 mM. The linear regression equation is: i = -0.1985 + 1.7499 C_{gly} (correlation coefficient is 0.999).

3.5. Stability and Reproducibility

Table 1 shows that the response current a little increased when the electrode was renewed while a slightly decreased after storage for 3 months. The possible reason is Fe NPs on the surface of the electrode was partly oxidized, resulting decrease of the complexation center of AAs. Five Fe–MWCNTs electrodes were fabricated, and the RSD for the individual determination of 0.5 mM glycine was 3.8%. The long-term stability of the modified electrode was tested after being stored in dry conditions at room temperature for 60 days, and no significant change in current responses was observed. Thus, the Fe–MWCNTs electrode exhibited acceptable stability and reproducibility.

Table 1. Measurement precision for the sensor in the buffer solution (pH 8.0) including 0.5 mM of glycine (n = 11). [a]R.S.D.: relative standard deviation.

Times	Response/μA	Response/μA After renew	Response/μA After storage for 3 months
1	2.31	2.69	2.20
2	2.44	2.56	2.10
3	2.23	2.51	2.25
4	2.34	2.60	2.27
5	2.38	2.63	2.22
6	2.40	2.50	2.15
7	2.24	2.63	2.13
8	2.22	2.47	2.09
9	2.41	2.55	2.30
10	2.39	2.67	2.15
11	2.25	2.52	2.12
Average/μA	2.33	2.58	2.18
RSD[a]	3.5%	2.8%	3.3%

3.6. Optimal experiment conditions of FIA

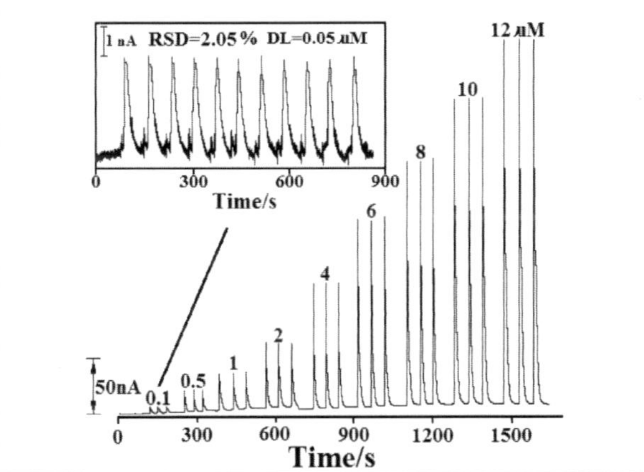

Figure 8. When FIA system with the electrode proposed as a detector, responses of a series of glycine concentration (0.1-12 μM) with 250 μl/min of flow rates in pH 8.0 of PBS carrier solution. A set of 11 replicate measurements for 0.1 μM glycine listed in inset, yields a relative standard deviation (RSD) of 2.05%, the detection limit (S/N=3) is 0.05 μM. A set of 11 replicate measurements for 0.5 mM glycine yielded a relative standard deviation (RSD) of 3.5%.

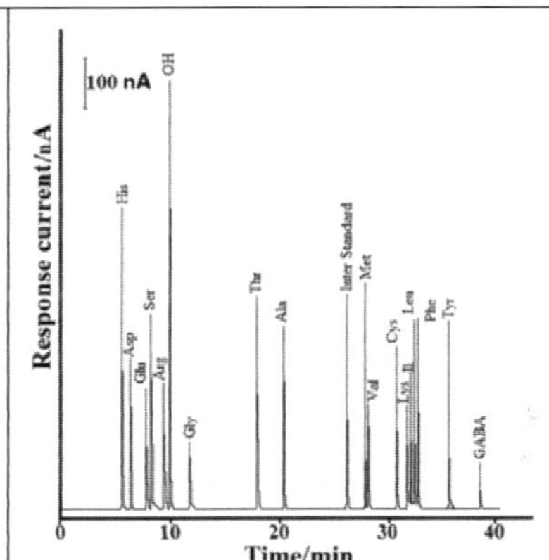

Figure 9. Chromatogram of a mixture of 17 amino acids (5 μM) containing an internal standard of 6-aminocaproic acid. A silica-based HPLC separation column was used for separation of the mixture of AAs before the samples were entranced into FIA with the electrode proposed as the detector.

The optimized experimental condition including 25% ratio of W_{Fe}/W_{Fe+CNT}, pH 8.0, 0.6 V of the electrode potential are adopted in FIA measurements. Considering 2.5 s of response time of the electrode as the detector and the effect of flow rates for the carrier on peak height and peak area, 250 μl/min of flow rates was applied. The obtanied results are shown in Figure 8. Besides glycine, there are many kinds of amino acids in the dialysis solution, all of which can produce the sensor response. Combined with FIA, calibration curve of glycine was obtained. Its linear range is from 0.1 to 12 μm with a regression equation $I = 29.999\ C_{gly} + 5.9999$(R=0.9994), sensitivity 30.0 nA and detection limit

IOP Conf. Series: Materials Science and Engineering **167** (2017) 012025 doi:10.1088/1757-899X/167/1/012025

(DL) 0.05 μm. Recovery from 96 to 110% of glycine in the assay was listed in Table 2. Using same procedures, linear range, R, sensitivity, and DL of determination of other 16 amino acids were obtained and list in Table 3. As a comparison, the measurement results from the typical methods [32] were also listed in Table 3.

Table 2. Recovery of glycine in the FIA assay after silica column separation

Spiked (μM)	Detected (μM) (n=5)	Recovery (%)
0.10	0.11	110
0.50	0.48	96
1.00	1.05	105
5.00	4.91	98
10.00	10.1	101

Table 3. Comparison of Linear range, regression (R), sensitivity, and detection limit (DL) of determination of 17 amino acids using standard solution combined with FIA with Reference [11]

Amino acids	Linear range μM		R		Sensitivity nA/μM		*DL* (*S*/*N* = 3) μM	
	This study	Ref. [11]	This study	Ref. [11]	This study	Ref. [11]	This study	Ref. [11]
Ala	0.60-2.42	5-500	0.9985	0.9990	42.5	7.29	0.30	24
Arg	0.05-6.60	5-300	0.9891	0.9960	29.4	5.97	0.03	587
	-	5-500	-	0.9894	-	5.05	-	695
Asn	-	5-300	-	0.9998	-	11.68	-	94
	-	5-500	-	0.9901	-	9.78	-	112
Asp	1.40-9.20	5-500	0.9949	0.9951	34.9	5.98	0.50	2693
Cys	0.10-9.00	10-300	0.9952	0.9955	37.4	6.41	0.05	875
Glu	0.10-14.0	5-500	0.9903	0.9901	28.6	4.90	0.10	624
Gln	-	5-500	-	0.9962	-	7.40	-	221
Gly	0.10-12.0	5-500	0.9994	0.9996	30.0	2.83	0.05	161
His	1.90-19.8	5-500	0.9958	0.9961	70.3	12.07	0.30	44
Iso-Leu	0.20-8.00	5-500	0.9974	0.9976	25.7	4.40	0.07	297
Leu	0.20-12.6	5-300	0.9976	0.9978	43.4	7.44	0.08	231
Lys	2.00-6.40	5-500	0.9988	0.9994	25.0	4.30	0.70	1890
Met	0.20-6.00	5-500	0.9919	0.9929	52.4	8.99	0.10	102
Phe	0.15-8.60	5-500	0.9987	0.9991	44.4	7.62	0.10	472
Pro	-	5-300	-	0.9988	-	3.70	-	2545
Ser	0.20-8.00	5-500	0.9911	0.9914	44.2	7.58	0.10	1393
Thr	0.06-15.0	5-500	0.9952	0.9954	49.3	8.46	0.03	429
Trp	-	5-500	-	0.9993	-	6.66	-	389
Tyr	1.90-11.0	5-500	0.9998	0.9999	41.8	7.17	0.80	548
Val	0.30-9.20	5-500	0.9990	0.9992	24.5	4.21	0.15	574
GABA	0.10-7.50	-	0.9982	-	30.6	-	0.05	-

3.7. Application

Figure 9 shows the chromatogram of a mixture of 17 amino acids (5 μM) containing an internal standard, 6-aminocaproic acid, using a silica-based separation column with the electrode detector. Furthermore, level of various amino acids detected in hypothalamus dialysis solution of Guinea pigs were listed in Table 4. The results obtained from both two methods are consistent. It confirms that the sensor reaches the needs of amino acids determination in the animals.

Table 4. Comparison of levels of 17 amino acids detected in microdialysis fluid of Guinea Pigs Hypothalamus with two methods

Amino acids	This study (μM)	HPLC (μM)	Amino acids	This study (μM)	HPLC (μM)
Ala	1.83±0.15	1.95±0.17	Lys	1.14±0.09	1.25±0.13
Arg	0.81±0.06	1.01±0.07	Met	0.13±0.01	0.24±0.02
Asp	1.41±0.11	1.53±0.12	Phe	0.66±0.05	0.78±0.08
Cys	1.05±0.08	1.12±0.09	Ser	0.52±0.04	0.67±0.05
Glu	0.46±0.05	0.61±0.06	Thr	1.69±0.14	1.82±0.13
Gly	2.66±0.21	2.48±0.28	Tyr	0.63±0.06	0.59±0.05
His	1.33±0.12	1.55±0.14	Val	1.56±0.12	1.44±0.10
Iso-Leu	0.63±0.05	0.78±0.06	GABA	0.26±0.02	0.23±0.02
Leu	1.21±0.10	1.37±0.17	/	/	/

4. Conclusion

The proposed electrode combined with FIA and HPLC exhibits good performances such as linear response, sensitivity, The determination of 17 amino acids in the hypothalamus microdialysis fluids of guinea pigs, illustrates that it is a powerful tool to investigate physiology and pathology mechanisms. Compared with HPLC for determination of amino acids, the biggest advantage this study is that derivatization steps can be saved, save time, save money.

References

[1] Zhen C H, Sun S G, Fan C J, Chen S P, Mao B Wand Fan Y J, 2004, In situ FTIRS and EQCM studies of glycine adsorption and oxidation on Au (111) electrode in alkaline solutions, *Electrochim. Acta* **49,** 1249–1255

[2] Yang J Q, Chen J W and Zhou Y K, 2011, A nano-copper electrochemical sensor for sensitive detection of chemical oxygen demand, *Sens. Actuatos* B-Chemical, **153**(1)78-82.

[3] Zhang J, Xiao X X and He Q Q, 2014, A Nonenzymatic Glucose Sensor Based on a Copper Nanoparticle-Zinc Oxide Nanorod Array, *Anal. Lett.* **47**(7) 1147-1161.

[4] Mho S and Johnson D C, 2001, Electrocatalytic response of amino acids at Cu–Mn alloy electrodes, *J. Electroanal. Chem.* **495,** 152–159.

[5] Sivakumar V, Suresh R and Giribabu K, 2014, Copper vanadate nanoparticles: synthesis, characterization and its electrochemical sensing property, *J. Mat. Sci.-Mat. Electronics* **25**(3) 1485-1491.

[6] Yeo I H and Johnson D C, 2001, Electrochemical response of small organic molecules at nickel–copper alloy electrodes, *J. Electroanal. Chem.* **495,** 110–119.

[7] Arribas A S, Bermejo E, Chicharro M, Zapardiel A, Luque G L, Ferreyra N F and Rivas G A, 2006, Analytical applications of a carbon nanotubes composite modified with copper microparticles as detector in flow systems, *Anal. Chim. Acta* **577,** 183–189.

[8] Majzlikova P, Prasek J and Chomoucka J, 2013, Cu_2O based electrochemical sensor for direct glucose detection 2013 *IEEE Sensors*, Baltimore, MD, USA, *IE Sensors*, pp.4.

[9] Vidotti M, Torresi S I C and Kubota L T, 2008, Electrochemical oxidation of glycine by doped nickel hydroxide modified electrode, *Sens. Actuat.* B **135,** 245 – 249.

[10] Majdi S, Jabbari A, Heli H and Moosavi-Movahedi A A, 2007, Electrocatalytic oxidation of some amino acids on a nickel–curcumin complex modified glassy carbon electrode, *Electrochim. Acta* **52,** 4622–4629.

[11] Zen J M, Hsu C T, Kumar A S, Lyuu H J and Lin K Y, 2004, Amino acid analysis using disposable copper nanoparticle plated electrodes, *The analyst* **129,** 841-845.

IOP Conf. Series: Materials Science and Engineering **167** (2017) 012026 doi:10.1088/1757-899X/167/1/012026

Synthesis of functionalized CNTs/hyperbranched polyester nanocomposites

Yufeng Pan, Xiaokun Cui, Yue Zhang[1]

Institute of Materials Science and Engineering, Ocean University of China, Qingdao 266101, China

E-mail: zhangyue@ouc.edu.cn

Abstract. Carbon nanotubes (CNTs) were unzipped using the modified Hummer method to prepare the CNTs-GO microstructure (see Fig. 1). A new type of CNTs-GO-H20 nanocomposite has been synthesized by grafting hyperbranched (HB) polyester (Boltorn H20) brushes on the CNTs-GO by coupling agent (KH560). The morphology of CNTs-GO-H20 was characterized by FTIR, TEM, XPS and TGA. The FT-IR data and XPS data evidenced that CNTs-GO-H20 nanocomposites were synthesized successfully. The addition of CNTs improved the thermal stability of the nanocomposites. The TEM data showed that the CNTs-GO microstructure was also prepared. These electrochemical measurements results indicated that coatings provided greater protection against corrosion behavior. Moreover, the nanocomposite material improved corrosion resistance of the coating.

1. Introduction

The research of mature corrosion protection technology generally adopted the way of adding fillers into the liquid coating. The introduction of nanoparticles has always been considered to increase the crosslinking density and make the coating more dense. Among those nanoparticles, CNT fillers can improved the mechanical properties of the coatings owing to its high strength and modulus of elasticity [1]; GO was considered to have the anticorrosive properties owing to its impermeability to oxygen and water, its very high surface area, nanometric thickness and plate-like structure[2]. However, both CNTs and GO have the agglomeration problem which would influence the compatibility between the coating matrix and fillers. Therefore, the aggregation problem of GO and CNTs has to be resolved in order to move the field forward.

Figure 1. CNTs-GO structure;

In recent years, it was found that the dispersion of CNTs/GO hybrid in solution media can be improved by ultrasonication [3,4]. Moreover, the chemical functionalization was proved to provide better long-term stability and better dispersibility [5].

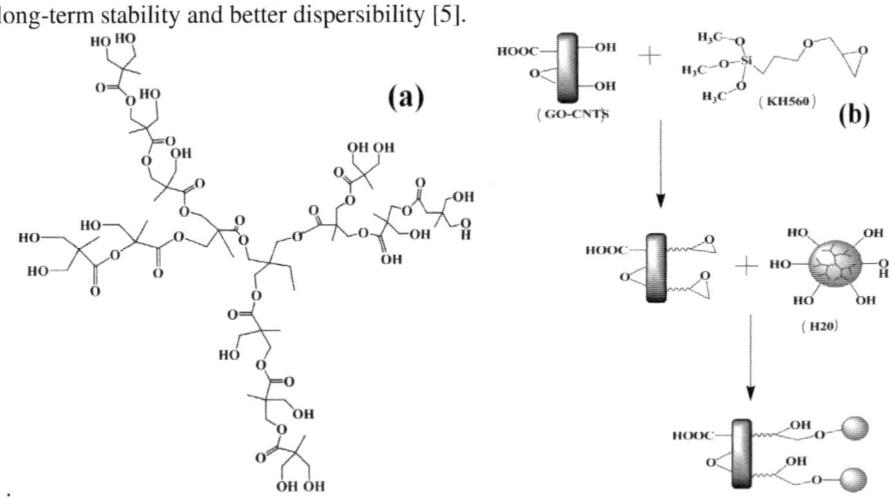

Figure 2. (a) Molecular formula of H20; (b) scheme of the synthesis route for CNTs-GO-H20;

HB polyester H20 (see Fig. 2a) is characterized by its spherical molecular shape, large number of functional end groups, good film forming ability, high solubility and low viscosity at melt and in solution [6]. In spite of the fact that H20 is widely used in the fields of composite materials and catalysts, but its application in the metal protection field is rarely studied. Anticorrosive coatings are prepared in solution all the time while commercial H20 is only available in powder form, the primary problem is to prepare a composite coating solution that can be stored stably. In this paper, the new liquid coating solution has been prepared for the application of copper corrosion protection.

2. Experimental section

2.1. Materials and reagent

MWCNTs used in this study (97% purity, average ED 20-40 nm) were purchased from Shenzhen Nanotech Port Co., Ltd. Boltorn H20 (Hydroxyl number: 565.1mg KOH/g, molecular weight: 2100) was supplied by Shanghai Seebio Biotech,Inc. KH560 was purchased from Aladdin Industrial Corporation. Chemical reagents were obtained from sinopharm chemical reagent Co.,Ltd. The Cu foil.

2.2. Preparation of CNTS-GO/H20 nanocomposite

The preparation of CNTs-GO employed a modified Hummer method [7]. The CNTs-GO was functionalized by the coupling agent (KH560). Equivalent mass of functionalized CNTs-GO and H20 were dissolved in 50ml water respectively. After ultrasonication for 30min, 5M NaOH was then added to the suspension of CNTs-GO as a catalyst. Then the mixture was stirred for 6h at 80°C in a round-bottom flask equipped with a condenser. After the reaction was complete, nonreactive H20 was removed by successive filtration using ethanol and acetone. Residual NaOH was washed out by distilled water until the pH of the filtrate became neutral. The solid was then dried under vacuum to obtain CNTs-GO-H20 nanocomposite (see Fig. 2b).

2.3. Characterization of samples

Fourier transfer infrared spectroscopy (FT-IR) was recorded between 4000-400cm^{-1} on a Nicolet in10 FT-IR spectrometer, U.S.A and using KBr pellets for sample. The size and morphologies of the resultant samples were characterized by JEM-1200EX transmission electron microscopy (TEM).

Thermogravimetric analysis (TGA) was performed under a nitrogen flow using SDT Q600 from room temperature to 800°C at a heating rate of 10°C /min. X-ray photoelectron spectroscopy measurements were carried out with a monochromatic Al Kα X-ray source by the Thermo ESCALAB. The properties of the viscosity of composite coating were investigated by ubbelohde viscometer. Electrochemical measurements were recorded in 3.5 wt% NaCl solution by a CHI660C electrochemical workstation (Shanghai CH Instruments) at room temperature. The Cu foil was washed several times with acetone, NaOH solution and deionized water, respectively.

3. Results and discussion

3.1. FT-IR analysis of samples

As shown in Fig. 3, the pristine CNTs have only three obvious peaks including C=C stretching vibration (1570cm^{-1}), C-O stretching vibration (1100cm^{-1}) and -OH stretching vibration (3378cm^{-1}). After pretreated by the modified Hummer method, three new peaks emerged in the CNTs-GO. The signals at 1710cm^{-1} and 1388cm^{-1} were absorbed by C=O and C-OH. The band observed at 1100cm^{-1} has become wider due to the coincidence of -CH(O)CH- vibration peaks and C-O vibration peaks. The values confirmed that the functional groups of -COOH, -OH, -CH(O)CH- were introduced to CNTs successfully. Compared with CNTs-GO, some new absorption bands appeared at functionalized CNTS-GO. The absorption at 2920cm^{-1} and 2852cm^{-1} were due to -CH_2- stretching, the band at 1210cm^{-1} and 909cm^{-1} assigned to the epoxy group.

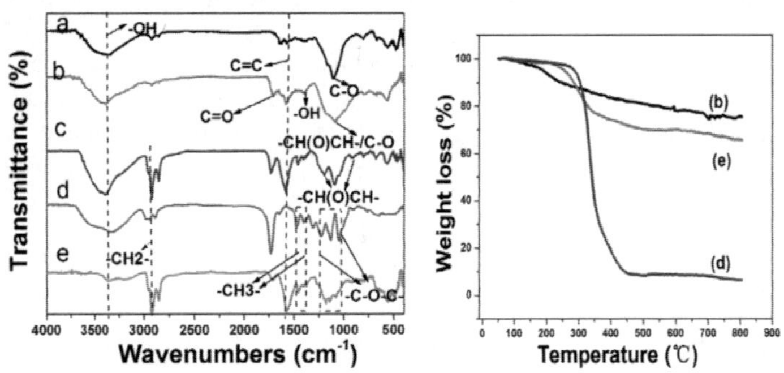

Figure 3. FT-IR (left) spectra and TGA (right) curve of samples
(a) CNTS; (b) CNTs-GO; (c) functionalized CNTs-GO; (d) H20; (e) CNTs-GO-H20;

The results confirmed that the surface of CNTs-GO was successfully grafted by KH560 [8]. In the pure H20 (d), there were two strong peaks, -CH_3 stretching vibration (1450cm^{-1}, 1380cm^{-1}) and -C-O-C- stretching vibration in -COOR (1300~1050cm^{-1}). Compared with pure H20 and functionalized CNTs-GO, the signals at wavenumbers 1300~1050cm^{-1}, 1380cm^{-1}, 1450cm^{-1} appeared in the resultant products, which proved that CNTS-GO-H20 was obtained successfully.

3.2. TG analysis of samples

As shown in Fig. 3, for CNTs-GO (b), the first weight loss of about 4% was attributed to the evaporation of the physically absorbed water. In addition, the 8% weight loss between 168-250°C registered for GO was due to the decomposition of labile oxygen functional groups. The slight decline occurred at 300-800°C when decomposition of the carbon skeleton structure occurs with the residual mass of 75.52% obtained at 800°C. In the case of functionalized CNTS-GO-H20 (e), the onset degradation temperature of functional groups was increased from 168°C to 209°C, which showed that a part of the functional groups reacted with the coupling agent. And from 265°C, the weight loss of the functionalized CNTs-GO-H20 was more significant than CNTs-GO. This may be mainly caused by the pyrolysis of the polyester grafted on the surface of CNTS-GO [9]. And the residual mass of functionalized CNTS-GO-H20 was 65.88%. This means that the weight loss of the CNTs-GO-H20 was almost 10% greater than CNTs-GO, which can be related to the amount of H20 grafted on the CNTS-GO surface. As regards the CNTs-GO-H20 nanocomposite a 10% weight loss was observed between 260°C and 544°C while the commercial H20 (d) sample thermal degradation only begins at 280°C and the commercial H20 sample was completely degraded at 460°C. The increase in thermal stability for the CNTs-GO-H20 nanocomposite may be ascribed to the ability of the outer graphene surface of CNTs-GO to capture free radicals generated during thermal decomposition [10].

3.3. XPS analysis of samples

The changes of C and O elements content in the three samples were shown in Table 1. C1s XPS spectra before and after modification were analyzed as shown in Figs. 4a, b and c.

Table1. Atomic percentages in XPS spectra of samples

Element(x%)	C	O	Si
CNTS	97	3	0
CNTs-GO	79.43	20.57	0
functionalized CNTs-GO	79.36	19.24	1.4
CNTs-GO-H20	83.73	15.38	0.89

As shown in Table1, the content of the C and O atoms in the pristine CNTs was 97% and 3%. After pretreated by $KMnO_4$, the content of the C and O atoms was changed to 79.43% and 20.57%. The signals in Fig.4a at 284.6eV, 285.1eV, 287eV and 288.2eV correspond to C-C, C-OH, C-O-C and C=O, respectively [11]. It was believed that the CNTs introduced oxygen functional groups, which indirectly proved that CNTs were opened [12]. After CNTs-GO was functionalized with KH560, the content of Si element was increased to 1.4%, which indicated the successful attachment of KH560 on the surface of CNTs-GO. And the new peak at 285.5eV appeared in Figs. 4b and 4c was contributed by C-O-Si, confirming the successful attachment of KH560.

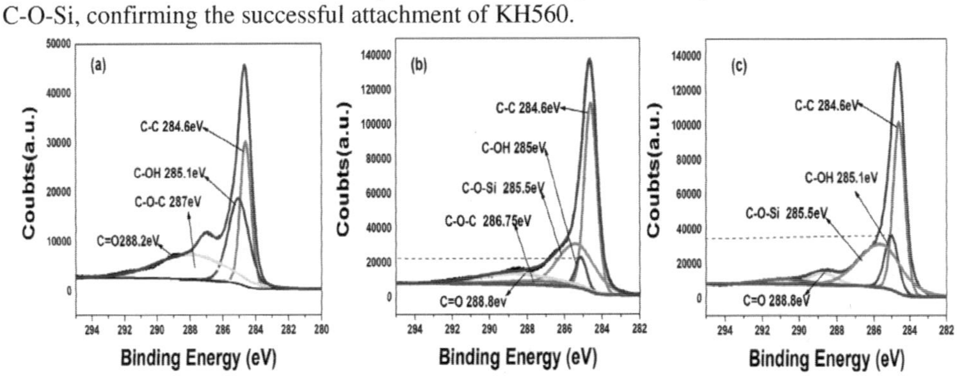

Figure 4. XPS spectra C1s of samples
(a) CNTs-GO; (b) functionalized CNTs-GO; (c) functionalized CNTs-GO-H20

Comparing Figs. 4b and Fig. 4c, the C1s showed two significant intensity changes that the increase at 285eV and the decrease at 286.75eV, which indicated the addition of hydroxyl groups and loss of epoxy groups. Those changes were caused by the successful grafting of hyperbranched polyester.

3.4. TEM analysis of samples
As shown in Fig. 5 TEM images of the CNTs revealed the presence of hollow tubes with 25nm in mean diameter (Fig. 5a).

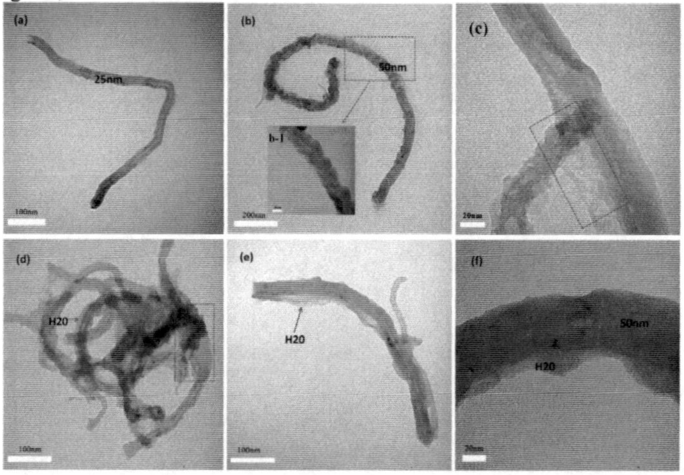

Figure 5. TEM images of (a) pristine CNTS; (b, c) CNTS-GO; (d, e, f) CNTS-GO-H20

After the treatment by Hummer method, the diameter increased to 50nm (Fig. 5b). At higher magnification (20nm), it can be observed that CNTS-GO exhibits serrated border (Fig. 5b-1). As shown in Fig. 5c, we found that the side walls of CNTs unravelled to form the nanoribbons. All those phenomena indicated that the outer structure of CNTs was successfully opened. From the TEM images of CNTs-GO-H20 (Figs. 5d, e, f), it can be found that the wall of CNTs became rough. In Fig. 4f, the mean diameter of the CNTs is 50nm and that of the outer polymer H20 layer is about 10nm. The values indicated that CNTs core were effectively functionalized with a soft organic shell through the grated reaction.

3.5. Dispersion analysis of nanocomposite materials-sediment experiment
Fig. 6 shows results of the sedimentation experiment. The pristine CNTs (Fig. 6A) quickly precipitated in the water solution even after sonication owing to the van der Waals interactions. For the CNTs-GO (Figs. 6B and 6C), the π-π stacking interactions between the outer "GO" structure and inner CNTs structure bind the hybrid, while the hydrophilic surfaces of the CNTs-GO improved the dispersion in solution [13]. So the dispersion can be stable for six months. As shown in Figs. 6D and 6E, CNTs-GO-H20 can prevent the entanglement of CNTs due to the larger static hindrance and the more terminal hydrophilic groups of H20. The nanocomposites /H20 suspension in Fig. 6G was uniform and stable more for than five days. Physical effects may also come into play: the sphere-like superstructure of H20 can allow the CNTs to permeate into the polymer to improve the dispersion [14].

Figure 6. Sedimentation experiments of various solution samples
(A) pristine CNTs/water solution, (B) CNTs-GO/water solution, (C) CNTs-GO/DMF solution, (D) CNTs-GO-H20/water solution, (E) CNTS-GO-H20/DMF solution, (G)CNTS-GO-H20/H20 solution

3.6. The electrochemical measurements of the composite coating

Fig. 7 shows results of the electrochemical measurements of samples. The addition of nanocomposite materials in the composite coating was one percent of the HB polyester H20. As shown in Fig. 7A, it is found corrosion potentials (Ecorr) positively increases from -0.358 V (a) to -0.270 V (b) and -0.209 V (c). In addition, corrosion current densities (icorr) of H20 coating and composite coating are 1.0718E-5 and 3.1707E-7 A/cm^2. And the icorr of bare Cu is 3.9703E-5 A/cm^2. More positive Ecorr and lower icorr of coating indicate that the film to be served as an effective barrier to diffusion of corrosive species. In Fig. 7B, we can observe that the bare Cu includes a capacitive loop at high frequencies and a straight line (Warburg impedance) at low frequencies, indicating that the diffusion process is controlled by the cathodic diffusion of dissolved oxygen from bulk solution to the electrode surface and the anodic diffusion of soluble Cu species from electrode surface to bulk solution. The diameter of the Nyquist loop in the coating is significantly larger than that of bare copper, indicating the polarization resistance has been greatly enhanced with the presence of coating.

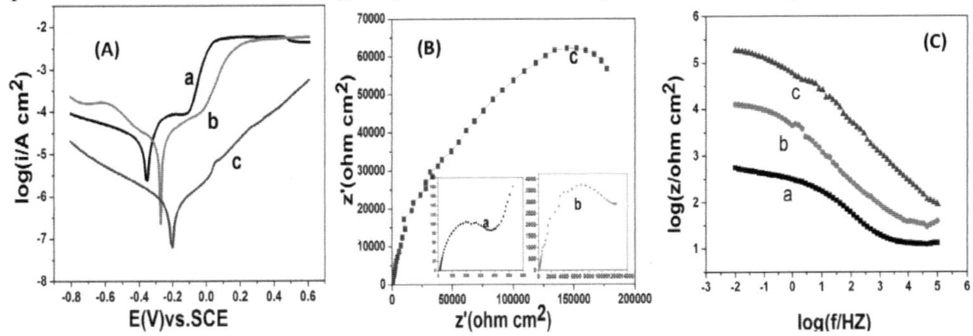

Figure 7. Potentiodynamic polarization curves

(A), Nyquist plots (B) and Bode plots (C) of bare Cu (a), H20 coating (b), composite coating (c)

And in Fig. 7C, it can be clearly seen that the impedance modulus at low frequency limit (10 mHz) of coating sample is greater than untreated sample. These phenomena indicate that the coating provides high resistance to corrosive medium [15].

4. Conclusions

The CNTs-GO-H20 nanocomposites were synthesized successfully as evidenced by FTIR and XPS spectra. The reduced weight loss of the nanocomposites is attributed to the amount of H20 grafted on the CNTs-GO surface, which indicates better thermal stability. The nanocomposite/H20 solution can be stable for a few days, which shows promise in liquid composite coating. At the same time, the coating samples exhibited better corrosion inhibitive properties in comparison with bare Cu substrate, as demonstrated by measurement results from potentiodynamic polarization curves and EIS techniques in 3.5 wt% aqueous NaCl solution. Compared with the traditional liquid coating, it has the following advantages: convenient for storage, better stability and less pollution for environment.

Acknowledgements

This work was financially supported by National Nature Science Foundation of China (51572248).

References

[1] Naik R B, Jagtap S B and Ratna D 2015 Effect of carbon nanofillers on anticorrosive and physico-mechanical properties of hyperbranched urethane alkyd coatings *Progress in Organic Coatings.* **87** 28-35

[2] Ramezanzadeh B, Ghasemi E, Mahdavian M, Changizi E and Moghadam M H M 2015

Covalently-grafted graphene oxide nanosheets to improve barrier and corrosion protection properties of polyurethane coatings *Carbon.* **93** 73-555

[3] Yen M Y, Hsiao M C, Liao S H, Liu P I, Tsai H M, Ma C C M, Pu N W and Ger M D 2011 Preparation of graphene/multi-walled carbon nanotube hybrid and its use as photoanodes of dye-sensitized solar cells *Carbon.* **49(11)** 606-3597

[4] Zhang C, Ren L, Wang X and Liu T 2010 Graphene oxide-assisted dispersion of pristine multiwalled carbon nanotubes in aqueous media *The Journal of Physical Chemistry C.* **114(26)** 11435-40

[5] Rana S, Karak N, Cho JW and Kim YH 2008 Enhanced dispersion of carbon nanotubes in hyperbranched polyurethane and properties of nanocomposites *Nanotechnology.* **19(49)** 495707

[6] Zhao Y, Wang F, Fu Q and Shi W 2007 Synthesis and characterization of ZnS/hyperbranched polyester nanocomposite and its optical properties *Polymer.* **48** 2853-9

[7] Higginbotham A L, Kosynkin D V, Sinitskii A, Sun Z and Tour J M 2010 Lower-defect graphene oxide nanoribbons from multiwalled carbon nanotubes *ACS nano.* **4(4)** 2059-69

[8] Qi B, Yuan Z, Lu S, Liu K, Li S, Yang L and Yu J 2014 Mechanical and thermal properties of epoxy composites containing graphene oxide and liquid crystalline epoxy *Fibers and Polymers.* **15** 326-33

[9] Bao C, Guo Y, Song L, Kan Y, Qian X and Hu Y 2011 In situ preparation of functionalized graphene oxide/epoxy nanocomposites with effective reinforcements *Journal of Materials Chemistry.* **21(35)** 13290-8

[10] Gonçalves G, Marques P A A P, Barros-Timmons A, Bdkin I, Singh M K, Emami N and Grácio J 2010 Graphene oxide modified with PMMA via ATRP as a reinforcement filler *Journal of Materials Chemistry.* **20(44)** 9927-34

[11] Ma W S, Li J and Zhao X S 2013 Improving the thermal and mechanical properties of silicone polymer by incorporating functionalized graphene oxide *Journal of Materials Science.* **48** 5287-94

[12] Wang Y, Shi Z X and Yin J 2010 Unzipped multiwalled carbon nanotubes for mechanical reinforcement of polymer composites *The Journal of Physical Chemistry C.* **114(46)** 19621-8

[13] Dong X, Xing G, Chan-Park M B, Shi W, Xiao N, Wang J, Yan Q, Sum T, Huang W and Chen P 2011 The formation of a carbon nanotube–graphene oxide core–shell structure and its possible applications *Carbon.* **49(15)** 5071-8

[14] Mahapatra S S, Yadav S K, Yoo H J, Cho J W and Park J S 2013 Highly branched polyurethane: Synthesis, characterization and effects of branching on dispersion of carbon nanotubes *Composites Part B: Engineering.* **45(1)** 165-71

[15] Liu L, Chen R, Liu W, Zhang Y, Shi X and Pan Q 2015 Fabrication of superhydrophobic copper sulfide film for corrosion protection of copper *Surface and Coatings Technology.* **272** 221-8

Fabrication of Micro/Nano optical Fiber by Electrospinning Direct-writing

Liu Yifang

School of Aerospace Engineering, Xiamen University, Xiamen 361005, China

E-mail: Yfliu@xmu.edu.cn

Abstract. Because of the excellent performance, Micro/Nano optical fiber has been more and more widely applied in passive photonic devices, micro-optical sensors, field of atomic manipulation, etc. Currently the main manufacturing mode of Micro/Nano optical fiber is the stretching method which is susceptible to air impact, vulnerable to contaminant and has poor reproducibility. In order to solve these problems, the fabrication of the Micro/Nano optical fiber by electrospinning direct-writing is researched in this article. The experimental platform is set up after the scheme of electrospinning direct-writing is designed. A series of comparative experiments are carried out with changing three experimental variables. The PMMA Micro/Nano optical fiber of controllable diameter is fabricated by regulating the distance between the sprinkler head and collecting plate, flow rate and concentration of PMMA solution. The testing results indicate that the light transmission power loss rate of the PMMA Micro/Nano optical fiberis 0.41dB/mmexcited by a 633-nm-wavelength light. The problem is expected to be solved by further optimization of the experimental process and parameters.

1. Introduction

Because of the low optical waveguiding losses, tight optical confinement, strong evanescent fields, small mass or weight, Micro/Nano optical fibers[1] will be the one of best optical waveguide. They can be used as air-clad wire-waveguides with subwavelength-diameter cores, and in building blocks in the future Micro/Nano photonic devices. They renewed research interests in optical MNFs as potential building blocks for applications in high density and miniaturized PICs[2-5]. So far, Micro/Nano optical fibers are mostly fabricated from amorphous materials including glasses and polymers. For these materials, the top-down physical drawing techniques[6-8] include flame-heated taper drawing of glass fibers, laser-heated taper drawing of glass fibers, flame-heated drawing of bulk glass and electrically heated taper drawing of glass fibers are researched. Though the fibers fabricated by these drawing techniques have uniform diameter, smooth surface, large length and low cost, they are vulnerable to air pollution and poor to repeat.

Electrospinning is a popular technique for fabricating polymer Micro/Nano optical fibers from a wide range of polymer materials[9-10]. Electrospinning direct-writing is a near-field electrospinning system, in which the distance between the spray-head and collector is a few millimeters. The near field will make the jet stability to obtain orderly Micro/Nano fiber. The precisely and orderly positioning of the Micro/Nano fiber will be implemented by controlling the trajectory and velocity of the collector. And the collection of single Micro/Nano fiber will be realized by the stability of the jet in the direct-writing system. Both these two characters provide the feasibility for the preparation of Micro/Nano optical fiber.

2. The design of the electrospinning direct-writing scheme

The electrospinning direct-writing platform is shown in Fig.1, including high-voltage DC power supply, solution storage device, solution injection mechanism and fiber collection system. The solution is poured into the transparent plastic tube between injection syringe and needle. The generation of air bubbles must be avoided in the solution injection process. The collector is silicon or tin foil which is placed on the X-Y motion stage. The anode of the dc high voltage power supply is connected to the syringe, and the ground wire is connected to the collector. By setting the voltage and the velocity of the fluid for predetermined value and controlling the X-Y motion stage to move along the default path, the Micro/Nano optical fiber is collected to be inspected and measured by electron microscope and SEM.

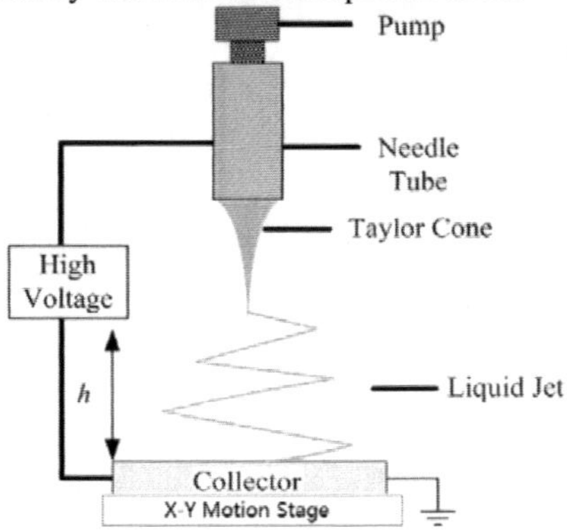

Figure 1. The electrospinning direct-writing platform

3. Experimental details

In the experimental process, multiple sets of experimental results are compared by changing one of the three experimental parameters (the distance between the nozzle and collector, PMMA solution concentration and flow rate). And the influence of the morphology and diameter of the Micro/Nano optical fiber by the experimental variables is analyzed qualitatively and quantitatively. By analyzing the mechanism of the Micro/Nano optical fiber surface defects such as porosity, distortion, and flatness, the universal laws of fabricating Micro/Nano optical fiber by electrospinning direct-writing from PMMA solution is summarized.

Firstly, PMMA is dissolved in mixed solvent (CHCl3: DMF =2:1) and solution concentration is 25%. Flow rate and the diameter of the spinning nozzle are 200 μ l/h and 0.21mm, respectively. The different samples have been prepared, changing the spinneret-collector distance: (a) 2 mm, (b) 3 mm, (c) 4 mm, (d) 5 mm, and (e) 6 mm. The applied voltage is the one which just can form stable jet. The morphology and varying tendency of the diameter of Micro/Nano optical fibers are attained in Fig.2. and Fig.3, respectively.

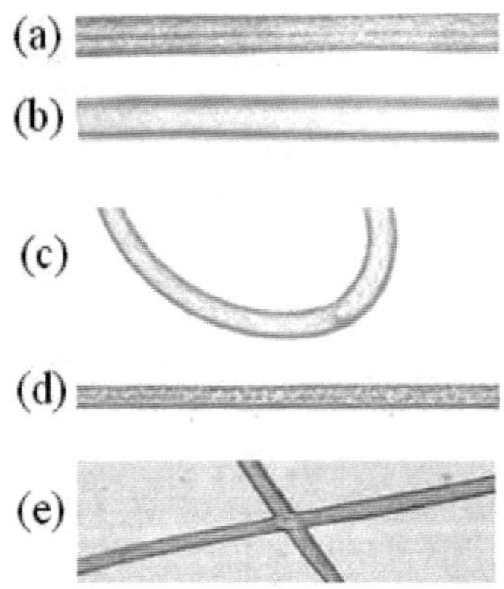

Figure 2. The morphology of Micro/Nano optical fibers at the different spinneret-collector distance: (a) 2 mm; (b) 3 mm; (c) 4mm; (d) 5 mm; (e) 6 mm;

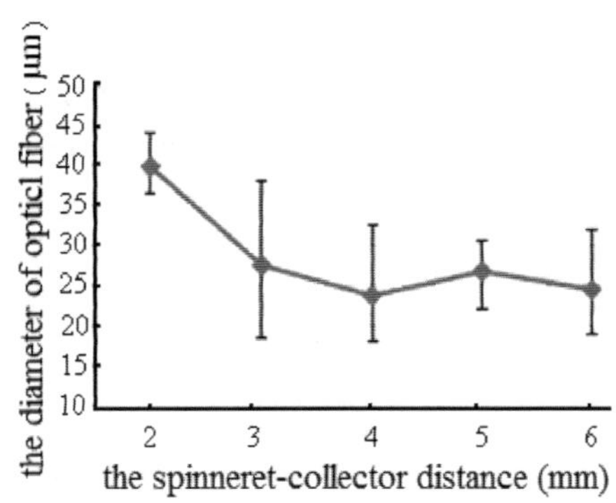

Figure 3. The varying tendency of the diameter of optical fibers with the spinneret-collector distance

From Fig.2 and Fig.3 we can attain that the diameter of the PMMA optical fiber decreases with the increasing of the spinneret-collector distance. The reason is that the flight time in space of spinning jet gets longer because of the lager spinneret-collector distance to make the volatilization of the solvent in the jet more fully.

Secondly, the flow rate, the diameter of the spinning nozzle and the spinneret-collector distance are 200 μ l/h, 0.21mm and 2 mm, respectively. The mixed solvent is CHCl3： DMF =2:1. The different samples have been prepared, changing the PMMA concentration (a) 8%, (b) 12%, and (c) 16%. The applied voltage is the one which just can form stable jet. The morphology of Micro/Nano optical fibers is attained in Fig.4. It can be deduced that though the solution concentration gradient is large, all the optical fibers are continuous. The optical fiber electrospinned from 8% PMMA concentration is finer than the ones from 12% and 16% PMMA concentration, because the more volatile solvent makes the fiber volatile fully in the electrospinning process.

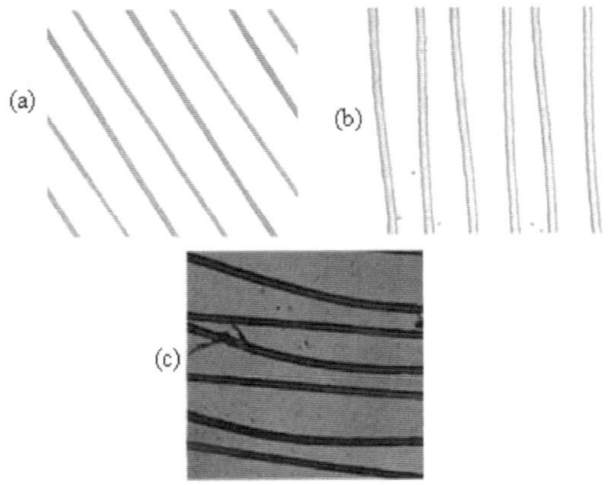

Figure 4. The morphology of fibers at the different solution concentration: (a)8%; (b)12%; (c)16%.

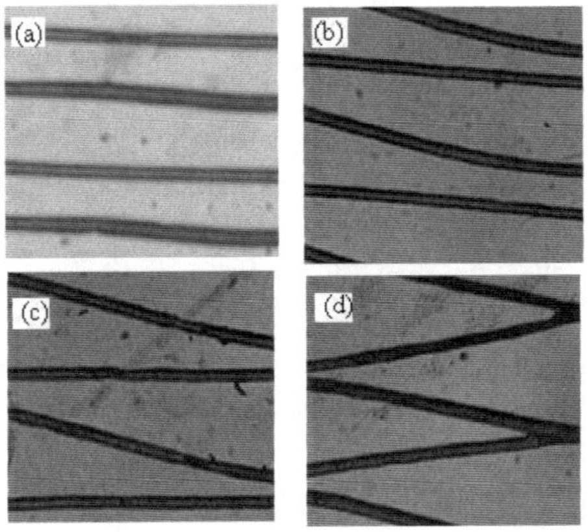

Figure 5. The morphology of fibers at the different flow rate: (a) 100μl/h; (b) 200μl/h; (c) 300μl/h; (d) 400μl/h

Thirdly, the PMMA concentration, the diameter of the spinning nozzle and the spinneret-collector distance is 16%, 0.21mm and 2 mm, respectively. The mixed solvent is CHCl3： DMF =2:1. The different samples have been prepared, changing the flow rate (a) 100μl/h, (b) 200μl/h, (c) 300μl/h, and (d) 400μl/h. The morphology of Micro/Nano optical fibers is attained in Fig.5. It can be observed that the flow rate does a little influence on the diameter of the optical fiber, but significantly affect the distribution of the fiber. When the flow rate is 100μl/h, the straightness and parallelism of fiber is good. With the increasing of the flow rate, fibers become distorted to form an angle. The extrusion speed of the solution from the needlepoint is controlled by the flow rate. So the greater the feed flow rate, the faster the solution out of needlepoint and the faster the jet scream to the collector which makes the deviation between the position of fiber deposition and the pre-set position.

4. Light loss tests and discussion

The light loss is one of the main parameters of characterization of optical fiber transmission performance. We test the optical power loss of the PMMA Micro/Nano optical fiber for two times, which is shown in the Fig.6. It can be calculated that the optical power loss of the PMMA fiber is 0.41dB/mm excited by a 633-nm-wavelength light, which is much greater than the common optical fiber in the market. The reason is that PMMA fiber has poor morphology, rough surface, sparse holes in the surface and inside the fiber, flat rather than round cross section, etc.

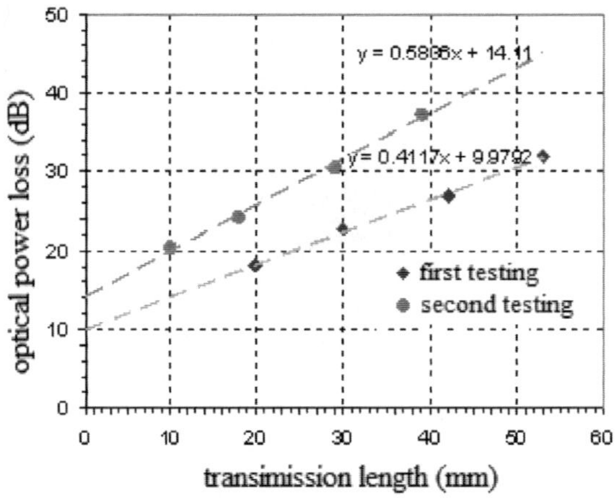

Figure 6. The optical power loss of the PMMA Micao/Nano fiber

5. Conclusions

The manufacturing of Minco/Nano optical fibers by electrospinning direct-writing using PMMA solution is feasible. The optimal design of the experimental scheme guarantees the collection of single and ordered fibers, which satisfies the basic requirements of Micro/Nano fiber. By controlling the distance between the sprinkler head and collecting plate, flow rate and concentration of PMMA solution, the spinning voltage, the PMMA Micro/Nano fiber with controllable diameter can be obtained. The controllable range of diameter is from ten to several hundred micrometers. At the same time, we know that the PMMA fiber made by electrospinning has the function of basic optical transmission.

Acknowledgements

This work was financially supported by National Natural Science Foundation of China (61404111) and Aviation Science funds (Aviation Key Laboratory of Science and Technology on Inertia)(20130868002).

References

[1] Maric D M, Meier P F and Estreicher S K 1992 Mater. *Sci. Forum* **83-87** 119.
[2] Zhu H, Wang Y Q, and Li B J 2009 Tunable refractive index sensor with ultracompactstructure twisted by poly(trimethylene terephthalate) nanowires *ACS Nano* **3** 3110-3114.
[3] Guo M L, Shi J C, and Li B J 2008 Polymer-based micro/nanowire structures for threedimensionalphotonic integration *Opt. Lett.* **33** 2104-2106.
[4] Wang Y Q, Zhu H, and Li B J 2009 Cascaded Mach–Zehnder interferometers assembledby submicron PTT wires *Photon. Technol. Lett.* **21** 1115-1117.
[5] Yu H, Liao D, Johnston M B, and Li B 2011 All-optical full-color displays using polymernanofibers *ACS Nano* **5** 2020-2025.
[6] Tong L M and Pan X Y 2007 Subwavelength-diameter optical fibers: waveguiding properties and photonic applications *Physics* **36(08)** 626-630.
[7] Sumetsky M, Dulashko Y and Hale A 2004 Fabrication and study of bent and coiled free silica nanowires: Self-coupling microloop optical interferometer *Opt. Express* **12(15)** 3521-3531.
[8] Shi L, Chen X, Liu H, Chen Y, Ye Z, Liao W and Xia Y 2006 Fabrication of submicron-diameter silica fibers using electric strip heater *Opt. Express 14(12)* 5055-5060.
[9] Bellan L M and Craighead H G 2009 Nanomanufacturing using electrospinning *Journal of manufacturing science and engineering* **131(3)** 034001.1-034001.4.
[10] Xing X B, Yu H Q, Zhu D B, Zheng J P, Chen H, Chen W and Cai J 2012 Subwavelength and Nanometer Diameter Optical Polymer Fibers as Building Blocks for Miniaturized Photonics Integration *Optical Communication* 289-319.

Flower-like Rh/Cu superhydrophobic coating via electrodeposition and its self-cleaning property

N Zhao, S X Lu[1], W G Xu, X L Cheng

School of Chemistry and Chemical Engineering, Beijing Institute of Technology, Beijing 100081, P.R. China

E-mail: shixianglu@bit.edu.cn

Abstract. Rh/Cu superhydrophobic coatings with textured flower-like dual micro-nano structures were fabricated on copper substrates via electrodesposition of Rh followed by annealing treatment. The scanning electron microscopy and X-ray diffraction pattern were conducted to analyse the morphology and chemical composition of the coating surfaces, respectively. The water drops exhibited a water contact angle of 168±1° and a sliding angle of 0±1° on the coating surfaces which were obtained by etching in concentrated hydrochloric acid and nitric acid, electrodeposited at -1.0 V for 30 min and annealing in an oven at 180°C for 60 min. The prepared superhydrophobic coatings showed excellent water repellence and self-cleaning property due to its extremely high water contact angle.

1. Introduction

A superhydrophobic surface with a water contact angle (CA) above 150° and a sliding angle below 10° has a low affinity to water, which lead to the limitation of various phenomena on such a surface, for example, icing, contamination or oxidation, and corrosion [1]. The most typical superhydrophobic surface is the lotus leaf surface in nature. Water droplets on a lotus leaf can roll freely, thus remove contaminants on the blade. This natural phenomenon has aroused tremendous interests in revealing the essence of the superhydrophobicity and exploring methods of preparing artificial superhydrophobic materials [2-3]. The fundamental reason for the superhydrophobicity is the micro-nano dual rough structure. This kind of rough structure can effectively adsorb a large amount of air on the coating surface to reduce the contact area between the solid and the water, resulting in the superhydrophobicity. At the same time, the low surface energy of material is one of the reasons that lead to the superhydrophobicity.

Inspired by the lotus leaf, an extensive research on designing functional artificial superhydrophobic coatings has been carried out on a vast range of materials through a combination of rough surface roughening and low surface energy materials [4]. In this way, many efforts including electrodeposition [5, 6], chemical vapour deposition [7], layer-by-layer self-assembly [8], chemical replacement [9], sol–gel methods [10] and others have been devoted to obtaining the various artificial devices or coatings with special wettability.

It is well known that electrodeposition is an effective, easily controlled and reproducible technology, which has been widely used for manufacturing uniform coatings. Copper is a common life and industrial material with high thermal and electrical conductivity, mechanical workability and malleability. In this paper, we exploited a facile electrodeposition method to manufacture superhydrophobic coatings on copper substrates, which were fabricated by etching in 6 mol·L^{-1}

IOP Conf. Series: Materials Science and Engineering **167** (2017) 012028 doi:10.1088/1757-899X/167/1/012028

hydrochloric acid and 8 wt. % nitric acid for 8 min, electrodeposition of Rh at -1.0 V for 30 min and annealing in an oven at 180°C for 60 min without any modification of low surface energy substance. The wettability, surface morphology, chemical composition of the superhydrophobic coatings were investigated in detail. The prepared superhydrophobic coatings showed excellent water repellency and self-cleaning property.

2. Experimental

2.1. Reagents and Materials

Rhodium (III) chloride hydrate ($RhCl_3 \cdot 3H_2O$, Rh 38.8-39.2%), nitric acid (HNO_3, 65%), hydrochloric acid (HCl, 36%-38%), acetone (CH_3COCH_3, 99.5%) and ethanol (CH_3CH_2OH, 99.7%) were of analytical grade and used without further purification and were purchased from Sinopharm Chemical Reagent Co. Ltd. Copper sheets (2.0cm×1.0cm×0.1cm) were obtained from Beijing Cuibolin Nonferrous Metal Research Institute.

2.2. Pretreatment of the copper substrates and electrochemical preparation of superhydrophobic coatings

The copper substrates were cleaned under sonication in ethanol, acetone and distilled water respectively for 10 min to remove grease on the surfaces. Then the substrates were etched vertically in a mixture of 6 $mol \cdot L^{-1}$ HCl and 8 wt. % HNO_3 solution for 8 min to remove oxide layer and roughen the surfaces, followed by rinsing with distilled water and drying in air.

The electrodeposition was conducted in a typical three electrodes cell with a copper substrate as working electrode, a saturated calomel electrode (SCE) as reference electrode, and platinum foil as counter electrode. The electrolyte is a mixture of 0.005 $mol \cdot L^{-1}$ $RhCl_3$ and 0.1 $mol \cdot L^{-1}$ KCl solution. Rh films were electrodeposited on copper substrate at a constant voltage of-1.0 V for 1800 s. After that, the surface was taken out from solution, rinsed slightly with distilled water and dried in air. Then the sample was placed in a petri dish covered with filter papers on both sides and annealed at 180°C for 1 h.

2.3. Sample characterization

All characterizations were conducted under an ambient atmosphere at room temperature. The water contact angle (CA) was measured with a water drop volume of 8 μL using a contact-angle meter (FAT200, Dataphysics Inc. USA) at five different locations on the surface, and averaged to the value, and the error was in the range of ±1°. The chemical composition information of the surface was acquired by X-ray powder diffractometer (XRD, D8 ADVANCE, Bruker, Germany) at a continuous scanning mode between a 2Theta range of 35 to 100 and a scanning rate of 5°/min. The surface morphology of each sample was characterized by scanning electron microscope (FEI SEM Nova NanoSEM430).

2.4. Self-cleaning property test

Cover the chalk powder with a thickness of 1 mm on the copper substrate, the coating after electrodeposition without annealing and the coating after annealing, and observe whether the water drop can take away the chalk from these surfaces with a slight lean.

3. Results and discussion

3.1. Wettability and morphology of each sample.

Figure 1 shows the water CAs of surfaces obtained under different treatment conditions, reflecting the obvious change of wettability with different treatments. As seen in figure 1a, the water CA of bare Cu substrate is 63±1°. After etching, the CA reaches 106±1° as seen in figure 1b, which decreases to 76±1° after electrodeposition as shown in figure 1c. While the water CA increases to 168±1° exhibiting superhydrophobicity after annealing at 180°C for 1 h as shown in figure 1d. All of these phenomena suggest that the diverse treatments especially annealing on surface wettability are important. The sample after annealing exhibits complete water repellence which is in line with the Cassie–Baxter model [11]. The Cassie–Baxter model states:

$$\cos\theta' = f_1\cos\theta - f_2 \qquad (1)$$

where f_1 and f_2 represent the area fractions of the solid-liquid interface and air-liquid interface, respectively; θ' and θ are apparent water CA on rough composite surface and intrinsic water CA on smooth surface, respectively. Here, f_1 is 0.015 and f_2 is 0.985 considering that $\theta = 63\pm1°$, $\theta' = 168\pm1°$ and $f_1 + f_2 = 1$, which means that the contact area of the solid and water droplet occupies about 1.5% of the total composite area. Thereby the coating exhibits water repellence.

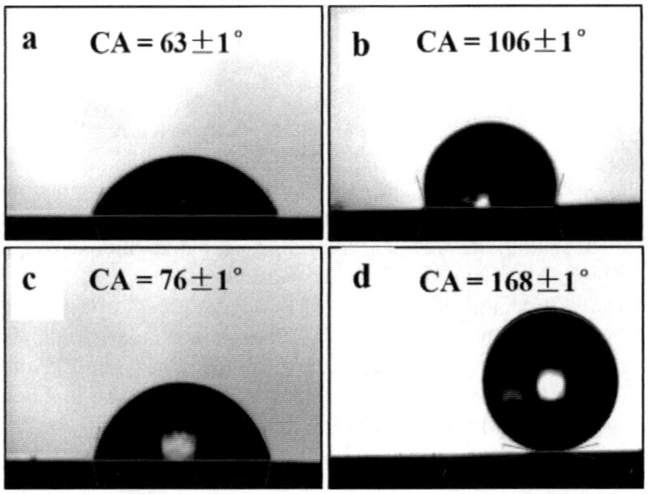

Figure 1. Images of the water CAs on the surfaces: (a) bare copper substrate; (b) after etching with a mixture of 6 mol·L^{-1} HCl and 8 wt. % HNO_3 solution for 8 min; (c) electrodeposited at −1.0 V for 1800 s;(d) after annealing at 180°C for 1 h

Figure 2 shows the SEM images of samples treated with different conditions at different magnifications. From the SEM image of figure 2a, the bare copper substrate is relatively smooth. While etching in a mixture of 6 mol·L^{-1} HCl and 8 wt. % HNO_3 solution, as shown in figure 2b, the surface becomes rough. After electrodeposition, it can be seen in figure 2c_1 that the surface develops into many ravines and densely packed spherical particles which grows upward and have an average diameter of 300 nm in figure 2c_2. After annealing treatment at 180 °C for 1h,it can been clearly seen in figure 2d_1 that every original sphere turns into smaller flower-like nanospheres, which have diameters ranging from 40 to 100 nm as shown in figure 2d_2. This change of morphology before and after annealing makes the surface form more grooves, which can adsorb more air, thereby reducing the contact area between the droplet and the solid surface and exhibiting excellent superhydrophobicitity.

3.2. Chemical compositions of the superhydrophobic coating

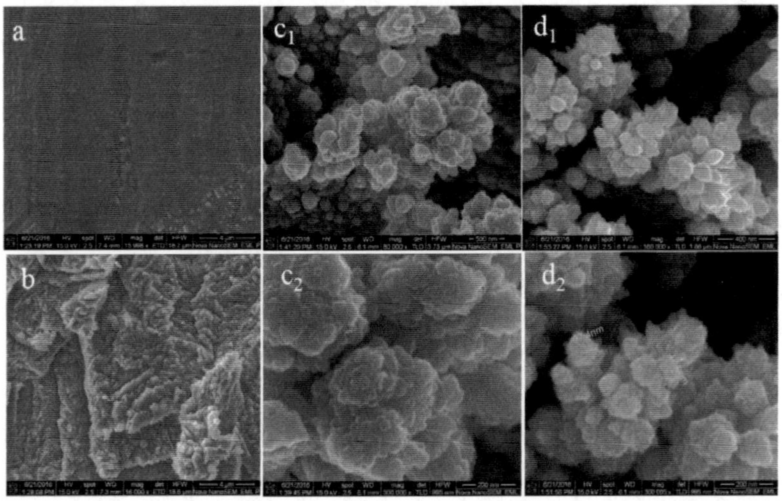

Figure 2.SEM images of each surface: (a) bare copper substrate; (b) after etching with a mixture of 6 mol·L^{-1} HCl and 8 wt. % HNO_3 for 8 min; (c_1 and c_2) electrodeposited at −1.0 V for 1800 s; (d_1 and d_2) after annealing at 180°C for 1 h.

3.2.1 XRD analysis. X-ray powder diffractometer (XRD) was applied to ascertain the elemental composition of the surfaces which could affect the surface wettability. The XRD patterns of bare copper substrate, electrodeposited coating before and after annealing treatment were analysed as shown in figure 3. The symbols of♦ and* present the peaks of Cu and Rh, respectively. The sharp diffraction peaks in figure 3a belong to pure metal copper substrate (JCPDS Card No. 04-0836), namely Cu (111), Cu (200), Cu (220) and Cu (311). Compared with the XRD patterns of the bare copper substrate, new diffraction peaks appear for the surface after electrodeposition at about 41.0°, 69.8°, 84.3° as shown in figure 3b, which are indexed to (111), (220), and (311) planes of cubic Rh (JCPDS Card No. 05-0685), suggesting that Rh was deposited onto the Cu surface during the electrochemical process. From figure 3c, there are no new peaks illustrating no new substance is formed after annealing treatment.

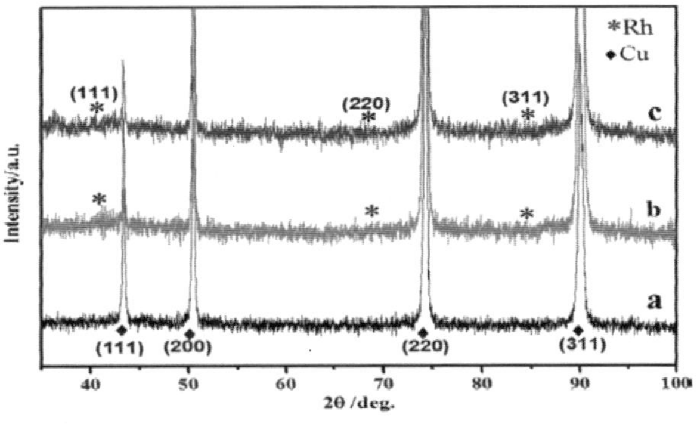

Figure 3. XRD patterns of the samples: (a) bare copper substrate; (b) electrodeposited at −1.0 V for 1800 s;(c) after annealing at 180°C for 1 h.

3.2.2 EDS analysis. Energy-dispersive spectroscopy (EDS) was applied to further confirm the chemical composition. The EDS spectra of bare copper substrate and superhydrophobic coating after annealing are shown in figure 4a and figure 4b respectively. For bare copper substrate, there is only Cu element as shown in figure 4a. As seen in figure 4b, the Rh element appear after electrodeposition and annealing treatment, further indicating that Rh was deposited onto the Cu surface successfully.

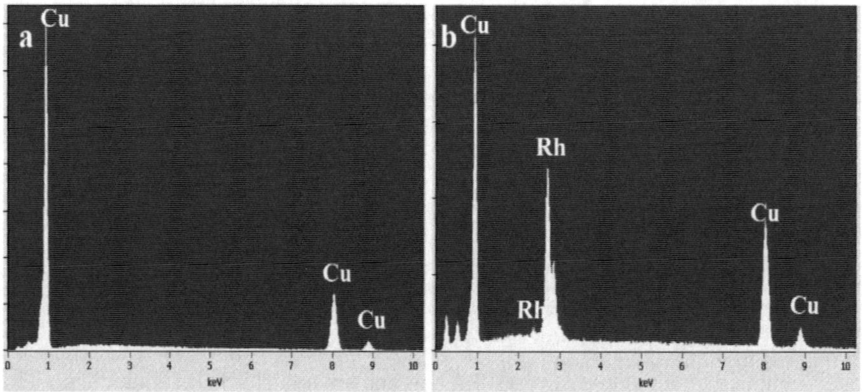

Figure 4. EDS spectra of the samples: (a) bare copper substrate; (b) superhydrophobic coating after annealing at 180°C for 1 h

3.3. Self-cleaning property of the superhydrophobic coating

Self-cleaning property of superhydrophobic coating has been a hot spot in the field of surface technology. The behaviours of water droplets on the bare copper substrate, the coating before and after annealing all covered with chalk powder are shown in figure 5. It can be seen in figure 5a that the water droplets detached from the syringe and chalk powder remained and adhered to the copper surface. Form figure 5b, the water droplets on the coating before annealing cannot take away chalk powder, which was same as the situation of figure 5a. As the water droplets detached from the syringe rolled down on the superhydrophobic coating after annealing, chalk powder were removed adequately and the surface remained dry as shown in figure 5c illustrating the prepared superhydrophobic coating has self-cleaning ability. The self-cleaning ability of superhydrophobic coating might have important potential and practical applications in industrial field.

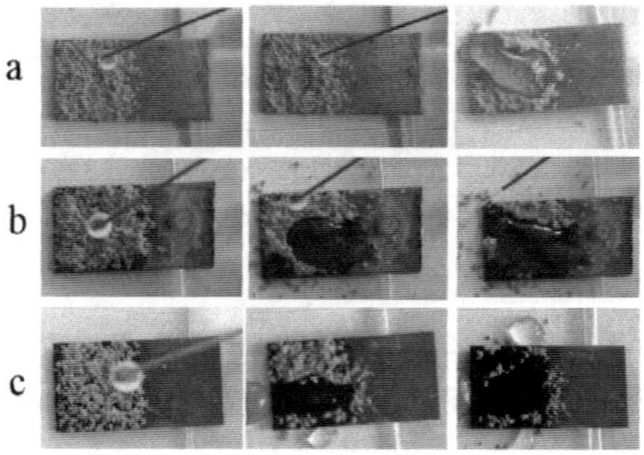

Figure 5. Self-cleaning property test: (a) bare copper substrate; (b) surface before annealing; (c) superhydrophobic coating after annealing.

4. Conclusion

In summary, Rh/Cu superhydrophobic coatings with flower-like structures were manufactured by electrodeposition method without modification of organic low surface energy materials. The micro-nano structures play important roles on the superhydrophobicity. The wettability, surface morphology, and chemical composition of the prepared superhydrophobic coatings were evaluated in detail. The fabricated superhydrophobic coatings showed self-cleaning property, which have potential industrial applications.

Acknowledgements
We gratefully thank the National Natural Science Foundation of China (no. 21271027) for this work.

References

[1] Nakajim A, Fujishima A, Hashimoto K and Watanabe T 1999 *Adv. Mater.* **11** 1365-1368.
[2] Neinhuis C.; Barthlott W 1997 *Ann. Bot.* **79** 667-677
[3] Gao XF and Jiang L 2004 *Nature* **432** 36-36.
[4] Gao L K, GanW T.; Xiao S L, Zhan X X.and Li J 2016 *Ceram. Int.* **42** 2170-2179.
[5] Li W and Kang ZX 2014 *Surf. Coat. Technol.* **253** 205-213.
[6] He G, Lu S X, Xu W G, Szunerits S, Boukherroub R.and Zhang H F 2015 *Phys. Chem. Chem. Phys.* **17** 10871-10880.
[7] Ishizaki T, Hieda J, Saito N, Saito N and Takai O 2010 *Electrochim. Acta.* **55** 7094–7101.
[8] Sun Z, Liao T, Liu K, Jiang L, Kim J and Dou S X 2013 *Nano Research* **6** 726–735
[9] Cheng Y Y, Lu S X, Xu W G, Wen H D and Wang J 2015 *J. Mater. Chem. A.* **3** 16774-16784.
[10] Han J T, Zheng, Y, Cho J H, Xu X and Cho K J 2010 *J. Phys.Chem. B.* **109** 20773–20778
[11] Cassie A B D and BaxterS 1944 *Transactions of the faraday society* **40** 546

An investigation on the fracture of linkage connecting wing flap of aircraft

Yanpeng Feng[1], Haijun Tang, Chun Wang, Ping Shu, Xiaoming Ma, Chunguang Li

Analysis and Appraisal Lab of Civil Aviation Safety, China Academy of Civil Aviation Science &Technology, Beijing, China

E-mail: Fengyp@mail.castc.org.cn

Abstract. Linkage that connected wing flap and supports was found broken several times during from 2011 to 2014. Flights data was check, no heavy landing or exceed limiting speed warning were found. To investigate the root cause of the failure process, macroscopic and micrograph of the fracture surface were research with optical micrograph and scanning electron micrograph. Results show that the surface of fracture was dimples and shearing fracture structure which mean transient breaking features. Nearby the fracture, a lot of scratch and paint marks were found, that indicated that the rod may collapsed by impacted with some mechanical components. But the distance to the nearest component is within the tolerance, during ground inspection. Basing on stability analysis of column, the linkage will be bending deflection with elastic deformation, when compress exceed a critical value, which decrease the distance. But it will be recovered, during ground inspection for lower compress. Finite Element methods were used to demonstrate the bending deformation too. Basing on the reports and analysis, enhanced linkages were provided and substituted for older versions, and the relevant incidents were never found.

1. Introduction

Trailing-edge flap is the activity component that installed in the tailing edge wing surface, usually close to under the wing surface during level flight. During the approach and take-off process, the flap is gradually extension, the bending of cross-section drawn is increase and the windward area of main wing is increase too, so the lift force of aircraft is improved 85%~95%[1,2]. So it is one of the systems to adjust and control the lift coefficient, the disable and failure of the system is dangerous for passengers and aircrafts [3, 4].

The extension mechanism of flaps is shown in figure 1a, a mechanical component with thread screw fixed at main wing, allowing movement in two directions used in the flap extension and retraction system. The installation location of flap connecting rod is shown in figure 1a with red circle, one side of linkage connected to the body with universal connector by simply supported, the other side connected to slide rail by simply supported too. So the condition of loading is compression stress along axial direction.

The connecting linkage was found broken or bend three times in two years for some aircraft, and no heavy landing signs were found for the failure process. The broken linkage is shown in figure 1b comparing with normal linkage in figure 1c. The bending or broken position close to the middle of the connecting rod and no obstacle components was found during the extension and retraction process.

2. Microstructure analysis

Fractured and bending linkages were shown in figure 2, the rod pipe was fabricated with 2024 aluminium alloy, the surface of pipe was coating with white colour paint, no corrosion products alone axial of the connecting rod and no broken at the connected side were found. The angle of bending is about 12 degree for failed rod. A universal connector at the end of each rod in figure 2 is flexible rotation without any jamming condition, and no damage and impacting marks were found on the connector.

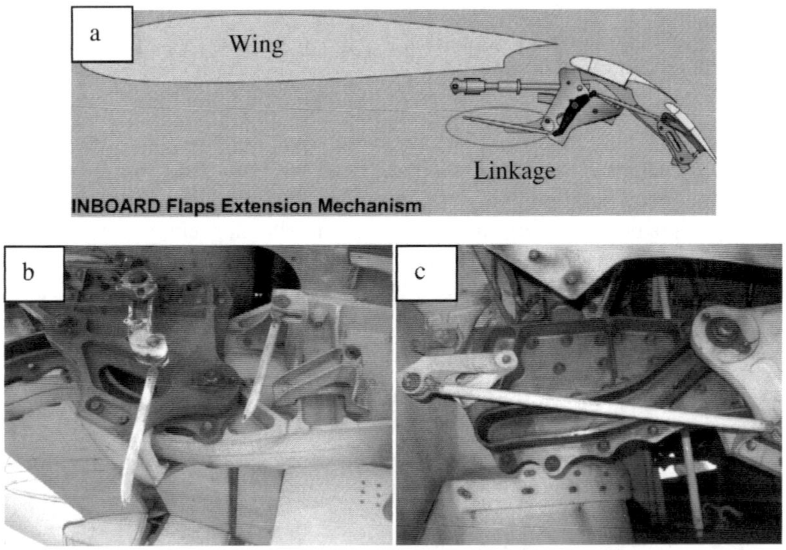

Figure 1. Assemble sketch map of linkage rod and unfolded mechanism of open the flap

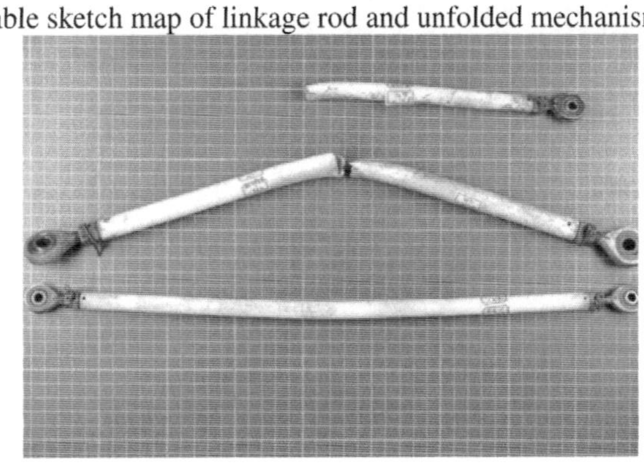

Figure 2. Fractured and bending connecting linkage

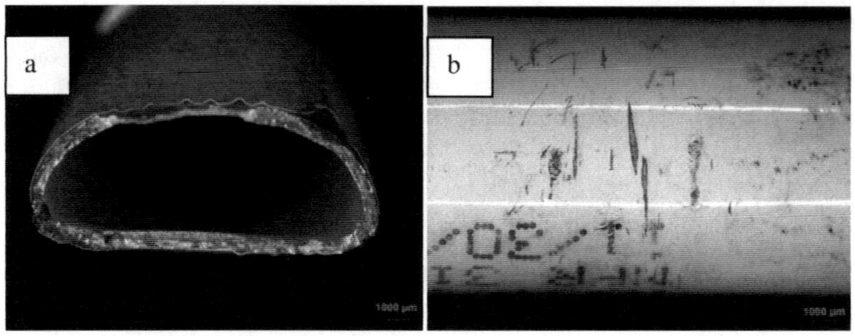

Figure 3. Macroscopic fracture surface and side wall of the broken linkage

Macroscopic fracture surface of broken linkage were shown in figure 3, cross section of normal linkage is circle, but the damaged one is ellipse. Flank wear and impacted marks are show in figure 3 b, surface coating was cut off from the linkage. From the trend of the impacted marks, it may instantly contact with some components. But after inspection in land situation, no object was found alone the travel path of the linkage.

The microstructure of fracture is shown in figure 4, the surface of fracture is type oblique and fluctuation, full of dimples and quasi-cleavage tear ridges without any fatigue lines, which mean that the process of the break is transient behaviour never inoculation process.

Figure 4. Scanning electron micrograph of fracture morphology

3. Stress analysis

Base the above analysis, the broken of linkage is instantly process, and no objects hinder the track, while the failed is not frequently found for the aircraft. So to clarify that what happy to the linkage, a stress analysis for the linkage is necessary. Basing the installation position and stress situation of the linkage, it is typical compression stability of column [5, 6]. The moment of section inertia for thin-walled circular tube is express as:

$$I = \frac{\pi(D^4 - d^4)}{64} = 4.33\times10^{-10}\,(\mathrm{m}^4) \tag{1}$$

Where D is the outside diameter for the cross section of the tube, which is 12.32 mm in this case, and d is 10.92 mm for the inside diameter. Inertia radius of rod is:

$$\mathrm{i} = \sqrt{\frac{I}{A}} = 0.0041\,(\mathrm{m}) \tag{2}$$

Where A is the cross-sectional area for the circular tube, the flexibility of the rod is express as:

$$\lambda = \frac{\mu L}{i} = 89.9 \tag{3}$$

Here μ is the length coefficient of compressed rod, μ is 1 for simply supported beam. And L is the length of the rod, is 0.37 m. Base the critical stress checking the stability.

$$\lambda_p = \pi\sqrt{\frac{E}{\sigma_p}} = 62.8 \tag{4}$$

E is the elasticity modulus of the aluminium alloy, as 72 GPa. And σP is the yield stress, is 180 MPa. Because of λ > λp, and according to euler formula, the critical stress Pcr to bending the rod is express as follow:

$$P_{cr} = \frac{\pi^2 EI}{(ul)^2} = 2246.9 \tag{5}$$

So, if the load of linkage over 2246.9 N, even the normal stress of it never greater than the yield stress, bending deformation will also come up. And with the increasing of deformation, the middle side of the linkage will experience plastic deformation.

4. Finite element analogy

Software of finite element analysis is adopted with ANSYS 15.0, to simulate the stability of the compressed rod. The geometric dimension of the rod was the above mentioned, and both side of the rod are fixed with simple supported. The compressive stress situation was forced on one side of the rod. The material parameter was adopted with 2024 aluminium alloy, elasticity modulus is 72 GPa, and poisson ratio is 0.3.

Boundary conditions setting as follow, according to the practical situation of flap linkage. 1) Displacement constraint of X, Y, Z directions were adopted on side 'C' (in figure 5), and rotation of three directions were free. 2) Displacement constraint of X and Z directions were adopted on side 'A' (in figure 5). 3) Preload of 100 N was adopted on Y direction on side 'A'.

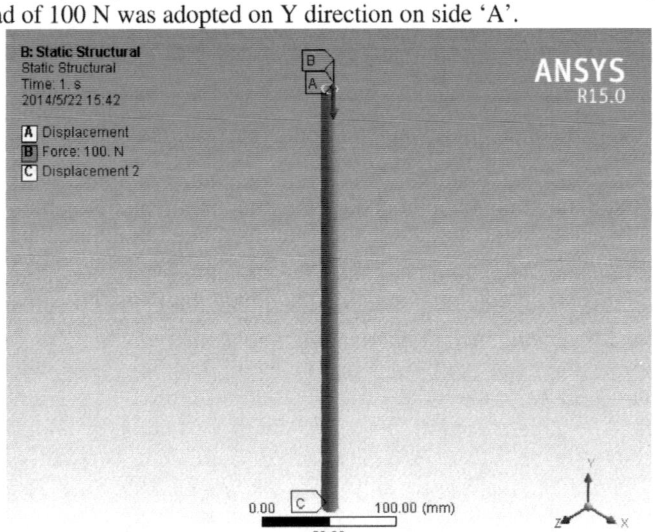

Figure 5. Boundary conditions setting of compressive linkage

Deformation cloud of simulation result is shown in figure 6, bulking deformation in the middle of the linkage. That provides lateral displacement about 1.1 mm, decrease the gap between linkage and objects. The critical load for bulking deformation is 2.2 KN, approach to calculated result.

When unload, the linkage recover in elastic deformation. So maintenance and service engineers never found any objects could impact the linkage. The bulking deformation occurs on release and retraction flap in the air condition, with air load on the wing flap. In the land condition, without any load bring to the linkage, bulking deformation never emergence.

5. Results

The failure module of linkage on wing flap is impacted with objects and shear fracture in the middle side. Because of instability on bulking deformation under compressive load, the middle of the rod move toward perpendicular to the axial direction.

The solution of the problem is increase the strength of linkage, that resistant the bulking deformation. So one selected alternative component of linkage with more strength flexural rigidity is provided, and replaced the previous versions. Failure of linkage never found in the past two years.

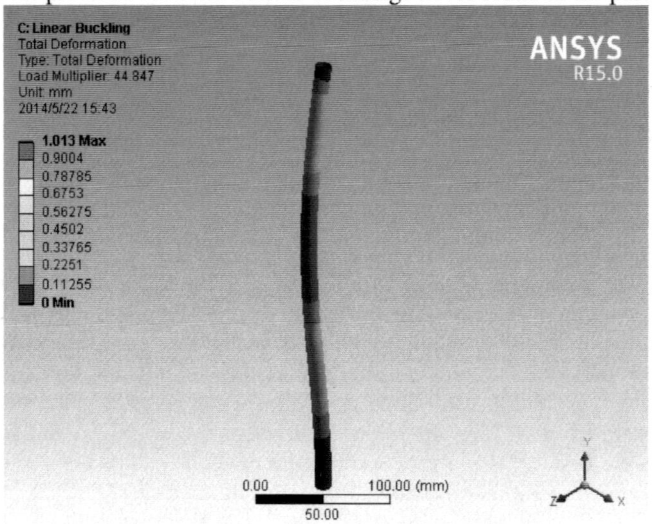

Figure 6. Bulking deformation of compressive linkage

Acknowledgments

Authors wishing to acknowledge financial support from civil aviation safety project 152146903107 and nature science funds U1333124 and U1533102 for financial supporting the project.

References

[1] Han S Hua D 2016 Logic of Protection for Flap's A symmetry Movement and Flap Skew Advances in Aeronautic Science and Engineering Vol 7 P 62
[2] Meng W 2007 Brief talks on the flap system of EMB145 regional jet airplanes *Aviation Maintenance &Engineering* Vol **1** P 51
[3] Cheng K 2015 Aircraft Control System Condition Monitoring and Fault Forecast Method Research Nanjing: Nanjing University of Aeronautics and Astronautics
[4] GE A 2003 Fracture Analysis of Joint Pole for Front Edge Wing on Boeing 735-500 plane *Materials Engineering* P 114
[5] Yizhi S He J 2010 Mechanical Design and Research of Trailing Edge Flap for Civil Aircraft Aircraft Design Vol **30** P 43
[6] Huangwei G, Li C, Zhongkui Z 2013 On the Stability and Axial Vibration of Compressive Bars Based on Nonlocal Elasticity Theory *Journal of Vibration and Shock* Vol **32** P 154

Preparation and magnetic properties of nickel nanowires by reduction in ethylene glycol medium under the influence of magnetic field

Wanshuo Sun[1], Junsheng Cheng, Lankai Li, Shunzhong Chen, Kun Chang

Institute of Electrical Engineering, Chinese Academy of Science, Beijing 100190, China

E-mail: sunwanshuo@mail.iee.ac.cn

Abstract. Nickel nanowires have successfully been fabricated through a simple liquid reduction in ethylene glycol medium with a 0.3T magnetic field applied. The effect of uniform magnetic field and solvent on the morphology and the crystal structure of magnetic nickel were studied. Scanning electron microscope images and transmission electron scope images s how that the effect of the external magnetic field on the morphology of nickel nanowires. X-ray diffraction shows the crystal structure of as-prepared products. And a energy disperse spectroscopy and a vibrating sample magnetometer are used to analyze the composition and static magnetic properties. The results show that the straight wires with an average diameter of about 100 nm and a length of several microns were obtained and mainly composed by fcc structure in the solvent of ethylene glycol. Magnetic measurements show that the saturation magnetization of the as-obtained products in a 0.3 T external magnetic field is 36 emu/g, less than that of bulk nickel crystal, and the coercivity of them is 186 emu/g, larger than that of bulk crystal with the mole ratio of sodium borohydride to nickel sulfate is 1:1000. This kind of nanowires array has potential applications with the special one-dimensional structures.

1. Introduction

In recent years, nanoscale magnetic metal materials have attracted intensive interest because of their unique physical and chemical characteristics including catalytic properties and magnetic shape anisotropy and electronic properties. Among them because of their one-dimension structures magnetic metal nanowires are a topic of growing interest. Thus they have many promising applications, for example, high density magnetic storage media[1], optoelectronic devices[2], sensors, catalysts and conductive fillers. To date, different techniques have been developed to synthesize one-dimensional nanostructured magnetic metal materials such as electrodeposition into templates[3-6] and reduction in solution without the presence of a template agent[7-9].

This work reports a simple approach for the fabrication of nickel nanowires on a large scale through liquid reduction method at room temperature, assisted by external magnetic field. The method has the merits of simples process, low cost and good performance in which $NiSO_4 \cdot 6H_2O$ acted as nickel source, $NaBH_4$ and $N_2H_4 \cdot H_2O$ acted as reducing agent, NaOH acted as pH-modifier. We present here the morphology of the product as-prepared and characterize their structure and magnetic properties in our effort to investigate the formation mechanism of nanowires.

2. Experiment

2.1. Preparation of Nickel Nanowires
In a typical experiment, all the reagents are of analytical grade and used as received without further purification.

First, nickel sulphate (Ni concentration of 0.5 M) was dissolved in ethylene glycol under mechanical stirring for 60 min at 40 ℃. 6.5 ml 80% hydrazine was mixed with 10 g sodium hydroxide with stirring fast in order to prevent them from agglomeration and the slurry was dissolved to 6.5 ml deionized water. Sodium borohydride was dissolved in 5 ml de-ionized water. In the second step, the reactor containing the nickel sulphate was placed in a magnetic field. Then the sodium borohydride was added to the above solution and some dark grey precipitates appeared immediately. After that, the slurry was quickly added to the reactor. Then a large amount of dark grey precipitates appeared accompanied a large number of bubbles. The precipitates were separated from the solution by placing a magnet under the container and then washed several times with de-ionized water and ethanol to remove any reaction residuals. At last, the as-prepared products were obtained by drying wet precipitates in a vacuum system at 30 ℃ for 12 h.

2.2. Instruments and measurements
The morphology of the as-synthesized product was recorded by scanning electron microscope (SEM). Transmission electron microscope was used to determine the size of the samples as-prepared. The samples prepared were characterized by X-ray diffraction (XRD) at 50 kV and 30 mA using Cu Kα radiation on Bruker D8X-ray diffractometer. Magnetic hysteresis loops were measured using a Lake Shore 7410 vibrating sample magnetometer (VSM).

3. Results and discussion
In order to examine the influence of the solvent to the reactor, we performed the reaction in deionized water and ethylene glycol when the mole ratio of the sodium borohydride to nickel sulfate is 1:1000 without external magnetic field. Figure 1 shows the transmission electron microscope image of as-prepared products obtained in deionized water and ethylene glycol. It is found that the average particle size of the as-prepared products in ethylene glycol was 86 nm, less than in deionized water (215 nm). In ethylene glycol, its high viscosity compared with deionized water to some extent hindered the spread of nickel ion. At the same time, the ethylene glycol hydroxyl absorbed on the surface of the nickel particles and in a certain extent prevented the growth and conglomeration of the nickel particles.

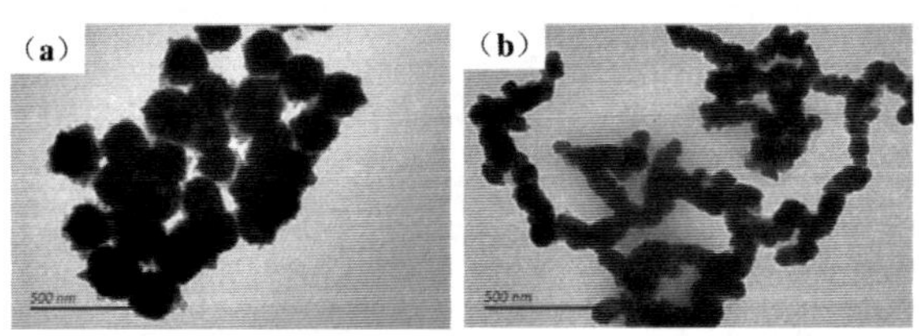

Figure 1. TEM image of the nickel particles in different solvent
(a) deionized water (b) ethylene glycol.

To further reveal the crystal structure of the products shown in Figure 1, X-ray diffraction analysis was carried out. Figure 2 shows the X-ray diffraction pattern of the as-prepared products obtained in deionized water and ethylene glycol at room temperature. It demonstrates that the crystal structure the as-prepared products in deionized water is the mixture of hcp structure and fcc structure. Compared with the peak positions of standard Ni (JCPDS, 04-0850) the diffraction peaks located at 2θ= 44.507°, 51.846°, 76.370° correspond to the (111), (200), (220) planes of fcc-Ni, respectively. Besides, hcp-NiOH (001), (100), (101), (102), (110), (111) have also been indicated. On the other hand, when the solvent is ethylene glycol the typical result for the as-obtained nickel array has been indicated in

Figure 2b. all the reflection peaks can be well indexed with fcc-Ni (space group Fm-3m(225); JCPDS card :04-0850), without impurity peaks.

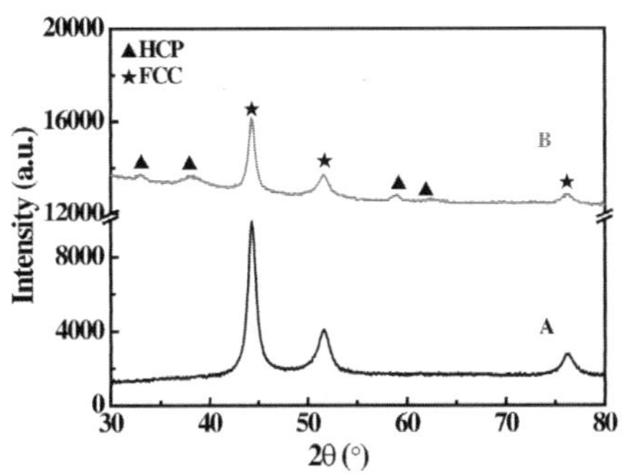

Figure 2. The XRD pattern of the nickel particles in different solvents (a) deionized water (b) ethylene glycol.

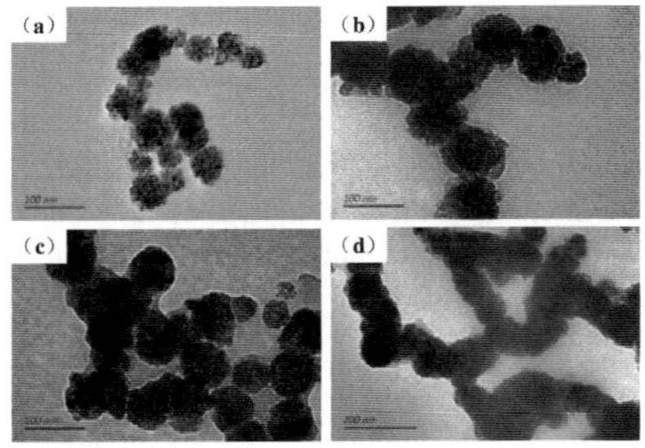

Figure 3. TEM image of the nickel particles with different mole ratio of sodium borohydride to nickel sulfate of (a) 1:10 (b) 1:100 (c) 1:500 (d) 1:1000.

Figure 3 shows the transmission electron microscope images of the as-obtained products in the absence of an external magnetic field when the mole ratio of the sodium borohydride to nickel sulfate is 1:10, 1:100, 1:500 and 1:1000, respectively. It is clearly seen that the average particle size of the nickel particles increased from 47 nm to 86 nm with the mole ratio of the sodium borohydride to nickel sulfate changed from 1:10 to 1:1000.

Nickel irons acted with a strong reducing agent sodium borohydride immediately. And nickel irons were reduced by sodium borohydride into Ni-B nano-particles that catalyzed the following reaction between nickel irons and hydrazine. When the mole ratio of sodium borohydride to sulfate nickel is 1:10, the larger amount of nickel irons were reduced by sodium borohydride. The as-obtained nano-particles acted as catalytic cores to catalyze the remaining smaller amount of nickel irons reduced by hydrazine. Thus the growth of the nickel particles in the last reaction was not sufficient and the size of nickel particles was smaller.

In order to further study if the morphology of the product as-prepared is related to the presence of the magnetic field, a sample has been prepared under similar conditions but in the absence of an external magnetic field. Figure 4 shows the morphology of the as-obtained products with and without an external magnetic field applied. It is clearly seen that the as-prepared products were spherical nickel particles and the nanowires were no longer observed in the absence of an external magnetic field while the nickel nanowires were obtained assisted by external magnetic field of 0.3 T. The nanowires were made up of spherical particles with averaging 100 nm in diameter. The length of the as-prepared nanowires is mostly in microscale. Thus it can be concluded that the external magnetic field is the key factor which control the morphology of products prepared. Figure 5 show the elemental composition of nickel nanowires was carried out on EDS. The EDS spectra show the presence of nickel without other impurities, indicating that the as-prepared products were pure.

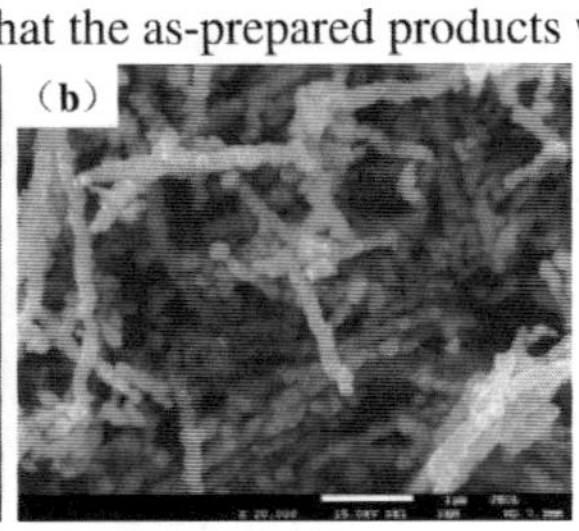

Figure 4. The morphology of the products obtained in the presence of a magnetic field of (a) 0 T (b) 0.3 T.

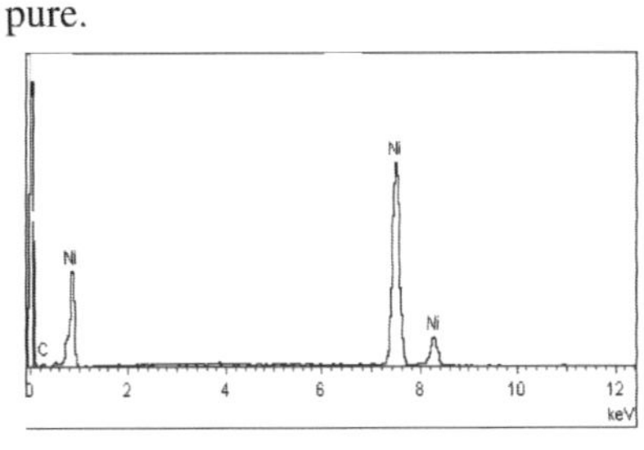

Figure 5. The EDS spectra of the nickel nanowires.

Figure 6 and Figure 7 show the scanning electron microscope image and the transmission electron microscope image of the nanowires prepared by reduction in ethylene glycol in the presence of a magnetic field of 0.3T when the mole ratio of sodium borohydride to nickel sulfate is 1:10, 1:100, 1:500 and 1:1000, respectively. It is clearly seen that the morphology of the products were straight wires. As shown in Figure 3 and Figure 7, the average diameter of the nickel particles changed from 47 nm to 86 nm after by adding a external magnetic field of 0.3 T with the mole ratio of sodium borohydride to nickel sulfate of 1:10. In contrast the average diameter of as-obtained nickel nanowires with different mole ratio of sodium borohydride to nickel sulfate remained about the same.

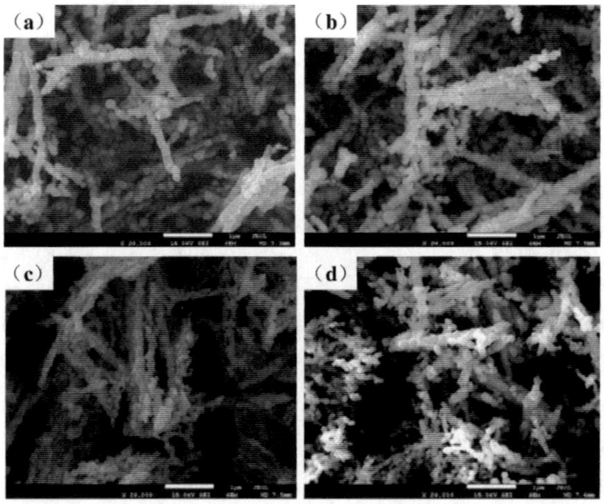

Figure 6. SEM image of nickel nanowires when an external magnetic field of 0.3 T is applied with the mole ratio of sodium borohydride to nickel sulfate of (a)1:10 (b)1:100 (c)1:500 (d)1:1000.

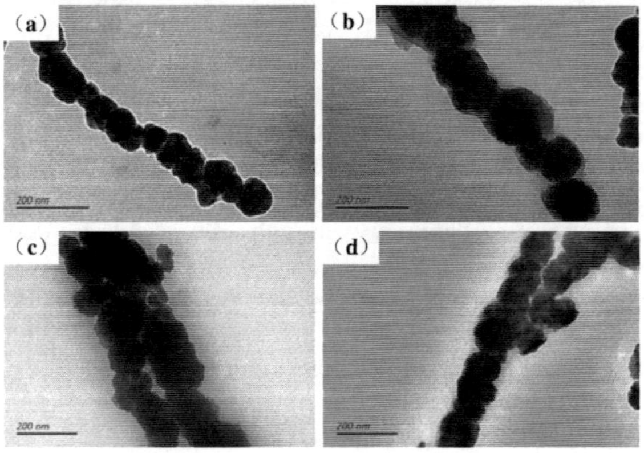

Figure 7. TEM image of nickel nanowires when an external magnetic field of 0.3 T is applied with the mole ratio of sodium borohydride to nickel sulfate of (a)1:10 (b)1:100 (c)1:500 (d)1:1000.

Magnetization makes nickel particles aggregate along the magnetic line of force by magnetic field-induction whilst the nickel particles distributed in whole reaction system without an external magnetic field. Their self-assembly behavior along the magnetic line forms the final morphology of nanowires.

Figure 8 shows the X-ray diffraction pattern of the nickel nanowires at room temperature. The XRD pattern of as-prepared products with different mole ratio of sodium borohydride to nickel sulfate are similar. It demonstrates that the crystal structure of the nickel nanowires is fcc structure. As shown in Figure 8, the diffraction peaks locate at 2θ= 44.507°, 51.846°, 76.370°, correspond to the （111），

（200），（220）planes of fcc-Ni, respectively(space group Fm-3m(225); JCPDS card :04-0850). It can be seen that the three main diffraction peaks of the crystal faces of (111), (200), (220) are strong and acute.

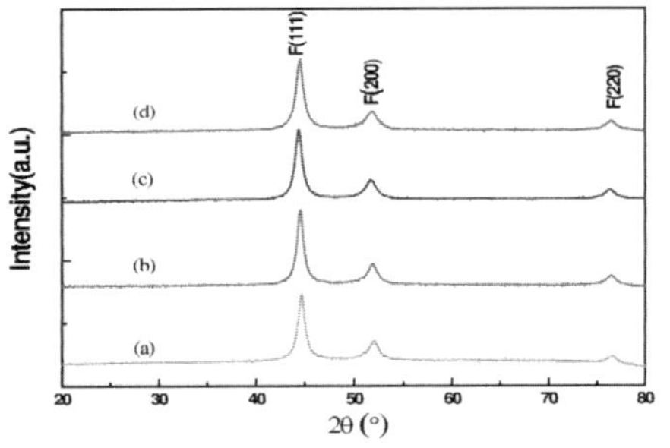

Figure 8. The XRD pattern of the nickel nanowires with the mole ratio of sodium borohydride to nickel sulfate of (a)1:10（b）1:100（c）1:500（d）1:1000.

Figure 9 shows the M-H hysteresis loops of the as-synthesized products. It can be seen that the saturation and magnetization value M_s of the nickel nanowires formed with the mole ratio of sodium borohydride to nickel sulfate of 1:10 and 1:1000 is 39 and 36 emu/g, respectively. And the coercivity value H_c for the nickel nanowires is 184 and 186 Oe, respectively. The saturation value of bulk nickel is 55 emu/g. The saturation magnetization of the nickel nanowires at room temperature is decreased compared to bulk nickel. The decrease in M_s for the nickel nanowires is usually ascribed to the formation of oxides on the particle's surface and a decrease in crystallinity[10] .In addition, it also might be partly due to a very small quantity of impurities on the products. The coercivity increased greatly, compared to the value of bulk nickel (100 Oe)[11]. Some literatures have reported that the coercivity of the nanowires depends on the magnetization reversal mechanism that is related to many factors such as the diameter, the presence of the crystalline defects and dipolar interaction[9].

According to the chains of spheres of model, the coercivity H_c under the fanning mechanism [12]is

$$H_{C,n} = \frac{\mu}{R^3}(6K_n - 4L_n) \qquad (1)$$

Where μ and R are the dipole moment and diameter of the spherical particles, respectively, and K_n and L_n are the constant interaction between magnetic particles. It is clearly seen that the coercivity H_c varies approximately as the reciprocal of the third power of the diameter, which means the coercivity H_c decreases with the diameter of the nickel particles decreasing.

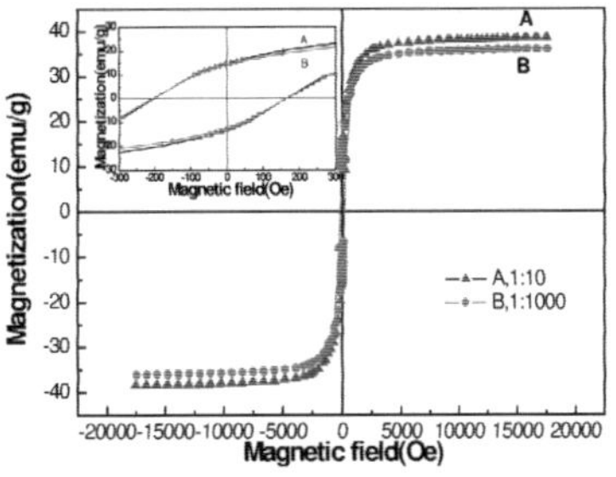

Figure 9. Hysteresis loops of the cobalt particles for different dosage of sodium borohydride.

4. Conclusion

Nickel nanowires have successfully been fabricated through a simple liquid reduction in ethylene glycol medium with a 0.3T magnetic field applied. The average size of the nickel nanowires prepared is about 100 nm in diameter and several microns in length. They crystallize with fcc structure in ethylene glycol medium. A less saturation magnetization and a larger coercivity for the nickel nanowires are achieved compared to the bulk nickel crystal with the mole ratio of sodium borohydride to nickel sulfate is 1:1000 by magnetic induction. It is suggested that this method could be a promising approach to obtain magnetic metal materials with one-dimensional structure and excellent magnetic properties.

Reference

[1] Wu M, Xiong Y, Jia Y, Niu H, Qi H, Ye J and Chen Q 2005 Magnetic Field-assisted Hydrothermal Growth of Chain-like Nanostructure of Magnetite *Chemical Physics Letters* **401** 374-379

[2] Shi J B, Chen Y C, Lee C W, Lin Y T, Wu C and Chen C J 2008 Optical and Magnetic Properties of 30 and 60nm Ni Nanowires *Materials Letters* **62** 15-18

[3] Pirota K R, Navas D, Hernández-Vélez M, Nielsch K and Vázquez M 2004 Novel Magnetic Materials Prepared by Electrodeposition Techniques: Arrays of Nanowires and Multi-layered Microwires *Journal of Alloys and Compounds* **369** 18-26

[4] Imran M M A 2008 Structural and Magnetic Properties of Electrodeposited Ni Nanowires *Journal of Alloys and Compounds* **455** 17-20

[5] Xu C L, Li H, Zhao G Y and Li H L 2006 Electrodeposition of Ferromagnetic Nanowire Arrays on AAO/Ti/Si Substrate for Ultrahigh-density Magnetic Storage Devices *Materials Letters* **60** 2335-2338

[6] Nielsch K, Wehrspohn R B, Barthel J, Kirschner J, Fischer S F, Kronmüller H, Schweinböck T, Weiss D and Gösele U 2002 High Density Hexagonal Nickel NanowireArray *Journal of Magnetism and Magnetic Materials* **249** 234-240

[7] Athanassiou E K, Grossmann P, Grass R N and Stark W J 2007 Template Free, Large Scale Synthesis of Cobalt Nanowires Using Magnetic Fields for Alignment *Nanotechnology* **18** 695-700

[8] Niu H, Chen Q, Ning M, Jia Y and Wang X 2004 Synthesis and One-dimensional Self-assembly of Acicular Nickel Nanocrystallites under Magnetic Fields *Journal of Physical Chemistry B* **108** 3996-3999

[9] Soumare Y, Dakhlaoui-Omrani A, Schoenstein F, Mercone S, Viau G and Jouini N 2011 Nickel Nanofibers and Nanowires: Elaboration by Reduction in Polyol Medium Assisted by External Magnetic Field *Solid State Communications* **151** 284-288

[10] Li X G, Murai T, Saito T and Takahashi S 1998 Thermal Stability, Oxidation Behavior and Magnetic Properties of Fe–co Ultrafine Particles Prepared by Hydrogen Plasma–metal Reaction *Journal of magnetism and magnetic materials* **190** 277-288

[11] Hwang J H, Dravid V P, Teng M H, Host J J, Elliott B R and Johnson D L 1997 Magnetic Properties of Graphitically Encapsulated Nickel Nanocrystals *Journal of materials research* **12** 1076-1082

[12] Zhou W, Zheng K, He L, Wang R, Guo L, Chen C, Han X and Zhang Z 2008 Ni/Ni3C Core–shell Nanochains and Its Magnetic Properties: One-step Synthesis at Low Temperature *Nano letters* **8** 1147-1152

Analysis of flow field characteristics in IC equipment chamber based on orthogonal design

W F Liu, Y Y Yang[1], C N Wang

School of Engineering and Technology, China University of Geosciences (Beijing), Beijing 10083, China

E-mail: yangyy@cugb.edu.cn

Abstract. This paper aims to study the influence of the configuration of processing chamber as a part of IC equipment on flow field characteristics. Four parameters, including chamber height, chamber diameter, inlet mass flow rate and outlet area, are arranged using orthogonally design method to study their influence on flow distribution in the processing chamber with the commercial software-Fluent. The velocity, pressure and temperature distribution above the holder were analysed respectively. The velocity difference value of the gas flow above the holder is defined as the evaluation criteria to evaluate the uniformity of the gas flow. The quantitative relationship between key parameters and the uniformity of gas flow was found through analysis of experimental results. According to our study, the chamber height is the most significant factor, and then follows the outlet area, chamber diameter and inlet mass flow rate. This research can provide insights into the study and design of configuration of etcher, plasma enhanced chemical vapor deposition (PECVD) equipment, and other systems with similar configuration and processing condition.

1. Introduction

The vigorous development of electronic information industry depends on the improvement of IC (integrated circuit) equipment design and manufacturing technology. The processing chamber, a quite common part of IC equipment, has not attracted enough attention to be optimized during the equipment research process. Meanwhile, more emphasises were put on single structure parameter during the research process. Furthermore, the distribution of temperature and flow among the processing chamber are the most important factors determining the process homogeneity. Consequently, it is of great significance to research the influence of structure characteristics and processing parameters on the uniformity of temperature and flow distribution in the chamber. Actually, researchers have laid a solid foundation of research jobs regarding the temperature and flow distribution through experiment [1-3] and numerical simulation [4-7].

As the influence of inlet location and mass flow rate on etcher process were analysed. Moreover, the impact of inlet location and mass flow, vacuum pump outlet area and location were also the key parameters should be optimized to improve the uniformity of flow filed[8,10].The influence of showerhead structure on RF-PECVD chamber were researched through finite volume method(FVM) and SIMPLE algorithm[5]. Moreover, using a 12 inch PECVD device, the simulation of flow velocity, pressure and temperature distribution based on four types of steady flow chambers were studied [11].During the past few years, investigations concerned with the flow field or other parameters also have been mentioned. However, most experiments focused on the influence of only a single parameter on flow distribution inside processing chamber, while analysis of the overall impact of the relationship among various parameters was absence.

This present study informs a variable structure plasma chamber and builds the 3-D symmetrical simulation models of the chamber. Besides, the parameters of the chamber, such as chamber height, chamber diameter, inlet flow mass rate and outlet area were arranged by using orthogonal design method. In addition, every simulation corresponding to the orthogonal design was computed by a numerical code-Fluent. After the simulation results of velocity, pressure and temperature above the holder of chamber were analyzed respectively, the velocity difference value between the maximum and minimum velocity value was selected as the evaluation criteria to analyze the influence of these parameters on the uniformity of the gas flow distribution in chamber. Finally, the quantitative effect of these parameters to gas flow distribution were obtained.

2. Modeling and Orthogonal Experimental Design

2.1. Chamber model

There are various kinds of processing chambers used in integrated circuit manufacturing processing technology, and as the process and application are different, so are the structure of chambers. However, those parameters affecting the flow distribution are similar, and which also are the important factors of chamber system, such as inlet system, gas distribution system, chamber height, chamber diameter and outlet system. This paper presents a chamber structure, whose gas flows into the chamber from the inlet employed at the center of top cover of the chamber. And next, the gas flow through two-stage showerhead following, by which makes the gas distributes more evenly. The gas flow out of chamber through the outlet, which area could be changed, arranged around the bottom of chamber radially. The schematic of the chamber structure is shown in Figure 1. As figure 1 shows, the distance between the second showerhead and holder is H, the diameter of chamber is D, and the outlet section is a rectangular section (not shown in figure 1) with which area is S ($a \times$ b). Due to the processing chamber with symmetrical structure, about one-eighth of the three-dimensional structure was taken to model and mesh. The one-eighth 3-D grid model is shown in figure 2.

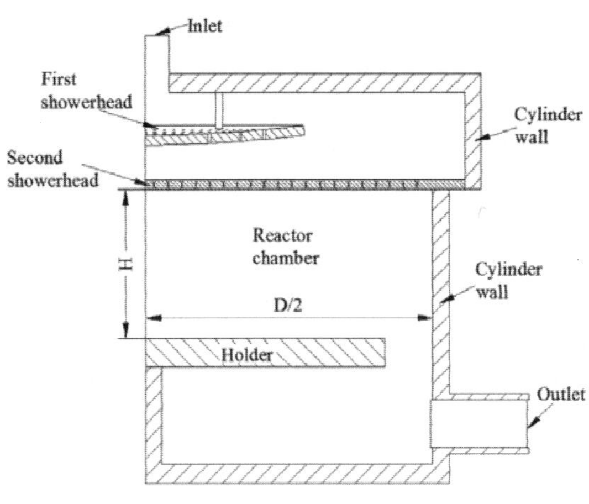

Figure 1. Schematic of the Chamber Structure

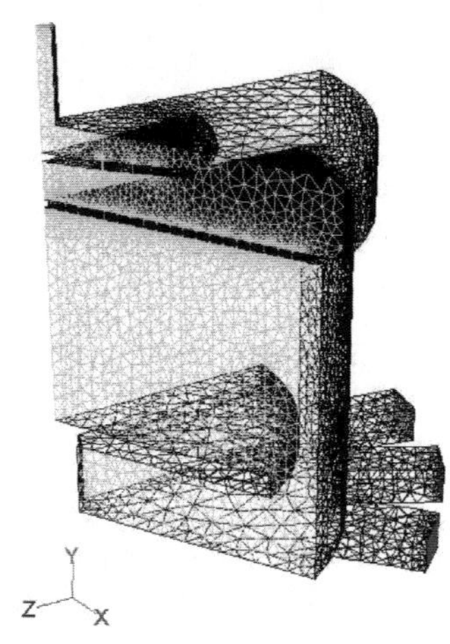

Figure 2. One-eighth 3-D grid model

2.2. The Flow Field Numerical Model

The flow field of processing chamber is proceeded by a commercial software-Fluent, which uses finite-volume method (FVM) to discrete the control equation of module. And the control equation in terms of the gas flow field in chamber should follow continuity equation, momentum and energy conservation equation.

The boundary conditions are defined as follows: 1.Inlet: the inlet sets as a mass-flow inlet; 2.Outlet: to make the outlet flow divergent completely, the model outlets are extended than their actual length which would not affect the simulation results. The outlet sets as a pressure outlet; 3.Symmetry Wall: as the simulation model is one eighth of the whole model, the two sides are set as symmetry wall; 4.Cylinder Wall: besides the symmetry wall, the cylinder walls set as no-slip wall; 5.Holder Surface: as the holder is the heat source of the chamber space, the holder surface temperature is set as constant as473.16k.

2.3. Orthogonal Experimental Design

The inlet mass rate *Q*, holder height(instead of chamber height) *H*, chamber diameter *D*, outlet section area *S* are taken as variable factors, the orthogonal experimental design methods is adopted and $L9(3^4)$ table is chosen for the experiment. And every variable factor has three levels, such as the inlet mass rate is 90sccm (outlet pressure is 0.5Pa), 325sccm (outlet pressure is 30 Pa) and 625sccm (outlet pressure is 50Pa); holder height is 65mm, 125mm and 185mm; chamber diameter is 360mm, 470mm and 560mm; outlet section area is 5 mm × 25 mm, 15 mm × 25 mm and 30 mm × 25 mm. The orthogonal experimental tableis shown in Table 1.

Table 1. Orthogonal experimental table.

No.	Column No.	1	2	3	4
	Factor	Inlet Mass Rate sccm	Holder Height mm	Chamber Diameter mm	Outlet Section Area mm×mm
	Symbol	Q	H	D	$S(a\times b)$
1		$1(Q_1)$	$1(H_1)$	$1(H_1)$	$1(a_1\times b)$
2		$1(Q_1)$	$2(H_2)$	$2(H_2)$	$2(a_2\times b)$
3		$1(Q_1)$	$3(H_3)$	$3(H_3)$	$3(a_3\times b)$
4		$2(Q_2)$	$1(H_1)$	$1(H_1)$	$3(a_3\times b)$
5		$2(Q_2)$	$2(H_2)$	$2(H_2)$	$1(a_1\times b)$
6		$2(Q_2)$	$3(H_3)$	$3(H_3)$	$2(a_2\times b)$
7		$3(Q_3)$	$1(H_1)$	$1(H_1)$	$2(a_2\times b)$
8		$3(Q_3)$	$2(H_2)$	$2(H_2)$	$3(a_3\times b)$
9		$3(Q_3)$	$3(H_3)$	$3(H_3)$	$1(a_1\times b)$

3. Simulation results and discussion

Every experiment from No.1 to No.9 in Table1 is simulated, and the distribution of flow velocity, pressure and temperature in the area above 10mm [14] of the holder are analyzed respectively.

3.1. Velocity Distribution

The velocity distribution lines in the chamber of every experiment are shown in figure 3 as follow:

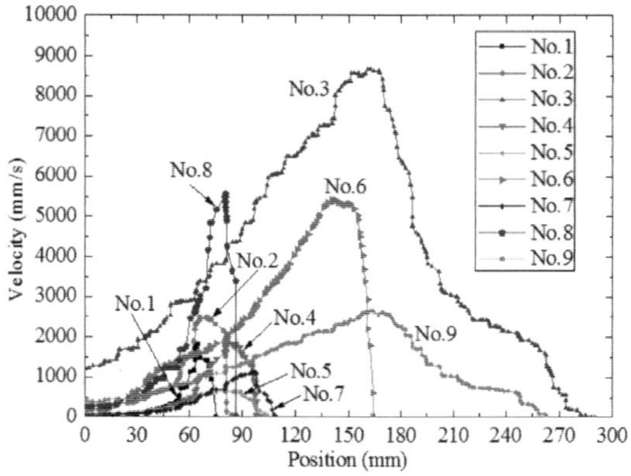

Figure 3. Flow Velocity Distribution

It can be concluded from figure 3 that while changing the chamber characteristic parameters, the velocity distribution shows regularity. The velocity value has been increasing from the center of holder to a certain position at which the value reaches maximum. Then, from the certain position to the chamber wall, the velocity value decrease monotonously. That is because when the gas flows through the second showerhead into chamber, then flows out of chamber from the outlets radial arranged around the bottom part of chamber. At the center position of holder, as the movement of gas in all directions is same, the velocity is zero. In every experiment model, the diameter of holder is 300mm (radius is 150mm). In No.1, No.6 and No.8 experiments, the climax velocity valuesare at the edge of holder, at which the position is 150mm and all of them have the same parameter of chamber diameter. Likewise, in No.3, No.6 and No.9 experiments, all of which have same holder height, the maximum velocity values also emerge at the edge of holder and all velocity lines have a similar trend on the left side. Although No.2, No.4 and No.9 experiments have same chamber diameter, the holder height of No.9 experiment is too high to make the holder closer to the second showerhead, which makes the right-side of velocity line of No.9 experiment different from No.2 and No.4 experiments. As the holder height of No.3 experiment is higher than the others, the velocity line of No.3 experiment is not similar to No.5 and No.7 experiments, despite of their same chamber diameter. Nevertheless, the velocity lines of No.5 and No.7 experiments have the same trend. From the lines of No.1, No.2 and No.3 experiments, we can obtain that the peak values of velocity shift left gradually, and end up with the middle of the lines.

3.2. Pressure Distribution

The pressure distribution lines in the chamber of every experiment are shown in figure 4 as follow:

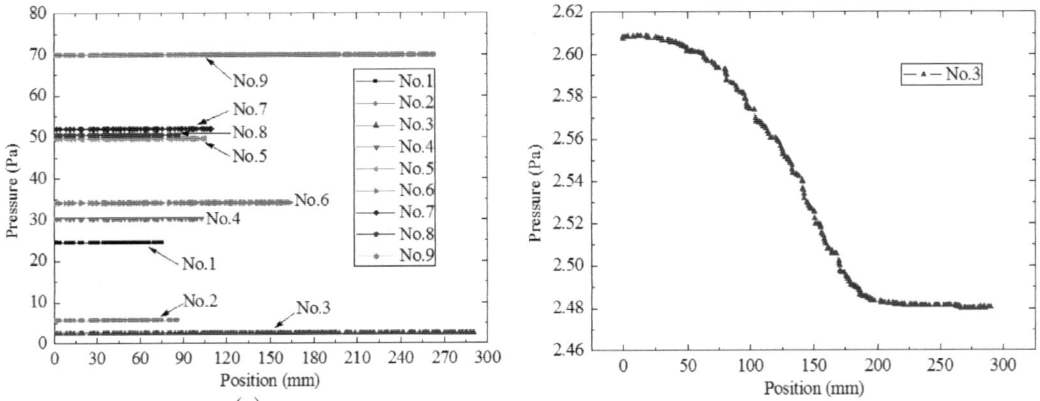

(a) (b)

Figure 4. Gas Flow Pressure Distribution

It can be concluded from figure 4 (a) that while changing the chamber characteristic parameters, the pressure distribution shows regularity. The pressure above the holder almost unchanged. During every experiment, all the pressure values in the area above 10mm of the holder decrease slowly, from the center of holder to the chamber wall. The figure 4 (b) shows the pressure distribution of No.3experiment. In addition,it shows that the pressure value is invariable fundamentally at the beginning, and then decreases from the position of 50mm to 175mm,then trends to a stable value. Similarly, other experiments also have a stable distribution segment at the initial stage, after then the pressure value decrease slowly to a certain value (the other experiments'results not show as figures). Though pressure values in every experiment have a decreasing trend, the change of every pressure value is subtle. Consequently, the process and structure parameter have negligible effect on the pressure distribution.

3.3. Temperature Distribution

The temperature distribution lines in the chamber of every experiment are shown in figure 5 as follow:

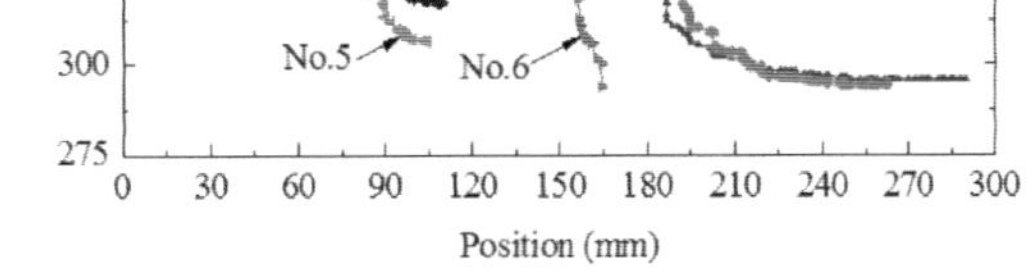

Figure 5. Gas Flow Temperature Distribution

It can be concluded from figure 5 that while changing the chamber characteristic parameters, the pressure distribution show regularity. As the position value increases, the temperatures above 10mm of holder area keep a constant temperature at a segment. When reaching at a certain position, the temperatures decrease rapidly. Moreover, the reason lays in the fact that the holder is the main heat source that distribute uniformly, so the temperature distribution above holder area has uniform distribution. However, on the edge of holder, along with the gas flow and heat source disappearing, the temperatures at this position change acutely. As every experiment from No.1 to No.9 has the same holder diameter. Therefore, every experiment has the same temperature distribution. While the chamber diameter varies, the temperature decreases until it equals to the temperature of wall. It is thus clear that when the holder is the main heat source, change of process and structure parameters barely affect the temperature.

3.4. Data Analysis

Through above analyses, it can be concluded that when the process and structure parameters changed, they had little influence on pressure and temperature distribution uniformity. However, the velocity distribution varied observably while the process and structure parameters changed. Therefore, the velocity difference value between the maximum and minimum velocity in the area above 10mm of holder could be used as the analytical factor. In addition, the velocity difference value can show the gas distribution uniformity above the holder. The fewer of the velocity difference value, the more uniformity of the gas distribution. The analysis of velocity difference value ΔV between the maximum and minimum velocity is shown in Table 2.

Table 2. Orthogonal Experiment Table.

No.	Column No.	1	2	3	4	5
	Factor	Inlet Mass Rate sccm	Holder Height mm	Chamber Diameter mm	Outlet Section Area mm×mm	Velocity Difference Value m/s
	Symbol	Q	H	D	S (a×b)	ΔV
1		1(Q_1)	1(H_1)	1(D_1)	1(a_1×b)	1.5
2		1(Q_1)	2(H_2)	2(D_2)	2(a_2×b)	0.55
3		1(Q_1)	3(H_3)	3(D_3)	3(a_3×b)	6.7
4		2(Q_2)	1(H_1)	2(D_2)	3(a_3×b)	0.36
5		2(Q_2)	2(H_2)	3(D_3)	1(a_1×b)	0.26
6		2(Q_2)	3(H_3)	1(D_1)	2(a_2×b)	4.2
7		3(Q_3)	1(H_1)	3(D_3)	2(a_2×b)	0.35
8		3(Q_3)	2(H_2)	1(D_1)	3(a_3×b)	4.4
9		3(Q_3)	3(H_3)	2(D_2)	1(a_1×b)	2

Table 3. Results Analysis

	1	2	3	4
K1	8.75	2.21	7.7	3.76
K2	4.82	5.21	2.91	5.1
K3	6.75	12.9	7.31	11.46
k1	2.19	0.55	1.93	0.94
k2	1.20	1.30	0.73	1.28
k3	1.69	3.23	1.20	2.87
Range	0.99	2.68	1.20	1.93
Important Order	$H>S>d>Q$			

The velocity difference value between the maximum and minimum values reflects velocity changes of the airflow above holder. Moreover, the velocity distribution also plays a pivotal role in the process of practical devices, such as thin film deposition or etching. Uniform gas flow means a uniform distribution process gas, whose velocity distribution affects the practical process substantially. From the results of orthogonally designed experiment in Table 3, the sequence of factors affecting the gas distribution is holder height, outlet section area, chamber height and inlet mass. In addition, the holder height, corresponding to chamber height or the distance between up and down electrode, is the most significant factor affecting the uniformity of gas distribution in practical devices. In conclusion, at the initial stage of device design, the chamber height and electrode distance should be confirmed firstly. Furthermore, various outlet section areas still affect the gas distribution.Since the outlet section area is different, the outlet gas flow under a certain vacuum pumping speed also differs. In other words, the changes in outlet gas flow affect gas distribution above the holder area in chamber. Hence, under circumstances of satisfying structure performance and cost control requirements with similar cases, install valves can adjust sizes of outlet section, so as to suffice various process requirements. In actual process devices, we can increase chamber diameter under cost control limitations, enhancing the capability of handling chips simultaneously and process efficiency as well.

Conclusion

The parameters of processing chamber such as inlet mass rate$Q$, holder height(instead of chamber height) H, chamber diameter D, outlet section area S, all of which have three levels, have been modelled and analysed by the orthogonal experimental design. Firstly, the orthogonal experiment design table was built with numbers of factors and levels of every factor. After analysing the velocity, pressure and temperature distribution in the area of 10mm above the holder, the velocity difference value between the maximum and minimum velocity is chosen as the analytical index. Furthermore, from the results of orthogonally designed experiment, the sequence of indexes affecting the gas distribution uniformity is holder height, outlet section area, chamber height and inlet mass. Finally, it comes to the conclusion that the experiment results could be a reference for IC processing chamber design and manufacture.

As the IC process is a complicated physical chemical process, under the coupling of physical field and multiphase, further works and theoretical study are needed to simulate the comprehensive process by considering the coupling of flow field, thermal field, electromagnetic field, and plasma and chemistry reaction simultaneously. Meanwhile, the simulation results also should be verified with experimental data. The research in this paper is a guide for development and application of chamber structure design. Hence further research about the uniformity of plasma distribution is needed, which will be our main research work in future.

Acknowledgments

This work was financially supported by National Science and Technology Major Project (2011ZX02403-4-3).

References

[1] Yabe H, Yuuki A and Matsui Y 1991 *Japan. J. Appl. Phys.* **30** 2873-77
[2] Alayo M I, Pereyra I and Carreno M N P 1998 *Thin Solid Films* **323** 40-5
[3] Komiyama H, Shimogaki Y and Egashira Y 1995 *Chem. Eng. Sci.* **54** 1941-57
[4] Sansonens L, Bondkowski J and Mousel S 2003 *Thin Solid Films* **42** 721-6
[5] Kim Y, Boo J and Hong B 2005 *Surf. Coat. Tech.* **193** 88-9
[6] Capuineau H, Enache I and Gherardi N 2009 *J. Phys. D Appl. Phys.* **42** 125201-7
[7] Li Z M, Hao Y and Zhang J C 2010 *Chinese J. Synth. Cryst.* **39** 226-31
[8] Economou D, Bartel T 1996 *IEEE Trans Plasma Sci.* **24** 131
[9] Khater M H, Overzet L J and Cherrington B E 1998 *J. Vac. Sci. Technol. B.* **16** 490
[10] Ra Y, Bradley S and Chen C 1994 *J. Vac. Sci. Technol. A.* **12** 1328
[11] Xia H X, Xiang D and Mou P 2012 *Chinese J. Synth. Cryst.* **41** 1030-6
[12] Versteeg H K, Malalasekera W 1995 *An Introduction to Computational Fluid Dynamics-The Finite Volume Method* (New York: Longman Scientific & Technical)
[13] Patankar S V 1980 *Numerical Heat Transfer and Fluid Flow* (London: Hemisphere Publishing Corporation)
[14] Cheng J, Zhu Y, Duan G H and Wang C H 2008 *Chinese Journal of Semiconductors* **29** 780-4

Sonochemical fabrication of Ag_2CO_3 nanomaterial and influencing factors on photocatalytic properties

Li Zhou, Liuying Liang, Dilinuer Talifu, Abulikemu Abulizi[1]

Key Laboratory of Coal Conversion & Chemical Engineering Process (Xinjiang Uyghur Autonomous Region), College of Chemistry and Chemical Engineering, Xinjiang University, Urumqi 830046, PR China.

E-mail: aablek@163.com

Abstract. Visible-light-active Ag_2CO_3 nanoparticles were synthesized by a simple sonochemical method (ambient temperature for only 1 h). The effects of ultrasound pulse mode, ultrasound time and pH value on the photodegradation properties of Ag_2CO_3 were investigated. The samples were characterized by X-ray diffraction (XRD), scanning electron microscopy (SEM), energy dispersive X-ray (EDS), UV-vis diffuse reflectance spectroscopy (DRS) and nitrogen adsorption–desorption techniques. The photocatalytic activity was evaluated by degrading methylene blue (MB) under visible light irradiation (λ>420nm). According to obtained results, it was found that MB degradation (about 99.4% after 20 min of irradiation) was higher compare to conventional stirring method, which might be due to the higher specific surface areas and active sites. The method described here can be used to improve the efficiency of visible-light sources and develop an ecologically friendly, practical wastewater treatment for pollutants.

1. Introduction

At present, the environmental pollution and energy shortage become more and more serious [1]. Semiconductor photocatalysts have attracted great attention worldwide by high photocatalytic activity and strong stability under visible light [2]. Semiconductor photocatalysts have been regarded as a promising candidate in solar energy conversion and water pollutant treatment [3-7]. As a typical photocatalyst, Ag-based semiconductor photocatalysts have narrow band gap, which can absorb visible light and usually show good catalytic activity [8-9]. In recent years, Ag-based photocatalysts such as AgX (X=Cl, Br, I) [10-11], Ag_3PO_4 [12], $Ag_8W_4O_{16}$ [13], Ag_2O [14], β-$AgAl_{1-x}Ga_xO_2$ [15], $Ag@Ag_2SO_4$ [16] etc. have been demonstrated as an efficient photocatalytic materials under visible light irradiation for the photodegradation of organic pollutants. Dong [9] successfully prepared Ag_2CO_3 semiconductor photocatalyst by a simple ion-exchange method based on a strategy incorporating of p-block C element into a narrow bandgap Ag_2O. Liu [17] and co-workers reported the microwave hydrothermal synthesis of Ag_2CrO_4 photocatalyst for fast degradation of PCP-Na under visible light irradiation. Lin [18] synthesized PVB/CdS composite materials via a microwave-assisted electrospinning process. However, in the process of photocatalysis, the above methods offered some defects such as low yield, undesired morphology and complex procedure.

In contrast, the sonochemical processing as a one-step process offers very important features such as low temperature processing and low cost [19]. Sonochemistry uses the ultrasonic irradiation to induce the formation of particles with smaller size and higher surface area in contrast to other synthesis method. Sonochemistry has been employed extensively for the synthesis of nanostructured materials due to its rapid reaction rate, controllable reaction conditions, simplicity and safety. Moreover, powder particles synthesized through this method normally have uniform shape with narrow size distribution. Owing to its advantages, sonochemistry has experienced a large promotion in various fields concerning the production of novel nanostructured materials with improved properties in the recent years. The tailored preparation and improvement of chemical and physical properties of the synthesized materials are now the main objects of the recent researches [20].

However, to the best of our knowledge, reports on the preparation of visible light-driven Ag_2CO_3 photocatalysts by sonochemical route remain rare. Herein, Ag_2CO_3 was successfully synthesized via a simple ultrasonic method. In particular, we systematically investigated the influencing factors of Ag_2CO_3 and proved that ultrasonic method can offer higher surface areas and photocatalytic activity. This work may provide a more green and efficient strategy for exploring the excellent Ag-based composite photocatalysts.

2. Experimental

2.1. Materials
Sodium carbonate (Na_2CO_3), silver nitrate ($AgNO_3$), sodium hydroxide (NaOH), nitric acid (HNO_3), all the chemicals used in this study were analytical grade and purchased from Tianjin Yong Sheng Chemical Factory of China without further purification. Distilled water was used in all experiments.

2.2. Synthesis of Ag_2CO_3
The Ag_2CO_3 samples were prepared by a typically simple ion-exchange method under the effect of ultrasound at room temperature. In a typical procedure, 20 mL $AgNO_3$ solution (0.36 M) was added to a 100 mL beaker, 20 mL Na_2CO_3 solution (0.18 M) which pH was adjusted with HNO_3 and NaOH was added dropwise to the above suspension in sequence under ultrasonic irradiation for 60 min. The ultrasonic irradiation was performed with a high intensity ultrasonic probe (Xin zhi Co, China, JY92-2D, 10 mm diameter, Ti-horn, 20 kHz, 60 W cm^{-2}) immersed in the reaction system. The obtained precipitate was collected by centrifugation, washed several times with distilled water and ethanol, and dried at 50°C for 12 h.

2.3. Characterization
The phase of the Ag_2CO_3 samples was characterized by power X-ray diffractometer (Shimadzu XRD-6000) equipped with a Cu-Kα radiation at a scanning rate of 5°/min in the 2θ range of 10~80°. The morphological characteristics and microstructure of the nanoparticles were observed using scanning electron microscopy (SEM; Hitachi S-4100, Tokyo, Japan). The chemical composition of the samples was determined by EDS. The UV–vis diffuse reflectance spectra (DRS) were obtained on UV–vis spectrophotometer (U-3010, Hitachi, Japan) by using $BaSO_4$ as the matrix. Nitrogen adsorption-desorption isotherms were collected on an Autosorb-iQ sorption analyzer and analyzed using the Brunauer-Emmett-Teller (BET) equation and Barret-Joyner-Halenda (BJH) model.

2.4. Photocatalytic activity measurement
In a typical measurement, 0.1 g samples were added into 100 ml of MB (20 mg/L). Prior to irradiation, the suspensions were magnetically stirred for 0.5h in the dark. This was to ensure that the dye could reach the absorption–desorption equilibrium on the photocatalyst surface. Subsequently, the suspension was continuously magnetically stirred under visible light irradiation supplied by a 280 W Xe lamp (Color temperature 6000K, PHILIPS) with a 420 nm cut-off filter. The degradation of MB was monitored by UV-vis spectrophotometer (PG, TU-1901) every 10 min. Before measurement, the photocatalysts were removed from the photocatalytic reaction system by centrifugation.

IOP Conf. Series: Materials Science and Engineering **167** (2017) 012032 doi:10.1088/1757-899X/167/1/012032

3. Results and discussion

3.1. Effects of ultrasonic pulse mode and reaction time

The degradation efficiency of MB by photocatalysts prepared at different ultrasonic pulse mode was illustrated in Figure 1a. The photocatalytic activity of Ag_2CO_3 changed with different pulse mode [21]. The photocatalysts prepared at pulse mode of "1 s off / 9 s on" displayed the highest degradation efficiency. The results show that degradation efficiency increases with increasing ultrasound power. The pulse mode "1 s off / 9 s on" was selected as the optimal condition for the following section. Moreover, the degradation efficiency of MB by photocatalysts prepared at different ultrasonic time was illustrated in Figure 1b. The photocatalytic activity of Ag_2CO_3 increased with the increase of ultrasonic time. Therefore, the ultrasonic time for 60 min was also selected as the optimal condition for the following section. The increase of photocatalytic efficiency might be due to the advantages of sonochemical method for smaller size and higher surface areas.

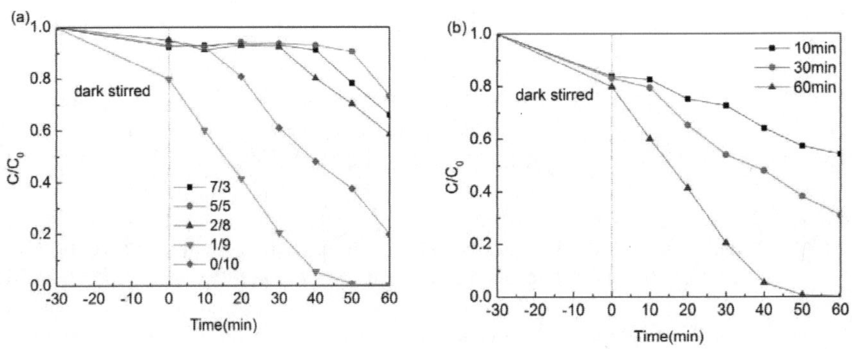

Figure 1. (a) Effect of ultrasonic pulse mode (pH: 11, ultrasonic time: 60 min); (b) Effect of ultrasonic time (pH: 11, ultrasonic pulse mode: 1 s off / 9 s on) on the degradation efficiency of MB

3.2. The effect of pH

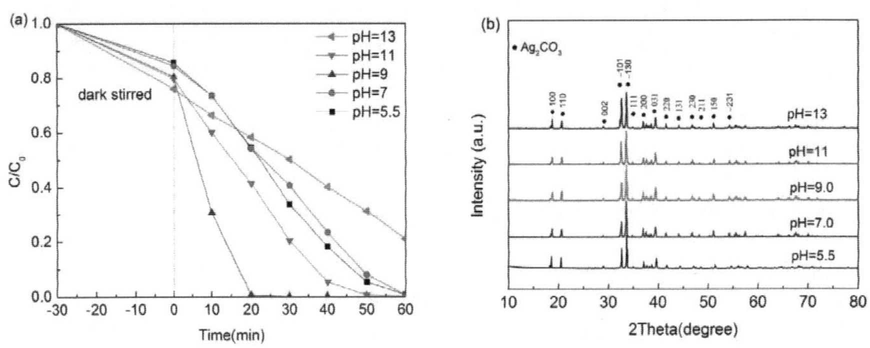

Figure 2 The photocatalytic degradation of MB (a) and XRD patterns (b) of Ag_2CO_3 prepared at different pH (ultrasonic pulse mode: 1 s off / 9 s on, ultrasonic time: 60 min)

The degradation efficiency and XRD patterns of Ag_2CO_3 prepared at different pH were showed in Figure 2. Figure 2(b) presents XRD patterns of the as-prepared Ag_2CO_3 by ultrasonic method. All the diffraction peaks of Ag_2CO_3 could be well indexed with monoclinic Ag_2CO_3 (JCPDS No. 70-2184), and no other impurities were found, which incidates that pH does not influence the purity of samples. However, the photocatalysts prepared at pH 9.0 displayed the highest degradation efficiency (Figure 2(a)). This suggests that pH will influence the degradation efficiency of Ag_2CO_3. The pH of the raw material solution can neither too high nor too low, because the acidic conditions lead to low yield of samples and alkaline conditions can lead to poor degradation ability. Thus, pH played an important role in the formation and photocatalytic degradation of pure Ag_2CO_3. And the Ag_2CO_3 prepared at pH 9.0 was selected as the optimum catalyst for the following sections. According to the above experimental results, the optimum conditions of preparation of Ag_2CO_3 samples are pulse mode for 1 s off / 9 s on, pH for 9.0 and ultrasonic time for 60 min.

3.3. The effect of methods

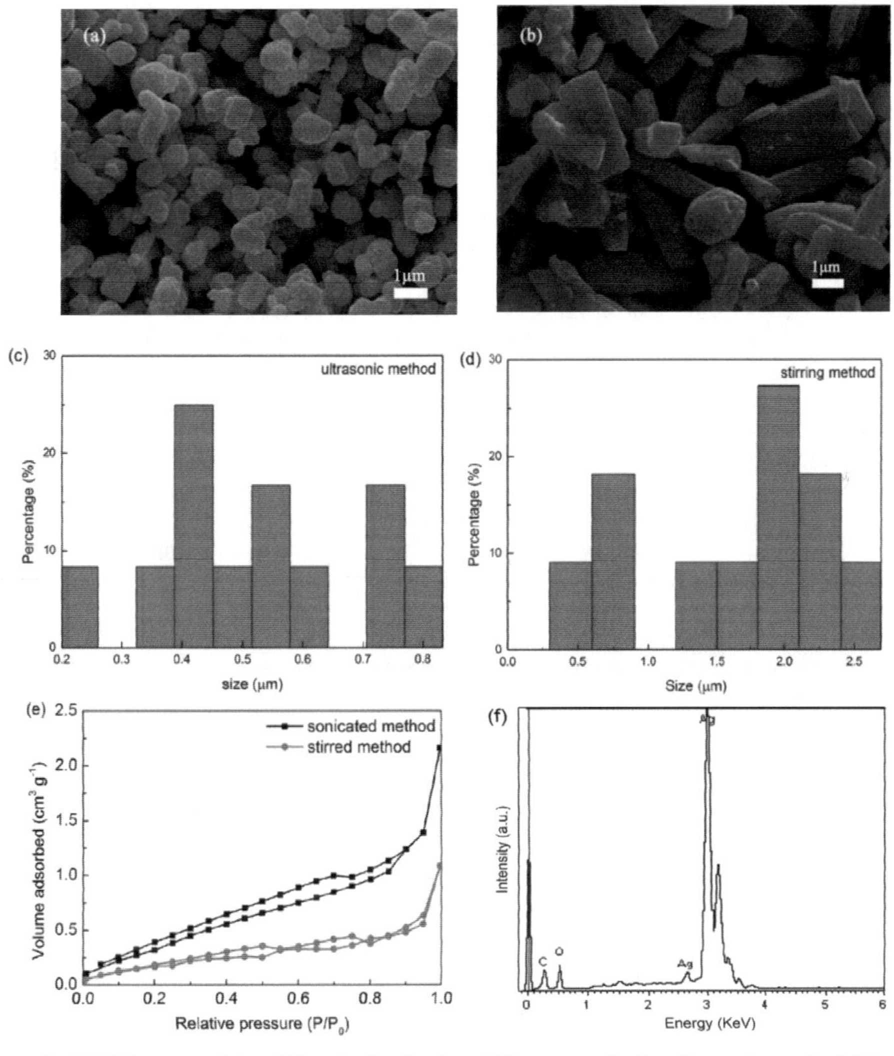

Figure 3. SEM images of Ag_2CO_3 obtained using different methods: ultrasonic method (a), stirring method (b); Size distribution histogram of Ag_2CO_3 nanoparticles over ultrasonic (c) and stirring (d) methods calculated from the corresponding SEM image; N_2 adsorption and desorption isotherms of the as-prepared Ag_2CO_3 (e); EDS spectrum of Ag_2CO_3 obtained with ultrasonic method (f)

Figure 3(a) and 3(b) show the SEM images of Ag_2CO_3 synthesized through ultrasonic and stirring method. The uniform and small size of Ag_2CO_3 nanomaterials can be clearly observed (Figure 3(a) and 3(c)). Figure 3(f) shows the EDS spectrum of the Ag_2CO_3 sample that contains C, O and Ag element, which indicates that the preparation process does not bring any impurities. To consider the effect of ultrasound, comparative experiments such as conventional stirring instead of the ultrasonic treatment were performed, as shown in Figure 3(b). Compared with Ag_2CO_3 formed by ultrasonic treatment (Figure 3(a)), the Ag_2CO_3 formed by conventional stirring did not show uniform structure, the main form of Ag_2CO_3 consist of irregular nanorods 1 2.5 μm (Figure 3(b) and (d)). N_2 adsorption and desorption isotherms and the Brunauer-Emmett-Teller (BET) specific surface area of the samples was measured (Figure 3(e)). Specific surface area 5.265 m^2 g^{-1} of Ag_2CO_3 was bigger than 2.166 m^2 g^{-1} of stirring method. And the porous structure of Ag_2CO_3 prepared via ultrasonic method is illustrated. The above results indicate that a more uniform small size of Ag_2CO_3 nanostructures have been formed by the sonochemical route without the use of any surfactants or templates, which means that high-intensity ultrasonic irradiation is necessary for the shape-controlled synthesis [22].

Figure 4(a) and 4(b) show the absorption spectral changes of MB by different samples. After stirring for 30 min in dark, the adsorption capacities of ultrasonic method (19.5%) is higher than stirring method (10.4%), which might be due to the bigger specific surface areas. The large specific surface areas are useful for the better adsorption of organic compounds and also provide a greater number of reactive sites for the photocatalytic process, thereby enhancing the photocatalytic activity [23]. In addition, the samples formed by stirring method exhibits obvious photocatalytic activity with degradation rate of 59.4% after visible-light irradiation of 20 min (Figure 4(c)). Dong [9] reported that all of MB (100 mL 10 mg/L) was degraded after 30 min by 0.05 g samples prepared with stirring method. Figure 4(d) indicated that MB solution was degraded within 30 min under the same amount samples and solution, which means pH will strongly influence the photocatalytic performance. Song [24] reported that about 65% of MB (100 mL 20mg/L) was degraded in 60 min by 0.075 g samples prepared with stirring method. After changed the method to ultrasonic, the photocatalytic activity has been further improved. Specifically, the degradation efficiency is 99.4% (0.1g samples) after visible-light irradiation of 20 min (Figure 4(c)). These results confirmed the excellent photocatalytic activity of Ag_2CO_3 prepared through an ultrasonic-assisted route. The DRS of the samples prepared using different method are examined and the results are showed in Figure 4(e). The two samples showed its fundamental absorption edge around 470 nm [25]. However, the transition band-gaps of two samples estimated from inset of the curve edges are about 1.58 and 1.98 eV. The absorption range of the samples using ultrasonic method apparently increased in the visible light and did have obvious red shifts to the long wavelength, which may be due to the narrow band-gap (1.58 eV) that caused by preparation condition and method. The absorption intensity of the samples in the visible light region became more prominent, which was in accordance with the color change. All above results suggested that ultrasonic cavitation played a vital factor in formation of Ag_2CO_3. Ultrasonic cavitation could provide high temperature, pressure and cooling rate, which lead to the smaller size, larger specific surface area, more active sites exposing and porous structure formation [26]. The smaller size and larger specific surface area can offer better adsorption capacities that make pollutants degrade in time. More active sites provide high activity of photocatalysts and porous structure makes catalysts good stability. Figure 4(f) shows the XRD patterns of Ag_2CO_3 after reactions. The new peak with 2θ values of 38.1 corresponding to (111) crystal plane of Ag (JCPDS 65-2871) is observed, which confirmed that Ag_2CO_3 prepared by stirring method is partially decomposed to Ag. However, the peak does not appear in samples that synthesized by ultrasonic method. The results indicate the better photocatalytic stability of Ag_2CO_3, which might be due to porous structure and reaction times of ultrasonic method for only 20 min and stirring method for 50 min.

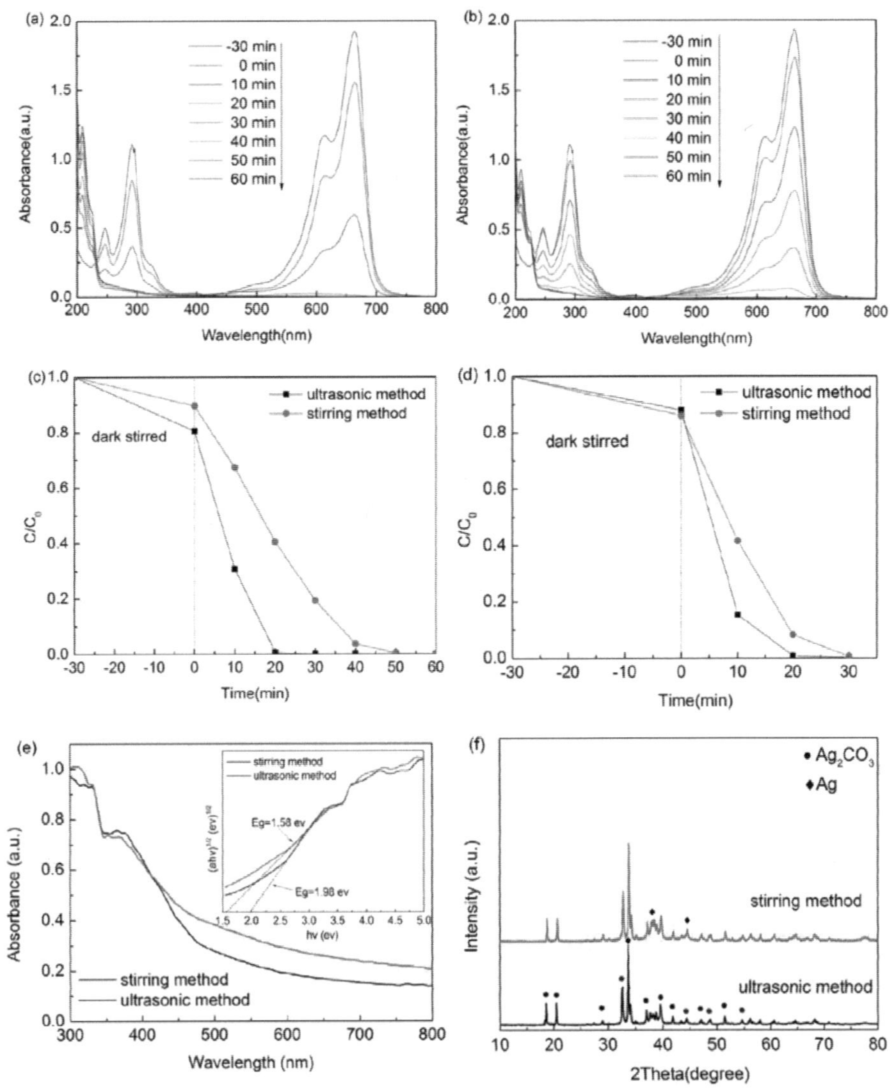

Figure 4. The absorption spectral changes of MB by the samples using ultrasonic method (a) and conventional stirring method (b); The degradation of MB (c) by the samples; The degradation of MB (100 mL 10 mg/L) by the 0.05 g samples (d); The UV-vis DRS and the bandgap (inset) by the samples (e); The XRD patterns of Ag_2CO_3 after reactions (f).

4. Conclusion

In summary, Ag_2CO_3 was successfully prepared via an easy and convenient sonochemical method. The effects of influencing factors on photodegradation ability were investigated at the first time. The optimum conditions are 9.0 value of pH, 60min of total sonication time and "1 s off / 9 s on" of sonication pulse mode. When Ag_2CO_3 nanoparticles were used as photocatalyst, the MB degradation percentage was 99.4% after 20 min irradiation of visible light. The DRS results and XRD patterns of

Ag_2CO_3 after reactions indicated that the Ag_2CO_3 nanoparticles synthesized by these conditions have high activity and good stability.

Acknowledgements
This work was supported by the Dr. Startup funds of Xinjiang University (BS150223) and National Natural Science Foundation of China (No. 21465022).

References

[1] H. Tong, S.X. Ouyang, Y.P. Bi, N. Umezawa, M. Oshikiri, J.H. Ye, *Adv. Mater.* **24** (2012) 229-251
[2] C. Cui, Y. Wang, D. Liang, W. Cui, H. Hu, B. Lu, S. Xu, X. Li, C. Wang, Y. Yang, *Appl. Catal. B-Environ.* **158-159** (2014) 150-160
[3] J.G. Yu, Q. Li, S.W. Liu, M. Jaroniec, *Chem. Eur. J.* **19** (2013) 2433-2441.
[4] K.F. Wu, H.M. Zhu, T.Q. Lian, *Acc. Chem. Res.* **48** (2015) 851-859.
[5] K. Wu, J. Chen, J.R. Mcbride, T. Lian, *Science* **349** (2015) 632 - 635.
[6] Q.J. Xiang, B. Cheng, J.G. Yu, *Angew. Chem. Int. Ed.* **54** (2015) 11350 - 11366.
[7] M. Soylu, A.A. Al-Ghamdi, W.A. Farooq, F. Yakuphanoglu, *Microelectron. Eng.* **154** (2016) 26-37.
[8] J. Wang, C. Dong, B. B. Jiang, K. L. Wu, J. Sun, X. Z. Li, W. J. Zhang, B. Zhang X. W. Wei, *Mater. Lett.* **131** (2014) 108-111
[9] H. Dong, G. Chen, J. Sun, C. Li, Y. Yu, D. Chen, *Appl. Catal. B-Environ.* **134-135** (2013) 46-54
[10] C.Y. Zeng, B.Z. Tian, J.L. Zhang, *J. Colloid, Interf. Sci.* **405** (2013) 17–21
[11] L. Han, P. Wang, C.Z. Zhu, Y.M. Zhai, S.J. Dong, *Nanoscale.* **3** (2011) 2931–2935
[12] J.H. Liu, X. Li, F. Liu, L.H. Lu, L. Xu, L.W. Liu, W. Chen, L.M. Duan, Z.R. Liu, *Catal. Commun.* **46** (2014) 138–141
[13] X.F. Wang, S. Zhan, Y. Wang, P. Wang, H.G. Yu, J.G. Yu, C.Z. Hu, *J. Colloid Interface Sci.* **422** (2014) 30–37
[14] Y.J. Jin, Z.Y. Dai, F. Liu, H.J. Kim, M.P. Tong, Y.L. Hou, *Water Res.* **47** (2013) 1837–1847
[15] J. Li, Q. Lu, J. Zhang, S. Zhao, *Mater. Sci. Semicond. Process.* **49** (2016) 54-60
[16] Z. Zarghami, M. Maddahfar, M. Ramezani, *J. Mater. Sci: Mater. Electron.* **26** (2015) 6339-6343
[17] Y. Liu, H. Yu, M. Cai, J. Sun, *Catal. Commun.* **26** (2012) 63-67
[18] P.Y. Lin, Z.S. Wu, Y.D. Juang, Y.S Fu, T.F. Guo, *Microelectron. Eng.* **149** (2016) 73-77
[19] J. Xie, C. Fang, J. Zou, H. Lu, C. Tian, C. Han, D. Zhao, *Mater. Lett.* **170** (2016) 62-66
[20] J. H. Bang, K. S. Suslick, *Adv. Mater.* **22.** (2010) 1039-1059
[21] A. Abulikemu, K. Okitsu, J.J. Zhu, *Ultrason. Sonochem.* **21** (2015) 1174-1181
[22] T.F. Yeh, J.M. Syu, C. Cheng, T.H. Chang, H. Teng, *Adv. Funct. Mater.* **20** (2010) 2255–2262
[23] J.S. Xie, L.S. Li, Y.C. Guan, H.D. Lu, C.L. Han, D.F. Zhao, C.A. Tian, *Mater. Lett.* **120** (2014) 54-57
[24] Y. Song, J. Zhu, H. Xu, C. Wang, Y. Xu, H. Ji, K. Wang, Q. Zhang, H. Li, *J. Alloys Compd* **592** (2014) 258-265
[25] J. Bao, S. Guo, J. Gao, T. Hu, L. Yang, C. Liu, J. Peng, C. Jiang, *RSC. Adv.* **118.** (2015) 97195-97204
[26] L. Xu, L.P. Jiang, J.J. Jun, *Nanotechnology.* **20** (2009) 045605

Experimental study on extinction performance of carbon nanotubes smoke to infrared radiation

Hongxia Wang[1], Lianfen Wang, Bin Xu and Haifei Zhu

Xi'an Research Inst. of Hi-Tech, Hongqing Town, Xi'an 710025, China

E-mail: redlightw@163.com

Abstract. Carbon nanotubes are one of the most typical materials in the nanoscale world. In order to study the IR interference performance of carbon nanotubes as smoke agent, using indoor large smoke box, the infrared extinction performance of three kinds of carbon nanotubes were measured in 8μm~12μm band. The smoke forming performance of carbon nanotubes were obtained by means of the testing of smoke mass concentration. Based on the experimental data, the dynamic mass extinction coefficients of three kinds of carbon nanotubes were calculated. The results show that carbon nanotubes smoke have good extinction performance to infrared radiation.

1. Introduction

Since the discovery of carbon nanotubes in 1991, the excellent physical and chemical properties of carbon nanotubes have aroused great interest and research upsurge. At present, the synthesis technology of carbon nanotubes has matured, especially the carbon nanotubes can be produced by chemical vapor flow catalytic method[1-3]. Both theoretical and experimental results show that the carbon nanotubes have specific properties, and the application prospect is immeasurable. Carbon nanotubes not only have the characteristics of small size effect, surface effect, quantum size effect and so on, and the unique spiral, tubular structure, high conductivity and electromagnetic loss tangent angle, making it also has than other nano materials more superior and unique wave absorbing properties and has a very broad application prospects. Preliminary study shows that the carbon nanotubes not only have a good absorption property for microwave, but also have a good absorption property in the infrared band[4-5]. The absorbing material, which is made of carbon nanotubes as absorbent, has the characteristics of wide frequency bandwidth, controllable, good compatibility and so on. It is a new generation of absorbing material with great potential. Preliminary results have been achieved in the application of the coating stealth field, but the study of carbon nanotubes as smoke agent to interfere infrared radiation, has rarely been reported in the literature.

In view of the carbon nanotubes have many advantages such as small scale, light weight, excellent levitation characteristics in the air and large specific surface area, absorption and scattering of electromagnetic wave in a wide band, etc. We carried out carbon nanotubes as smoke agent interfering materials research on anti infrared radiation. The extinction characteristics of several kinds of carbon nanotubes smoke were tested in the 8μm ~12μm band.

2. Measurement principle of mass extinction coefficient

Scattering and absorbing of smoke agent on infrared radiation totally called extinction. The extinction properties commonly descripted using mass extinction coefficientα (m^2/g), which is defined as the extinction cross section per unit mass of smoke particles. The greater the extinction coefficient of the material becomes, the better the attenuation effect of infrared radiation will be. Define I_0 as infrared radiation incident intensity, through the smoke the strength is I, according to the Lambert-Beer law:

$$T = \frac{I}{I_0} = I_0 exp(-\alpha C_m L) \quad (1)$$

Where, T is transmission rate, C_m is smoke mass concentration(g/m^3), L is optical path(m). From the above expression, the smoke concentration and path under certain circumstances, for a certain intensity of infrared radiation, the transmission rate is determined by the mass extinction coefficientα, so, α is an important parameter for evaluating smoke extinction performance. According to formula (1), the calculation formula of mass extinction coefficient can be obtained:

$$\alpha = \frac{1}{C_m L} ln\frac{1}{T} \quad (2)$$

Therefore, in the case of the known optical path, through the measurement of infrared radiation transmittance and smoke mass concentration, the values of smoke material mass extinction coefficientα can be obtained.

3. Test system

Test system shown in figure 1. The main test equipment is as follows:

Smoke box. The volume is 20m^3(6.1m×2m×1.8m), optical path L= 6.1m, two built-in mixing fan.

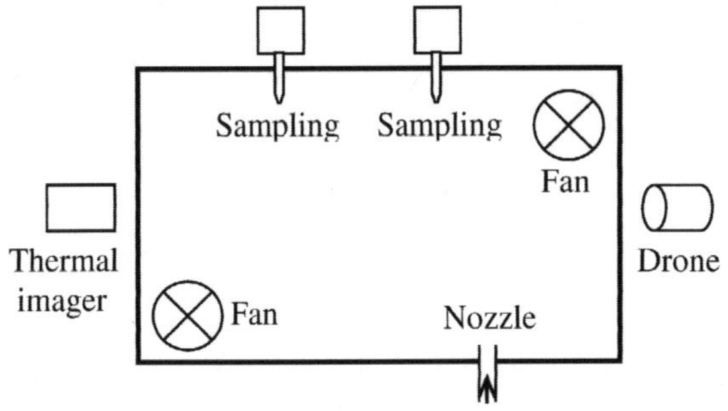

Figure 1. The test system of smoke box

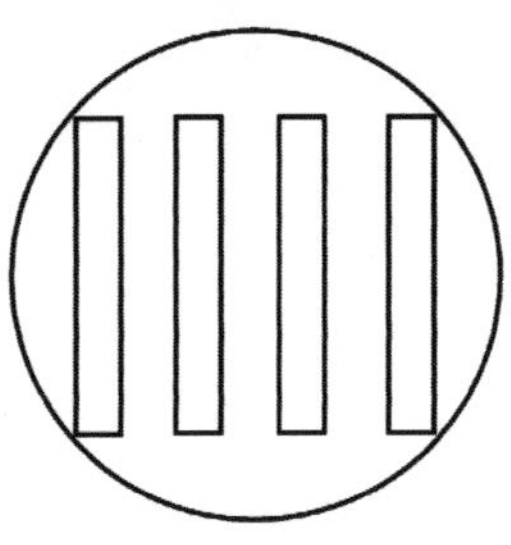

Figure 2. The structure of drone

Filter membrane concentration sampling system. The mass concentration of smoke is weighed by membrane weighing method. Extracting a volume of smoke in the smoke box, making it through the glass fiber filters, the smoke particles are blocked in the filter paper, according to the ratio of membrane added mass to the sampling smoke volume, calculating the mass concentration of smoke solid particles. The calculation formula is as follows:

$$C_m = (m_1 - m_2) / Q \quad (3)$$

Where, m_1 is the value of the filtration membrane quality after sampling, m_2 is the value of the filtration membrane quality before sampling, Q is sampling flow, its value is 40L/min. In order to make the mass sampling concentration representative, in the different positions of smoke box testing light path set two sampling pump, at the same time sampling, and the average value is the smoke mass concentration. Sampling time is 1 minutes, a total of three times of sampling, the interval is 2 minutes.

Drone. Drone provides an infrared heat source with a constant temperature difference of 30°C at room temperature, the temperature control precision is ±0.01°C, the drone shape is shown in Figure 2.The infrared heat radiated by drone pass through the smoke box and received by the infrared thermal imager. Finally, infrared thermal image of drone are formed on the monitor.

Thermal Imager. Working waveband 8μm ~12μm, external video recorder and monitor.

Drone and infrared thermal imager are arranged at both ends of the smoke box, the infrared thermal image of the drone in the absence of interference, after being accepted by the thermal imager, a clear infrared image of black and white of the drone on the monitor. After spraying smoke material, using the absorption and scattering of smoke particles to drone infrared radiation energy, eliminate difference between drone and background infrared radiation, so that the infrared thermal imager can't form normal infrared drone image, the shielding condition of smoke to infrared drone image can be visually observed on the monitor, according to the drone image gray value at different times. The transmittance curve can be obtained through the quantitative calculation of smoke transmittance.

4. Test result

There are three kinds of carbon nanotubes used in the test, the size and the performance parameters are shown in Table 1.

Table 1. The performance parameters of carbon nanotubes.

Sample number	Material name	Diameter(nm)	Length(μm)	specific surface area ($m^2 \cdot g^{-1}$)
①	carbon nanotubes	10-30	1-2	40-300
②	carbon nanotubes	40-60	1-2	40-300
③	carbon nanotubes	60-100	5-15	40-300

4.1. *Smoke mass concentration*

Weighing each size of materials 20g, using the high-pressure airflow sprayed into the smoke box, through sampling, weighing, the smoke mass concentration at different time is calculated according to equation (3), as shown in Table 2.

Table 2. The test results of smoke mass concentration.

Sample number	Sampling sequence	mass concentration C_m ($g \cdot m^3$)		$\overline{C}_m$($g \cdot m^3$)
		C_1	C_2	
①	1	0.392	0.516	0.454
	2	0.279	0.335	0.307
	3	0.178	0.253	0.215
②	1	0.443	0.559	0.501
	2	0.215	0.202	0.208
	3	0.106	0.150	0.128
③	1	0.315	0.429	0.372
	2	0.197	0.266	0.232
	3	0.141	0.168	0.154

For some smoke interference materials, the effect of interference is more obvious when the smoke mass concentration becomes higher, which the number of smoke particles is more in the unit volume. So, to improve the smoke mass concentration can improve the overall effect of the smoke extinction. As shown in table 2, the smoke mass concentration formed by ① and ② carbon nanotubes higher than the ③ carbon nanotube. It is concluded that the thin and short carbon nanotubes have good performance of smoke forming.

4.2. Shielding of infrared drone

When the carbon nanotubes are not sprayed, the infrared target is clearly visible on the monitor, as shown in Figure 3 (a). After spraying 20g carbon nanotube materials in the smoke box, as a result of scattering and absorption of infrared radiation by carbon nanotubes, the thermal imager to receive the infrared drone image is disturbed, the image becomes fuzzy straight completely disappear, figure 3 (b) for the drone image is completely covered. With the passage of time, due to the smoke particles of natural sedimentation, so that the smoke concentration decreased gradually in the smoke chamber, and the shadowing effect weakening, the drone image on the monitor gradually emerge and clear, figure 3 (c) is the drone image after the appearance.

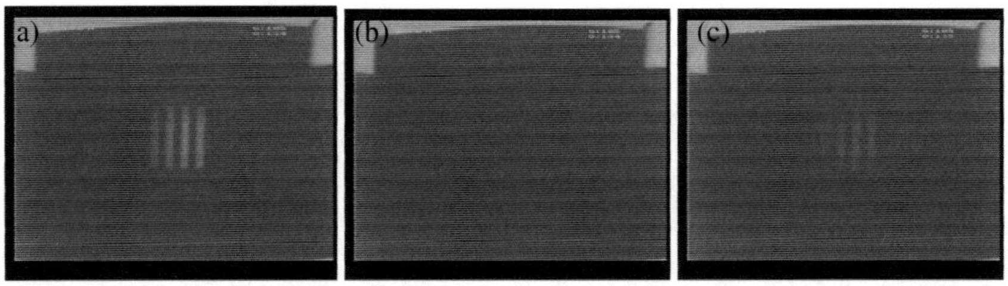

Figure 3. The shielding of carbon nanotubes smoke to IR drone.(a) Original drone image;(b) The image when smoke completely obscured drone; (c) The image after appearance of drone

4.3. Transmission rate

The thermal imager infrared detector output signal is proportional to the received irradiance. Therefore, ray intensity on the CRT screen is also proportional to the irradiance received by detector, strong radiation at high temperature parts, corresponding to the infrared thermal images bright parts, low-temperature parts corresponding to grey area. Therefore, according to the gray image formed by thermal imager, the smoke transmission on infrared radiation can be calculated by the following formula:

$$T = \frac{h_2(t) - h_1(t)}{h_2(0) - h_1(0)} \times 100\% \tag{4}$$

Where, $h_1(0)$ and $h_1(t)$ respectively represent the gray value of the dark zone of infrared thermal images before and after throwing up a smokescreen, $h_2(0)$ and $h_2(t)$ respectively represent the gray value of the bright zone of infrared thermal images before and after throwing up a smokescreen. Select an infrared thermal image interval of 10s and to calculate the arithmetic mean value of the corresponding transmission rate of each part of the drone as the transmittance. The smoke transmission curve of three kinds of carbon nanotubes is shown in Figure 4.

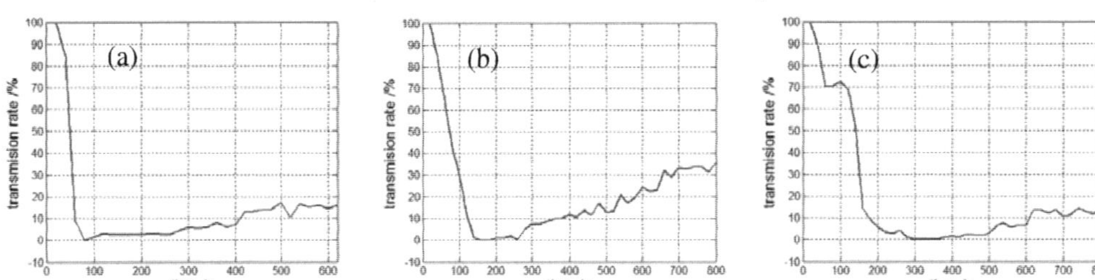

Figure 4. The transmission rate curve. (a) sample ①; (b) sample ②; (c) sample ③

4.4. Mass extinction coefficients

According to the smoke mass concentration shown in Table 2 and the corresponding average transmission rate, the mass extinction coefficients can be calculated based on equation (2).Table 3 is the mass extinction coefficient of three kinds of carbon nanotubes in 8μm~12μm wave band at different time. As can be noted from table 3, the mass extinction coefficient of the three carbon nanotube materials are all greater than 1 $m^2 \cdot g^{-1}$. The mass extinction coefficient of Sample ③ is greater than those sample ① and sample ②.

Table 3. Infrared extinction coefficients of carbon nanotubes smoke to 8μm~12μm.

Sample number	Time /s	$C_m/g \cdot m^{-3}$	T / %	α / $m^2 \cdot g^{-1}$	$\overline{\alpha}$ / $m^2 \cdot g^{-1}$
①	1~2	0.454	2.6813	1.3067	1.4086
	4~5	0.307	6.8382	1.4325	
	7~8	0.215	14.2310	1.4867	
②	1~2	0.501	1.2970	1.4218	1.4755
	4~5	0.208	14.3384	1.5308	
	7~8	0.128	31.8129	1.4668	
③	1~2	0.372	1.3186	1.9075	1.9519
	4~5	0.232	6.6938	1.9107	
	7~8	0.154	14.2613	2.0733	

5. Conclusions

Based on the Lambert Beer law, infrared interference extinction performance of three kinds of carbon nanotubes in 8μm~12μm wave band have been tested using indoor smoke box. The mass extinction coefficients in this wave band are 1.4086 m^2/g、1.4755 m^2/g and 1.9519m^2/g respectively. It shows that the carbon nanotube material has a good infrared extinction performance. Although the thin and short carbon nanotubes sample ① and sample ② have better performance in forming smoke than sample ③, but the mass extinction coefficient is lower than sample ③. Although the mass extinction coefficient of carbon nanotube sample ③ is big, the mass concentration in the smoke box is not high, the maximum value is 0.372$g \cdot m^{-3}$, and with the increase of time decreased significantly. It shows that the smoke forming and suspension performance of carbon nanotube material need to be further improved.

References

[1] Colomer J F, Sterhan C and Lefrant S 2000 Large-scale Synthesis of Single-wall Carbon Nanotubes by Catalytic Chemical Vapor Deposition (CCVD) Method *Chem Phys Lett* **317**:83-89
[2] Endo M, Hayashi T and Kim Y A 2006 Large-scale production of carbon nanotubes and their applications *Pure and applied chemistry* **78**: 1703-1713.
[3] Zheng G B, Kouda K and Sano H 2004 A model for the structure and growth of carbon nanofibers synthesized by the CVD method using nickel as a catalyst *Carbon* **42**: 635-640.
[4] Sfeir M Y, Beetz T and Wang F 2006 Optical spectroscopy of individual single-walled carbon nanotubes of defined chiral structure *Science* **312**: 554-556.
[5] Wang H X, Song Z B and Liu D Z 2013 Study on Attenuation Characteristic of Carbon Nanomaterials Smoke to Laser *Advanced Materials Research* **712**: 215-219.

Enhanced electrochromic properties of TiO_2 nanoporous film prepared based on an assistance of polyethylene glycol

Shunjian Xu[1], Xiaorui Luo, Zonghu Xiao, Yongping Luo, Wei Zhong, Hui Ou, and Yinshuai Li

Xinyu Institute of New Energy, Xinyu University, Xinyu 338004, China

E-mail: xushunjian@126.com

Abstract. Polyethylene glycol (PEG) was employed as pore-forming agent to prepare TiO_2 nanoporous film based on spin-coating a TiO_2 nanoparticle mixed paste on fluorine doped tin oxide (FTO) glass. The electrochromic and optical properties of the obtained TiO_2 film were investigated by cyclic voltammetry (CV), chronoamperometry (CA) and UV–Vis spectrophotometer. The results show that the PEG in the mixed paste endows the TiO_2 film with well-developed porous structure and improves the uniformity of the TiO_2 film, which are helpful for the rapid intercalation and extraction of lithium ions within the TiO_2 film and the strengthening of the diffuse reflection of visible light in the TiO_2 film. As a result, the TiO_2 film derived from the mixed paste with PEG displays higher electrochemical activity and more excellent electrochromic performances compared with the TiO_2 film derived from the mixed paste without PEG. The switching times of coloration/bleaching are respectively 10.16/5.65 and 12.77/6.13 s for the TiO_2 films with PEG and without PEG. The maximum value of the optical contrast of the TiO_2 film with PEG is 21.2% while that of the optical contrast of the TiO_2 film without PEG is 14.9%. Furthermore, the TiO_2 film with PEG has better stability of the colored state than the TiO_2 film without PEG.

1. Introduction

Electrochromic materials, showing a reversible color and transparency change during electrochemical oxidation and reduction, have displayed a great potential for applications, such as smart windows, low-power displays, auto-mobile anti-glare rearview mirror and e-papers [1-3]. In general, electrochromic materials can be divided into organic and inorganic ones according to the material components. In 1961, the pioneered work performed by Piatt [4] marks the investigation start of the inorganic electrochromic materials. Recently, motivated by the practical requirements of electrochromic devices, much effort has been dedicated to improve the electrochromic properties of the inorganic electrochromic materials, such as the optical contrast, coloration efficiency and response time.

As an important type of the inorganic electrochromic materials, many transition-metal oxides with variable oxidation state, including nickel oxide (NiO), niobium pentoxide (Nb_2O_5), molybdenum trioxide (MoO_3) and titanium oxide (TiO_2), were reported due to their high refractive indexes, good chemical and thermal stabilities and wide bandgaps [5-7]. Among them, the TiO_2 has exhibited particular potential as an electro-active material, because of its high activity, strong oxidation capability and superior chemical stability. However, there still exist many shortcomings for the TiO_2 electrochromic film, such as slow switching speed and poor optical contrast. Recently, it is recognized that the fabrication of the TiO_2 electrochromic film with ordered structure or nanoporous structure provides a useful technique for enhancing the electrochromic properties [8, 9]. In this work TiO_2 nanoporous film was synthesized via an assistance of polyethylene glycol (PEG) as pore-forming agent based on spin-coating a TiO_2 nanoparticle mixed paste on fluorine doped tin oxide (FTO) glass. The PEG in the mixed paste not only endowed the TiO_2 film with well-developed nanoporous structure, but also improved the uniformity of the TiO_2 film. Therefore, the electrochromic performances of the obtained TiO_2 film were enhanced.

2. Experimental

2.1. Preparation of TiO_2 nanoporous film

The TiO_2 nanoparticles (particle size: ~30 nm) were obtained from Alfa Aesar chemical Co. LTD, and used without further purification. First, 2 g of TiO_2 nanoparticles was ground in a mortar with 0.8 g of PEG (MW: 20000), 15 mL of deionized water, 0.2 mL of acetylacetone, 1 mL of alcohol and 1 mL of 10% Triton X−100 aqueous solution to obtain a TiO_2 mixed paste with PEG. For comparison, a TiO_2 mixed paste without PEG was achieved by the same process. Then, both of the mixed pastes were coated on FTO glass by spin-coating. After being sintered at 450 °C for 30 min, the film was treated with 40 mM $TiCl_4$ aqueous solution followed by heat treatment at 450 °C for 30 min. The TiO_2 films derived from the mixed pastes with PEG and without PEG are labeled as NTF-1 and NTF-2, respectively.

2.2. Characterizations and measurements

The morphology of the TiO_2 film was characterized by an optical microscope (OM) and a field emission scanning electron microscopy (FE-SEM; JSM-6701F, JEOL, Japan). The transmittance of the TiO_2 film on FTO at different wavelengths was obtained by using a UV–Vis spectrophotometer (TU-1901, China). A three-electrode cell was used for both of the cyclic voltammetry (CV) and chronoamperometry (CA) measurements. The cell included a platinum counter electrode, an Ag/AgCl reference electrode and a 1 M $LiClO_4$ dissolved in propylene carbonate as electrolyte. The TiO_2 film under study was employed as a working electrode. The CV measurement was realized at 50 mV/s scan rate between –2 and 2 V potential range. A coloration potential of –2 V and a bleaching potential of +2.0 V were applied for 30 s one after another in CA measurement.

3. Results and discussion

The electrochromic performances of the TiO_2 films of NTF-1 and NTF-2 were measured using a three-electrode electrochemical cell in 1 M $LiClO_4$ in PC electrolyte solution. Figure 1(a) shows the transmittance of both the TiO_2 films on FTO glasses in the bleached and colored states with applying potentials of +2 and –2 V, respectively. In the bleached state, the transmittance of NTF-1 is lower than that of NTF-2 in a wavelength range between 400 and 700 nm, indicating the existence of more well-developed porous structure in the former TiO_2 film owing to its stronger diffuse reflection of visible light. The transmittance of NTF-1 and NTF-2 in the bleached state at a wavelength of 550 nm is 69.3% and 73.5%. When the potential is switched to -2 V, the color of both the TiO_2 films quickly changes from colorless to blue. The transmittance of NTF-1 and NTF-2 in the colored state at a wavelength of 550 nm is 56.5% and 61.4%. The optical contrast (i.e. the transmittance difference between the film in the bleached state and in the colored state) of the TiO_2 films of NTF-1 and NTF-2 is displayed in Figure 1 (b). As a whole, NTF-1 derived from the mixed paste with PEG shows a larger transmittance modulation than NTF-2 derived from the mixed paste without PEG. The maximum value of optical contrast is 21.2% (at 400 nm) and 14.9% (at 700 nm) for NTF-1 and NTF-2, respectively.

Figure 2 exhibits the stability of the colored TiO_2 films of NTF-1 and NTF-2 under a condition without applying any potential. With the prolonged time, the color of both the TiO_2 films gradually changes from blue to colorless, as shown in the inserts in Figure 2. When the time is 25 min, the relative transmittance (T_t/T_0) (1.018) of NTF-1 is slightly lower than that (1.027) of NTF-2. It indicates that the TiO_2 film prepared via an assistance of PEG possesses stronger storage capacity of lithium (Li^+) ions and better stability.

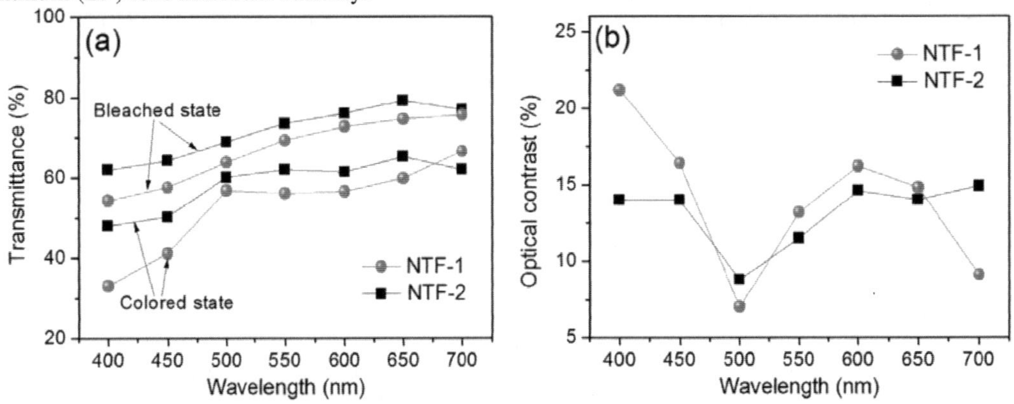

Figure 1. Transmittance (a) and optical contrast (b) of TiO2 films derived from mixed pastes with PEG (NTF-1) and without PEG (NFT-2).

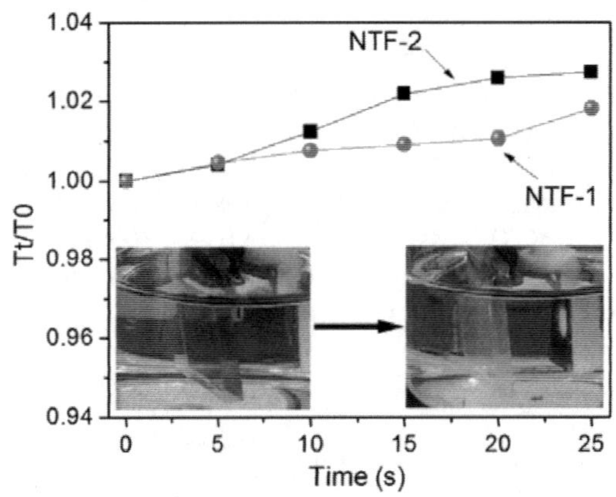

Figure 2. Stability of the coloredTiO2 films derived from mixed pastes with PEG (NTF-1) and without PEG (NFT-2) in the released state of potential. The inserts show the photos of NTF-1 at time of 0 and 25 min.

Response time is an important parameter to measure the speed that electrochromic film varies between colored and bleach states by applying different potentials. The switching kinetics is investigated by evaluating the times defined as the requirement for 90% of the total current density change. Figure 3(a) demonstrates the CA curves of the TiO_2 films of NTF-1 and NTF-2. From Figure 3(a), the switching times are 10.16 s for coloration and 5.65 s for bleaching for NTF-1, significantly quicker than those measured for NTF-2. The switching times of the coloration and bleaching of the latter are 12.77 and 6.13 s, respectively. These results indicate that a fast switching speed is achieved within the TiO_2 film prepared via an assistance of PEG (NTF-1) because of the rapid Li^+ ion intercalation and extraction, a thesis that could be supported by the results of the CV measurements. The CV curves of the TiO_2 films of NTF-1 and NTF-2 are revealed in Figure 3(b). The cathodic and anodic peaks of the TiO_2 film appear at around –1.5 and –0.5 V, respectively, which are assigned to Li^+ ion insertion into and extraction out of the TiO_2 framework. The electrochemical reaction can be expressed as $\mathrm{TiO_2} + x\mathrm{Li}^+ + xe^- \leftrightarrow \mathrm{Li}_x\mathrm{TiO_2}$ [10]. It is well known that the insertion and extraction of

Li^+ ions and electrons from a film result in coloration and bleaching, respectively. It is obvious from Figure 3(b) that the peak current density of NTF-1 is higher than that of NTF-2, suggesting that the TiO_2 film based on an assistance of PEG shows a higher electrochemical activity, which is associated with its superior electrochromic performances.

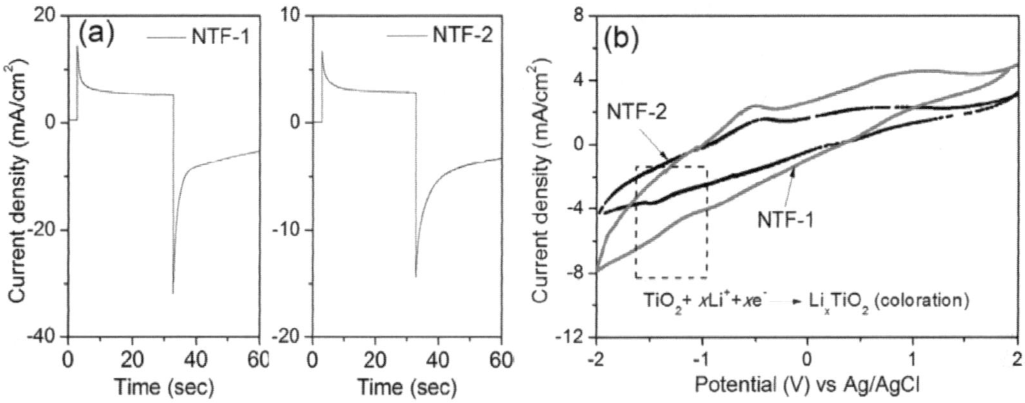

Figure 3. CA (a) and CV (b) curves of TiO2 films derived from mixed pastes with PEG (NTF-1) and without PEG (NFT-2).

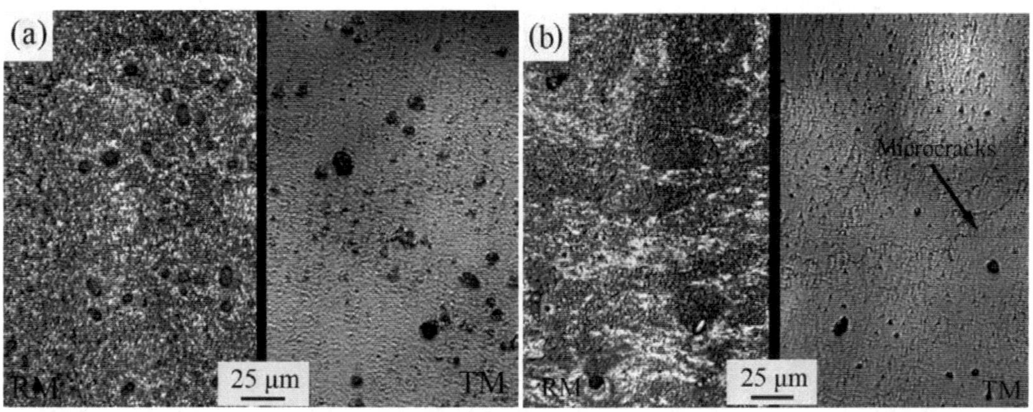

Figure 4. OM morphologies of TiO2 films derived from mixed pastes with PEG (a) and without PEG (b) in reflection mode (RM) and in transmittance mode (TM).

Figure 4 exhibits the OM morphologies of the TiO_2 films of NTF-1 and NTF-2 in reflection mode (RM) and in transmittance mode (TM). It is obvious from Figure 4 that the morphological difference of two TiO_2 films could be observed. NTF-1 based on an assistance of PEG is more uniform compared to NTF-2 in spite of the existence of slightly more macropores with a size of 1~10 μm (the black phases) in the former TiO_2 film. Moreover, the microcracks appeared on NTF-2 are not observed on NTF-1. These results suggest that the PEG with high viscosity in the mixed paste is beneficial for the uniformity of the TiO_2 film. The SEM image in Figure 5 further proves that the TiO_2 film of NTF-1 based on an assistance of PEG possesses well-developed nanoporous structure, which provides effective channels for Li^+ ion diffusion to further participate in the electrochemical reaction. From above analyses, we could carefully conclude that the excellent performances of the TiO_2 film based on an assistance of PEG are attributed to the well-developed nanoporous structure and high uniformity of the TiO_2 film induced by PEG to a great extent.

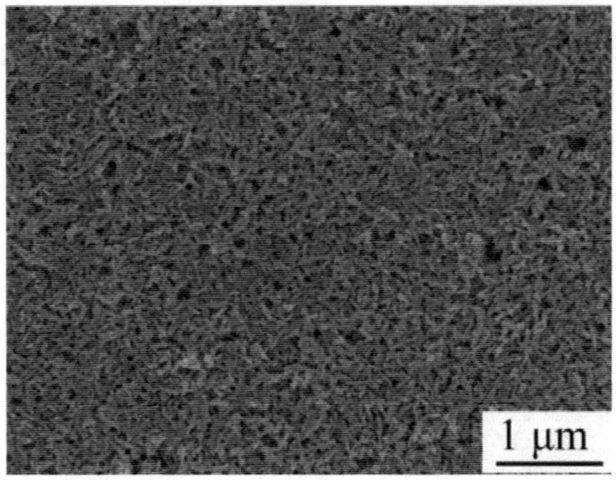

Figure 5. SEM image of TiO_2 film prepared via an assistance of PEG (NTF-1).

4. Conclusion

TiO_2 nanoporous film is synthesized using PEG as pore-forming agent based on spin-coating a TiO_2 nanoparticle mixed paste on FTO glass. The TiO_2 film derived from the mixed paste with PEG displays higher electrochemical activity and more excellent electrochromic performances compared with the TiO_2 film derived from the mixed paste without PEG. The switching times of the coloration/bleaching are respectively 10.16/5.65 and 12.77/6.13 s for the TiO_2 films with PEG and without PEG. The maximum value of the optical contrast of the TiO_2 film with PEG is 21.2% while that of the optical contrast of the TiO_2 film without PEG is 14.9%. Furthermore, the TiO_2 film with PEG possesses stronger storage capacity of lithium (Li^+) ions and better stability. The outstanding performances could be attributed to the well-developed nanoporous structure and the high uniformity of the TiO_2 film induced by PEG to a great extent.

Acknowledgements

This work was financially supported by National Natural Science Foundation of China (51462035), Training Programme Foundation for Young Scientist of Jiangxi, China (20133BCB23035), Natural Science Foundation of Jiangxi (20161BAB206106) and Educational Commission of Jiangxi, China (KJLD13100).

References

[1] Falahatgar S, Ghodsi F, Tepehan F, Tepehan G, Turhan L and Pishdadian S 2015 *J. Non-Cryst. Sol.* **427** 1
[2] Fu X, Jia C, Wan Z, Weng X, Xie J and Deng L 2014 *Org. Electron.* **15** 2702
[3] Zhang W, Ju W, Wu X, Wang Y, Wang Q, Zhou H, Wang S and Hu C 2016 *Appl. Surf. Sci.* **367** 542
[4] Platt J 1961 *J.Chem.Phys.* **34** 862
[5] Pehlivan E, Koc K, Tepehan F and Tepehan G 2016 *J. Sol-Gel Sci. Technol.* **77** 172
[6] Shi J, Lai L, Zhang P, Li H, Qin Y, Gao Y, Luo L and Lu J 2016 *J. Solid State Chem.* **241** 1
[7] Niu W, Wang G, Liu X, Tang J and Bi X 2015 *Int. J. Electrochem. Sci.* **10** 2613
[8] Lee K, Kim D, Berger S, Kirchgeorg R and Schmuki P 2012 *J. Mater. Chem.* **22** 9821
[9] Patil R, Devan R, Liou Y and Ma Y 2016 *Sol. Energ. Mat. Sol. C.* **147** 240
[10] Liu S, Zhang X, Sun P, Wang C, Wei Y and Liu Y 2014 *J. Mater. Chem.* C **2** 7891

Catalytic oxidation of dye waste water by biomass charcoal loaded multiple rare earth composite material

Suriga[1] and Liping CHEN[2,3]

[1]College of Chemistry and Environment Science, postgraduate, Inner Mongolia Normal University, Hohhot 010022, China
[2]College of Chemistry and Environment Science, Inner Mongolia Normal University, Hohhot 010022, China

E-mail: clp@imnu.edu.cn

Abstract. The main purpose of this study is to investigate the individual effect as well as the interactions of different influencing factors like catalyst dosage, aeration rate, temperature and pH on the removal of methylene blue (MB) using biomass charcoal loaded multiple rare earth composite material. Design-Expert 7.0 was used to design testing program and establish response surface model. The result showed that among the factors, catalyst dosage played the most important role, then pH value, aeration rate and temperature in turn. By the optimization of process parameters, the optimum experimental conditions were catalyst dosage of 2.50 g, aeration rate of 2.5 $L\cdot min^{-1}$, temperature of 21 ℃ and pH value of 12, under these optimum conditions, maximum predicted and observed decolorization rate were 100.00% and 99.61%, the observed value was well match with the predicted value.

1. Introduction

With growing urbanization and industrialization, the problem of the dye waste water has been of increasing concern in many parts of the world. China discharge 1.6 hundred million m^3 of dye waste water every year [1].They damage the environment and contain many chemicals that are toxic, mutagenic, hazardous and cancerogenic [2,3].They also block light penetration, destroy the water quality of the receiving streams, and deliver hazardous and toxic materials to organisms in the food chain[4].Therefore, dye waste water needs to be treated for the sake of the protection of human beings health and environmental safety. For the practical purpose of water treatment, it was necessary to investigate an efficient, practical method for application in industrial degradation. Dyes was very stable and difficult to be biodegraded because of the synthetic origin and complex aromatic molecular structures, chemical coagulation and biodegradation lead to secondary pollution [5,6]. Recently, the treatment of waste water through catalytic oxidation techniques has proven to be simple, cheap, efficient and nontoxic [7].

Biomass charcoal is a solid waste produced in the agriculture production process. In China, the biomass charcoal has discharged increasing numbers with the increase of sunflower planting, its output is 11.8 billion ton every year in china. Biomass charcoal is a kind of low cost and easily available catalyst and often used in wastewater treatment due to its large surface area, porosity, high adsorption capacity and good stability [8].

In order to study the optimal parameters for catalytic oxidation of MB, we used response surface methodology to design testing program and establish response surface model. RSM is a useful method which studying the effect of individual factors as well as their interaction by a limited number of experiments. So, the optimum conditions for desirable responses could be obtained.

In our previous study, biomass charcoal loaded multiple rare earth composite material has been prepared for the removal of MB and investigated the influences of catalyst dosage, aeration rate, temperature and pH on the catalytic oxidation process through single factor test, then obtained the optimal conditions by using RSM and then briefly discussed the experimental result. The results provides a theoretical basis for preparation of the biomass charcoal composite material and catalytic oxidation study.

2. Experimental

2.1. Materials and chemicals

Sunflower straw was taken from the Hetao region of Inner Mongolia; Mixed rare earth was obtained from Baotou of Inner Mongolia. The rare earth oxides and their contents in mixed rare earth used in this study are given in table 1. It was evident from the table that the mixed rare earth were rich in CeO_2 of 52% and La_2O_3 of 28%. That is, CeO_2 played a bigger role in mixed rare earth for catalytic oxidation of MB. The other chemicals were methylene blue, phosphoric acid (15 mol·L^{-1}), HCl (0.1 mol·L^{-1}), NaOH (0.1 mol·L^{-1}), sodium silicate, distilled water.

Table 1. Oxides and their contents in mixed rare earths.

Oxides	Contant / (wt %)
La_2O_3	28
CeO_2	52
Pr_6O_{11}	5
Nd_2O_3	15

2.2. Preparation of Biomass Charcoal

Select the sunflower straw before preparation and sheared to desired size fragment (1-2 cm), then fully washed with distilled water several times and oven dry. The samples were immersed in phosphate solution for 12 h with an impregnation rate of m (sunflower straw): m (phosphate) = l: 1.8, then placed the mixture into the oven and activation for 24 h at 150℃, remove the mixture and rinsed with distilled water until the pH was neutral, oven dry at 105 ℃, Then, the catalyst was ground and sieved to a particle size of 120 meshes for further studies.

2.3. Preparation of Multiple Rare Earth/Biomass Charcoal Composite Material

Activated carbon, mixed rare earth oxide powder and sodium silicate were mixed with the quality ratio of 40:1:1, appropriate amount of water added into the mixture and transferred into the sludge, heating to 70-80 ℃ and washing with distilled water until the pH value neutral, drying, grinding 1-2 hours, 120 mesh sieve, then get the composite materials.

2.4. Catalytic oxidation experiment

30 mL MB solution (500 mg·L^{-1}) was added into the beaker, adjusted the pH by the NaOH solution(0.1 mol·L^{-1})and HCl solution(0.1 mol·L^{-1}),then 0.1 g composite material was added into above solution to form a mixed system with surging for 3 h on a ED-85A thermalstat oscillator,1 mL filtrate was separated from the solution and measuring the concentration of MB by 722N visible spectrophotometer at 665 nm. The decolorization rate was calculated by the following equation (1):

$$\eta = (1 - \frac{c}{c_0}) \times 100\% \quad (1)$$

Where η (%) represented the decolorization rate of MB, c_0 (mg·L^{-1}) represented the initial concentration of MB, c (mg·L^{-1}) represented the concentration of MB after reaction.

2.5. *Box–Behnken experimental design*

According to single-factor experiment, a four factor, three level Box–Behnken experimental design was established. The experimental levels of independent variables considered in this study were presented in **Table 2.**The levels of independent factors of the Box–Behken experimental design were chosen based on the single factor experiment, the experimental result was analyzed by response surface methodology.

3. Results and discussions

3.1. *Box–Behnken experimental result and ANOVA analysis*

The relationship between decolorization rate and four influencing factors (catalyst dosage, aeration rate, temperature and pH) and the interaction between variables were analyzed based on the box–behnken experimental design. The results were tabulated in Table 3. Final equation in terms of coded factors was evaluated and given in Eq. (2).

$$\eta = 97.47+3.83x_1-0.44x_2+0.000x_3+2.83x_4+1.69x_1x_2+5.00\times10^{-3}x_1x_3-4.58x_1x_4+5.00\times10^{-3}x_2x_3-0.44x_2x_4+0.023x_3x_4-2.09x_1^2-1.60x_2^2+3.86x_3^2-0.56x_4^2 \quad (2)$$

The significance of the model was studied by analysis of variance (ANOVA). The ANOVA for response surface quadratic model for MB removal was listed in Table 4.

F-value indicated the significant level of the model and homologous variables, F-value >3 indicates that the model or homologous variables are significant. In this Model, F-value of 3.16 implies the model is significant.

The homologous variables or interaction effect would be more significant if the p-value< 0.05, values greater than 0.1000 indicate the model terms are not significant, p-value of 0.0197 in this model implies the model is significant.

In this case, x_1、x_4、x_1x_4 and $x_3^2$are significant model terms. The significant level of the four factors is $x_1>x_4>x_2>x_3$. "Adeq Precision" measures the signal to noise ratio, a ratio greater than 4 is desirable. In this model, ratio of 7.918 indicates an adequate signal. This model can be used to navigate the design space.

Table 2. Independent variables and their coded levels.

Variables	Symble	Range and level		
		-1	0	1
Catalyst dosage/(g)	x_1	0.20	0.25	0.30
Aeration rate /(L·min^{-1}))	x_2	0.31	1.41	2.50
Temperature/(℃)	x_3	20	25	30
pH	x_4	8	10	12

Table 3. Box-Behken experimental design and experimental data.

Run	Coded variables				η /%	Run	Coded variables				η /%
	x_1	x_2	x_3	x_4			x_1	x_2	x_3	x_4	
1	-1	0	0	1	99.83	16	0	1	0	1	99.77
2	-1	0	1	0	99.93	17	0	0	-1	1	99.72
3	0	0	0	0	97.47	18	0	0	0	0	97.47
4	0	-1	0	1	99.81	19	1	0	0	1	99.81
5	0	0	0	0	97.47	20	0	0	1	1	99.81
6	-1	0	0	-1	81.07	21	1	0	-1	0	99.96
7	0	0	1	-1	99.95	22	0	1	1	0	99.94
8	0	0	0	0	97.47	23	0	1	-1	0	99.96
9	1	-1	0	0	99.84	24	1	0	0	-1	99.37
10	0	-1	1	0	99.91	25	-1	-1	0	0	89.39
11	1	0	1	0	99.95	26	0	-1	0	-1	91.38
12	0	1	0	-1	93.09	27	0	-1	-1	0	99.95
13	-1	0	-1	0	99.95	28	1	1	0	0	99.71
14	0	0	0	0	97.47	29	0	0	-1	-1	99.95
15	-1	1	0	0	82.52						

Table 4. ANOVA for response surface quadratic model analysis of variance.

Source	Sum of square	DF	Mean square	*F*-value	*P* value
Model	547.64	14	39.12	3.16	0.0197
x_1	175.95	1	175.95	14.20	0.0021
x_2	2.33	1	2.33	0.19	0.6710
x_3	0.000	1	0.000	0.000	1.0000
x_4	95.99	1	95.99	7.75	0.0147
x_1x_2	11.36	1	11.36	0.92	0.3546
x_1x_3	2.500E-005	1	2.500E-005	2.018E-006	0.9989
x_1x_4	83.91	1	83.91	6.77	0.0209
x_2x_3	1.000E-004	1	1.000E-004	8.071E-006	0.9978
x_2x_4	0.77	1	0.77	0.062	0.8073
x_3x_4	2.025E-003	1	2.025E-003	1.634E-004	0.9900
x_1^2	28.40	1	28.40	2.29	0.1523
x_2^2	16.61	1	16.61	1.34	0.2664
x_3^2	96.83	1	96.83	7.81	0.0143
x_4^2	2.06	1	2.06	0.17	0.6895
Residual	173.47	14	12.39		
Lack of fit	173.47	10	17.35		

3.2. Response surface analysis and optimization by RSM

The R^2 and adjusted R^2 values were found to be 0.7594 and 0.5189, which indicates that there was a good agreement between the actual and the predicted values[9], as shown in **figure 1.** This suggests that the obtained regression model is adequate to explain most of the variability for MB removal under a given experimental condition.

Three dimensional response surface was estimated the interactive effect of two variables on decolorization rate of MB. Figure 2 showed the effect of catalyst dosage and pH on efficiency of composite material for decolorization of MB when the temperature and aeration rate were coded factor (0.000). The interaction between catalyst dosage and pH has significant influence on MB removal with an P value<0.05.The capacity of composite material for the removal of MB can be determined by the catalyst dosage at any given pH, the decolorization rate gradually increased with the increase of the catalyst dosage and pH, and then reached a stable level. During the initial reaction stage, with the increase of catalyst dosage, surface area and dispersity of catalyst has increased, so the reaction becomes faster, to a certain dosage, reaction basically completed and decolorization rate had no obvious change. With the increase of dissolved oxygen concentration, a series of free radical reactions occurring on the surface of the catalyst [10]: O_2= O·+O·, O·+H_2O= HO·+ HO· and RH+O_2= R·+HOO·(RH represents the organic matter), HO· and HOO· are free radicals with strong oxidizing property, attack the -C=N-,-C=S- and other heterocyclic structures of MB, so that the MB conversion to two methyl aniline and further oxidized to two methyl amine, then be degraded into CO_2, H_2O and other small molecular substances eventually [11,12], at alkaline condition, the intermediate organic acids were further reacted with alkali, so the MB was degraded [13].

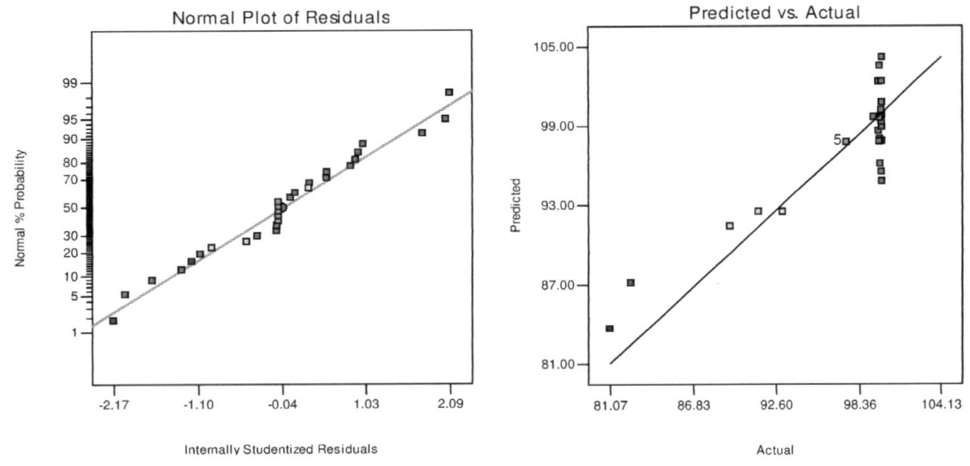

Figure 1. Plot of the actual response versus predicted response for decolorization rate.

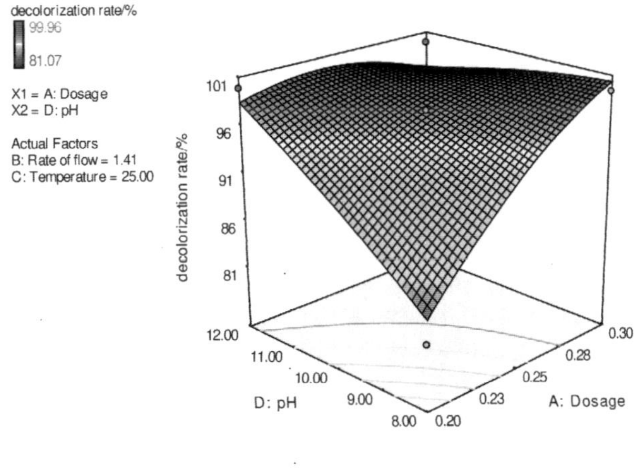

Figure 2. The three-dimensional response plot.

The model predicted a maximum decolorization rate of 100.00%, the optimum conditions were found to be as follows: catalyst dosage 2.50 g, aeration rate 2.5 L·min^{-1}, temperature 21 ℃ and pH value 12.

To confirm that the model is satisfactory in predicting the maximum decolorization rate of MB, an experiment was performed under the optimum conditions, the experimental result was 99.61%, which is in good agreement with the predicted value. It indicated that the response surface methodology could be an reliable and effective methods to optimizing the catalytic oxidation process of MB onto the biomass charcoal loaded multiple rare earth composite material.

4. Conclusion

The present investigation was carried to study the removal of MB using biomass charcoal loaded multiple rare earth composite material and to conduct process optimization using RSM for finding the optimum values of parameters affecting the process to achieve maximum removal efficiency. The significant level of the four factors is $x_1>x_4>x_2>x_3$, the interaction between catalyst dosage and pH was found to be a significant factor. The optimum conditions were found to be as follows: catalyst dosage 2.50 g, aeration rate 2.5 L· min^{-1}, temperature 21 ℃ and pH value 12, under these optimum conditions, the maximum decolorization rate of MB was 99.61%. The observed values were consistent with the theoretical values, leading support to the conclusion. The optimized result showed that the biomass charcoal loaded multiple rare earth composite material was supposed to be an environmentally friendly, economically and efficiency catalyst for the treatment of MB.

Acknowledgement

This research was financially supported by the Hohhot Natural Science Foundation (2014MS0221).

References

[1] Dong Li and Yin Guo-Jie 2008 J. Study on adsorption of methylene blue by modified coal gangue. Non-Metallic Mines 31 59-61

[2] Demirbas E and Nas M Z 2009 J. Batch kinetic and equilibrium studies of adsorption of Reactive Blue 21 by fly ash and sepiolite. *Desalination* **243** 8–21

[3] Hajati S, Ghaedi M and Yaghoubi S 2015 *J.* Local, cheep and nontoxic activated carbon as efficient adsorbent for the simultaneous removal of cadmium ions and malachite green: Optimization by surface response methodology. *Journal of Industrial and Engineering Chemistry* **21** 760–767

[4] Kah Aik Tan, Norhashimah Morad and Tjoon Tow Teng 2015 *J.* Synthesis of magnetic nanocomposites (AMMC-Fe_3O_4) for cationic dye removal: Optimization, kinetic, isotherm, and thermodynamics analysis. *Journal of the Taiwan Institute of Chemical Engineers* **54** 96–108

[5] Masoud Giahi , Roohan Rakhshaee and Mohammad A 2011 *J.* Removal of methylene blue by tea wastages from the synthesis waste waters. *Chinese chemical letters* **22** 225-228

[6] Abdul Halim, Abdullah and Jia Hui 2012 *J.* Response surface methodology analysis of the photocatalytic removal of methylene blue using bismuth vanadate prepared via polyol route. *Journal of environmental sciences* **24** 1694-1701

[7] Gao Jun-Jie, Qin Ye-Bo and Zhou Tao 2013 *J.* Adsorption of methylene blue onto activated carbon produced from tea (camellia sinensis L.) seed shells: Kinetics, equilibrium, and thermodynamics studies. *Zhejiang university-science B (Biomed & Biotechnol)* **14** 650-658

[8] Xu Nan-Nan, Lin Da-Song and Xu Ying-Ming 2014 *J.* Adsorption of Aquatic Cd^{2+} by biochar obtained from Corn Stover. *Journal of Agro-Environment Science* **33** 958-964

[9] Liu Jian-Yong, Qi Zhou and Chen Jun-Hua 2013 *J.* Phosphate adsorption on hydroxyl–iron–lanthanum doped activated carbon fiber. *Chemical Engineering Journal* , **216** 859–867

[10] Zhang Yong-Li 2009 *J.* Treatment of printing and dyeing wastewater with catalytic wet oxidation technology. *Chinese Journal of Environmental Engineering,* **06** 1011-1014

[11] Sadana A and Katzer J 1974 *J.* Involvement of Free Radicals in the Aqueous-Phase Catalytic Oxidation of Phenol over Copper Oxide. *Catalysis,* **35** 140-152

[12] Rivas F, Kolaczkowski S and Beltran F 1998 *J.* Development of a Model for Wet Air Oxidation of Based on a Free Radical.Mechanism . *Chemical Engineering Science,* **53** 2575-2586

[13] Liu Zhan-meng, Tang Chao-chun, Li Jing and Bao Dong-jie 2009 J. Catalytic ozonation with activated carbon for degradation of methylene blue in wastewater. *Journal of Zhejiang Forestry College,* **26** 406-410

Synthesis and characterization of structural and optical properties of single crystalline a-TiO_2 films on $MgAl_2O_4(111)$ substrate

Haisheng Xu, Xianjin Feng, Caina Luan, Jin Ma[1]

School of Microelectronics, Shandong University, Jinan 250100, China

E-mail: sduxhs@163.com

Abstract. Anatase phase TiO_2 (a-TiO_2) films have been deposited on $MgAl_2O_4(111)$ substrates by the metal organic chemical vapor deposition (MOCVD) method at the substrate temperatures of 500-650℃. The structural analyses showed that the films were highly (004) oriented with tetragonal anatase structure and the epitaxial relationship was given as a-TiO_2 (004)||$MgAl_2O_4$ (111). The sample prepared at 600℃ exhibited the best crystallization with a single-crystalline epitaxial film. The average transmittance of every TiO_2 film in the visible range exceeded 90% excluding the influence of the substrate. The morphology and composition of the TiO_2 films have also been studied in detail.

1. Introduction

The wide band gap oxide semiconductor material is an important material for the transparent electronic devices and optoelectronic devices. The wide band gap oxide semiconductor material has been widely used in the fields of flat panel display[1], transparent film transistors(TFT)[2], solar cells[3], light-emitting diodes (LED)[4] and UV detectors[5]. High quality epitaxial single crystal oxide films have attracted more and more attentions because of the advantages of the integrity of the structure, few defects and high mobility. The band gap of the titanium dioxide (TiO_2) is about 3.2eV[6], which is relatively close to those of the GaN[7], SiC[8] and ZnO[9] materials. Therefore, TiO_2 is a promising wide band gap oxide semiconductor material with excellent properties of stable physical and chemical properties, completely non-toxic and abundant storage.

Rutile and anatase are the common crystal structures of TiO_2. The rutile TiO_2 has poor adsorption capacity for oxygen, small surface area, easy combination of the electrons and holes and poor catalytic activity. Conversely, the anatase TiO_2 has larger surface area, strong adsorption capacity for oxygen and better photocatalytic activity. Therefore the anatase phase TiO_2 was widely used in the fields of photocatalytic catalysts and dye-sensitized solar cells. The MOCVD method has the properties of good control of the growth parameters and easy commercial production. The TiO_2 films deposited by the MOCVD method normally have good single crystal properties with complete structure. The TiO_2 films prepared on the substrates of Y-stabilized ZrO_2(YSZ)[10], $LaAlO_3$(LAO)[11], $SrTiO_3$(STO)[12], α-Al_2O_3[13] and LSAT[14] have been reported, but the a-TiO_2 film deposited on the $MgAl_2O_4$ substrate is seldom reported until now. The $MgAl_2O_4$ substrate is suitable for the growth of the a-TiO_2 film with low cost and good matching. In the present work, the high quality a-TiO_2 single crystalline films were deposited on $MgAl_2O_4(111)$ substrates by MOCVD. The structural

IOP Conf. Series: Materials Science and Engineering **167** (2017) 012036 doi:10.1088/1757-899X/167/1/012036

properties, surface morphology, optical properties and the growth mechanism of the films were investigated in detail.

2. Experimental

TiO_2 films were deposited on $MgAl_2O_4(111)$ substrates using a high vacuum MOCVD system with two different gas flows. The tetrakis-dimethylamino titanium (TDMAT) was used as the organometallic (OM) precursor and was stored in a stainless steel bubbler maintained at 20℃. The high purity oxygen (5N) is used as the oxidant. Before loaded into the reaction chamber, the substrates were cleaned with ethanol and then ultrasonic cleaned with de-ionized water, followed by compressed N_2 drying. The TDMAT and the oxidant were separately transported into the reaction chamber by ultra-high-purity N_2. During the deposition, the flows rates of OM vapor and O_2 were kept at 25sccm and 68sccm, respectively. The chamber pressure was fixed at 10Torr.

The epitaxial relationships of the obtained samples were examined by a Rigaku D/MAX 2200PC X-ray diffractometer (XRD) with Cu K□ radiation. The surface micrograph was obtained using a JSM-6700F scanning electron microscope (SEM). The chemical composition of the TiO2 film prepared at 600 ℃ was determined by the X-ray photoelectron spectroscopy (XPS) using an ESCALAB MK II multi-technique photoelectron spectrometer. High-resolution transmission electron microscopy (HRTEM) and selected area electron diffraction (SAED) were performed on the sample by a Tecnai F30 transmission electron microscope operated at 300 kV. The optical transmittance spectra were performed with a TU-1901 double-beam UV-vis-NIR spectrophotometer in the wavelength range from 200 to 800 nm.

3. Result and discussion

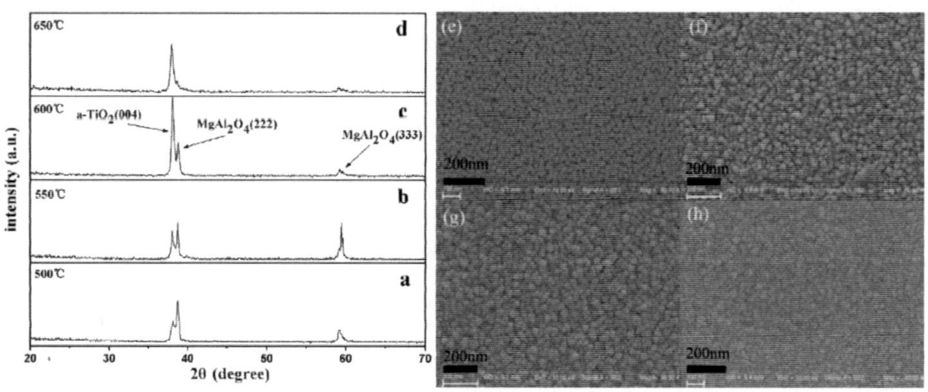

Figure 1. XRD spectra of the TiO_2 samples deposited at different substrate temperatures: (a)500, (b)550, (c)600 and (d)650℃. Plan view SEM micrographs of the TiO2 films prepared at (e) 500, (f) 550, (g) 600 and (h) 650℃.

XRD patterns of θ-2θ scan for TiO_2 films deposited on $MgAl_2O_4(111)$ substrates at different substrate temperatures of 500, 550, 600 and 650℃ are shown in Figure 1(a), (b), (c) and (d), respectively. The diffraction peak of the $MgAl_2O_4(222)$ is located at about 38.5° and that of the $MgAl_2O_4(333)$ is located at about 59.3° (JCPDS No.21-1152), which can be seen in Figure 1(a)-(d). Only one diffraction peak located at about 37.8° corresponding to a-$TiO_2(004)$ (JCPDS No.21-1272) is observed besides the peaks of the substrates for all the samples, which indicates that the films obtained at all temperatures are single crystalline a-TiO_2 structure. In addition, the full width at half maximum (FWHM) of the peak of the $MgAl_2O_4(222)$ deposited at 500-650℃ are 0.651°, 0.430°, 0.411° and 0.487° obtained by the Jade, respectively. Therefore the a-TiO_2 film deposited at 600℃ exhibits the best crystal quality with the smallest FWHM. The analyses of XRD θ-2θ patterns show that the crystal quality of the TiO_2 films prepared by MOCVD is obviously influenced by the substrate temperature. The out-of-plane growth relationship of the film is a-TiO_2(004) || $MgAl_2O_4$(111).

Figure 1(e)-(h) show the SEM images of TiO_2 films deposited at 500, 550, 600 and 650℃, respectively. The SEM images show that the grains become bigger as the substrate temperature increases from 500 to 600℃, which indicates the crystal quality of the film becomes better. In Figure 1(e), compact grains can be seen, but the grains are small and there is no evident boundary between the grains. The bigger grains are obviously seen in Figure 1(f), but the sizes of the grains are uneven. Figure 1(g) shows uniform grains with evident boundaries, which indicates the film obtained at 600℃ has a excellent crystallization. The surface morphology becomes smooth with unsystematic grains in Figure 1(h), indicating a degradation in the film crystallinity. The analysis of the SEM result shows that the film deposited at 600℃ exhibits the best crystallization, which is consistent with the XRD analyses.

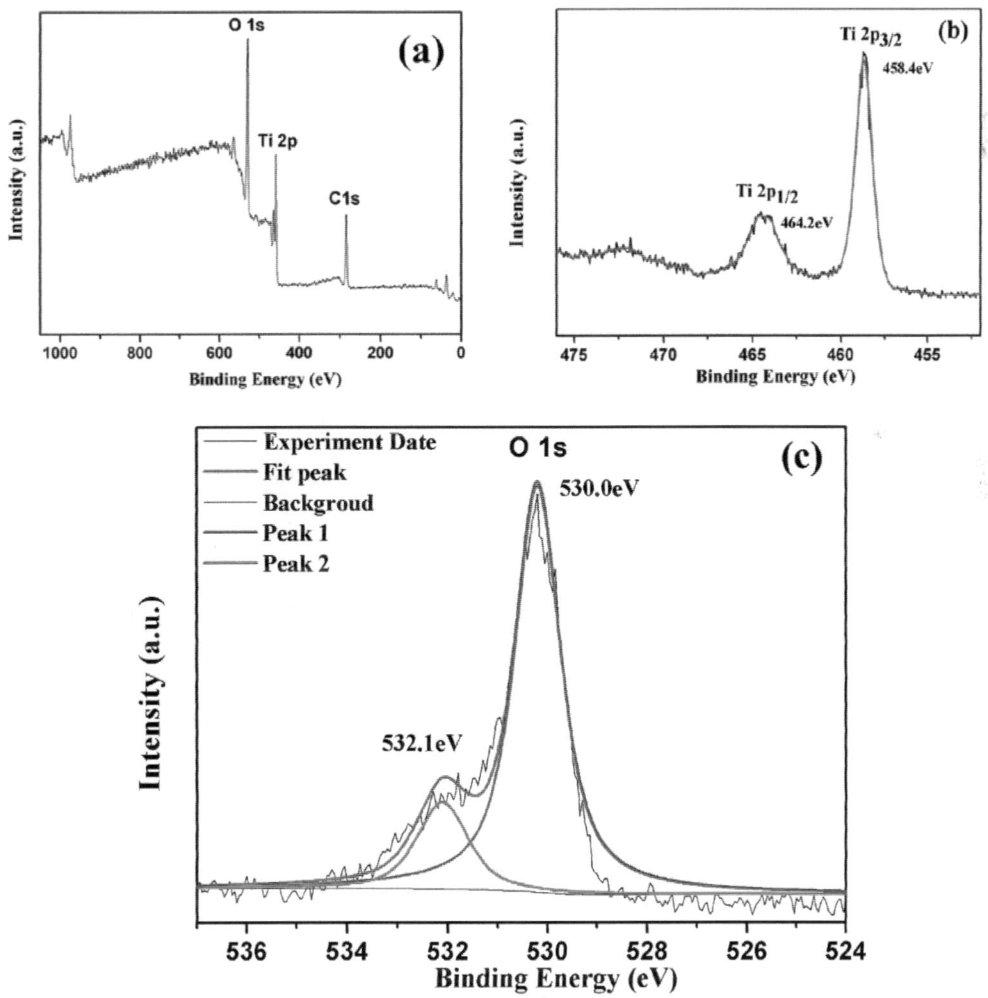

Figure 2. XPS spectra of the TiO_2 film grown at 600℃: (a) survey, (b) Ti2p and (c) O1s core level, respectively.

The X-ray photoelectron spectroscopy (XPS) technique was performed to research the chemical compositions of the sample deposited at 600 °C. Figure 2(a) shows a full scan in the energy ranging from 0-1200ev, from which the Ti2p, C1s and O1s core levels can be seen without the perks of any other elements detected, indicating the formation of titanium oxide. The C1s peak is located at about 285.0eV, which is used as a reference for the calibration of Ti2p and O1s core levels. The presence of C1s peak is related to surface pollution of the sample when it is exposed to the air before the XPS experiments. The Ti2p spectra of the TiO_2 film deposited at 600℃ is shown in Figure 2(b). The peak of $Ti2p_{1/2}$ is located at about 464.2eV and that of $Ti2p_{3/2}$ is located at about 458.4eV, which is the characteristic of TiO_2 with a separation of 5.8eV[15]. No Ti^{3+} signal located at 457eV is observed on the surface of the film[16]. In Figure 2(c), the O1s core level located at about 532.1eV and 530.0eV after gaussian fitting were assigned to oxygen from OH^- adsorption and TiO_2, respectively[17]. The stoichiometry of the film can also be calculated by the relative areas of the total Ti2p and O1s XPS peaks corrected by the relative sensitivity factors, the result of which is about 2.0. The analyses of the XPS indicates the film deposited at 600℃ is high-purity TiO_2.

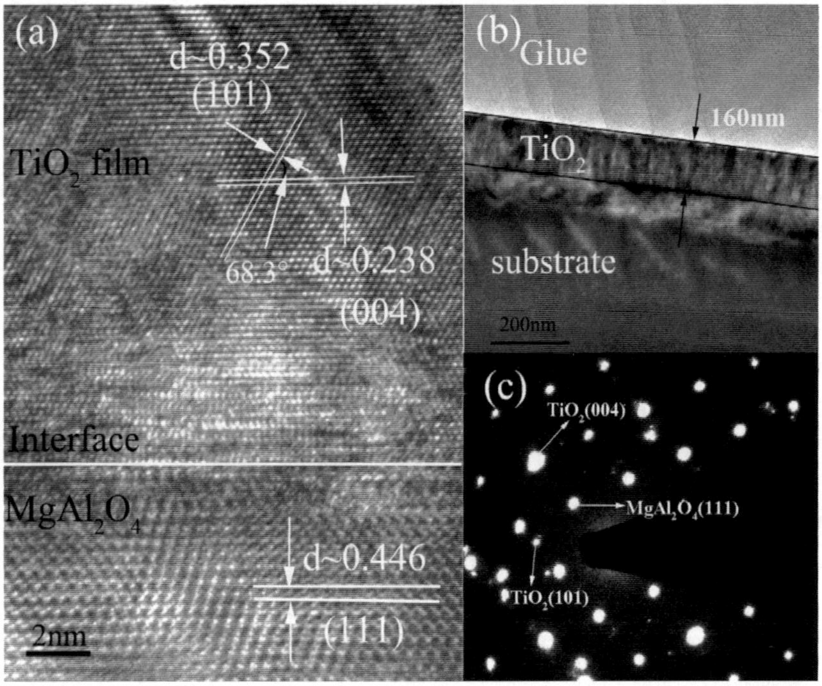

Figure 3. The cross-sectional (a) HRTEM micrographs, (b) low magnification TEM and (c) SAED micrographs of the interface between TiO_2 film grown at 600℃ and $MgAl_2O_4$ substrate.

The heteroepitaxial growth mechanism of the TiO_2 film grown on $MgAl_2O_4(111)$ substrate was further studied using the cross-section TEM measurements. The HRTEM, low resolution TEM images and the SAED pattern of the interfacial region between the $MgAl_2O_4$ substrate and the TiO_2 film grown at 600℃ are shown in Figure 3 (a)–(c), respectively. From Figure 3(a), the crystal lattice of both the TiO_2 film and $MgAl_2O_4$ substrate can be seen clearly with a well-defined interface. For the film, the as-marked interplane spacings are 0.352nm and 0.238nm, corresponding to a-TiO_2 (101) and (004) planes, respectively. The as-marked interplane spacings for the substrate is 0.466nm, which is corresponding to $MgAl_2O_4(111)$ planes. The low resolution TEM image shows that the thickness of the film is about 160nm as is shown in Figure 3(b). The SAED pattern of the corresponding interface area is shown in Figure 3(c), from which the diffraction spots of a-TiO_2 (101) and (004) as well as $MgAl_2O_4$ (111) and (222) can be seen clearly. The regular diffraction spots array implies good single crystallinity. From the HRTEM image and SAED pattern, the epitaxial relationship between a-TiO_2 film and $MgAl_2O_4$ substrate can be given as a-TiO_2(004)||$MgAl_2O_4$(111), which is consistent with the XRD analyses.

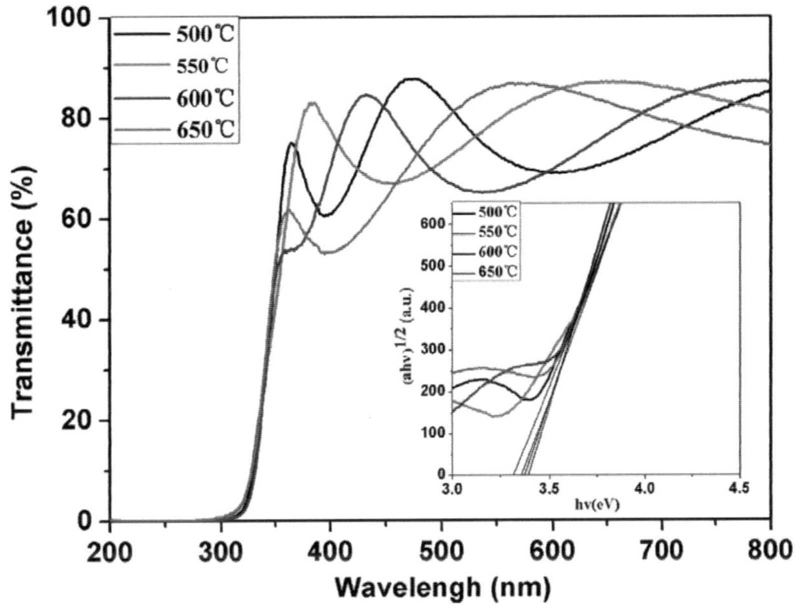

Figure 4. The optical transmittance spectra of a-TiO_2 samples prepared at different temperatures with the plot of $(\alpha hv)^{1/2}$ vs. hv shown in the inset.

The optical transmittance spectra of the TiO_2 samples deposited at the temperatures ranging from 500 to 650℃ are shown in the Figure 4. Deducting the influence of the substrate, the average transmittance of the TiO_2 films in the visible range exceeds 90%. The a-TiO_2 film is an indirect-band gap semiconductor, so the optical band gap(E_g) of the a-TiO_2 can be estimated by the equation $\alpha hv=A(hv-E_g)^2$ [18], where v is the frequency of the incident photon, h is the Planck's constant, A is a material dependent constant and α is the adsorption coefficient (α=(-1/d)lnT, where d and T are the thickness and transmittance of the film, respectively). The energy gap of the a-TiO2 films can be estimated by plotting $(\alpha hv)^{1/2}$ vs. hv. The optical band gaps of the TiO_2 films deposited at 500-650℃ are in the range of 3.2-3.38eV.

4. Conclusions

Anatase TiO_2 films have been deposited on $MgAl_2O_4$(111) substrates at the substrate temperatures of 500-650℃ by MOCVD. The structural analyses showed that the film deposited at 600℃ was single crystalline a-TiO_2 with the epitaxial relationship of a-TiO_2(004)||$MgAl_2O_4$(111). The HRTEM and SAED measurements further confirmed the single crystal structure and epitaxial relationship of the film prepared at 600 ℃. The average transmittance of all the samples prepared at different temperatures exceeded 90% in the visible range. The optical band gap of theTiO_2 film deposited at 600℃ was calculated as about 3.3eV. Due to the excellent structural and optical properties, the TiO_2 films prepared on $MgAl_2O_4$(111) will be used widely in the fields of photovoltaic devices after doping Niobium or Tantalum element.

Acknowledgment
This work was financially supported by the National Natural Science Foundation of China, China(Grant no.51472149).

References

[1] Lewis B G and Paine D C 2000 *Mrs Bulletin* **25(08)**: 22-27.

[2] Zhang H, Uochi H and Takayama T 1996 *U.S. Patent 5,563,426[P]* **10-8**.

[3] Wan J, Liu R and Tong Y 2016 *Nanoscale Res. Lett.* **11** 1-9.

[4] Mont F W, Kim J K and Schubert M F 2008 *J. Appl. Phys.* **103** .

[5] Basak D, Amin G and Mallik B 2003 *J. Cryst. Growth* **256** 73-77.

[6] Tang H, Prasad K and Sanjines H 1994 *J. Appl. Physi.* **75** 2042-2047.

[7] Matsuoka T, Okamoto H and Nakao M 2002 *Appl. Phys. Lett.* **81** 1246-1248.

[8] Mattausch A and Pankratov O 2007 *Phys. Rev. Lett.* **99** 076802.

[9] Srikant V and Clarke D R 1998 *J. Appl. Phys.* **83** 5447-5451

[10] Wang W, Ma J and Wang W, 2015 *Mater. Lett.* **161** 9-12.

[11] Kurita D, Ohta S and Ohta H 2006 *J. Appl. Physi.* **100** 096105-1.

[12] Hsieh C C, Wu K H and Juang J Y 2002 *J. Appl. Physi.* **92** 2518-2523.

[13] Chang H L M, You H and Gao Y 1992 *J. Mater. RES.* **7** 2495.

[14] Wang M, Wang W and Li Z 2016 *RSC Adv.* **6** 4867-4871.

[15] Stefanov P, Shipochka M and Stefchev P 2008 *J. Phys. Conference Series. IOP Publishing* **100.**

[16] Sandell A, Anderson M P and Alfredsson Y 2002 *J. Appl. Physi.* **92** 3381-3387.

[17] Lu Z, Yip C T and Wang L 2002 *ChemPlusChem* **77** 991.

[18] Bernardi M I B and Lee E J H 2001 *Mat. Res.* **4** 223.

Fe_2O_3/TiO_2 nanocomposite photocatalyst prepared by supercritical fluid combination technique and its application in degradation of acrylic acid

J Wei[1] and J C Zhang

College of Architecture and Civil Engineering, Beijing University of Technology, Beijing 100124, China

E-mail: weij@bjut.edu.cn

Abstract. Fe_2O_3/TiO_2 nanocomposite photocatalysts were synthesized by supercritical fluid combination technique, consisting of sol-gel method and supercritical fluid drying. The photocatalytic activity of the samples was evaluated by the degradation of acrylic acid. The results indicated that the Fe_2O_3/TiO_2 nanocomposite catalysts prepared by this novel technique showed significant improvement in catalytic activity compared with pure TiO_2 or Fe_2O_3/TiO_2 catalysts prepared by traditional drying. Both infrared and ultraviolet spectrum of Fe_2O_3/TiO_2 nanocomposite photocatalysts shift a little to lower wavelength indicating that the absorption threshold of Fe doped nanocomposite photocatalysts shift into the visible light region. This phenomenon was also attested by the photocatalytic degradation test under visible light.

1. Introduction

While acrylic acid and acrylic esters[1] are playing important roles in chemical industry, much attention is being given to their production because of a huge amount of wastewater released in the production.

Semiconductor photocatalysis appears as an emerging destructive technology leading to the total mineralization of most of organic pollutants[2]. It offers various advantages such as low cost, no secondary pollution, photoreaction occurring at room temperature and atmospheric pressure[3]. Although significant progress has been made on the waste water treatment, photocatalytic degradation of industrial wastewater with acrylic acid remains an untouched area.

Although TiO_2 is widely used in photocatalyst due to its good optical and electronic properties, chemical stability and non-toxicity and low cost[4], several factors limit its practical application. The first factor is that TiO_2 has a wide band gap of 3.2 eV which can only be excited by high energy UV irradiation with wavelength no longer than 387.5 nm. The second factor is that TiO_2 has a high recombination of photogenerated electron/hole pairs[5, 6]. One of these materials that has aroused much research interest is the so-called composite semiconductor photocatalyst. It could extend light absorption to the visible light range[7]. A possibility to achieve this goal is to dope the titanium dioxide with transition metal ions. It has been shown that Fe (III) doping can increase the photocatalytic activity of TiO_2[8].

In this paper, the nanocomposite photocatalyst Fe_2O_3/TiO_2 was prepared by supercritical fluid combination technique (SCFCT). Its photocatalytic activity for the degradation of acrylic acid under ultraviolet light and visible light was also assessed. The success of the present study should shed a light on the feasible application of acrylic acid in the waste water treatment.

2. Materials and Methods

2.1. Preparation of Catalyst

Surfactant (0.3v%) was added to 0.3 mol·L^{-1} $TiCl_4$, stirring vigorously at room temperature. The pH of the solution was adjusted to 8-9. After aging for 16 h, the sol was filtered and washed until no presence of Cl^{-1}. The water in the sol was replaced by anhydrous ethanol. The alcogel obtained was transferred into an autoclave and dried under the supercritical condition of ethanol (262 °C, 8.5 MPa) to produce aerogel. After calcination at different temperature for 1 h, pure nano-TiO_2 was obtained.

0.1 mol·L^{-1} $Fe(NO_3)_3$ and surfactant (0.3v%) was added to 0.3 mol·L^{-1} $TiCl_4$ in sequence, stirring vigorously at room temperature. The following processes were carried out as mentioned in above. Finally, different molar percentage of Fe_2O_3/TiO_2 nanocomposite catalysts was obtained.

The composite alcogel was also dried at room temperature and then calcined at different temperature for 1 h in air atmosphere. Then (0.03mol%) Fe_2O_3/TiO_2 composite catalysts were obtained.

2.2. Characterizations

The UV irradiation of the high pressure mercury lamp was measured with the UV radiometer (Beijing Normal University photo electric company). The content of Fe was determined by atomic absorption spectrophotometer (Z-8000). The surface morphologies and particle size were determined from transmission electron micrographs made with Hitach-800 TEM. X-ray diffraction measurements were performed using a Shimadzu HR6000X (Cu target X tube, voltage 40.0 KV, current 30.0 mA). The IR spectra were recorded with a Shimadzu IRPrestige-21 spectrophotometer. Diffuse UV-Vis spectra were obtained with a UV-2501PC Ultraviolet spectrophotometer. The UV irradiation of the high pressure mercury lamp was measured with the UV radiometer (Beijing Normal University photo electric company).

2.3. Evaluation of Photocatalytic Activity

The photocatalytic activities of the synthesized catalysts were examined using a self-designed photocatalytic reaction apparatus. It consists of double-layered cylindrical quartz flask with an ultraviolet lamp (9 W, wavelength: 254 nm, intensity: 5210 μw·cm^{-2}) or a solar lamp (18 W) positioned in the middle. Acrylic acid aqueous solution (300 mL, 300 mg·L^{-1}) with the chemical oxygen demand (COD) value of 400~500 mg·L^{-1} was used as degradation solution. The concentration of catalyst was 1g·L^{-1}. The air was introduced from the bottom of the reaction solution at the rate of 30 mL·min^{-1}. The temperature was maintained at 25±1 °C. Stirring was kept to make sure that catalyst was dispersed evenly in the reaction solution. A small volume of the reaction solution was obtained periodically. After centrifugation, the absorbance of these solutions was measured at 210 nm by B-UV 752 spectrophotometer. The COD value of the reaction solution was determined by potassium dichromate method. The photocatalytic activities were estimated from the degradation rate of acrylic acid.

3. Results and discussion

3.1. Results of AAS and XRD measurements

The results of AAS showed that the content of Fe in the composite photocatalysts was 0.03%, 0.1% and 0.15% (molar percentage), respectively.

Figure1 shows the XRD patterns of selected samples (0.03mol% Fe_2O_3/TiO_2) prepared by SCFCT with calcination at 500 ℃, 600 ℃, 700 ℃ and without calcination. The observed patterns can be attributed to the TiO_2 anatase as unique phase, which has diffraction peaks at d_1=0.35256 nm, d_2=0.18945 nm and d_3=0.23768 nm. No transformation from anatase type to rutile type was observed after heat treatment. It confirms the conclusion that TiO_2 prepared by SCFCT exists only as anatase type crystal, that is the drying and crystallizing could be accomplished in a single step in supercritical fluid drying process[9]. It also can be found that the crystallinity was enhanced by the calcination temperature. There is no peak corresponding to Fe_2O_3 present in the XRD patterns. This result could be attributed to that there is no independent Fe_2O_3 crystalline phase accumulated on the surface of TiO_2, or Fe_2O_3 disperses uniformly in the bulk of TiO_2 particle as very small clusters.

The size of domain crystallite estimated from Scherrer equation was 8.9 nm (without calcination), 10.9 nm (500 °C), 15.8 nm (600 °C) and 27.3 nm (700 °C). These results indicate that the size of the particles increases with the increase of the calcination temperature.

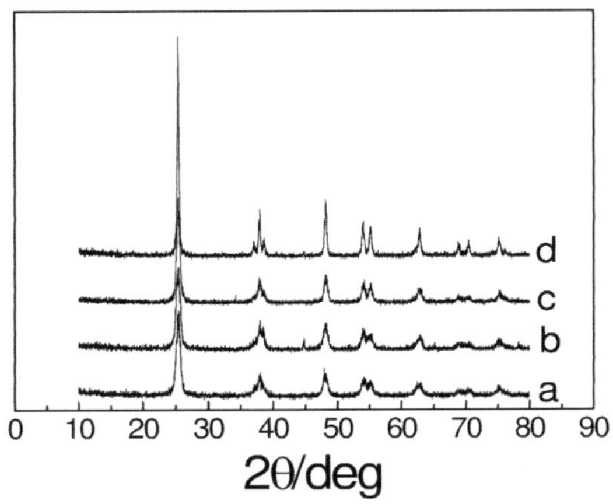

Figure 1. XRD pattern of Fe_2O_3/TiO_2 (0.03mol%) samples with different calcinations temperature.
(a) Fe_2O_3/TiO_2 (SCFCT, without calcinations)
(b) $Fe_2O_3/$ TiO_2 (SCFCT, 500 °C)
(c) Fe_2O_3/TiO_2 (SCFCT, 600 °C)
(d) $Fe_2O_3/$ TiO_2 (SCFCT, 700 °C)

3.2. Results of TEM Measurement

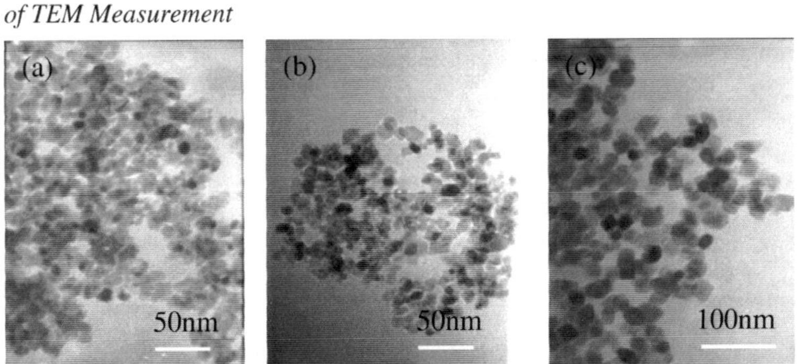

(a) $Fe_2O_3/$ TiO_2 (TD,600 °C) (b) $Fe_2O_3/$ TiO_2 (SCFCT) (c) Fe_2O_3/TiO_2 (SCFCT, 600 °C)

Figure 2. TEM photographs of Fe_2O_3/TiO_2 nanocomposite particles prepared by different methods.

The TEM photographs of 0.03mol% Fe_2O_3/TiO_2 nanocomposite particles prepared with different methods are shown in figure 2. For the powders obtained using traditional drying (TD) method, the crystals are in the form of aggregates and the composite particles are observed to be 17-23 nm in size (figure 2a). In contrast, the samples prepared by SCFCT have good dispersity. The particle size determined from the TEM graph is 8-10 nm (figure 2b). After calcination at 600 °C, the particles size is up to 14-18 nm (figure 2c), as is consistent with the size derived from the Scherrer equation.

3.3. Results of IR and UV-Vis

The IR spectra of Fe_2O_3, TiO_2 and 0.03mol% Fe_2O_3/TiO_2 samples are shown in figure 3. In the spectrum of Fe_2O_3/TiO_2, the intensified absorption at about 480 cm^{-1} is different from that of TiO_2 and Fe_2O_3, which indicates that the formation of a new chemical bond Fe-O-Ti. Figure 3 also shows that the red-shift takes place in the spectra of Fe_2O_3/TiO_2 in comparison with TiO_2, indicating that the absorption threshold of Fe-doped catalyst shift into the visible light region.

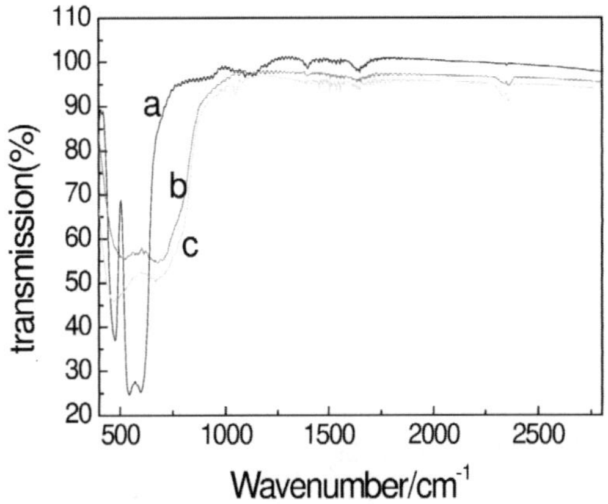

Figure3.FTIR spectrum of different samples (SCFCT, 600°C)
(a) Fe_2O_3; (b) TiO_2; (c) Fe_2O_3/TiO_2 (0.03%)

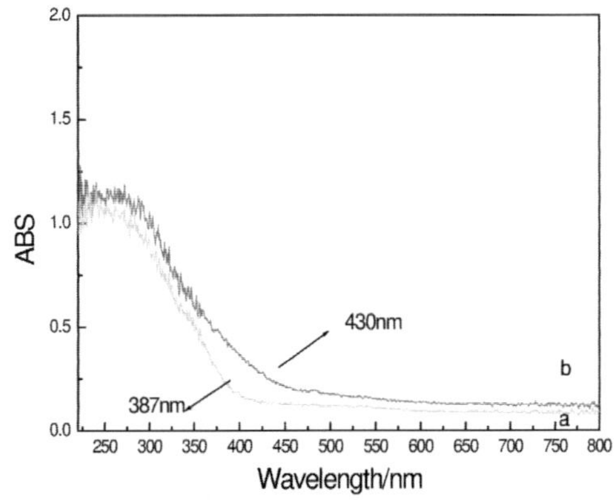

Figure 4.UV-Vis spectrum:
(a) TiO_2; (b) 0.03%Fe_2O_3/TiO_2.

The UV-Vis spectra for TiO_2 and 0.03mol% Fe_2O_3/TiO_2 are shown on the figure 4. From the curves we can see that UV-Vis absorbance of 0.03mol% Fe_2O_3/TiO_2 is shifted to longer wavelength (the red shift) compared with TiO_2. The band gap energy (Eg) of anatase TiO_2 extrapolates the absorption edge onto the energy 3.2 ev (387.5 nm)[10]. The electron-hole pairs can only be created by the light whose wavelength is shorter than 387.5 nm. In this experiment, the value of band gap energy of pure TiO_2 shown in the figure 4 is fairly consistent with the above. The band gap energy of 0.03mol% Fe doped on TiO_2 is reduced to 2.8 ev (430.0 nm), showing that red shift occurred. It is thought that additional band gap energy was created by Fe doping so that electron-hole pairs can be excited in the visible light wavelength region and red shift occults is possible.

3.4. Effect of Fe_2O_3 dosage on catalytic activity

The catalytic activities of Fe_2O_3/TiO_2 nanocomposite catalysts prepared under the same conditions (containing 0.00mol%, 0.03mol%, 0.10mol% and 0.15mol% Fe_2O_3 respectively) are shown on figure 5. It can be seen that 0.03mol% Fe_2O_3 doped TiO_2 is more photoactive than any other samples. 0.1mol% Fe_2O_3 doped TiO_2 has high photoactivity in comparison with TiO_2, but 0.15mol% Fe_2O_3 doped TiO_2 has lower Fe_2O_3 doped TiO_2 than pure TiO_2.

The beneficial effect of Fe^{3+} may be explained as follows: 1) The ion radius of Fe^{3+} (0.063 nm)[11] is much closer to that of Ti^{4+} (0.068 nm)[12], so the Ti^{4+} sites could be easily substituted by Fe^{3+} ions; 2) The Fe^{2+}/Fe^{3+} energy level lies close to Ti^{3+}/Ti^{4+} level (figure 6), similar to anatase, thus photogenerated electrons are transferred from TiO_2 to Fe^{2+}/Fe^{3+} energy level ($Fe^{3+} + e^- \rightarrow Fe^{2+}$). Fe^{3+} can be an effective electron trap. Meanwhile, due to the energy level of Fe^{3+}/Fe^{4+} being above the valence band edge of anatase TiO_2[13], Fe^{3+} can also be served as vacancy trap: $Fe^{3+} + h^+ \rightarrow Fe^{4+}$. However, high doping concentrations are not desirable, because of the high possibility that both charge carriers (e^-/h^+ pair) will be trapped, and on the other hand, more doping agents reduce the number of active centers on the surface of TiO_2 catalysts.

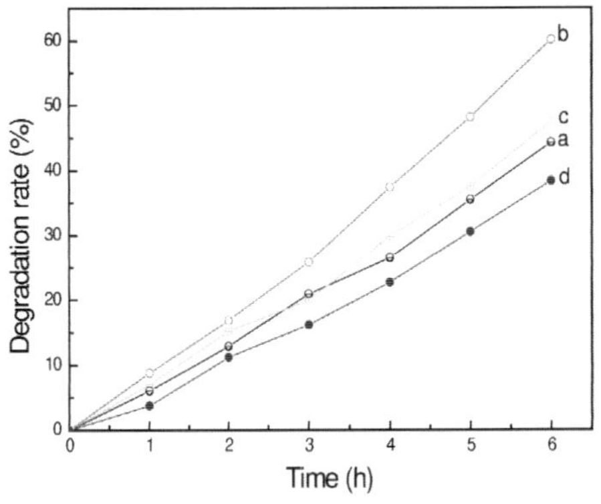

Figure 5. Effect of Fe doping concentration ondegradation of acrylic acid (SCFCT, °C).
(a) 0.00% Fe_2O_3/ TiO_2; (b) 0.03% Fe_2O_3/ TiO_2
(c) 0.10% Fe_2O_3/ TiO_2; (d) 0.15% Fe_2O_3/ TiO_2

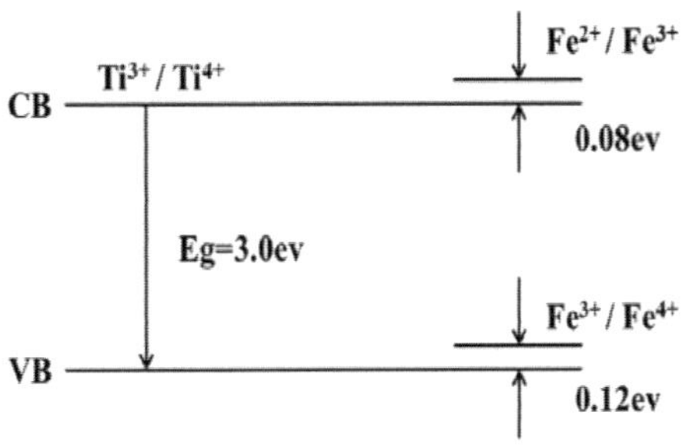

Figure 6. Energy level diagram of Fe^{3+} in rutile.

It can be seen from table 1 that 0.03mol% Fe_2O_3/TiO_2 nanocomposite photocatalyst has higher photoactivity than TiO_2 and Fe_2O_3, both under the ultraviolet light and visible light. This result is constant with the result of Uv-Vis. It is also suggested that light condition plays a vital importance in the degradation of acrylic acid. The efficient application of solar spectra still need to be further investigated.

Table 1. Effect of different light conditions on photocatalytic degradation of acrylic acid.

Light condition	no light/%	ultraviolet light/%	visible light/%
Fe_2O_3	0	30.13	19.34
TiO_2	0	44.25	14.30
0.03mol% Fe_2O_3/TiO_2	0	60.13	37.78

3.5. Effect of calcination temperature on the catalytic activities
0.03mol% Fe_2O_3/TiO_2 samples prepared by SCFCT were calcined at 500, 600 and 700 ℃ for 1 h respectively. The catalytic activities of the obtained catalysts were determined with photocatalytic degradation tests including 300 $mg{\cdot}L^{-1}$ acrylic acid. The degradation rates are listed in table 2.

Table 2. Effect of calcination temperature on the photo-catalytic activities[a, b, c]

Calcination temp. (°C)	Calcination time (h)	Degradation rate (%)
Without calcination	/	55.8
500	1	60.13
600	1	73.0
700	1	46.75

(a) Concentration of acrylic acid: 300 $mg{\cdot}L^{-1}$.
(b) Addition of catalyst: 1 $g{\cdot}L^{-1}$.
(c) Reaction temperature: 25±1 °C.

It can also be found that the degradation rate increases with the increase of calcination temperature. Sample calcined at 600 °C has the highest activity. When the temperature raises up to 700 ℃ the activity decreases to 46.75%. Low catalytic activity at low calcination temperature is attributed to the covered active centers by some organic compounds such as alcohols and surfactants. With the increase of calcination temperature, decomposition of the impurities will expose more active center on the surface, thus leading to improvement in the activity of the catalysts. However, when the calcination temperature is increased to 700 °C, a rapid particle growth from 15.8 nm to 27.3 nm is occurred. The growth of the powder particles leads to the decrease of the surface area and photoatalytic activity.

3.6. Effect of preparation methods on the catalytic activities
It is obvious that the catalyst prepared by SCFCT shows higher photocatalytic activity than that of prepared by TD as shown in figure 7. It is due to the interfacial force of the solvent, which leads to collapse of the cavity structure and aggregation of particles. And therefore, the particle size increases and cavity volume decreases. Correspondingly, the surface area and the photocatalytic activity are reduced. This phenomenon could be avoided in the preparation of aerogels obtained by SCFCT. At supercritical condition of alcohol, the solvent in the pores expands rapidly without interfacial force. The composite particles with good dispersity and small size lead to quantum size effect and improve the carriers transfer efficiency. Thus, the catalysts prepared through SCFCT show higher catalytic activity.

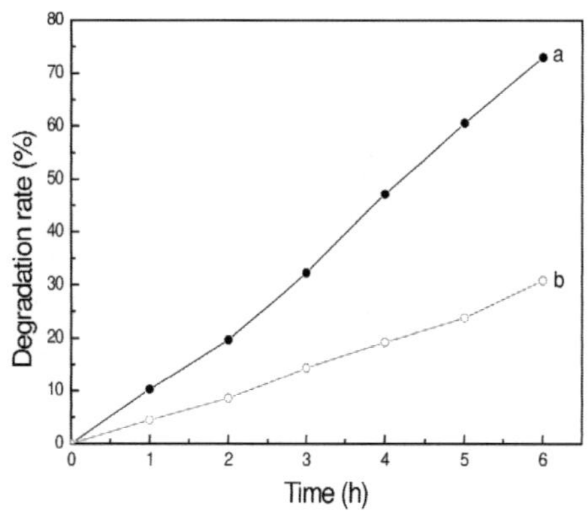

Figure7. Effect of different preparation methods of Fe_2O_3/ TiO_2 (0.03%) on photocatalytic property: (a) SCFCT ; (b) TD.

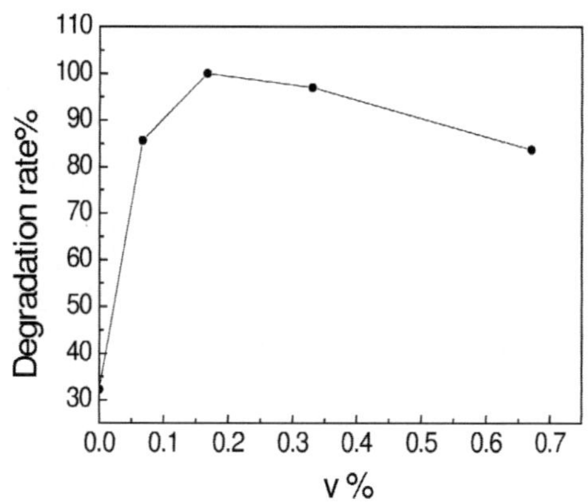

Figure 8. Effect of H_2O_2 concentrationon the acrylic acid degradation rate.

3.7. Effect of H_2O_2
Effect of amount (different volume percentages 0.00v%, 0.033v%, 0.167v%, 0.33v%, 0.66v%) of H_2O_2 on the photocatalytic degradation of acrylic acid suspensions is shown in figure 8. It is clear that the addition of H_2O_2 greatly enhance the degradation in comparison with nanocomposite photocatalyst alone. The mixture of 0.03mol% Fe_2O_3/TiO_2 and 0.167v% H_2O_2 can completely degrade the acrylic suspensions in 3 h. Adding H_2O_2 (lower than 0.167v%) could increase the degradation rate. This may be attributed to the enhancement of the initial concentration of hydroxyl radicals. At the same time H_2O_2 can capture one electron of the conduction band and produce the hydroxyl radical (figure 9), which hinders the electron–vacancy recombination rate and increase the rate of the photocatalytic process[14].

$$H_2O_2 + hv \rightarrow 2\cdot OH$$
$$H_2O_2 + e^- \rightarrow \cdot OH + OH$$

Figure 9. Formation of hydroxyl radicals from H_2O_2.

At high concentrations of hydrogen peroxide, the reactions shown in figure 10 will occur.

$$H_2O_2 + \cdot OH \rightarrow H_2O + HO_2\cdot$$
$$HO_2\cdot + \cdot OH \rightarrow H_2O + O_2$$

Figure10. Consumption of hydroxyl radicals in excess.

3.8. Results for COD_{Cr}

As can be seen from table 3, after 3 h treatment with 0.03mol% Fe_2O_3/TiO_2 and 0.167v% H_2O_2, the COD_{Cr} (η_{COD}) value of acrylic acid suspensions is reduced from 448.4 mg·L^{-1} to 35.57 mg·L^{-1} (or by 92.07%). The degradation rate determined by absorbency method is 100%. Both results indicate that almost the whole amount of acrylic acid is decomposed into H_2O and CO_2.

Table 3. The value of COD and absorbency on photocatalytic degradation of acrylic acid.

Photocatalytic degradation time (h)	0	3
The COD value (addition of H_2O_2 0.167%) (mg·L^{-1})	448.4	35.57
Degradation rate (η_{COD}/%)	/	92.07
Absorbency (A)	1.243	0.00
Degradation rate (η_{COD}/%)	/	100.00

4. Conclusions

Anatase phase Fe_2O_3/TiO_2 nanocomposite photocatalysts can be obtained directly by SCFCT, indicating that drying and crystallizing process can be accomplished in one step. Moreover, SCFCT can inhibit effectively the phase transformation from anatase to rutile. The Fe_2O_3/TiO_2 nanocomposite photocatalysts prepared posses good dispersity, crystallinity and high catalytic activity both under the ultraviolet light and visible light. In addition, H_2O_2 plays a vital role in improving the efficiency of photocatalysts.

References

[1] Wang H, Xu L, Li R, Pang L, Hu J, Wang M and Wu G, Highly hydrophilic ultra-high molecular weight polyethylene powder and film prepared by radiation grafting of acrylic acid 2016, *Appl. Surf. Sci.***382** 162-9.

[2] Ahmad M A A, Kamaruzzaman M R and Chin S Y, New method for acrylic acid recovery fromindustrial waste water via esterification with 2-ethyl hexanol 2014, *Process Saf. Environ.***92** 522-531.

[3] Kabir M S, Namjoshi O A, Verma R, Polanowski R, Krueger S M, Sherman D, Rott M A, Schwan W R, Monte A and Cook J M, A new class of potential anti-tuberculosis agents: Synthesis and preliminary evaluation of novel acrylic acid ethyl ester derivatives 2010, *Bioorg. Med. Chem.***18**4178-4186.

[4] Ângelo J, Magalhães P, Andrade L and Mendes A, Characterization of TiO_2-based semiconductors for photocatalysis by electrochemical impedance spectroscopy 2016, *Appl. Surf. Sci.***387** 183-9.

[5] Zhu Y, Wang Y, Chen Z, Qin L, Yang L, Zhu L, Tang P, Gao T, Huang Y, Sha Z and Tang G, Visible light induced photocatalysis on CdS quantum dots decorated TiO_2 nanotube arrays 2015, *Appl. Catal. A- Gen.***498** 159-166.

[6] Lin L, Wang H, Luo H and Xu P, Enhanced photocatalysis using side-glowing optical fibers coated with Fe-doped TiO_2 nanocomposite thin films 2015, *J Photoch.Photobio. A: Chem.***307–308** 88-98.

[7] Kalantari K, Kalbasi M, Sohrabi M and Royaee S J, Synthesis and characterization of N-doped TiO_2 nanoparticles and their application in photocatalytic oxidation of dibenzothiophene under visible light 2016, *Ceram. Int.***42**14834-42.

[8] Kumar R, Rashid J and Barakat M A, Zero valent Ag deposited TiO_2 for the efficient

photocatalysis of methylene blue under UV-C light irradiation 2015, *Colloids Interf. Sci.Commun.***5** 1-4.

[9] Han Z, Yu Y, Zhang J, Yang X and Cao W, The chemical interaction of poly(3-octylthiophene)/titanium dioxide composite materials 2012, *Synthetic Met.***162** 212-6.

[10] Fujihara K, Izumi S, Ohno T, Matsumura M, Time-resolved photoluminescence of particulate TiO_2photocatalysts suspended in aqueous solutions 2000, *J Photoch. Photobio. A: Chem.***132** 99-104.

[11] Zhu J, Zheng W, He B, Zhang J and Anpo M, Characterization of Fe–TiO_2 photocatalysts synthesized by hydrothermal method and their photocatalytic reactivity for photodegradation of XRG dye diluted in water 2004, *J Mol Catal. A: Chem.***216** 35-43.

[12] Garcoa J C and Takashima K, Photocatalytic degradation of imazaquin in an aqueous suspension of titanium dioxide 2003, *J Photoch. Photobio. A: Chem.* **155** 215-226.

[13] Choi W and Termin A, Hoffmann M R, The role of metal ion dopants in quantum-sized TiO_2: correlation between photoreactivity and charge carrier recombination dynamics 1994, *J Phy. Chem.***98** 13669-79.

[14] Tang J W, Wu P X, Zeng S Y and Liu Y, Research progress of titanium dioxide visible-light photocatalysts 2005, *Mod. Chem. Ind.***25** 25-8.

Preparation, characterization and photocatalytic activities of TiO_2-$SrTiO_3$ composites

Yan Wang, Lianjie Zhu[1], Fubo Gao and Hanjie Xie

School of Chemistry & Chemical Engineering, Tianjin Key Laboratory of Organic Solar Cells and Photochemical Conversion, Tianjin University of Technology, Tianjin 300384, PR China.

E-mail: zhulj@tjut.edu.cn

Abstract. Series of TiO_2-$SrTiO_3$ composites were synthesized by hydrothermal method, using TiO_2 nanotube array as a precursor and $Sr(OH)_2$ as a Sr source material. TiO_2-$SrTiO_3$ products with various composition were obtained by simply changing the reaction time. The as-synthesized products were characterized by scanning electron microscopy (SEM) and X-ray diffraction (XRD). The optical properties were studied by means of UV-Vis absorption spectroscopy and photoluminescence (PL) spectra. Their photocatalytic activities were assessed by photodegradation of rhodamine B (RhB) solution and the photocatalytic reaction mechanism was discussed. The TiO_2-$SrTiO_3$ composites obtained at 2 h exhibits the highest activity for photodegradation of RhB.

1. Introduction

Nanostructured TiO_2 have attracted wide attention due to their high photocatalytic activity, low cost and excellent chemical stability [1-3]. Particularly, TiO_2 nanotube array film is a promising catalyst for photodegradations of organic pollutants because of its outstanding advantage in easy reclamation from a slurry system after photocatalytic reaction. However, pure TiO_2 (Eg=3.2 eV) can only absorb UV light and very little visible light, which limits the practical applications owing to the poor utilization of solar light [4,5]. Although cubic perovskite structure of $SrTiO_3$ (Eg=3.2 eV) also has large band energy, the TiO_2-$SrTiO_3$ composites could perform higher photocatalytic activity because of the difference in conduction and valence bands which could change migration process of photo-generated carriers, effectively promoting photo-generated electron-hole separation [6].

Many efforts have been devoted to the synthesis of TiO_2-$SrTiO_3$ heterostructure using hydrothermal method. For instance, TiO_2 powder was taken as a precursor to synthesize TiO_2-$SrTiO_3$ nanoparticles [7-10]. The other researchers [6,11-18] made use of TiO_2 nanotube array films as precursors to fabricate TiO_2-$SrTiO_3$ heterostructures. They mostly focus on the length and pore diameter of the TiO_2-$SrTiO_3$ composite nanotubes because longer nanotubes have larger surface areas enabling more light harvesting and more active sites accessible to pollutants in the aqueous environment [6]. However, the relations between reaction time and product morphology, composition and photodegradation activity have not been reported.

The present paper reports successful syntheses of the TiO_2-$SrTiO_3$ composites with different morphology and composition. The effects of reaction time on morphology, composition, crystal structure, optical property and photodegradation activity of the TiO_2-$SrTiO_3$ composites were investigated systematically. The results show that photocatalytic activity of the TiO_2-$SrTiO_3$ composites are closely related with morphology and composition.

2. Experiment

TiO_2 nanotube array (NTA) precursor was synthesized as follows. In a typical procedure, a cleaned Ti foil (70 × 50 × 0.25 mm^3, 99.6% purity) was put into the electrolyte containing 1.0 g of NH_4F in glycol solution with 3 vol.% H_2O and 7 vol.% formamide, whose pH was adjusted to 4 by drop-wise addition of H_2SO_4. TiO_2 NTAs were obtained by potentiostatic anodization in a constant stirring two-electrode cell at 40 V for 2 h at room temperature with graphite as a counter electrode placed at 3 cm from the Ti foil. Then the as-obtained TiO_2 films were rinsed with ethanol and water for several times, dried at 60℃ in an oven. The TiO_2-$SrTiO_3$ composites were fabricated by a hydrothermal method. The anodized TiO_2 nanotube array films (25 × 13 mm^2) and 0.03 g strontium hydroxide ($Sr(OH)_2\cdot 8H_2O$) was added to 20 mL deionized water in a Teflon stainless steel autoclave. The autoclave was sealed and heated to 150℃ in an oven for several hours. After it was cooled down to room temperature, the sample was removed from the autoclave, soaked in a 0.1 M HCl solution for 1 min, and rinsed with deionized water. Then the sample was calcined at 400℃ for 3 h in a muffle furnace with a heating rate of 5℃ min^{-1}. Consequently, series of TiO_2-$SrTiO_3$ composites were obtained.

X-ray diffraction (XRD) measurements were carried out on a Rigaku Ultima IV diffractometer) with a Cu K radiation source (λ = 0.15418 nm) operated at 40 kV and 40 mA. The scanning electron microscope (SEM) were taken on Carl Zeiss MERLIN Compact scanning electron microscope and transmission electron microscope (Germany). The UV-vis diffuse reflectance absorption spectra were recorded on a Hitachi/U-3900 UV-visible spectrophotometer. The PL emission spectra were obtained from a F-7000 steady and time resolved spectrometer excited at 330 nm.

The photodegradation activities of the series of TiO_2-$SrTiO_3$ composites on RhB solutions were studied under simulated sunlight illumination. The TiO_2-$SrTiO_3$ composites was horizontally placed at the bottom of a 10 mL of beaker with the nanotubes upwards. Then 8 mL of 0.01 g L^{-1} RhB aqueous solution was put into above beaker, which was in dark sonicated for 30 min and air bubbled for another 30 min to reach adsorption-desorption equilibrium. The catalytic system was then exposed to the simulated sunlight irradiation from a 400 W halogen light (5% UV light) located 15 cm above the solution and kept air bubbling during the whole photodegradation reaction. The concentration of the RhB solution at certain time interval was determined by measuring its absorbance at 552 nm on a UV-visible spectrophotometer, from which the photodegradation rate was expressed by $(A_0-A)/A_0$, where A_0 is the initial absorbance of the RhB solution and A is the absorbance at certain reaction time.

3. Results and discussion

3.1. Morphology and crystal structure

The surface and lateral SEM images of the as-anodized TiO_2 films and TiO_2-$SrTiO_3$ composites obtained at different reaction time show that they are mostly well-aligned nanotube arrays, grown vertically from Ti substrates (figure.1). As shown in figure.1(A), the pure TiO_2 film was composed of ordered nanotubes with an average diameter and tube length of about 106.8 nm and 3.1 μm, respectively. The formation of TiO_2 NTAs in fluorinated electrolyte was the result of three simultaneously occurring processes [19]: (1) field-assisted oxidation of Ti metal to TiO_2 ($Ti + H_2O \rightarrow TiO_2+H_2$), (2) field-assisted dissolution of Ti metal into electrolyte ($Ti + 6HF + 2H^+ \rightarrow [TiF_6]^{2-}+2H_2$) and (3) chemical etching of TiO_2 to TiF_6^{2-} ($TiO_2 + 6F^- + 4H^+ \rightarrow [TiF_6]^{2-} + 2H_2O$).

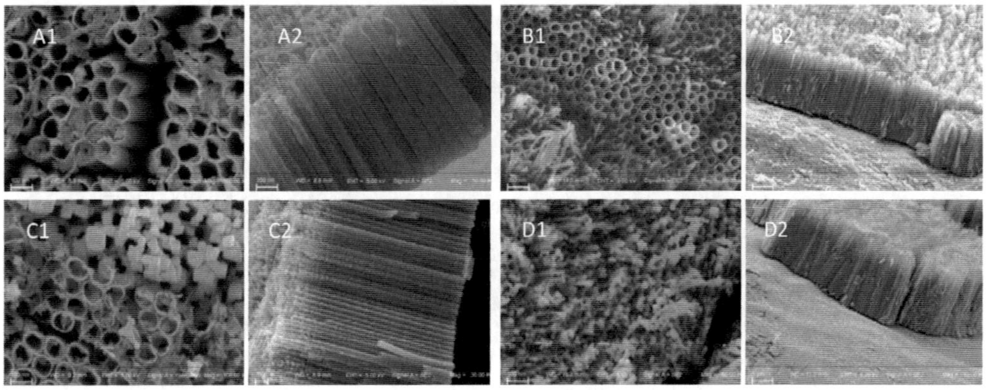

Figure 1. SEM images of TiO_2-$SrTiO_3$ composites obtained at different reaction times: (A) 0 h, (B) 2 h, (C) 3 h and (D) 6 h

It is noticed that the morphology of the TiO_2-$SrTiO_3$ composites change with prolonging reaction time (especially the surface morphology) although the tubular structure of the TiO_2 precursor was remained mostly (figure. 1B-D). When the reaction time is 2 hours (figure. 1B), the nano-tubular structure of the product was well preserved and mouths of the tubes became clear, which may be due to the transformation reaction from TiO_2 to $SrTiO_3$. This implies the specific surface area of the sample obtained at 2 h remained high. After 3 h reactions, $SrTiO_3$ nanocubes were partly dispersed on the nanotube surface and interior, as shown in figure.1(C), suggesting that the specific surface area of this TiO_2-$SrTiO_3$ composite was decreased. Further prolonging the reaction time to 6 h led to formation of more nanocubic $SrTiO_3$ which covered the whole surface of TiO_2 nanotubes. Thus, the tubular structure of the product became obscure from top surface view and the specific surface area was further decreased. Thus, the TiO_2-$SrTiO_3$ composite obtained at 2 h may possess the largest specific surface area which would lead to more active sites available for adsorption of dyes on their surfaces [20].

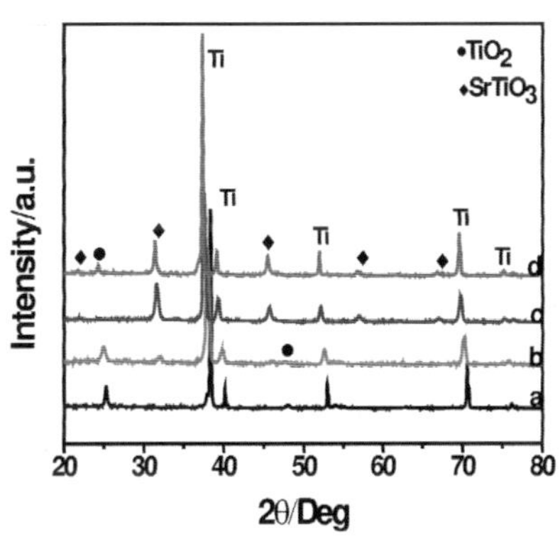

Figure 2. XRD patterns of TiO_2-$SrTiO_3$ composites prepared at different reaction times: (a) 0 h, (b) 2 h, (c)3 h and (d) 6 h

On the basis of these results and literature [13], we propose a formation mechanism of TiO_2-$SrTiO_3$ composites, which can be hypothetically described as follows: (i) TiO_2 molecules were dissolved by hydroxyl radicals (OH^-) to generate $[Ti(OH)_6]^{2-}$ ions ($TiO_2 + 2OH^- + 2H_2O \rightarrow [Ti(OH)_6]^{2-}$), (ii) more and more as-formed $[Ti(OH)_6]^{2-}$ ions further reacted with Sr^{2+} to generate

$SrTiO_3$ nanoparticles on the surface of TiO_2 nanotubes (Sr^{2+} + $[Ti(OH)_6]^{2-}$ → $SrTiO_3$ + $3H_2O$). With the reaction proceeding, more and more TiO_2 were converted to $SrTiO_3$ and nanocube morphology occurred on the surface layer, so that the nanotube structure can no longer be maintained at the surface layer. The SEM results also support that the transformation reaction from TiO_2 nanotubes to $SrTiO_3$ nanostructures were proceeded from surface to Ti substrate. Therefore, only top layers of TiO_2 nanotubes were converted to $SrTiO_3$ nanostructures and down layers of TiO_2 nanotubes closed to the Ti substrate were remained. Consequently, the as-obtained products from 2-6 h are all TiO_2-$SrTiO_3$ composites which are proven by the following XRD results.

The XRD pattern of the as-obtained pure TiO_2 nanotube array film (figure. 2a) shows that except the diffraction peaks of the Ti substrate, two peaks of a tetragonal anatase TiO_2 at 2θ = 25.3° and 48.0° are observed, corresponding to (101) and (200) planes (JPCDS: 21-1272, space group: I41/amd (141), cell: a = 3.785 Å, b = 3.785 Å, c = 9.514 Å). For the TiO_2-$SrTiO_3$ composites obtained at different reaction times (Fig. 2b-d), however, other five peaks of a tausonite $SrTiO_3$ at 2θ = 22.8°, 32.4°, 46.5°, 57.8° and 67.8° are observed, corresponding to (100), (110), (200), (211) and (220) planes (JPCDS: 05-0634, space group: Pm-31[221], cell: a = b = c =3.905 Å). All diffraction lines of TiO_2 and $SrTiO_3$ are sharp and narrow, indicating a high crystallinity of the TiO_2-$SrTiO_3$ composites. With prolonging the reaction time, the peak intensities of the $SrTiO_3$ phase gradually increased, whereas the peaks of TiO_2 gradually weakened. These results indicate that the TiO_2 nanotubes were gradually converted to $SrTiO_3$ and the phase ratio of $SrTiO_3$:TiO_2 gradually increased. $SrTiO_3$ became dominant phase after 3 h's reaction. However, when the reaction time is 2 h, the TiO_2 phase was still predominant in the presence of TiO_2-$SrTiO_3$ heterostructure, which may result in the higher photocatalytic activity.

3.2. Optical peoperties of TiO_2-$SrTiO_3$ composites

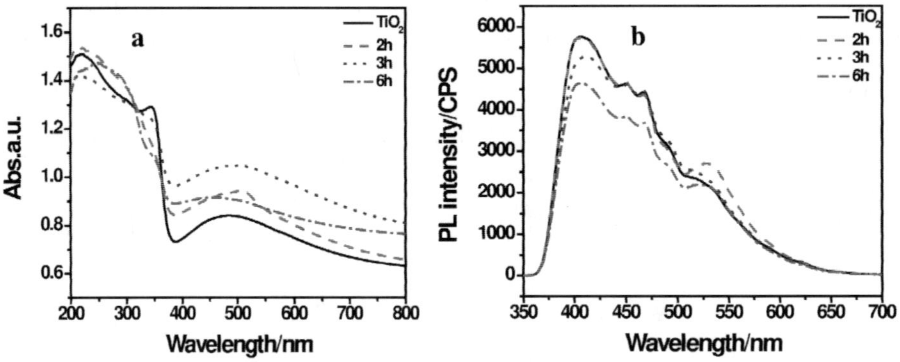

Figure 3. (a) UV-Vis diffuse reflection absorption spectra and (b) Photoluminescence spectra of TiO_2-$SrTiO_3$ composites prepared by hydrothermal method for different reaction time

The optical properties of the TiO_2-$SrTiO_3$ composites obtained at different reaction time were investigated by UV-vis diffuse reflectance absorption spectra and PL spectra. As shown in figure.3(a), a strong absorption band in UV light region and a relatively weaker absorption band in visible light region were present for each TiO_2-$SrTiO_3$ composite sample, which are similar to that of the pure TiO_2 nanotube array film except that the absorptions to visible light are enhanced significantly. With increasing reaction time, the absorption peak intensities of the TiO_2-$SrTiO_3$ composites in the visible light region firstly increase and then decrease. These results indicate that the TiO_2-$SrTiO_3$ heterostructures could greatly enhance the absorption to visible light and thus may effectively increase the photocatalytic activity of the catalyst. For TiO_2-$SrTiO_3$ heterostructures, the photo-generated electron could transfer from conduction band (CB) of $SrTiO_3$ to conduction band (CB) of TiO_2. At the same time, the holes on valence band of TiO_2 could transit to valence band of $SrTiO_3$. This maybe make TiO_2-$SrTiO_3$ composites absorb light in longer wavelength, resulting in enhanced absorption of TiO_2-$SrTiO_3$ composites on visible light.

The PL spectra in figure. 3(b) demonstrate that the emission bands of the TiO_2 NTAs and TiO_2-$SrTiO_3$ composites are all in the range of 380-600 nm with peak values at around 410 nm. The PL peak intensity of the TiO_2-$SrTiO_3$ composite obtained at 2 h is similar to that of the pure TiO_2 NTAs, except the peak at around 530 nm, which may be caused by the TiO_2-$SrTiO_3$ heterostructure, suggesting that TiO_2-$SrTiO_3$ heterostructure with suitable phase ratio is beneficial to separation of photo-generated electrons and holes, which may result in higher photocatalytic activity.

3.3. Photodegradation activities on RhB and reaction mechanism

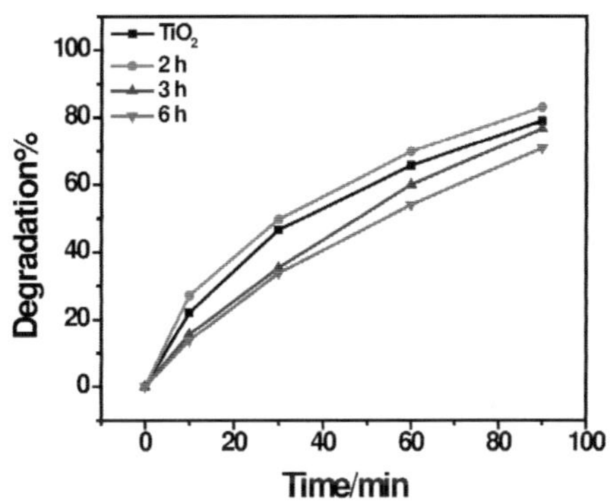

Figure 4. Photocatalytic performance of TiO_2-$SrTiO_3$ composites obtained at different reaction time on rhodamine B

Figure.4 demonstrates the photocatalytic performance of the TiO_2 NTA films and TiO_2-$SrTiO_3$ composites on RhB dye. With prolonging the illumination time, the photodegradation percentages of RhB over the series of catalysts increase gradually and reach 83% at 90 min for the best catalyst. With increasing the hydrothermal time, the photodegradation rate of RhB over the TiO_2 NTAs and TiO_2-$SrTiO_3$ composites firstly increased, reaching maximum and then decreased gradually. The TiO_2-$SrTiO_3$ composite obtained at 2 h possesses the highest photocatalytic activity, which may be attributed to its high specific surface area, TiO_2-$SrTiO_3$ heterostructure with mostly TiO_2 phase and enhanced visible light absorption ability.

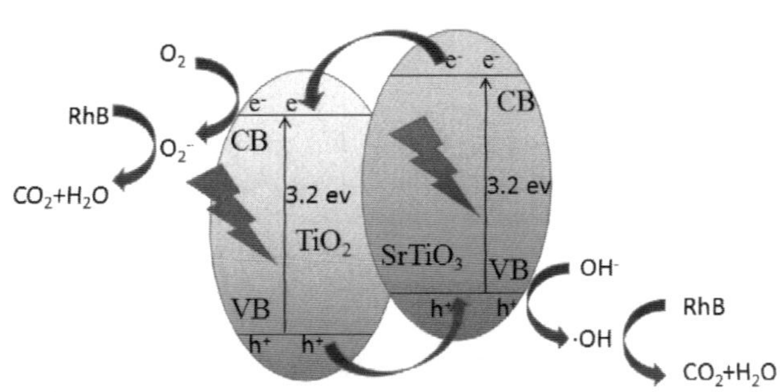

Figure 5. Schematic illustration of the reaction mechanism of photodegradation of RhB over the TiO_2-$SrTiO_3$ composites.

The possible photocatalytic reaction mechanism could be as follows. As shown in figure.5, under illumination, TiO_2 and $SrTiO_3$ generate photo-generated electron and holes. And then the photo-generated electron (e^-) could transfer from conduction band (CB) of $SrTiO_3$ to conduction band (CB) of TiO_2. At the same time, the holes (h^+) on valence band of TiO_2 could transit to valence band of $SrTiO_3$. The holes can oxidize RhB or react with OH^- or water to form hydroxyl radicals (·OH). Electrons (e^-) in the conduction band (CB) reduces dissolved molecular oxygen to create superoxide (O^{2-}). RhB in the solution then are degraded by hydroxyl radicals (·OH) and superoxide anions (O^{2-}) to inorganic compounds of CO_2 and H_2O [6]. In order to achieve maximum photodegradation efficiency, high visible light absorption ability and rapid charge separation, originated from TiO_2-$SrTiO_3$ heterostructure with suitable phase ratio, is desired. Besides, a large active surface area is also an important factor because more active sites are available for photocatalytic reactions. The TiO_2-$SrTiO_3$ composites obtained at 2 h has more surface active sites, strong visible light absorption capability and good photo-generated electrons/holes separation due to suitable composition of $SrTiO_3$, resulting in the highest activity for photodegradation of RhB.

4. Conclusion

Series of TiO_2 NTA-$SrTiO_3$ composites with different phase ratio have been successfully prepared by a facile hydrothermal method. With prolonging hydrothermal time, TiO_2 nanotubes are firstly converted to $SrTiO_3$ nanotubes with remained high specific surface area. Then $SrTiO_3$ nanocubes appear on the surface layer of the TiO_2 nanotube and gradually expand toward Ti substrate, accompanying with phase transformation from TiO_2 to $SrTiO_3$. When the hydrothermal time is 2 h, the obtained TiO_2 NTA-$SrTiO_3$ composite possesses high surface area (more active sites) with TiO_2 as the main phase which exhibits the highest activity for photodegradation of RhB. Optical studies of the series of samples indicate the high photocatalytic activity may be also related with its high visible light absorption and good separation of the photo-generated electrons/holes originated from suitable TiO_2-$SrTiO_3$ heterostructure.

References

[1] Liu Y, Wang Z, Wang W and Huang W 2014 *J. Catal.* **310** 16
[2] Yang X, Qin J, Li Y, Zhang R and Tang H 2013 *J. Hazard. Mater.* **261** 342
[3] Zheng W, Liu X, Yan Z and Zhu L 2009 *ACS Nano* **3** 115
[4] Cesano F, Bertarione S, Damin A, Agostini G, Usseglio S and Vitillo J G 2008 *Adv. Mater.* **20** 3342
[5] Kitano M, Matsuoka M, Ueshima M and Anpo M 2007 *Appl. Catal. A* **325** 1
[6] Zhang X, Huo K, Hu L, Wu Z and Chu P 2010 *J. Am. Ceram. Soc.* **93** 2771
[7] Ruzimuradov O, Nurmanov S, Hojamberdiev M, Prasad R M, Gurlo A, Broetz J, Nakanishi K and Riedel R 2014 *Mater. Lett.* **116** 353
[8] Yue X, Zhang J, Yan F, Wang X and Huang F 2014 *Appl. Surf. Sci.* **319** 68
[9] Yan J, Zhu Y, Tang Y and Zheng S 2009 *J. Alloys Compd.* **472** 429
[10] Wang T, Liu S and Chen J 2011 *Powder Technol.* **205** 289
[11] Zhu Y, Xu L, Hu J, Zhang J, Du R and Lin C *2014 Electrochim Acta* **121** 361
[12] Yang Y, Lee K Y, Kado Y and Schmuki P 2012 *Electrochem Commun.* **17** 56
[13] Huang J, Tan X, Yu T, Zhao L and Hu W 2014 *Electrochim Acta* **146** 278
[14] Zhou J, Yin L, Li H, Liu Z, Wang J, Duan K, Qu S, Weng J and Feng B 2015 *Mater.Sci. Semicond.Process.* **40** 107
[15] Zhang Y, Bu Y, Yu J and Li P 2013 *J Nanopart Res* **15** 1717
[16] Jiao Z, Chen T, Yu H, Wang T, Lu G and Bi Y 2014 *J. Colloid Interface Sci.* **419** 95
[17] Wu Z, Su Y, Yu J, Xiao W, Sun L and Lin C 2015 *Int. J. Hydrogen Energy* **40** 9704
[18] Guo L, Wang X, Zhang H and Li L 2013 *Ceram. Int.* **39** S633
[19] Lai C W and Sreekantan S 2011 *J. Nanomater.* **2011** 99
[20] Alsawat M, Altalhi T, Shapter J and Losic D 2014 *Catal Sci. Technol.* **4** 2091

IOP Conf. Series: Materials Science and Engineering **167** (2017) 012039 doi:10.1088/1757-899X/167/1/012039

Ce-doped titania nanoparticles: The effects of doped amount and calcination temperature on photocatalytic activity

Jianwen Shi[1], Yajun Zou and Dandan Ma

Center of Nanomaterials for Renewable Energy, State Key Laboratory of Electrical Insulation and Power Equipment, School of Electrical Engineering, Xi'an Jiaotong University, Xi'an 710049, China.

E-mail: shijnwn@163.com

Abstract. A series of Ce-doped TiO_2 nanoparticles with different doped amount and calcination temperature were prepared by sol-gel method. These obtained samples were characterized with X-ray diffraction (XRD), transmission electron microscope (TEM) and ultraviolet-visible diffuse reflectance spectra (DRS), and their photocatalytic activities were evaluated by the photocatalytic degradation of methyl orange. Results showed that Ce doping inhibits the growth of crystal size and the phase transformation from anatase to rutile, leads to lattice distortion and expansion of TiO_2. Furthermore, Ce doping brings the red-shift of absorption profile and the increase of photons absorption in the range of 400-600 nm. Photocatalytic degradation of methyl orange shows that Ce doping improves the photocatalytic activity of TiO_2. The optimal doped amount is 0.05 mol% and the optimal calcined temperature is 600 ^{o}C for the maximum photocatalytic degradation efficiency in our experiment.

1. Introduction

As a promising photocatalyst, TiO_2 has been intensely researched for the treatment of environmental pollution in recent decades. However, there are still some problems which restrict its application in practical industry, such as restrictive light absorption due to its large band gap, the fast recombination of photo-generated electrons and holes [1]. Therefore, it is of great interest to develop new photocatalysts with visible-light response and effective separation of electron-hole pairs.

Recently, some reports have demonstrated that doping TiO_2 with suitable metal or nonmetal ions is a useful strategy to improve the mentioned above two performances of TiO_2 [2]. Due to incompletely occupied *4f* and empty *5d* orbitals, rare earth metals are often used as catalyst or promote catalysis. Some results have shown that doping TiO_2 with rare earth metals or their oxides is an effective method to improve the photocatalytic activity of TiO_2 [3]. As one of the rare earth metals, Ce has been used as a dopant to enhance the photocatalytic activity of the TiO_2 [4]. However, there is an example that Ce-doping brought negative effect on the photocatalytic performance of TiO_2 [5].

In order to clarify the effects of Ce-doping, in current work, a series of Ce-doped TiO_2 nanoparticles with different doped amount and calcination temperature were prepared by sol-gel method. The obtained samples were characterized by X-ray diffraction (XRD), transmission electron microscope (TEM) and ultraviolet-visible diffuse reflectance spectra (DRS), and their photocatalytic activities were investigated by photocatalytic degradation methyl orange (MO) in water.

2. Experimental

2.1. The preparation of photocatalysts

All reagents used in this experiment are analytical grade without further purification. Ce-doped TiO_2 nanoparticles were prepared by a sol-gel route. In a typical procedure, 17.0 mL tetra-n-butyl titanate was firstly dissolved in 30.0 mL ethanol under vigorous stirring. Another solution containing 28.3 mL ethanol, 7.2 mL distilled water, 20.0 mL acetic acid, and a certain weight of $Ce(NO_3)_4 \cdot 6H_2O$ was slowly added into the above solution under magnetic stirring at room temperature. The mixture was hydrolyzed for 60 min under vigorous stirring, and then the transparent sol was obtained. The sol was aged for 6 h at 60°C to obtain gel, and then the gel was dried at 80 °C in a vacuum oven to remove the solvents. The dry gel was milled into powders in a mortar, and then was calcined in a muffle furnace at different temperature for 2 h, resulting Ce-doped TiO_2 nanoparticles. Un-doped TiO_2 was prepared by the same procedure without $Ce(NO_3)_4 \cdot 6H_2O$ addition. For convenience, Ce-doped TiO_2 samples were labelled as TiCe(X)Y and un-doped TiO_2 sample was labelled as TiY, where X stood for the molar percentage of Ce in theoretical product and Y referred to the calcined temperature. For example, a sample doped with 0.05 % Ce and calcined at 600 °C, we labelled it as TiCe(0.05)600.

2.2. Characterizations

XRD patterns of all samples were obtained at room temperature with a Holand X'pert PROMPD diffractometer (Cu K_{a1} radiation, λ= 1.5406 Å). The crystal size was estimated by applying the scherrer equation $D = K\lambda/(\beta cos\theta)$ to the full width at half maximum (FWHM) of the (101) peak of anatase, where K is a constant (shape factor, about 0.89), λ is the X-ray wavelength, β is the FWHM of the diffraction line, and θ is the diffraction angle. Transmission electron microscope (TEM) was carried out on JEM-2012. A Shimadzu spectrophotometer (TU1901, China) equipped with an integrating sphere was used to record DRS of the samples. The baseline correction was done using a calibrated sample of barium sulfate, and the spectra was recorded at room temperature in air in the range of 240-600 nm.

2.3. Photocatalytic activity testing

The experiment of the photo degradation was carried out in a photoreaction system. Ultraviolet irradiation was offered by a 500-W high-pressure mercury lamp with major emission at 365 nm. The reactive bottle was a 250-mL cylindrical vessel with a water-cooled quartz jacket, and the lamp was located in the center of the quartz jacket. A magnetic stirrer was equipped at the bottom of the reactor to achieve effective dispersion. Air was bubbled through the reaction solution from the bottom to ensure a constant dissolved O_2 concentration. The amount of catalysts was 2 g/L. The initial MO concentration was 40 mg/L, and 250 mL MO solution was used for the photocatalytic degradation every time. Some aqueous samples were withdrawn at different intervals. The residual concentration of MO was measured at 465 nm with a spectrophotometer (DR/2500, America HACH Company).

3. Results and discussions

3.1. The effects of Ce-doped amount

In order to investigate the effects of doped amount, Ce-doped TiO_2 samples with different doped amount (i.e. 0.05%, 0.10%, 0.50% and 2.00%) and calcined at 500 °C were prepared. Figure 1A shows the XRD patterns of these samples. Only the single anatase TiO_2 was detected in all samples. With the increase in Ce-doped amount, the diffraction peaks of anatase phase become broader in width and weaker in intensity, which implies that Ce doping inhibited the growth of nanocrystallite size. In addition, the mean crystalline sizes calculated by the Scherrer equation were summarized in Table 1, from which it is observed that the crystalline size decreased with the increase in Ce-doped amount, which can be ascribed to that a given amount of Ce ions has entered into the lattice of TiO_2 and the Ti-O-Ce bonds have formed on the surface of TiO_2 during calcination, which retarded the contact of particles, the transfer and rearrangement of Ti and O in particles [6]. Smaller crystal size means more powerful redox ability due to the quantum size effect, and smaller crystal size is in favour of the shifting of photogenerated carriers to the surface of photocatalyst and reacting with reactant. That is, Ce doping improved the transfer efficiency of photogenerated carriers to surrounding supports, and then, enhanced the photocatalytic activity of TiO_2. From Table.1, it can be seen that unit cell volumes and lattice distortion of Ce-doped TiO_2 particles were larger than that of un-doped TiO_2, meaning that the entering of Ce ions into the TiO_2 lattice resulted in the distortion and expansion of crystal lattice. It is well known that oxygen molecules are easy to be adsorbed as electron trap on the sites of structure defects brought by distortion and expansion of crystal lattice. Therefore, the recombination rate of electron and hole pairs can be reduced, which has a positive influence on the photocatalytic activity of TiO_2 [7].

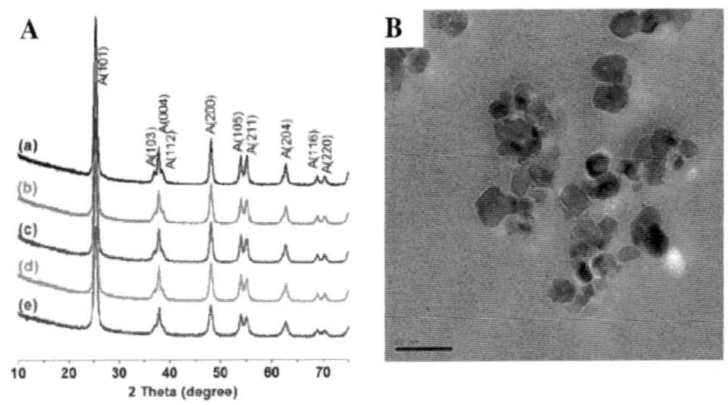

Figure 1. (A) XRD patterns of samples (A: Anatase): (a) Ti500, (b) TiCe(0.05)500; (c) TiCe(0.10)500; (d) TiCe(0. 50)500; (e) TiCe(2.00)500; (B) TEM image of TiCe(0.05)500.

Table 1. XRD analysis results of all samples.

Sample	Crystal size (nm)	Matrix distortion (%)	Crystal parameter		
			a (nm)	c (nm)	V (nm^3)
Ti500	18.2	0.8595	0.3892	0.8084	0.1225
Ti600	51.9	0.3012	0.3888	0.8166	0.1234
Ti700	79.1	0.1976	0.3898	0.8127	0.1235
Ti800	101.8	0.1534	0.3884	0.8221	0.1240
TiCe(0.05)500	16.3	0.9610	0.3892	0.8119	0.1230
TiCe(0.10)500	16.2	0.9669	0.3883	0.8302	0.1252
TiCe(0.50)500	15.4	1.0719	0.3897	0.8197	0.1245
TiCe(2.00)500	13.6	1.1481	0.3891	0.8154	0.1234
TiCe(2.00)600	13.2	1.1838	0.3886	0.8235	0.1244
TiCe(2.00)700	15.3	1.0208	0.3878	0.8286	0.1246
TiCe(2.00)800	41.0	0.3816	0.3888	0.8276	0.1251

Figure 2A shows the evolution of DRS of TiO_2 doped with different Ce amount and calcined at 500 ^{o}C. Compared with the spectrum of un-doped TiO_2, a red-shift of the absorption profile and photons absorption increase in the range of 400-600 nm in the Ce-doped TiO_2 were observed, and the red-shift extent increased with the increase in Ce-doped amount. This red-shift can be attributed to the charge-transfer transition between the rare earth ion *f* electrons and the TiO_2 conduction or valence band [8]. During the photocatalytic process, electrons are excited and transferred from valence band to conduction band when photons are absorbed by photocatalyst, thus electron-hole pairs generate. The electrons in the conduction band are captured by oxygen molecules dissolved in the suspension, and the holes in the valence band are captured by OH^- or H_2O species adsorbed on the surface of catalysts, to produce the hydroxyl radical. So the catalysis activity is related with the number of electron-hole pairs in TiO_2. The more electron-hole pairs exist on the surface of TiO_2, the more highly efficiency obtains the catalysis activity in decoloration [9]. The number of electron-hole pairs depends on optical property of samples. The longer wavelength light can be absorbed, under the same light irradiation, the more electron-hole pairs will be formed on the surface of TiO_2. Ce doping leads to red-shift of the absorption profile and photons absorption increase in the range of 400-600 nm, which are advantaged to generating of electron and hole under light irradiation.

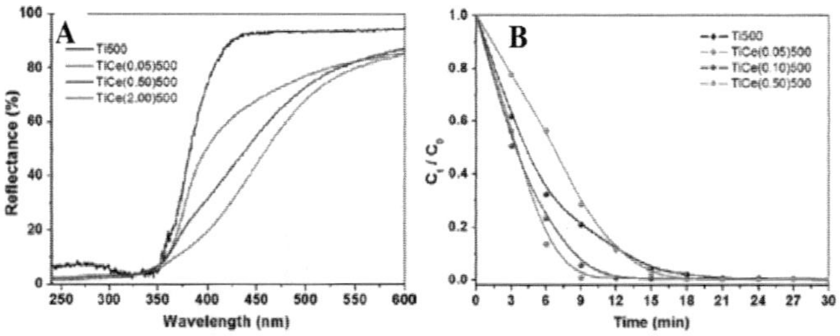

Figure 2. (A) DRS spectra of samples; (B) Degradation curves of MO over samples.

The photocatalytic activities of the samples were measured by the degradation MO solution without concerning the degradation intermediates in detail. The residual concentration ratios C_t/C_0 of MO versus time under mercury lamp irradiation are obtained and shown in Figure 2B. Where, C_t and C_0 (mg/L) are signed as t time and the initial concentrations of MO in solutions, respectively. It can be seen that doping with proper Ce ions can improve the photocatalytic activity of TiO_2, and the reasons may be attributed to the smaller crystallite size, lattice distortion, the red shift of the absorption profile, and photons absorption increase in the range of 400-600 nm because of Ce doping. The optimal doped amount is 0.05 mol% for the maximum photocatalytic degradation efficiency.

3.2. The effects of calcination temperature

Figure 3 shows the XRD patterns of un-doped TiO_2 and Ce-doped TiO_2 calcined at different temperature. It can be seen that all samples presented anatase structure except for Ti800, in which 11 wt% rutile structures appeared, indicating that phase transition from anatase to rutile was inhibited due to Ce doping. It can be seen from Table. 1 that the crystal size increased with the increase in calcination temperature, and the increased velocity of un-doped TiO_2 was more rapid than that of Ce-doped TiO_2, implying that Ce doping inhibited the growth of crystal size.

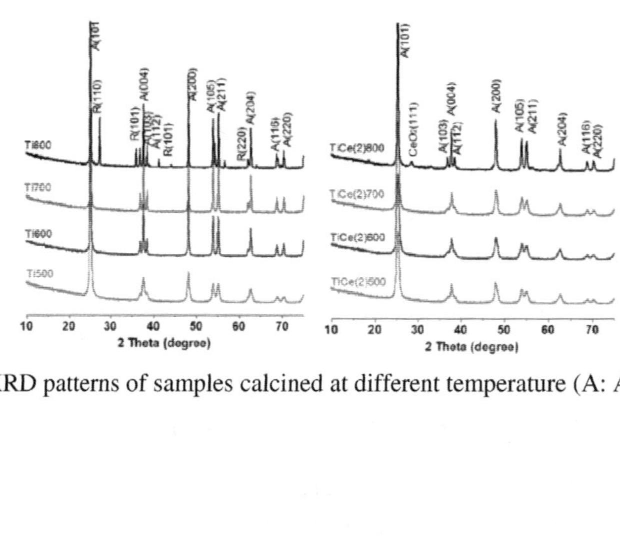

Figure 3. XRD patterns of samples calcined at different temperature (A: Anatase; R: Rutile).

Figures 4A and B shows the effect of calcination temperature on the DRS results of un-doped and Ce-doped TiO_2 (doped amount is 0.05%), respectively. Some red-shift of the absorption profile was observed with calcination temperature increase from Figure 4A. The red-shift extent of the sample calcined at 800 °C was the biggest among the four samples, which can be attributed to the appearance of rutile structure confirmed by XRD pattern. Compared with the spectrum of un-doped TiO_2, the red-shift of the absorption profile and photons absorption increase in the range of 400-600 nm of Ce-doped TiO_2 were also observed from Figure 4B.

Figure 4C shows the photocatalytic degradation results of Ce-doped and un-doped TiO_2 calcined at different temperature. In comparison with un-doped TiO_2, Ce doping improved the photocatalytic activity of TiO_2 markedly. The effects of calcination temperature on photocatalytic activity of un-doped TiO_2 and Ce-doped TiO_2 were very obvious. The optimal calcined temperature was 600 °C. Furthermore, the Ce-doped sample revealed its higher catalytic activity. The reason may be ascribed to the restraining of crystal size increase, the lattice distortion and expansion, the red-shift of the absorption profile and photons absorption increase in the range of 400-600 nm.

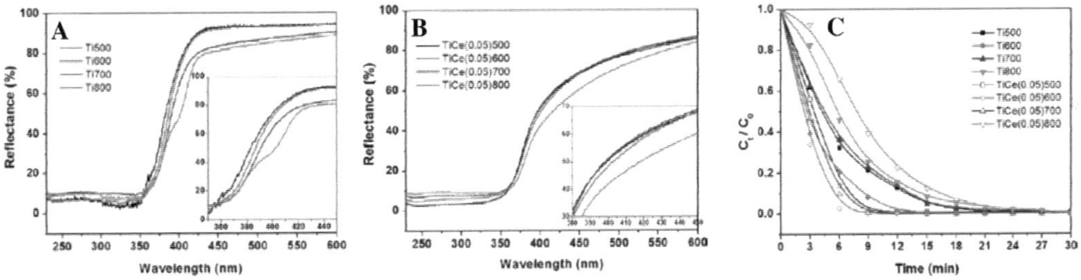

Figure 4. (A) DRS spectra of TiO_2 calcined at different temperature; (B) DRS spectra of Ce-doped TiO_2 calcined at different temperature; (C) Degradation curves of MO over samples calcined at different temperature.

4. Conclusions

Ce doping decreases the crystal size of TiO_2 nanoparticles, which is in favour of the shifting of photogenerated carriers to the surface of photocatalyst and reacting with reactant. Ce doping leads to lattice distortion and expansion of TiO_2, which reduces the recombination of electron and hole, and then produces more hydroxyl radical. Furthermore, Ce doping brings a red-shift of the absorption profile and photons absorption increase in the range of 400-600 nm, which is advantaged of the generation of electron and hole under light irradiation. Therefore, Ce doping is an effective mean to improve the photocatalystic activity of TiO_2 for degradation organic pollutants in water.

Acknowledgements

This work was sponsored by the Fundamental Research Funds for the Central Universities (2015gjhz06).

References

[1] Chen X and Mao SS, 2007 *Chem. Rev.* **107** 2891
[2] Liu G, Wang L, Yang HG, Cheng H-M and Lu GQ, 2010 *J. Mater. Chem.* **20** 831
[3] Shi J-w, Zheng J-t and Wu P, 2009 *J. Hazard. Mater.* **161** 416
[4] Xie YB and Yuan CW, 2003 *Appl. Catal. B* **46** 251
[5] Xiao J, Peng T, Lia R, Peng Z and Yan C, 2006 *J. Solid State Chem.* **179** 1161
[6] Xu Z, Yang Q, Xie C, Yan W and Du Y, 2005 *J. Mater. Sci.* **40** 1539
[7] Wu X, Qin W, Ding X, Wen Y, Liu H and Jiang Z, 2007 *J. Phys. Chem. Solids* **68** 2387
[8] Xu AW, Gao Y and Liu HQ, 2002 *J. Catal.* **207** 151
[9] Gou YQ, Chen DY and Su ZX, 2004 *Appl. Catal. A* **261** 15

Spin electronic manipulation based on zigzag-edgegraphene nanojunction with a line defect

Haidong Li[1], Lili Zheng and Ruixue Li

Henan University of Engineering, College of Science, No. 1, Xianghe Road, Xinzheng, 451191, Zhengzhou, China

E-mail: lihd06@mails.jlu.edu.cn

Abstract. We investigate the transport of spin electron through the zigzag-edge graphene nanojunction with a line defect. When the magnetization is zero, the conductance spectra exhibits a well-defined insulating band around the point far away from Dirac point. And the width of the insulating band is exactly equal to the energy splitting between the lowest conduction band of the left lead and the new state of the device region. For the parallel configuration, with the enhancement of the magnetization, the conductance value will be reduced by half around the Dirac point. For the antiparallel configuration, the width of the well-defined insulating band becomes larger with the rising of the magnetization.

1. Introduction

Graphene is the first experimental realization of truly two-dimensional crystal[1], which have exhibited many unusual physical properties in comparison with the conventional semiconductor materials[2,3], rendering it attractive for use in electronics. And the long spin relaxation times[4,5] and the small spin-orbit coupling, further promoting graphene as an interesting material for spintronics applications[2, 6]. Meanwhile, the spin injection from ferromagnetic (FM) metal contacts into graphene has been experimentally achieved[7,8]. The studies have predicted that the spin-polarized carriers can be realized by growing the graphene on a FM insulator[9], others have reported spin current injection in graphene using dynamical and thermal methods[10,11], the nonlocal magneto-resistance measurements have performed, and the graphene-based spin-valve device could have a high magnetoresistance[12,13], so the graphene have attracted a lot of attentions in spin-based quantum computing and spintronics applications. The experimental discovery a peculiar topological line defect in graphene, which can be created by alternating the Stone-Thrower-Wales defect and divacancies[14], extensive attention has been given to this field. The first-principles calculation and experimental observations show that this line defect acts as a one dimensional metallic wire, which motivates researchers to further discuss the electron properties or design the schemes of valley polarization [14,15].

Inspired by possible potential device applications, in this work, based on the Landauer-Butikker formula and the tight binding model, we investigate the transport of spin electron through the zigzag-edge graphene nanojunctions (ZGNJs) with a line defect.

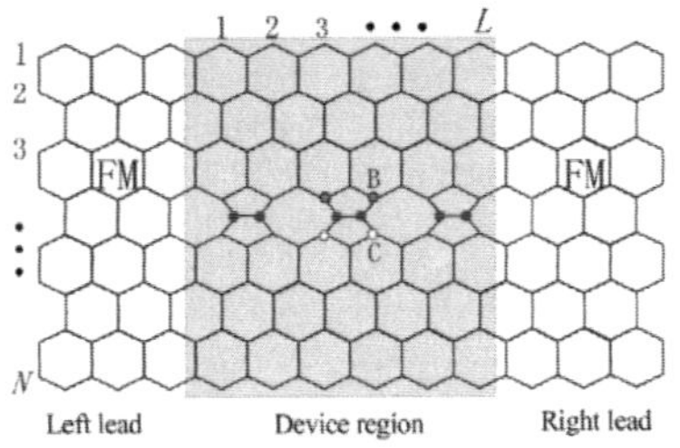

Figure 1. The schematic of the ZGNJs with a line defect.

2. Model and theory

The structure of graphene nanojunction with zigzag-edge is depicted in figure 1, the structure has three regions: the two semi-infinite FM graphene nanoribbons at the left and the right sides serve as two leads for electronic transmission, which are needed to characterize a clear junction. The central region with a line defect constitutes the device region (shaded), where the electron tunneling is scattered. We use the parameters N and L as the latitudinal and longitudinal size, respectively. We adopt the tight-binding Hamiltonian with nearest-neighbor in-plane interaction to describe the low-energy electronic properties of the graphene nanoribbons. In the tight binding approximation, the single layer of graphene can be described by the following Hamiltonian:

$$H_T = H_d + H_\alpha + H_T \quad (1)$$

the first term H_d is the tight-binding Hamiltonian of the π band electron in the device region.

$$H_C = \sum_i d_i^\dagger(\varepsilon_\alpha \sigma_I + \sigma \cdot M_\alpha)d_i + t\sum_{\langle i,j \rangle}(d_i^\dagger \sigma_I d_j + H.c) \quad (2)$$

where d_{is}^{+} (d_{is}) is the electron creation(annihilation) operator associated with spin on the local atomic orbits i in graphene, and $d_{is}^{+} = (d_{i\uparrow}^{+}, d_{i\downarrow}^{+})$. The notation <i,j> means that the summation is restricted between the pairs of the nearest neighbor carbon atoms. t is the corresponding hopping energy. The second term H_α is the Hamiltonian of the two semiinfinite leads.

$$H_\alpha = \sum_i d_i^\dagger(\varepsilon_\alpha \sigma_I + \sigma \cdot M_\alpha)d_i + t\sum_{\langle i,j \rangle}(d_i^\dagger \sigma_I d_j + H.c) \quad (3)$$

$\sigma = (\sigma_x, \sigma_y, \sigma_z)$ are the Pauli matrices and I is a 2 × 2 unit matrix. ε_d, ε_L and ε_R are the on-site energies in the center region, left and right leads, respectively. M_L and M_R are the magnetizations of the left and right FM leads, which can be along arbitrary direction. The magnetization value in the left and right FM leads are assumed to be equal $M_L = M_R = M$. The last term H_T in the Hamiltonian describes the electron tunneling between the leads and graphene.

$$H_T = t \sum_{\langle i,j \rangle (i \in C, j \in \alpha)} (d_i^\dagger \sigma_I d_j + H.c) \quad (4)$$

with the help of the Landauer-Butikker formula in the discrete lattice representation[16], the linear conductance $\mathcal{G}$(E) as a function of the incident electron energy E can be evaluated. It gives

$$\mathcal{G}(E) = \frac{e^2}{h} Tr(\Gamma_L G^r \Gamma_R G^a) \quad (5)$$

where G^r is a retarded Green function, $\Gamma_{L(R)}$ is the coupling function. ε_M is the on-site energies of the line defect. We define the magneto-resistance as $R_M = (\mathcal{G}_P - \mathcal{G}_A)/\mathcal{G}_P \times 100\%$, where $\mathcal{G}_P$ and $\mathcal{G}_A$ are the conductances of the parallel and antiparallel configurations, respectively.

3. Results and discussion

Following the above theory, in the numerical calculation, we will select a unified constant t as the corresponding hopping energy, the lattice constant a is used as the units of the length, the zero point of the energies as the Fermi energy level. When the magnetization M = 0, in figure 2 we plot the conductance spectra ($\mathcal{G}$ vs E) for the ZGNJs. First, we fixed the width of N = 12 and the center region longitudinal size L=20 of the ZGNJs. And applying the gate bias ε_d to the device region. The gate bias can be simulated by shifting the on-site energy of all the lattice points. We can see that the conductance drops to zero and presents a well-defined insulating band in the vicinity of the Dirac point. And with the rising of ε_d further, the width of the well-defined insulating band becomes larger. In this regard, we can give an explanation by the sub-band selective rule[17]. Briefly the energy-band of the ZGNJs on the up and down side of the Fermi surface is symmetric and comply with sub-band selective rule, or the electrons belonging to the even (odd) parity subbands in the left ZGNJs are transported only into the even (odd) parity subbands of the right ZGNJs. In addition, we can see that the conductance zero gradually disappear when the ε_d becomes very larger. However when we only change the on-site energy ε_m of the line defect, from the figure 2 we can see that the conductance value remains no change ($\mathcal{G}$=2 e^2 /h) in the vicinity of the Dirac point. But the width of the conductance valley gradually narrows with the rising of ε_m at the point far away from Dirac point. We can infer that the conductance insulating band arises from the linear defect, which leads to a new zero conductance at the point far away from Dirac point.

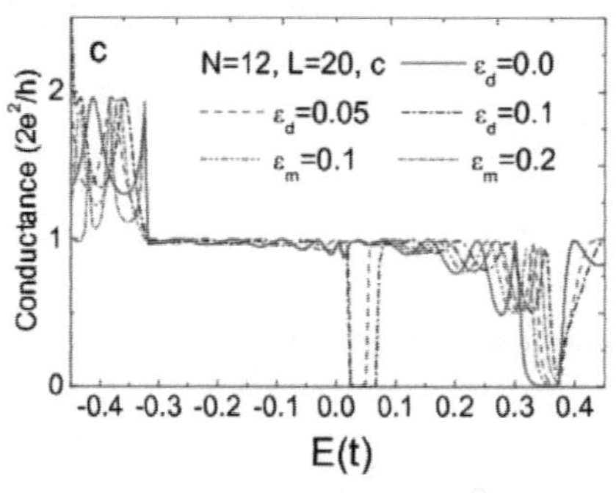

Figure 2. The comparison of the conductance spectra of the ZGNJs, nc representative does not contain line defect, and c represent contains line defect. Change the on-site energy ε_d and ε_m.

In order to further study the influence of the line defect on the conductance spectra of the system, we plot the sub-band structures of the pristine ZGNJs (N = 12, nc) and the ZGNJs with a line defect (N = 12, c). For a charge neutral junction, from the figure 3, we can see that the Dirac points of the two ribbons are aligned with each other, and two additional states (red line) are included in contrast to the band structure of the pristine ZGNJs. Here we use Δ to label the energy splitting between the lowest conduction band and the Fermi energy of the left lead. And use Δ_m to denote the new state of the device region above the Fermi energy. From the figure 3(b) we also can find that the new state could be tuned by the on-site energy "m of the line defect. Comparing with the conductance spectrum (show in figure 2), we can see that the width of the well-defined insulating band at the point far away from Dirac point is exactly equal to the energy splitting $|\Delta-\Delta_m|$. Accordingly, we found a theoretical explanation that the on-site energy of the line defect can change the width of the well-defined insulating band.

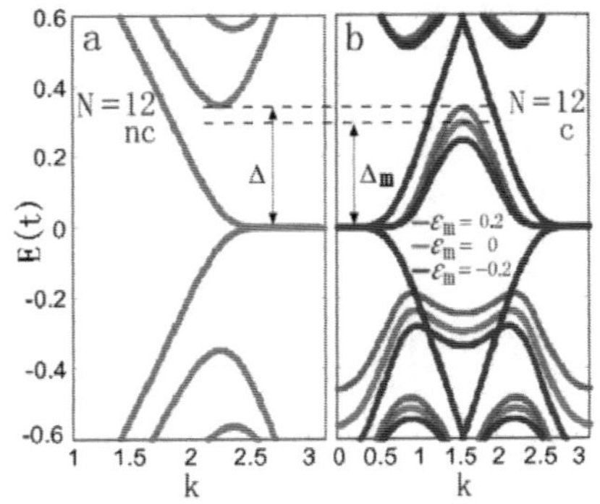

Figure 3. [(a) and (b)] The sub-band structures of the pristine ZGNJs and the ZGNJs with a line defect. Δ and Δ_m denote the energy splitting.

In figure 4(a) and (b), we plot the conductance spectrum of the magnetization $M \neq 0$. In this way we can observe the spin valve effect in the ZGNJs with a line defect. According to the energy band structure of the pristine ZGNJs, we can calculate the energy splitting Δ between the lowest conduction band and the Fermi energy $E_F = 0$, where Δ=0.34 and Δ=0.27 for N=12 and N=16, respectively. In this work, we focus on the conductance behavior in the low-energy region. This is because at this level range, only the lowest electronic conductance channels are involved in electron transport. Therefore, we consider only the case of $M< \Delta$. In addition, for the direction of magnetization, we only consider the parallel $\theta = 0$ and antiparallel $\theta = \pi$ configurations. Firstly, in the parallel configuration, the results show in the figure 4(a). We can see that the conductance value decreased from G=2 e^2/h to G= e^2/h with the enhancement of M in the Dirac point. But the conductance insulating band gradually disappears at the point far away from Dirac point. Secondly, for the antiparallel configuration, from the figure 4(b), we can see that the conductance value decreased from G=2e^2/h to G=0 in the vicinity of the Dirac point and generating the conductance insulating band, and the width of insulating band becomes larger with the rising of the magnetization M. At the same time, we also can find that the width of insulating band becomes larger at the point far away from Dirac point. Similarly, when we change the size of the central region, we also can find the same conclusion, which suggests that the impact of the magnetization is universal to the conductance spectra.

For these characteristics of the conductance when the magnetization $M \neq 0$, we can give explanation according to the energy band structure. Firstly, for the parallel configurations, the zigzag-edge energy band structure will split and no longer degenerate together with the enhancement of M, which will cause the spin-up electrons and the spin-down electrons have only a single electron conductive channel, respectively. So the conductance value decreased from G=2e^2/h to G=e^2/h in the vicinity of the Dirac point. Similarly, the splitting can make the energy difference between the lowest conduction band of the left lead and the new state (red line) of the device region decrease. Furthermore it can cause the conductance insulating band gradually disappear at the point far away from Dirac point. Secondly, for the antiparallel configurations, the energy band of the spin-up and the spin-down electrons in the left and the right lead will split along the opposite direction with the enhancement of M, according to the band-selective rule, the spin-up electrons and the spin-down electrons in the center region is no electron conductive channel when the incident electron energy around the Dirac point. Which make the conductance value decreased from G=2e^2/h to G=0 in the Dirac point and at the point far away from Dirac point.

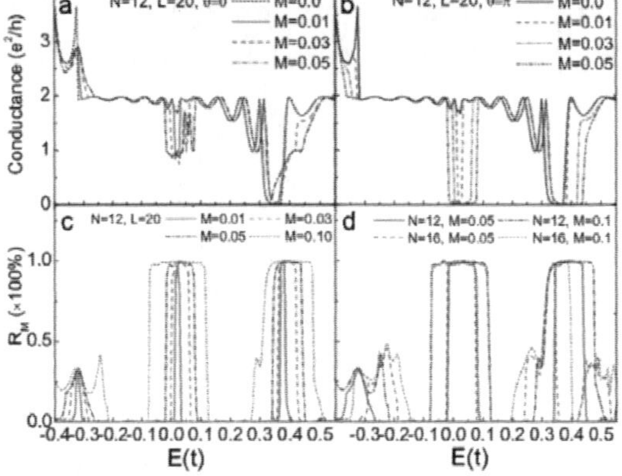

Figure 4. [(a) and (b)] A comparison of the conductance spectrum in parallel configuration and in antiparallel configuration, respectively. (c) The magneto-resistance R_M for different the magnetization M. (d) A comparison of the magneto-resistance by changed the width N and the magnetization M.

Based on the conductances of the parallel and antiparallel configurations, we can plot the magneto resistance RM as a function of the incident electron energy E for different magnetization M. From the figure 4(c) we can see that the magneto-resistance is very large as far as R_M =100% at the Dirac point and at the point far away from Dirac point. In this work, we define the R_M =100% as the unit magneto-resistance (UMR). Firstly, by contrast with the band structure, we can find that the width of the UMR at the Dirac point exactly equal to 2M, and these results for the pristine ZGNJs are also suitable. Secondly, when we fixed the magnetization M and change the width N of the ZGNJs (N=12 and N=16), from the figure 4(d) we can see that the width of the UMR remains unchanged around the Dirac point. But for the UMR at the point far away from Dirac point, we can see that the plateau width becomes narrow with the increase of the width N. Similarly, when we fixed the width N and change the magnetization M (M=0.05 and M=0.1), we can find that the width of the two UMR increases with the enhancement of the magnetization M. Meanwhile, we also find the width of the UMR is not affected by the longitudinal size L. By comparing conclusions of figure 4(a) and (b), we can easily find the UMR is derived from the zero conductance in antiparallel configurations. It is worth noting that the plateau width of the UMR (at the point far away from Dirac point) is determined by magnetization M and the size of the center region N. This indicates that it can be considered as a promising candidate for realizing the spin electron transport manipulation.

4. Conclusion

In this work, we have performed a theoretical investigation about the spin electronic manipulation based on ZGNJs with a line defect. The results show that the conductance spectra present a well-defined insulating band at the point far away from Dirac point when the magnetization M=0. We find that the width of the insulating band is exactly equal to the energy splitting $|\Delta-\Delta_m|$, and the width of the insulating band could be tuned by the onsite energy ε_m of the line defect. For the parallel configurations, with the enhancement of M, the conductance value decreased from $2e^2/h$ to e^2/h around the Dirac point, and the insulating band gradually disappears at the point far away from Dirac point because of the energy band splitting. For the antiparallel configuration, the band-selective rule makes the width of the insulating band becomes larger with the rising of the magnetization. Based on the above results, we can find that the magneto-resistance is very large as far as 100%. Compared with the energy band structure, we find that the plateau width of the UMR at the Dirac point is exactly equal to 2M. And find the plateau width can be readily tuned by the magnetization M and the size N.

Acknowledgments
This work was supported by the Special Funds of the National Natural Science Foundation of China (Grant no.11147139), Key scientific research projects in colleges and universities in Henan Province (Grant no.16A140042), and supported by the Henan Provincial CollegeYouth backbone teachers (Grant no.2012GGJS-186).

References

[1] Novoselov K S, Geim A K, Morozov S V, Jiang D, Zhang Y, Dubonos S V, Grigorieva I V and Firsov A A 2004 *Science* **306** 666
[2] Zhang Y, Tan Y W, Stormer H L and Kim P 2005 *Nature* **438** 201
[3] Morozov S V, Novoselov K S, Katsnelson M I,Schedin F, Ponomarenko L A, Jiang D and Geim A K 2006 *Phys. Rev. Lett.* **97** 016801
[4] Min H, Hill J E,Sinitsyn N A,Sahu B R,Kleinman L and MacDonald A H 2006 *Phys. Rev. Lett.* **74** 165310
[5] Huertas-Hernando D, Guinea F and Brataas A 2006 *Phys. Rev. B* **74** 155426
[6] Novoselov K S, Geim A K, Morozov S V, Jiang D, Katsnelson M I, Grigorieva I V, Dubonos S V and Firsov A A 2005 *Nature* **438** 197
[7] Tombros N, Jozsa C, Popinciuc M, Jonkman H T and van Wees B J 2007 *Nature* **448** 571.
[8] Cho S, Chen Y F and Fuhrer M S 2007 *Appl. Phys. Lett.* **91** 123105
[9] Haugen H, Huertas-Hernando D and Brataas A 2008 *Phys. Rev. B* **77** 115406
[10] Tang Z, Shikoh E, Ago H, Kawahara K, Ando Y, Shinjo T and Shiraishi M 2013 *Phys. Rev. B* 87 140401
[11] Torres A, Lima M P, Fazzio A and da Silva A J R 2014 *Appl. Phys. Lett.* **104** 072412
[12] Jozsa C, Popinciuc M, Tombros N, Jonkman H T and van Wees B J 2009 *Phys. Rev. B* **79** 081402
[13] Han W, Wang W H, Pi K, McCreary K M, Bao W, Li Y, Miao F, Lau C N and Kawakami R K 2009 *Phys. Rev. Lett.* **102** 137205
[14] Lahiri J, Lin Y, Bozkurt P, Oleynik I I and Batzill M 2010 *Nat. Nanotech.* **5** 326
[15] Liu Y, Song J, Li Y, Liu Y and Sun Q F 2013 *Phys. Rev. B* **87** 195445
[16] Datta S 1995 Electronic Transport in Mesoscopic Systems (Cambridge University Press, Cambridge, England)
[17] Chen J, Cheng S, Shen S and Sun Q 2010 *J. Phys.: Condens. Matter* **22** 035301

A novel composite hydrogen sensor based on Pd nanoclusters/TiO_2 nanotube arrays

X B Wei[1,3], X H Yang[1], T Wu[1], H Y Pang[1], G D Lv[1], S H Wu[1], S B Li[1] and Z Chen[1,2]

[1] School of Optoelectronic Information, University of Electronic Science and Technology of China, Chengdu, 610054, China

[2] Department of Electrical & Computer Engineering and Center for Nanoscale Science &Engineering, University of Kentucky, Lexington, Kentucky 40506, USA

E-mail: weixiongbang@uestc.edu.cn

Abstract. A novel composite hydrogen sensor, consisting of Pd nanoclusters/TiO_2 nanotube arrays, was fabricated and evaluated at room temperature. The Pd nanoclusters layer was deposited on top surface of the nanotube arrays by using a direct current (DC) magnetron sputtering method. Resistive response of the composite sensors to 0.5% hydrogen was measured. Experimental results indicated that the Pd nanoclusters can quickly and continually form or break multiple passages by absorbing or desorbing hydrogen, so that the composite hydrogen sensors have promising hydrogen sensitivity at room temperature.

1. Introduction

As an important reductive gas and clean energy, hydrogen is being widely used in chemical, aviation, medical, petrochemical, transportation and energy [1-4]. However, several security issues should be taken into account because of the low spark ignition energy and explosive with wide flammable range (4%–75%) of H_2 in air. Hence, a sensor with high sensitivity, fast response and recovery times, long time stability and low cost is needed to detect the leakage of hydrogen at room temperature. In recent years, various structures and mechanisms of hydrogen sensor have been reported[5-7]. Nanostructured metal oxide films, such as SnO_2, ZnO and WO_3, have been found to be the active materials of hydrogen sensors, duo to their high surface-to-volume ratio, simple working principle and portability[8-10].

Recently, one-dimensional titanium dioxide nanotubes have attracted significant interest as an attractive sensor material, because of its unique structure and special physical/chemical properties. Varghese and co-workers studied hydrogen sensing properties of titania nanotubes made through anodization. The results showed the nanotubes with smaller pore diameter (46 nm) have promising sensitivity to hydrogen at 290℃[11]. Erdem Sxennik et al. reported that sensors based on TiO_2 nanotubes synthesized through anodic oxidation of a titanium foil in an aqueous solution showed a good sensitivity, but relatively poor response/recover behaviours at room temperature[12]. Moreover, the high operating temperature (200℃ to 500℃) limited their applications.

In this work, we present a simple method to fabricate a novel "network" of Pd nanoclusters on the surface of TiO_2 nanotube arrays by using DC magnetron sputtering, with the amount of the Pd to be controllable by controlling the sputtering parameters. We will demonstrate that the Pd nanoclusters/TiO_2 nanotubes composite has promising hydrogen sensitivity at room temperature.

2. Proposition and design of the composite structure

Compared with TiO_2 powder, TiO_2 nanotubes have larger specific surface area and stronger adsorption capacity, and thus showed higher hydrogen sensitivity. Resistance of the TiO_2 nanotube arrays will be significantly decreased after they absorb hydrogen gas, which will recover to original state when the hydrogen gas is absent. TiO_2 nanotube surface has a strong chemical adsorption to the dissociated H atom, so as to make partial charge transfer from H to the TiO_2 conduction band, which produces an electron accumulation layer on surface of the TiO_2 nanotubes, resulting in an increase in the nanotube conductance. When the hydrogen gas is absent, the electron is transferred back to the H atom, so that the TiO_2 nanotubes return to their original resistance. Therefore, TiO_2 nanotubes have a sensitive resistance response to hydrogen. Although the TiO_2 nanotube arrays have good sensitivity and selectivity, high operating temperature limit their applications as hydrogen sensors.

Under normal circumstances, resistance of metallic Pd membrane will become larger after hydrogen adsorption. However, when the metal Pd membrane made of nanoclusters structure, in the non-absorbed hydrogen state, the binding between nanoclusters is not sufficient, so it has a higher resistance. As each nanoparticle absorbs hydrogen gas and the expansion of the phase shift from the alpha phase to the beta phase, the bonding between the neighboring particles will reduce the resistance. Therefore, this Pd nanoclusters based hydrogen sensor have a good response at room temperature. Before the hydrogen absorption, the mesoscopic Pd nanoclusters are in the off state, resistance value is very large, in the uptake of hydrogen, Pd nanoclusters expanded and connected to each other, resulting in resistance decreases, which is named "break-junction" mechanism. The resistance change of this structure showed the opposite trend with the metallic palladium (Pd) membrane

The results of previous studies showed that the response time of the hydrogen sensor prepared by Pd thin film deposited on anodic aluminium oxide (AAO) and other porous substrates is much faster than that of the sensor prepared by the corresponding flat substrate[13-16]. In order to prepare the sensor with faster response, it is needed to prepare a kind of good near discontinuity network structure to isolate the Pd nanometer point, and to prepare the highly sensitive hydrogen sensor using the "break-junction" principle of Pd nanoclusters. Based on Chen's former work on " break-junction " principle of Pd nanoclusters, TiO_2 nanotube arrays was determined as substrate to deposited Pd nanoclusters. Using the network of TiO_2 nanotube arrays structure, good process controllability and large hole and thin wall can make the Pd films become quasi isolated nano-islands, and also can make full use of the hydrogen sensing properties of TiO_2 nanotubes. At the same time, it also make use of the self-cleaning function of TiO_2 nanotubes.

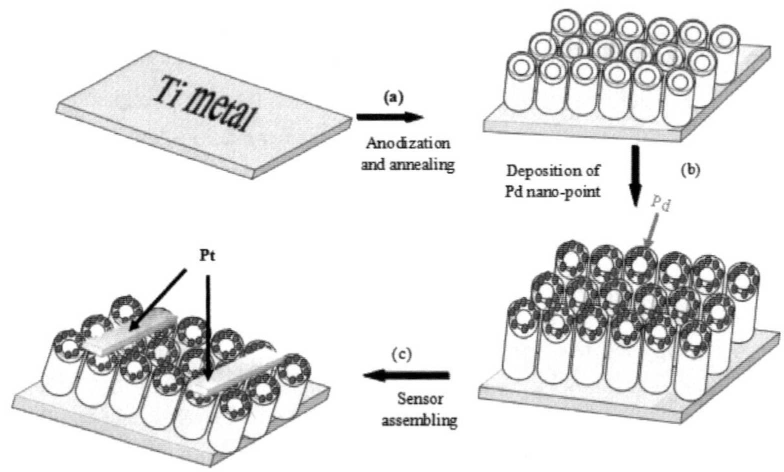

Figure 1. The Pd/TiO_2/Ti multilayer structure preparation process.

Thus the Pd/TiO_2/Ti multilayer and it preparation process are illustrated in Figure 1. Conductive copper wires were connected to the Pt electrode with silver conductive paint. The morphology of TiO_2 nanotubes and the surface composition of Pd were characterized using scanning electron microscope (SEM; FEI-Inspect F50, Holland). X-ray diffractometer (XRD; Bede D1 with Cu Kα radiation) was used to investigate the crystalline phase of the prepared TiO_2 nanotubes. The hydrogen sensitivity detection was conducted in a home-made chamber with a volume of 60 ml. Before each test, ambient air was circulated inside the chamber. All of the sensor testing was carried out at room-temperature. A Keithley 2700 multimeter (Tektronix (China) Co., Ltd, Chengdu, China) was used as a data collector to record the variation of the sensors in the atmospheres of H_2 in/out.

3. Preparation and hydrogen sensitivities of the Pd nanoclusters/ TiO_2 nanotube arrays composite structure

3.1. Preparation of TiO_2 nanotube arrays

For the growth of titania nanotubes, a 0.1-mm-thick Ti foil (99.8% purity, HeFei Ke Jing Materials Technology Co., LTD, Hefei, China) was cut into 20×25 mm^2 pieces. The titanium foil was then sonicated in acetone, ethanol, deionized water for 20 min and dried at 65℃ in drying cabinet. TiO_2 nanotubes were fabricated on the Ti substrate by electrochemical anodization[17,18]. TiO_2 nanotube arrays were first prepared by anodic oxidation of NH_4F/ glycol organic electrolyte aqueous solution in 0.25wt% for 3 hours, and then were annealed at 450 ℃ in atmosphere for 3 hours. After annealing treatment, the amorphous TiO_2 nanotubes were transformed into anatase (A) crystal phases with A (101), A (004), A (105), (200), A (211) strong characteristic peaks. Figure 2 (a)(b) showed the XRD diagrams of TiO_2 nanotubes before and after annealing. Figure 3 demonstrated the surface morphologies of TiO_2 nanotube arrays. The TiO_2 nanotube arrays prepared were very regular. The triangle area between the pipe and the pipe is very obvious.

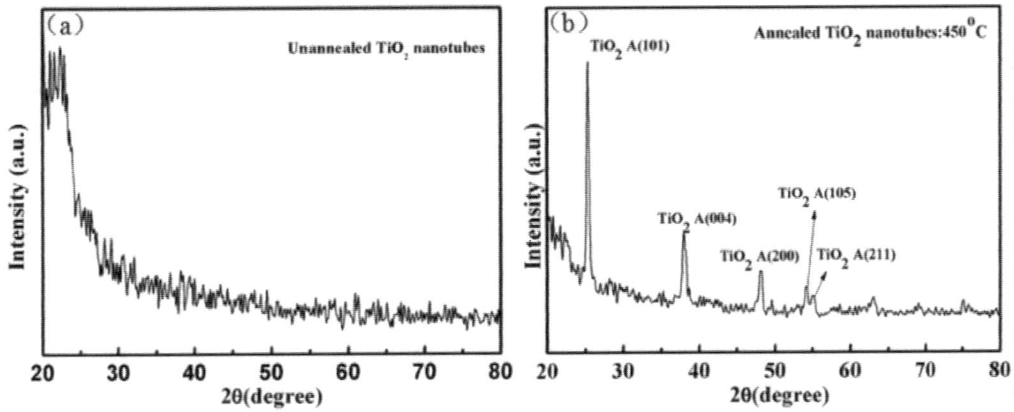

Figure 2. XRD diagrams of TiO2 nanotubes before and after annealing.

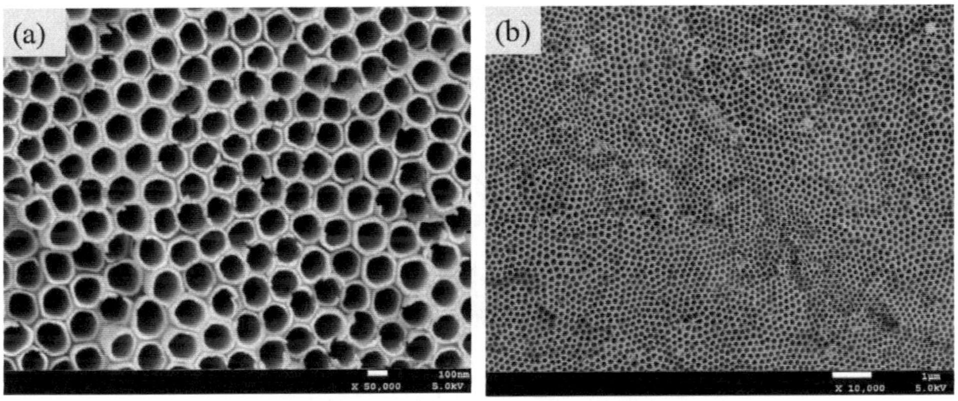

Figure 3. Surface images of TiO2 nanotube arrays with magnification of (a)50000, and (b)10000.

3.2. Preparation of the Pd nanoclusters and measurements of the Pd nanoclusters/TiO_2 nanotube arrays composite structure

The Pd nanoclusters were isolated from the surface of TiO_2 nanotubes by magnetron sputtering, and the parameters such as vacuum degree, sputtering power and sputtering time were controlled. Figure 4 was a typical hydrogen response-recovery curves of Pd nanoclusters/ TiO_2 nanotube arrays composite structure. In this structure, the Pd nanoclusters was 30 nm-thickness and the TiO_2 nanotube arrays pore diameter was about 90 nm. The response and recovery time of the sensor were 21 s and 23 s at room temperature and 0.5% hydrogen concentration.

Figure 4 is a typical 30 nm-thickness Pd nanocluster/90 nm pore diameter TiO_2 nanotubes composite structure surface SEM diagram. As can be seen from the figure, Pd nanoclusters show an obvious quasi-isolated state. The resistance change of the composite before and after the injection of hydrogen proved this point, as shown in Figure 5.

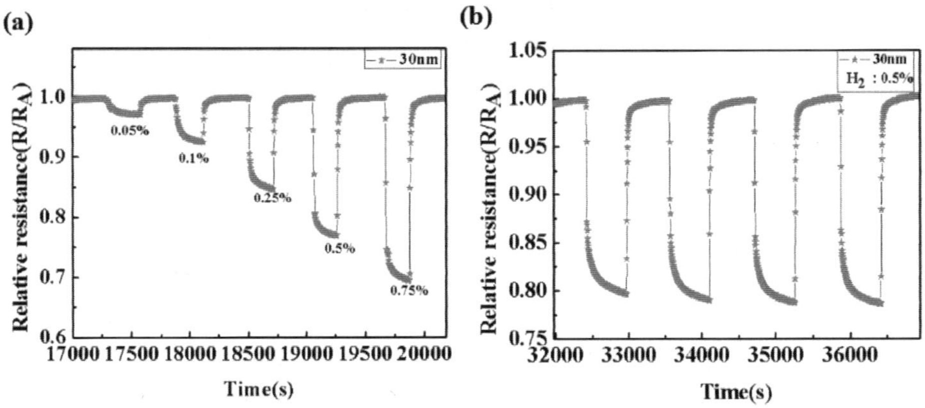

Figure 4. Hydrogen response-recovery curves of the Pd nanoclusters/TiO_2 nanotube arrays composite structure with Pd nanoclusters 30 nm-thickness and TiO_2 nanotube arrays pore diameter about 90 nm (a) response-recovery curves at different concentrations; (b) device performance repeatability test with hydrogen concentration of 0.5%.

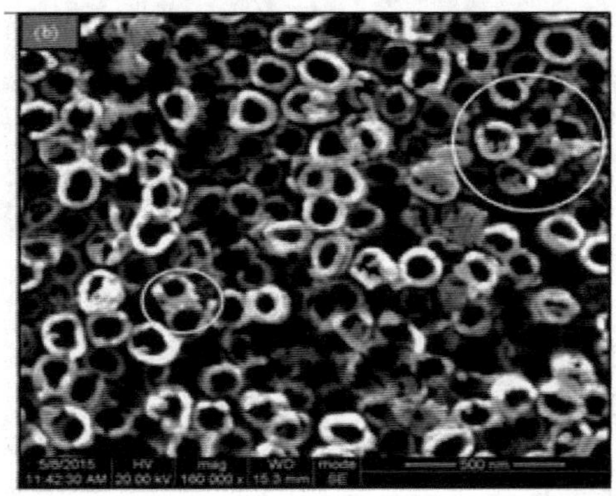

Figure 5. SEM image of a typical 30 nm-thickness Pd nanocluster/90 nm pore diameter TiO_2 nanotubes composite structure surface.

As we all know, the Pd nanoclusters can absorb a large amount of H_2 and lead to rapid volume expansion. In the case, the expansion of Pd nanoclusters will cause the separated nanoclusters connected with each other and then one or more conducting passage between the electrodes were created, which will result resistance decrease. When the Pd deposited get a certain thickness, the separated Pd nanoclusters could connect with each other and create more conducting passages between the electrodes, which will result rapid resistance decrease. In this case, the ideal isolated Pd nanoclusters can quickly and continually form or break multiple passages by absorbing or desorbing hydrogen, and the hydrogen sensors have the optimum response and recovery characteristics.

4. Conclusions

A novel composite structural hydrogen sensor of Pd nanoclusters/TiO_2 nanotube arrays was fabricated and tested. This novel composite structure is based on the "break-junction" principle of Pd nanoclusters and make full use of the quasi-isolated areas on top of the TiO_2 nanotube arrays. Combined with the advantages of the structure, hydrogen sensing properties of the materials themselves, the potential room temperature hydrogen sensing sensor was obtained. Hydrogen sensor was fabricated by sputtering network Pd nanoclusters on TiO_2 nanotube arrays, which is capable of working at room temperature. Well-dispersed Pd nanoclusters was obtained on the TiO_2 nanotube arrays. The composite structure sensor has better hydrogen sensing performance when the diameter is about 90 nm for TiO_2 nanotube and the Pd nanoclusters is about 30 nm-thickness. A typical better hydrogen sensing performance was obtained that the response time and recovery time is 21 s and 23 s at 0.5% H_2, respectively.

Acknowledgements
This work was supported by National Natural Science Foundation of China under Grant Nos. 61474016 and 61405026, 61371046.

References
[1] Goltsov V A and Veziroglu T N 2002 *Int. J. Hydrogen energ.* **27** 719-23
[2] Ahmed S and Krumpelt M 2001 *Int. J. Hydrogen energ.* **26** 291-301
[3] Brown L F 2001 *Int. J. Hydrogen energ.* **26** 381-97
[4] Ohsawa I, Ishikawa M, Takahashi K, Watanabe M, Nishimaki K, Yamagata K, Katsura K I, Katayama Y, Asoh S and Ohta S 2007 *Nat. Med.* **13** 688-94
[5] Maffei N and Kuriakose A K 2004 *Sens. Actuators* B **98** 73-76
[6] Nedrygailov I I and Karpov E G 2010 *Sens. Actuators* B **148** 388-91
[7] Chiu S Y, Tsai J H, Huang H W, Liang K C, Huang T H, Liu K P, Tsai T M, Hsu K Y and Lour W S 2009 *Sens. Actuators* B **141** 532-37
[8] Kadir R A, Li Z Y, Sadek A Z, Rani R A, Zoolfakar A S, Field M R, Ou J Z, Chrimes A F and Kourosh K Z 2014 *J. Phys. Chem.* C **118** 3129-39
[9] Horprathum M, Srichaiyaperk T, Samransuksamer B, Wisitsoraat A, Eiamchai P and Limwichean S 2014 *ACS Appl. Mater. Interfaces* **6** 22051-60
[10] Mondal B, Basumatari B, Das J, Roychaudhury C, Saha H and Mukherjee N 2014 *Sens. Actuators* B **194** 389-96
[11] Varghese O K, Gong D, Paulose M, Ong K G and Grimes C A 2003 *Sens. Actuators* B **93** 338-44
[12] Sxennik E, Colak Z, Klnc N and Ozturk Z Z 2010 *Int. J. Hydrogen energ.* **35** 4420-27
[13] Ding D Y and Chen Z 2007 *Adv. Mater.* **19** 1996-99
[14] Ding D Y, Chen Z, Rajaputra S and Singh V 2007 *Sens. Actuators* B **124** 12-7
[15] Ding D Y and Chen Z 2006 *Sens. Lett.* **4** 331-3
[16] Ding D Y, Chen Z and Lu C 2006 *Sens. Actuators* B **120** 182-6
[17] Yang X H, Wei X B, Wu S H, Wu T, Chen Z and Li S B 2015 *J. Mater. Sci. Mater. Electron.* **26** 7081-5
[18] Wu T, Wei X B, Yang X H, Quan Y, Lv G D, Shi Y, Wu S H, Li S B, Chen Z 2016 *J. Mater. Sci. Mater. Electron.* **27** 2481-5

Liquid phase synthesis: preparation of co microspheres by hydrazine reducing method

Y Deng[1,3], Z Q Lu[1], F Xie[1], H Liang[1],Y J Wang[1] and X H Du[2]

[1]Chongqing University of Arts and Science, Chongqing 402160, China
[2]College of Materials Science and Engineering, Sichuan University, Chengdu 610065, China

E-mail: 397194810@qq.com

Abstract. The study synthesizes Co microspheres with perfect spherical shape by use hydrazine hydrate as reducing agent to reduce $CoCl_2$ in alkaline conditions at room temperature, the influence of NaOH addition amount on crystal morphology and structure of cobalt particle were studied. The results show that, under the condition of concentration of NaOH, 0.030g ml^{-1} and having ethylene glycol as solvent, are the best choice to synthesize Co microspheres, and the grain is smaller, suitable for industrial production. At the condition of less amount of NaOH, cobalt particle's surface is smooth spherical, and with the increase of NaOH, particles will deviate from the spherical morphology and the surface become rough. The first section in your paper

1. Introduction

Cobalt is a kind of strategic metals, which is the main material of metal composite material with its ductility property. [1] Studies show that spherical morphology cobalt particle is more suitable for metal composite materials due to its small contact surface, which make particles not easy adhering and disperse well. [2]Anymore, it can significantly reduce the aggregation, reduce cobalt pool of composite, therefore the spherical cobalt particle is the best shape of metal composite material application. [3]There are many methods to synthesize spherical cobalt particles, such as the adding of nodulizing agent and surfactant, spray, plasma method and so on. [4] But neither chemical nor physical method, the spheroidization rate is lower, making the shape come to spherical-like, resulting the particle contact area increasing, unfavorable to disperse. [5] In recent years, it was found that hydrazine hydrate reduction (HHR) method can synthesize perfect spherical metal particles,[6] based on it, this study discusses the condition of hydrazine hydrate method to synthesize cobalt microspheres, in order to contribute to the batch production of spherical cobalt particle.

2. Experiments

2.1. Reagents

Cobalt dichloride ($CoCl_2$), absolute ethyl alcohol, ethylene glycol, glycerol, Sodium hydroxide, polyvinylpyrrolidone (PVP), hydrazine hydrate. All the chemical reagents used in this study were of analytical grade, and used as received without further purification.

2.2. *Experimental Procedures*

Adding an appropriate amount of sodium hydroxide and surfactant PVP to 30 ml of hydrazine hydrate, made PVP, sodium hydroxide dissolved in hydrazine hydrate by ultrasonic stirring, so got a uniform mixed solution; Weighing 2.9g $CoCl_2$ and dissolved it in four 30ml solvent separately (water, alcohol, glycol, glycerol) to get B which is Co ions solution; Ultrasonic dispersing solution A, meanwhile, dropped the solution B to the solution A at the rate of 20 drops /min. after dropping, continued ultrasonic dispersing the mixing solution 20 min and then static. Finally, the mixing solution was divided into two layers, of which the upper was transparent liquid, and the precipitation of bottom were black particles sample. Then separated the precipitation by centrifuge, and vacuum dried the sample to obtain the target particles.

3. Results and Discussion

3.1. *The influence of PH value*

We can observe from the HHR reaction experiment, with the different of PH values and solvents using, the morphology and the size of cobalt particles are different, but the crystal structure of them are similar. Studies show that, [7] with the increase of PH, hydrazine hydrate reduction ability will be enhanced at alkaline conditions.Smoluchwski's Particle Collision Theory pointed out that the collision frequency of primary grains and other particles, will increase with the increase of the number of crystal nucleus of per unit volume, it lead the primary grains size smaller. [8]

Figure 1 shows the SEM imagines of cobalt particles from HHR with different NaOH adding. From the figure, it can be observed that, with the increase of PH value added, the particles size decreased, and the surface was more and more rough. This is because of the HHR ability will increase with the PH value adding, while the number of nuclei of Co increases. This will results that the collision and aggregation of grains occurs easier, and limit the free growth of grains, finally get smaller cobalt particle particles with rough surface.

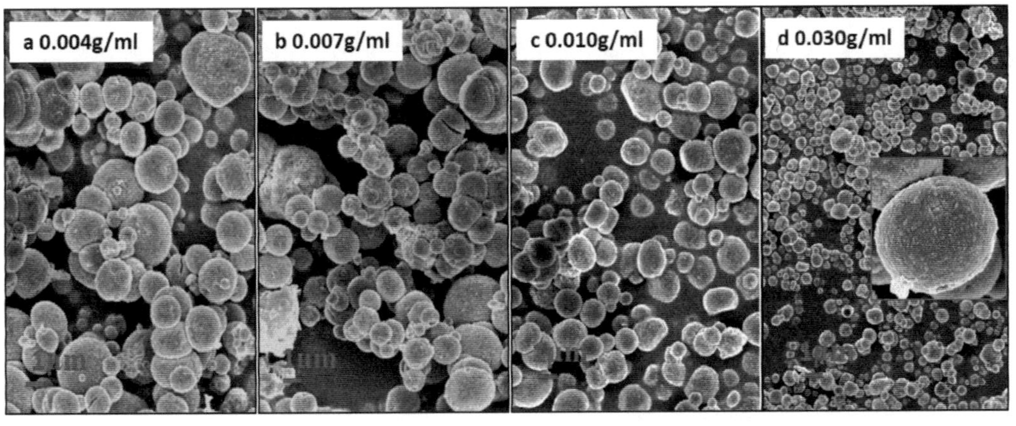

Figure 1. SEM images of Co prepared by HHR with the different NaOH adding

3.2. *The influence of the solvent*

Studies have shown that, Co tends to priority form small pieces shape in the process of liquid phase synthesis. [9] Adding polyols in HHR course, it will gather together to form micro spherical reactor, and the reaction of HHR will be limited to a small space, small cobalt pieces will gather to form spherical particles. The formation of micro reactor requires a sufficient number of hydroxyl (OH), and hydroxyl numbers will influence the sphericity of particles. In the solvent of HHR, the largest number of hydroxyl is glycerol, followed is ethylene glycol, ethanol and water. Figure 2 shows the SEM imagines of Co particles from different solvent systems of HHR. It can be clearly seen from the figure that, the best sphericity degree of particles is from the glycol, followed is glycerol, and the shapes of Co particles are irregular which from water and ethanol solvent systems, especially the water solvent makes the Co shape to pieces. This indicates the role of the hydroxyl reactor, which is only proper quantity of hydroxyl numbers can form complete spherical cobalt particles, such as ethylene glycol.

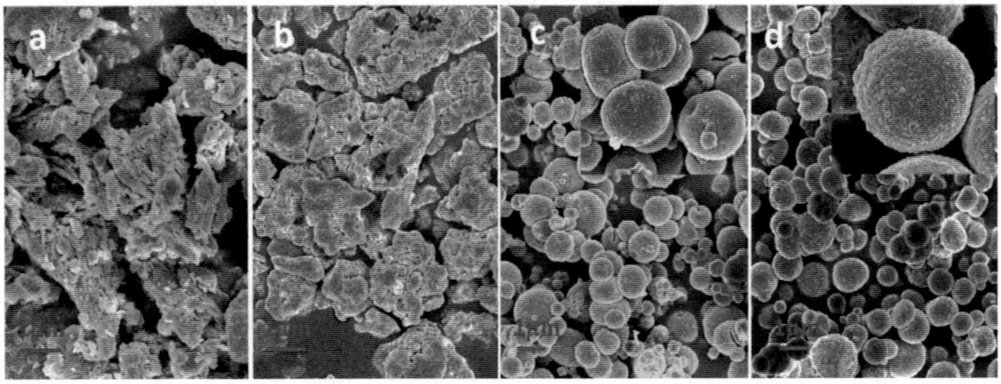

Figure 2. SEM images of Co prepared by HHR from the different solvent systems
a. water b. alcohol c. glycerol d. glycol

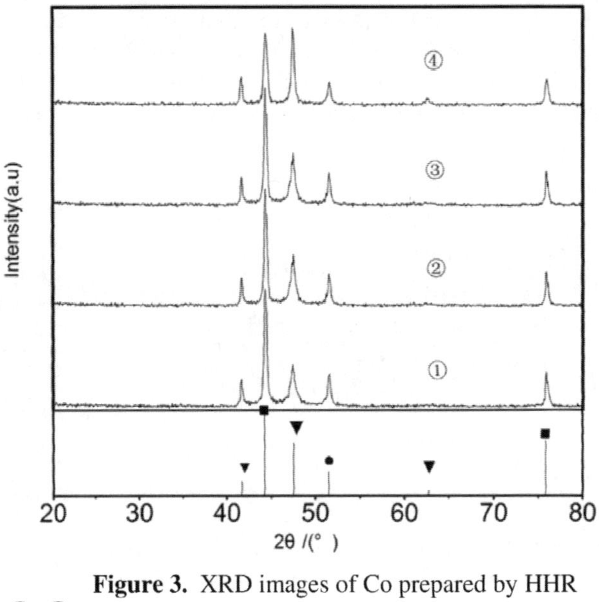

Figure 3. XRD images of Co prepared by HHR
①-②water and alcohol as the solvent, NaOH conc. 0.030g·ml^{-1}
③-④ NaOH conc. 0.030/0.004g·ml^{-1}, glycolic as the solvent

3.3. Structure of Co crystal

Figure 3 shows the XRD patterns for Co particles from HHR course. ①-② of the figure are the samples from water and glycol separately, and the concentrate of NaOH is 0.030g·ml^{-1}, while ③-④ are from glycerol and the concentrate of NaOH is 0.030/0.004g·ml^{-1} separately. Compared the patterns of figure 3 to standard PDF card of Co, we can conclude that the Co particles from all HHR course are pure Co with fcc and hcp strctures, the reaction of HHR can synthesize pure Co particles with perfect spherical shape.

4. Conclusions

(1)Hydrazine hydrate reduction can be used to synthesize Co particles with perfect spherical shape by using glycol as solvent with 0.030g·ml^{-1} NaOH adding. The particle from it is greater than 1 μm, and suitable for the batch production of metal composite material;
(2) NaOH adding can influence the morphology of the cobalt powdes: lower amount of NaOH will produce spherical Co particle particles with smooth surface; with the increase of NaOH, the particle of Co deviates from the spherical shape and become irregular, and its surface become coarse;
(3)Only proper quantity of hydroxyl numbers can form complete hydroxyl reactors, by using glycol as solvent, it can produce suitable spherical reactors and results spherical Co particles forming.

Acknowledgments

This work was supported by the scientific research project of Chongqing University of Arts and Science Y2014CJ26, the cooperation project "Development of high strength and toughness micro//nano cermet materials" between Chongqing University of Arts and Science and Chongqing academy science and technology.

References

[1] Bl'anda M, Duszová A, Csanádi T, Hvizdo˘s P, Lofaj F and J Dusza 2014 Indentation fatigue of WC grains in WC-Co composite *J. Eur.Ceram. Soc.* **34** 3407.
[2] Li XS 1994 Ceramic cutting tools: an introduction *J. Key Eng Mater* **96**:1.
[3] Rajabi A, Ghazali M, Syarif J and Daud AR 2014 Development and application of tool wear: A review of the characterization of TiC-based cermets with different binders *J. Chem. Eng.* **255**:445.
[4] Carles VB, Arnal A, Poquillon D and Tailhades P 2008 Correlation between the morphology of cobalt oxalate precursors and the microstructure of metal cobalt powders and compacts *J. Powder Technology.* **185** 231.
[5] Wang XD, Graugnard E , King JS , Zhong LW and Summers CJ. 2004 Large-Scale Fabrication of Ordered Nanobowl Arrays *J. Nano Lett.* **4** 2223.
[6] Wang YM, Li JQ and Liu PC 2005 Preparation of Ultrafine Cobalt Powder by Glycol Reduction *J. Nonferrous metal* **4**: 37.
[7] Li ZY, Han ZH and Shen JY. 2006 Reaction Kinetics of Reduction of Co2＋ by Hydrazine in Solution for the Preparation of Cobalt Nanoparticles *J. acta chimica sinica* **64** (4): 295.
[8] Xu R, Xie T, Zhao YG and Li YD. 2007 Single-crystal metal nano-platelets: Cobalt, nickel, copper, and silver *J. Cryst Growth Des* **7 (9)** 1904.
[9] Hou YL, Kondoh H and Ohta T. 2005 Self-assembly of Co nanoplatelets into spheres: Synthesis and characterization *J. Chem Mater* **17 (15)** 3994.

IOP Conf. Series: Materials Science and Engineering **167** (2017) 012043 doi:10.1088/1757-899X/167/1/012043

Study on the compounds of $Mg_{12}La$ and $(Mg,Zn)_{11.5}La$

M L Huang [1], Z Y Si and H X Li

School of Materials Science and Engineering, Northeastern University, Shenyang 110819, China

E-mail: mingli_huang@126.com

Abstract: The composition, crystal structure and phase relationship of the main intermetallics of the Mg-Zn-La system on the low La side have been studied. The results show that there exists a binary solid solutions of $Mg_{12}La$ as $(Mg,Zn)La_{12}$ and a linear ternary compound $(Mg,Zn)_{11.5}La$. Though the composition formula of the two phases are similar, but the composition of Zn in them are different. The maximum Zn content of $(Mg,Zn)_{12}La$ is 7.2at% and the minimum Zn content of $(Mg,Zn)_{11.5}La$ is 8.5at%. What's more, the crystal structures of $(Mg,Zn)_{12}La$ and $(Mg,Zn)_{11.5}La$ phase are also totally different. $(Mg,Zn)La_{12}$ have the body centered tetragonal lattice structure, but the crystal structure of $(Mg,Zn)_{11.5}La$ is C-centered orthorhombic lattice structure.

1. Introduction

Magnesium alloys are one of the lightest structural metal materials, and its application potential in automobile industry, aviation industry and electron industry has been focused [1-6]. Mg-Zn system is the basic system for Mg alloys, but the system exhibits a low eutectic temperature. For this reason, the mechanical properties of the Mg-Zn alloys are poor at elevated temperature. It is known that the addition of rare earth element can not only refine the grains of the Mg-Zn system alloys, but can also improve the creep resistance at elevated temperature[7-12], because of the formation of high melting-point compounds.

According to the literature [13], the phases in Mg-Zn-La system are Mg based solid solution, binary phases and ternary phases. The metallic phases in this system often play an important role on elevating the mechanical properties especially on the creep resistant property. From this point of view, the information of the intermetallic phases in Mg-Zn-La system must be the important for alloying.

2. Experimental

The experimental alloys with different composition were prepared in this work. The pure alloying elements of Mg (purity about 99.99wt.%), Zn (purity about 99.99wt.%) and La(purity about 9.98wt.%) were mixed up and melted in an induction furnace under Ar (purity about 99.9wt.%) atmosphere, then cooled to room temperature with the furnace. In order to achieve a relative equilibrium state, the as-cast alloys were annealed at 400℃ for 1800 h, and then quenched in ice water.

Microstructure of each alloy was studied by scanning electron microscopy (SEM). The compositions of equilibrium phases in alloys were studied by electron probe microanalyzer (EPMA) with a beam size of 1μm and a voltage of 15kV. The high pure Mg, Zn, and La were served as standards to revise the characteristic radiations with ZAF4 correction. The crystal structures of phases were studied by X-ray diffraction (XRD) and transmission electron microscopy (TEM) respectively. Thin foil specimens for TEM were prepared by ion polishing.

3. Results and discussion

3.1. The analysis of (Mg,Zn) $_{12}$La

The microstructure and XRD pattern of $Mg_{93}La_7$ alloy are shown in Fig. 1. The results suggest that $Mg_{93}La_7$ is a two-phase alloy with $Mg_{12}La$ and Mg solid solution. As shown in Fig. 2b, the diffraction peaks of $Mg_{12}La$ can be indexed as a body-centered tetragonal lattice structure. According to the result of EPMA, the large light area in Fig. 2a is pro-eutectic $Mg_{12}La$, and the region having the alternating light and dark lamellar structure is ($Mg+Mg_{12}La$) eutectic structure.

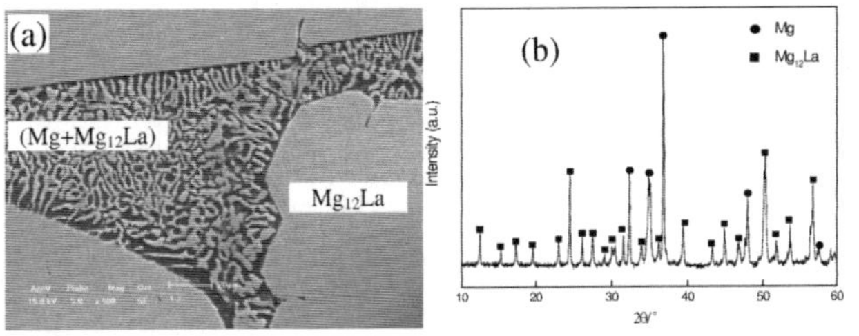

Figure 1. The microstructure (a) and X-ray diffraction pattern (b) of $Mg_{93}La_7$ alloy

3.2. The analysis of (Mg,Zn)$_{11.5}$La

Fig.2 shows the microstructures of alloys of $Mg_{85}Zn_{10}La_5$, $Mg_{67}Zn_{25}La_8$, $Mg_{88}Zn_{10}La_2$ and $Mg_{46}Zn_{46}La_8$ at 400℃. As shown in Fig.2, the alloys $Mg_{85}Zn_{10}La_5$ and $Mg_{88}Zn_{10}La_2$ all consist of two phases. According to the results of Table 1, the black region is the solid solution of Mg, and the white phase of each alloy is a ternary compound. $Mg_{67}Zn_{25}La_8$ and $Mg_{46}Zn_{46}La_8$ alloys are nearly the single phase alloys. The single phase in each alloy is a ternary compound as shown in Table 1. It is not difficult to find that the compositions of the four ternary compounds in the key alloys all contain about 8at% La, and the Zn content in them are increasing with the decreasing of the Mg contents from 16at% to 44.3at%. The compounds in $Mg_{85}Zn_{10}La_5$, $Mg_{67}Zn_{25}La_8$, $Mg_{88}Zn_{10}La_2$ and $Mg_{46}Zn_{46}La_8$ were identified asτ_1, τ_2, τ_3, τ_4 respectively. The results show that the formation of the ternary compounds is on the substitution of Mg atom by Zn atom.

Table 1. Phase compositions of the alloys in Figure 2. by EPMA (at%)

Alloys		Ternary compound (T-phase)			Mg solid solution		
		Mg	Zn	La	Mg	Zn	La
$Mg_{85}Zn_{10}La_5$	τ_1	76.2	16.0	7.8	99.2	0.8	0.0
$Mg_{67}Zn_{25}La_8$	τ_2	66.9	25.2	7.9		—	
$Mg_{88}Zn_{10}La_2$	τ_3	54.3	37.8	7.9	98.4	1.6	0.0
$Mg_{46}Zn_{46}La_8$	τ_4	48.7	43.3	8.0		—	

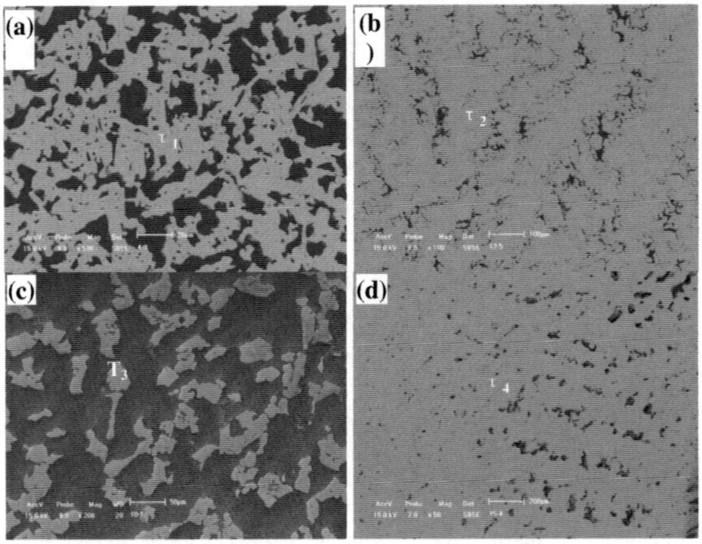

Figure 2. The microstructures of alloys: (a) $Mg_{85}Zn_{10}La_5$; (b) $Mg_{67}Zn_{25}La_8$; (c) $Mg_{88}Zn_{10}La_2$; (d) $Mg_{46}Zn_{46}La_8$

Fig.3 is the XRD patterns of $Mg_{85}Zn_{10}La_5$ alloy, $Mg_{67}Zn_{25}La_8$ alloy, $Mg_{88}Zn_{10}La_2$ alloy and $Mg_{46}Zn_{46}La_8$ alloy. It shows that the law of the appearance of the diffraction patterns of the ternary compounds and the intensity of the diffraction peaks of the other ternary compoundsτ_1, τ_2, τ_3, τ_4 are coherent. That is to say, the four ternary compounds have the same crystal structure. According to the analysis of the single phase alloy of $Mg_{67}Zn_{25}La_8$, the crystal structure ofτ_2 have be identified as the C-centered orthorhombic crystal structure. The results of Table 1 and Fig. 3a suggest that all the ternary compounds belong to one phase, and it has the C-centered orthorhombic crystal structure. According to the result in Table 1, The ternary compounds fromτ_1 toτ_4 have the characterization of the linear compound in composition, and the chemical formula is about $(Mg,Zn)_{11.5}La$, it is a typical replacement solid solution.

Fig.3b is enlargement of Fig.3a, which shows that the diffraction angles for the same (*h k l*) triplet of the ternary compound shift to a little higher with the increase of Zn content in it. It is well known that the radias of Zn is shorter than that of Mg. Therefore, the phenomenon must be due to the substitution of Mg by Zn according to Bragg Equation and Crystal plane spacing equation.

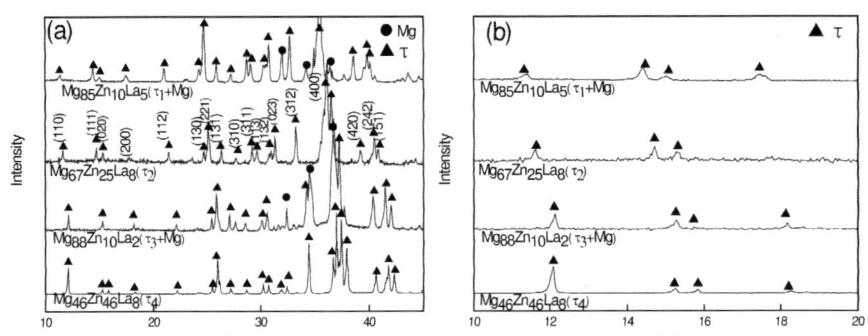

Figure 3. The XRD patterns of the alloys $Mg_{85}Zn_{10}La$, $Mg_{67}Zn_{25}La_8$, $Mg_{88}Zn_{10}La_2$ and $Mg_{46}Zn_{46}La_8$:
(a) 2θ from10°to 45°; (b) 2θ from10°to 20°

The composition and the crystal structure characteristic of the ternary compounds from τ_1 to τ_4 suggests that all the compounds belong to one ternary linear phase which identified as $(Mg,Zn)_{11.5}La$ here.

3.3. Discustion about $Mg_{12}La$ and $(Mg,Zn)_{11.5}La$

Fig.4 is the microstructures of alloys as $Mg_{93}Zn_2La_5$, $Mg_{90}Zn_5La_5$ and $Mg_{85}Zn_{10}La_5$.All the alloys contain two areas, the black and the light. According to the results of composition shown in Table 2, the black area is Mg solid solution, and the light one is a ternary compound. The TEM microstructures in Fig.5 show that the light region in $Mg_{93}Zn_2La_5$ and $Mg_{85}Zn_{10}La_5$ contains the big complete grain, with straight boundaries. But that in $Mg_{90}Zn_5La_5$ contains large numbers of fine grains. The fine grains in light regions of $Mg_{90}Zn_5La_5$ are formed from one another. But the compositions of the fine grains ternary compounds have slight difference as shown in Table 2.

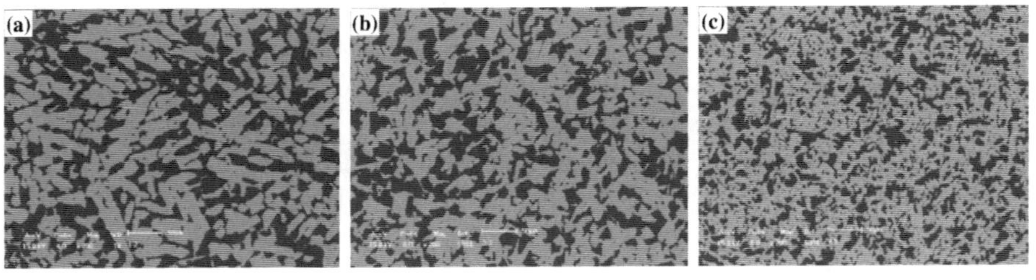

Figure 4. The microstructures of $Mg_{93}Zn_2La_5$ (a), $Mg_{90}Zn_5La_5$ (b) and $Mg_{85}Zn_{10}La_5$ (c)

Table 2. The phase composition of alloys as $Mg_{93}La_5Zn_2$, $Mg_{90}La_5Zn_5$ and $Mg_{85}La_5Zn_{10}$ (at %)

Composition of alloys	Compounds			Mg solid solution		
	Mg	Zn	La	Mg	Zn	La
$Mg_{93}La_5Zn_2$	88.7	3.1	8.2	99.8	0.2	ignored
$Mg_{90}La_5Zn_5$	84.5	7.2	8.3	99.5	0.5	ignored
	83.3	8.5	8.2			
$Mg_{85}La_5Zn_{10}$	76.2	16.0	7.8	99.2	0.8	ignored

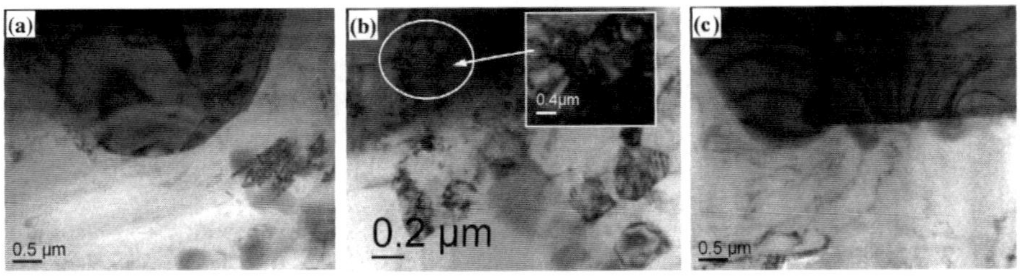

Figure 5. The TEM microstructures of $Mg_{93}La_5Zn_2$ (a), $Mg_{90}La_5Zn_5$ (b) and $Mg_{85}Zn_{10}La_5$ (c)

Fig.6 is the X-ray diffraction patterns of the alloys in Table 2. Fig.5a and Fig.6c shows that the diffraction patterns of the compounds in $Mg_{93}Zn_2La_5$ and $Mg_{85}Zn_{10}La_5$ are different to each other, though the microstructures and phase compositions of the two compounds are resemble. That is to say, the two ternary compounds are not the same phase. Through analysis of the diffraction patterns (Fig.6), the compound in $Mg_{93}Zn_2La_5$ alloy is $Mg_{12}La$ with body-centered tetragonal lattice structure and that in $Mg_{85}Zn_{10}La_5$ alloy is $(Mg,Zn)_{11.5}La$ with C-centered orthorhombic lattice structure.

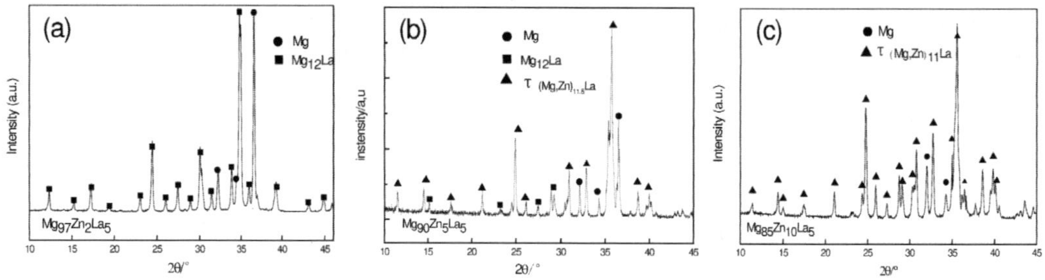

Figure 6. The XRD patterns of $Mg_{93}La_5Zn_2$ (a), $Mg_{90}La_5Zn_5$ (b) and $Mg_{85}Zn_{10}La_5$ (c)

Fig.6b shows that, $Mg_{90}Zn_5La_5$ alloy consists two compounds and a Mg solid solution. One of the compounds is a binary solid solution of $Mg_{12}La$, the other being a ternary compound $(Mg,Zn)_{11.5}La$. Because the composition of the two compounds are close to each other as shown in Table 2, so no phase interface between them has been observed in the black electron images as shown in Fig.4b. However, according to the TEM microstructrues (Fig.5), the compound bulk in $Mg_{85}Zn_{10}La_5$ alloy is composed by so many small grains which is different from those in $Mg_{93}Zn_2La_5$ and $Mg_{85}Zn_{10}La_5$. The result in Fig.6b also suggests that the small fine grains consist two phases, such as $Mg_{12}La$ with maximum Zn content and $(Mg,Zn)_{11.5}La$ with minimum Zn content.

4. Conclusions

- There is a binary substitutional solid solution solid of $Mg_{12}La$ as $(Mg,Zn)_{12}La$, the maximum Zn must be Less than or equal to 7.2at%.
- A ternary linear compound $(Mg,Zn)_{11.5}La$ has been identified, the Zn content of it must be Greater than or equal to 8.5at%, and the Zn content in this compound at least up to 43.3at%.
- The crystal structure of $Mg_{12}La$ is the body-centered tetragonal lattice structure, and that of $(Mg,Zn)_{11.5}La$ is C-centered orthorhombic lattice structure.

Acknowledgements

This work was financially supported by the National Natural Science Foundation of China (51301032), the Natural Science Foundation of Hebei Province (E2014501102) and the Fundamental Research Foundation of Central Universities (N152304002).

References

[1] Bondarev I B and Bessonov A M. 2005 *Tsvetn. Met.* 5-6 131-134
[2] Schumann S and Friedrich H 2003 Mater. Sci. Forum. 2003;419-422:51-56.
[3] Gröbner J, Hampl M, Schmid F R, Easton M A Zhu S Gibson M A and Nie J F 2012 *Intermetallics* **28** 92-101
[4] Pérez P, Garcés G, Maeso M and Adeva P 2012 *Metall. Mater. Trans. A* **43** 4383-4396
[5] Yu L N, Nakata K and Liao J S 2009 *J. Alloys Compd.* **480** 340-346
[6] Eddahbi M, Perez P, Monge M A, Garces G and Pareja R 2009 *J. Alloys Compd.* **473** 79-486
[7] Zhang K, Li X G, Li Y J and Ma M L 2008 *Trans. Nonferrous Met. Soc.* China 18: s12-s16
[8] Omori G, Matsuo S and Asada H 1975 Trans. *Jpn. Inst. Met.* **16** 247-255
[9] Yu K, Li W X and Wang R C 2005 *J. Cent. S. Univ. Technol.* **12** 499-502
[10] Wei L Y, Dunlop G L and Westengen H 1995 *Metall. Mater. Trans. A* **26** 1705-1716.
[11] Li Q, Wang Q D, Wang Y X, Zeng X Q and Ding W J 2007 *J. Alloys Compd.* **427** 115-123
[12] Jun J, H Park B K, Kim J M, Kim K and Ung W J 2006 *Mater. Sci. Forum* 510-511 214-217
[13] Dobatkina T V and Muratova E V 1978 *Russ. Metall.* **1** 205-208

IOP Conf. Series: Materials Science and Engineering **167** (2017) 012044 doi:10.1088/1757-899X/167/1/012044

Negative temperature dependence of recrystallized grain size: analytical formulation and experimental confirmation

M Elmasry, F Liu, Y Jiang, Z N Mao, Y Liu[1] and J T Wang

School of Material Science and Engineering, Nanjing University of Science and Technology, Nanjing 210094, PR China

E-mail: liuying517@njust.edu.cn

Abstract. The catalyzing effect on nucleation of recrystallization from pre-existing grains is analyzed, analogy to the foreign nucleus size effect in heterogeneous nucleation. Analytical formulation of the effective nucleation site for recrystallization leads to a negative temperature dependence of recrystallized grain size. Non-isochronal annealing, where annealing time is set just enough for the completion of recrystallization at different temperature, is conducted on pure copper after severe plastic deformation. More homogeneous and smaller grains are obtained at higher annealing temperature. The good fitting between analytical and experimental results unveils the intrinsic feature of this negative temperature dependence of recrystallized grain size.

1. Introduction

The history of metalworking, including the procedures of deformation and heating, can be traced back to the early days of Neolithic Age, when the first metalsmiths began working with native copper [1]. The birth of "Recrystallization" was declared by a paper of 1887, where Sobey reported the emergence of new equiaxed grain when heating the deformed iron with elongated structure [2]. Establishing quantitative models based on the physics of recrystallization, and using them to improve, optimize and control microstructure, has long been the crucial mission of modern metal industry, and also the frontier researches of nanomaterials [3].

In summarizing the "laws of recrystallization", it is stated by Burke and Turnbull that [4, 5] the final grain size depends chiefly on the degree of deformation and to a lesser degree on the annealing temperature, normally being smaller the greater the degree of deformation and the lower the annealing temperature. Experimental observations show that final grain size increases significantly with the increase of annealing temperature after pre-deformation in steel [6,7], Cu–Zn alloys [8,9], magnesium alloy [10], molybdenum alloy [11] and so on. However, on the other side, Eastwood et al. reported that recrystallized grain size is insensitive to annealing temperature and they disclosed that different annealing temperatures yield a similar grain size after recrystallization for brass [12].

In addition to these contradictive experimental results, a basic gap in our knowledge for many years has been the lack of a quantitative model to account for the grain size after recrystallization [13]. One principle reason for the existence of this basic gap is the complexity of recrystallization process, especially the complex nucleation behavior of the pre-deformed materials. In order to reach a quantitative formulation of the recrystallized grain size, and meantime to avoid the complex problem related to nucleation process, it will be helpful to estimate quantitatively the number of effective nucleation site, which serves as an essential base for calculating the nucleation rate for recrystallization, and is always proportional to the final number of recrystallized grains after the completion of recrystallization.

The existing high angle grain boundaries (HAGB) in the pre-deformed materials play an important role in promoting nucleation for recrystallization, through the mechanism, for example, strain-induced boundary migration (SIBM) [14, 15]. Severe plastic deformation introduces high density of non-equilibrium HAGBs, which serves as preferential nucleation site for recrystallization in the annealing subsequent to the deformation, and significantly promotes the nucleation rate which leads to much smaller recrystallized grain [16, 17]. Analogy to the catalyzed nucleation by "foreign particles" during phase transformation, the catalyzing effect on nucleation of recrystallization from HAGB can be analyzed following the foreign nucleus in Heterogeneous Nucleation [18,19].

2. Analytical formulation of catalyzed nucleation by HAGB in recrystallization

Consider the catalyzing effectiveness on nucleation of recrystallization from an existing grain with a diameter of d in the as-deformed microstructure with HAGBs, and the catalyzed embryo has a critical radius of r^*, which can be calculated from the classical nucleation theory [18]:

$$r^* = \frac{2\gamma}{\Delta G_V} \tag{1}$$

where γ is the interfacial energy; ΔG_V, the difference of Gibbs free energies per unit volume of the parent and new phases; in the present case of recrystallization, this Gibbs free energy difference equals to stored energy from pre-deformation, and is basically constant upon change of annealing temperature.

Fletcher [18] showed that the free energy of formation of a critical embryo with a radius of r^* can be written as

$$\Delta G^* = \frac{8\pi\gamma^3}{3(\Delta G_V)^2} f(m,x) \tag{2}$$

in which $f(m, x)$, known as the Fletcher shape factor for heterogeneous nucleation on a convex spherical catalyst particle, is given by

$$f(m,x) = 1 + \left(\frac{1-mx}{g}\right)^3 + x^3\left[2 - 3\left(\frac{x-m}{g}\right) + \left(\frac{x-m}{g}\right)^3\right] + 3mx^2\left(\frac{x-m}{g} - 1\right) \tag{3}$$

Where, $g = (1 + x^2 - 2mx)^{1/2}$; $x = d/(2r^*)$ is the dimensionless diameter of the spherical catalyst particle, normalized by the radius of critical embryo r^*, and d is the actual dimeter of the catalyst particle; m, in the case of melt solidification, is the cosine of the contact angle of the new embryo on the catalyst surface, with $0 \leq m \leq 1$; In the present case of recrystallization, m resembles the catalyzing efficiency of the pre-existing grain boundary on the forming of new embryo for recrystallization, and could be considered constant in the case of high angle grain boundary misorientation. Thus $f(m, x)$ is a function depends solely on x, the diameter of existing grains with HAGBs catalyzing the nucleation of recrystallization in subsequent annealing.

Generally, in dealing with the effect of foreign particle size on heterogeneous nucleation [20], and in dealing with the nucleation issue of a pre-deformed material during recrystallization, probability has been used to relate nucleation activities with structure parameters in the parent phases (such as grain boundary misorientation etc.) [13, 21]. Hence, the probability p for an existing grain in the pre-deformed matrix to act as an effective nucleation site for recrystallization is considered currently. It is reasonable to assume p to be proportional to a thermal activation factor with ΔG^* as the energy barrier, and to f_{HAGB}, the fraction of HAGBs which are evidenced to be preferential nucleation site. Thus

$$p \approx C \cdot \exp\left(-\frac{\Delta G^*}{k_B T}\right) \cdot f_{HAGB} \tag{4}$$

Where C is a coefficient, and k_B the Boltzmann constant.

In considering the catalyzed nucleation by "foreign particles" during phase transformation, it is clear that the larger the diameter of the "foreign particle", the higher its efficiency in catalyzing the nucleation. Analogy to this tendency, suppose at $x \geq x_c$, $p \approx 1$, which means grains in the pre-deformed matrix with HAGB and normalized dimensionless diameter x larger than x_c will act as effective nucleation sites upon annealing at temperature T with 100% possibility; and considering equations (2) and (3),

$$p \approx 1 \approx C \exp\left[-\frac{8\pi\gamma^3}{3\Delta G_V{}^2 k_B T} f(m, x_c)\right] f_{HAGB} \tag{5}$$

This leads to a relation between the annealing temperature T and pre-existing grain size x_c:

$$\frac{8\pi\gamma^3}{3\Delta G_V{}^2 k_B} f(m, x_c) = T \cdot \ln\left(C \cdot f_{HAGB}\right) \tag{6}$$

Using typical parameters of f_{HAGB}=0.67 [22], ΔG_V = 57 J/mol $\approx$ 8x10^6 J/m^3 [23] after severe plastic deformation, and γ = 0.625 J/m^2 [2] for copper, equation (6) gives a drastic dependence of x_c upon T, through f(m, x_c). As shown in Figure. 1, all the curves describing the dependence of x_c upon T at different catalyzing efficiency m indicate that, the higher the annealing temperature T, the smaller the critical grain size x_c, and thus more grains in the pre-deformed matrix will act as effective nucleation sites. This means that higher annealing temperatures lead to smaller recrystallized grain size, or an negative temperature dependence of recrystallized grain size, which is not only against the "laws of recrystallization" summarized by Burke and Turnbull that [4, 5], but also contradictive to the experimental results for copper [24-28], , as well as other metals and alloys [6-11].

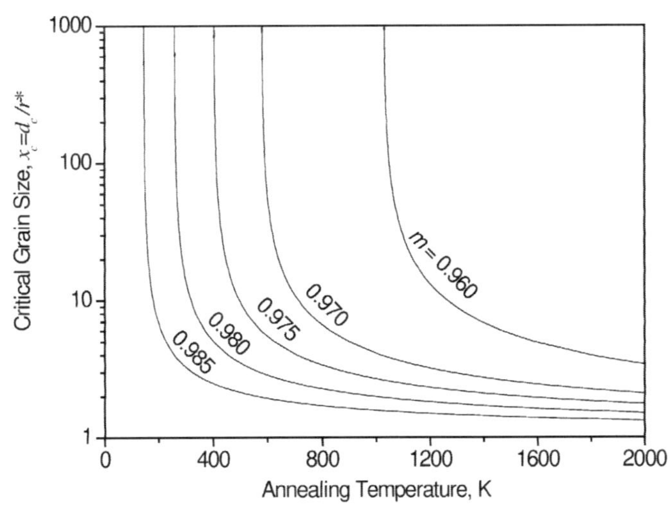

Figure 1. Dependence of critical grain size x_c upon annealing temperature T at different level of catalyzing efficiency m, which is large enough to sufficiently catalyze nucleation of recrystallization.

In studying the recrystallization behavior of a two-phase brass, a negative temperature dependence of recrystallized grain size was observed by Naether et al. [9], where the reduction in the average grain size with increasing temperature was observed accompanied by a concurrent increase of the volume fraction of β phase in the sample. This case did not indicate the intrinsic negative temperature dependence of recrystallized grain size itself, but the increasing effect of β phase stabilization from the concurrent increase of the volume fraction of β phase in the sample upon increase of annealing temperature [9].

The key to unlock this contradiction between the analytical formulation given in equations (2) to (6) and the experimental data in the literature might be the annealing time. Isochronal annealing at different temperatures is usually used for the annealing and recrystallization to obtain temperature dependence of grain size in the literature [24-28]. When the annealing time is set enough for completion of recrystallization at a lower temperature, this isochronal annealing time is far more than that for recrystallization to complete at higher temperatures, and the grain size obtained in this case for high temperature annealing inevitably includes the effect of grain growth subsequent to the completion of recrystallization, and this leads to the resulting "pseudo-morph" of positive temperature dependence of grain size after "recrystallization".

To avoid the influence of this isochronal annealing issue and to obtain a true annealing temperature dependence of recrystallized grain size, kinetics of recrystallization process could be characterized first, and the time just enough for the completion of recrystallization at each annealing temperature could thus be determined, which could then be used for non-isochronal annealing at different temperatures to obtain grain size immediately after the completion of recrystallization process. This non-isochronal annealing may remove the effect of grain growth, and obtain the intrinsic annealing temperature dependence of recrystallized grain size.

3. Experimental confirmation

To confirm the negative temperature dependence of recrystallized grain size unveiled from the above analytical formulation, cold rolled (CR) high purity copper (99.97 wt%) after equal channel angular pressing (ECAP) was used.

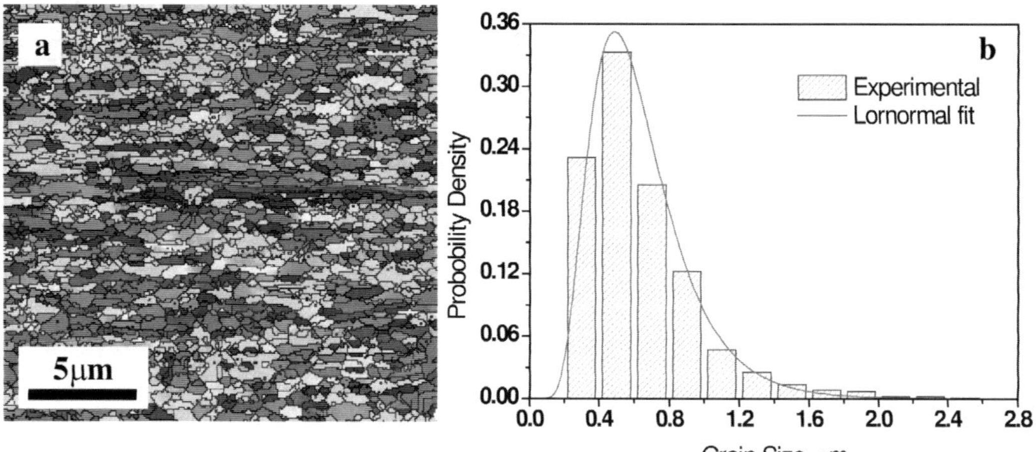

Figure 2. An EBSD image of the cross section (a) and its grain size distribution (b) of high purity (99.97 wt%) copper pre-deformed by ECAP+CR.

Hot rolled high purity copper, with initial heat treatment at 873 K (600 °C) for 1 h to achieve a homogenous microstructure with an average grain size of ~100 μm, is used in this study. Billets with size of 32 × 32 × 160 mm are processed by ECAP at room temperature with a die angle of 90° for 8

IOP Conf. Series: Materials Science and Engineering **167** (2017) 012044 doi:10.1088/1757-899X/167/1/012044

passes using route B_C. The sample is then cold rolled (CR) at room temperature to a thickness reduction of ~75% . After pre-deformation through ECAP+CR, annealing is conducted at temperatures ranging from 373 K (100 °C) to 573 K (300 °C), and for times ranging from 3 to 48000 s. Vickers microhardness is measured at room temperature with a load of 100 g for 10 s. More than 20 random indentations were made to obtain a representative bulk hardness value. Electron back-scattering diffraction (EBSD) technique is used to characterize microstructural evolution on the cross-sections of the ECAP+CR samples during annealing treatment. Figure. 2 shows the EBSD microstructure of the sample after cold rolling (Figure. 2(a)) and its grain size distribution (Figure. 2(b)). The HAGB fraction of the materials is ~68%.

The recrystallization kinetics of the ECAP+CR samples upon annealing is characterized by the microhardness variations versus annealing time for different isothermal annealing temperatures, as shown in Figure. 3, which indicates when the annealing temperature increases from 473 K (200 °C) to 573 K (300 °C), the time for the completion of the recrystallization is reduced from 25 mins to 35 s.

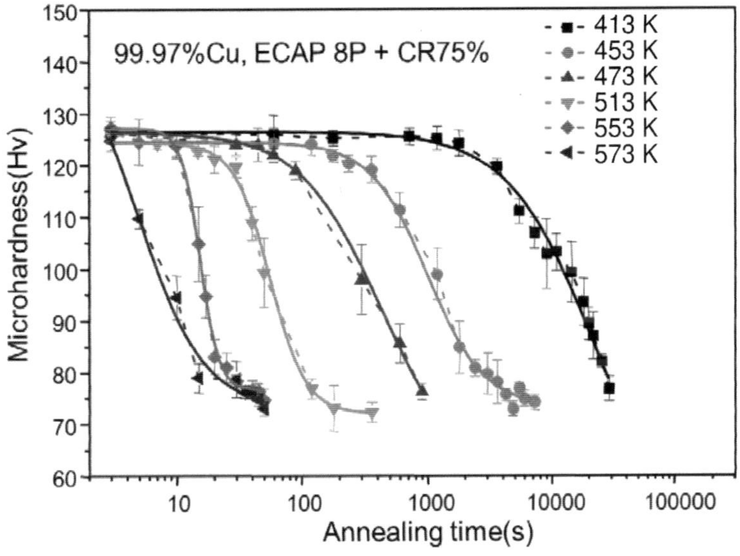

Figure 3. Variation of the microhardness vs. time at different annealing temperatures for ECAP+CR samples.

Figure. 4(a, b) show the microstructure of the fully (~95% volume fraction) recrystallized specimens annealed at 473 K (200 °C) and 573 K (300 °C) respectively. The recrystallized grain size after 473 K/25 mins annealing ranges from 1 to 14 μm with an average grain size of 3.0± 0.6 μm; while that after the 573 K/35 s annealing ranges from 1 to 7.5 μm with an average value of about 1.9 ± 0.5 μm.

The microstructures clearly show that with increasing annealing temperature, the microstructure becomes more homogeneous with smaller average grain size, which means that the recrystallized grain size has a negative dependence on the annealing temperature. Grain size data from specimen annealed at other temperatures also confirms this negative dependence, as shown in Figure. 4(c). It is clear from this figure, that the recrystallized grain size at ~95% completion of recrystallization decreases, even the distribution span of grain size at each annealing temperature also decreases, upon the increase of annealing temperature.

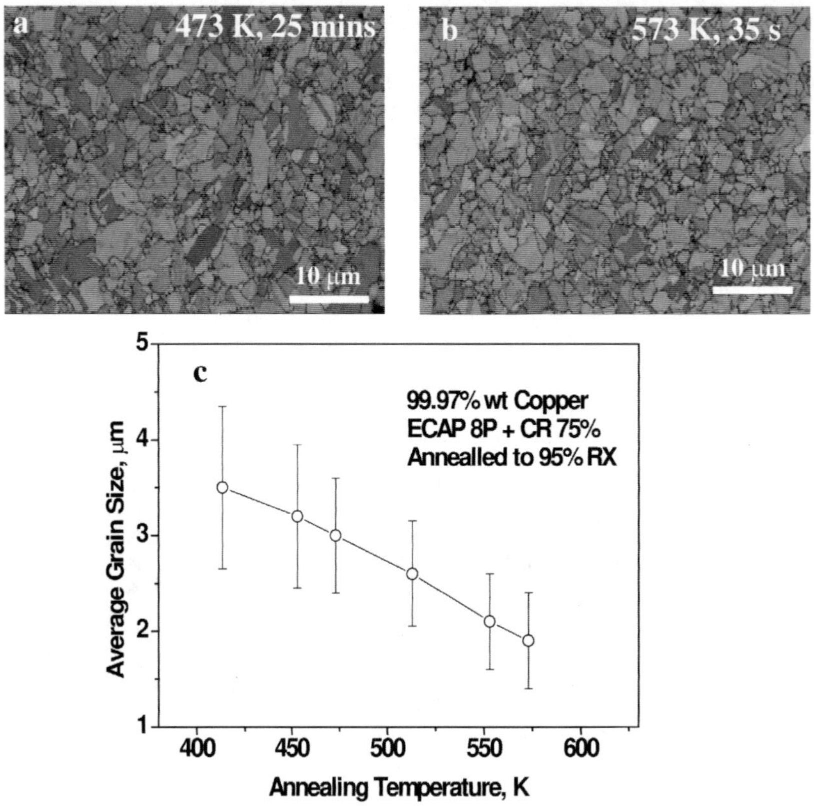

Figure 4. EBSD images after ~95% recrystallization of the specimens upon annealing at (a) 473 K (200 °C), 25 mins (b) 573 K (300 °C) , 35 s, and (c) the dependence of the recrystallized grain size at 95% recrystallization (RX) on annealing temperature.

To compare these experimental results with the prediction of equation (6), grain size distribution in the as-deformed matrix is needed. It has been demonstrated that the grain size d distribution after severe plastic deformation follows the log normal distribution [29, 30]

$$f(d;\mu,\sigma) = \frac{1}{d\sigma\sqrt{2\pi}}\exp\left[-\frac{(\ln d - \mu)^2}{2\sigma^2}\right] \tag{7}$$

where the two distribution parameters μ and σ are the mean and standard deviation of ln$d$ respectively. The number fraction of grains with a diameter $\geq d_c$ ($d_c = 2\, r^* x_c$) is:

$$F_N = 1 - \int_0^{dc} f(x;\mu,\sigma)dx = \frac{1}{2} - \frac{1}{2}erf\left(\frac{\ln d_c - \mu}{\sigma\sqrt{2}}\right) \tag{8}$$

On the other hand, from the experimental data, the number of grains in the recrystallized microstructure annealed at temperature T is inversely proportional to its average grain size D_T, and the number of grains in the pre-deformed matrix is inversely proportional to its average grain size d_{av}. Thus the number fraction of grains in the pre-deformed matrix which transformed into recrystallized grains after annealing at temperature T is:

$$F_{NT} = (\, d_{av}/D_T)^3 \tag{9}$$

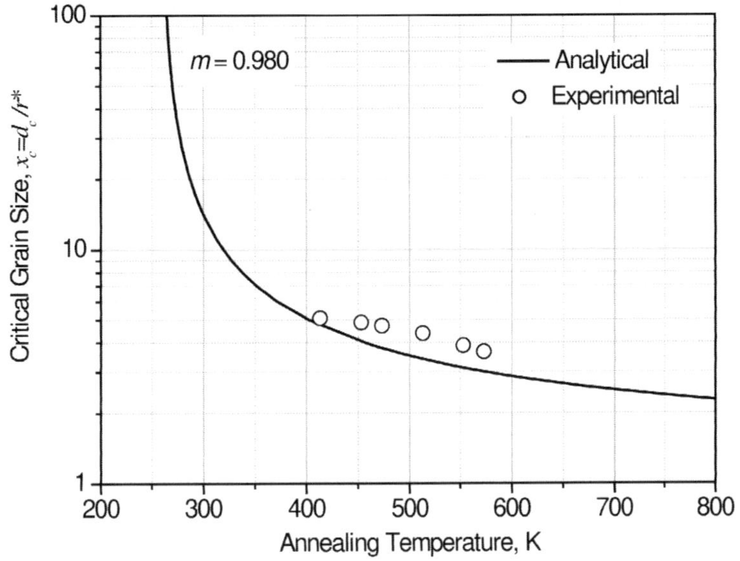

Figure 5. Comparison of analytically and experimentally obtained dependence of critical grain size for catalyzing nucleation site (x_c=d_c/2r*) upon annealing temperature T.

Combination of equations, (8) and (9) gives:

$$erf\left(\frac{\ln d_c - \mu}{\sigma\sqrt{2}}\right) = 1 - 2\left(\frac{d_{av}}{D_T}\right)^3 \tag{10}$$

d_{av} can be calculated from Figure.2(b), together with μ = -0.538 and σ = 0.424 for the lognormal distribution of grain size, and D_T is given in Figure. 4c. Thus F_{NT} and d_c for recrystallized microstructure annealed at temperature T can be calculated according to equations (9) and (10). An experimental relation of x_c ($x_c = d_c/(2r^*)$) and T can thus be obtained and is shown in Figure. 5 by the hollow circles. A partial of the m=0.980 curve from Fig. 1, which is analytical prediction from equation (6), is also shown in Figure. 5. It can be seen that the data points from experiments fit fairly well the analytical prediction for m=0.980. The high value of m (= 0.980), which yields good fitting between analytical prediction and experimental data, reflects the high catalyzing efficiency of the pre-existing HAGB on the forming of new embryo for recrystallization.

4. Conclusion

In summary, the catalyzing effect on nucleation of recrystallization from pre-existing grains with HAGB is analyzed, analogy to the foreign nucleus size effect in heterogeneous nucleation. Analytical formulation of effective nucleation site leads to a negative temperature dependence of recrystallized grain size, where a critical grain size d_c pre-existing in the as-deformed matrix, is formulated to be negatively dependent on the annealing temperature T; Grains with diameter larger than this critical d_c in the as-deformed matrix will act as effective nucleation site for recrystallization in the subsequent annealing at T. The higher the T, the smaller the d_c, and the higher the fraction of grains in the as-deformed matrix can act as effective nucleation sites, and thus smaller recrystallized grain size is formulated. Non-isochronal annealing, where annealing time is set just enough for the completion of recrystallization at different temperatures, is conducted on a high purity (99.97 wt. %) copper after room temperature severe plastic deformation. Negative temperature dependence of recrystallized grain size is confirmed, where more homogeneous and smaller grains are obtained experimentally at higher annealing temperature. The good fitting between analytical prediction and experimental results unveils the intrinsic feature of the negative temperature dependence of recrystallized grain size.

Acknowledgements

This work was supported by the Chinese Ministry of Science and Technology of China under Grant No. 2012CB932203 of the National Key Basic Research Program, the Natural Science Foundation of China under Grant No. 51304123, and the Key grant Project of Chinese Ministry of Education under Grant No. 311035.

References

[1] Cathro R J 2005 Tin Deposits and the Early History of Bronze. *CIM Bulletin* **98** p 1088
[2] Humphery F J and Hatherly M 2004 *Recrystallization and related annealing phenomena*, 2nd edit. (Oxford:Elsevier)
[3] Wang Y, Chen M, Zhou F and Ma E 2002 *Nature* **419** p 912
[4] Burke J E and Turnbull D 1952 Recrystallization and grain growth, *Progress in Metal Physics*, (London: Pergamon Press) vol.3 p 220
[5] Cahn R W and Haasen P 1996 *Physical metallurgy* (North-Holland) vol.3 p 2419
[6] Rodrigues D L, Silveira J R F and Landgraf F J G 2011 *IEEE Trans* **47** p 2179
[7] Sohn S S, Lee B J , Lee S, Kim N J and Kwak J -H 2013 *Acta Mater* **61** p 5050
[8] Oishi K, Sasaki I and Otani J 2003 *Mater Lett* **57** p 2280
[9] Naether G, Krumnacker M and Pysz G 1980 *Metals Technology* **7** p 345
[10] Mishra S K, Tiwari S M, Carter J T and Tewari A 2014 *Mater Sci and Eng A* **A599** p 1
[11] Marinchenko G E 2005 *Proceedings of NAU* **24** p 59
[12] Eastwood L W, Bousu A E and Eddy C T 1935 *Trans Metall Soc AIME* **117** p 246
[13] Wang S, Holm E A, Suni J, Alvi M H, Kalu P N and Rollett A D 2011 *Acta Mater* **59** p 3872
[14] Beck P A, Sperry P R 1949 *Trans. AIME* **180** p 240
[15] Beck P A, Sperry P R 1950 *J Appl Phys* **21** p 150
[16] Molodova X, Gottstein G, Winning M and Hellmig R J 2007 *Mater Sci and Eng A* **A460–461** p 204
[17] Wang Y L, Lapovok R, Wang J T, Qi Y S and Estrin Y 2015 *Mater Sci and Eng A* **A628** p 21
[18] Fletcher N H 1958 *J Chem Phys* **29** p 572
[19] Perepezko J H and Wilde G 2016 *Curr Opin Solid State Mater Sci* **20** p 3
[20] Weinberg M C 1992 *J Chem Phys* **96** p 9144
[21] Piękoś K, Tarasiuk J, Wierzbanowski K and Bacroix B 2008 *Comput Mater Sci* **42** p 36
[22] Wang J T, Zhang Y and Liu J Q 2008 *Mater Sci Forum* **584-586** p 929
[23] Zhang Y, Wang J T, Cheng C and Liu J Q 2008 *J Mater Sci* **3** p 7326
[24] Li Y S, Zhang Y, Tao N R and Lu K 2008 *Scr Mater* **59** p 475
[25] Takayama A, Yang X, Miura H and, Sakai T 2008 *Mater Sci and Eng A* **A478** p 221
[26] Daly R, Zghal S and Njeh N 2009 *Physics Procedia* **2** p 677
[27] Blum W, Li Y J and Durst K 2009 *Acta Mater* **57** p 5207
[28] Flinn J E, Field D P, Korth G E, Lillo T M and Macheret J 2001 *Acta Mater* **49** p 2065
[29] Islamgaliev R K , Chmelik F, Kuzel R 1997 *Mater Sci and Eng A* **A237** p 43
[30] Gendelman O V, Shapiro M, Estrin Y, Hellmig R J and Lekhtmakher S 2006 *Mater Sci and Eng A* **A434** p 88

Porosity evolution in additively manufactured aluminium alloy during high temperature exposure

J Bai[1, 2,4], H L Ding[2], J L Gu[3], X S Wang [2] and H Qiu[2]

[1]School of materials science and engineering, Northeastern University, Shenyang 110819, China

[2] School of Resources and Materials, Northeastern University at Qinhuangdao, Qinhuangdao 066004, China

[3] College of Materials Science and Engineering, Yanshan University, Qinhuangdao 066004, China

E-mail:baij@neuq.edu.cn

Abstract. A 2319 aluminum alloy is deposited by the Wire+Arc Additive Manufacturing technology with Cold Metal Transfer process. Porosity that are both existing in the as-deposited and as-heat treated state metal are revealed by optical microscopy and quantitatively analyzed. It explains the reason why the newly initiated pores are easily tend to gather between each layer around the fusion line zone for the WAAM metal after heat treatment. The inner morphology of the pores are demonstrated by Scanning Electron Microscopy. Porosity evolution and distribution during high temperature exposure are demonstrated. Thus two porosity growth and number increase mechanisms are proposed eventually, providing theoretical basis for related material design and process optimization.

1. Introduction

Wire+Arc Additive Manufacture (WAAM) is a promising manufacturing process which offers near-net shape fabrication of complex components without dies or substantial machining, resulting in a reduction in lead-time, waste, and cost, as well as the ability to produce complex geometries which allows the manufacture of novel, light-weight structures from a low density alloy[1]. Nowadays, industry applications such as aerospaceandautomotive are normally applying aluminium alloys instead of steel due to their low density and low cost[2]. The production of such components is of great interest for large scale metallic parts[3]. However the implementation of additively manufactured aluminium is limited by its inferior physical propertiesand the hydrogen pores defect. It is reported by H. Toda[4]that the existing of micro pores will further reduce the mechanical properties of the alloy.

Hydrogen porosity are always hard to control during WAAM depositing process for two reasons. One reason is that aluminum alloys are the most susceptible to hydrogen due to the great difference in solubility between liquid and solid metal (0.036 against 0.69 $cm^3/100$ g at a melting point of 660℃) [5]. Thus even small amount of hydrogen may excess the solubility limit in solid state. The other causeof porosity is that the moisture, grease and other hydrocarbon contaminants on the wire, which is the raw material for WAAM process, are converted into atomic hydrogen which is then available for absorption into the molten pool[6].

During practical industrial manufacturing process, the metallic parts are always strengthened by solution treatment and subsequent aging process technology. However, some researches have been carried out on casting and wrought aluminium alloys and have shown that the secondary porosity problem would be observed during high-temperature exposure[7]. Although the porosity number increase, formation and growth during homogenization were shown in these studies, the factors affecting those behaviors and the physical properties of those secondary pores in solid phase are not yet to be fully understood. Especially the porosity evolution of additively manufactured aluminium alloy have not been investigated before. The objective of the present research is to experimentally study the evolution of micro pores during high temperature exposure, thus to provide theoretical basis for related material design and process optimization.

2. Experimental

The compositions of ER2319 wire, which was applied in the present study as raw material for WAAM metal, is Al-6.3%Cu-0.3%Mn. 2219-T851aluminium plate was used as substrate metal in this trail. The 1.2mm wire was feed at a constant wire feed speed (6m/min) by a wire feeder machine to deposit a wall dimensions in 230mm×50mm×6.5mm.The deposits were built by pulse advanced Cold Metal Transfer (CMT) process supplied by a Fronius CMT Advanced 4000R power source,connecting to an ABB robot IRB2400 at a fixed travel speed of 0.6m/min. Flow rate of the shielding gas of pure argon (99.99%) was kept at 25 L/min during the whole process. Half of the wallwas solution treated in a furnace with a ramping speed at 200℃/h to 535℃ and then held for 90min, followed by quenching in water at 20℃. Specimens for porosity observation were mechanically sectioned from the central region (20mm from the top surface) of the wall along the longitudinal-direction. They were prepared in epoxy mount and polished to a mirror finish. Microstructure of the alloy was etched and revealed by Keller reagent. Pictures for porosity observation were taken by anOlympus optical microscopy and quantitativelyanalyzed by Image-Pro Plus (IPP) software.The inner morphology of the pores was examined by Scanning Electron Microscopy (SEM).

3. Results and discussion

Figure1(a) and (b) show the metallographic images of micro pores in as-deposited and as-heat treated additively manufactured 2319 aluminium alloy. There are various sized micro pores distributed in the two states of the WAAM alloys. In addition, it clearly shows that the pore numbers in the as-heat treated sample are greatly increased compared with the as-deposited one. The results analyzed by IPP have demonstrated that the number density of pores greater than 5 μm increases from 512 to 1667 per $100mm^{-2}$, which means that the number of pores increases by three times during high temperature exposure. Meanwhile the mean diameter of the pores just increases from 13.5mm to 15.5mm. Microstructure of the as-heat treated sample in figure 1(c) shows that the boundary between the dendrites and equiaxed grains is the fusion line between each layer. In other words, the newly initiated and grown pores (black points) are mainly distributed around the fusion line zone (partially melted zone).

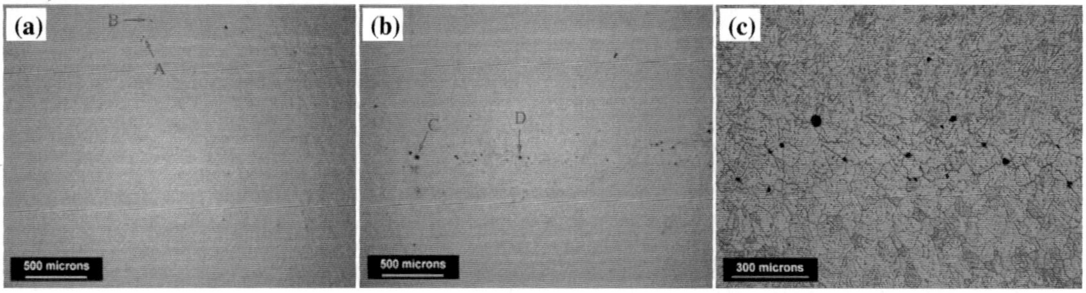

Figure 1. Porosity in WAAM aluminum metal both in (a) as-deposited and (b) as-heat treated state, (c) is the microstructure of as-heat treated state

During the additive depositing process, when the liquid metal get into the molten pool, solidification begins to process as the dissolved atomic hydrogen are rejected from the newly formed solid into the surrounding liquids[8] due to the big gap of solubility between the two existing states. Along with the process of solidification, the continuously incremental hydrogen in the liquid will be gathered to surpass the solubility limit[9]. Thus pores will be nucleated and grew along the dendrite grain boundaries. If the newly initiated pore size and buoyancy force grow big enough, it may float to the upper part surface of the liquid metal[10]. Ultimately, the gas pores are usually distributed at the top of the solidified molten pool because they are always entrappedat the top layer of the fusion zone. In addition, the quickly formed oxide thin films are very easily absorbing molecular hydrogen and moisture in the air, which will further increase the amount of hydrogen in the top region of each layer[11]. Thus around the fusion line zone of each layer, there is always a higher amount of entrapped hydrogen and fine micro pores, which will be got together and grown to be bigger ones during high temperature exposure. That is why the greatly increased observable pores are mostly distributed around the fusion line zone between each layer.

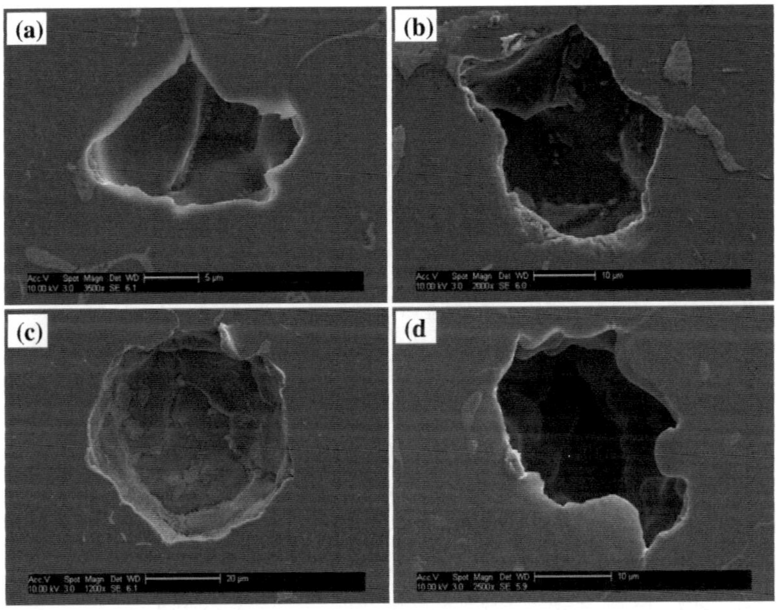

Figure 2. Inner morphology of micro pores distributed in the WAAM 2319 alloy in (a) and(b) as-deposited state, (c) and (d) as-heat treated state

Porosity of point A and point B in figure 1(a) are further magnified by SEM as shown in figure 2(a) and (b), as well as pore C and pore D in as-heat treated state are indicated in figure 2(c) and (d), respectively. There are distinctly different characteristics of inner morphology for these pores. Although the pores may be nucleated as spheres initially, as analyzed above and shown in figure 2(a) and (b), they will grow into various sized irregular hydrogen pores together with the primary eutectic phases along with the inter-dendritic grain boundaries and distributed in the dendritic arms and cells during the solidification process.

During heat treatment process, the pores evolution is obviously controlled by classicalOstwald Ripening effect[12] as characterized in that the pores have grown to be bigger and the sphericity of some individual pores are improvedas shown in figure 2(c). During this process, hydrogen pores are governed by inter-pores coalescence in solid aluminum both from molecular hydrogen dominating in micro pores or hydrogen entrapped in the other surrounding trap sites, such as precipitates, grain boundaries or lattice interstices. Intra-pore growth and coarsening is a local ripening of individual pores, which is controlled by increasing internal hydrogen pressure comparing with the surface tension.

However, another mechanism is also dominating the growth and initiation of gas pores, because there is one contrary observed phenomenon comparing with the classical Ostwald Ripening in that a prominent increase of number density of porosity is observed. It may be caused by the migration and combination of hydrogen and vacancies that originally existed after depositing or are induced by dissolving of primary eutectic phases (Al_2Cu). In addition, the ample dispersion particles and abundant grain boundaries around the inter-layer region provide energetically favorablesites for the heterogeneous nucleation of hydrogenwhen supersaturated atomic hydrogen is precipitated asmolecules. The hypothesis is in great accordance with the inner morphology as indicated in figure 2(d), where the eutectic phases are completely dissolved into the matrix metal. Besides, irregular pores are formed along the wall of inter-dendrites, exhibiting non-spherical shape characters.

4. Conclusions

Irregular shaped hydrogen pores are observed in the as-deposited 2319 aluminum alloy due to the solubility difference of the liquid and solid aluminum metal and the unavoidable hydrogen contaminants existing on the wire surface. During high temperature exposure, there is a great increase of number density as well as a slight increase of the mean size of porosity. The newly initiated pores are mostly gathered between each layer around the fusion line zone. In addition, the inner morphology of the pores both generated before and after heat treatment are demonstrated. It is illustrated that there are two related mechanisms in controlling the growth and number increase of porosity after heat treatment. They are the classical Ostwald Ripening effect and combination of hydrogen and vacancies effect, respectively.

Acknowledgements

This work is supported by the National Natural Science Foundation of China (Grant No. 51301036), the Fundamental Research Funds for the central Universities of China (Grant No. N130523001), the Research Fund for the Doctoral Program of Higher Education of China(Grant No. 20120042110021).

References

[1] Zhang Y, Chen Y, Li P and Male A T 2003 *J. Mater. Process. Tech.* **135** 347-357
[2] Li J, Zhang M and Zhao Y 2004 *Journal of East China Shipbuilding Institute* **18** 78-81
[3] Ding J and Colegrove P 2011 *J Comp. Mater. Sci.* **50** 3315-3322
[4] Toda H, Oogo H and Horikawa K 2014 *Metall. Mater. Trans. A* **45A** 765-776
[5] Devletian J H and Wood W E 1983 Welding Research Council of the Engineering Foundation in: Factors affecting porosity in aluminum welds - a review NY 1-18
[6] Gu J, Ding J and Cong B 2015 *Advanced Materials Research* **1081** 210-214
[7] Toda H, Hidaka T and Kobayashi M 2009 *Acta Mater.* **57** 2277-2290
[8] Lee P D and Hunt J D 2001 *Acta Mater.* **49** 1383-1398
[9] Fruehan R J and Anyalebechi P N 2008 *ASM Handbook* **15** 64-73
[10] Hwang L R, Gung C H and Shih T S 2001 *J. Mater. Process Tech.* **116** 101-113
[11] Wu Z, Qu L and Lu L 2014 *Welding Technology* **43** 59-61
[12] Toda H, Yamaguchi T and Nakazawa M 2010 *Mater. Trans.* **51** 1288-1295

Effect of QPQ nitriding time on microstructure and wear resistance of SAF2906 duplex stainless steel

D Liu[1], G X Wu and L X Shen

School of Mechanical Engineering and Automation, Fuzhou University, Fuzhou 350108, China.

E-mail: ldfzu@fzu.edu.cn

Abstract. QPQ salt bath treatment of SAF2906 duplex stainless steel was conducted at 570 ℃ for 60 min, 90 min,120 min,150 min and 180 min, followed by post-oxidation process with heating temperature of 400℃ and holding duration of 30 min. The effect of QPQ nitriding time on microstructure and wear resistance of SAF2906 duplex stainless steel was investigated by means of OM, SEM, XRD, microhardness test, adhesion strength test and wear resistance test. Microstructure observation showed outer layer was composed of Fe_3O_4. The main phase of the intermediate layer was CrN, αN and $Fe_{2\text{-}3}N$. The main phase of the inner layer was CrN and S. The adhesion strength test of the surface layer-substrate showed the QPQ treated samples have favorable adhesion strength of HF-1 level. With the increase of nitriding time, the growth rate of the compound layer gradually slowed down and the surface hardness first increased and then decreased, and the maximum hardness was 1283 $HV_{0.2}$ at 150 min. The dry siliding results showed that the wear resistance of the QPQ treated samples was at least 20 times than that of the substrate, and the optimum nitriding time to obtain the best wear resistance is 150 min. The worn surface morphology observation showed the main wear mechanism of the substrate was plough wear, while micro-cutting is the main wear mechanism that causes the damage of the QPQ treated samples.

1. Introduction

Duplex stainless steels which have a two-phase microstructure consisting of ferrite and austenite and possess the excellent performance of ferritic and austenitic stainless steel are widely used in petroleum-refining, chemical and military industries, etc. [1-3]. The research and application of duplex stainless steels were conducted since 1930 s. With the progress of metallurgical technology, duplex stainless steels have been developed to the third-generation of super duplex stainless steel, whose pitting resistance equivalent number is over 40 [4]. SAF2906 super duplex stainless steel has better corrosion resistance, and it is suitable for deep water or seawater. However, deposited in joint action of multiple factors (e.g. corrosion, wear and overloading), SAF2906 duplex stainless steel, of whose low hardness and poor wear resistance, failed early. In industrial production, there are some enterprises using quench-polish-quench(QPQ) technology to improve wear resistance of stainless steels [5]. For example, the has been used QPQ technology treating engine valve materials of 5Cr21Mn9Ni4N austenitic stainless steel to increase its wear resistance under high temperature condition. The German textile machinery manufacturers also adopt QPQ technology to treat guide rail materials of 0Cr18Ni12Mo2Ti austenitic stainless steel for improving its wear resistance. Current opinion on the mechanisms of stainless steels possessing excellent wear resistance after QPQ treatment that has reached a unified is the formation of high-hardness and a certain thickness of compound layer in its surface. The thicker the compound layer, the better the wear resistance. And some scholars are specialized in investigating the influence of nitriding time on surface layer thickness and wear resistance of stainless steel [6-7], the results show that when nitriding time is short, the compound layer is too thin to meet the requirement of the actual operating condition. Yet long nitriding time makes the growth rate of the compound layer slow down and the loose layer appear, and the nitrides in the compound layer grow and coarsen, which leads to the decrease of surface hardness. As a consequence, choosing the reasonable nitriding time is important for stainless steels to improve its wear resistance. At present, considerable researches on the effect of QPQ nitriding time on microstructure and wear resistance of stainless steels is mainly about single phase stainless steels such as austenite and martensite, and the research on SAF2906 duplex stainless steel is rarely done. Therefore, it is necessary to study the effect of QPQ nitriding time on the microstructure and wear resistance of SAF2906 duplex stainless steel.

2. Materials and methods

The material used in this investigation was SAF2906 duplex stainless steel, whose preparation method is as follows: first, the raw materias such as 316L stainless steel, chromium metal, iron- molybdenum, electrolytic nickel and nitride alloy were melted in 300 KW/30 Kg medium frequency induction furnace, the chemical compositions of the alloy was obtained by OES as shown in table 1. The as-cast sample was solid solution treated by KSS-1700 box-type resistance furnace at 1050 ℃ for 2h, then followed by water quenching. The heat treatment samples were cut into two groups: 10 mm×10 mm×10 mm and 30 mm×7 mm×6 mm. The former were used for optical microscopy and micro-hardness test, and the latter were used for adhesion strength and wear resistance test. Before QPQ treatment, all the flat surfaces of each sample were ground with various grades of SiC sandpaper up to 2500 grit and polished to a mirror fish using Cr_2O_3 solution, then followed by ultrasonically rinsing in acetone and alcohol for 10 min, and finally dried in air.

Table 1. Chemical compositions of SAF2906 DSS samples (wt.%).

C	Si	Mn	P	S	Ni	Cr	Mo	Cu	N	Fe
0.03	0.72	0.81	0.02	0.02	6.61	29.19	2.11	0.79	0.26	Bal.

The samples were nitrided at 570 ℃ for 60 min, 90 min,120 min,150 min and 180 min, then followed by the same post-oxidation process with heating temperature of 400℃ for 30 min. The CNO- concentration was kept at 33 %.

The cross-section microstructure of the QPQ treated samples after inlaying, grinding, polishing and etching was observed by XJP-300 optical microscopy and S-3000N scanning electron microscopy, and measured the compound layer thickness by image analysis software of Image-Pro Plus. The phase compositions of the surface layer were identitied by X'Pert Pro MPD X-ray diffractometer with Co Kα radiation in the range of glancing angles from 20° to 100°. The adhesion strength test of the surface layer-substrate adopted HRC indentation method commonly used in engineering. The test load was 150 kgf. The indent morphologies of the treated samples were observed under 100 times of optical microscopy, then compared with criteria morphologies of evaluating adhesion strength [8-9], and the adhesion level of the surface layer was obtained. The surface hardness of the samples was measured by MVC-1000D1 micro-hardness tester with 200 g load for 15 s, taking five readings to determine the arithmetic mean value as the surface hardness. Dry-sliding tribological tests were carried out on MMS-2A wear test machine. Before test, the substrate and treated samples were cleaned by alcohol and weighed. During test, a 40 mm diameter GCr15 steel ball rotated at a speed of 200 rpm on the surface of the samples. The test load was 50 N. After 60 min, the samples were removed from the wear test machine. They were cleaned and dried, and measured by FA2014 photoelectron analytical balance with an accuracy of 0.1 mg, then calculated mass loss to characterize its wear resistance. In order to reduce the influence of test error on the results, the average mass loss was calculated by three parallel experiments carried out under the same condition.

3. Results and discussions

3.1. Metallographic observations

Figure 1 shows the cross-sectional microstructure of SAF2906 duplex stainless steel nitrided at 570 ℃ for different time (60 min, 90 min, 120 min, 150 min and 180 min). From the outmost surface to the core,a thin outer layer,intermediate layer,inner layer were formed. The outer layer is too thin to differentiate by optical microscopy,while the intermediate layer and inner layer can be distinguished from color after etching, but the appearance is still unclear. So as to clearly observe the morphology of different layers, the sample nitrided for 150 min was selected for SEM observation, the result is shown in figure 1(f). It can be seen from the figure that the outer layer is a black banding layer, whose thickness is about 2 μm. In order to furtherly determine the phases in the outer layer, the samples of nitrided only and QPQ treated were selected (nitriding time of 150 min) for XRD test, and the results were depicted in figure 2. The main phases of the surface layer nitrided only was CrN, αN (expanded martensite), $Fe_{2-3}N$ and S (expended austenite). In addition to CrN, αN, $Fe_{2-3}N$ and S, the surface layer QPQ treated also has some Fe_3O_4. Therefore, it can be inferred that the outer layer is composed of Fe_3O_4, which was produced in the post-oxidation process, and this finding is consistent with previous reports [10-11], namely, the outer layer is an oxide layer. Nitriding treatment is the most important procedure of QPQ treatment, and its purpose is to form a certain thickness and high-hardness of nitrided layer. As shown in figure 2a, the nitrided layer consists of CrN, αN, $Fe_{2-3}N$ and S. Due to the nitrogen concentration in the surface layer gradually declines from the outmost surface to the core, the hardness of the surface layer also drops. Besides, the nitrogen concentration in the inner layer has a significant drop, which leads to the hardness of the inner layer, whose effect on improving wear resistance is small, is far lower than that of the intermediate layer. Therefore, the intermediate layer also called compound layer, is the expected microstructure by nitriding treatment, and its main phases are high-hardness nitrides of CrN and $Fe_{2-3}N$, which plays a leading role in wear resistance. The inner layer, also called diffusion layer, is a black zone between the intermediate layer and the substrate, and its main phases are low-hardness phase of S and fine precipitate of CrN. Compared with common microstructure of diffusion layer, the diffusion layer of SAF2906 duplex stainless steel is discontinuous and composed of two parts: black region "1" and white region "2", which is closely related to the substrate phases of ferritie and austenite. This may be attributed to that the original microstructure of "1" region is ferrite, and its solubility of nitrogen is small, which results in nitrogen diffusing into ferrite combining extremely easily with high concentration of chromium to form chromium nitride and subsequently leads to Cr-depleted areas in the matrix and deteriorates its corrosion resistance, after reagent etching, the region appears "dark". The original microstructure of "2" region is austenite, and its solubility of nitrogen is higher than that of ferritie, more nitrogen atoms can be furtherly dissolved into austenite, which increases its corrosion resistance. Therefore, the region becomes "bright" after reagent etching. It is also obtained from figure 1 that when nitriding time is 60 min, the thickness of the compound layer is thin. With the extension of nitriding time, its thickness increases significantly. When nitriding time is over 120 min, its thickness increases slowly. The thickness of the compound layer is shown in figure 1, it was approximately 9.35 μm, 12.15 μm, 14.20 μm, 15.71 μm and 16.53 μm, respectively. Obviously, with the increase of nitriding time, the growth rate of the compound layer gradually slows down. The reason for the phenomenon is due to the extension of nitriding time, the nitrides formed on the surface of SAF2906 duplex stainless steel increases, which results in diffusion resistance of nitrogen increasing and the formation ability of nitrides reducing.

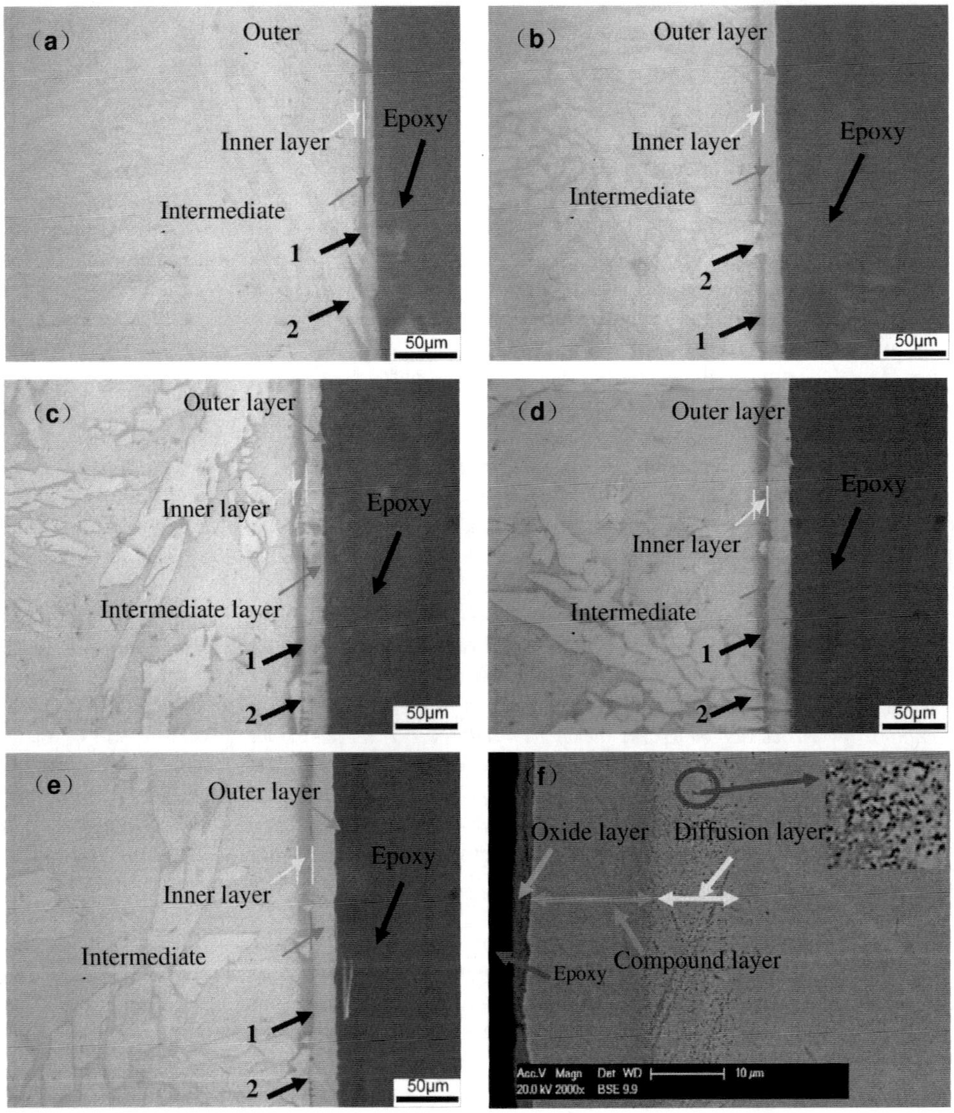

Figure 1. Cross-sectional metallographic microstructure of 2906 duplex stainless steel treated at 570 ℃ for different time.
(a)60min (b)90min (c)120min (d)150min (e)180min (f)150min.

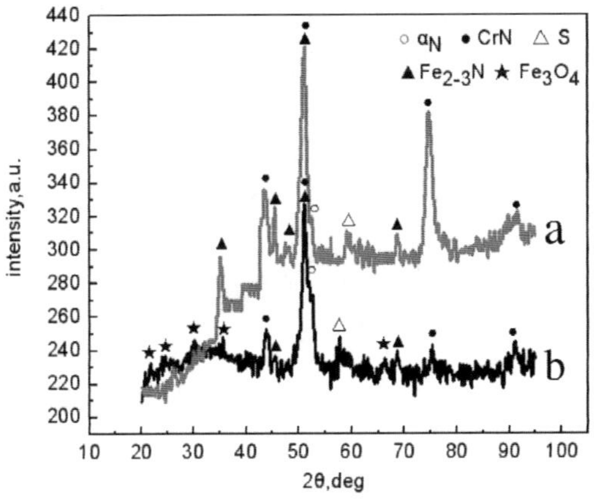

Figure 2. X-ray diffraction patterns of the QPQ treated sample and nitrided only sample.
(a) nitrided only sample (b) QPQ treated sample.

3.2. Adhesion strength analysis

Indentation method is one of the commonly uesd for evaluating the adhesion strength of hard layer by using large loading, and make layer-substrate system occur significant plastic deformation, thereby qualitatively evaluating adhesion strength of layer-substrate system [12]. Figure 3 shows indent morphologies of the treated samples under the action of 150 kgf. It can be obtained from the figure 3 that the treated samples have favorable adhesion strength of HF-1 level as compared with criteria morphologies of evaluating adhesion strength, which is due to distribution characteristics of nitrogen in the surface layer displays a continuous gradient variation, making the hardness of the surface layer gradually decline from the outmost surface to the core and the surface layer exhibits good ability of coordination deformation, thus greatly guaranteeing the treated samples have excellent adhesion strength of surface layer-substrate. Moreover, nitrogen atoms combine with Mo、Cr and Ti, etc., to form stable nitrides, which distribute dispersely in surface layer- substrate interface, thus froming a certain thickness of transition layer between the surface layer and substrate, which is beneficial to improve adhesion strength of layer- substrate. It is also found that there is no cracks occurred on the surface layer of the treated samples with nitriding times of 60 min, 90 min and 120 min. Yet when nitriding time is 150 min and 180 min, there is some cracks appeared on the surface layer of the treated samples, and the longer the nitriding time is, the more the cracks are. This indicates when nitriding time is over 120 min, the adhesion strength of surface layer-substrate declines, and the longer the nitriding time is, the more obvious the decline is. This may be attributed to that the brittle phase of CrN increses and coarsens with nitriding time increasing, which causes a decrease in toughness of the surface layer. Besides, the structural differences of the phases increase the residual stress in the surface layer.

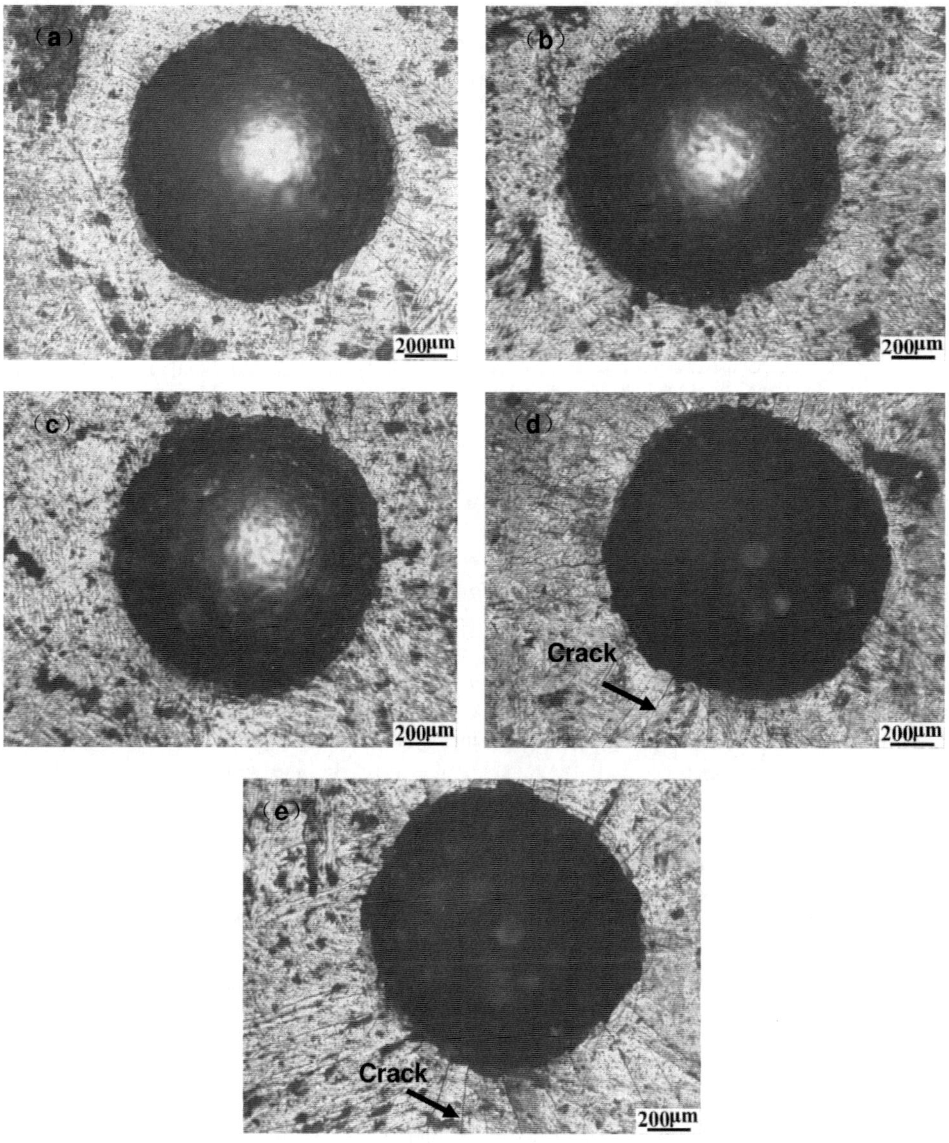

Figure 3. Indent morphologies of the QPQ treated samples by indentation method.
(a) 60min (b)90min (c)120min (d)150min (e)180min.

3.3. Hardness analysis

Figure 4 shows the relationship between the surface hardness of the treated samples and nitriding time. It can be obtained that the surface hardness of the samples after QPQ treatment has a significant increase and is over 3 times of the substrate. The significant increase in hardness is due to the surface microstructure of the substrate changes from low-hardness phases of α and γ into high-hardness phases of CrN, $Fe_{2\text{-}3}N$ and αN etc, which makes the surface hardness notably reinforce. It can be also obtained from the figure that with the increase of nitriding time, the surface hardness of the treated samples gradually increases, when nitriding time is 150 min, its hardness reachs the maximum of 1283 $HV_{0.2}$. To furtherly increase nitriding time, the surface hardness of the treated samples decreases, which is related to the change of surface microstructure of SAF2906 duplex stainless steel. At an initial stage of nitriding treatment, nitrogen atoms is dissolve into the substrate of α and γ, causing its lattice expansion, which improves the surface hardness of SAF2906 duplex stainless steel to some extent. Besides, nitrogen atoms combines with alloy elements or carbides to form nitrides (e.g. CrN and $Fe_{2-3}N$), which significantly enhances its surface hardness. When the surface layer reaches a certain thickness, the diffusion resistance of nitrogen atoms increases, and the degree of lattice distortion and the amount of nitrides have unconspicuous increase, the surface hardness of the treated sample reaches the maximum. With prolonged nitriding time, nitrides grow and coarsen, which causes the decrease of surface hardness instead.

3.4. Wear resistance analysis

Figure 5 shows the mass loss of the samples nitrided at 570 ℃ for different time. It can be seen from the figure that the mass loss of the substrate is about 36.5 mg, that of the treated samples is below 1.5 mg, that is, the wear resistance of the latter is over 20 times of the former. Besides, the surface hardness of the treated samples fist increased and then decreased with nitriding time increasing. When nitriding time is 60 min, its mass loss has a maximum of 1.5 mg, with nitriding time extended to 150 min, its mass loss reaches the minimum value of 0.8 mg, indicating that the optimum nitriding time to obtain the best wear resistance is 150 min, which is in good agreement with the surface hardness changing regularity.

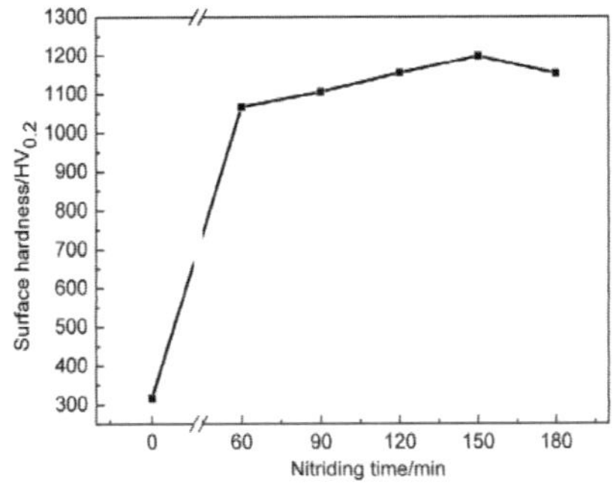

Figure 4. Variation of surface hardness versus nitriding time

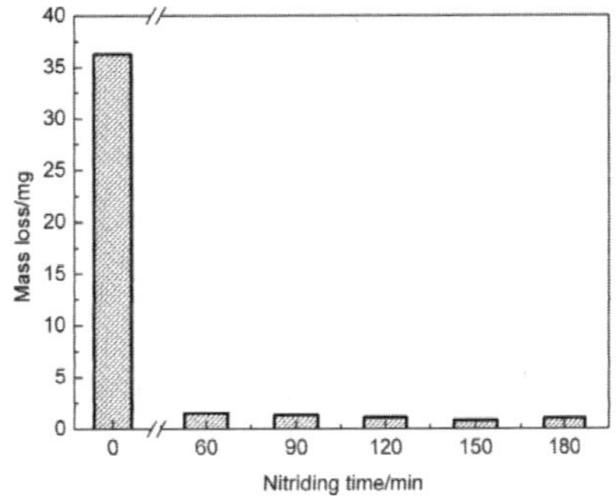

Figure 5. The mass loss of the samples nitrided at 570 ℃ for different time.

Figure 6 shows the surface SEM micrographs of the substrate and the sample nitrided for 150 min after wear test. It can be seen from figure 6a that a very rough metallic surface with severe abrasion grooves, peeling-off and plastic deformation is seen on the surface of the substrate. As for the sample nitrided for 150 min, a basically smooth appearance with very mild abrasion grooves is observed, nearly no peeling-off and plastic deformation occurred on the surface are found, and only some wear debris are seen from the surface. Therefore, plough wear is the main wear mechanism that causes the damage of the substrate, the main wear mechanism observed at the surface of the treated sample is micro-cutting. The reason why the substrate and the treated sample show different wear mechanism, which has a great relationship with the microstructure and property of material. This can be explained that due to the hardness of the substrate is far lower than that of mating material, its surface is prone to produce plastic deformation under the action of loads. In addition, some weak areas appear crack and fall off, then form tiny wear debris. In the wear process, wear debris have "plough" effect on the substrate surface, thus producing severe abrasion grooves on the surface. As for the treated sample, the surface of which consists of high-hardness and dispersed nitrides (e.g. CrN and $Fe_{2-3}N$), is not easy to produce wear debris compared with the substrate, and wear debris is also not easy to insert into the substrate. So only some shallow scratches occur on the treated sample surface.

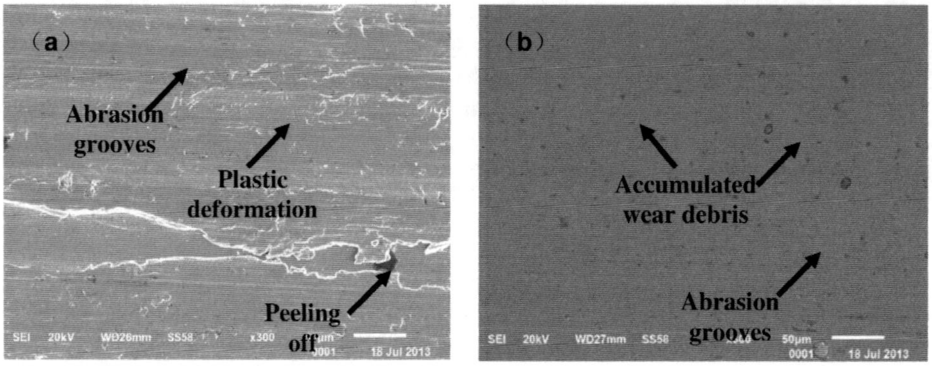

Figure 6. Worn surface morphology of the substrate and the sample treated at 570℃ for 150min (a) Substrate (b) QPQ treated sample

4. Conclusions

The surface layer of the samples after QPQ treatment was composed of outer layer, intermediate layer and inner layer from the outmost surface to the core. The outer layer was composed of Fe_3O_4. The main phase of the intermediate layer was CrN, αN and $Fe_{2\text{-}3}N$. The main phase of the inner layer was CrN and S. The QPQ treated samples have favorable adhesion strength of HF-1 level. With the increase of nitriding time, the growth rate of the compound layer gradually slowed down and the surface hardness first increased and then decreased, and the maximum hardness was 1283 $HV_{0.2}$ at 150 min.

The wear resistance of the QPQ treated samples was at least 20 times than that of the substrate, and the optimum nitriding time to obtain the best wear resistance is 150 min. The worn surface morphology observation showed the main wear mechanism of the substrate was plough wear, while micro-cutting is the main wear mechanism that causes the damage of the QPQ treated samples.

Acknowledgements

This work was financially supported by the Key Project of Sinence and Technology Department of Fujian Province(No:2014H003); The Development of Science and Ttechnology Fund of Fuzhou university(2014XQ13); Education Scientific Research Project Plan of Young and Middle-aged Teachers in FuJian Province(No.JAT160066).

References

[1] Wu J 1999 Duplex stainless steel (Beijing: China Metallurgical industry press) p 50-52
[2] Koztowski R H 1995 *J. J. Mater. Process. Technol.* **53** 239–46
[3] Jimenez J A, Carsi M and Ruano O A 2000 *J. J. Mater. Sci.* **35** 907–15
[4] Tan H, Jiang Y and Deng B 2009 *J. Mater. Charact.* **60** 1049-54
[5] Li H Y, Luo D F and Wu S X 1997 Principle and application of QPQ technology (Beijing: China Machine Press) p 30-32
[6] Cai C, Wei G F and Meng F N 2012 *J. Transaction of materials and heat treatment.* **34** 153-6
[7] Cai W W and Luo D F 2012 *J. Material and heat treatment.* **41** 176-9
[8] Heinke W, Leyland A and Matthews A1995 *J. Thin Solid Films.* **270** 431-8
[9] Burnett D J and Rickerby D S 1987 *J. Thin Solid Films.* **154** 403-16
[10] Li H Y, Luo D F and Yeung C F 1997 *J. J. Mater. Process. Technol.* **69** 45-9
[11] F Y C, Lau K H and Li H Y 1997 *J. J. Mater. Process. Technol.* **66** 249-52
[12] Fu T, Wang C P, Hou B, Zhu X D and Ma S G 2010 *J. China surface engineering.* **23** 64-7

Crystalline electric field schemes of Ce_2NiGe_6 and Ce_2NiGe_3

Yushan Li[1], Hongyan Liu

Department of Physics, Heze University, Heze 274015, China

E-mail: lysh507@163.com

Abstract. The splitting levels and corresponding wave functions of rare earth compounds Ce_2NiGe_6 and Ce_2NiGe_3 have been presented by taking into account the crystalline electric field. The influences of Ge content on rare earth ion ground state splitting also have been analysed. The first excited energy and overall splitting energy of Ce_2NiGe_3 coincides with the ones calculated by simulating Schottky specific heat.

1. Introduction

The investigation of crystalline electric field (CEF) effect for 3d-4f rare-earth inter-metallic compounds has become an important subject in the field of magnetism and magnetic materials from both theoretical and application point of view in recent years [1-6]. Although the 4f electron is firmly bound in the depth of atoms, many rare earth compounds appear various electronic interactions related to 4f electron, including magnetic order, Kondo effect, RKKY interaction, heavy electron effect and CEF effect. The multiple ground state of local 4f electron in rare earth compounds under the CEF actions of surrounding ions will be split or excited. The degeneracy may partially or completely disappear, and then producing a great impact on magnetic susceptibility, specific heat, resistivity and magnetization. Therefore, we can stimulate the temperature dependence of the inverse magnetic susceptibility, specific heat, magnetization to obtain the CEF schemes on the bases of CEF theories. For example, the CEF splitting levels of $PrNi_2Al_5$ [3], $Ce_2Pd_3Si_5$ [4] and $Ce(Pd_{1-x}M_x)_2Al_3$ [5] have been reported by means of simulating their respective inverse magnetic susceptibility-temperature curves.

Recently, Huo [4] and Im [5] have measured the magnetic susceptibility, magnetization and electrical resistivity of Ce_2NiGe_6 and Ce_2NiGe_3, respectively. In this paper, we use the CEF program to simulate inverse magnetic susceptibility-temperature curves for Ce_2NiGe_6 and Ce_2NiGe_3 in order to obtain CEF splitting levels. It is of interest to make an attempt to see some systematic behavior by comparing the same type compounds with different crystallographic structures.

2. Model and theory

We suppose the charge distribution of the surrounding environment is $\rho(R)$. Then electrostatic interaction potential between metal ion and i-valence electrons can be expressed by

$$V(\vec{r}) = \int \frac{\rho(\vec{R})}{\left|\vec{R} - \vec{r}\right|} d\vec{R} \cdot \qquad (1)$$

$\vec{R}$ and $\vec{r}$ denote the coordinate vectors of around ion and i-valence electrons, respectively.

The CEF Hamiltonian of rare ion is generally written as

$$H_{CEF} = -e\sum_i V(\vec{r}_i). \tag{2}$$

The summation includes all 4f electrons.
We can use the following formula

$$\frac{1}{|\vec{R}-\vec{r}|} = \sum_{k=0}^{k=\infty} \frac{r^k}{R^{k+1}} P_k(\cos(\vec{R},\vec{r})) \tag{3}$$

and

$$P_k(\cos(\vec{R},\vec{r})) = \frac{4\pi}{2k+1}\sum_q (-1)^q Y_{kq}(\theta_i,\phi_i) Y_{k-q}(\theta,\varphi) \tag{4}$$

to obtain the concrete form of CEF Hamiltonian. Meanwhile, we also introduce tensor operator

$$C_q^k(\theta,\varphi) = \left(\frac{4\pi}{2k+1}\right)^{1/2} Y_{kq}(\theta,\varphi). \tag{5}$$

Here, θ_i and φ_i represent the polar angle and azimuth angle of $\vec{r}$ which characterize the position of 4f electron. θ and ϕ denote the polar angle and azimuth angle of $\vec{R}$ which characterizes the positions of surrounding ions.

Then, we can obtain the CEF Hamiltonian H_{CEF} on the base of simple point-charge model with

$$H_{CEF} = \sum_{k,q} B_q^k C_q^k. \tag{6}$$

B_q^k and C_q^k are the CEF parameters and Stevens equivalent operators, respectively. The CEF Hamiltonian can be expressed as various polynomials of B_q^k and C_q^k within different symmetries.

Following the Hutchings notation, the CEF Hamiltonian for the orthogonal symmetry Ce_2NiGe_6 can be expressed in general form as

$$H_{CEF} = B_2^0 C_2^0 + B_2^2 C_2^2 + B_4^0 C_4^0 + B_4^2 C_4^2 + B_4^4 C_4^4. \tag{7}$$

The compound Ce_2NiGe_3 belongs to hexagon crystal system, only two nonzero CEF parameters are needed

$$H_{CEF} = B_2^0 C_2^0 + B_4^0 C_4^0, \tag{8}$$

In these Hamiltonians, B_2^0 coefficient is dominant since the others are smaller by an order of magnitude. The sixth order term is often neglected. The direction of the rare earth magnetic moment shows correlation with the sign of B_2^0 coefficient. For orthorhombic point symmetry, a negative sign of B_2^0 parameter implies that the magnetic moment aligns along the b-axis, whereas a positive sign is connected with the moment lying in the a-c plane, while for hexagon, a negative sign implies that the magnetic moments align along the hexagonal axis, a positive sign reveals that they are normal to it. We next try a CEF analysis for a deeper understanding of this behaviour.

3. Results and discussion

It can be seen that the inverse magnetic susceptibility-temperature curves of Ce_2NiGe_6 and Ce_2NiGe_3 in Refs. [4] and [5] significant deviate from Curie-Weiss law fits below 50 K, owing to that the strong

CEF effects which results in the splitting of the multiple ground states degeneracy for rare ion Ce^{3+}. And for this we calculate the eigen values and corresponding wave functions of Ce_2NiGe_6 and Ce_2NiGe_3.

The first excited energy, total splitting energy and wave functions for Ce_2NiGe_6 and Ce_2NiGe_3 derived from CEF calculation are listed in table 1. The analytical results of the CEF schemes show that the ground state for Kramers ion Ce^{3+} is doublet. The six-fold degenerated ground state $^2F_{5/2}$ of orthogonal symmetry Ce_2NiGe_6 splits into three mixed doublets. The six-fold degenerated ground state $^2F_{5/2}$ of hexagon symmetry Ce_2NiGe_3 splits into three pure doublets. This is consistent with the Kramers law and the new wave functions are linear combinations of function with one J but different M in the presence of CEF. It is also worth mentioning that the difference of adjacent splitting energy in Ce_2NiGe_6 and Ce_2NiGe_3 is rather small, meaning that the influences of Ge may be ignored, but the influences of different crystal system seem enormous [7, 8].

Figure 1 and Figure 2 show the calculated and experimental results of the inverse magnetic susceptibility for Ce_2NiGe_6 and Ce_2NiGe_3, respectively. We can read off that the calculated magnetic susceptibility at higher temperature is consistent with the experiment results above 100K but the discrepancy becomes larger at lower temperatures, as shown in Figure 1 and Figure 2. There are mainly several possible reasons that may be leading to this discrepancy between the experiments and the calculation. First, we use isotropic field parameters in the calculation, but in fact there is a strong anisotropic effect especially in the case of low symmetry. Second, there is a strong hybridization between the f-electrons of the rare atoms and the d-electrons of the transition metal. Third, the appearance of spin glass freezing has caused the spin reorientation and fluctuation. Additionally, the other factors such as Kondo effect and RKKY interaction may have an influence on the magnetic moments [9, 10]. It is also worth mentioning that in some compounds CEF can change the magnetic ordering temperature in some compounds. Note that if we replace the simulation values of inverse magnetic susceptibility by magnetic susceptibility, then the discrepancy between CEF calculated and experiment data will become very small. Therefore, although the fitting of susceptibility gives only approximate values of splitting energies, the CEF calculation is rather reasonable within the experimental accuracy. In the case of Ce_2NiGe_3, the first excited energy 138K and overall splitting energy 496K coincides with 135K and 498K calculated by Huo [4] group through simulating the Schottky specific heat.

Table1. Eigen values E/K and corresponding wave functions of Ce_2NiGe_6 and Ce_2NiGe_3.

Eigen value	Wave function
Ce_2NiGe_6	
0	$-0.04418\|\mp 2.5\rangle + 0.65751\|\mp 0.5\rangle + 0.75215\|\pm 1.5\rangle$
78	$0.11092\|\mp 2.5\rangle + 0.75146\|\mp 0.5\rangle - 0.65039\|\pm 1.5\rangle$
557	$0.99285\|\mp 2.5\rangle - 0.05469\|\mp 0.5\rangle + 0.10613\|\pm 1.5\rangle$
Ce_2NiGe_3	
0	$\|\pm 1.5\rangle$
138	$\|\pm 2.5\rangle$
496	$\|\pm 0.5\rangle$

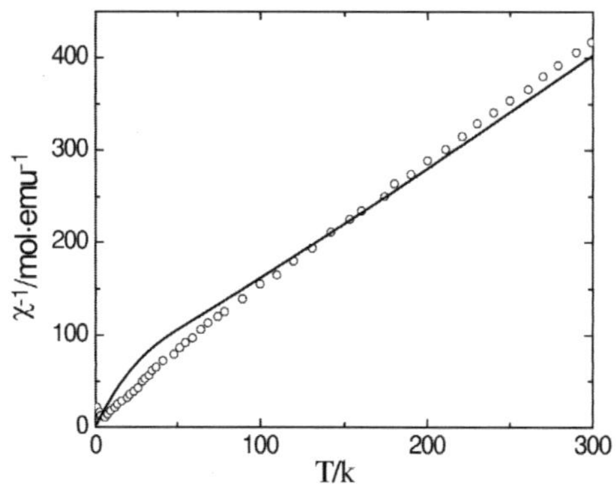

Figure 1. Temperature dependence of inverse magnetic susceptibility for Ce_2NiGe_6. Solid line shows the CEF calculated curve, and the open circles are the experiment curves.

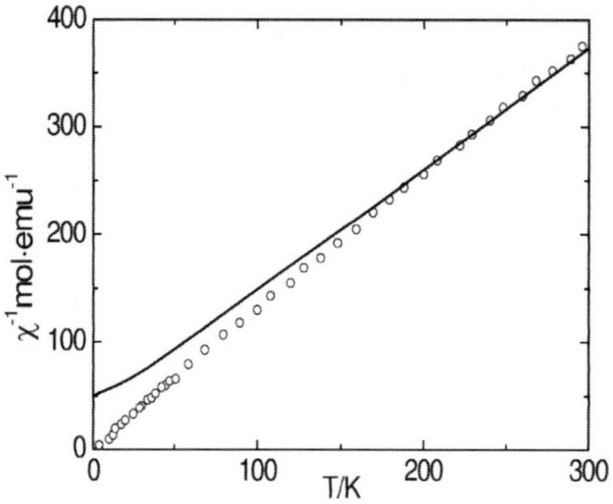

Figure 2. Temperature dependence of inverse magnetic susceptibility for Ce_2NiGe_3.

4. Conclusions

We succeed in simulating the temperature dependence of inverse magnetic susceptibility for Ce_2NiGe_6 and Ce_2NiGe_3, and obtaining the CEF splitting levels and corresponding wave functions of Ce^{3+}. We analyze the influences of Ge content on CEF-scheme theoretically. Because of the complexity and sensitivity of magnetic susceptibility in the lower temperature, it is difficult to simulate exactly the lower temperature part. Therefore, the CEF calculated results give only approximate values of the splitting energies. The more exactly neutron scattering experiments are desired to determine the magnetic structure and confirm CEF splitting scheme.

Acknowledgements

This work is supported by the science and technology research program for colleges and universities of Shandong province (No. J15LJ55) and the technology program for doctor of Heze University (Nos. XY17BS11 and XY17BS12).

References

[1] Saha S R, Sugawara H, Matsuda T D, Aoki Y and Sato H 2000 Magnetic, thermal, and transport properties of single crystals of antiferromagnetic Kondo-lattice Ce_2PdSi_3 *Phys Rev B* **62** 425-429

[2] Huo D, Skurai J, Kuwai T, Isikawa Y and Lu Q 2001 Electric, magnetic, and thermal properties of Ce_2NiGe_3: A Kondo lattice compound showing spin glass behavior *Phys Rev B* **64** 224405

[3] Akamaru S, Isikawa Y and Sakurai J 2001 Magnetic properties and de Hass-van Alphen effect of $PrNi_2Al_5$ *J Phys Soc Jpn* **70** 2049-2053

[4] Huo D, Sakurai J, Kuwai T, Mizushima T and Isikawa Y 2002 Antiferromagnetic in the Kondo lattice compound $Ce_2Pd_3Si_5$ *Phys Rev B* **65** 144450

[5] Im H J, Min B H, Kwon Y S, Kim D L and H C Ri 2002 Magnetic and transport properties of the antiferromagnetic dense Kondo lattice system $Ce_2Cu_xNi_{1-x}Ge_6$ *Physica B* **312-313** 197-199

[6] Sun P, Lu Q, Kobayashi K, Kuwai T and Isikawa Y 2004 Schottky peaks in specific heat induced by electronic change in the antiferromagnetic-ferromagnetic alternating Kondo systems $Ce(Pd_{1-x}M_x)_2Al_3$ (m=Cu or Ag) *Phys Rev B* **70** 174429

[7] Deng D and Li Y 2007 Theoretical calculation of Ce^{3+} ground state for orthorhombic crystalline electric field *Chin J Low Temp Phys* **29** 237-240 (in Chinese)

[8] Li Y, Du B, Wang H, Zhang X and Lu Q 2007 Theoretical calculation of crystalline electric field for rare earth compound CeRhGe and CeRhSn *J At Mol Phys* **24** 40-44 (in Chinese)

[9] Tetsuya T, Arumugam T, Tomoyuki O, Mineko Y, Noriko, Takeshi Y, Yoshihiko I, Kiyohiro S, Andrei G, Etsuji Y, Koichi K, Takao E and Yoshichika O 2003 Anisotropic, thermal, and magnetic Properties of $CeAgSb_2$: Explanation via a crystalline electric field scheme *Phys Rev B* **67** 064403

[10] Tran V H 2004 Magnetic, electritic transport and thermodynamic properties of Ce_5CuSn_3 *J Alloy Compd* **383** 281

IOP Conf. Series: Materials Science and Engineering **167** (2017) 012048 doi:10.1088/1757-899X/167/1/012048

Particles control in selective laser melting in-situ oxide dispersion strengthened method

Xin Zhou[1], Zhibin An[1,4], Zhijian Shen[2], Wei Liu[2], Chenguang Yao[3]

[1]Science and Technology on Plasma Dynamics Laboratory, 710038, Xi'an, China
[2]School of Materials Science and Engineering, Tsinghua University, 100084, Beijing, China
[3]China Women's University, 100101, Beijing, China

E-mial: azb1999@163.com

Abstract. Stainless steel selective laser melting (SLM) can be considered as a new possible approach for in-situ formation of oxide dispersion strengthened (ODS) steels because of the dispersion of amorphous oxide nano-particles due to the trace amounts of laser chamber oxygen and in-situ internal oxidation of reactive elements. In this paper it is demonstrated that the particle sizes and distributions can be adjusted by choosing different chamber oxygen level through controlling the quantity of initial reactive cores and the available reactive solutes of each core.

1. Introduction

The in-situ formation of oxide nano-particles during 316L stainless steel selective laser melting (SLM) has been reported in [1-3]. It is considered as a possible approach for in-situ formation of ODS-steels (Oxide Dispersion Strengthened) where the constituents of Si and Cr readily react with oxygen presented inside the 316L steel powder and the laser chamber. Saeidi reported that the ODS SLM 316L steel achieved a higher tensile strength (703 MPa) and a higher yield strength (456 MPa) compared to that of the conventionally casted 316L steel (620MPa and 310 MPa, accordingly), and a higher ductility with 46% elongation compared to that of HIPed ODS 304 steel (20~25%) and cast 316L (30%) [1].

A uniform and controlled dispersion of second phase particles is very critical in the material performance, ODS alloys commonly employ a high energy ball milling technique but it is time-consuming [4]. The internal oxidation process is another possible ODS method involving selective reactions of a less noble solute with oxygen [5, 6]. The in-situ formation of ODS-steels through SLM belongs to the category of internal oxidation and includes two basic physical steps: one is the formation and growth of oxide particles and the second is the dispersion of formed particles under laser driven strong convections. The fluid flow characteristics and particle migration behaviors within the melt pool are well reported in SLM and/or laser welding [7-11], then uniform dispersions of second phase particles can be expected under proper laser parameters.

The growth kinetics and size control of oxide nano-particles are another research hot spots. The particles growth is controlled by the mechanism of "reactive crystallization", "diffusion-controlled precipitation" and "collisional growth" in melt pool [12]. The growth rate is proportional to the initial oxygen concentration, the melt flow field intensity and the particle collision probability, therefore the sizes and distributions of nano-particles can be adjusted by manipulating the chamber oxygen level. In this paper, 316L selective laser melting experiments are conducted with different chamber atmosphere (different oxygen levels) then the sizes, distributions and constituents of oxide nano-particles are analyzed, the forming mechanism and growth behavior are also discussed.

2. Material and methods

316L stainless steel powder granules with an overall chemical composition of 17 wt.% Cr, 10.6 wt.% Ni, 2.3 wt.% Mo, 0.98 wt.% Mn, 0.5 wt.% Si, 0.5 wt.% Ti, and Fe being the balance was used as precursor, supplied by Renishaw (Renishaw, Shanghai, China). SLM experiments were conducted at a Renishaw AM250 facility (Renishaw AMPD, Stone, UK) with laser parameters of power 190W, scan speed 700 mm/s, line spacing 0.05 mm, and "zigzag" scan strategy. The chamber residual oxygen contents in Ar atmosphere were set as extreme low (<50ppm), medium (300~500ppm), and high (1000~1200ppm) respectively. TEM observations were conducted on a JEM-2100 microscope (JEOL, Tokyo, Japan) and the sample preparation followed the procedures in [2].

3. Results and discussion

As predicted, no oxide particles can be found in the microstructure in extreme low oxygen (less than 50ppm) chamber environment and the dislocations are non-uniform, Figure 1 (a ~ b). Increasing the chamber oxygen level to 300~500ppm, some large oxide particles can be formed with size around 500 nm~2 μm, Figure 1 (d) and Figure 2 (a ~ b). The quantity of oxide particles is limited at this oxygen level but plenty of dislocations and the pinning effects of particles can be clearly observed. With further increase of the chamber oxygen level to 1000~1200 ppm, oxide nano-particles of size 50~100 nm are dispersive and distributed homogeneously, Figure 1 (c). The particles in 1000 ppm oxygen level are fairly accorded with results reported in former literatures. Based on EDS analysis, the inclusion constituents are actually Fe, Ni, Cr, Ti and Si oxides, Figure 2 (c), the Cr, Ti, Si and O elements are relatively high but the Ni element is lower compared with the matrix Ni content, Figure 2 (d). This implies the oxide particles have similar chemical constitutions and amorphous structures (presented in [1]) with stainless steel oxide (passive) films which are not the harmful brittle phases, such as σ (tetragonal), $M_{23}C_6$ (FCC) or χ phase (BCC). Thus the following conclusions can be made: the formation of nano-particles is closely related to the oxygen environment and with an increase in the chamber oxygen level, the particle size reduces but the particle distribution is more dispersed. The particle size discrepancy can be as high as 40 times in different chamber oxygen content, 50 nm in 1000 ppm while 2 μm in 300~500 ppm, Figure 1(c, d).

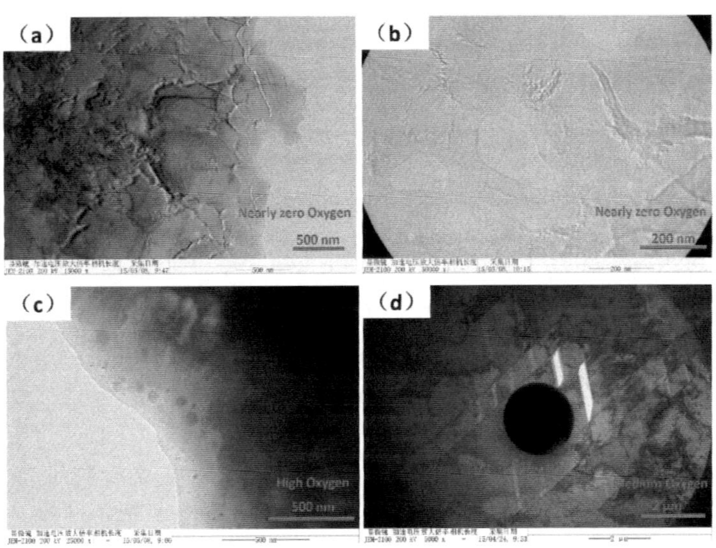

Figure 1. The distribution of oxide particles, chamber oxygen level of less than 50ppm (a ~ b), 1000~1200ppm (c), and 300~500ppm (d).

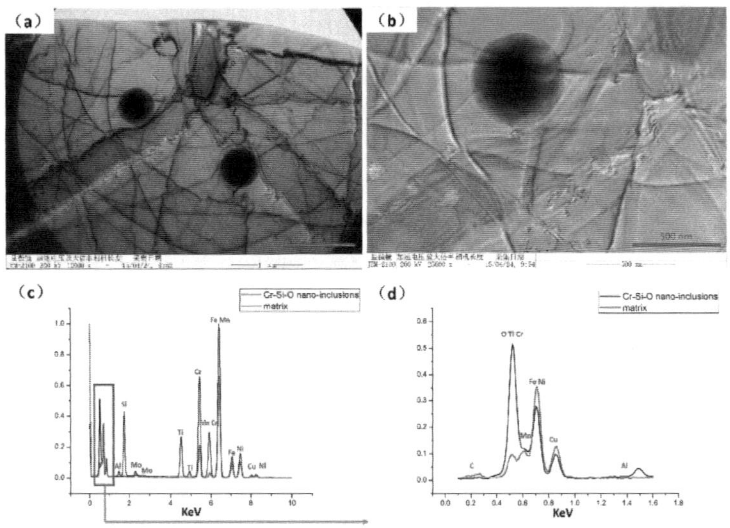

Figure 2. The oxide particles formed in the chamber oxygen level of 300~500ppm (a, b). EDS spectrums of a nano-inclusion and the matrix (c), the partial enlarged view of 0~1 KeV (d).

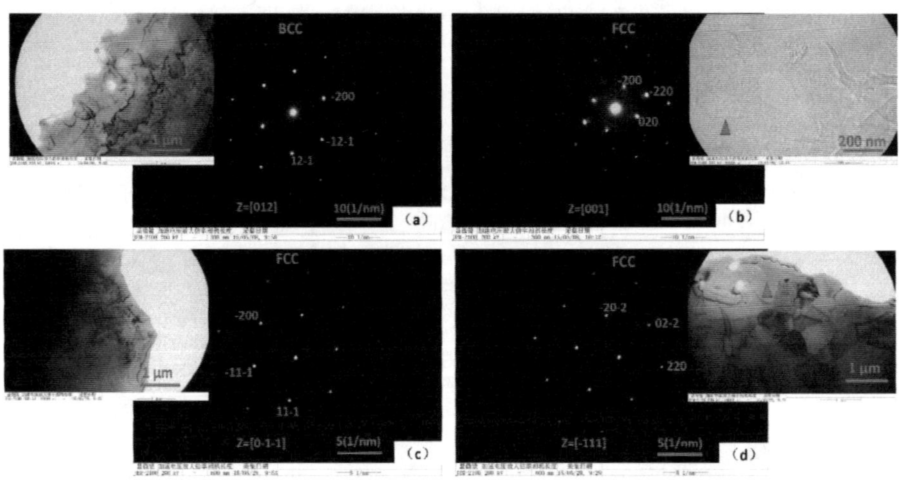

Figure 3. TEM analysis and corresponding electron diffraction patterns.

(a) and (b) with extreme low oxygen less than 50ppm, (c) and (d) with high oxygen 1000~1200 ppm. The diffraction positions taken in (a) and (c) are with high density dislocations, while in (b) and (d) are with nearly no dislocations.

The chamber oxygen level can affect not only the inclusion formations but also the matrix microstructures. As shown in Figure 3 the TEM analysis and corresponding electron diffraction patterns in extreme low oxygen chamber environment, most areas can be identified as the FCC structures (austenite) but some areas with high density dislocation tangles can be identified as the BCC structures demonstrating the ferrite nucleation. In a high oxygen environment (1000~1200 ppm), all the diffraction regions regardless of dislocation densities are all identified as the FCC structures (austenite).

316L stainless steel solidifies as the main austenite but a small amount of ferrite can be generated through eutectic reaction in the cellular/dendritic interface according to its constituents and phase diagram. Austenite stabilizing elements (e.g. Ni) will be consumed during laser melting and the re-solidifying process, while ferrite stabilizing elements (e.g. Si, Cr, Ti, and Mo) will be enriched in the solidification front and dendritic interface. With low oxygen chamber atmosphere, in this local area of dendritic interface the equivalent proportion of Cr and Ni (Cr_{eq}/Ni_{eq}) will be increased which will then form ferrite nucleation and high density dislocations, Figure 3 (a). With a certain amount of oxygen in the chamber, a thin oxide film on the melt pool surface will be generated rapidly by oxidation of the rejected Ti, Si and Cr for the high laser irradiation temperature and shallow melt pool (100~200 μm depth). Moreover, melt convection in the laser melt pool is another significant feature [8, 13] and under this strong stirring action the formed oxide film will be scattered and then dispersively distributed in the microstructure as the discovered oxide nano-particles. For Ni the element is consumed by austenitic solidification, so it's counting is a bit lower in oxide-particles than in the matrix. Cr, Ti and Si are excessively consumed by oxidation, so ferrite nucleation seems impossible in high oxygen chamber atmosphere. Mo element (solute) is heavier than Fe melt (solvent) and does not participate in the formation of oxide film [14], so it will be distributed in the cellular boundaries.

There still remains difficulty in calculating the particle sizes and distributions quantitatively because of the lack of some thermophysical data, but the qualitative analysis based on growth kinetics of particles can be still involved. Nano-particles in laser melt pool can be considered as a polydisperse system with a radius of a continuous distribution, the density distribution function of particles $n(r)$ is defined as a steady-state exponential distribution in particle radius range $(r, r+dr)$ [15]:

$$n(r)=n_0\, exp(-\lambda r),\ r_0 < r < r_1 \qquad (1)$$

The parameters n_0, λ, r_0, r_1 are decided by specific experimental conditions, where

$$n_0 = \frac{3\rho_f[o\%]}{4\pi\rho_p X_o \int_{r_0}^{r_1} r^3 e^{-\lambda r} dr} \qquad (2)$$

$[O\%]$ is the oxygen content in the melt, X_o is the oxygen mass fraction in particles. Therefore the particle number density is directly correlated with oxygen level, the higher oxygen level in laser chamber and the higher dissolved oxygen in the melt ($[O\%]$), means more amounts of oxide particles (n_0 and $n(r)$) can be generated.

Based on sphere diffusion model [12], the maximum radius R_∞ of oxide-particles growth is defined as:

$$R_\infty = R_d \frac{(c_0 - c_p)^{1/3}}{c_p^{\ 1/3}} \qquad (3)$$

Where R_d is the radius of sphere diffusion (Figure 4 b ~ c), C_o is initial concentration of reactive elements (Cr, Si, Ti, O, etc.) in the melt, C_p is mass fraction of reactive elements in the particles. Therefore, the maximum radius R_∞ is controlled by the number of reactive cores and diffusion radius R_d. Higher density of initial reactive cores ($n(r)$) decides the shorter diffusion radius R_d which means the available solute for each reactive core is limited.

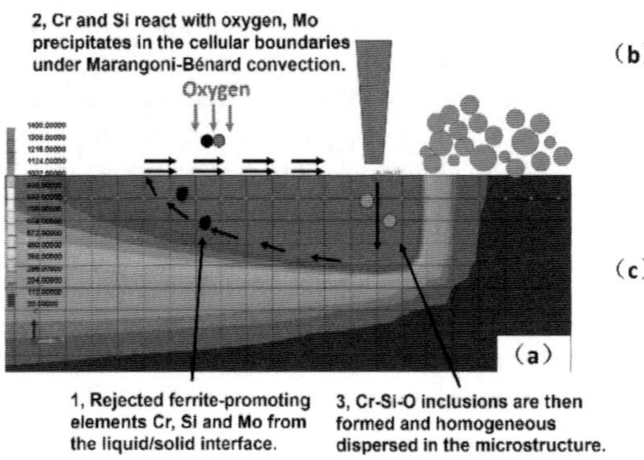

Figure 4. The formation mechanism of oxide particles (a). (b) and (c) exhibit the scenarios of different oxygen level, 300~500ppm and 1000~1200ppm separately.

4. Conclusions

Based on these analyses, oxygen level can be used as a regulating method. Therefore, the quantity of initial reactive cores is limited and oxide-particle growth is sufficient in the condition of 300~500 ppm chamber oxygen. On the contrary, the initial reactive cores are numerous and the final radius is limited in the condition of 1000~1200 ppm chamber oxygen. Higher oxygen content may further reduce the particle size but high oxygen level (over 1500ppm) will deteriorate SLM densities because of the balling phenomenon [16]. Adding small amounts of Ti, Al and Y and fabricating in an atmosphere of 300~500ppm oxygen can be predicted as a new possible process for in-situ ODS 316L steel.

References

[1] K. Saeidi, L. Kvetkova, F. Lofajc, Z. Shen, Austenitic stainless steel strengthened by the in situ formation of oxide nanoinclusions, *RSC ADV* **5**(2015) 20747-20750.
[2] K. Saeidi, X. Gao, Y. Zhong, Z.J. Shen, Hardened austenite steel with columnar sub-grain structure formed by laser melting, *MAT SCI ENG A-STRUCT* **625**(2015) 221-229.
[3] K. Saeidi, X. Gao, F. Lofaj, L. Kvetkova, Z.J. Shen, Transformation of austenite to duplex austenite-ferrite assembly in annealed stainless steel 316L consolidated by laser melting, *J. ALLOY COMPD* **633**(2015) 463-469.
[4] A. Chauhan, D. Litvinov, J. Aktaa, High temperature tensile properties and fracture characteristics of bimodal 12Cr-ODS steel, *J. NUCL MATER* **468**(2016) 1-8.
[5] M. Hansel, C.A. Boddington, D.J. Young, Internal oxidation and carburisation of heat-resistant alloys, *CORROS SCI* **45**(2003) 967-981.
[6] M. Udyavar, D.J. Young, Precipitate morphologies and growth kinetics in the internal carburisation and nitridation of Fe-Ni-Cr alloys, *CORROS SCI* **42**(2000) 861-883.
[7] P. Yuan, D. Gu, D. Dai, Particulate migration behavior and its mechanism during selective laser melting of TiC reinforced Al matrix nanocomposites, *MATER DESIGN* **82**(2015) 46-55.
[8] C. Panwisawas, C.L. Qiu, Y. Sovani, J.W. Brooks, M.M. Attallah, H.C. Basoalto, On the role of thermal fluid dynamics into the evolution of porosity during selective laser melting, *SCRIPTA MATER* **105**(2015) 14-17.
[9] D. Dai, D. Gu, Tailoring surface quality through mass and momentum transfer modeling using a volume of fluid method in selective laser melting of TiC/AlSi10Mg powder, *INT J. MACH TOOL MANU* **88**(2015) 95-107.
[10] S. Tammas-Williams, H. Zhao, F. Léonard, F. Derguti, I. Todd, P.B. Prangnell, XCT analysis of the influence of melt strategies on defect population in Ti–6Al–4V components manufactured by Selective Electron Beam Melting, *MATER CHARACT* **102**(2015) 47-61.

[11] S. Pang, W. Chen, J. Zhou, D. Liao, Self-consistent modeling of keyhole and weld pool dynamics in tandem dual beam laser welding of aluminum alloy, *J. MATER PROCESS TECH* **217**(2015) 131-143.

[12] B. Zhang, B. Li, Y. He, Growth kinetics of single inclusion in molten metal, *Journal of Iron and Steel Research* **17**(2005) 19-25.

[13] X. Zhou, D. Wang, X. Liu, D. Zhang, S. Qu, J. Ma, G. London, Z. Shen, W. Liu, 3D-imaging of selective laser melting defects in a Co–Cr–Mo alloy by synchrotron radiation micro-CT, *ACTA MATER* **98**(2015) 1-16.

[14] L.J. OBLONSKY, T.M. DEVINE, A SURFACE-ENHANCED RAMAN-SPECTROSCOPIC STUDY OF THE PASSIVE FILMS FORMED IN BORATE BUFFER ON IRON, NICKEL, CHROMIUM AND STAINLESS-STEEL, *CORROS SCI* **37**(1995) 17-41.

[15] L.F. Zhang, S. Taniguchi, K.K. Cai, Fluid flow and inclusion removal in continuous casting tundish, *METALL MATER TRANS B* **31**(2000) 253-266.

[16] X. Zhou, X. Liu, D. Zhang, Z. Shen, W. Liu, Balling Phenomena in Selective Laser Melted Tungsten, *J. MATER PROCESS TECH* **222**(2015) 33-42.

First-principles studies of the influences mechanism of Y,Hf elements on the high temperature oxidation of γ -TiAl alloys

G Y Zhang[1,3], C M Liu[2], and G L Fang[1]

[1] College of Physics Science and Technology, Shenyang Normal University, Shenyang 110034, China
[2] School of Materials and Metallurgy, Northeastern University, Shenyang 110809, China

E-mail: gyzhang1965@sina.com

Abstract. The influences mechanisms of Y, Hf elements on the high temperature oxidation of γ-TiAl alloys were studied by using a first-principle plane wave pseudopotential method within the density functional theory. It is shown that Y, Hf atoms segregate to the γ-TiAl surface by substituting Ti atoms, and Y atom is easier to segregate than Hf. Y, Hf atoms segregated at γ-TiAl surface decrease the adhesion of O and γ-TiAl surface, which decreases the oxidation rate of γ-TiAl alloys and thereby hinders the growth of oxide films. The electron structure calculations suggest that the interaction between O and Ti, Al, Y, Hf atom exists both ionic and covalent binding characteristics, the decrease of the adhesion of O and γ-TiAl surface with the doped Y, Hf is mainly attributed to the weaken covalent interactions. As a result, the results obtained by first principles can make us get a deeper understanding of the mechanism of the segregation of Y, Hf to the γ-TiAl surface and the influence mechanism of surface active elements Y, Hf on the initial oxidation of γ-TiAl alloys.

1. Introduction

The γ-TiAl based alloys exhibit a large number of outstanding properties such as high melting point, low density and high specific strength. Therefore they are being considered as prospective structural materials for applications in aerospace, automobile and energy industries. However, its oxidation resistance can not satisfy the demands for increased operating temperatures, which limits the practical application of TiAl alloys [1]. Similar activities of Ti and Al result in the formation of titania and alumina simultaneously on TiAl alloys. Compared with alumina, titania has high diffusivity and poor protectiveness. This results in the poor oxidation resistance of TiAl alloys at temperatures higher than 1073 K[2]. Therefore the selective oxidationof Al on TiAl alloys surface for improving the oxidation resistance of γ-TiAl was in focus of both experimental and theoretical researches all the time. In 1992, Becker et al.[3] found that the small amounts of addition of Nb to TiAl leads to the formation of a protective Al_2O_3 layer at the metal/scale interface which is very resistant to thermal cycling. The improvement of oxidation resistance by Si is mainly attributed to promoting the formation of continnous and dense Al_2O_3 layer in the oxide scale on TiAl alloys during the initial stages of oxidation[4]. The addition of W element can significantly reduce the oxygen solubility in the TiAl alloys. Owing to this effect, the internal oxidation is inhibited in the alloy[5]. Taniguchi et al [6] found that the addition of Hf is very effective in decreasing the overall oxidation rate at temperature up to 1300k. The excellent oxidation resistance is attributable to the formation of alumina-rich layers in the scale. Wu et al [7] found that the addition of small amounts of Y was effective in enhancing the oxidation resistance of TiAl-1.4Mn-2Mo-0.3C alloy. They thought that this improvement was due partly to the improved adhesion of the scale and due partly to the formation of a continuous Al_2O_3 layer in the outer scale.

On the theoretical side, Li et al [8] revealed that the abilities of interstitial oxygen and Ti vacancy formation are lowered by Nb addition, thus the diffusion of oxygen and Ti is reduced correspondingly, which enhances the oxidation resistance of TiAl alloy. Wu et al [9] studied the effect of Nb on the bonding characteristics of TiAl alloys. The results suggest that Nb reduces the interactions between Ti and O, and enhances the interactions between Al and O. Meanwhile, with the introduction of Nb, the formation energy barrier of TiO_2 is increased, while the generation of Al_2O_3 is facilitated due to the decrease of the formation energy barrier. Liu et al [10] found that the surface segregation of Si makes the interactions between O and Al atoms stronger whereas those between O and Ti weaker, which is benefit to the nucleation of Al_2O_3.

However, to the best of our knowledge, there is few theoretical studies reported dealing with the influence of the reactive elements (e.g. Y and Hf) on the high temperature oxidation of TiAl alloys. Thus in this paper, we presents a first principles study on the effect of Hf, Y on O adsorption on the surface of γ-TiAl alloys to reveal the effect mechanism of Hf, Y on the high temperature oxidation of γ-TiAl alloys.

2. Calculation Method and Models

2.1. Calculation Method

In this study, we utilized the CASTEP[11] based on the density function theory (DFT) in MS6.0. The generalized gradient approximation (GGA) in the Perdew-Burke-Ernzerhof (PBE) functional form was used to describe the exchange-correlation energy [12]. The electron-ion interaction was described with ultra-soft pseudo-potential [13]. The structure models have been relaxed according to the BFGS algorithm [14]. The geometry optimization is completed with a convergence tolerance of energy of 2.0×10^{-5} eV/atom, a maximum force of 0.05 eV/nm, a maximum displacement of 0.02nm and a maximum stress of 0.1GPa. A kinetic energy cut off of 300 eV and a 4×4×1 K-point mesh for all the electronic structure calculations.

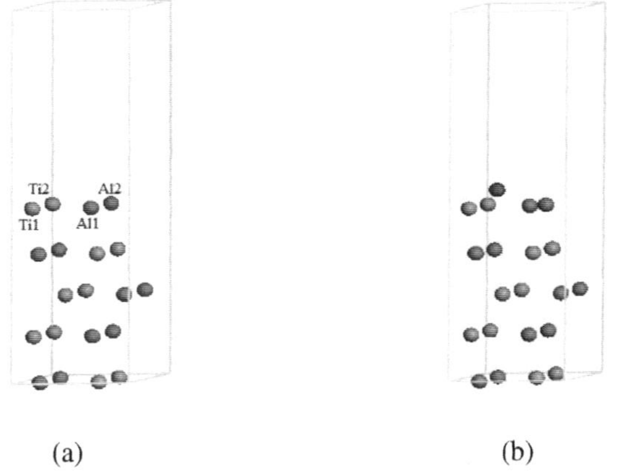

Figure 1. Models of clean (a) and O adsorbed (b) (2×2) γ-TiAl(111) surfaces (Red, pink and gray balls represent O, Ti and Al atoms, respectively).

2.2. *Models*

The γ-TiAl (111) surface was modeled by a (2×2) supercell of a slab of 5 γ-TiAl (111) layer separated by a 1.1nm-thick vacuum layer. In Ref.15, we found that the most stable sites for O adsorption on γ-TiAl alloy surface are the hcp-Al adsorption sites. In the following, we just discuss the hcp-Al adsorption sites. Figure1(a) shows the clean γ-TiAl (111) (2×2) surface, and figure1(b) represents the γ-TiAl (111) (2×2) surface with O adsorption at hcp-Al sites. When considering the doping of the reactive elements, Y, Hf atom substitutes Ti1 or Al1 atom at the γ-TiAl (111) surface. In our calculations, the three metal layers below the surface were allowed to relax.

3. Results and discussion

3.1. *Surface segregation of Hf, Y.*

The formation energy of a defect (dopant) is a crucial factor in determining its doping concentration. The lower the formation energy of a defect is, the higher the doping concentration is. One Hf or Y atom substitutes one Ti (or Al) atom in γ-TiAl or at γ-TiAl(111), the impurity formation energy is defined as:

$$E_{imp}^{X} = E_{t}^{X-Y\,or\,Hf} - E_{t}^{X} + E_{Ti\,or\,Al}^{bulk} - E_{Y\,or\,Hf}^{bulk} \quad (1)$$

Where X represents the γ-TiAl(111) surface or the γ-TiAl bulk; E_T^X and $E_T^{X-Hf\,or\,Y}$ represent the total energies of X and Hf or Y-doped X; $E_{Ti\,or\,Al}^{bulk}$ and $E_{Y\,or\,Hf}^{bulk}$ are the total energies of one Ti or Al and one Hf or Y atom in their bulk states, respectively.

The tendency of an impurity atom to segregate to the surface can be characterized by the surface segregation energy E_{seg} defined as [8]

$$E_{seg}^{sur} = E_{imp}^{sur} - E_{imp}^{bulk} \quad (2)$$

Where E_{imp}^{sur} and E_{imp}^{bulk} donate the impurity formation energy of RE (Hf and Y) at the surface and in the bulk, respectively. A negative E_{seg} indicates that impurity atom is energetically favourable to segregate to the surface. The lower the segregation energy, the stronger the degree of an impurity atom to segregate to the surface is.

The impurity formation energies of Hf and Y in the bulk and at the surface are calculated by equation (1), and the surface segregation energies of Hf and Y are obtained by equation (2). The results are listed in table 1.

Table 1. The calculated impurity formation energies E_{imp} (eV) of Hf, Y at the γ-TiAl(111) surface and bulk γ-TiAl and surface segregation energy E_{seg} (eV) of Hf, Y at the γ-TiAl(111) surface.

substituted		Bulk Ti	Bulk Al	Surface Ti	Surface Al
Y	E_{imp}	1.569	2.172	0.146	0.288
	E_{seg}			-1.423	
Hf	E_{imp}	-0.022	4.058	-0.403	0.0225
	E_{seg}			-0.381	

From table 1, it can be seen that the impurity formation energies of Hf or Y substituting for Ti atom, no matter in the bulk or on the surface, are lower than that of Hf or Y substituting for Al atom. This indicates that only Ti atoms in the bulk or on the surface can be substituted by Hf, Y atoms. The surface segregation energies of Y, Hf calculated with equation (2) are negative, which indicates that Y, Hf atom can segregate to the surface. The surface segregation energy of Y is more negative than that of Hf, which suggests that Y has a stronger segregation tendency to the surface than Hf.

The surface segregation of Y, Hf will influence the composition of the oxide scale and the phase formation. Due to the strong affinity among Y and O atoms, the formation of Y_2O_3 and Y-rich multiphase (Y, Al) O oxides would impede the oxygen transport and the growth rate of the oxide scale.

3.2. The effect of Hf, Y on O adsorption on the γ-TiAl surface

The adsorption energy is an important physical quantity in the surface problem, the higher the adsorption energy, the more stable the surface adsorption is. The adsorption energy of O atom is defined as [8]

$$E_{ads} = -\frac{1}{N}(E_{O/TiAl(111)} - E_{TiAl(111)} - \frac{N}{2}E_{O_2}) \tag{3}$$

Where $E_{O/TiAl(111)}$ is the total energy of a given surface slab containing the pure γ-TiA surface or the Hf,Y alloyed γ-TiAl surface, $E_{TiAlM(111)}$ is the corresponding energy of the bare slab. E_{O_2} is the total energy of an oxygen molecule, which is determined by the optimization of a 1nm cube with O_2 at eight corners. N is the number of adsorbed O atoms on the γ-TiAl(111) surface(here, N=1). According to the defination by equation (3), a positive adsorption energy indicates an exothermic chemisorption of oxygen onto the alloy surfaces. The larger the value of the adsorption energy, the stronger the interaction between O and γ-TiAl(111) surface is, the more possible the adsorption of O will happen, and the more stable the surface adsorption is. The adsorption energies of O atoms adsorbed on clean and Y, Hf segregated surfaces were calculated by equation (3) to be 6.040, 5.237, 5.543eV respectively, thus it can be seen that the segregation of Y or Hf to γ-TiAl surface lowers the binding energy of O to the γ-TiAl(111) surface. As a result, the initial oxidation rate of γ-TiAl alloy decreases, and the growth the oxide scale is impeded.

3.3. Electronic structure analysis

In order to further analyze the influence mechanism of the surface active elements to the high temperature oxidation of γ-TiAl alloys, the density of states (DOS) of O, Al1, Ti1, Ti2(Ti1 is replaced with the doped atom), Y, Hf atoms of the clean and the alloying element segregated O/γ-TiAl(111) surface systems are calculated and shown in figure 2. For the clean O/γ-TiAl(111) surface system(see figure.2(a)), the s and p orbitals of Al hybridize with s, p and d of Ti in the nearly whole energy range, indicating a strong covalent bonding combined with the metal-metal valence characteristics. It is clearly that a peak hybridized with O_{2s} orbital and Al_{3p} orbital appear near energy level of -20 eV. The 2p orbital of O hybridize with 3s, 3p of Al in the -10 to -4 eV energy range, while the hybridization of p orbital of O with 4s, 4p, 3d of Ti in the -10 to -4 eV energy range is weak. All these reflect that oxygen adsorption results in the formation of new Ti-O and Al-O bonds, but the Ti-O bond is weaker than and Al-O bond. The existence of Ti-O and Al-O bonds weakens the original Ti-Al bonds.

For the Y segregated O/γ-TiAl(111) surface systems (Ti1is replaced with Y) shown in figure 2(b), the interactions between O_{2p} and Al_{3p} electrons are also strong, but the interactions between O_{2p} and Y_{4d} electrons are weak(the peak of Y_{4d} energy level of -5 eV is very small), thus it can be seen that a weaker covalent bonding forms between Y and O. After the substitution of Ti1 with Hf, it is obviously seen from figure 2(c) that the density of states of Hf_{5d} is very different from that of $Ti1_{3d}$(compared with figure 2(a)), the density of states of d orbitals Hf is obviously lower than that of Ti1 and Y, indicating a large change of the bonding behavior. The numerical discussion will be given in next section.

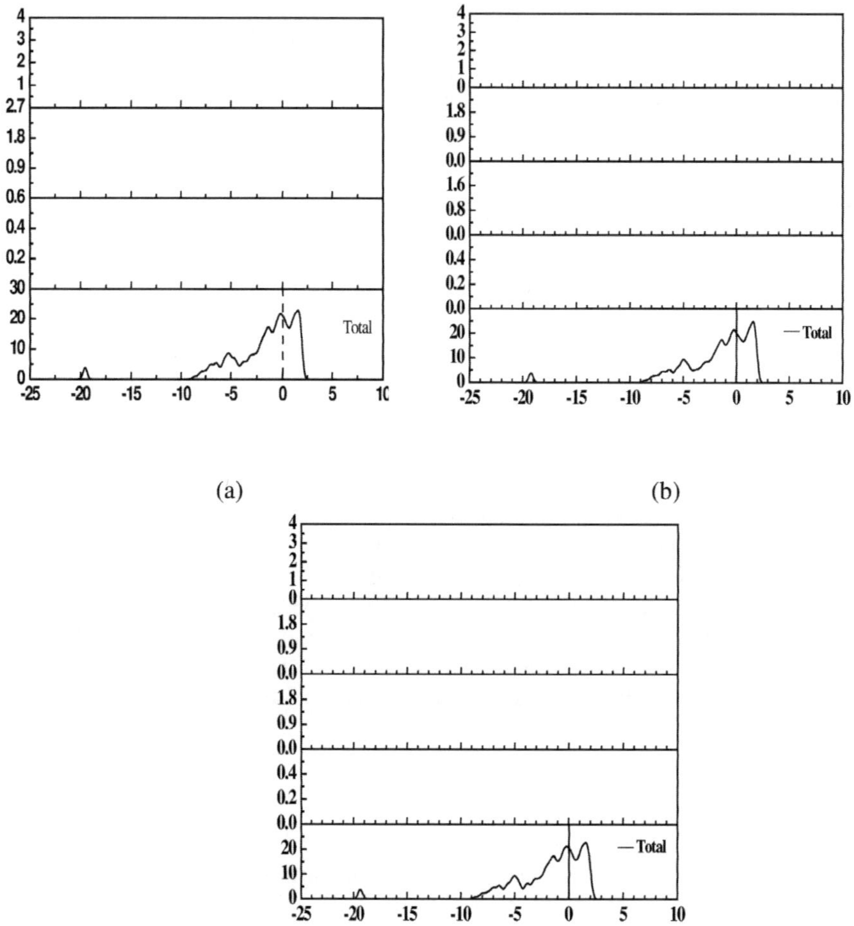

(a) (b)

(c)

Figure 2. The PDOS of atoms at the clean(a) and the Y(b), Hf (c) segregated O/γ-TiAl(111) surface systems, and the TDOS of each system.

3.4. Population analysis

The mulliken overlap populations per unit bond length indicate the overlap degree of electron clouds of two bonding atoms and can be used to quantify the covalent bond strengths [14]. The results of Mulliken population between O and neighbor atoms on the clean and Y, Hf segregated O/γ-TiAl(111) surface systems are given in table 2. The atomic labeling is shown in figure 1, Y, Hf replaces Ti1.

For the pure O/γ-TiAl(111) surface system, the overlap population per unit bond length values of O-Ti1、O-Ti2 bonds are smaller than that of O-Al1, indicating stronger covalent interaction between O and Al1. For the Y segregated O/γ-TiAl(111) surface systems, the population values of O-Y, O-Al1 bonds decrease obviously, while that of O-Ti2 increases. In general, Y substitution for Ti1 reduces the adhesion of O to the γ-TiAl(111). For the Hf segregated O/γ-TiAl(111) surface systems, the population value of O-Hf bond decreases obviously, the population value of O-Al1 basically does not change, but that of O-Ti2 increases. In general, Hf substitution for Ti1 also reduces the adhesion of O to the γ-TiAl(111) surface. In addition, it can be seen from table 2 that the influence of Y on the adhesion of O to the γ-TiAl(111) surface is stronger than that of Hf. All these results are consistent with the above discussions of the O adsorption energy and DOS. The surface segregation of Y, Hf reduces the adhesion of O to the surface, thereby reducing the initial oxidation rate of γ-TiAl alloy and improving the oxidation resistance of γ-TiAl alloys.

Table 2. The overlap population per unit bond length of each bond （P_B/L_B） at the pure and segregated O/γ-TiAl(111) surface systems.

O/γ-TiAl（111）		Y segregated O/γ-TiAl（111）		Hf segregated O/γ-TiAl（111）	
Bond	$P_B/L_B/Å^{-1}$	Bond	$P_B/L_B/Å^{-1}$	Bond	$P_B/L_B/Å^{-1}$
O-Ti 1	0.170	O-Y	0.103	O-Hf	0.131
O-Ti 2	0.170	O-Ti 2	0.216	O-Ti 5	0.178
O-Al1	0.206	O-Al 1	0.173	O-Al 1	0.223

The strength of the ionic bond between atoms is generally determined by the amount of transferred charge. So we performed the calculations of the loss or gain of charge of atoms at the clean and Y, Hf segregated O/γ-TiAl(111) surface systems. All calculated values are given in table 3. The atom numbering is shown in figure.1.

Table 3. The loss or gain of charge of atoms at the pure and Y, Hf segregated O/γ-TiAl(111) surface systems (e).

O/γ-TiAl（111）		Y segregated O/γ-TiAl（111）		Hf segregated O/γ-TiAl（111）	
atom	net charge	atom	net charge	atom	net charge
O	-0.78	O	-0.76	O	-0.78
Ti 1	0.40	Y	0.28	Hf	0.08
Ti 2	0.41	Ti 2	0.64	Ti 2	0.63
Al 1	0.28	Al 1	0.15	Al 1	0.27

For the pure O/γ-TiAl(111) surface system, as seen from table 3, the adsorbed O atom on the surface gains charges, while those metal atoms at the surface lose charges, so the bonding of the adsorbed O to the surface has a partial ionic component. For the Hf segregated O/γ-TiAl(111) surface systems, the amount of charge obtained O is almost not change. For the Y segregated O/γ-TiAl(111) surface systems, the amount of charge obtained O is reduce slightly. Thus it can be seen that Y,Hf has little influence on the ionic interaction of the adsorbed O atom with the γ-TiAl(111) surface.

4. Conclusions

We have investigated the influences mechanisms of Y and Hf surface reactive elements on the high temperature oxidation of γ-TiAl alloys by using first-principles calculations. Through the studying, we get the following conclusions:

- The calculations of the impurity formation energies and the surface segregation energies show that Y, Hf atoms segregate to the γ-TiAl surface by substituting Ti atoms, and Y atom is easier to segregate to the γ-TiAl surface than Hf.
- The results of the adsorption energy of O atom on the γ-TiAl surface show that the segregation of Y, Hf atom at γ-TiAl surface decreases the adhesion of O to the γ-TiAl surface, which decreases the oxidation rate of γ-TiAl alloys and thereby hinders the growth of oxide films. The influence of Y on the adhesion of O to the γ-TiAl surface is stronger than that of Hf.
- The calculations of DOS, the overlap population and the loss or gain of charge suggest that the interaction between O and Ti, Al, Y, Hf atom exists both ionic and covalent binding

characteristics, but the decrease of the adhesion of O to γ-TiAl surface with the segregated Y, Hf is mainly attributed to the weaken covalent interactions between O and γ-TiAl surface, and there is no direct relationship between the decrease of the adhesion of O to γ-TiAl surface and the ionic binding of O to γ-TiAl surface.

Acknowledgements
This work was financially supported by National Natural Science Foundation of China under Grant No.51371049

References

[1] Becker S, Rahmal A, Schorr M and Schüte M 1992 *Oxid. Met.* **38** 425-64.
[2] Rahmal A, Schütze M and Quadakkers W J 1995 *Mater. and Corros.* **46** 271-85
[3] Becker S, Schütze M and Rahmel A 1993 *Oxid. Met.* **39** 93-106
[4] Dong L M, Cui Y Y, Yang R and Wang F H 2004 *Acta Metall. Sin.*, **40** 383-7
[5] Shid Y and Anada H 1994 *Mater. Trans. JIM*, **35** 623-31
[6] Taniguchi S, Shibata T, Juso H and Katoh N 1994 *Oxid. Met.* **42** 205-22
[7] Wu Y, Hwang S K and Morris Jr J W 2003 *Metell. Mater. Trans.* A, **34A** 2077-87
[8] Li H, Wang S Q and Ye H Q 2009 *Acta Phys. Sin.* **63** s224-9 (in Chinese)
[9] Wu H L, Zhang W and Gong S K 2008 *Acta Chim. Sin.* **66** 1669-75 (in Chinese)
[10] Liu S Y, Shang J X, Wang F H and Zhang Y 2009 *J. Phys: Condens. Matter.* **21** 225005-12
[11] Segall M D, Lndan P L D, Probert M J, Pickard C J, Hasnip P J, Clark S J and Payne M C 2002 *J. Phys. Condens. Matter.*, **14**, 2717-44
[12] Marlo M and Milman V 2000 *Phys. Rev. B*, **62** 2899-907
[13] Vanderbilt D 1990 *Phys. Rev. B*, **41** 7892-5
[14] Pfrommer B G, Cote M, Louie S G and Cohen M L 1997 *J. Comput. Phys.* **131** 233-40
[15] Zhang G Y, Liu G L, Bao J S, Liu C M and Zhang H 2014 *Advan. Mater. Resear.* **1052** 28-33

A facile and effective strategy to synthesize orthorhombic $Sr_2Al_6O_{11}:Eu^{2+},Dy^{3+}$ with blue-green persistent luminescence

Juan Han[1], Ziqiu Jiang[1], Wenyan Zhang[1,4], Lingyun Hao[1], Yaru Ni[2,3], Chunhua Lu[2,3], and Zhongzi Xu[2,3]

[1] College of Material Engineering, Jinling Institute of technology, Nanjing, China
[2]Jiangsu Collaborative Innovation Center for Advanced Inorganic Function Composites, Nanjing Tech University, Nanjing, PR China
[3]State Key Laboratory of Materials-Oriented Chemical Engineering, College of Materials Science and Engineering, Nanjing Tech University, Nanjing, PR China

E-mail: wiseyanyan@jit.edu.cn

Abstract. $Sr_2Al_6O_{11}:Eu^{2+},Dy^{3+}$ is known as a high efficient material for generating persistent luminescence. Due to its low structural stability, it is a challenge to prepare such orthorhombic material in large scale. In this work, a facile and effective strategy was designed for the preparation of $Sr_2Al_6O_{11}:Eu^{2+},Dy^{3+}$ with high purity by combining the advantages of solid state reaction and chemical vapor deposition method. The prepared $Sr_2Al_6O_{11}:Eu^{2+},Dy^{3+}$ could effectively store the UV light energy and emit blue-green luminescence for 240 min by slow liberation of photo-excited electrons. Its blue-green afterglow was composed of two luminescent emissions which released from the Eu centers located in different crystal fields.

1. Introduction

Afterglow materials have attracted a great deal of interest since ancient years. Owning to its special ability to store and release light energy, such material could be applied widely for the rewritable optical memory media, security signs, traffic signage, dial markers and medical diagnostics[1, 2]. Persistent luminescence of materials depends not only on their crystal structure but on the distribution and concentration of luminescent centers inside the materials. Consequently, the development of afterglow material was rather slow until the discovery of $SrAl_2O_4:Eu^{2+},Dy^{2+}$ in 1996, which made a significant breakthrough in this field as it extended the afterglow duration from minutes to tens of hours[3]. From then on, people found it the key to enhance the persistent luminescence is to construct appropriate trap levels inside the materials to realize effective capturing, storing and slowly releasing of excited carriers (electron and hole). After that, extensive research was triggered on inventing new persistent phosphors as well as bringing out reasonable mechanism models of long-lasting luminescence. Silicates, germinate, gallate and titanate materials were also found to be promising hosts for storing and releasing photo-generated carriers with high efficiency[1-6]. However, the aluminates phosphors are still recognized to be the most effective afterglow material at present due to its high brightness, long phosphorescence time, suitable chemical stability, non-toxicity, and low cost.

$Sr_2Al_6O_{11}:Eu^{2+},Dy^{3+}$ is a "mysterious" species in the family of aluminates afterglow phosphors. Due to its low structural stability, such material could only be prepared in narrow and strict synthesis conditions. Smets et al. firstly reported the synthesis of $Sr_2Al_6O_{11}:Eu^{2+}$ and suggested that it was a single phase with similar crystal structure like $Sr_{1.33}Pb_{0.67}Al_6O_{11}$, which could yield high efficient luminescence [7]. Wang et al. named such material as "so-called $Sr_2Al_6O_{11}:Eu^{2+}$" because their work indicated that the $Sr_2Al_6O_{11}:Eu^{2+}$ was not a single-phase crystal but a mixture of $SrAl_2O_4$ and $Sr_4Al_{14}O_{25}$ [8]. Furthermore, it is ambiguous that whether $Sr_2Al_6O_{11}$ is an appropriate matrix for the generation of persistent luminescence. Takashi et al. studied the optical properties of $Sr_2Al_6O_{11}:Eu^{2+},Dy^{3+}$ and found that the material could not yield persistent luminescence[9]. Song et al. also reported that the $Sr_2Al_6O_{11}:Eu^{2+}$ didn't show obvious persistent luminescence whereas the $Sr_2Al_6O_{11}:Eu^{2+},Er^{3+}$ could emit blue phosphorescence longer than 14 h [10]. Xue et al. investigated the effects of sintering temperature on the phase formation and luminescent properties of $Sr_2Al_6O_{11}:Eu^{2+},Dy^{3+}$ [11]. However, as the phosphor they synthesized was a mixture of $SrAl_2O_4:Eu^{2+},Dy^{3+}$, $Sr_2Al_6O_{11}:Eu^{2+},Dy^{3+}$ and $Sr_4Al_{14}O_{25}:Eu^{2+},Dy^{3+}$. Though $Sr_2Al_6O_{11}$ was main phase in the mixture, the afterglow luminescence they reported was actually a mingled emission of those three phosphors [11]. After that, little progress was got about the structure and optical property of $Sr_2Al_6O_{11}:Eu^{2+},Dy^{3+}$.

To make clear about the optical property of $Sr_2Al_6O_{11}:Eu^{2+},Dy^{3+}$, it is necessary to obtain such "mysterious" species with high purity and favorable quality. In this work, an effective and moderate strategy was proposed for the synthesis of such phosphor. $Sr_2Al_6O_{11}:Eu^{2+},Dy^{3+}$ with high purity could be obtained by that strategy which combined the advantages of solid state reaction and chemical vapor deposition method. The prepared $Sr_2Al_6O_{11}:Eu^{2+},Dy^{3+}$ exhibited favorable persistent luminescence at room temperature.

2. Material and Experiments

Al_2O_3, Eu_2O_3 (99.9%), Dy_2O_3 (99.9%), H_3BO_3 (99.5%), $SrCO_3$ (99.5%), urea (99.5%) and active carbon (99.5%) were purchased from Guoyao Corporation. Reactant ratio for $Sr_2Al_6O_{11}:Eu^{2+},Dy^{3+}$ were $0.01Eu_2O_3+0.02Dy_2O_3+0.97SrCO_3+1.5Al_2O_3+0.4H_3BO_3+1urea$ (mol%). A blank sample, the $Sr_2Al_6O_{11}$, was prepared by the same reactant ratio of $Sr_2Al_6O_{11}:Eu^{2+},Dy^{3+}$ but without the addition of Eu_2O_3 and Dy_2O_3. Then the reactants were ground in an agate mortar to mix them together, and pressed into pellets.

As shown in Figure 1, solid state reaction was carried out in a tubular oven. The tubular hearth is made up of two regions which were regulated by independent temperature-control system. Active carbon particles were placed at the first region and heated at 700 °C to produce large amount of thermal CO gas ($2C+O_2 \rightarrow CO$). The pellets were located at the second region and the second region was then heated at 1000 °C for 2h. Ar gas flew from the first region to the second region continuously to deliver the CO gas to the pellets.

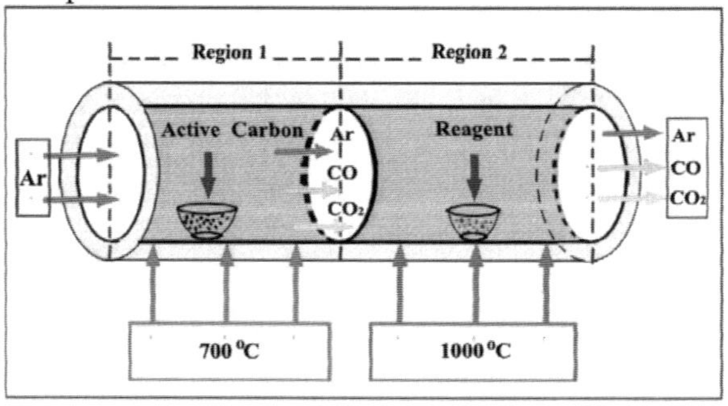

Figure 1. Scheme of synthesis process.

Material structures were investigated by X-ray diffraction (Thermal ARL X'TRA) and Fourier transforming infrared spectra (FTIR, Nicolet IS-10, 4000-400 cm^{-1}). Photo-luminescence (PL) and afterglow spectra were analyzed by fluorescence spectrometer (Edinburghinstruments FS5). Optical properties were also studied by incident reflection (UV-Vis-NIR spectrometer, Shimadzu, UV 3101PC). Thermal behavior was analyzed by the DSC-TG characterization (NETZSCH STA 409 PC).

3. Results and discussion

DSC-TG curves indicated the $Sr_2Al_6O_{11}$:Eu^{2+},Dy^{3+} could be obtained at 1000°C as there were neither weight loss nor thermal variation above that temperature (Figure 2(a)). XRD pattern shows the prepared $Sr_2Al_6O_{11}$ has high purity as all the diffraction peaks matched well with the crystal planes of orthorhombic $Sr_2Al_6O_{11}$ (JCPDS#72-7796, Figure 2(b)). The orthorhombic $Sr_2Al_6O_{11}$ structure was made up by two kinds of planes: the AlO_4 plane which was connected by corner-shared AlO_4-tetrahedra groups, and the AlO_6 plane which was connected by edge-shared AlO_6-octahedra groups (Figure 2 (c))[7, 9]. The AlO_4 and AlO_6 planes are cross-linked by AlO_4-tetrahedra. FTIR spectra reveal the existence of the two planes, as shown in Figure 2(d). The FTIR bands at 850, 780 and 643 cm^{-1} corresponds to the stretching vibration of AlO_4-tetrahedra. The 523, 476 and 430 cm^{-1} bands are assigned to the vibration of AlO_6-octahedra[12].

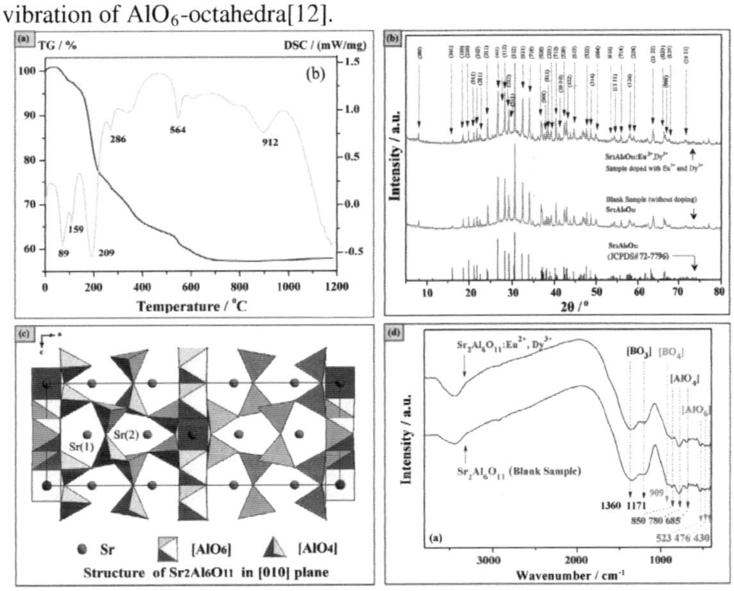

Figure 2. (a) DSC-TG curves of reactants; (b) XRD patterns of products; (c) Scheme of $Sr_2Al_6O_{11}$ structure; (d) FTIR spectra of products

In fact, the orthorhombic $Sr_2Al_6O_{11}$ is fairly unstable as its zigzag AlO_4 planes tends to transform spontaneously to release its structural stress[9]. That enhanced the difficulty in stabilizing the $Sr_2Al_6O_{11}$ structure during the synthesis process. So the $Sr_2Al_6O_{11}$ structure could only be synthesized in narrow and strict conditions. Herein, as the CO gas was produced from the oxidation of carbon, they provided the pellets with not only a reductive atmosphere but high thermal energy to accelerate the crystallization of $Sr_2Al_6O_{11}$ phase (Figure 1). Intermediate products, such as CO_2 and volatile boric acid, have little influence on the crystallization of $Sr_2Al_6O_{11}$ as they were removed from the reaction system with the flow of CO gas. Thus the preparation strategy in this work could facilitate the formation of high purity $Sr_2Al_6O_{11}$ by taking advantages of solid state reactions and CVD method. The 909 cm^{-1} band is assigned to the vibration of residual amorphous boron oxygen compounds, indicating that the boric acid could help to prevent the structure transformation of orthorhombic $Sr_2Al_6O_{11}$ and improve its stability during the reaction.

Under 365 nm excitation, $Sr_2Al_6O_{11}$:Eu^{2+},Dy^{3+} exhibited a broad emission band in 375-650 nm (Figure 2(a, b)). The blue-green emission band centered at 492 nm is mainly attributed to the $4f^65d$-$4f^7$ electron transition of Eu^{2+} ions. Excitation band of $Sr_2Al_6O_{11}$:Eu^{2+},Dy^{3+} overlapped with the emission band at 375-450 nm, indicating the emission light at 375-450 nm could be applied by the $Sr_2Al_6O_{11}$:Eu^{2+},Dy^{3+} to excite second emission lights (Figure 2(a)). That was supported by the reflection spectra, which show that the $Sr_2Al_6O_{11}$:Eu^{2+},Dy^{3+} could absorb 340-475 nm light whereas the $Sr_2Al_6O_{11}$ without doping have no absorption at that range (Figure 2(c)). Those absorbed light was captured by the Eu^{2+} ions to generate its $4f^65d$-$4f^7$ electron transitions. In addition, the weak absorption at 797-1655 nm are assigned to the Dy^{3+} f-f electron transitions and confirms the doping of Dy^{3+} ions (Figure 2(c,d)).

The 492 nm emission band is asymmetric and could be de-convoluted into two Gaussian bands at 485 and 521 nm (Figure 2(b)). The splitting of 492 nm band originated from the variation of crystal fields around the Eu(1) and Eu(2) sites, as the $4f^65d$-$4f^7$ electron transitions of Eu^{2+} ions are very sensitive to coordinate environments. The 492 nm emission of $Sr_2Al_6O_{11}:Eu^{2+},Dy^{3+}$ could still be detected after 240 min (Figure 2(e)) [7,9,13]. The profile of 492 nm emission did not change with the decay of emission intensity, indicating the Eu(1) and Eu(2) sites both contribute to the blue-green persistent luminescence (Figure 2(f, g)).

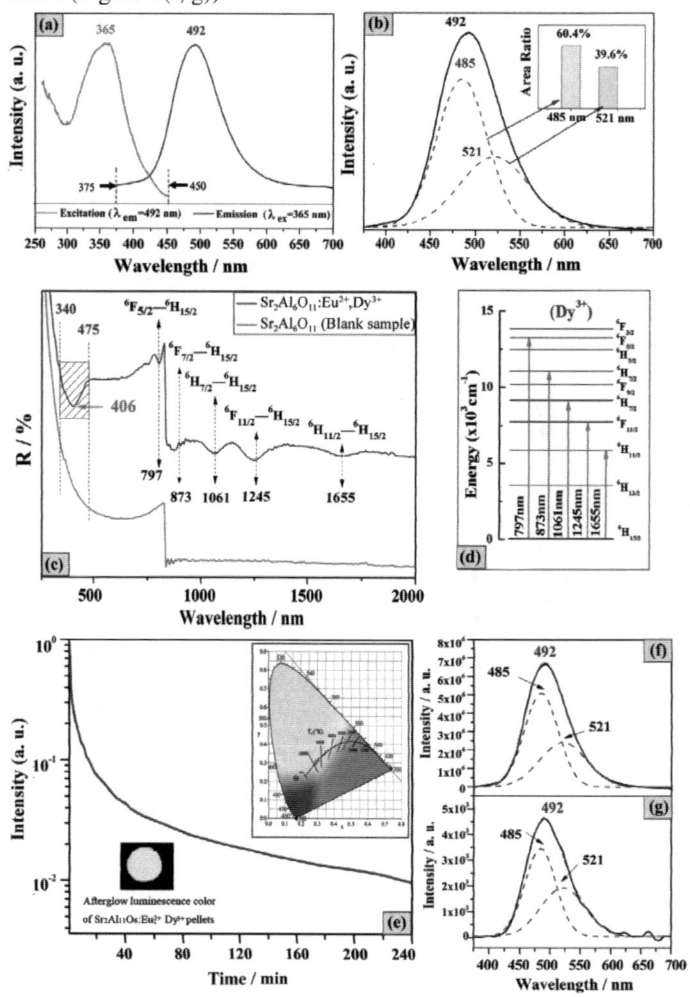

Figure 3. Excitation and emission spectra of $Sr_2Al_6O_{11}:Eu^{2+},Dy^{3+}$ (λex=365nm, λem=492nm); (b) Gaussian fitting of 492 nm emission band; (c) UV-vis-NIR spectra of $Sr_2Al_6O_{11}:Eu^{2+},Dy^{3+}$ and $Sr_2Al_6O_{11}$; (d) Scheme for the electron levels of Dy ions; (e) Persistent luminescence as a function of time for $Sr_2Al_6O_{11}:Eu^{2+},Dy^{3+}$ (Inset, CIE chromaticity diagram of $Sr_2Al_6O_{11}:Eu^{2+},Dy^{3+}$); (f) Emission spectrum of $Sr_2Al_6O_{11}:Eu^{2+},Dy^{3+}$ (1 min after excitation); (g) Emission spectrum of $Sr_2Al_6O_{11}:Eu^{2+},Dy^{3+}$ (240 min after excitation)

Modified Clabau theory is recognized to be a reasonable model to explain the afterglow mechanism of Eu^{2+},Dy^{3+} doped aluminates[14-16]. According to that theory, the afterglow emission process of $Sr_2Al_6O_{11}$:Eu^{2+},Dy^{3+} could be speculated as follows (Figure 4): (1) the doping of Eu^{2+} and Dy^{3+} ions generated V_o and V_{Sr} defects; (2) Under UV irradiation, the electrons excited from Eu^{2+} $4f^7$ to $4f^65d^1$ level could move into conductive band (CB), while some electrons were excited from valence band (VB) to Eu^{3+} $4f^7$ level; (3) V_o defects trapped the electrons released from CB, while V_{Sr} defects trapped the holes from VB; (4) The trapped electrons and holes were released slowly from V_o and V_{Sr} to yield the blue-green persistent luminescence.

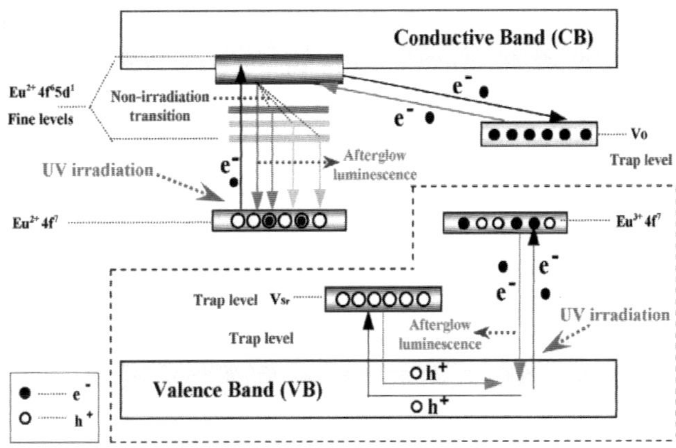

Figure 4. Proposed afterglow mechanism for $Sr_2Al_6O_{11}$:Eu^{2+},Dy^{3+}

4. Conclusion

In this work, we report the synthesis and optical properties of a distinctive afterglow material $Sr_2Al_6O_{11}$:Eu^{2+},Dy^{3+}. It was simple and effective to synthesize the orthorhombic $Sr_2Al_6O_{11}$:Eu^{2+},Dy^{3+} with high purity. Under UV excitation, the special aluminates exhibits broad emissions at 375–650 nm and the 375-450 nm emission lights could be utilized by the material to excite second emission. The Eu(1) and Eu(2) sites in $Sr_2Al_6O_{11}$ lattice both contribute to the blue-green persistent luminescence.

Acknowledgments

This work is supported by Jiangsu Provincial Natural Science Foundation for Youths (No. BK20130095), Educational Commission of Jiangsu Province (15KJD430014), National Natural Science Foundation of China (No.51502143), Innovation Foundation for scientific research (2014-jit-n-06), and Key University Science Research Project of Jiangsu Province (No.15KJB430022).

References

[1] Li Y, Gecevicius M, Qiu JR 2016 *Chem. Soc. Rev.* **45** 2090.
[2] Pan ZW, Lu YY, Liu F 2012 Nat. Mater. **11** 58.
[3] Matsuzawa T, Aoki Y, Takeuchi N, et al. 1996 *J. Electrochem. Soc.* **143** 2670.
[4] Zhang X, Xu XH, He QL, et al. 2013 *J. Solid State Sci. Technol.* **2** R225.
[5] Ueda J, Shinoda T, Tanabe S 2013*Opt. Mater. Express* **3** () 787.
[6] Zhuang YX, Katayama Y, Ueda 2014J *Opt. Mater.* **36** 1907.
[7] Smets B, Rutten J, Hokeks G, et al. 1989 *J. Electrochem. Soc.* **136** 2119.
[8] Wang D, Wang MQ 1999 *J. Mater. Sci. Lett.* **18** 1433.
[9] Takeda T, Kasori T, Uheda K, et al. 2002 *J. Japan Soc. Power and Power Metall.* **49** 1128.

[10] Song HH, Liu WF 2008 *Rare Metal Mater. Eng.* **37** 1167.
[11] Xue HW, Cui C, Huang P, et al. 2009 *J. Synth. Cryst.* **38** 1384.
[12] Tarte P 1967 *Spectrochim. Acta Part A* **23** 2127.
[13] Huang P, Cui CE, Hao HZ 2009 *J. Sol-Gel Sci. Technol.* **50** 308.
[14] Clabau F, Rocquefelte X, Jobic S, et al. 2005 *Chem. Mater.* **17** 3904.
[15] Clabaua F, Rocquefeltea X, Jobica S, et al. 2007 *Solid State Sci.* **9** 608.
[16] Ngaruiyaa JM, Nieuwoudt S, Ntwaeaborwa OM, et al. 2008 *Mater. Lett.* **62** 3192.

IOP Conf. Series: Materials Science and Engineering **167** (2017) 012051 doi:10.1088/1757-899X/167/1/012051

Accelerated lifetime test of vibration isolator made of Metal Rubber material

Hongrui Ao [1], Yong Ma, Xianbiao Wang, Jianye Chen, Hongyuan Jiang

School of Mechatronics Engineering, Harbin Institute of Technology, Harbin 150001 P.R. China

E-mail: Hongrui_ao@hit.edu.cn

Abstract. The Metal Rubber material (MR) is a kind of material with nonlinear damping characteristics for its application in the field of aerospace, petrochemical industry and so on. The study on the lifetime of MR material is impendent to its application in engineering. Based on the dynamic characteristic of MR, the accelerated lifetime experiments of vibration isolators made of MR working under random vibration load were conducted. The effects of structural parameters of MR components on the lifetime of isolators were studied and modelled with the fitting curves of degradation data. The lifetime prediction methods were proposed based on the models.

1. Introduction

The Metal Rubber (MR) material is a kind of porous material with the good environmental suitability, long working life-span, and adjustable stiffness. MR material is manufactured by using a special cold pressing technology to metal spiral wires. When it is applied in the vibration condition, it shows nonlinear damping characteristics for its internal dry friction between adjacent metal wires. The features of the vibration isolation of MR material comprise good damping capacity, strong environmental suitability, adjustable stiffness, and so on. Since it is the alternative of the rubber damping material under special working condition, it gains wide application in the fields such as spacecraft, weaponry, and petrochemical engineering [1]. The research on this material concentrated on its vibration performance, mathematical description, and application [2,3]. The researches on the vibration isolation and vibration damping performance of MR mainly comprise the studies in the property of nonlinear deformation, energy dissipation mechanism, and fatigue damage. Li *et al* testified the effects of shape factor on the mechanical property of material structure, obtained the variation relation of the mechanical property of MR with the changing of structure parameter, and acquired the theoretical equation of the loss factor changing with the material height and balance weight respectively [4]. Jiang *et al* [5] took the broadband random vibration experiment on the MR vibration isolation system and discussed the influence factors of the vibration isolation system according to the result. Wang et al [6] considered the comprehensive damage factor as the failure criteria of MR components. Zhao et al [7] discussed the relation between the fatigue circulation circles and the stiffness of the parts with soft characteristic, the static loss factor, and the dynamic loss factor respectively through the cyclic loading experiment on MR material.

Usually the MR components work under the random vibration condition. Therefore, for the popularization of metal rubber material, it is meaningful to study the lifetime and reliability of MR vibration isolator based on the random load. The statistical objects of traditional lifetime research and reliability analysis are the failure time which needs lots of lifetime and accelerated lifetime experiments to obtain the failure data of products. But for the MR isolators with long lifetime and high reliability, the many needed experiments and long time to obtain these data restrict the application of traditional reliability assessment method. Therefore, in this study the analytical method based on the performance degradation is adopted. As a necessary parameter for reliability analysis, the performance degradation can make up the deficiencies of the traditional reliability analysis in analyzing the products with long lifetime and high reliability.

In the experiments, a high level stress is put on the products than the requirement of their normal works to accelerate the failure process and rapidly acquire the failure data without changing the failure mechanism of the products. At present, the accelerated lifetime experiment has already been an important method of reliability assessment. According to the researches, during the disabling process, some important performance may change solely as the time changing, so the disabling process of products can be seen as the process that the performance parameter degrades constantly. For the reliability analysis of products with long lifetime and high reliability, it has been an important method to do the reliability assessment by building up the appropriate model after extracting the failure information from the degradation data.

The reliability model based on the performance degradation amount assumes that the performance degradation data of samples correspond a certain distribution law at the same moment, estimate the distribution parameter by the degradation data, and calculate the reliability of products at each moment according to the definition of reliability function.

Generally, the basic researches mentioned above are helpful to understand the vibration damping principle, the damping effect, and the fatigue damage mechanism. However, currently, the researches on the lifetime forecast, reliability, and other features of MR material are still immature, and the insufficient data accumulation also restricts the wider application of the technology. In this paper, the lifetime of MR isolator was tested experimentally and the mathematic models by the performance degradation data to forecast its lifetime were established.

2. The accelerated lifetime test of Metal Rubber isolator

The fatigue lifetime experiment includes two steps: the sine sweep frequency experiment and the random vibration experiment. The sine sweep frequency experiment can measure the resonant frequency and resonant transmissibility of the MR vibration isolator. Through the random vibration experiment, the MR elements will bear a certain random load, the amount of which is determined by the acceleration stress. The dynamic stress generated inside the metal rubber elements during the random vibration is expressed as

$$\sigma_d = \frac{ma}{S} = \frac{m}{S}\sqrt{\frac{\pi A f_0 \eta_0}{2}} \tag{1}$$

where σ_d is the dynamic stress inside the MR elements (MPa), m is the mass of the system (kg), S is the equivalent contact area of the MR element (mm^2), a is the accelerated speed response of the random vibration (g), f_0 is the resonant frequency of the system (Hz), and η_0 is the resonant transmissibility of the system.

According to the equation (1), after the stress condition of the accelerated lifetime experiment is determined, the power spectral density of the random vibration can be calculated as

$$A = \frac{2\sigma_d^2 S^2}{\pi m^2 f_0 \eta_0} \tag{2}$$

It is noted that the Metal Rubber isolators need to be stabilized before the sine sweep frequency experiment.

After the stabilization is completed, the resonant frequency f_0 and resonant transmissibility η_0 measured from the sine sweep frequency experiment are taken as the original value of the performance degradation data. The power spectrum density of the random vibration experiment is calculated according to the equation (2), and conduct the fatigue lifetime experiment for a while under the random vibration condition. The resonant frequency and transmissibility were measured through the sine sweep frequency experiment every other certain time to compare with the initial value. When the performance degradation data exceeds the threshold value of initial value, the fatigue lifetime experiment can be stopped.

During the fatigue lifetime experiment, the variation quantity of the threshold value of stopping the experiment need to be 12%, which is determined by the requirements of the customers. The prediction accuracy and the rationality of the models are verified by the repetition of the experiments.

3. The experiment results and analysis

The control variable method is adopted in the fatigue experiment which means to successively change influential factors of MR elements including the relative density, metal wire diameter, and height, and obtain the performance degradation data of different stress condition under each influential factor. By the lifetime forecast model based on the Bayes theory mentioned to deal with the performance degradation data, the lifetime forecast value of the MR vibration isolation under the single influential factor can be obtained.

Figure 1 shows the fatigue lifetime curve of the MR vibration isolation with different metal wire diameter under the condition that the relative density, $\bar{\rho}$, defined as the ratio of the density of MR components and the density of the steel, is equal to 0.23 and the height of MR component is h=1.5mm. It can be seen from the Figure 1 that the magnitude of load of random vibration experiment has great influence on the fatigue lifetime of MR vibration isolator. Under the different stress condition, the fatigue lifetime differs a lot. The fatigue lifetime is expressed in a logarithm way and fitted linearly as

$$\lg(T) = -2.377\sigma + 5.408 \quad (3)$$

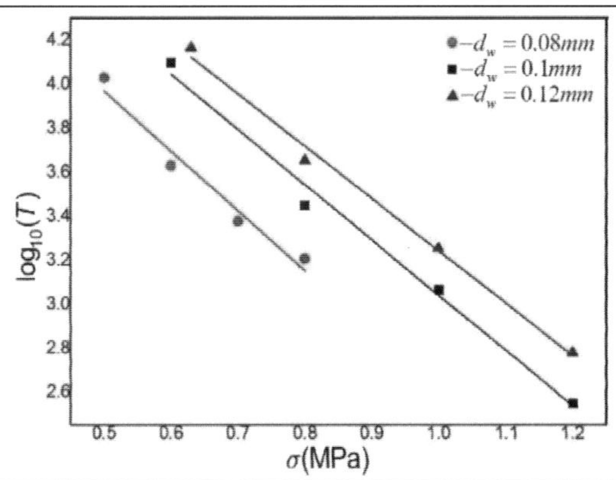

Figure 1. The fatigue lifetime curve of the MR elements with different metal wire diameters.

In a similar way, the relation between the logarithm value of fatigue lifetime and the stress value received by the MR with different diameters of metal wire, relative densities, or heights of the MR can be obtained. It can be seen from Figure 1 that the fatigue lifetime of MR elements increased with the increasing of the metal wire diameter under the same stress condition.

Then, the coefficient function is adopted to deduce the relation between the fatigue lifetime and metal wire diameter of MR vibration isolation. Taking the lifetime curve of the MR elements of $\bar{\rho} = 0.23, d_w = 0.1\text{mm}, h = 1.5\text{mm}$ as the standard, the forecast model is established as following:

$$\lg(T) = -2.377\alpha_1(d_w)\sigma + 5.408\alpha_2(d_w) \tag{4}$$

In the formula, $\alpha_1(d_w)$ and $\alpha_2(d_w)$ are the coefficient function related to the metal wire diameter. Because the standard of lifetime forecast model is the lifetime curve of MR elements of $d_w = 0.1$mm, so $\alpha_1(0.1)=1$, $\alpha_2(0.1)=1$. For the coefficient function value of d_w=0.08mm and d_w=0.12mm, the calculation method corresponds with the coefficient specific value between fitting formula and standard formula. It is:

$$\alpha_1(0.08) = -2.72/(-2.337) = 1.144, \alpha_2(0.08) = 5.329/5.408 = 0.985,$$

$$\alpha_1(0.12) = -2.229/(-2.377) = 0.936, \alpha_2(0.12) = 5.458/5.408 = 1.01$$

According to the above coefficient function value, the coefficient function obtained from the least square fitting is as follows

$$\begin{cases} \alpha_1(d_w) = 100d_w^{\ 2} - 25.2d_w + 2.52 \\ \alpha_2(d_w) = -6.25d_w^{\ 2} + 1.875d_w + 0.875 \end{cases} \tag{5}$$

Then, the lifetime forecast formula about the metal wire diameter of the fatigue lifetime can be obtained as

$$\lg(T) = -2.377(100d_w^{\ 2} - 25.2d_w + 2.52)\sigma + 5.408(-6.25d_w^{\ 2} + 1.875d_w + 0.875) \tag{6}$$

Figure 2 shows the relationship between the fatigue lifetime of MR vibration isolators and different relative density under the condition that d_w=0.1mm and h=1.5mm. Similarly, based on the lifetime curve of MR elements of $\bar{\rho}$=0.23, d_w=0.1mm, h=1.5mm, the fatigue lifetime about the relative density can be predicted as follows

$$\lg(T) = -2.377(-83.2\bar{\rho}^2 + 41.37\bar{\rho} - 4.114)\sigma + 5.408(-40.8\bar{\rho}^2 + 21.11\bar{\rho} - 1.697) \tag{7}$$

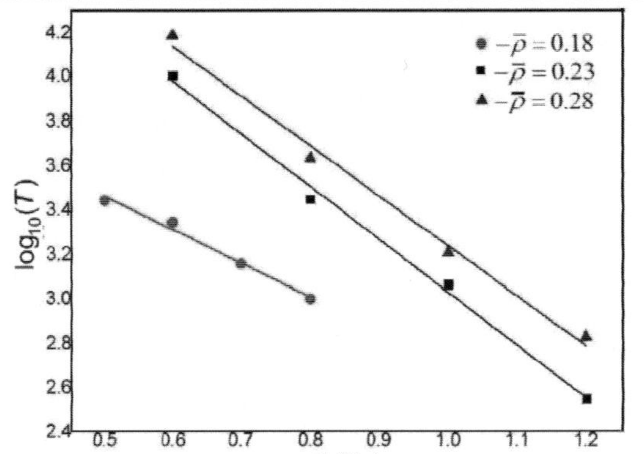

Figure 2. The fatigue lifetime curve of MR elements with different relative density.

Figure 3 is the fatigue lifetime curve of MR vibration isolators with different heights under the condition that the metal wire diameter of MR elements is d_w=0.1mm and the relative density is $\bar{\rho}$ =0.23. Under the same stress condition, the fatigue lifetime of MR isolators increases with the increase of the MR component heights.

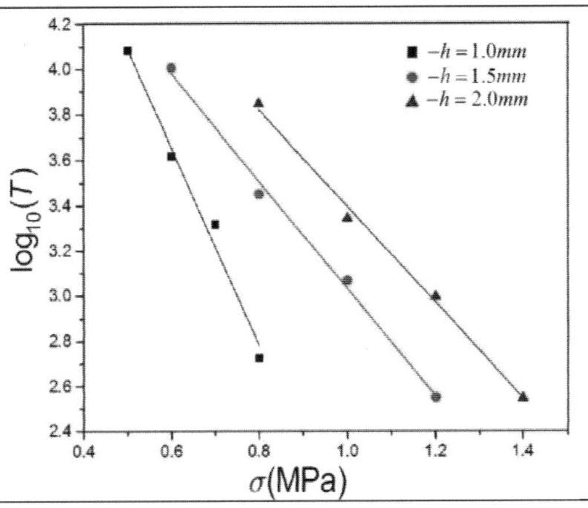

Figure 3. The fatigue lifetime curve of MR elements with different heights.

Using the same method discussed above, the forecast formula of fatigue lifetime about the height of MR elements can be obtained

$$\lg(T) = -2.377\left(1.798h^2 - 6.429h + 6.598\right)\sigma + 5.408\left(0.446h^2 - 1.519h + 2.275\right) \quad (8)$$

The above three formulas are respectively the lifetime forecast formulas obtained under the condition of single influential factor. But in the practical usage, the three factors must be used together and affect the lifetime of MR. After the further exploration in that if the three factors have dependency, the lifetime forecast formula of multi-factor coupling can be obtained on this foundation.

4. Conclusion

Dynamic experiments under random vibration loads are conducted to test the lifetime of MR isolator for the description of its performance degradation. An exponential mathematic model was put forward based on the degradation data. The influencing factors, including the relative density, the diameter of metal wires, and the preloaded deformation of MR components, on the lifetime of isolator were studied. The fitting method for the lifetime fitting formula of the MR vibration isolators based on the single factor was discussed. These approach are beneficial to the further theoretical study and practical application of Metal Rubber vibration isolation system.

References

[1] A.M. Ulanov. Proceedings in Samara Scientific Center of Russia Scientifc Academy. 2009, 11(5): 88~94. (in Russian)

[2] A.M.Ulanov, U. K. Ponomarev. Proceedings of vibration protection design with elastic elements made of MR material. Russia: Samara State Aerospace University, 2008. 10(3):853-857. (in Russia)

[3] Li Shengbo, Yan Hui, Jiang Hongyuan, et al. *Acta Physica Sinica*, 2012, 61(1):010702.

[4] Li Tuo, Bai Hongbai, Lu Chunhong, et al. *Journal of Vibraiton and Shock*, 2014, 33(3):142-147.

[5] Yan Hui, Jiang Hongyuan, Liu Wenjian, et al. *ACTA PHISICA SINICA*, 2009,58(8):5238-5243.

[6] Wang Youyan, Bai Hongbai, Hou Junfang. *Chinese Journal of Mechanical Engineering*, 2011,47(2): 65-71.

[7] Zhao Cheng, Luo Kun, Liu Changhong, *Noise and Vibration Control*, 2010,01:106-108+111.

Microstructure and mechanical properties of zirconium doped NiAl/Cr(Mo) hypoeutectic alloy prepared by injection casting

L Y Sheng[1,3], B N Du[1] and J T Guo[2]

[1] Shenzhen Institute, Peking University, southern 7 street, Hi-Tech Park, Shenzhen, China

[2] Institute of Metal Research, Chinese Academy of Sciences, Wenhua Road 72, Shenyang, China

E-mail: lysheng@yeah.net

Abstract. NiAl based materials has been considered as most potential candidate of turbine blade, due to its excellent high-temperature properties. However the bad room-temperature properties handicap its application. In the present paper, the zirconium doped NiAl/Cr(Mo) hypoeutectic alloy is fabricated by conventional casting and injection casting technology to improve its room-temperature properties. The microstructure and compressive properties at different temperatures of the conventionally-cast and injection-cast were investigated. The results exhibit that the conventionally-cast alloy comprises coarse primary NiAl phase and eutectic cell, which is dotted with irregular Ni_2AlZr Heusler phase. Compared with the conventionally-cast alloy, the injection-cast alloy possesses refined the primary NiAl, eutectic cell and eutectic lamella. In addition, the Ni_2AlZr Heusler phase become smaller and distribute uniformly. Moreover, the injection casting decrease the area fraction of primary NiAl phase at the cell interior or cell boundaries. The compressive ductility and yield strength of the injection-cast alloy at room temperature increase by about 100% and 35% over those of conventionally-cast alloy, which should be ascribed to the microstructure optimization.

1. Introduction

Compared with many NiAl-based alloys, NiAl-28Cr-6Mo (NiAl-Cr(Mo) for short) eutectic alloy is regarded as the most logical choice of the multielement system because of it good balance between elevated temperature creep resistance and room temperature fracture toughness [1-3]. But, the high temperature strength of the alloy still needs improve further [4-5]. The recent researches [6-8] reveal that the addition of Heusler phase forming elements (Ti, Zr, Hf) could increase the strength of the NiAl based eutectic alloy. However, the Heusler phase prefer to form along the eutectic cell boundary and is detrimental to the room-temperature mechanical properties [9,10]. Then improvement on the size and distribution of the strengthening precipitates is important for the alloy. It is well known that rapid solidification is a feasible way to refine the microstructure and decrease the segregation, which is helpful to adjust the morphology and distribution of precipitates [11-13]. Moreover, the refined eutectic structure is also beneficial to the strength and ductility of the eutectic alloy [14]. Additionally, the researchers [15-16] have found that appropriate amount of primary NiAl is helpful to the wear property. Then in the present paper, the zirconium (Zr) doped NiAl-24Cr-4.6Mo-0.4Zr alloy (at.%) hypoeutectic alloy is fabricated by conventional casting and injection casting. Its microstructural evolution and mechanical properties were investigated as well.

2. Experimental

Alloy ingots with a composition of NiAl-24Cr-4.6Mo-0.4Zr were prepared in a vacuum induction furnace with starting materials of 99.9% Ni, 99.9% Al, 99.8% Zr, 99.5% Cr and 99.9% Mo. The melted alloy was cast into the ceramic shell mould to obtain the conventionally-cast alloy with 40 mm in diameter. The slices cut from the conventionally-cast alloy ingot for the injection casting. The slices were re-melted in quartz tube and injected through a nozzle into Cu mold to obtain the injection-cast alloy with 10mm in diameter.

Specimens for microstructure characterization and mechanical properties test were cut from the conventionally-cast and injection-cast alloy. The S-3400 scanning electron microscopy (SEM) was employed to perform the microstructure characterization. Constitute phase composition in the composite was analyzed by the EPMA-1610 electronic probe microanalysis. The specimen for transmission electron microscope (TEM) characterization was cut from composite with thickness of 0.4 mm. Then the specimen was polished to 30 μm and shaped into φ3 mm in size followed by twin-jet electropolishing. JEOL-2100 was employed to perform the TEM observation and analyses. Compression sample (4×4×6 mm^3) was cut from the alloys and mechanical grounded by 800-grit SiC abrasive. Gleeble 3800 was employed to carry out the compression test at the initial strain rate of 1×10^{-3}/s.

3. Results and discussion

3.1. Microstructural characteristics

The SEM observations on conventionally-cast NiAl-Cr(Mo)-Zr alloy are shown in Figure 1. Clearly, the alloy mainly consists of black NiAl phase, gray Cr(Mo) phase and white Heusler phase, as shown in Figure 1 (a). The NiAl and Cr(Mo) phases form the eutectic cell with an average size of 50μm, as shown in Figure 1 (a). Analysis on the conventionally-cast alloy exhibits the eutectic lamella has an average thickness of 4μm and the size of primary NiAl phase is about 20μm. In the eutectic cell, black NiAl and gray Cr(Mo) plates emanate radially from the cell center to boundary, as shown in Figure 1 (b). Coarser Cr(Mo), primary NiAl phase and white Heusler phase mainly distribute at the intercellular zone. Furthermore, it also can find many Cr(Mo) precipitates in the NiAl phase and NiAl precipitates in the Cr(Mo) phase. TEM observation on the Heusler phase reveals that it contains cubic crystal structure (a=b=c=0.6123 nm) and Fm3m space group, as shown in Figure 1 (c).

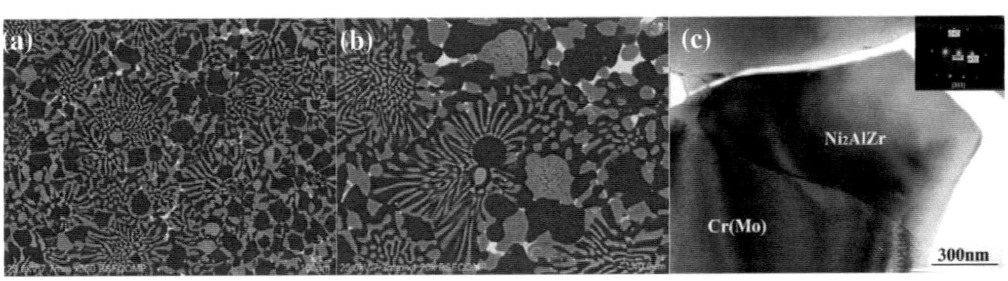

Figure 1. (a) SEM image of conventionally-cast alloy, (b) Morphology of precipitates and eutectic structure, (c) TEM image of Ni_2AlZr Heusler phase

The typical microstructure of injection-cast NiAl-Cr(Mo)-Zr alloy is shown in Figure 2. Apparently, the microstructure of injection-cast alloy is quite different from that of conventionally-cast alloy as a result of high cooling rate during injection casting. The morphology of primary NiAl phase exhibits dendritic crystal characteristic and its amount decreases a little, as shown in Figure 2 (a). The eutectic cell size in injection-cast alloy is about 25-35μm, which is smaller than that of conventionally-cast alloy, as shown in Figure 2 (a). The primary NiAl dendrites in injection-cast alloy with an average size of 15μm mainly distribute at the eutectic cell boundary. Moreover, the thickness of the NiAl/Cr(Mo) eutectic lamella in eutectic cell is less than 1μm, but that near the eutectic boundary is about 2μm, as shown in Figure 2 (c). In addition, TEM observation also reveals plentiful nano-scale NiAl and Cr(Mo) precipitates. By comparing Figure 1 (a) with Figure 2 (a), it can be specific that the Heusler phase has been distinctly fined and evenly distributes at cell boundaries.

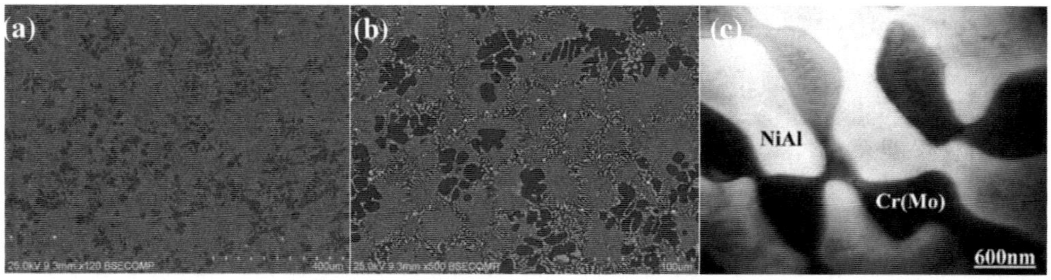

Figure 2. (a) SEM image of the injection-cast alloy, (b) Morphology of the precipitates and eutectic structure, (c) TEM image of the ultrafine NiAl/Cr(Mo) eutectic structure

According to the former research [9], the eutectic cell size and lamellar thickness of NiAl-Cr(Mo) eutectic alloy would decrease with the growth rate increasing. In the present research, the cooling rate of the injection casting is higher than that of the conventional casting, so the NiAl/Cr(Mo) eutectic structure and eutectic cell become finer. Compared with the conventionally-cast alloy, the precipitation in the injection-cast alloy is smaller and less. The EPMA analyses of different phase in conventionally-cast and injection-cast alloys are enumerated in Table 1. For injection-cast alloy, the solubility of Cr, Mo and Zr in primary NiAl phase, Ni, Al in Cr(Mo) phase and Cr in Heusler phase are higher than that of conventionally-cast alloy. Although 7.67% Cr was detected in primary NiAl phase of injection-cast alloy by EPMA, the solid solution of Cr in NiAl phase should be less than the measured result, because the precipitation in NiAl and Cr(Mo) phase could result in some deviation. However, the results still exhibit that the rapid solidification could restrict the element diffusion and increase the solid solution of the matrix. Moreover, it can be understood that why the NiAl and Cr(Mo) phase near the eutectic cell boundary are coarser than those in the center. Because the alloy element concentrating along liquid/solid interface could increase the constitutional supercooling, that would result in the rapid growing of the NiAl and Cr(Mo) phase.

Table 1. Composition of different phases in conventionally-cast and injection-cast alloys (at.%)

Alloys	Phase	Ni	Al	Cr	Mo	Zr
	Primary NiAl	47.32	50.08	2.50	0.20	0.20
Conventionally- cast	Cr(Mo)	3.30	4.90	80.18	11.62	-
	Heusler	50.78	23.37	3.20	-	22.25
	Primary NiAl	44.66	45.82	7.67	0.85	0.36
Injection-cast	Cr(Mo)	9.20	10.08	69.88	10.84	-
	Heusler	47.07	26.38	6.27	-	20.28

3.2. Mechanical properties

The true stress-strain curves and mechanical test data at RT of conventionally-cast and injection-cast alloys were shown in Figure 3 and Table 2, respectively. It can be seen that the alloys fabricated by different technologies have the similar stress-strain curves, which exhibit continuous work hardening. The injection-cast alloy attains yield strength of 1360 MPa and compressive strength of 1920 MPa, both of which are about 25% higher than those of conventionally-cast alloy. Compressive strain values at RT are calculated based on practical plastic deformation in order to eliminate the contribution from the compliance of the testing system. The injection-cast alloy has a better RT compressive ductility with about 18% than about 8% for conventionally-cast alloy. It means that the RT ductility and strength of injection-cast alloy have been improved at the same time. The compression tests at 1273 K and 1373 K reveal that the yield strength and compressive strength of the injection-cast alloy are almost similar as that of the conventionally-cast alloy, which indicates that the rapid solidification could improve the room temperature mechanical properties obviously but no influence on the high temperature mechanical properties.

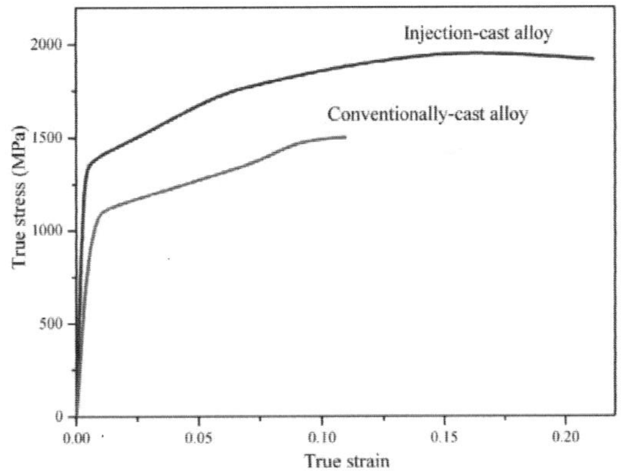

Figure 3. True stress-strain compression curves c of conventionally-cast and injection-cast alloys at room temperature

Table 2. Results of compressive tests under the nominal strain rate of 2x10-3 s-1

Alloy	Test temperature(K)	Yield strength (MPa)	Compressive strength (MPa)	Compressive strain (%)
	RT	1010	1510	8
Conventionally-cast	1273K	324	395	>30
	1373K	205	225	>30
	RT	1360	1920	18
Injection-cast	1273K	340	420	>30
	1373K	205	265	>30

Figure 4 shows the typical RT compressive fractographies of conventionally-cast alloy and injection-cast alloy, both of which exhibit similar fracture morphologies, i.e. typical stripping of NiAl/Cr(Mo) interface and cleaving of primary NiAl. However, many dimple-like cavities can be observed on the fracture surface of SC alloy. These cavities were formed by thin Cr(Mo) phase and NiAl phase in eutectic cell pulling out from each other, which are propitious to attain good compressive ductility and strength.

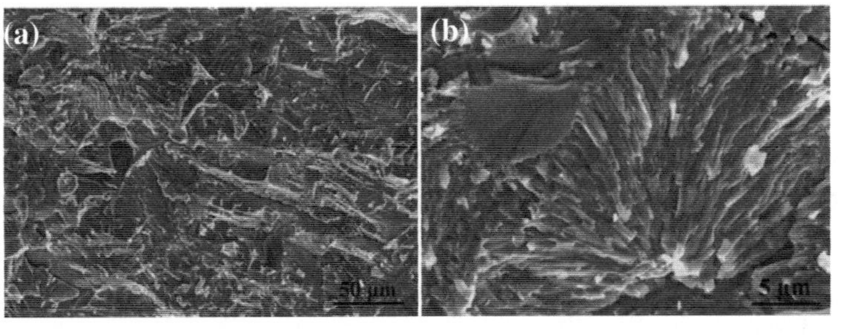

Figure 4. Typical RT fracture surface of (a) conventionally-cast alloy and (b) injection-cast alloy

The refinement of eutectic cells, the decrease of lamellar spacing, the extended solubility and the increased area fraction of cell eutectic as well as the refinement of Heusler phase particles may be the main factors that are relevant to the improved strength and ductility of injection-cast alloy at RT. The decrease of lamellar spacing and increase of cell eutectic zone produce more interfaces between NiAl and Cr(Mo) phases. Investigations [17-19] demonstrated that the dislocation network between NiAl and Cr(Mo) phase interface plays an important role on the strength of NiAl based alloys. The more NiAl/Cr(Mo) interfaces result in more dislocation network at the interface, so the strength of alloys can be enhanced accordingly. The increscent solubility of alloying elements in NiAl and Cr(Mo) is beneficial to the room temperature strength by solution strengthening. Besides, the remarkable increase in the total area of cell boundaries or phase boundaries by injection casting also induces a significant decrease in the segregated concentration of Heusler phase per unit area of cell boundaries that is beneficial to the strength improvement by particle strengthening. The improvement of ductility can be attributed to the increased area fraction of eutectic cell as well as the fine NiAl and Cr(Mo) plates.

The mechanical test data by compressive test at 1273K and 1373K of conventionally-cast and injection-cast alloys (Table 2) show that both alloys have the similar high temperature strength. For injection-cast NiAl-Cr(Mo)-Zr alloy, the high solid solution of alloying element in NiAl and Cr(Mo) phase, the great area fraction of eutectic cell and fine lamellar spacing should be propitious to the improvement of the strength at 1273K and 1373K. However, it exhibits similar high temperature strength to the conventionally-cast alloy. The causation can be ascribed to the weak intercellular zone. For NiAl-Cr(Mo)-Zr lamellar eutectic alloy, crack is easy to pass along the cell boundary when stress is applied to the eutectic alloy due to the coarser NiAl/Cr(Mo) plate and primary NiAl at intercellular zone, as shown in Figure 5. During injection casting process, the eutectic cell fined due to high cooling rate and the total area of cell boundary increases distinctly. The failure of the boundaries may become the dominant factor in determining the high temperature strength. Then it can be deduced that the strengthening effect of dislocation networks and solid solution was counteracted by the weaken effect from the weak cell boundary.

4. Conclusions

- The Zr doped NiAl-Cr(Mo) hypoeutectic alloy is mainly composed of coarse NiAl/Cr(Mo) eutectic cell and primary NiAl phase. The Zr addition results in the Ni_2AlZr Heusler phase, which prefers to segregate along eutectic cell boundary.
- The microstructure of injection-cast alloy presents a fine microstructure including the refinement of eutectic cell size, interlamellar spacing and intercellular zone as well as primary NiAl phase and Ni_2AlZr Heusler phase. In addition, the extension of solid solubility occurred in the alloy.
- The injection-cast alloy attains room temperature yield strength of 1445 MPa and compressive ductility of about 14%, which are higher than 1168 MPa and 7% for conventionally-cast alloy respectively. Both alloys possess the similar high temperature strength.

5. Acknowledgements

The authors are grateful to the Shenzhen International Cooperative Research Project(GJHZ20140419114548516), the Shenzhen Technology Innovation Plan (CXZZ20140731091722497 and CXZZ20140419114548507), China Postdoctoral Science Foundation (2015M580923) and Shenzhen Basic Research Project (JCYJ20140419114548515, JCYJ20150625155931806 and JCYJ20160407090231002) for financial support.

References

[1] MiracleDB 1993 Overview No. 104 the physical and mechanical properties of NiAl Acta Metal. Mater. **41** 649-684
[2] Sheng L Y, Zhang W, Guo J T, Yang F, Liang Y C and Ye H Q 2010 Effect of Au addition on

the microstructure and mechanical properties of NiAl intermetallic compound Intermetallics **18** 740-744

[3] Sheng L Y, Zhang W, Guo J T, Wang Z S and Ye H Q 2009 Microstructure evolution and elevated temperature compressive properties of a rapidly solidified NiAl-Cr(Nb)/Dy alloy Mater.Design **30** 2752-2755

[4] Wu Q H, Zhang Jand Luo Y S 2014 Composition and Mechanical Property of DD6 Superalloy Revert Mater. Sci. Forum **788** 488-492

[5] Sheng L Y, Yang F, Guo J T and Xi T F 2014 Anomalous yield and intermediate temperature brittleness behaviors of directionally solidified nickel-based superalloy Trans. Nonferrous Met. Soc. China **24** 673 - 681.

[6] Sheng L Y, Xie Y, Xi T F, Guo J T, Zheng Y F and Ye H Q 2011 Microstructure characteristics and compressive properties of NiAl-based multiphase alloy during heat treatments Mater. Sci. Eng. A **528** 8324-8331.

[7] Sheng L Y, Guo J T, Lai C and Xi T F 2015 Effect of Zr addition on microstructure and mechanical properties of NiAl/Cr(Mo) base eutectic alloy Acta Metall. Sin. **51** 828-834.

[8] Sheng L Y, Zhang W, Guo J T and Ye H Q 2009 Microstructure and mechanical properties of Hf and Ho doped NiAl–Cr (Mo) near eutectic alloy prepared by suction casting Mater. Charact. **60** 1311-1316

[9] Johnson D R, Chen X F, Oliver B F, Noebe R D and Whitterberger J D 1995 Processing and mechanical properties of in-situ composites from the NiAl-Cr and the NiAl-(Cr,Mo) eutectic systems Intermetallics **3** 99-113.

[10] Sheng L Y, Nan L, Zhang W, Guo J T and Ye H Q 2010 Microstructure and mechanical properties determined in compressive tests of quasi-rapidly solidified NiAl-Cr(Mo)-Hf eutectic alloy after hot isostatic pressure and high temperature treatments J. Mater. Eng. Perform. **19** 732-736

[11] Sheng L Y 2016 Microstructure and wear properties of the quasi-rapidly solidified NiAl/Cr(Mo,Dy) hypoeutectic alloy Strength Mater. **48**107-112.

[12] Sheng L Y, Yang F, Xi T F, Zheng Y F, and Guo J T 2013Microstructure and room temperature mechanical properties of NiAl-Cr(Mo)-(Hf, Dy) hypoeutectic alloy prepared by injection casting Trans. Nonferrous Met. Soc. China **23** 983–990

[13] Sheng L Y, Zhang W, Guo J T, Wang Z S, OvcharenkoV E, Zhou L Z and Ye H Q 2009 Microstructure and mechanical properties of Ni_3Al fabricated by thermal explosion and hot extrusion Intermetallics **17** 572-577

[14] BarabashRI, LiuW, TischlerJZ, BeiH and BudaiJD 2012 Phase-specific elastic/plastic interface interactions in layered NiAl–Cr(Mo) structures Acta Mater.**60**(8) 3279 – 3286

[15] Yang J-M,Jeng S M, Bain K andAmato R A 1997 Microstructure and mechanical behavior of in-situ directional solidified NiAl/Cr(Mo) eutectic composite Acta Mater. **45** 295-308

[16] Sheng L Y, Lai C, Yang F, Wang Q L and Xi T F 2014 Microstructure and wear behaviour of ceramic particles strengthening NiAl based composite Mater. Res. Innov. **18** 544-549

[17] Sheng L Y, Guo J T, Zhou L Z and Ye H Q 2010 Microstructure and compressive properties of NiAl-Cr(Mo)-Dy near eutectic alloy prepared by suction casting Mater. Sci. Tech. **26** 164-168

[18] Wang L, Shen J, Zhang Y P and Fu H Z 2016 Microstructure, fracture toughness and compressive property of as-cast and directionally solidified NiAl-based eutectic composite Mater. Sci. Eng. A **664** 188-194

[19] Sheng L Y, Yang F, Xi T F, Zheng Y F and Guo J T 2012 Improvement of compressive strength and ductility in NiAl–Cr(Nb)/Dy alloy by rapid solidification and HIP treatment Intermetallics **27** 14-20

Propane dehydrogenation over PtSnMg/Cr_2O_3·Al_2O_3 catalysts: effect of the amount of Mg loading

Peng Zhao[1], Yunfei Guan[1], Yanmei Wang[1], Xianzhi Guo[1],Jingya Zhang[1], Zongjie Du[1], Shoumin Zhang[1], Qinxing Xie[2] and Shihua Wu[1,3]

[1] Inorganic Chemistry Department, Nankai University, Tianjin, 300071, China
[2] Tianjin Key Laboratory of Fiber Modification and Functional Fibers, School of Materials Science and Engineering, Tianjin Polytechnic University, Tianjin, 300387, China

E-mail: wushh@nankai.edu.cn

Abstract. A series of PtSnMg/Cr_2O_3·Al_2O_3 catalysts with different Mg loading amounts were synthetized by chemical co-deposition method and their propane dehydrogenation catalytic performances were tested. The catalysts were characterized by XRD, BET, TG, XPS. It was found that alkali metal Mg can reduce the surface acidity of carrier, improve the catalytic performance. Propane conversion peaks at Mg loading of 0.6wt.%, increasing Mg loading further than this saw a decline in conversion.

1. Introduction

Propene is one of the most important bulk chemicals in petrochemical industry and mainly used in the synthesis of polypropylene, acrylonitrile, propylene oxide and so on[1], in particular, its derivative polypropylene, which is widely used in the daily life. In the recent years, with the rapid development of China's real estate, packaging, textile industry, automobile industry, the traditional propene production process has been unable to meet the growing demand, in addition, due to the discover in abundance of shale gas in the USA and the uncertain price of oil in future, the catalytic propane dehydrogenation (PDH) has received much attention as an alternative way for producing propene. Now there are three companies owning the technology of industrial production of propane dehydrogenation[2,3]: UOP's Oleflex, Lummus's Catofin and Krupp Uhde's Star. A large number of companies chose UOP's Oleflex and the core part of its PDH device is the catalyst. They used catalyst PtSn/Al_2O_3 (produced by UOP), the propane conversion was about 35% and the propene selectivity was 84% in the reaction (600-700°C). Lummus's Catofin used Cr-based catalysts (produced by Süd-Chemie AG). Cr-based catalysts[4,5] have the advantage of price, so does the carrier Al_2O_3, in addition Al_2O_3 has a good mechanical stability[6,7,8]. Krupp Uhde's Star technology was applied in only one device in Egypt, the catalyst was PtSn/$ZnAl_2O_4$ (produced by Uhde). For the purpose of getting a cheap and stable carrier, we used chemical co-deposition method to synthetized a composite oxide carrier Cr_2O_3·Al_2O_3. After loading the traditional PtSn[9], we got the standard catalyst - PtSn/Cr_2O_3·Al_2O_3. Mg is an alkali metal, so it may reduce the surface acidity of carrier[10]. We do the experiments to test whether Mg could improve the catalytic performance and what is the appropriate content of Mg.

2. Experimental

2.1. Catalyst preparation
$Cr(NO_3)_3 \cdot 9H_2O$, $Al(NO_3)_3 \cdot 9H_2O$ and 150ml deionized water were added into a 250ml beaker, after stirring for 5min, drops of $NH_3 \cdot H_2O$ was added until the pH of solution is between 4.8 to 5.1, then using the centrifuge (3800r/min) 10min to separate the precipitate. Washed the precipitate several times, dried at 80°C for 8h, calcined at 550 °C for 4 h, then composite oxide $Cr_2O_3 \cdot Al_2O_3$ was obtained. Loading 0.4wt.% Pt: Added H_2PtCl_6 and $Cr_2O_3 \cdot Al_2O_3$ into a round flask, stirring 12h, also dried at 80°C for 8h, calcined at 550 °C for 4 h. The steps of loading 0.6 wt.% Sn and (0.2wt.%, 0.4wt.%, 0.6wt.%, 0.8wt.% and 1.0wt.%) Mg was the same as above steps.

2.2. Catalytic evaluation
Propane dehydrogenation of the catalyst (mass 0.3g) was carried out in quartz tube fixed-bed reactor. The reaction conditions: the fed gas mixture was N_2:H_2:C_3H_8=35:8:7(molar ratio); the propane weight hourly space velocity (WHSV) is 3.0 h^{-1}; the reaction temperature is 590°C; the data was directly analysed by N2000 chromatography data workstation of Zhejiang University through GC-7800(FID, Al_2O_3 packed column). The following formulas were used to measure the propane conversion and propene selectivity.

Content of propane in feedstock = (C2 in product × 2/3+C1 in product × 1/3+ propane in product + propene in product)

Propane Conversion = (content of propane in feedstock- content of propane in product) × (content of propane in feedstock)$^{-1}$ × 100%

Propene Selectivity = content of propene in feedstock × (content of propane in feedstock – content of propane in product)$^{-1}$ × 100%

2.3. Catalyst characterization
Wide-angle XRD patterns were measured by a powder X-ray diffractometer (Rigaku D/max-2500) using Cu Kα radiation, the anode was operated at 40 kV and 100mA, the scanning rate is 8°/min from 3° to 100 °. Micromeritics' ASAP 2020/Tristar 3000 was used to test the specific surface area, pore size and distribution. TG was in the charge of NETZSCH's TG209 to determine the amount of coke deposition the catalyst after the reaction, the sample was heated from room temperature to 1000°C at a rate of 10°C/min in air atmosphere. XPS was recorded by Kratos Analytical' Axis Ultra DLD, the test condition: X-ray source is Al Kα (hv= 1486.6 eV), using C1s line(284.6 eV) as the internal standard to calibrated the binding energies.

3. Results and discussion

3.1. Propane dehydrogenation catalytic performances
Figure 1. shows the catalytic performances of different Mg loading catalysts, z is the standard catalysts (initial propane conversion was 36.69%, final conversion was 28.94%; initial propene selectivity was 94.09%, final selectivity was 96.35%). After loading Mg, the propane conversion increased obviously, no matter the initial conversion or the final conversion. Especially catalyst C, it has the highest conversion, the initial conversion improved from 36.69% to 44.97%, the final conversion was 5 point higher than z. The amount of Mg loading is not the higher the better, the 0.6wt.% is the top point, loading more Mg will reduce the propane conversion. For propene selectivity of the catalysts, they were all nearly or higher than 90% in the whole reaction.

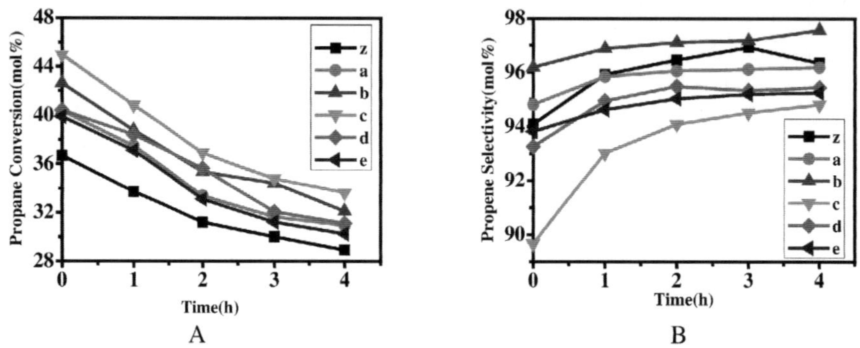

Figure 1. (A) Propane Conversion and (B) Propene Selectivity of the catalysts
z: Pt(0.4%)-Sn(0.6%)/$Cr_2O_3 \cdot Al_2O_3$ a: Pt-Sn-Mg(0.2%)/$Cr_2O_3 \cdot Al_2O_3$
b: Pt-Sn-Mg(0.4%)/$Cr_2O_3 \cdot Al_2O_3$ c: Pt-Sn-Mg(0.6%)/$Cr_2O_3 \cdot Al_2O_3$
d: Pt-Sn-Mg(0.8%)/$Cr_2O_3 \cdot Al_2O_3$ e: Pt-Sn-Mg(1.0%)/$Cr_2O_3 \cdot Al_2O_3$

3.2. Catalyst characterization

3.2.1. XRD results Several characteristic peak positions of Cr_2O_3 (24.7°, 33.9°, 36.5°, 55.4°, PDF Number: 84-0315) and Al_2O_3 (25.6°, 35.2°, 37.8°, 43.4°, PDF Number: 75-1863) are given in Figure 2. (A), which proved that the synthesized compound is composite oxides: $Cr_2O_3 \cdot Al_2O_3$. Due to the low loading amount of Pt, Sn and Mg and highly dispersed on the surface of the carrier, it was difficult to find their own characteristic peak as shown in Figure 2. (B).

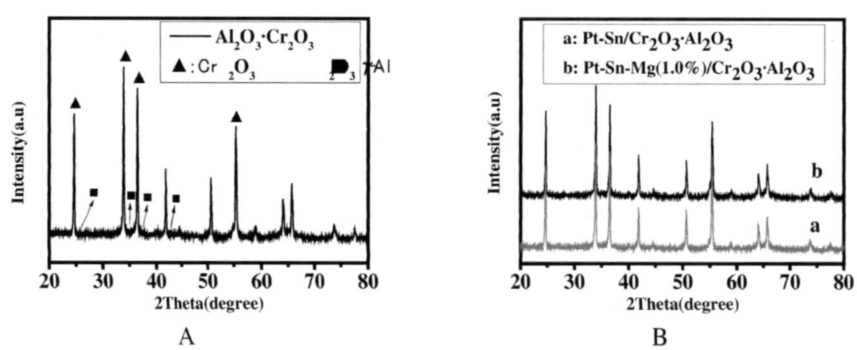

Figure 2. XRD spectrums of the catalysts

3.2.2. BET results and N2 adsorption-desorption results. From Table 1. it can be seen that the BET Surface area of the carrier $Cr_2O_3 \cdot Al_2O_3$ was up to 123.82 m^2/g and the catalyst PtSnMg/$Cr_2O_3 \cdot Al_2O_3$ was 100.52 m^2/g still higher than 100 m^2/g, which means more active Pt sites would be exposed in the reaction. Figure 3. (A) the N_2 adsorption-desorption isotherms exhibits the type IV isotherm hysteresis loops of type H1, indicated the catalyst was a kind of mesoporous material with a narrow pore size distribution. After loading Mg, a small part was into the holes and large part was on the surface so the pore diameter was rise slightly from 5.65nm to 5.77nm.

IOP Conf. Series: Materials Science and Engineering **167** (2017) 012053 doi:10.1088/1757-899X/167/1/012053

Table 1. BET of the catalysts

Catalyst	Surface area（m^2/g）
$Cr_2O_3 \cdot Al_2O_3$	123.82
Pt-Sn-Mg(0.6%)/$Cr_2O_3 \cdot Al_2O_3$	100.52

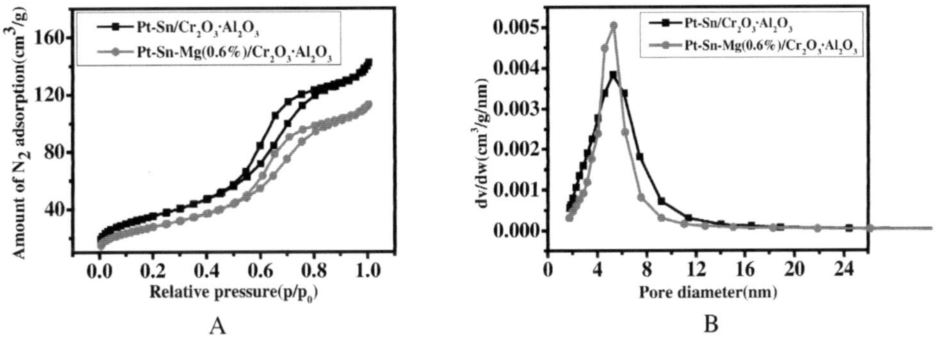

Figure 3. N_2 adsorption-desorption isotherms (A) and Pore size distribution curve of catalysts (B)

3.2.3. TG results Coke is one of the reasons that make the catalysts deactivation. TG was used to analysis the coke. As shown in Table 2, compared to catalyst PtSn/$Cr_2O_3 \cdot Al_2O_3$, after loading Mg, the coke of PtSnMg/$Cr_2O_3 \cdot Al_2O_3$ reduced clearly, from 2.89wt.% to 1.40wt.%, indicated Mg can improve the ability of anti-coke deposition.

Table 2. Coke of the catalysts

Catalysts	Coke(wt.%)
Pt-Sn/$Cr_2O_3 \cdot Al_2O_3$	2.89
Pt-Sn-Mg(0.6%)/$Cr_2O_3 \cdot Al_2O_3$	1.40

3.2.4. XPS results Figure 4. shows the XPS corresponding to the Sn3$d_{5/2}$ binding energy of the catalysts[11]. A slight move to the high binding energy can be seen from PtSn/$Cr_2O_3 \cdot Al_2O_3$ to PtSnMg/$Cr_2O_3 \cdot Al_2O_3$, it indicated the alloying degree between Pt and Sn was changed, Sn could keep at a high valence: Sn^{2+}, Sn^{4+}. That means there would be more active catalytic sites.

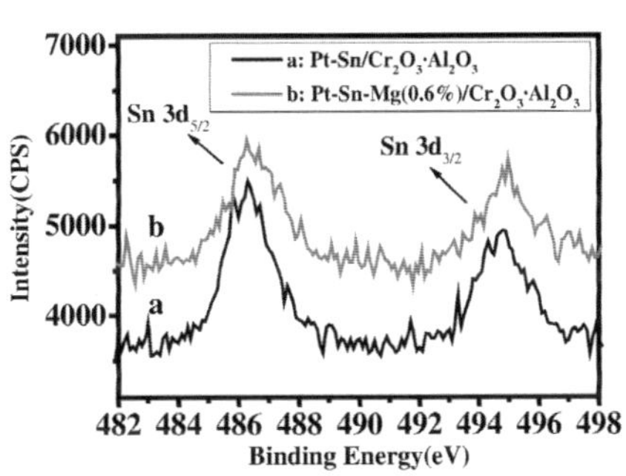

Figure 4. XPS spectra of the catalysts

4. Conclusion

A series of PtSnMg/$Cr_2O_3 \cdot Al_2O_3$ catalysts with different Mg loading amounts were prepared by chemical co-deposition method and the catalysts showed a prominent activity for propane dehydrogenation, proving Mg could improve the catalytic ability. With the increasing of Mg, the propane conversion was rising to the top at first and then fall down. The catalyst PtSnMg(0.6wt.%)/$Cr_2O_3 \cdot Al_2O_3$ had the highest propane conversion: initial conversion was 44.97%, final conversion was 33.65% and propene selectivity was above 90% for the time-on-stream under study. These catalysts exhibited some promising results and could pave the way for commercial dehydrogenation catalysts with improved performance and durability.

Acknowledgements

This research was supported by National Natural Science Foundation of China (21271107).

References

[1] Lobera M P, Tellez C, Herguido J and Menendez M 2008 Transient kinetic modelling of propane dehydrogenation over a Pt–Sn–K/Al_2O_3 catalyst *Applied Catalysis A: General* **349** 156-64.

[2] Aitani A M 2004 Advances in propylene production routes *Erdöl, Erdgas, Kohle* **120** 36-9.

[3] ZINGER S 2005 On-purpose propylene production *Petroleum technology quarterly* **10** 125-35.

[4] Mentasty L R, Gorriz O F and Cadus L E 1999 Chromium oxide supported on different Al2O3 supports: catalytic propane dehydrogenation *Industrial & engineering chemistry research* **38** 396-404.

[5] Llorca J, Homs N, Leon J, Sales J, Fierro J L G and De La Piscina P R 1999 Supported Pt–Sn catalysts highly selective for isobutane dehydrogenation: preparation, characterization and catalytic behavior *Applied Catalysis A: General* **189** 77-86.

[6] Fan X, Li J, Zhao Z, Wei Y, Liu J, Duan A and Jiang G 2015 Dehydrogenation of propane over PtSnAl/SBA-15 catalysts: Al addition effect and coke formation analysis *Catalysis Science & Technology* **5** 339-50.

[7] Sattler J J H B, Beale A M and Weckhuysen B M 2013 Operando Raman spectroscopy study on the deactivation of Pt/Al_2O_3 and Pt–Sn/Al_2O_3 propane dehydrogenation catalysts *Physical Chemistry Chemical Physics* **15** 12095-103.

[8] Gascón J, Téllez C, Herguido J and Menéndez M 2003 Propane dehydrogenation over a Cr_2O_3/Al_2O_3 catalyst: transient kinetic modeling of propene and coke formation *Applied Catalysis A: General* **248** 105-16.

[9] Bariås O A, Holmen A and Blekkan E A 1996 Propane dehydrogenation over supported Pt and Pt–Sn catalysts: catalyst preparation, characterization, and activity measurements *Journal of Catalysis* **158** 1-12.

[10] Bai L, Zhou Y, Zhang Y, Liu H and Sheng X 2009 Effect of magnesium addition to PtSnNa/ZSM-5 on the catalytic properties in the dehydrogenation of propane *Industrial & Engineering Chemistry Research* **48** 9885-91.

[11] Morgan W E and Van Wazer J R 1973 Binding energy shifts in the x-ray photoelectron spectra of a series of related Group IVa compounds *The Journal of Physical Chemistry* **77** 964-9.

Aging life evaluation of a new carbon fiber composite core wire

Jiangming Cao, Zhihua Wang[1], Yongbo Zhang and Huimin Fu

School of Aeronautic Science and Engineering, Beihang University, Beijing 100191, China

E-mail: wangzhihua@buaa.edu.cn

Abstract. The carbon fiber composite core wire is the key component for new overhead transmission lines in the power industry. As the primary load-bearing member, the composite mandrel will produce long-term creep because of the tensile during operation. Therefore, designers need to understand the long-term deformation behaviour of the composite mandrel to assess its durability. To this end, an accelerated creep test on the composite core rod under different temperatures and stresses has been conducted in the current study. According to the time- temperature- stress superposition principle, the shift factors are calculated, and the creep under normal working condition is predicted. Results show that the carbon fiber composite core wire can satisfy the design requirements under normal working condition.

1. Introduction

The carbon fiber composite core wire is made of a highly conductive soft aluminum stranding in carbon fiber composite core. The new wire is a kind of energy saving and environmental protection new wire because of its high strength, light weight, low sag and other significant characteristics, and it has extensive application prospect in high voltage overhead conductors [1-5]. As the key load- bearing component of the wire, the carbon fiber composite core has marked advantages in short-term mechanical properties and thermal properties. During operation, the wire's weight and wind vibration will act as load putting on the composite core. This generates creep behavior for long-term of power transmission, and the process will also be affected by environment factors such as wet, heat etc [6-7].

With the rapid development of rubber and plastic industry, researchers have gradually realized the importance of the study on the viscoelasticity of substrate material. The main feature of the mechanical behavior of viscoelastic materials is time dependence. The essence is the internal clock of the material. Consequently, the analysis and the prediction of long-term mechanical behavior from short-term experiment are possible. In the literature, the time- temperature- stress superposition principle has been applied to forecasting the long-term mechanical properties of polymers by considering the effects of time, temperature and stress. The results have shown the effectiveness.

Creep of resin matrix composites is a kind of macroscopic deformation, and the creep mechanism of the composite materials is very complicated [8-13]. According to the study of Johnson [8], the creep of composite materials is caused by the viscoelasticity of the substrate material. Then the viscoelasticity and performance prediction can be connected by the time-temperature superposition principle, which states that the same mechanical behavior of polymer can be observed not only under an acute higher temperature, but also under a lower temperature condition for a much longer term. This principle has been widely adopted in corresponding polymer researches since the Ferry's pioneering work [9]. Ferry and his colleagues proposed the Williams-Landel-Ferry (WLF) formula for shift factors to apply the time-temperature superposition principle to the study of the long-term properties of polymers.

Akinay [14] studied the creep of PLC by applying the time - stress equivalence principle. Firstly, the creep curves under 9 stress conditions were obtained by means of the time-stress superposition principle. Then, the parameters of the curve model and the shift factors were calculated. Finally, he studied how to get more accurate results with less data (2 groups and 3 groups of experimental data), and concluded that the results became more accurate when the interval between different stress levels increased. Jazouli [15] conducted an accelerated creep test on polycarbonate and applied the time-stress superposition principle to dealing with the experimental results. Brostow [16-17] proposed a new formula based on free volume and the chain relaxation capability model to calculate the temperature shift factors. In his study, a stress relaxation test involving four materials was conducted, and the characteristic parameters of the shift factor formula were obtained. Then, he applied the new formula to fitting the shift factors, and got a result with a higher precision.

Furthermore, a comprehensive time-temperature-stress superposition principle, which combines the previously mentioned time-temperature and time-stress superposition principles, has been proved effective in viscoelastic behavior analysis [18-19]. Consequently, the principle is adopted in this paper to study the long-term creep behavior of the carbon fiber composite core wire. And an accelerated creep test has been conducted on its key load-bearing component (the carbon fiber composite core) via the fatigue testing machine. Thereafter, a creep curve under normal conditions which is of engineering concerns is determined. The evaluation of aging and long-term creep properties for the carbon fiber composite core is also presented.

The remainder of the paper is organized as follows. The experiments including the material selection and the experimental methods are introduced in Section 2. The long-term creep property analysis and evaluation process are discussed in Section 3. A two-step method is presented to treat the accelerated test (short-term) data by shifting the creep curves for two times. First, 25 ℃ is set as a reference temperature state. The creep strain-logarithmic time curves under other temperatures are shifted according to the reference curve, and the time-temperature shift factors are calculated. Second, 500 MPa is supposed as a reference stress state. The creep strain-logarithmic time curves under other stresses are shifted according to the reference curve, and the time-stress shift factors are calculated. Then the main creep curve of the carbon fiber composite core is obtained. Finally, the aging life and long-term creep properties are evaluated and predicted. Then a summary and conclusion is given in Section 4.

2. Experiments

2.1. *Material selection*

As the key load- bearing component for the carbon fiber composite core wire, creep performance of the carbon fiber composite core is studied through a comprehensive test. Carbon fiber composite core rods have been produced as test sample, as shown in figure 1. Its glass transition temperature (T_g) is 196 ℃. The diameter of the rod for testing is 8.1 mm and the length is 450 mm.

2.2. *Characterization methods*

An accelerated tensile creep test considering higher temperature and higher load conditions has been conducted on an Instron fatigue test machine to test the tensile creep properties for the carbon fiber composite core rod. According to the time-temperature-stress superposition principle, the short-term test data in accelerated conditions can be utilized to analyze and predict the long-term behavior in normal condition. The severe temperatures and stresses are chosen based on the principle of both maintaining the same physical failure mechanism and obtaining more life information in a limited test interval.

For the carbon fiber composite core wire, the normal operating condition is given as 25℃ and 500 MPa. Through testing, the intensity of the carbon fiber composite core is 2000Mpa, and as we know that its physical failure mechanism can be maintained at any temperature lower than its glass transition temperature (T_g). Consequently, the accelerated stress levels are preset as 1800 MPa and 1000 MPa, and the higher temperatures are taken as 160 ℃ and 140 ℃. The normal operating condition of 25℃ and 500 MPa are also considered as a reference in the test. Thereafter, the fatigue machine is set to conduct a 24-hour creep testing. Then creep strain data sets are collected under the preset temperatures and stress levels, as shown in Figure 2.

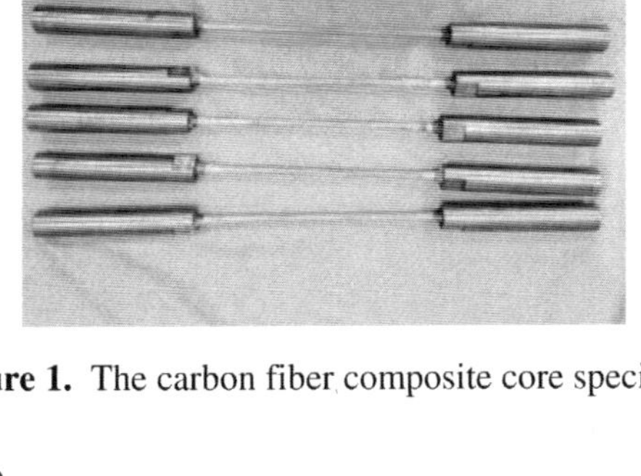

Figure 1. The carbon fiber composite core specimen.

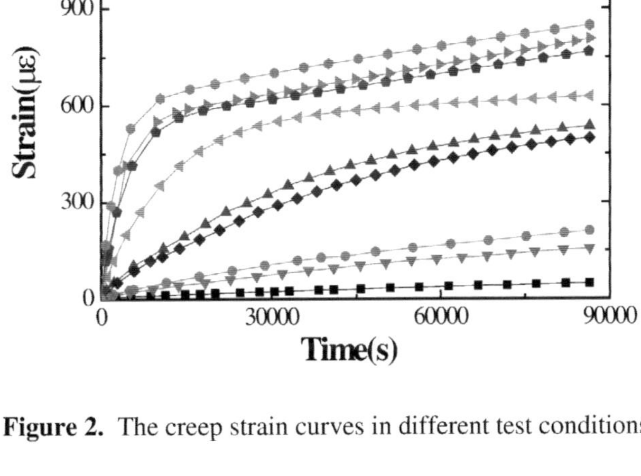

Figure 2. The creep strain curves in different test conditions.

3. Results and discussion

Creep process can generally be divided into three stages: 1) the attenuation creep, where the strain rate decreases with the increase of time; 2) the steady creep, where the strain rate can almost be assumed as a constant value; 3) the accelerate creep, where the strain rate increases gradually over time, and finally results in creep rupture. Since the test in the current study regards short-term creep, only the first two stages can be observed.

3.1. The influence of temperature

The time-temperature superposition principle establishes an equivalent relationship between temperature and the effective time of creep response by a conversion factor. According to its basic theory, mechanical properties of viscoelastic materials are equivalent under a high temperature for a short period of time and a low temperature for a long time span. That is, the increase of temperature is equivalence to the decrease of the time scale. This kind of equivalence can be accomplished by translating logarithmic coordinate axis. Then creep compliance function[10] can be represented as:

$$J(T,t) = J\left(T/_{\phi_T^\sigma}, t_0\right) \tag{1}$$

where ϕ_T^σ is the time-temperature shift factor at a fixed stress. The creep- log time curves for different conditions are obtained and shown in Figure 3(a).

In the current study, the reference temperature state is taken as 25 ℃, the creep strain - log time curves under other temperatures shown in figure 3(a) are shifted to the reference curve, and the shift distance is the corresponding temperature shift factor $\log\phi_T^\sigma$. The obtained creep strain - log time curves are shown in figure 3(b), and the shift factors are listed in Table 1.

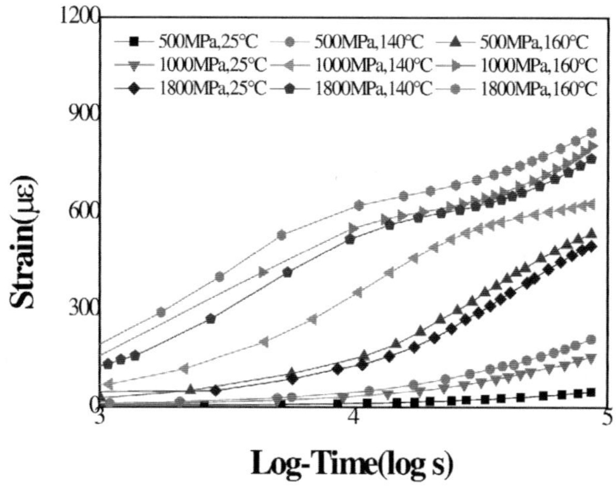

(a)

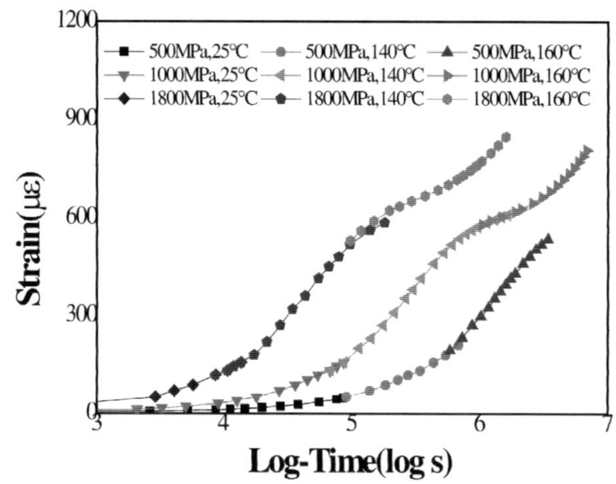

(b)

Figure3. The creep strain- log time curves of composite rods at different temperatures(a) and the superposed creep curves at T_{ref}=25℃(b).

Table 1. the time-temperature shift factors under different stresses.

Temperature (℃)	Stress（MPa） 500	1000	1800
25	Ref.	Ref.	Ref.
140	-0.18	-1.7	-1
160	-1.58	-2.1	-1.04

Relative to the reference temperature of 25 ℃, the temperature shift factors under 160 ℃ are $\log\phi_T^{(500)} = -1.58$, $\log\phi_T^{(1000)} = -2.1$, $\log\phi_T^{(1800)} = -1.04$, respectively. Since the temperature shift factor differs from stresses, superscripts for shift factors are used to represent the stress conditions. It can be seen that the equivalent test time span shown by the master curve increases significantly compared to the initial 24 hour accelerated test time. According to the time-temperature superposition principle, the 24 hour creep data under 160 ℃ contains almost 4 year creep information under 25 ℃. Then the time-temperature superposition principle provides an accelerated characterization method for long-term mechanical properties of materials.

3.2. The influence of stress

Similar with temperature, stress also has an impact on material internal clock. Mechanical properties of viscoelastic materials under high stress in a short time period are equivalent to that under low stress in a long time span. According to the time-stress superposition principle, the long-term performance under low stress can be accessed from short-term experimental data by calculating the shift factor ϕ_σ.

The reference stress state is taken as 500 MPa. The creep strain - log time curves shown in figure 3(b) can be shifted to the reference curve, and the shift distance is the corresponding stress shift factor $\log\phi_\sigma$. The creep strain - log time curves before and after shifting are shown in Figure 4, and the shift factors are summarized in Table 2.

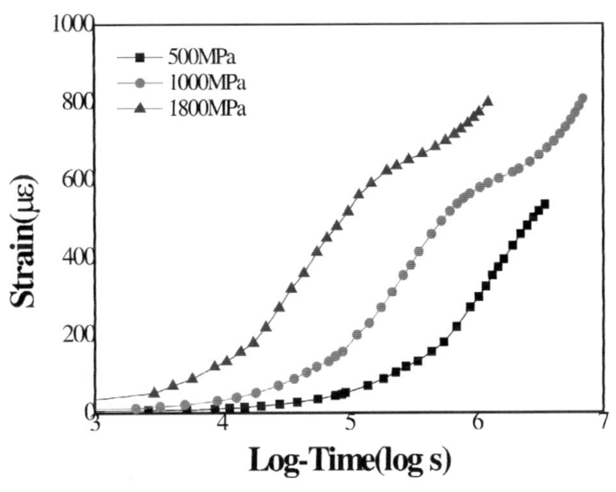

(a)

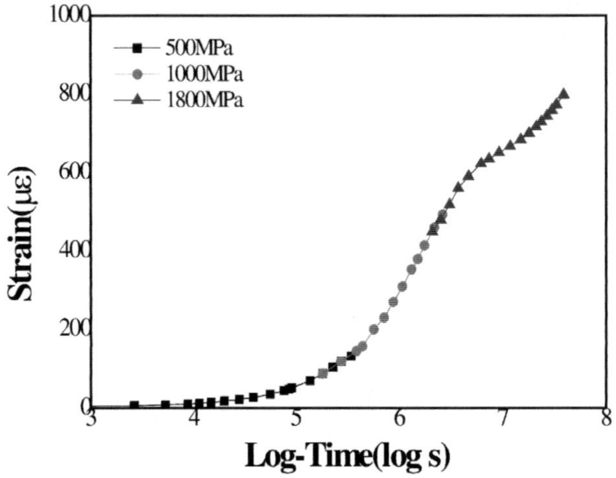

(b)

Figure 4. The creep strain- log time curves of composite rods under different stresses(a) and the superposed creep curves under σ_{ref}=500MPa(b).

Table 2. the time-stress shift factors at reference temperature.

Stress（MPa）	Time-stress shift factor
500	Ref.
1000	-0.78
1800	-1.52

3.3. Time - temperature - stress superposition principle

Time - temperature - stress superposition principle concludes that level of stress and temperature have an effect on the material free volume. An assumption can be given as:

$$\eta(T,\sigma)=\eta(T_0,\sigma_0)\phi_{T\sigma} \tag{2}$$

Then according to Doolittle equation, one can get

$$\log\phi_{T\sigma}=\frac{-B}{2.303f_0}\left[\frac{\alpha_T(T-T_0)+\alpha_\sigma(\sigma-\sigma_0)}{f_0+\alpha_T(T-T_0)+\alpha_\sigma(\sigma-\sigma_0)}\right]=-C_1\left[\frac{C_3(T-T_0)+C_2(\sigma-\sigma_0)}{C_2C_3+C_3(T-T_0)+C_2(\sigma-\sigma_0)}\right] \tag{3}$$

Supposing there are stress shift factor under constant temperature ϕ_σ^T and temperature shift factor under constant stress ϕ_T^σ, which can hold:

$$\eta(T,\sigma)=\eta(T,\sigma_0)\phi_\sigma^T=\eta(T_0,\sigma_0)\phi_T^{\sigma_0}\phi_\sigma^T=\eta(T_0,\sigma)\phi_T^\sigma=\eta(T_0,\sigma_0)\phi_\sigma^{T_0}\phi_T^\sigma \tag{4}$$

Then,

$$\log\phi_\sigma^T=-C_1\frac{C_2}{C_2+(T-T_0)}\bullet\left[\frac{C_2(\sigma-\sigma_0)}{C_2C_3+C_3(T-T_0)+C_2(\sigma-\sigma_0)}\right] \tag{5}$$

$$\log\phi_T^\sigma=-C_1\frac{C_3}{C_3+(\sigma-\sigma_0)}\bullet\left[\frac{C_3(T-T_0)}{C_2C_3+C_3(T-T_0)+C_2(\sigma-\sigma_0)}\right] \tag{6}$$

The following relationship can be given according to Eqs. (5)-(6).

$$\phi_{T\sigma} = \phi_T^{\sigma_0} \phi_\sigma^T = \phi_\sigma^{T_0} \phi_T^\sigma \tag{7}$$

Then from the results in Table 1 and Table 2, the temperature stress joint shift factors for the reference temperature T_0 and stress σ_0 are shown in Table 3. And the following results for the unknown parameters can be determined according to Equation. (3).

$$\begin{cases} C_1 = -0.083 \\ C_2 = -162.83 \\ C_3 = -4.97 \times 10^3 \end{cases} \tag{8}$$

Table 3. the temperature-stress joint shift factors.

Temperature (℃)	Stress（MPa）		
	500	1000	1800
25	Ref.	-0.78	-1.52
140	-0.18	-2.48	-2.52
160	-1.58	-3.62	-2.56

There are two ways to accomplish the shift of material's mechanical properties from a certain condition of temperature T and stress σ to the reference condition of temperature T_0 and stress σ_0. One is a one-step method, which can be realized by a temperature stress joint shift factor. The other one is a two-step approach, which is accomplished by a stress shift factor under constant temperature and a temperature shift factor under constant stress.

3.4. Long-term creep prediction

The carbon fiber composite core wire is designed to serve 30 years. During this long time span, various external forces and environmental factors will influence the creep process. Therefore, it is necessary to explore the long-term creep performance. The master creep strain curve obtained based on the previous procedure is adopted to evaluate the long-term creep behaviour.

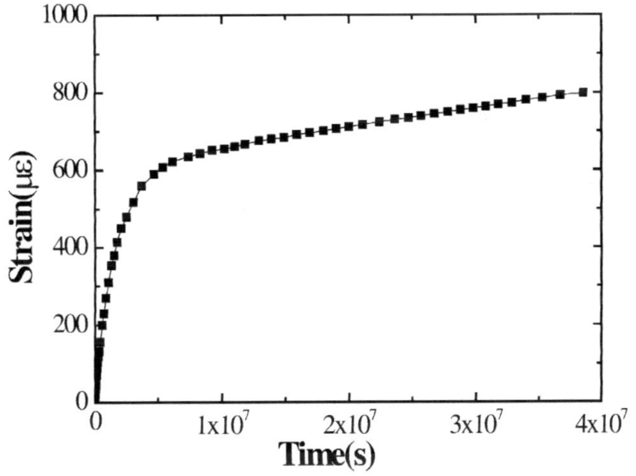

Figure 5. The creep strain-time curves of the composite rods.

The creep strain-time curve under 25 ℃ and 500 MPa is shown in Figure 5. The improved Findley model is adopted for fitting.

$$\varepsilon = a_1 + a_2(1 - e^{-\frac{t}{a_3}}) + a_4 t \tag{9}$$

The results of the parameters are as follows:

$$\begin{cases} a_1 = 9.62 \\ a_2 = 584.93 \\ a_3 = 1.03 \times 10^7 \\ a_4 = 3.58 \times 10^{-6} \end{cases} \tag{10}$$

According to experimental results, creep rupture will appear when the creep strain develops to 20000με for the carbon fiber composite core. From the creep strain-time curve and the corresponding fitting function in Equation (1) under normal condition in Figure 5, it can be known that the creep strain of the material will reach 20000με in about 175 years.

As we know, the dispersion has to be considered in engineering for materials and components. In the current study, the dispersion is considered via a safety factor for the carbon fiber composite core, which is predefined as 1.5. Then the extreme strain changes to 14000με. Therefore, the service life of the carbon fiber composite core is about 117 years for the normal condition.

In addition, the long-term creep behaviour and service life can also be assessed in other conditions with different temperatures and stress levels. Results for several conditions are analysed and summarized in Table 4. Accordingly, it can be concluded that the carbon fiber composite core wire can still satisfy the 30-year design requirement under conditions more severe than normal service.

Table 4. the service life of the composite rods under different working conditions (unit: year)

Temperature (℃)	Stress level 25%	30%	35%	40%
70	109	108	107	106
150	63	59	54	48
160	47	43	36	32

4. Conclusion

- To study the long-term creep behaviour of a new carbon fiber composite core wire, an accelerated creep test is conducted on their key load-bearing component, the carbon fiber composite core. 9 data sets regarding different conditions are obtained.
- The time-temperature-stress superposition principle is adopted to analyse and predict the long-term creep behaviour. A 24-hour creep test span for accelerated condition is equivalent to a nearly 4-year period for normal service condition according to the theory.
- The relationship between creep and temperature, stress, time is determined, and the service life of the carbon fiber composite core under normal service condition and several conditions are evaluated. The results show that the service life of the carbon fiber composite core can satisfy the design requirements.

References

[1] Zhang G 2008 Application of Aluminum Conductor Composite Core in Electric Power Transmission Line *Shanxi. Elec. Pwr.* **4** 56-7

[2] Liu H, Chen X and Wang Y 2013 Research on the Resins for High Temperature Carbon Fiber Composite in Power Transmission and Distribution *Fiber Reinf. Plast. Compo.* **3** 22-5

[3] He Z, Chen X and Wang Q 2010 Research and Application Overview of ACCC Conductor in China *Elec. Pwr. Const.* **31** 90-3
[4] Ju Y, Li Q and Meng Y 2011 Comparative Research on Aluminum Conductor Composite Core and Traditional Conductor *E. Chin. Elec. Pwr.* **7** 1191-4
[5] Lan F, Chen X and Wang Y 2014 Creep Behavior Prediction for Composite Core of Novel Electric Conductor *Fiber Reinf. Plast. Compo.* **3** 41-4
[6] Qiao H, Zeng J and Du G 2007 Research on Hygrothermal Aging Properties of Cable Cores Made from Hybrid Fiber Reinforced Epoxy Composites *Fiber Reinf. Plast. Compos.* **1** 42-5
[7] Wang L, Ji Y and Wang J 2012 the Prediction of Service Life of Composites at Hydrothermal Environment *Fiber Reinf. Plast. Compos.* 43-5
[8] Hollaway L C and Leeming M 1999 *Strengthening of Reinforced Concrete Structures* (London: Woodhead) pp 183-203
[9] Ferry J D 1980 *Viscoelastic Properties of Polymers* (New York: Wiley)
[10] Brostow W 2000 *Performance of Plastics* (Munich: Hanser) p 84
[11] Akinay A E and Brostow W 2001 Long-term service performance of polymeric materials from short-term tests: prediction of the stress shift factor from a minimum of data *Polymer* **42** 4527-32
[12] Starkova O, Yang J and Zhang Z 2007 Application of time–stress superposition to nonlinear creep of polyamide 66 filled with nanoparticles of various sizes *Compo. Sci. Technol.* **67** 2691–8
[13] Akinay A E, Brostow W and Castaño V M 2002 Time-temperature correspondence prediction of stress relaxation of polymeric materials from a minimum of data *Polymer* **43** 3593-600
[14] Akinay A E and Brostow W 2001 Long-term service performance of polymeric materials from short-term tests: prediction of the stress shift factor from a minimum of data *Polymer* **42** 4527-32.
[15] Jazouli S, Luo W and Bremand F 2005 Application of time–stress equivalence to nonlinear creep of polycarbonate *Polym. Testing* **24** 463-7
[16] Burrows H, Weir R and Stohner J 2009 Reliability and prediction of long-term performance of polymer-based materials : Pure and Applied Chemistry *Int. Uni. Pure Appl. Chem.* **81** 417-32
[17] Akinay A E, Brostow W and Castaño V M 2002 Time–temperature correspondence prediction of stress relaxation of polymeric materials from a minimum of data *Polymer* **43** 3593-600
[18] Luo W, Yang T and An Q 2001 Time-temperature-stress equivalence and its application to nonlinear viscoelastic materials *Acta Mech. Sol. Sin.* **14** 195-9
[19] Yen S C and Williamson F L 1990 Accelerated characterization of creep response of an off-axis composite material *Compo. Sci. Technol.* **38** 103-18

IOP Conf. Series: Materials Science and Engineering **167** (2017) 012055 doi:10.1088/1757-899X/167/1/012055

Research on compressive strength of recycled cement mortar after high temperature

Xianggang Zhang[1,2,3], Jianhui Yang[1,2,4] and Dapeng Deng[1,2]

[1] Henan Polytechnic University, Henan Province Engineering Laboratory of Eco-architecture and the Built Environment, Jiaozuo 454000, China;
[2] Henan Polytechnic University, School of Civil Engineering, Jiaozuo 454000, China;
[3] Guangxi Key Lab of Road structure and materials, Nanning 530000, China.

E-mail: xgzhang@hpu.edu.cn

Abstract: In order to study cube compressive strength of recycled fine aggregate cement mortar after different temperatures, with the affect parameters between replacement rate of recycled fine aggregate and temperature, 45 standard cube test blocks were designed and produced to carry out compressive strength test. The failure process and failure mode of test blocks were observed. Ultimate compressive strength of cube blocks were measured, the relations between cube compressive strength and the replacement rates of recycled fine aggregate under different temperatures as well as the relations between cube compressive strength and temperatures under different replacement rates were all analyzed, the influence change parameters made on cube compressive strength was discussed. The results showed: the failure process and the failure mode of recycled fine aggregate cement mortar and the failure process and the failure mode of nature is similar; when the temperature reached 400℃, the block has no burst phenomenon, but the colour of block into a dark pink; with the increase of recycled fine aggregate, the mass lose rate of block is increase; effect different temperature make on cube compressive strength of test block is not obvious when temperature keeps same for 3h.

1. Introduction

At present, resource waste is very serious in our country, according to statistics [1], China's construction waste has occupied 30% ~ 40% of municipal waste, of which 50% ~ 60% is waste concrete. Due to the rapid development of urbanization, a large number of construction waste, produce great harm to the environment, to protect environment and save resources, realize sustainable development, the scientific use of construction waste is urgent[2-3]. Our country building sand have increased in recent years, natural resources are close to run out, continued exploitation will inevitably lead to environment, cost, security and other issues, etc[4]. Make the construction waste to be recycled, then after the program processing such as crushing, screening, cleaning, drying and part or all of the aggregate instead of natural aggregate to make recycled building materials[5-6]. After the mixture of crushing fine aggregate instead of natural fine aggregate and cement and water to make new building materials is called recycled cement mortar[7]. Construction waste recycling can reduce the consumption of natural aggregate, relieve the pressure on the per capita possession of resources, improve the economic and environmental benefit will have a positive effect[8-10].

In recent years, the studies of the recycled cement mortar performance have a certain size. Yang Z S et al[11] have studied construction waste's recycled mortar mechanical performance, the results have showed that the block strength of recycled aggregate can meet the requirement, grain composition has app on the strength. But they didn't explain the high temperature's effect on the block. Xu M et al[12] studied the stress-strain relationship of the uniaxial compression for the recycled cement mortar prism specimen(70.7mm×70.7mm×216mm) after high temperature, the results showed that different recycled fine aggregate replacement rate of mortar at high temperature after surface color changes obviously, more than 400 degrees, the quality tends to be stable. Under same temperature, with the increase of replacement ratio of recycled fine aggregate, the peak stress and elastic modulus decreases, peak strain and ultimate strain increases gradually. However, there are few reports on the standard cube test block after high temperature.

In this paper, forty five cement mortar with recycled fine aggregate of the standard cube test block were made in 3 different temperatures, the recycled cement mortar cube test block of uniaxial compressive strength test. Analysis of the same substitution rate under the condition of cube compressive strength changing with temperature, and at the same temperature as the cube compressive strength of recycled fine aggregate replace rate of change, in order to provide reference for the preparation of recycled cement mortar, and disaster control.

2. Test profiles

2.1. Test raw materials

The four components of cement, recycled fine aggregate, sand and water according to a certain proportion was made recycled fine aggregate cement mortar. The cement is 42.5 grade P.O cement of Jiaozuo strong production; reinforced concrete beams with recycled fine aggregate from the Henan Polytechnic University structure test hall has been destroyed, crushing, screening, grading, cleaning, drying and other processing procedures, the particle size in the range of 0.5~5mm, the bulk density is 1321kg/m^3, as shown in Fig. 1 (a); the sand to the natural river sand, its bulk density is 623kg/m^3, as shown in Figure 1 (b); by mixing water into ordinary tap water.

(a) Recycled fine aggregate (b) Natural river sand

Figure 1. Fine aggregate

Table 1. Mix ratio of recycled mortar

Test block type	cement (kg/m^3)	Water (kg/m^3)	Aggregate(kg/m^3)	
			Light sand	Ordinary sand
N	370.0	310.0	0.0	1321.0
RM25	370.0	310.0	155.8	990.8
RM50	370.0	310.0	311.5	660.5
RM75	370.0	310.0	467.3	330.3
RM100	370.0	310.0	623.0	0.0

2.2. Mix ratio

The standard test block of cement mortar with strength grade of M15, this experiment has produced 5 kinds of standard cube recycled fine aggregate cement mortar test block with different substitution rate. Among them, the cement mortar specimen was seem as reference block (N) when the replacement is 0%, 25%, 50%, 75%, 100% and four kinds of different substitution rate test blocks (RM25, RM50, RM75 and RM100) as contrast test block, according to JGJT 98-2010 "with mortar than the" design rules of mix design, mix design test as shown in table 1.

2.3. Test block fabrication, high temperature test and loading
45 block size made in this experiment are standard cube 70.7mm×70.7mm×70.7mm block, in 3 groups of temperature (20℃, 200℃, 400℃) conditions, each replacement rate (0%, 25%, 50%, 75%, 100%) has 3 block. The mold is used in cube steel mold standard, 24h after stripping, standard curing 28d, respectively, corresponding test block is heated to 200℃, 400℃, and keep in the constant temperature 3 hours then take out, natural cooling to room temperature, the heating device is TDL-1400F high temperature furnace, as shown in figure 2. The heating system at 200℃ and 400℃ are shown in Figure 3 and Figure 4 . According to JGJ-T70-2009 "mortar test method", the test blocks under different temperature conditions are uniformly loaded, and the compressive strength is measured. Load device is the SYE-2000 type pressure testing machine, the maximum range is 2000kN, as shown in Figure 5, the test using the loading rate is 3~5kN/s.

Figure 2. Heating device

Figure 3. Loading device

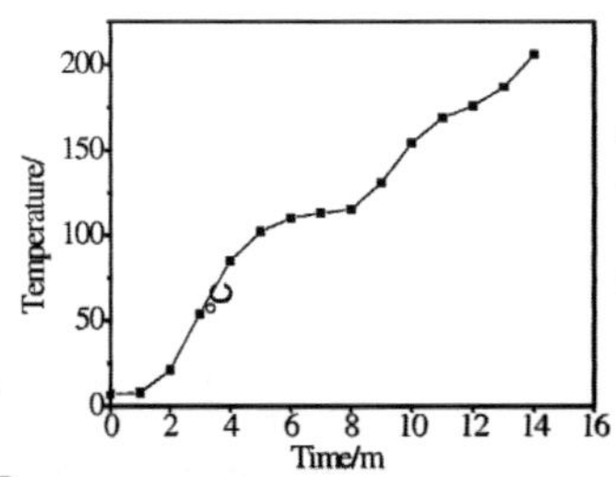

Figure 4. 200℃ heating system

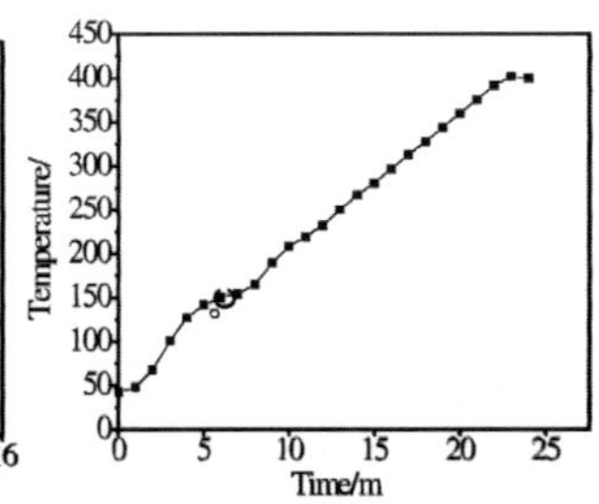

Figure 5. 400℃ heating system

3. Test results and analysis

3.1. Test block high temperature phenomenon
During the heating process, with the increase of the temperature, the moisture inside the test block has been evaporating, and the mortar test block has a great physical change in the process of high temperature. After the high temperature test, the two groups did not have a burst phenomenon. As shown in Figure 6, when the temperature reached 200 degrees, the color of the test block and the color of the test block at room temperature is basically no different. But when the temperature reached 400 degrees, the color of the test block had a significant change, was dark pink.

(a) (b)

Figure 6. Colour change after high temperature

3.2. The failure process and failure mode of the test block
At the beginning of the load, test block appearance has not changed significantly, as the load increases, the occasional faint sound of colloid cracking, with increasing load, block surface produced some vertical and slightly tilted cracks parallel to the axial force direction, when the load is close to the limit load, the micro crack develop quickly, through the formation of the main crack, four vertical free

IOP Conf. Series: Materials Science and Engineering **167** (2017) 012055 doi:10.1088/1757-899X/167/1/012055

surface began to drum, final failure. Recycled cement mortar specimens under normal temperature and ordinary cement mortar damage process and form block is basically similar to that of recycled cement mortar under different high temperature after the failure process and failure form of block are similar, showing two connected four Inverted pyramid. At different temperatures, the failure modes of all replacement rate test blocks are shown in Figure 7 (a), (b), (c). At different temperatures, the failure pattern of the 100% replacement rate test block are shown in Figure 6 (d), (e), (f).

(a) 20℃ (b) 200℃ (c) 400℃

(d) 20℃ 100% replacement rate (e) 200℃ 100% replacement rate (f) 400℃ 100% replacement rate

Figure 7. Destruction form

4. Influence factor analysis

4.1. Mass loss rate

The mass loss of the test block includes the evaporation of water, the melting of the fibre, the decomposition of calcium silicate hydrate and the decomposition of calcium carbonate[13], etc. Mass loss ratio is the ratio of the mass to the high temperature and the quality of the test block after high temperature[14]. Curve of mass loss rate η of recycled cement mortar and the temperature T are shown in Figure 8, the relation curve between the mass loss rate η and recycled fine aggregate replacement rate is r as shown in figure 9.

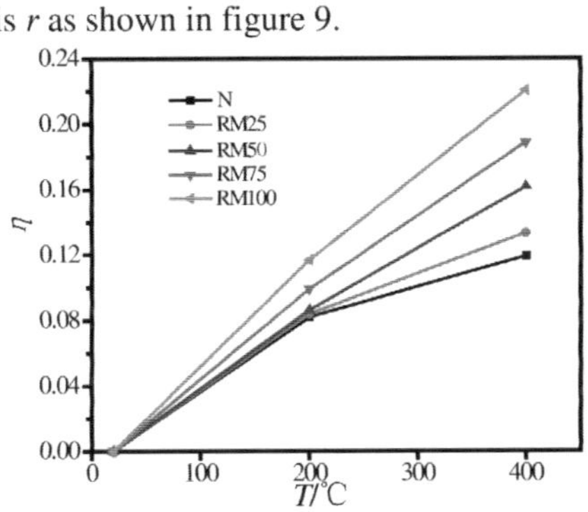

Figure 8. Mass loss rate and temperature curve

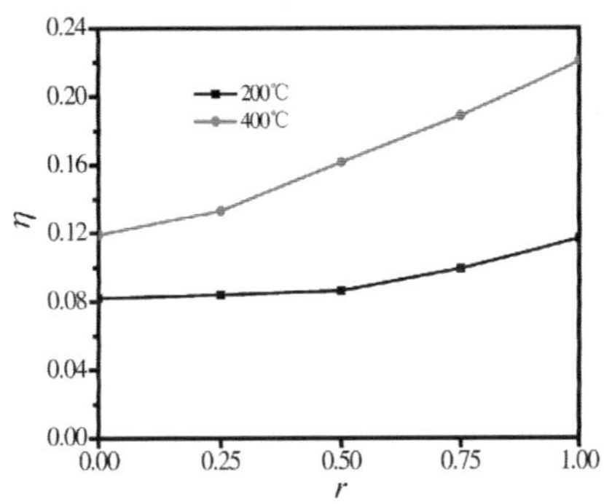

Figure 9. Mass loss rate and replacement rate curve

Figure 8 shows that with the increase of temperature, the mass loss rate of N type test block gradually increased, RM25, RM50, RM75, RM100 test block mass loss rate increased. Related to 20℃, at 200℃, N and RM25, RM50, RM75, RM100 block mass loss rate increases were 8.2%, 8.4%, 8.6%, 9.9%, 11.7%; related to 20℃, at 200℃, RM25, RM50, RM75, RM100 type specimen mass loss rate increases were 3.70%, 4.90%, 7.60%, 9%, 10.30%. The increases of mass loss rate of each replacement rate test block in the range of 20~200℃ are significantly higher than that in the range of 200~400℃. For the reason, this is because most of the water at the 20~200℃ has been evaporated, only a small part of the water evaporated at 200~400℃.

Figure 9 shows that when the replacement rates was 0%, 25%, 50%, 75%, 100%, 200 degrees corresponding block mass loss rates were 8.2%, 8.4%, 8.6%, 9.9%, 11.7%, 2.4%, increased by 2.4%, 15.1%, 18.2%; at 200℃, the corresponding block mass loss rate was 11.9% 13.3%, 16.2%, 18.9%, 22%, 21.8%, increased by 11.8%, 16.7%, 16.4%. With the increases of replacement rate of recycled fine aggregate, the mass loss rate increases gradually. The reason is that with the increase of the replacement ratio of recycled fine aggregate increased gradually, the number of test block within the pore and micro cracks increases gradually, the free water that did not participate in the hydration of cement became more, at the same temperature, free water evaporated and lost[15-16].

4.2. Cube compressive strength

According to JGJ-T70-2009 "mortar test method", the failure load of recycled cement mortar specimens with different replacement rates at different temperatures was obtained in 3 groups. Calculated and obtained the cube compressive strength of each specimen, as shown in table 2

Table 2. Cube compressive strength (Mpa)

Specimen category			Replacement rate				
			0%	25%	50%	75%	100%
Temperature	20℃	Measured strength	17.47	16.50	13.32	9.15	14.30
			17.34	11.51	12.15	14.22	15.11
			17.29	11.36	11.51	15.08	11.02
		Error	<15%	>15%	<15%	>15%	>15%
		Represent strength	17.36	11.51	12.33	14.22	14.30
		Cube compressive strength	22.57	15.00	16.03	18.49	18.59
	200℃	Measured strength	11.9	15.04	14.93	13.12	11.45
			11.6	9.79	13.06	10.02	10.08
			13.46	13.29	15.48	9.00	9.06
		Error	<15%	>15%	<15%	>15%	<15%
		Represent strength	12.32	13.29	14.49	10.02	10.20
		Cube compressive strength	16.02	17.28	18.84	13.03	13.26
	400℃	Measured strength	11.09	15.92	18.81	10.84	11.17
			18.17	16.75	16.52	12.46	13.03
			18.13	16.42	11.97	13.52	11.63
		Error	>15%	<15%	>15%	<15%	<15%
		Represent strength	18.13	16.36	16.52	12.28	11.94
		Cube compressive strength	23.57	21.27	21.48	15.96	15.52

Note: Cube compressive strength calculation refer to JGJ-T70-2009 "mortar test method". When the difference between the maximum value or minimum value of strength in group with middle one less than 15% of the median value, take the average of three; when the difference between the maximum value or minimum value of strength in group with middle one more than 15% of the median value, take the median value; when the difference between the maximum value, minimum value of strength in group with middle one more than 15% of the median value, the group data is not valid.

Table 2 shows that under each replacement rate, the cube mortar test block withstand different temperatures, the compressive strength reached the design strength. For all kinds of temperature, under different recycled fine aggregate replacement rate, the cube compressive strengths of recycled cement mortar are shown in figure 10. For various replacement rates, with the change of temperature, the cubic compressive strength of recycled cement mortar is shown in Figure 11.

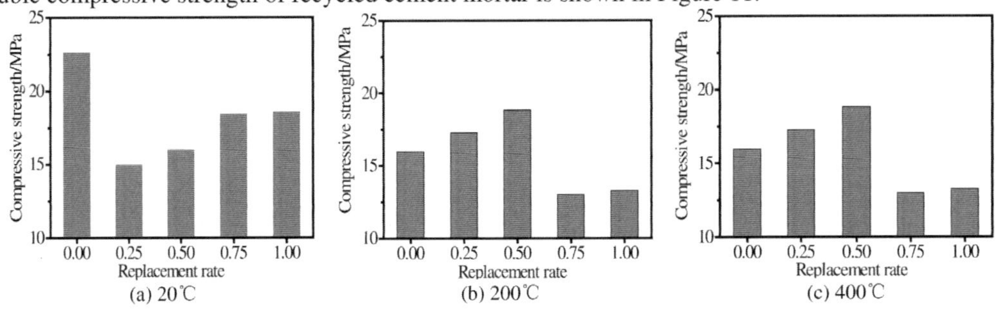

Figure 10. Cube compressive strength under different substitution rates

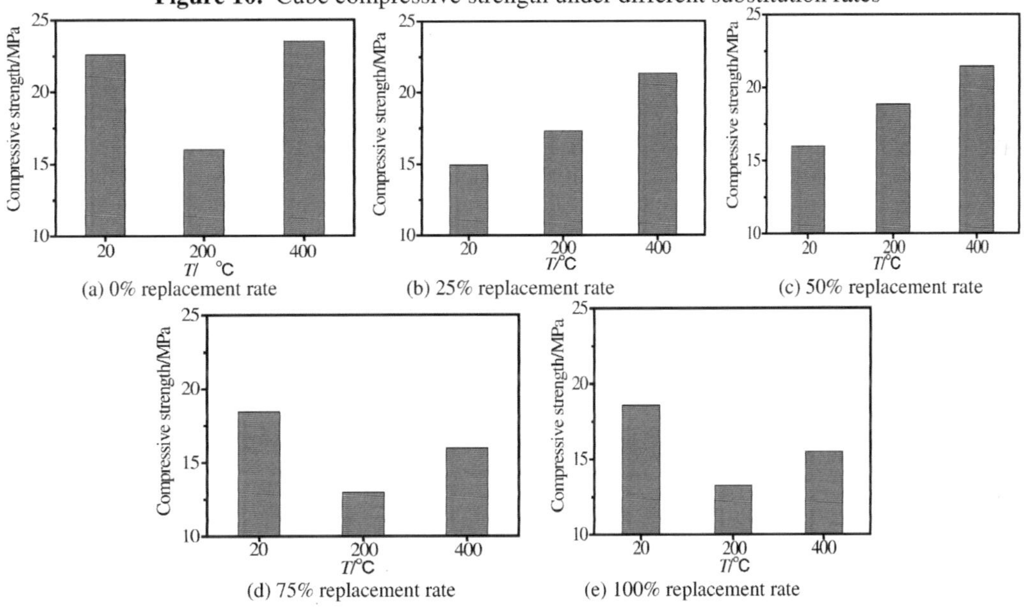

Figure 11. Cube compressive strength at different temperatures

Figure 10 shows that when the replacement rate was 0%, the compressive strength of cube test block at 20℃ and 400℃ was not significant different, and are significant higher than the recycled cement mortar cube compressive strength. But at different temperature, there is no good regularity between the compressive strength of recycled cement mortar and the replacement ratio of recycled fine aggregate. One reason is that discrete type of cement mortar; the two is increasing the recycled fine aggregate will reduce the compressive strength, big water binder ratio made the compressive strength decreased, mutual coupling, cross effect at cubic compressive strength; three is the temperature and time insufficient let the internal structure no obvious change.

Figure 11 shows that in the range of 20~200℃, when replacement ratio is 0%, 75% and 100%, with the increase of temperature, compressive strength of test block reduced, when the replacement ratio is 25% and 50%, with the increase of temperature, compressive strength of test block improved. In the range of 200~400℃, with the gradual increase of temperature, the compressive strength of recycled cement mortar test block has a certain improvement, the smallest increase is 2.26%, the maximum increase of the amplitude can reach 7.54%. In the range of 20~400℃, with the increase of temperature, the compressive strength of the test block is not obvious. One reason is that the temperature is insufficient and the constant temperature time is short, the internal structure of the test block is not destroyed; the two is due to the large discrete type of recycled cement mortar.

5. Conclusions

With increasing temperature, the change of surface features of recycled fine aggregate cement mortar block in the early stage was not obvious, but when the temperature increased to 400℃, block color became dark pink from gray, blocks did not have burst phenomenon;

Recycled cement mortar and ordinary cement mortar blocks damaging process and form are similar, showing two connected four Inverted pyramid;

With the increase of the replacement rate of recycled fine aggregate, the mass loss rate of recycled cement mortar increased. At different replacement rate, the mass loss rate of the test block at 400℃ increased to that at 200℃ in different degrees. When the replacement rate was 100%, the mass loss rate increased to 10.30%;

In the range of 20~400℃ and 3 hours under the conditions of constant temperature, with the increase of temperature, the law of compressive strength of recycled cement mortar test block is not obvious. It is suggested that further research should be done before the extension and application of recycled cement mortar. For example, the temperature range extended to 20~800℃, or the constant temperature time extended to 8 hours, or each group produced 6 or even 9 of the same parameters of the test pieces, and so on.

Acknowledgments

The authors would like to thank 2017 Science and Technology Research Projects of Henan Province, 2016 Science and Technology Project of Prevention and Control of the Key Technology on Safety Production Major Accident (henan-0006-2016AQ), 2015 Jiaozuo Science and Technology Plan Project, Key Scientific Research Project of Colleges and Universities in Henan province (15A560008), Open Topic of Guangxi Key Lab of Road structure and materials(2015gxjgclkf-004), for their generous support to the research projects.

References:

[1] Zhao J, Liu Q X and Lin L Q et al. 2013 Evolution and comparison of construction waste of large cities in China *J. Journal of Central South University(Science and Technology)*, **44(4)**: 1297-1304.

[2] Monalisa B, Bhattacharyya S K and Minocha A K, et al. 2014 Recycled aggregate from C&D waste & its use in concrete-A breakthrough towards sustainability in construction sector: A review[J]. *Construction and Building Materials*, **68**: 501-516.

[3] Liu S H and Yan P. 2010 Mechanical properties of recycled aggregate concrete *J. Engineering Journal of Wuhan University*, **43(1)**: 85-101.

[4] Zhou W J, Chen J L and Lu H B. 2009 Research on the green recycled mortar[J]. *Journal of Wuhan University of Technology* **31(7)**: 15-18.

[5] Xiao J Z. 2008 Recycled concret[M]. *Beijing:China Architecture & Building Press*,**2008**: 3-4.

[6] Zhang X G, Chen Z P and Xue J Y. 2015 Physical and mechanical performance of recycled aggregate concrete *J. Bulletin of the Chinese Ceramic Society*, **34(6)**: 1684-1688.

[7] Laneyrie C, Beaucour A L, Green M F, Hebert R L, Ledesert B and Noumowe A. 2016 Influence of recycled coarse aggregates on normal and high performance concrete subjected to elevated temperatures *J. Construction and Building Materials*, **111**: 368-378.

[8] Guelmine L, Hadjaband H and Benazzouk A. 2016 Effect of elevated temperatures on physical and mechanical properties of recycled rubber mortar *J. Construction and Building Materials*, **126**: 77-85.

[9] Sadatiand S and Khayat K H. 2016 Field performance of concrete pavement incorporating recycled concrete aggregate *J. Construction and Building Materials*, **126**: 691-700.

[10] Xu Y D, Zhou S Q and Xiao J. 2004 Experimental study of recycled concrete aggregate[J]. *Journal of Building Materials* **7(4)**: 447-450.
[11] Yang Z S, Wang A Q and Zhu Y Z. 2015 Study on the mechanical properties of construction waste recycled mortar *J. Journal of Zhongyuan University of Technology* **26(3)**: 72-75.
[12] Xu M, Wang T and Chen Z F. 2015 Experimental research on uniaxial compressive stress-strain relationship of recycled mortar after high temperature *J. Journal of Building Structures*, **36(2)**: 151-157.
[13] Li D, He R and Wang S, et al. 2015 High temperature performance of cementitious composites reinforced with PVA fibers *J. Bulletin of the Chinese Ceramic Society*, **34(6)**: 1604-1609.
[14] Zhang M X, Luo Z T and Bao W Z, et al. 2013 Effects of mixed with fly ash and slag powder on volume deformation of cement mortar under the environment of invariable temperature and variable humidity[J]. *China Concrete and Cement Products*, **3** 22-24.
[15] Zhang Y M, Qin H G and Sun W, et al. 2002 Preliminary study on the proportion design of recycled aggregate concrete *J. China Concrete and Cement Products* **1** 7-9.
[16] Chen Z P, Zhan D H and Xu J J. 2015 Research on mechanical properties of recycled concrete using different recycled coarse aggregate replacement *J. Industrial Construction* **45(1)** 130-135.

Research on Waste FRP Fiber Reinforced Adhesive Mortar

Y C Feng[1] and F Q Zhao[2,3]

[1] Hebei University of Science &Technology, Shijiazhuang 050018,PR China
[2] Hebei Engineering and Research Center of Solid Waste Utilization, Shijiazhuang050018, PR China

E-mail:zhaofq3366@126.com

Abstract. The use of FRP in industry results in large amount of waste FRP. If not treated properly, it will pollute the environment. In our study, waste FRP powder was used in mortar to substitute part of sand in mortar. The use of short waste FRP fibre can further increase the mechanical strength. The Waste FRP fibre reinforced adhesive mortar was prepared with the material proportioning: cement-sand ratio 1:2, polymer 8% (cement based), waste FRP fibre 3% (based on the total solids). The performance of the mortar conforms to JCT 547-2005 with excellent adhesive characteristics.

1. Introduction

Glass fiber reinforced plastic (GFRP) is a kind of light and high strength material, which is widely used in various fields of life. In 2013, the total production of GFRP reached 4 million tons, in which thermosetting (FRP) was 2.73 million tons and thermoplastic glass steel (FRTP) was 1.37 million tons. Waste glass fiber reinforced plastic has some characteristics such as high strength and strong corrosion resistance, which make it difficult for recycling and utilization. Recycling and utilization of waste glass fiber is becoming an important factor hindering the development of glass fiber industry [1-2]. How to deal with the wasted glass fiber has become a pressing issue in glass fiber industry in China and in the world.

The research on the recycling of waste glass fiber reinforced plastics is still in the primary stage, mostly going for incineration and landfill in practice. Both will pollute the environment, have high processing cost and cannot meet the increasing requirements of processing the wasted glass fiber [3-4]. According to the present recycling methods of GFRP waste materials, it can be divided into physical recycling, chemical recycling and energy recovery [5-6]. Among the three methods, physical recycling is the most used because of operatability and cost-effectiveness. Xiaosong et al. [7] treated waste powder GFRP material with organic surface treatment, and effectively improved the interfacial adhesion with the matrix resin to improve the mechanical properties of the prepared composites. The properties of waste powder FRP make it possible substitute of sand in mortars [8-9]. Shaojie et al. [10] studied that the core-shell emulsion polymerization styrene-acrylic polymer prepared by P (St-DVB-HEMA) can be used for adhesive mortar. Results showed that the core-shell cross-linked poly (styrene-hydroxyehyl methacrylate) could substantially improve the modified-mortar's bond strength compared to the mortar with commercial EVA as modifier.

The purpose of this paper is to formulate a waste FRP fiber reinforced adhesive mortar on the basis of selecting polymers, cement-sand ratio and different size of waste FRP fiber. Another purpose is to find a proper method to utilize waste FRP.

2. Materials and methods

2.1. Materials

Cement-sand ratio, fineness modulus of 2.5; Portland cement: P. 42.5. The physical properties are listed in Table 1. Fibers: waste glass fiber reinforced plastic fiber (WGFP), size 1~3 mm), supplied by Jizhou Hebei Machinery Equipment Technology Co., Ltd (Figure 1). Polymers: EVA powder (ethylene vinyl acetate copolymer), VAE emulsion (ethylene vinyl acetate copolymer emulsion), styrene butadiene latex, styrene acrylic emulsion, rubber latex. Water retaining agent: hydroxypropyl methyl cellulose ether.

Table 1. Mechanical properties of cement and cement mortar.

Specific surface area of cement (m2/kg)	Normal Consistency (%)	Setting time(min)		Compressive strength (MPa)		Bending strength (MPa)		Stability (boiling method)
		Initial setting	Final setting	3d	28 d	3d	28 d	
370	26.2	164	252	32.4	57.7	6.7	8.28	Qualified

Figure 1.WGFP fiber with different sizes (by mechanical crushing).

2.2. Methods

For preparation and testing of the samples.Portland cement, sand, Waste FRP fiber, re-dispersible emulsion and additives are mixed to form adhesive mortar, which is then mixed with water to make the test specimens (40 mm × 40 mm × 10 mm). The specimens are cured in the mold for 24 h, then keep in 23 (± 2) °C temperature & 50 (± 5) % relative humidity (RH) for the 27 days. The adhesive strength is tested according to JCT 547-2005 (China professional standard, adhesives for ceramic wall and floor tile). All the values collected were the average of six separate tests. SEM test was done using HITACHI S-570 Scanning Electron Microscope.

3. Results and discussion

3.1. Orthogonal experiment

On the basis of initial tests, VAE emulsion was selected for modification of mortar with the proper content of 8 % (based on the cement mass). The water retaining agent used is hydro-xypropyl methyl cellulose ether with the content 0.2% (based on the total solids). Three factors and three levels, the amount of cement sand ratio (A), polymer content (B) and waste FRP fiber (C) were selected for orthogonal experiment to determine the optimal proportioning of mortar. See Table 2.

Table 2. Factors and levels.

Factor	A（Cement-sand ratio）	B（Polymer content）	C（Short fiber content）
Level 1	1：2	4%	1%
Level 2	1:2.5	6%	2%
Level 3	1:3	8%	3%

3.2. Result analysis

The response data collected from the experiments were listed in Table 3.

Table 3. Range analysis.

Number	1	2	3	Bonding strength(MPa)
	A	B	C	
1	1	1	1	1.09
2	1	2	2	1.04
3	1	3	3	1.15
4	2	1	2	0.98
5	2	2	3	1.01
6	2	3	1	1.05
7	3	1	3	0.9
8	3	2	1	0.88
9	3	3	2	0.96
Range	0.180	0.076	0.023	

With the bonding strength of the samples as objective, the effects of VAE emulsion, cement-sand ratio and waste FRP fiber contents were studied. The range analysis of the results was performed. The influence order of the factor is: A>B>C. Among the three factors, cement-sand ratio shows the most remarkable impact.

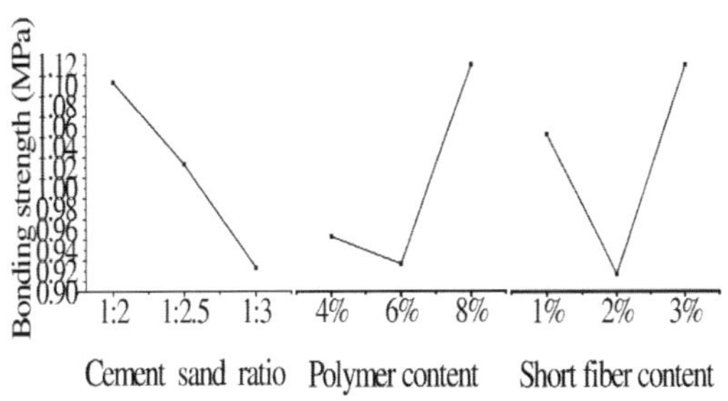

Figure 2. Factor-Effect diagram.

Figure 2 gives factor-effect diagram of the experiment. With bonding strength value of the samples as objective, then the optimal combination of adhesive mortar is A1B3C3, that is to say, when the cement-sand ratio is 1: 2, the polymer is 8% (cement based), and waste FRP fiber is 3% (based on the total solids).

3.3. Verification test

Based on the orthogonal experiment, the obtained optimal combination of factors and levels are A1B3C3. The verification test was performed. The bonding strength values of the mortar with the optimal combination is 1.19MPa. The product conforms to JCT 547-2005.

3.4. Microstructure of waste FRP fiber reinforced adhesive mortar

Figure 3 (a) and (b) show the SEM photos of the waste FRP fiber reinforced adhesive mortar (with and without polymer modification). Just as reinforced concrete, the needle-like fiber strengthens the structure. Cement paste wrapped around, and uniformly dispersed in the mortar and matrix interface, which increases the bonding strength. The polymer further improves the bonding of cement mortar as well as bonding with latex. In this case, cement paste and aggregate is connected closely, so that mortar adhesive properties are improved. It also shows that the adhesive properties of modified mortars are greatly influenced by VAE latex. Waste FRP fiber are evenly scattered in cement composite system so that adhesive strength of hardened cement composites is significantly improved. At the same time, the stress concentration on crack tip is effective alleviated, which offers great help to the performance of adhesive strength.

(a) Mortar with polymer emulsion (b) Mortar without polymer emulsion

Figure 3. SEM photos of FRP fiber reinforced adhesive mortar.

4. Conclusions

- Waste FRP fiber can be used in functional mortar with enhanced adhesion properties. The results show that the moderate content of fiber powder can greatly improve the adhesive strength.
- Different kinds of polymer were used to modify mortar; VAE emulsion exhibits excellent properties when used in the formulation. When the content of VAE emulsion is 8 %, the bonding strength is better.
- Based on three factors & three levels orthogonal experiment, the Waste GFRP fiber reinforced adhesive mortar was prepared with the material proportioning: cement-sand ratio 1:2, polymer 8% (cement based), waste FRP fiber 3% (based on the total solids). The performance of the mortar conforms to JCT 547-2005 with excellent adhesive characteristics.
- The mechanism of adhesive strength improvement of waste FRP fiber reinforced mortar was investigated by using SEM. Waste FRP fiber, and VAE emulsion penetrated each other, forming dense system, which increased the bonding strength of the mortar.

References

[1] Duan Z J, Duan W C and Zhang R Q 2011 *China Plastics Industry.* **39** 14-8
[2] Liu J, Liu L F and Yu J Y 2011 *Industrial Textiles.* **29** 26-8
[3] Ribeiro M C S, Meira-Castro A C, Silva F G, Santos J, Meixedo J P, Fiúza A , Dinis M L and Alvim M R 2013*Resour. Conserv. Recy.* **29** 670-4
[4] Mazzoli A and Moriconi G 2014 *Micron*,**67** 169-178
[5] Ma H X 2010 *Thermosetting Resin.* **25** 38-42
[6] Chen X S and Hou W S 2012 *Thermosetting Resin.* **27** 75-9
[7] Chen X S, Zhang Z M, Li S S and Wang Z F 2012 *Thermosetting Resin.* **27** 61-4
[8] Feng Y C ,Zhao F Q and Xu H 2016 *MATEC Web of Conferences.* **67** 1-5
[9] Zhao F Q, Li H, Liu S J and Chen J B 2011 *Construction & Building Materials.* **25** 2635-38
[10] Liu S J, Liu H J, Li P X and Zhao F Q 2014 *Journal of Adhesion Science and Technology.* **28** 722-730

Mixing temperature design and properties evaluation for SMA-13 mixture

X Y Huang [1,2,3], **B Y Li** [1]

[1] Key Laboratory for Special Area Highway Engineering of Ministry of Education, Chang an University, Xi an 710064, Shaanxi, China;
[2]Guangxi Communications Investment Group CO., LTD.，Nanning 530022, Guangxi

E-mail: 709547714@qq.com

Abstract. The mixing temperature of hot asphalt mixture could be reduced by adding WMA additive Sasobit, as well as reducing smoke emissions and energy construction during the mixing construction and paving. The reasonable mixing temperature were investigated in this paper. In addition, high temperature stability, water stability and low-temperature performance of warm asphalt mixture were evaluaterd. The test results indicate that the mixing temperature of SMA-13 with WMA additive Sasobit may reduce 15-20℃ at the same energy (compaction times). the dynamic stability were improved after adding WMA additive Sasobit, and the Water stability and low-temperature performance of mixture decreased， while all kinds of asphalt mixture properties can meet the requirements.

1. Introduction

Asphalt concrete construction are characterized by the work environment to the high temperature, smoke more, poor visibility and air circulation characteristics such as slow, common risk factors, harm has three aspects: one is toxic or harmful gases [1]. This is because the channel tunnel space narrow, poor ventilation, plus the asphalt send out of high temperature homework gasses and smoke, and car, paver, roller and other kinds of machinery equipment exhaust gas not easy to send out, gathered in the narrow space in the homework, construction workers once the body absorbs excess, will cause poisoning accident; second is high temperature. The asphalt concrete paving as high as 110 ℃ temperature above, plus poor ventilation and cooling effect is poor, and workplace temperature is higher, easy to cause the homework personnel heatstroke, affect human health and safety [2-3]; Three is fire. There are mainly the high temperature of asphalt surface and wash the shovel scraper with diesel, and vehicle machine operation and the temporary fire hazard of electricity [4-5]. In order to guarantee the channel tunnel in the process of paving asphalt concrete construction safety, in addition to strictly control the construction technology measures, but also by reducing asphalt mixing in order to reduce the viscosity of the compaction temperature, construction environment improvement avoid a lot of asphalt smoke production, improve the working environment, to ensure the construction quality [6].

For those purposes, the paper carry warm mix asphalt mixing and temperature control study on road performance, the findings for warm mix asphalt SMA-13 design has a guiding value.

2. Materials

The addition of Sasobit is 0.3% of the dosage of HMA, the temperatures of stone and mixing are reduced by 15~30 °C on the basis of HMA whose temperatures are 185 °C and 190 °C respectively. Test graduation of SMA-13 can be seen in Table 1. The binder content was 5.9% by weight of mix. Lignin fibre was added at the rate of 12% by the mass of mixture.

Table 1. Gradation of SMA-13 for testing.

Mixtures	Passing(by Mass) under different sieve size (mm) , %										Binder content , %
	16	13.2	9.5	4.75	2.36	1.18	0.6	0.3	0.15	0.075	
PAC-13	100	90.5	61.4	26.9	22.9	17.9	15.2	12.6	11.1	8.5	5.9

3. WMA's mixing temperature

In order to determine the mixing temperature, following molding temperature molding specimens and test mix parameters were tested.

1) stone's temperature is 170 °C, the mixing temperature is 175 °C; 2) stone's temperature is 170 °C, the mixing temperature is 170 °C; 3) stone's temperature is 165 °C, the mixing temperature is 175 °C; 4) stone's temperature is 155 °C, the mixing temperature is 165 °C; 5) stone's temperature is 150 °C, the mixing temperature is 160 °C; The volume indexes of WMA at different temperatures are shown in Table 2.

Table 2. The volume indexes at different temperatures

temperatures (°C)	185+190 (HMA)	170 175	170 170	165 175	155 165	150 160
asphalt-aggregate ratio (%)	5.9	5.9	5.9	5.9	5.9	5.9
fiber content (%)	0.3	0.3	0.3	0.3	0.3	0.3
Sasobit content (%)	0.3	0.3	0.3	0.3	0.3	0.3
volume of air voids VV(%)	4.1	3.7	4.1	4.2	5.2	5.5
voids in mineral aggregate VMA(%)	16.5	16.2	16.5	16.6	17.4	17.7
Voids filled with asphalt VFA(%)	75.2	76.6	75.2	74.5	70.4	69
Bulk specific gravity(%)	2.465	2.477	2.467	2.463	2.439	2.431
VCAmix	38.4	38.1	38.3	38.5	39	39.2

As can be seen from the Table 2, when the compaction works (compaction times) of WMA and HMA are consistent, if the degree of compaction of WMA is supposed to be same to HMA's, the temperatures could be reduced by 15 °C ~20 °C. Finally, stone's temperature is determined to 170 °C, the mixing temperature is 170 °C.

4. Performance of WMA

4.1. High temperature stability

The specific test is rutting test in which the index is dynamic stability at 60℃, under the wheel pressure of 0.7 MPa, the test results are shown in Table 3.

Table 3. Results of rutting test

Asphalt-aggregate ratio (%)	5.9				
	1	2	3	WMA average value	Criteria
Dynamic stability (times/mm)	15750	13695.65	14653.74	14699.79	≥3000

It is observed that the addition of Sasobit has an important influence on WMA's dynamic stability (14699.79 times /mm) and it is much larger than HMA's (10962.265 times/mm). Because at the normal temperature of road surface, Sasobit forms the lattice structure in the binder, which improves the high temperature stability of road and enhances the resistance to rutting, WMA's dynamic stability has improved significantly compared to HMA, and doesn't cause rutting damage due to inadequate compaction by lower temperatures in the process of the mixing and compaction or the ease of asphalt's aging.

4.2. Water stability

4.2.1. Test of soaking Marshall Stability. WMA's soaking Marshall stability is shown in table 4.

Table 4. WMA- soaking Marshall stability

Groups	Theoretical maximum specific gravity /r_t	measured bulk specific gravity /r_r	VV (%)	VMA (%)	VCA_{mix}	VFA (%)	WMA Marshall stability (KN)	WMA soaking Marshall stability of residues (%)	Criteria
MS_1	2.572	2.469	4.0	16.4	38.3	75.6	11.98	100.503	≥80
MS	2.572	2.464	4.2	16.6	38.4	74.4	11.92		

The comparative results of soaking Marshall stability of HMA and WMA are displayed in Table 5.

Table 5. The comparative of soaking Marshall stability of HMA and WMA

Groups	theoretical maximum specific gravity /r_t	HMA Marshall stability (KN)	HMA soaking Marshall stability (%)	WMA Marshall stability (KN)	WMA soaking Marshall stability of residues (%)	Criteria
MS_1	2.572	12.70	98.83	11.98	100.50	≥80
MS	2.572	12.85		11.92		

Compared to the HMA's, the values of WMA's soaking Marshall stability- MS are reduced a little, the difference of soaking residual stability MS0 of them looks small relatively. This result shows that the addition of Sasobit has a little negative effect on the short-term stability of asphalt mixture, but the effect on the long-term stability is little. Obviously the values of the MS0 of WMA with Sasobit aren't less than 80%, meeting the requirement of code (JTGF40-2004) in wet zone where annual rainfall is greater than 1000 mm.

4.2.2. Test of freeze-thaw splitting. The results of freeze-thaw splitting test of WMA are shown in Table 6.

Table 6. The results of freeze-thaw splitting test

groups	theoretical maximum specific gravity /r_t	measured bulk specific gravity /r_r	VV (%)	VMA (%)	VCA_{mix}	VFA (%)	splitting tensile strength Rt(MPa)	freeze-thaw splitting tensile strength TSR(%)	code requirement T0729
no	2.572	2.466	4.1	16.5	38.4	75.0	0.917	83.95	≥80
freeze-thaw	2.572	2.467	4.1	16.4	38.35	75.4	0.770		

The comparative results of freeze-thaw splitting test of HMA and WMA are displayed in Table 7.

Table 7. The comparisons of **the results of** freeze-thaw splitting test

groups	theoretical maximum specific gravity /r_{at}	HMA splitting tensile strength Rt(MPa)	HMA freeze-thaw splitting tensile strength TSR(%)	WMA splitting tensile strength Rt(MPa))	WMA freeze-thaw splitting tensile strength TSR(%))	code requirement T0729
no	2.572	1.029	85.43.	0.917	83.95	≥80
freeze-thaw	2.572	0.879		0.770		

The data presented in the Table 7.suggests that TSR(83.95%) of WMA with Sasobit of 0.3% is lower than HMA's(85.43%), however, which meets the requirement(TSR≥80%) of code(JTGF40-2004) in wet zone where annual rainfall is more than 1000 mm.

4.3. Low-temperature performance

The results of asphalt mixture WMA with Sasobit of 0.3% are presented in Table 8.

Table 8. The results of low temperature bending test of WMA

bending-tensile strength R_B(MPa)	bending stiffness modulus S_B(MPa)	limiting bending-tensile stain $\varepsilon_B(\mu\varepsilon)$	code requirement $\varepsilon_B(\mu\varepsilon)$
12.77	3781.95	3401.88	≥3000

The result demonstrates that the gradation of SBS I-C modified asphalt SMA-13 meet the criterion's requirement to limiting bending-tensile stain ($\varepsilon_B \geq 3000\mu\varepsilon$).

The comparative results of low temperature bending test of HMA and WMA are shown in Table 9.

Table 9. The comparison results of low temperature bending test

Types	bending-tensile strength R_B(MPa)	bending stiffness modulus S_B(MPa)	limiting bending-tensile stain $\varepsilon_B(\mu\varepsilon)$	code requirement $\varepsilon_B(\mu\varepsilon)$
HMA	13.22	3620.54	3675.85	≥3000
WMA	12.77	3781.95	3401.88	≥3000

From the foregoing, we can know that the bending-tensile strength and limiting bending-tensile stain of HMA are higher than WMA's, which suggests that Sasobit is harmful to low-temperature performance of asphalt mixture, such as the ability to resist damage and contraction deformation.

5. Conclusions

- In order to get the same compactness with the hot mix asphalt, the mixing temperature of SMA-13 with Sasobit warm agent may reduce 15-20℃ at the same energy (compaction times).
- Considering guaranteeing VMA, high temperature, low temperature and water stability and other road performance of warm mix asphalt, the temperature should be controlled appropriately at 160 ~ 175 ℃, and compaction temperature at 135 ~ 145 ℃.

- It can be found that warm agent have bad effect on asphalt mixture properties, including low temperature and water stability, while all kinds of asphalt mixture properties can meet the requirements.

References

[1] Ma F, Sha A M, Lin R Y, Huang Y and Wang C 2016 Greenhouse Gas Emissions from Asphalt Pavement Construction: A Case Study in China *J. International Journal of Environmental Research and Public Health* **13(3)** 351-366.

[2] Punith V S, Xiao F P and Wingard D 2013 Performance Characterization of Half Warm Mix Asphalt Using Foaming Technology *J. Journal of Materials in Civil Engineering* **25(3)** 382-392.

[3] Grallam C H and Brian D P 2005 Evaluation of AsPhalt-min Zeolite for Use inWarm-Mix Asphalt *R. Ameriea:NCAT Rrport* 05-04 .

[4] Parker C M, Schreiner C A and Hallmark N 2011 Evaluation of reproductive/development and repeated dose (subchronic) toxicity and cytogenetic effects in rats of a roofing asphalt fume condensate by nose-only inhalation J. Regulatory Toxicology and Pharmacology **59(3)** 445–453.

[5] Abdullah M E , Zamhari K A, Hainin M R, Ebenezer Akin Oluwasola E A, Yusoff N I M and Hassan N A 2016 High temperature characteristics of warm mix asphalt mixtures with nanoclay and chemical warm mix asphalt modified binders *J . Journal of Cleaner Production*, **122** 326-334.

[6] Ji J，SUO Z，SHI Y F and Xi J X 2015 Analysis on Fatigue Behaviors of Hot and Warm Recycled SMA Aspahtl Mixture J. Journal of Building Materials **18**（**6**）1095-1098.

An efficient synthesis of 3,4-Dihydrofuro [3,4-B] [1,4] dioxepine-6,8-dicarboxylic Acid

D D DENG, L P WANG, H W YAN, X X SUN [1], Z WANG

Jiangxi Key Laboratory of Organic Chemistry, Jiangxi Science and Technology Normal University, Nanchang 330013, People's Republic of China

E-mail:sunxiaoxia77@126.com;

Abstract. As a class of important conjugated polymers, electroluminescent material of polyfuran possess a lot of merits, including good stability, structure is easy to be modified and controllable electrochemical properties. This study uses ethylene diglycol, diethyl oxalate, 1,3-dibromopropane and so on as raw materials to synthesize 3, 4 - ethylenedioxy furan monomer and similar derivatives via oxidation, acetylation, hydrolysis and elimination reaction, etc.

1. Introduction

Conjugated polymer materials, as a new type of light-emitting body, has aroused great concern in the material and chemical industry because of its great application prospects in lighting, display device and so on. Compared with the current widely used fluorescent lamp and LCD display device, polymer light-emitting-diode (PLED) has characteristics of low energy consumption, simple manufacturing process, flexible, low-cost, colorful and tremendous market potential [1-3].

One of the most important applications of organic electroluminescence is realization of full-color display device to replace the traditional display. Red, green and blue are the basic colors of full-color display [4]. Among them, green fluorescent polymer materials such as P-PPV has made great progress in brightness and efficiency, but there is still space for improvement in material color purity and efficiency of the device [5-7]. In recent years, organic conjugated small molecule compounds gradually become the focus of study, this kind of material with determination of molecular structure, controllable synthesis, good repeatability, monodisperse, stable glass state and good film-forming properties, these characteristics make it can be applied in organic electro luminescence device and become a potential material [8-9].

As a new generation of information functional materials, organic polymer semiconductor materials have attracted worldwide attention and become the object of research and development that more and more research institutions competing for [10]. The study on organic polymeric semiconductor materials is mainly focused on the small molecules which can be deposited by evaporation or the polymer which can be processed by solution. Due to the expensive equipment and complex process, more and more research is focused on the polymer. Because of its wide range of material selection, solution processing(can use some low-cost processing methods, such as spin coating, inkjet printing, screen printing, static pulling, etc.), easy fabrication of large area devices, and good flexibility, the polymer semiconductor materials have attracted extensive attention [11-13]. By spin coating although can get uniform large area devices, but as solvent volatilizing speed too fast in the film-forming process and the donor material P3HT can not form a sufficient crystallization in the process of precipitation, cause the P3HT in the mixed film of the photosensitive layer did not form a good hole transport channel, so the efficiency of the device is usually very low [14-15].

Most of the electrochromic materials are printing and spraying in the traditional organic solution, printing and spraying of low steam pressure in water is a big challenge [16]. EDOT of side chain hydroxyl or carboxyl extension is very stable in aqueous phase, with super high conductivity, strong side chain function, can be converted into other groups, strong biological coupling ability and electrochemical polymerization is very easy to occur in conventional electrolytes [17-19]. In 2010, John R. Reynolds reported that after the side chain has been functionalized, the electroluminescent material can be sprayed in the water solution, which can reduce the environmental pollution and cut the cost [20].

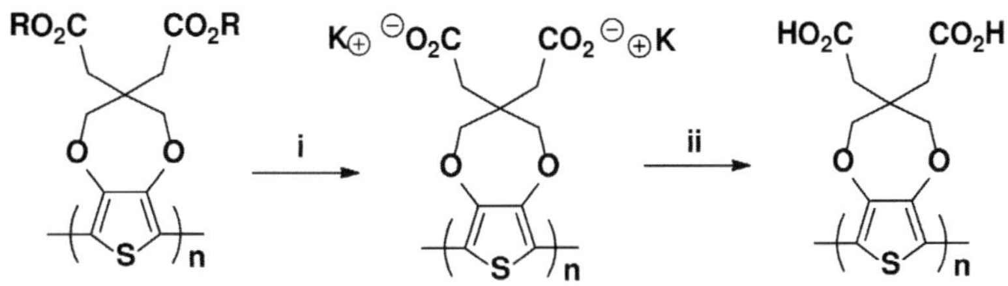

Figure 1. The structures of the materials

If we can further optimize the structure of the conjugated polymer and increase the molecular order or reduce the disorder to enhance the mobility of the carrier, the photoelectric conversion efficiency of polymer will be further improved. In view of D-A type polymers can regulate the intensity of intra molecular charge transfer by selecting appropriate donor and receptor, makes the D-A conjugated polymers exhibit properties such as narrow band gap, broad spectrum absorption, high electron affinity and low ionization energy, high migration rate and so on. In order to synthesize new type of optoelectronic materials, project team has synthesized three new compounds, namely, diethyl-3,4-dihydroxyfuran -2,5-dicarboxylate; 3,4-dihydrofuro[3,4-b][1,4]dioxepine-6,8-dicarboxylate; 3,4-dihydrofuro [3,4-B] [1,4] dioxepine-6,8-dicarboxylic acid

2. Experimental Section

2.1. Materials and Characterization

Methanol was dried over CaH_2 and then distilled prior to use in synthesis. Ethylene diglycol was purchased from Energy Chemical Co. and were used as received. Other solvents and reagents were analytical grade and purchased from Energy Chemical Co. and used without further purification. ^{1}H and ^{13}C NMR spectra were obtained with a mercury-300 spectrometer with tetramethylsilane (TMS) as the internal reference and chloroform-d ($CDCl_3$) or DMSO as solvent. The mass spectra were determined on a Daltonics Esquire 6000 spectrometer.

2.2. *Synthesis*

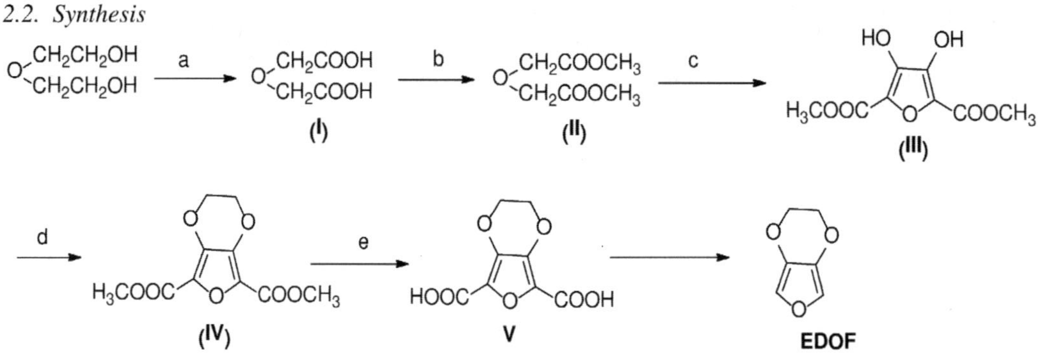

a) HNO_3, 90~100℃, 12h. b) CH_3OH, $SOCl_2$, RT, 24h. c) CH_3ONa, CH_3OH, $(COOEt)_2$, 70℃, 10h. d) $BrCH_2CH_2CH_2Br$, K_2CO_3, DMF 110℃, reflux 5h. e)NaOH)

Figure 2. The synthesis of 3,4-dihydrofuro[3,4-b][1,4]dioxepine-6,8-dicarboxylic acid

2.2.1. Diglycolic acid(I) Add 500 ml water and 500 ml concentrated nitric acid to an appropriate three-necked flask , the ethylene diglycol (111.8 g, 1.05mol) then was slowly added to the mixture, the resulting solution was stirred overnight at 100℃. After reaction, the mixture was concentrated by evaporation under reduced pressure to removal of the excess of nitric acid. Wait until the reaction liquid was evaporated to dryness, the resultant product was further dried by vacuum drying oven, and the yield is 95.3 g (67.7 %). ^{1}H NMR (400 MHz, DMSO) δ 12.79 (s, 2H), 4.09 (s, 4H).

2.2.2. Dimethyl 2,2'-oxydiacetate(II) Diglycolic acid (10.0 g, 0.075 mol) was added to a 250 ml three-necked flask, then $SOCl_2$ (21.6 ml, 0.30 mol) was added dropwise, hydrochloric acid can be produced in the process of dropping. After dropping, the mixture stirred at room temperature for 6h. Add an ice bath outside the reaction system, then the dry methanol (50 ml) was added drop by drop to the resulting mixture, when the system is no longer producing gas ,the mixture was warmed up to room temperature and stirred overnight. Evaporate the excess of $SOCl_2$ and methanol, the residual liquid was extracted by ethyl acetate and neutralized by saturated solution of $NaHCO_3$, the obtained organic phase was washed with water, dried over anhydrous $MgSO_4$. The solvent was then removed by rotary evaporation to give a white solid (yield: 9.27 g, 76.3 %). ^{1}H NMR (400 MHz, $CDCl_3$) δ 4.26 (s, 2H), 3.78 (s, 3H). ^{13}C NMR (101 MHz, $CDCl_3$) δ 170.11, 68.12, 51.99.

2.2.3. Diethyl-3,4-dihydroxyfuran-2,5-dicarboxylate(III) 200 ml of dry methanol was added into a 500 ml three-necked flask, and tiny sodium (2.13 g ,0.09 mol) was added, stirred until the sodium was dissolved entirely (the solution was milky white feculent liquid),add the dimethyl 2,2'-oxydiacetate (10.0 g,0.06 mol), the mixture was heated to 75℃ for 30min. Then the diethyl oxalate (6.76g, 0.05mol) was added, the solution becomes yellow feculent liquid, after that reflux for 3~4h, a great deal of yellow solid was precipitated. The yellow solid was washed with ethanol(dried over anhydrous $MgSO_4$),added a small amount of water to dissolve, then added slowly hydrochloric acid(which was dilute twice).The yellow solution transformed gradually into achromaticity and considerable white solid deposited from it, Then filtrated and dried the precipitation, the yield is 8.26 g (63.7%). ^{1}H NMR (400 MHz, d-DMSO, ppm): δ 3.76 (s, 6H), 10.26 (s, 2H); ^{13}C NMR (100 MHz, d-DMSO): δ 51.81, 127.76, 144.14, 158.96. LRMS (ESI-) M/z calcd for $C_8H_8O_7$. [M-H]- 215.1; found 215.2.

2.2.4. 3,4-dihydrofuro[3,4-b][1,4]dioxepine-6,8-dicarboxylate (IV) A 50 ml single-necked flask, diethyl-3,4-dihydroxyfuran-2,5-dicarboxylate (1.0 g ,4.6 mmol) was added, DMF (25 ml) was added as solvent and dry potassium carbonate (1.28 g ,9.2 mmol) was added. The mixture was heated to 110℃, after that 1,3-dibromopropane(1.87g , 9.2mmol) was added drop wise to reaction system, then the mixture was refluxed overnight. After completion, the reaction mixture was cooled to room temperature and add proper amount of water to the system, filtered to get a white solid and dried it, the yield is 0.69 g (58.5 %). ^{1}H NMR (400 MHz, $CDCl_3$) δ 4.41 - 4.29 (m, 4H), 3.91 (s, 6H), 2.36 (dt, J = 10.4, 5.1 Hz, 2H). ^{13}C NMR (101 MHz, $CDCl_3$) δ 157.75, 146.02, 129.87, 71.49, 51.47, 32.18; LRMS (ESI+) M/z calcd for $C_{11}H_{12}O_7$. [M+Na]+ 279.0; found 279.1.

2.2.5. 3,4-dihydrofuro[3,4-b][1,4]dioxepine-6,8-dicarboxylic acid (V) 3,4-dihydrofuro[3,4-b][1,4]dioxepine-6,8-dicarboxylate(0.5g ,2.0mmol) and 10 ml of 10 M sodium hydroxide solution were added into a 50 ml single-necked flask, the mixture was heated to 60°C. After the solid were dissolved entirely, the mixture was stirred for 3h. This mixture was washed three times with dichloromethane, and dilute hydrochloric acid was added drop by drop until no precipitation. The white solid was separated by filtration, and the yield is 0.39 g (85.5 %). ^{1}H NMR (400 MHz, $CDCl_3$) δ 13.24 (s, 2H), 4.32 - 4.09 (m, 5H), 2.32 - 2.02 (m, 2H). ^{13}C NMR (101 MHz, $CDCl_3$) δ 159.14 146.59, 130.98, 72.22, 33.19. LRMS (ESI+): M/z calcd for $C_9H_8O_7$. [M+H]+ 458.1; found 458.3.

3. Results and Discussion

In this study, the main routes of the reaction are: oxidation of the ethylene diglycol, acetylation of the diglycolic acid, synthesis of the furan ring, Williamson reaction, hydrolysis of esters, etc. In the esterification process of diglycolic acid, we adopt the method of first acetylation and then esterification to increase the rate and yield of the reaction. In order to get a better result, we adopt sodium hydroxide solution instead of ethanol solution of sodium hydroxide in the process of ester hydrolysis. The amount of methanol is the key to the synthesis of the furan ring and in further treatment the excess of the methanol must be evaporated.

4. Conclusions

In summary, we have successfully synthesized 3,4-dihydrofuro[3,4-b][1,4]dioxepine-6,8-dicarboxylic acid which can be utilized for synthesis of a new type of organic electroluminescent material by using low-cost and readily available ethylene diglycol as raw material. Compared to existing procedures, the high yield, operational simplicity, easy availability and inexpensive reagents make the procedure more attractive. 3,4-dihydrofuro[3,4-b][1,4]dioxepine-6,8-dicarboxylic acid has a special substituent group which can increase the solubility of the polymer and is conducive to the formation of noncoplanar structure, reduce the fluorescence quenching, improve the quantum efficiency and stability.

Acknowledgments

We are grateful for the financial support of the National Natural Science Foundation of China (No.21462018), the Science Fund of the Technology Office of Jiangxi, China (2009ZDS11100 and 20132BBE50024).

References

[1] Yu G, Gao J, Hummelen J, Wudl F and Heeger A 1995 *Science*, **270**, 1789-1791
[2] Hou J, Chen H, Zhang S, Li G and Yang Y 2008 *Journal of the American Chemical Society*, **130**, 16144-16145
[3] Wienk M, Turbiez M and Janssen R 2008 *Advanced Materials*, **20**, 2556-2560
[4] Kim J, Lee K, Coates N, Moses D, Nguyen T, Dante M and Heeger A 2007 *Science*, **317**, 222-225
[5] Gunes S, Neugebauer H and Sariciftci N 2007 *Chemical Reviews*, **107**,1324-1338
[6] Li Y and Zou Y 2008 *Advanced Materials*, **20**, 2952-2958
[7] Dennler G, Scharber M and Brabec C 2009 *Advanced Materials*, **21**,1323-1338
[8] Chen J and Cao Y 2009 *Accounts of Chemical Research*, **42**, 1709-1718
[9] Pierre M, Chad M and John R 2010 *Accounts of Chemical Research*, **43**, 1396-1407
[10] Wang J, Xiao Q and Pei J 2010 *Organic Letters*, **12**, 4164-4167

[11] Yang Y, Farley R, Steckler T, Eom S, Reynolds J, Schanze K and Xue J 2009 *Journal of Applied Physics*, **106**, 044509
[12] Liu G, Li A and Zhu X 2009 *Macromolecular Rapid Communications*, **30**, 1484-1491
[13] Yang R.; Garcia A.; Korystov D.; et al. 2006 *Journal of the American Chemical Society*, **128**, 16532-16539
[14] Bouffard, J.; Swager, T. M. 2008 *Macromolecules*, **41**, 5559-5562
[15] Tam T, Li H, Wei F, Tan K, Kloc C, Lam Y, Mhaisalkar S and Grimsdale A 2010 *Organic Letters*, **22**,3340-3343
[16] Kim S, Seo J and Jung H 2005 *Advanced Materials*,**17**, 2077-2082
[17] Shi P, Amb C, Knott E, Thompson E, Mei J, Dyer A and Reynolds J 2010 *Advanced Materials*, **22**, 4949-4953
[18] Lu G, Li L and Yang X 2007 *Advanced Materials*, **19,** 3594-3598
[19] Luo S, Ali E, Tansil N, Yu H, Kantchev E and Ying J 2008 *Langmuir*, **24**, 8071-8077
[20] Beaujuge P, Amb C and Reynolds J 2010 *Advanced Materials*, **22**, 5383-5387

The colour and micro-structure change of the aged madder-dyed silk fabric adding with consolidation material of EGDE/Ala

Liping Yang[1], Minjing Gu[1], Wenjian Li[1], Hailing Zheng[2], Zhiqin Peng[1,3], Bing Wang[1], Zhiwen Hu[1], Junmin Wan[1] and Yang Zhou[2]

[1]Institute of Textile Conservation, Zhejiang Sci-Tech University, Hangzhou, 310018, Zhejiang Province, China
[2]China National Silk Museum, Hangzhou, 310018, Zhejiang Province, China

E-mail: pengqiao6858@126.com

Abstract. A consolidated method by EGDE (ethylene glycol diglycidyl ether) and Ala (alanine) was used for fragile ancient silk fabrics protection. Therefore, an investigation has been carried out to figure out the effect of consolidation material in the system. Silk fibers were dyed with madder and then exposed to UV light beams for accelerated aging. The aged madder-dyed samples were treated with EGDE aqueous solution by spraying, and then treated with Ala aqueous solution by the same way on 10 minutes later. The samples were investigated by colour measurement and scanning electron microscopy (SEM). It has been observed that there is an obvious change of colour of samples after consolidation material treatment. The surface of the aged madder-dyed silk fabric were covered with EGDE and Ala thus protecting the silk fibers.

1. Introduction

Historic silk textiles as Chinese cultural heritages are very precious and of great value in research. For the ancient silk fabrics were deteriorated by water, light, heat and microorganisms [1-4], it is very important to protect historical silk fabric relics for further displays and researches. Many consolidation materials and methods were adopted in the protection of ancient silk fabrics. Such as weave [5], mounting [6], polymers [7], resin, adhesive [8] and so on. But the applications are limited by the problems, like high fabric stiffness, color difference or yellowing of silk fabrics and lead to potential harm to silks in the long term and so on, produced in the protection processes.

EGDE as an epoxide with bifunctional groups has high reactivity with amines, alcohols, phenols, carboxylic acids and thiols [9]. Consequently, EGDE was widely used in the preparation of three-dimensional polymers [10] and gels [11] as a cross-linking agent. Moreover, EGDE has been also used to modify polyrotaxane fibers [12], silk fibers [13] and silk fabrics [9, 14] to improve properties such as elongation at break, moisture regain, thermal stability and so on. In this study, EGDE was used as material to consolidate aged madder-dyed samples. Ala, which is homologous and compatible with silk, was used to consolidate fragile samples with the support of EGDE.

EGDE can be reacted with Ala and silk fabrics by simple spraying at low temperature, which is easy to perform and has significant effect on the improvements of mechanical properties compared with other materials [15, 16]. Therefore, EGDE was adopted as a material for treatment of very weak historic silk fabrics. Though we have some researches on properties of the aged silk fabric (undyed) adding with consolidation material of EGDG, there is still no work carried out to research the effect of dyed silk fabric on properties especially in color difference.

The consolidation materials of EGDE and Ala were added into artificial aged madder-dyed silk fabric by spraying at room temperature, and the effect of the aged madder-dyed silk fabric adding with consolidation material of EGDG/Ala on color and micro-structure were studied.

2. Experimental

2.1. *Materials*

Plain weaved reference silk fabrics (Hangzhou Fusi Textile Co.,Ltd) was used for experiments. Madder was purchased from Chinese medicinal herbs store, China. EGDE used in this study was supplied by Nagase ChemteX Co. Ltd. (Japan) as consolidation materials. Alanine was purchased from Aladdin Chemistry Co., Ltd. Distilled water was used in the experiments.

2.2. *Preparation of control and treated samples*

The madder were soaked in the water for 24 h. Then the water which the madder were soaked in was heated to boiling temperature for 90 min. Water was added during the heating process until the extract was cooled and then filtered two times to get the solution of Dye. The silk fabric was treated in water at 100℃ for 30 min, then rinsed and air dried. The dyeing procedure was performed in a beaker with the liquor ratio of 1:50 (For 1 g of silk fabric a dye solution of 50 mL is applied). 4g of silk fabric were used in this study. The silk fabrics were treated with the dye solution and then it was brought slowly to the 80℃ for 60 min. Then the samples were removed, squeezed, and air dried. After dyeing, the dyed samples were rinsed with hot water at 80℃ and then washed with hot water at 40℃. The unfixed dyestuff was removed by washing twice with cold water, and the dyed silk fabric were removed, squeezed, and air dried. The dyed samples were cut up into small pieces (2 × 10 cm). The small piece of dyed samples were put in UV Accelerated Weathering tester which equipped with 500 watt UV lamp, 70℃, 40% relative humidity (RH) for 30 days. Then the artificial aged madder-dyed silk fabric samples (control samples) were prepared. The aged madder-dyed silk samples were smoothly put on a net frame. The aged madder-dyed samples were treated with EGDE aqueous solution of 2.5 wt% by spraying, and Ala solution with the concentration of 1.5 wt% was gently sprayed on the samples which were treated with EGDE after 10 minutes. Then the samples were dried at 25℃ for tests.

2.3. *Characterization of the artificially aged madder-dyed silk fabric samples and the samples treated by consolidation materials*

The Colour differences of samples were measured using a Konica Minolta spectrophotometer (CM-700D, Japan) interfaced with a personal computer. The colour characteristics (L*a*b*) of the aged madder-dyed silk fabric were assessed for all samples treated in the presence and absence of consolidation materials. Colour differences between the samples were determined by the following equation:

$$\Delta E^*_{ab} = \sqrt{\left[(\Delta L^*)^2 + (\Delta a^*)^2 + (\Delta b^*)^2\right]} \tag{1}$$

Samples were cut into 5*5mm and mounted on aluminium stubs by using electrically conducting copper tape. Then the samples were sputtered with gold for 60 s at 15 mA. The morphology of the surface of the control samples in comparison to the two kinds of artificially aged samples were observed by using SEM (JSM-5610LV, Japan) under the voltage of 5kv and the working distance of 15mm .

3. Results and Discussion

3.1. *Colour difference*

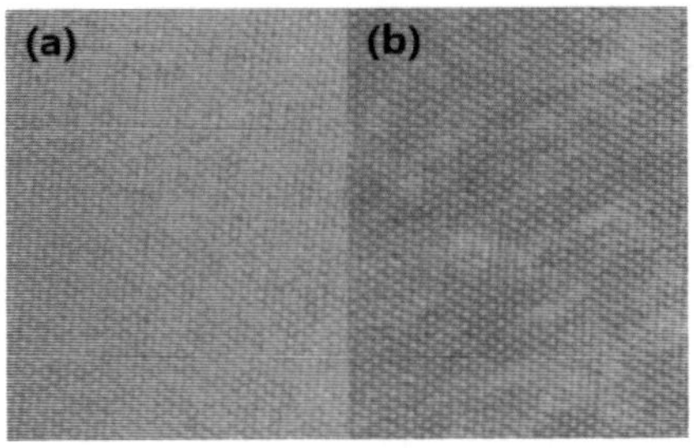

Figure 1. Images of the aged madder-dyed silk fabrics of (a) control, (b) EGDE\Ala treated

Figure 1 shows that the aged madder-dyed silk fabrics which treated with consolidation material appear in shades of red, and darker than the control sample. It has been observed that there is an obvious change of colour of samples after EGDE\Ala treatment.

Table 1. Colour difference of the aged madder-dyed silk fabrics and treated samples

	ΔL	Δa	Δb	ΔE
Colour difference	-5.38	2.14	-0.75	5.84

In order to exactly evaluate whether consolidation material of EGDE/Ala has an influence on colour, chromatic values of aged madder-dyed silk fabric before and after treatment were tested. The result in Table 1 shows that the total colour difference (△E) between control sample and treated sample is about 5.84. The value more than 1.0 suggests that the treatment is not acceptable in the protection of ancient silk fabrics [17]. The CIELAB system also provides a psychometric index of lightness L* which is about -5.38. This negative values indicated that the colour of treated sample distinctly darker than the untreated one. One colour coordinates is a* which is about 2.14 (reddish coloration), and the other is b* which is about -0.75 (bluish coloration). The greater of the absolute value shows the more obvious of the colour difference.

3.2. The morphology of the surface

Figure 2. SEM images of the aged madder-dyed silk fabrics ((a) and (c)) and EGDE\Ala treated samples ((b) and (d))

The samples were subjected to SEM analysis to evaluate morphological changes that may have occurred before and after EGDE\Ala treatment, which would led to changes in physical and mechanical properties. Scanning electron microscopy (SEM) images of fragile silk fabrics before and after treatment are shown in Fig.2. As shown in Fig.2 (a), the surface of the aged madder-dyed silk fabrics are rough and destroyed. The surface morphology of silk fiber wasn't changed significantly, as depicted in Fig.2 (b). These results suggest that treated with consolidation material of EGDG and Ala has obvious influence on its appearance with smooth the surface. EGDE and Ala might react with silk fabrics, and form a layer of film on the surface of silk fabrics (fig.2(c) and (d)), thus increasing the breaking strength and protect the silk fibers.

4. Conclusions

The aims of this paper is to study the effect of the aged madder-dyed silk fabric adding with consolidation material of EGDG/Ala on colour and micro-structure. The results concluded that the addition of consolidation material of EGDG/Ala affected the colour and micro-structure of the aged madder-dyed silk fabric. It has been obtained that there is an obvious change of colour of samples after EGDE\Ala treatment. EGDE and Ala were covered in the surface of silk fabrics to protect the silk fibers. The colour and micro-structure change probably results from the triadic interactions of EGDE, Ala and silk fabric. However, the detailed mechanism and property is unclear. Therefore, we need to do further research based on this work.

Acknowledgements

We thank to financial support from National Key Technology R&D Program (2013BAH58F01-02) and Special Funds from the Administration of Cultural Heritage of Zhejiang Province (2016011).

References

[1] Zhang X M, Berghe I V and Wyeth P 2011 *J. Cult. Heritage* **12** (4) 408–411
[2] Zhang X M, Yuan S and Chin X 2010 *J. Chem* **28** (4) 656–662
[3] Kotowa J S 2004 *Int. Biodeter. Biodegr* **53** 165–170
[4] Mang X M and Yuan S X 2010 *Spectrosc. Spectr. Anal* **30** 262–265
[5] Wang Z H and Ed 1995 The Protective Materails for Cultural Relics (Xian: Northwest University)
[6] Ahmed H E and Ziddan Y E 2011 *J. Cultural Heritage* **12** 412
[7] Halvorson B, Kerr N and H. S. F. R. T. Group 1992 Effect of Parylene C on Selected Properties of Silk (Washington: Harper's Ferry Regional Textile Group)
[8] Wu S Q, Li M Y, Fang B S and Tong H 2012 *Carbohyd. Polym.* **88** 496
[9] Freddi G, Shiozaki H, Allara G, Goto Y, Yasui H and Tsukada M 1996 *J. Soc. Dyers Colour.* **112** 88–94
[10] Lazaro J M, Chattah A K, Torres Sanchez R M, Buldain G Y and Dall'Orto V C 2012 *Polymer* **53** 1288–1297
[11] Topuz F and Okay O 2009 *Biomacromolecules* **10** 2652
[12] Katsuyama N, Shimizu K, Sato S, Araki J, Teramoto A, Abe K and Ito K 2010 *Text. Res. J.* **80** 1131–1137
[13] Tsukada M, Goto Y, Freddi G, Matsumura M, Shiozaki H and Ishikawa H 1992 *J. Appl. Polym. Sci.* **44** 2203–2211
[14] Tsukada M, Shiozaki H, Goto Y and Freddi G 1993 *J. Appl. Polym. Sci.* **50** 1841–1849
[15] Hu Z W, Huang X F, Zhang J, Cao X Y, Zheng H L, Zhou Y, Zhao F and Peng Z Q 2011 *Polym. Mater. Sci. Eng.* **27** 44
[16] Huang X F, Hu Z W, Peng Z Q, Zhang J, Cao X Y, Zhou Y and Zhao F 2010 Consolidation of Fragile Silk Fabrics with Fibroin Protein and Ehtylene Glycol Diglycidyl Ether by Spraying (Hang Zhou: Zhejiang University)
[17] Wu S Q, Li M Y, Fang B S and Tong H 2012 *Carbohyd.Polym.* **88** 496

IOP Conf. Series: Materials Science and Engineering **167** (2017) 012060 doi:10.1088/1757-899X/167/1/012060

High-Performance transparent Meta-surface for C-/X-Band Lens Antenna Application

Tong Cai[1], Guangming Wang

Air and Missile Defend College, Air force Engineering University, Xi' an 710051, China.

E-mail: caitong326@sina.cn

Abstract. A novel strategy of designing high-performance transparent meta-surfaces is proposed by using an ABBA system. The ABBA element provides a new freedom to largely enhance the transmission of a transmissive meta-surface and suppress the fluctuations in the transmission amplitude by tuning the coupling among cascaded layers. Dual-band operating property is straightforwardly performed by employing the polarization-control principle. A well-designed transparent meta-surface, operating at f_1=6.5 GHz and f_2=10.5 GHz, consists of 13×17 ABBA elements with parabolic phase distributions along different dimensions. Perfect focusing effects with the same focal length and small reflection are unambiguously observed at both frequencies under excitation of differently-polarized waves, indicating strong phase compensating capacities and high efficiencies of the meta-lens. For practical application, a dual-band lens antenna (LA) is implemented by launching the carefully-designed meta-lens with a self-made Vivaldi antenna. Numerical and experimental results coincide well, indicating that the proposed LA advances in many aspects such as high radiation gain of 18.7 dB at f_1 and 23 dB at f_2, competitive aperture efficiencies better than 30% and also a simple fabrication process based on a standard printed-circuit-board (PCB) technology. The finding opens up a new avenue to design high-performance meta-surfaces operating in multi-band or achieving integrated functionalities.

1. Introduction

Electromagnetic (EM) control plays an essential role in both science and technology, and has become a research spot with remarkable achievements. Recently, gradient meta-surfaces (GMSs) [1], proposed by YU et al, have provided a powerful solution to solve the mentioned issues due to their strong abilities to manipulate the reflections or refractions of incoming waves by locally controlling the phase and amplitude of the designed meta-atoms within a deep-subwavelength scale. As a result, different kinds of EM manipulation and a lot of fascinating phenomena have been discovered, including wavefront control (the generalized law of refraction/reflection [1], meta-holograms [2], polarization beam splitter [3], focusing [4]), functionality extension (multi-functional device design: bi-functional lens [5], bi-functional devices achieving focusing and vertex effects), polarization control (quarter-wave plate, polarizer) and also multi-frequency system realization [6-7]. Multi-band operating devices have attracted increasing attention for their capability of improving the communication quality, achieving EM integration and reducing device costs. Several techniques, controlling the EM responses by optimized meta-surfaces, have been reported very recently to realize multi-frequency operating devices [6]-[7]. Firstly, a straightforward method of achieving multi-band devices is to use several electric or magnetic resonators with different dimensions, and some relevant absorbers are designed and experimentally demonstrated [6]. Secondly, multi-mode resonators have been intensively investigated to realize multi-band devices [7]. However, the mentioned methods suffer from mutual coupling and sensitivity of the working frequencies. Especially, these techniques face huge challenges to design multi-band GMSs since it is uneasy to realize a complete phase-shift range over 360° at several separate frequencies without mutual interactions. For the multi-band reflective GMSs, multi-layer structures are independently controlled and optimized, which requires a large frequency interval and becomes invalid for a small frequency interval [8]. Transparent GMSs design is quite challenging for not only requiring a complete phase-shift range over 360° but also needing high transmission amplitude [9]. Therefore, conventional transparent GMSs is limited with low-efficiency, single operating frequency and complex configurations. To the best of our knowledge, high-efficiency transparent GMSs incorporating dual-band operating property, in addition to perfect focusing effects, are rarely studied in the open literature.

In this paper, we propose a new ABBA system to enhance the transmission efficiency of a transparent meta-surface. Compared with the conventional cascaded element (AAAA element), the proposed one increases another design freedom to modulate the coupling among different layers and to suppress the fluctuations in the transmission amplitude. Moreover, inspired by the method of manipulating the transmission beams through polarizations in our work [9], here, we control the operating frequencies of a transparent meta-surface through two orthogonal polarizations. For experimental demonstration, we conceived and realized a high-efficiency transparent meta-surface using the ABBA elements to realize a focusing lens operating in C and X bands with the same focal length. Finally, by carefully launching the lens with a wideband Vivaldi antenna at the focal point, we numerically and experimentally designed and optimized a dual-band lens antenna (LA).

2. Working mechanism of dual-band meta-surface and ABBA element design

We will provide a general strategy to achieve dual-band operating property and high-efficiency transmission of a transparent meta-surface, respectively. For an anisotropic element with mirror symmetries located at (x, y), the EM response can be described by the matrices **R** and **T**, where

$$\mathbf{R}(x,y)=\begin{pmatrix} r_{xx,f_1}(x,y) & 0 \\ 0 & r_{yy,f_2}(x,y) \end{pmatrix} \tag{1}$$

$$\mathbf{T}(x,y)=\begin{pmatrix} t_{xx,f_1}(x,y) & 0 \\ 0 & t_{yy,f_2}(x,y) \end{pmatrix} \tag{2}$$

For a perfect transparent element, the reflection coefficient satisfies **R**=0, while the transmission coefficient satisfies **T**=1. Here, under the condition of $|t_{xx}=1|$ and $|t_{yy}=1|$, we can control the transmission phases φ_{xx} and φ_{yy} by the local structures, respectively. From the angle of wavefront shaping, three fascinating properties can be found in Fig. 1. In the first case, with a different phase gradient along x and y directions for GMS1 ($\varphi_{xx}=\xi_1 x$, $\varphi_{yy}=\xi_2 x$), differently-polarized waves can split into two beams by passing through the meta-device, which results in an intriguing polarization beam splitter (PBS) [3], [9]. In the second case, referring to Figs. 1(c) and 1(d), beam deflector and focusing lens can be switched freely by illuminating GMS2 ($\varphi_{xx}=\xi x$, $\varphi_{yy}=k_0(\sqrt{F^2+x^2+y^2}-F)$) with differently-polarized waves, which is very interesting to satisfy the requirement of functional integration in the field of photonics. Finally, we can realize operating frequency control conveniently and precisely by modulating the phases in different dimensions of the transparent meta-surface. As shown in Figs. 1(e) and 1(f), with phase distributions of $\varphi_{xx}=k_1(\sqrt{F^2+x^2+y^2}-F)$ and $\varphi_{yy}=k_2(\sqrt{F^2+x^2+y^2}-F)$, GMS3 combines two focusing lenses with the same focal length F working at f_1 and f_2 under excitation of differently-polarized incident waves, respectively.

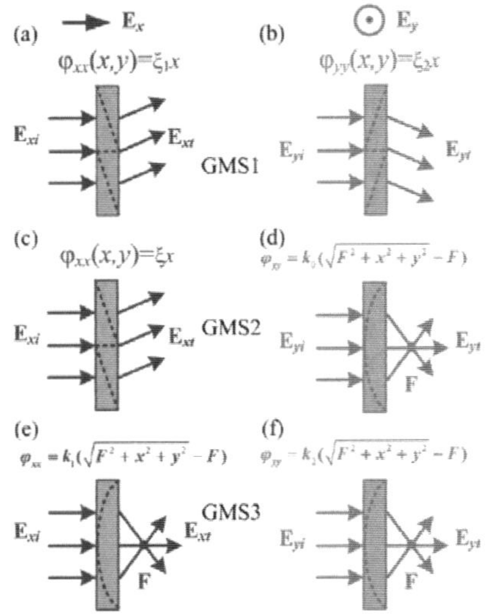

Figure 1. Schematics and working principles of the different types of transparent meta-surfaces.

GMS1 serves as a PBS with $\varphi_{xx}=\xi_1 x$, $\varphi_{yy}=\xi_2 x$ under excitation of (a) an x-polarized wave and (b) a y-polarized wave; GMS2 behaves as (c) a beam deflector and (d) a focusing lens with $\varphi_{xx}=\xi_1 x$, $\varphi_{yy}=k_0(\sqrt{F^2+x^2+y^2}-F)$; GMS3 operates at frequencies of (e) f_1 and (f) f_2 with $\varphi_{xx}=k_1(\sqrt{F^2+x^2+y^2}-F)$, $\varphi_{yy}=k_2(\sqrt{F^2+x^2+y^2}-F)$ under illuminating of different polarizations.

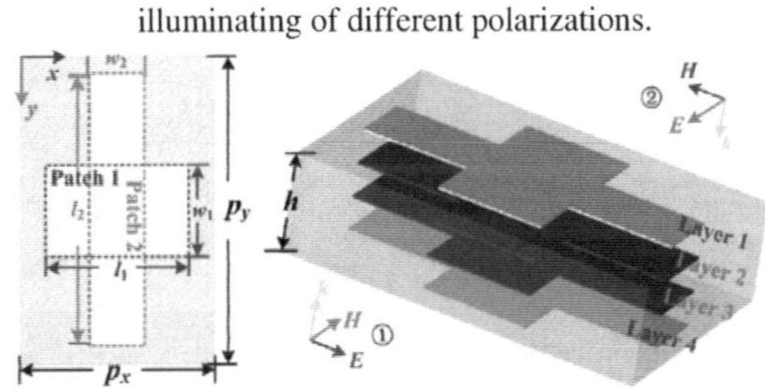

Figure 2. Topology and simulation setup of the proposed ABBA element.

(a) Top view of the basic element. The final geometrical parameters are listed as: p_x=11 mm, p_y=17 mm, w_1=5 mm, w_2=4 mm. (b) Perspective of the ABBA element and the simulation setup.

To design a high-performance transparent GMS, the first step is to find a proper element which is totally transparent but with transmission phases changeable via structural tuning. For a commonly used cascaded element, the variation in phase is typically linked with that in transmission amplitude, which generates fluctuations in the transmission spectrum. Here, we propose an ABBA system to improve the transmission amplitude by modulating the coupling among the cascaded layers, as shown in Fig. 2. In fact, the AAAA system is a special case of the ABBA system. The center frequencies for the proposed element are arranged as f_1=6.5 GHz and f_2=10.5 GHz. A basic patch structure can exhibit a perfect transmission at a particular frequency, induced by the Lorentz resonance. As seen in Fig. 2(a), a pair of orthogonal patches (parch 1 for x polarization, parch 2 for y polarization) are introduced to realize a dual-band operating property and a high isolation between different frequencies. The adopted substrate is the commonly used F4B material with h=4.5 mm, ε_r=2.65, tan δ=0.001. Then, we will make a fair comparison between two typical elements, composing of a conventional AAAA system and a proposed ABBA system, respectively, with the topologies shown in the inset of Fig. 3. In this special case, the I-shaped structures change the coupling among cascaded layers, resulting in a tunable transmission. It is worth noting that the transmission phase φ_{yy} is almost consistent from 4 to 8 GHz for the two elements. However, compared with AAAA element, the transmission coefficient has been significantly enhanced from 0.73 to 0.95 at the frequency of f_1. Straightforwardly, the transmission coefficients at other transmission phases can be manipulated with the same manner. Referring to Fig. 4(a), at frequency of f_1, within a 360° range for transmission phase, the variation of $|t_{yy}|$ is 0.73-1.0 for the conventional AAAA system, but shrinks to 0.88-1 for the proposed ABBA system. The frustrations in the transmission band have been largely suppressed. Therefore, adopting ABBA elements to design our transmissive meta-surface can surely enhance the overall working efficiencies of the device. At frequency of f_2, see Fig. 4(b), the transmission $|t_{xx}|$ of the AAAA system keeps higher than 0.85 with a changeable l_1, which is adopted for a design convenience. Moreover, the polarization-independent property is demonstrated by the consistent transmission phases φ_{xx} at f_1 and φ_{yy} at f_2, which provides a high isolation between two operating frequencies and makes the 2-D wavefront control possible [9].

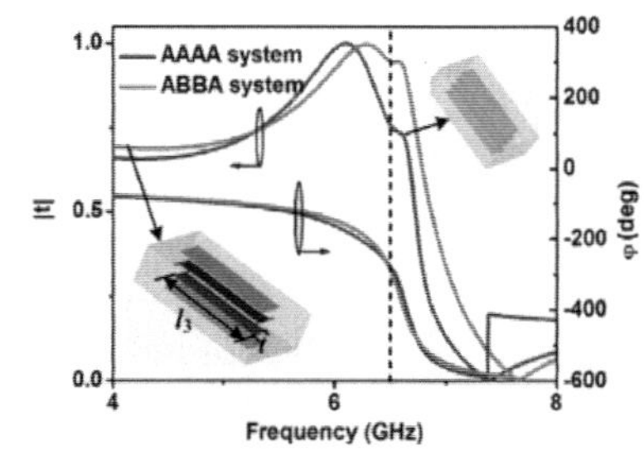

Figure 3. Transmission coefficients and transmission phases for different elements with the topology shown in the inset.

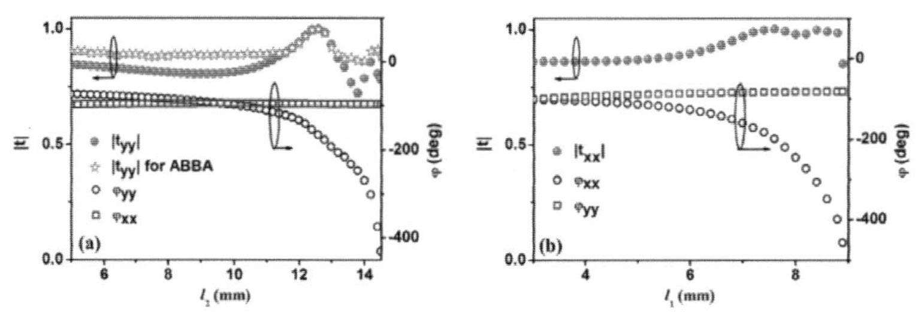

Figure 4. Final EM responses at frequencies of f_1 and f_2 under different polarizations.

(a) Transmission coefficients and transmission phases as a function of (a) l_2 at f_1 under excitation of an x-polarized wave and (b) l_1 at f_2 under excitation of a y-polarized wave for ABBA system.

3. Numerical results of high-efficiency dual-band focusing lens

The optimized ABBA system offers us the possibility to find a unit with an arbitrary transmission phase at dual operating frequencies. Here, we design a dual-band focusing lens with the parabolic transmission phase distributions satisfying $\varphi = k_i(\sqrt{F^2+x^2+y^2}-F)(i=1,2)$, where k_i and F denotes the wave vector at frequency of f_i and the focal length, respectively. Here, F=75 mm is arranged for both frequencies. Fig. 5(c) plots the calculated transmission phase distributions at f_2=10.5 GHz (the phase distributions at f_1=6.5 GHz is not given for brevity). Then, the dual-band focusing lens is designed with the pictures displayed in Figs. 5(a) and 5(b). To elucidate the key point of phase compensation, the electric field of E_x at f_2 along x axis (pink lines in Figs. 5(a) and 5(b)) are displayed clearly in Fig. 5(d) and the field is normalized to the maximum case (E_y at f_1 =6.5 GHz is not given for brevity). Here, two points should be highlighted. Firstly, high transmission is obtained for each element at both frequencies. Secondly, parabolic phase distributions are clearly demonstrated.

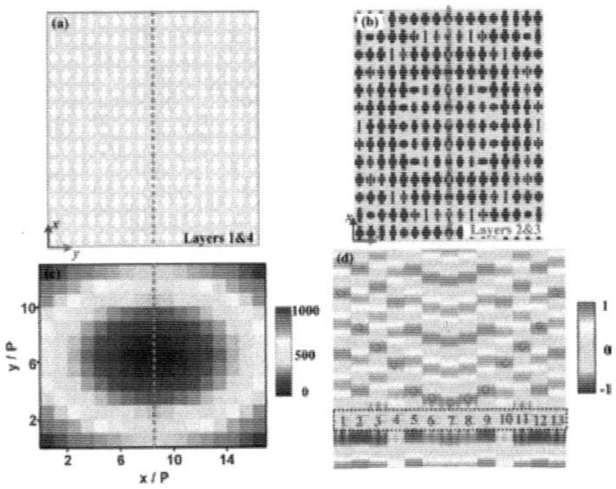

Figure 5. Design of the dual-band transparent focusing lens.

(a) Top view of the designed dual-band focusing lens (a) in layers 1 and 4 and (b) in layers 2 and 3; (c) The required phase distribution at f_2; (d) FDTD simulated E_x at f_2 along the red lines in (a) and (b).

To provide a physical insight into the focusing effects at both frequencies, we evaluate the *E*-field based on FDTD simulations. Fig. 6(a) depicts the simulated Re(E_y) distribution of a plane wave operating at f_1, and a continuous wavefront is unambiguously observed at *xoz* plane. Illuminating our dual-band focusing lens with an *x*-polarized plane wave at frequency of f_1, the Re(E_y) distribution at *xoz* plane is displayed in Fig. 6(b). As expected, perfect focusing effect is observed by the largely enhanced E_y around the focal point of the meta-surface. Moreover, an almost consistent wavefront at the input part demonstrates the minimal reflection and high efficiency of the focusing lens. We can identify the focal point by evaluating the *E*-field intensity distribution at *xoz* plane, as shown in Fig. 6(c). The maximum intensity at *z*=75 mm is found to be the focal point with *F*=75 mm, which coincides well with that of the designed value (*F*=75 mm). By rotating our meta-surface with 90°, a transparent focusing lens working at f_2 is pre designed. Compared with the plane wave shown in Fig. 6(e), the designed meta-surface enables the incident uniform wavefront to be converged tightly at a focal length of *F*=75 mm, shown in Figs. 6(f) and 6(g). The simulated focal length is consistent with that of the designed value (*F*=75 mm) for a second time. The decent focusing effects come from the strong phase compensating capacity of the meta-surface and also the exact design. Fig. 6(d) compares the normalized *E*-field intensity along *z* axis between the focusing lens and the referenced plane wave at both frequencies. A simple calculation indicates that the energy at the focal point is about 20 times larger than that of the plane wave at f_2 and 16 times at f_1. The enhanced field can also be validated by the *E*-field intensity along the line of *z*=75 mm, with the numerical results shown in Fig. 6(h).

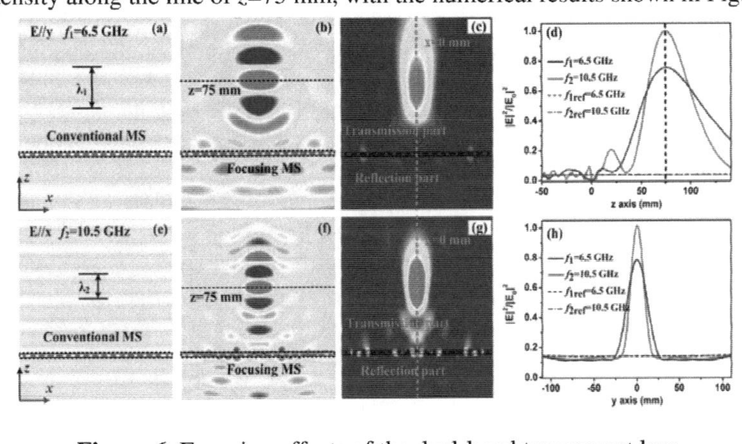

Figure 6. Focusing effects of the dual-band transparent lens.

E-field distributions of the referenced plane waves at (a) f_1 and (e) f_2 and of proposed dual-band focusing lens at (b) f_1 and (f) f_2; The power distributions for (c) f_1 and (g) f_2; Comparison of normalized *E*-field intensity along (d) *z* axis and (h) *z*=75 mm.

4. Dual-band lens antenna (LA) with improved radiation characteristics

Here, we try to improve the radiation characteristics of the dual-band LA by employing ABBA-designed focusing lens, which largely enhances the system transmission. The topology of the proposed dual-band LA is displayed in Fig. 7(a). Self-made Vivaldi antenna, operating in a wide bandwidth from 4 to 12 GHz, is applied as the feed antenna, which can generate quasi-spherical waves at both frequencies of f_1 and f_2. The parameter l=71 mm is optimized to obtain best radiation patterns at both frequencies. Then, we fabricate the focusing lens and the Vivaldi antenna with a standard printed-circuit-board (PCB) technology, with the pictures shown in Fig. 7(c).

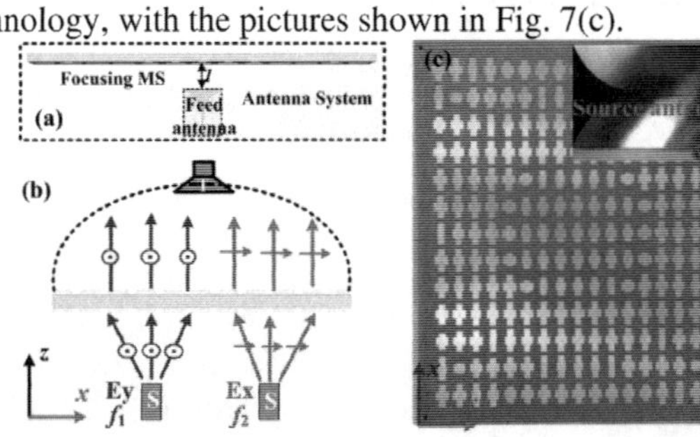

Figure 7. The (a) simulation model, (b) measurement schematic, and (c) the fabricated pictures for the dual-band LA.

For experimental demonstration, we measure the 2-D radiation patterns of the LA in two principal radiation planes through a far-field measurement system in an anechoic chamber. As seen in Fig. 8, the simulated and measured results are in reasonably consistent in all cases. Under excitation of a *y*-polarized wave at frequency of f_1, the measured (simulated) half power beam width (HPBW) is about 126.5° (127.3°) in *E* plane (*xoz* plane) and 91.2° (91.5°) in *H* plane (*yoz*) for the Vivaldi antenna, respectively. However, the HPBW reduces significantly to 13.5° (13.4°) and 14° (14.2°) in above mentioned planes for the LA, respectively. The measured (simulated) gain is about 18.7 dB (18.7 dB) for the LA, and a simple calculation shows that the gain has increased about 13.1 dB (13.2 dB) compared with that of the bare Vivaldi antenna. At frequency of f_2, the measured (simulated) HPBW is about 10.2° (10.4°) at *E* plane (*yoz* plane), 9.3° (9.4°) at *H* plane (*xoz* plane), respectively. A significant beam-width reduction of about 72.5° (68.8°) at *E* plane, 44.3° (42.5°) at *H* plane is achieved for the LA at f_2. The measured (simulated) radiation gain is about 23 dB (23.2 dB), 9.2 dB (9.4 dB) for the TA and bare Vivaldi antenna, respectively. Note that the aperture efficiency is evaluated with respect to the utmost directivity which can be calculated through the equation $\eta = G/D_{max} = G/(4\pi PQ/\lambda_0^2)\times 100\%$. The aperture efficiency of 30.3% (30.3%) is obtained at f_1, and 31.3% (32.8%) at f_2, respectively. Further analysis indicates that the F/B ratio is better than 15.5 dB in all cases. The side-lobe levels for all numerical results (experimental results) are about 16.3 dB (15.4 dB) lower than those of the mainlobe. The 1 dB gain bandwidth is evaluated as about 0.7 GHz (6.1-7 GHz), 1.2 GHz (9.8-11 GHz) at f_1 and f_2, respectively.

IOP Conf. Series: Materials Science and Engineering **167** (2017) 012060 doi:10.1088/1757-899X/167/1/012060

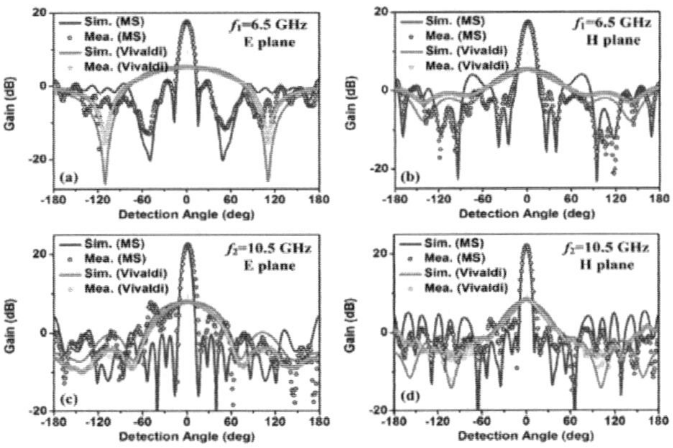

Figure 8. Simulated and measured radiation patterns at f_1 (a) *xoz* plane; (b) *yoz* plane and at f_2 (a) *yoz* plane; (b) *xoz* plane.

5. Conclusion

A novel LA operating in both C and X bands with improved radiation characteristics has been numerically and experimentally demonstrated by employing a high-performance transparent meta-surface. The designed LA shows a series of advantages such as high radiation gain of 18.7 dB at f_1 and 23 dB at f_2, competitive aperture efficiencies better than 30% at both frequencies and comparable wide bandwidth. Our numerical and experimental results can stimulate the realizations of high-performance transparent meta-surfaces working in other frequencies or achieving other integrated functionalities.

References

[1] Yu N, Genevet P and Kats M A 2011 Light propagation with phase discontinuities: generalized laws of reflection and refraction *Science* **334** 333-338

[2] Huang L, Chen X, Mühlenbernd H and Zhang H 2013 Three-dimensional optical holography using a plasmonic metasurface *Nat. Commun.* **4** 2808

[3] Ma H F, Wang G Z and Kong G S 2015 Independent controls of differently-polarized reflected waves by anisotropic metasurfaces *Sci. Rep.* **5** 9605

[4] Khorasaninejad M, Chen W T and Devlin R C 2016 Metalenses at visible wavelengths: diffraction-limited focusing and subwavelength resolution imaging *Science* **352** 1190

[5] Wan X, Shen X, Luo Y and Cui T J 2014 Planar bifunctional Luneburg-fisheye lens made of an anisotropic metasurface *Laser Photonics Rev.* **8(5)** 757-765

[6] Li H, Yuan L H and Zhou B Ultrathin multiband gigahertz metamaterial absorbers 2011 *J. Appl. Phys.***110** 014909

[7] Shen X, Cui T J and Zhao J 2011 Polarization-independent wide-angle triple-band metamaterial absorber *Opt. Express***19(10)** 9401-9407

[8] Huang C, Pan W, Ma X and Luo X 2016 Multi-spectral metasurface for different functional control of reflection waves *Sci. Rep.* **6** 23291

[9] Cai T, Wang G M and Zhang F X 2015 Ultra-thin polarization beam splitter using 2-D transmissive phase gradient metasurface *IEEE Trans. Antennas Propag.* **63(12)** 5629-5636

[10] Sun W, He Q, Sun S and Zhou L 2016 High-efficiency surface plasmon meta-couplers: concept and microwave-regime realizations *Light: Sci. Appl.* **5** e16003

[11] Pfeiffer C and Grbic A 2013 Metamaterial Huygens' surfaces: tailoring wave fronts with reflectionless sheets *Phys. Rev. Lett.* **110** 197401

Steer and split electro-magnetic waves by employing ultra-thin anisotropic meta-material

Wenlong Guo[1], Tangjing Li, Haipeng Li, Haisheng Hou

Air Force Engineering University, Xi'an 710051, People's Republic of China

E-mail: 13259461383@163.com

Abstract. As an ultra-thin developed form of meta-materials, metasurface does attract tremendous attention due to its flexibility in controlling wave front in different spectrums. In this paper, we propose a microwave metasurface with function of transforming spherical waves into plane waves and splitting waves according to polarized status of incident waves. By using a compact ultra-thin anisotropic element, we design a polarization beam splitter under spherical wave's illumination. With different phase gradients for *x/y*-polarized waves, the anisotropic metasurface has function of gain enhancement of 12.4 dB on average and reflecting *x/y*-polarized waves in *-x/+y* direction with the same reflected angel of 33° at the operating frequency of 15GHz. For verification of simulation results, the metasurface sample with size of $102\times102mm^2$ and cells of 17×17 is fabricated and measured. Experimental results are in excellent agreement with the simulated ones, in which the high isolation of 30dB at 15GHz between *x*- and *y*-polarized waves is obtained. Due to the great performance in beam steering and thickness reduction, our design provides a promising approach of beam splitting, steering and gain enhancement in microwave region.

1. Introduction

Metamaterails have gained much attention around the world since J. B. Pendry and D. Smith proposed the concept of negative permittivity [1] and permeability [2]. But multi-layered three-dimensional metamaterials suffer from a narrow usable bandwidth, low optical efficiency as well as complicated and costly fabrication. Recently, the concept of an abrupt phase change at the interface provides a powerful solution to overcome the limitations of metamaterials, which is also called metasurface [3-5]. A metasurface is usually introduced by applying a series of periodic or quasi-periodic subwavelength antennas on a surface from which novel optical properties are achieved [6], such as negative refraction [7], subwavelength focusing [8-10]. Based on the plasmonic response of the gold V-shaped nano-antenna, phase gradient metasurface (PGM) is first proposed by Yu et al. in 2011 [3]. By imposing a hyperbolical phase profile on the interface, planar focusing metasurface (PFM) has attracted attention since they can replace lossy and bulky curved conventional focusing elements [11-12].

Among the most researches, it's always a hot topic to manipulate polarized status of propagating waves by employing anisotropic metasurface [13-14]. For example, polarized converters as well as polarized beam splitters are commonly realized by anisotropic metasurfaces [15-16]. As for polarized converters, the study group of Shaobo Qu [16-17] and the group of Tiejun Cui [18-19] have done many in-depth researches on both cross-polarization conversion and linear-to-circular polarization conversion. But for beam splitters, as depicted in Ref. 19 and 20, only the condition of plane waves' illumination is considered in most research articles up to now. Therefore it's still in great desire to explore the beam splitters under the condition of spherical waves' illumination.

In this paper, we first propose a new kind of phase distribution which contributes to both anomalous reflection and gain enhancement. Then an ultra-thin anisotropic reflected metasurface element is designed to construct a beam splitter under spherical waves illumination according to the phase distribution we proposed. By putting different phase gradients for *x*/*y*-polarized waves on the anisotropic metasurface, it has function of gain enhancement of 12.4 dB on average and reflecting *x*/*y*-polarized waves in -*x*/+*y* direction with the same reflected angle of 33°. Besides the isolates degree above 30dB between *x*- and *y*-polarized waves is obtained in both simulated and measured setup at 15GHz. The metasurface is single-layered with thickness of 1mm (0.05λ), which makes contributions in thickness reduction of microwave devices.

2. Anisotropic metasurface cell design

The structure of the metasurface cell is demonstrated in Fig.1, in which the double improved orthogonally I-shaped metallic structures are employed to control the reflection phases of *x*/*y*-polarized electromagnetic waves respectively. The detailed dimensions of the structure shown in Fig. 1(b) are p=6 mm, b=0.4mm, d= 0.2 mm and the orthogonally I-shaped metallic structures are separated from a metal grounded plane by a dielectric spacer with thickness of h=1mm. The dielectric layer is with the relative permittivity ε_r=4.3 and loss tangent 0.001. In order to illustrate the polarization-independent property for differently-polarized incident waves, we fix a_y=2mm and vary a_x from 2.2 mm to 5.6 mm. Fig.1(c) depicts the reflection phases at 15 GHz for *x*/*y*-polarized incoming waves. A desirable phase variation from -133° to -473° is unambiguously observed for the *x*-polarization waves, while the reflection phase is not affected for the *y*-polarized waves. Similarly, if a_x is fixed at 2 mm and a_y is varied from 2.2 mm to 5.6 mm, the reflection phases of *y*-polarized waves are changed from -133° to -473°, while the reflection phases of the *x*-polarized waves are not affected. In addition, the phase range does not cover 360° despite of independent phase response for differently polarized waves, which will debase the precision of phase control. But fortunately the phase range covers 340°. When designing a metasurface sample, we will choose the cells with relative phase response of 340° and 360° (0°) instead of cells with phase ranging from 340° to 350° and 350° to 360°respectively. In this way the phase compensation error will be depressed to the least, which will contribute to the satisfying performance of the designed metasurface sample.

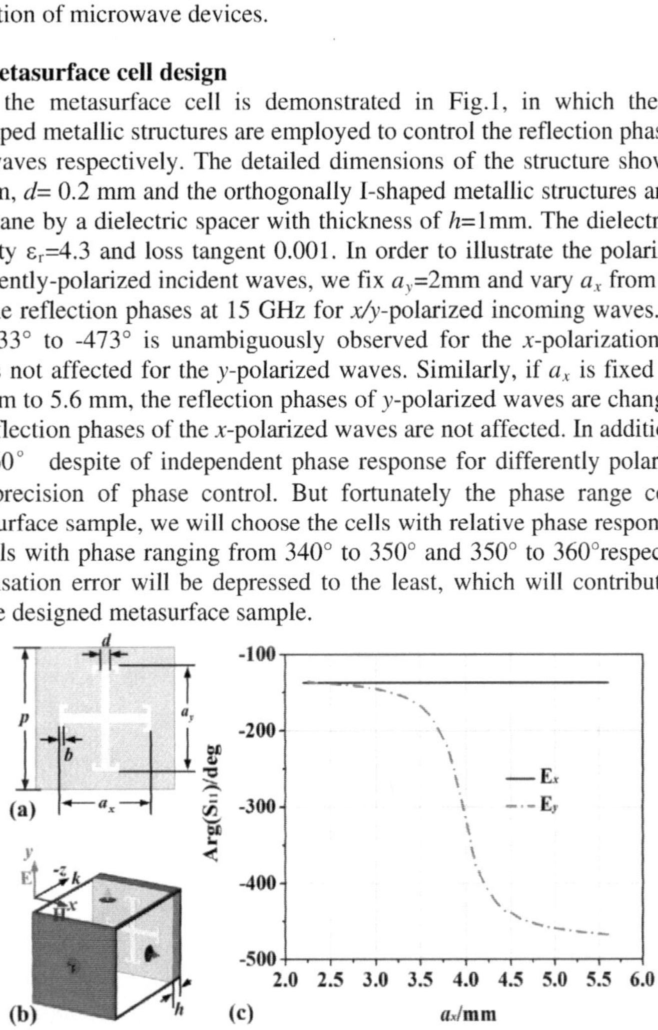

Figure 1. (a) Top view and (b) perspective view of the reflected MS element, (c) phase response for *x*- and *y*-polarzied incident waves with a_y=2 mm and a_x varying from 2.2 mm to 5.6 mm.

3. Beam Splitters under Plane Waves' Illumination

As we know the gradient phase distribution shown in Eq. (1) contributes to anomalous reflection according to the general reflection law as depicted in Eq. (2). Considering the element we proposed has independent phase response for orthogonally polarized waves, we can set the different phase

IOP Conf. Series: Materials Science and Engineering **167** (2017) 012061 doi:10.1088/1757-899X/167/1/012061

gradients for the orthogonally polarized beams. In this way, the designed metasurface sample will reflect the orthogonally polarized waves in different deflection angles.

$$\Phi(x) = \xi \cdot x + \Phi_0 \tag{1}$$

$$\sin\theta_r - \sin\theta_i = \frac{\lambda}{2\pi n_i}\frac{d\Phi}{dx} \tag{2}$$

Based on above designed procedure, we propose a one-dimensional phase gradient metasurface (a supercell) as shown in Fig. 2(a). The parameters of the cells along +*x* axis are demonstrated in table I, in which the phase response of the six distributed cells are also depicted. Moreover $\Phi_x(\Phi_y)$ in the table corresponds to the phase response for *x*(*y*)-polarized incident waves. Unanimously the phase gradient is fixed for 60° between adjacent cells. By fitting 3×3 supercells together as shown in Fig. 2(b), the constituted metasurface is simulated in CST Microwave Studio, in which periodic boundary condition is set in *y* direction while open boundary condition is set in *x* and *z* directions. Fig. 3(c) depicts the simulated reflection coefficients of the metasurface under *x*- and *y*-polarized waves' illumination. As shown, the valley lies at 15GHz, which demonstrate that the incident beams are mostly reflected anomalously at 15GHz. In order to detect the operating procedure of the anomalous reflection, the E-field distribution on *xoz* plane is calculated and shown in Fig. 3. Thereinto Fig. 3(a) and (b) depict the cases under *x*- and *y*-polarized waves' illumination respectively. In addition, the deflection angles for *x*- and *y*-polarized beams are denoted by 33.7° and 32.9° which are in good accordance with theoretic ones calculated by Eq. (2).

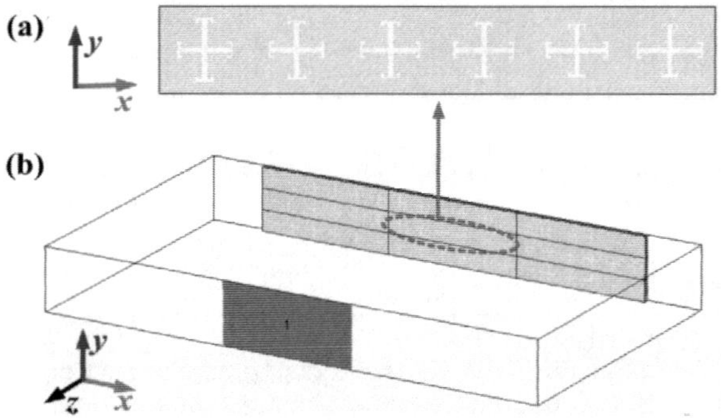

Figure 2. (a) The designed 1-dimensional phase gradient metasurface, (b) the simulated setup for the metasurface constituted by3×3 supercells.

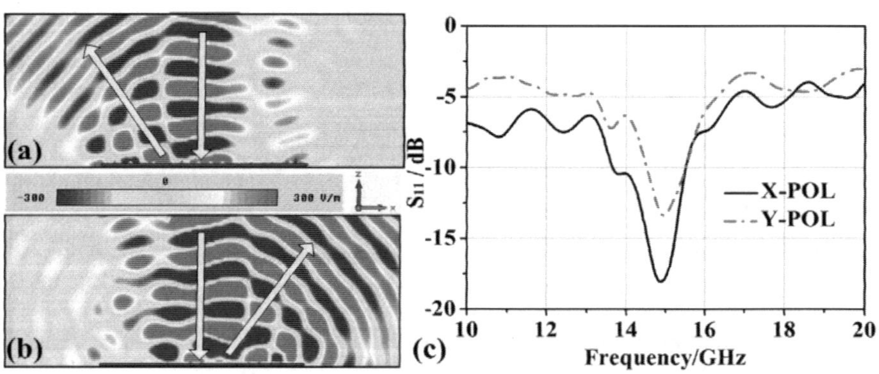

Figure 3. The E field distribution on *xoz* plane for (a) *x*-polarized and (b) *y*-polarized waves, (c) the reflection coefficients of the metasurface under *x*- and *y*-polarized waves' illumination.

Table 1. The sizes and absolute phase shifts of the six distributed cells

Index n	1	2	3	4	5	6
a_x (mm)	4.4	4.1	4	3.89	3.7	2.2
a_y (mm)	2.2	3.7	3.89	4	4.1	4.4
Φ_x(deg)	-435	-374	-314	-255	-197	-136
Φ_y(deg)	-136	-197	-255	-314	-374	-435

4. Beam Splitters under Spherical Waves' Illumination

As we know, the phase distribution as depicted in Eq. (3) is capable of converting spherical waves into plane waves and vice versa. Besides, it's the basic phase distribution for a planar focusing metasurface (PFM). According to the general reflection law as depicted in Eq. (2), the plane waves impinging on the PGM normally will be reflected to a certain angle in x direction. Moreover, seeing that the spherical waves will be converted into plane waves by PFM effectively, an additional gradient phase can be attached on the PFM lens to get the plane waves converted by spherical waves reflect to an anomalous angle. In addition, if different phase gradients are set for *x/y*-polarized waves, the spherical waves will be split into *x/y*-polarized plane waves with different deflection angles.

$$\Phi(x,y)=\frac{2\pi}{\lambda}(\sqrt{x^2+y^2+L^2}-L)+\Phi_0 \tag{3}$$

According to this point, the phase distributions for *x/y*-polarized incident waves should meet the phase relationship as described as below

$$\Phi_x(x,y)=\frac{2\pi}{\lambda}(\sqrt{x^2+y^2+L^2}-L)+\xi_1\cdot x+\Phi_0 \tag{4}$$

$$\Phi_y(x,y)=\frac{2\pi}{\lambda}(\sqrt{x^2+y^2+L^2}-L)+\xi_2\cdot y+\Phi_0 \tag{5}$$

where $\Phi_x(x,y)$ and $\Phi_x(x,y)$ represent the phase distributions for x- and y-polarized waves while ξ_1 and ξ_2 correspond to the extra linear phase gradients. In addition, λ is the wavelength in vacuum and L is the designed focal length. Deduced from above description, the reflection angles of the *x/y*-polarized waves are only related to the linear phase gradients ξ_1 and ξ_2, which can also be calculated according to generalized reflection law as below:

$$\theta_{r-x}=\sin^{-1}(\frac{\lambda}{2\pi}\xi_1) \tag{6}$$

$$\theta_{r-y}=\sin^{-1}(\frac{\lambda}{2\pi}\xi_2) \tag{7}$$

Besides, since the extra phase gradients ξ_1 and ξ_2 are appended in x and y direction, x- and y-polarized beams will be reflected along x and y direction with deflection angles of θ_{r-x} and θ_{r-x} respectively. Based on above designed principle, we propose a metasurface sample with size of 102×102 mm^2 and cells of 17×17 as shown in Fig. 4(a). With phase distributed as Eq. (4) and (5), Fig. 4(b) and (c) depict the relative phase distribution for x- and y-polarized waves on the metasurface. The focal length L is set for 40mm and the wavelength is denoted by 20mm (at 15GHz). In addition, the extra linear phase gradient ξ_1 and ξ_2 are set for $\pi/3p$ and $-\pi/3p$. The feed source of the designed metasurface is shown in Fig. 5, in which the patch antenna radiates omnidirectionally with an operating bandwidth from 14.4GHz to 15.5 GHz. With the feed source placed at the focal point of the metasurface, the constituted system is shown in Fig. 6(b), in which the patch antenna is set obliquely as shown in Fig. 6(a) in order to get both *x/y*-polarized incident waves at the same time. Furthermore, the polarized status of incident beam can be described as

$$E=e_x\cdot|E|\cos\varphi+e_y\cdot|E|\sin\varphi \tag{8}$$

Here we set φ=45° in order to keep the amplitudes of x- and y-polarized incident waves equal. And then we set the assembly metasurface sample simulated in CST Microwave Studio where open boundary conditions are set in all directions. In Fig. 6(c) is demonstrated the simulated far-field pattern of the system. Obviously, double pencil shaped beams with the maximum gain of 17.6 dB on

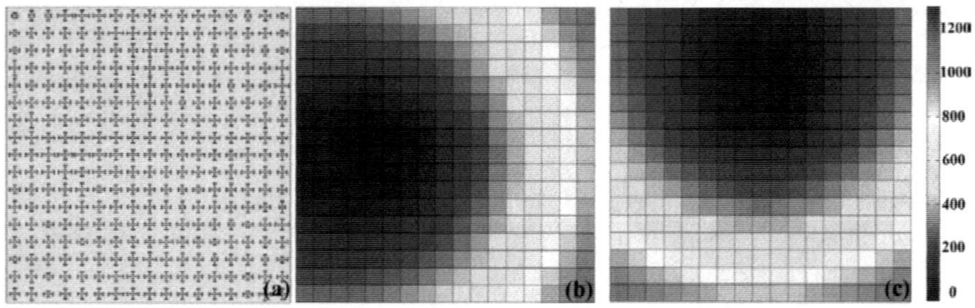

Figure 4. (a) The designed metasurface sample, (b) the relative phase distribution for *x*-polarized waves, (c) the relative phase distribution for *y*-polarized waves.

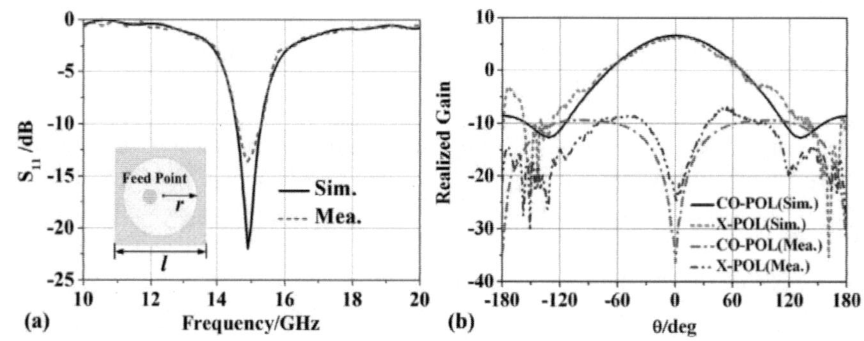

Figure 5. (a) The reflection coefficient and (b) the far-field pattern of the patch antenna

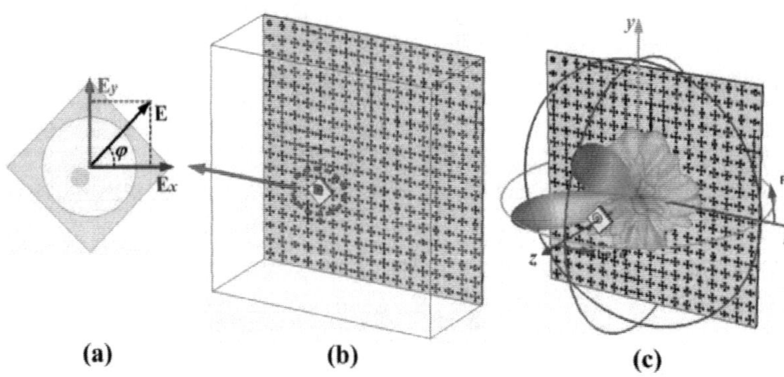

Figure 6. (a) The placement of the patch antenna; (b) the constituted metasurface sample, (c) the far-field pattern of the designed metasurface system.

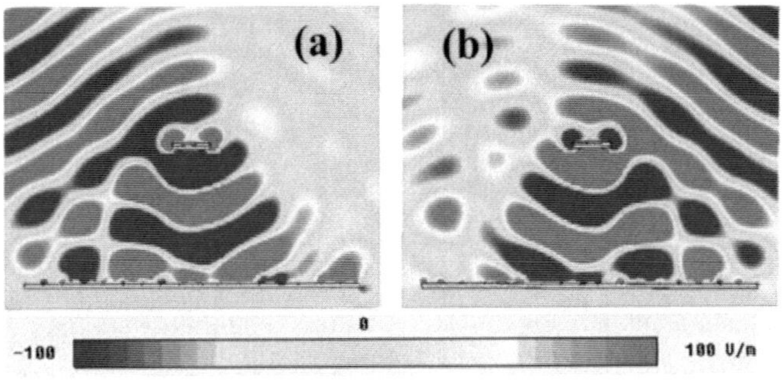

Figure 7. E-field distribution on (a) *xoz* plane and (b) *yoz* plane.

xoz and *yoz* plane are achieved. In order to illustrate the operating mechanism, the E-field distributions on *xoz* and *yoz* plane are calculated and shown in Fig. 7(a) and (b). Unanimously, the spherical waves emitted from the feed source have been converted into plane waves with reflection angle of 33.2°, which are in good accordance with the theoretical ones calculated by the general reflection law.

In order to further verify the design, the metasurface sample is fabricated and then measured in a microwave anechoic chamber as shown in Fig. 8. The test setup is consistent with the simulated one. The numerical and measured far-field patterns on *xoz* and *yoz* plane are plotted in Fig. 9. It's obvious that the *x*-polarized waves are reflected in $-x$ direction with deflection angle of 33.2° while *y*-polarized waves are reflected in $+y$ direction with reflection angle of 33.1°. Besides, the isolation degrees at the peak value between *x*- and *y*-polarized waves surpass 30dB, which is revealed about perfect beam splitting performance. Thus by designing and assigning a fire-new phase distribution on the reflected metasurface, the fabricated sample is capable of converting spherical waves to plane waves as well as splitting different polarized waves into diverse directions.

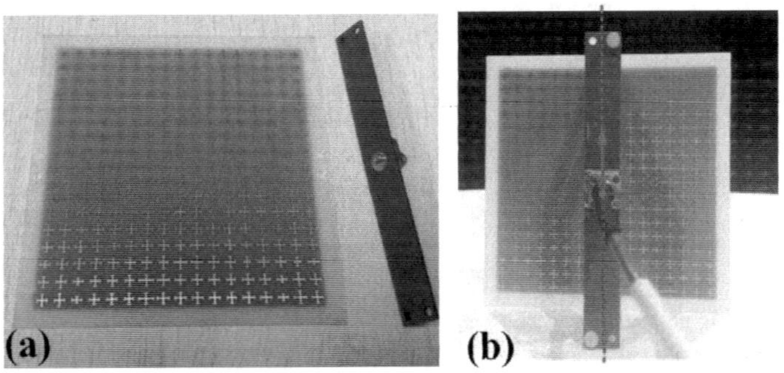

Figure 8. (a) The fabricated metasurface sample and (b) the measured setup

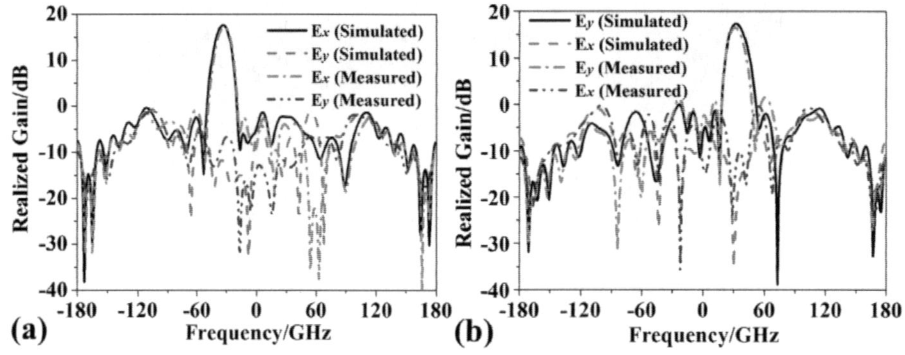

Figure 9. The simulated and measured gain on (a) *xoz* and (b) *yoz* plane.

5. Conclusion and perspectives

In conclusion, a new method to design beam splitter under spherical waves' illumination is proposed. By adding additional linear phase profile on hyperbolic phase distribution, the combined phase profile on surface contributes to both gain enhancement and anomalous reflection. With different phase profiles for orthogonally polarized waves, the accordingly designed meatsurface is capable of improving the patch antenna gain above 12.4dB and splitting *x/y*-polarized waves in *-x/+y* directions with deflection angle of 33.4° at 15GHz. With thickness reduction, the proposed metasurface has great application in the region of beam splitting, steering and gain enhancement in microwave spectrum.

Acknowledgment

The authors would like to express their gratitude to anonymous reviewers for their helpful comments and China North Electronic Engineering Research Institute for the fabrication. This work was supported by the National Natural Science Foundation of China (Grant Nos.61372034, 61501499)

References

[1] Pendry J. B., Holden A. J., et al. 1998 J. *Phys. Condens. Matter* **vol. 10** pp. 4785–4809
[2] Smith D. R., et al. 2000 *Phys. Rev. Lett.* **vol. 84** pp. 4184–4187
[3] Yu N. F., Capasso F., et al. 2011 *Science* **vol. 334** pp. 333-337
[4] Paul O., Reinhard B., et al. 2010 *Appl. Phys. Lett.* **vol. 96** p. 241110
[5] Lin Q., Cui T. J., et al. 2008 *Appl. Phys. Lett.* **vol. 92** p. 131904 2008.
[6] Zhang K., et al. 2014 *New J. Phys.* **vol. 16** p. 103020
[7] Ni X., et al. 2012 *Science* **vol. 335** pp. 427
[8] Aieta F. et al. 2012 *Nano Lett.* **vol. 12** 4932-4936
[9] Saeidi C., Weide D. van der 2014 *Appl. Phys. Lett.* **vol. 105** p. 053107
[10] Wan X., Li Y. B., Cai B. G., and Cui T. J. 2014 *Appl. Phys. Lett.* **vol. 105** p. 151604
[11] Li X., Cui T. J., and Zhou L., et al. 2012 *Opt. Lett.* **vol. 37** pp. 4940-4942
[12] Grady N. K., et al. 2013 *Science*, **vol. 340**, pp. 1304-1307
[13] Chen H. Y. et al. 2014 *J. Appl. Phys.*, **vol. 115**, p.154504
[14] Guo W. L., Wang G. M., et al. 2016 *Chinese Phys. B.* **vol. 25** no. 8
[15] Cai T. et al. 2015 *IEEE Trans. Antennas Propag.* **vol. 63 no. 12** pp. 5629-5636
[16] Li Y. F., et al. 2015 *J. Appl. Phys.* **vol. 117** p.044501
[17] Song K., Liu Y. H., Luo C. R., and Zhao X. P., 2014 *J. Phys. D: Appl. Phys.* **vol. 47** p.505104
[18] Ma H. F., Wang G. Z., Cui T. J. et al. 2014 *Opt. Matter. Express* **vol. 4, no. 8**, pp.1717-1724
[19] Ma H. F., and Cui T. J., et al. 2015 *Sci. Rep.* **vol. 5** p. 9605
[20] Jia S. L., Wan X. Cui T. J., et al. 2015 *Laser Photonics Rev.* **vol. 63 no. 5** pp. 545-553

Fabrication and characterization of non-volatile transistor memory based on polypeptide as gate dielectric

Lijuan Liang [1,3], LianFang Li[1], Xianfu Wei[1], Beiqing Huang[1], Yen Wei[2]

[1] Beijing Institute of Graphic Communication, 102600 Beijing, China
[2] Department of Chemistry, Tsinghua University, Beijing, 100084, China

E-mail: lianglijuan@bigc.edu.cn

Abstract. The organic thin film transistor (OTFT) fabricated with the polypeptide as a dielectric layer shows memory function. In order to investigate the effect of polypeptide structure on the performance of non-volatile transistor memory, the Fourier-transform IR (FT-IR) and Circular Dichiroism (CD) spectral of PMLG film has been applied, respectively. In conclusion, the memory transistor device fabricated with polypeptide as the ferroelectric exhibit promising behavior such as a large memory window, and the dipole moment of the amide group was considered as the main source of the memory function.

1. Introduction

Recently, the non-volatile transistor memory array has been investigated extensively because it is an essential element for flexible electronic application such as in the memory card, radio frequency identification tag, e-signage and sensor[1-2]. As the first polyvinylidene fluoride PVDF and its derivatives based memory device was reported [3], there are lots of conjugated molecules and solution processable ferroelectric materials such as DNA [4,5,6], nylon [7], poly(imide) [8] are expected to realize the low-cost and low temperature fabrication process.

Poly (γ-methyl-L-glutamate) (PMLG), a synthetic polypeptide, can exist in various solid state, the preparation method and solvents which were chosen has great effect on the conformation forming [9,10]. A β-sheet structure is observed in films swollen in formic acid or in crystals precipitated from diluted solution in formic acid [11]. In most other solvents, the PMLG exist in the α helix conformation. Twisted structures similar to the cholesteric liquid-crystal state of the concentrated solution are retained in films prepared from dichloroethane solution[12, 13]. A gel which contains fibrous aggregations can be formed when the N, N-dimethylformamide (DMF) was applied as the solvent [14, 15]. The poorly-crystalline films were obtained from dimethylformamide (DMF) solution. In contrast, the films prepared from solutions in chloroform, dichloromethane (DCM) or m-cresol are highly crystalline, and consist of α helices arranged in a hexagonal lattice with dimension a=1.20 nm[16,17]. Therefore, understanding the film state is important to improve the performance of memory device from the viewpoint of their practical applications.

In this study, we demonstrated a top contacted ferroelectric polymer non-volatile transistor memory with PMLG and pentacene as a gate dielectric and semiconductor, respectively. The structure of the PMLG and the electronic performance including the transfer characteristic and output characteristic has been investigated. It is clearly seen that the fabricate device exhibit the memory behavior. And the relationship between molecular structure and memory mechanism has been presented successfully.

2. Experiment

2.1. Material
The PMLG (degree of polymerization: 440) were provided by Kyowa Hakko Co., Ltd. 1,2-Dichloroethane (Kanto Kagaku Co., Inc.) was used as received. Pentacene (Nard Institute Ltd. 98% purity) was used without further purification. Au (Sigma-Aldrich) was applied as source and drain electrode.

2.2. Preparation of the TFT device
PMLG film was prepared by spin-coating method onto the indium thin oxide (ITO) glass substrate with a 1, 2-dichloroethane solution containing 3 wt. % PMLG. The thickness of the film was set to 800 nm. Pentacene film (thickness =50 nm) as an active layer was deposited on the PMLG film by vacuum deposition method under 2x10-3 Pa at a deposition rate of 0.2–0.4 Ås-1. Au layer as the source and drain electrodes (W/L=5 mm /20 μm) was deposited by successive vacuum deposition on this pentacene layer. The structure of the OTFT memory and polypeptide are shown in Figure 1.

2.3. Apparatus
Circular dichroism (CD) spectra were recorded with the CD spectrometer (JASCO, J-820). Transmission Fourier-transform IR (FT-IR) spectra was measured at room temperature using FT-IT-660 Plus (JUSCO) equipped with an MCT detector. All electric measurements were carried out using a Keithley 4200 semiconductor parameter analyzer under dark conditions in vacuo.

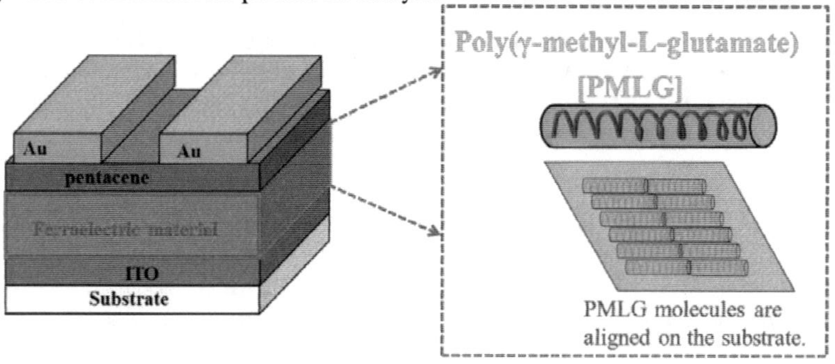

Figure 1. The configuration of the OTFT memory device using PMLG as the gate dielectric.

3. Results and Discussion
In order to investigate the conformation of poly (γ-methyl-L-glutamate) (PMLG) film employed in this study, the CD analysis has been carried out. Figure 2 shows the CD spectra of PMLG in film state which was spin-coated on calcium fluoride substrate. As shown in the figure, either positive or negative CD signals have been obtained. For instance, the negative Cotton effect at about 205 nm and 230 nm have been observed. In addition, the positive Cotton effect about 196 nm and 210 nm have also been obtained. All of these signals indicated that the polypeptide used in this study is similar in shape to α helix structure, which is consistent well with the previously reported papers. [7-8] In addition, the shape of the spectrum was also consistent with that in the case of a clockwise interaction of the dipole moment of the PMLG main chain. [11] This implied that the formation of the helical structure is in a tertiary order, similar to a cholesteric liquid crystal phase.

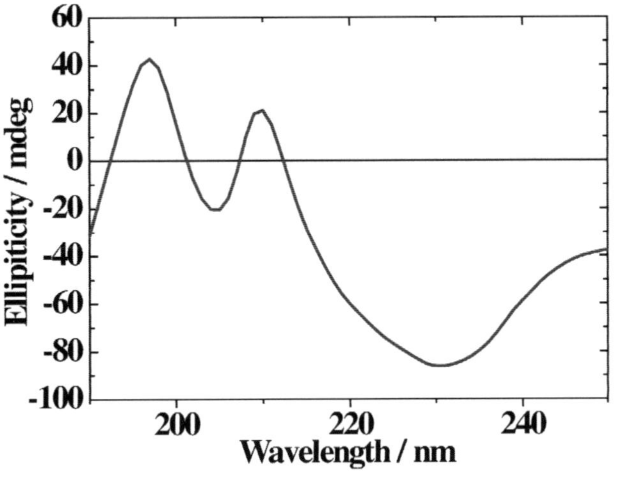

Figure 2. CD spectra of PMLG film.

Infrared spectroscopy (IR spectroscopy) which deals with the infrared region of the electromagnetic spectrum exhibits important information in the identification and investigation of the chemical materials. In this study, for the purpose of confirming the conformation of the PMLG film, IR spectroscopy has been carried out. In addition, through calculating the intensity ratio of amide1 to amide 2, the orientation of the molecular axis of PMLG can also be determined by IR spectra. For instance, if the ratio of amide 1 to amide 2 is nearly the same, the molecular axis of PMLG will be oriented upright on the film surface. In contrast, when the PMLG molecules lie on substrates, namely, which is parallel with the film surface, the ratio will be remarkably higher.

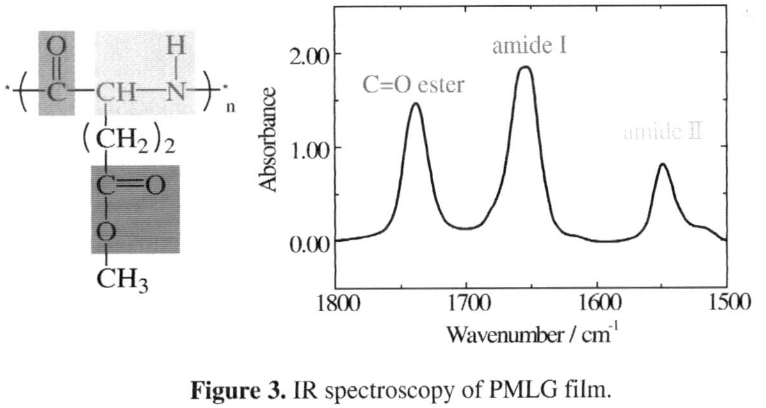

Figure 3. IR spectroscopy of PMLG film.

Figure 3 shows the IR spectroscopy of PMLG film which is spin-coated on calcium fluoride substrate and analyzed by FT-IR absorption over a range of 1800–1500 cm-1, and the left figure also illustrated the corresponding structure with the IR spectral. As the figure shows, the specific bands corresponding to the C=O ester has been well presented. In addition, the dipole groups with high dipole moment mainly including amide 1 and amide 2 groups also have been obtained, which is consistent well with its own structure. Further, based on the ratio of amide 1 to amide 2, it can be concluded that in the PMLG film, the molecules lie on the substrate.

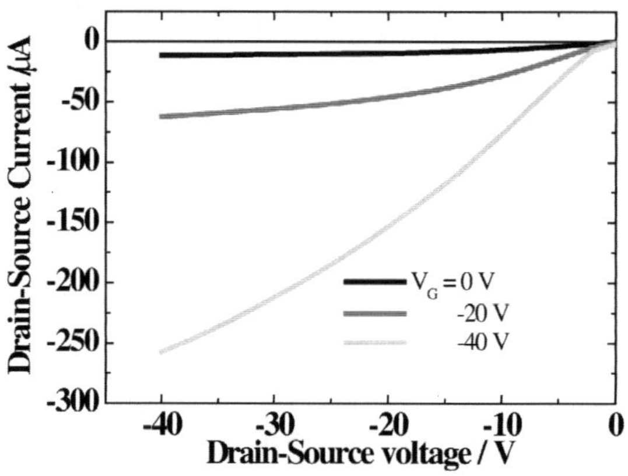

Figure 4. Output characteristic of memory device based on PMLG as gate dielectric.

Output characteristic is also one of the most important electronic characteristics of OTFT memory device, it can be obtained by measuring the I_{DS} as a function of the voltage applied to the drain source electrode with a fixed gate voltage (V_G), as shown in Figure 4. It shows that the dependence of the drain source current I_{DS} on the drain source voltage V_{DS} for the devices using PMLG film as the dielectric layer, and it exhibits clear field effect with operation in p-type accumulation mode through applying negative gate bias. Compared with the DNA-CTMA based device, PMLG based device exhibits 10 times higher current densities. At the same time, only the linear region can be observed. Namely, the device gives a non-ideal I_{DS} with less well-defined pinch off. It indicates that depletion layer in conducting channel was not formed at high V_{DS} region, suggesting that the induced charge at the conducting channel may be originated from the spontaneous polarization of main chain dipole of PMLG molecule.

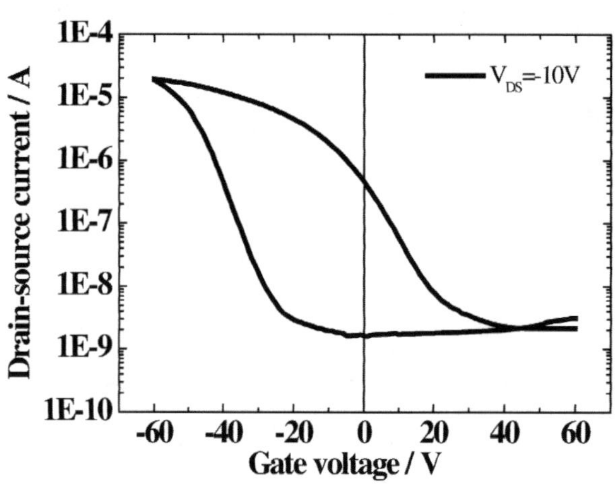

Figure 5. Transfer characteristic of memory device based on PMLG as gate dielectric.

Figure 5 presents the transfer characteristic which represents the transistor current I_{DS} plotted as a function of the gate voltage at constant V_{DS} of the OTFT memory fabricated with PMLG as a gate dielectric layer. The observation of sizable hysteresis indicated that PMLG film can work as the memory layer in the preparation of OTFT memories. The ON/OFF current ratio at V_G=0 V has been calculated as 2.79×10^2. At the same time, according to the plotting square root of I_{DS} against V_G, the threshold voltage of -36 V was obtained. Further, the field effect mobility was calculated by fitting the plot of the square root of I_{DS} versus V_G using the following equation [18, 19, 20].

$$I_{DS(lin)} = \frac{W}{L}\mu C_i(V_G - V_{TH})V_{DS} \quad (1)$$

Where I_{DS} is the drain source current; C is the capacitance of gate dielectric per unit area; V_G is the gate voltage; V_{TH} is the threshold voltage, W and L are the width and length of the channel of the device, respectively. The capacitance was measured to be 0.424 nF/cm^2. The carrier mobility of the PMLG based device was calculated to be 1.09 cm^2/V·s. To our best knowledge, it is the moderate mobility reported for a bottom gate top contact organic semiconductor TFT memory device based on PMLG gate dielectric.

4. Conclusion

In conclusion, we demonstrated a non-volatile transistor memory device based on PMLG as the ferroelectric dielectric layer. A high mobility up to 1.09 cm^2/V·s was obtained. Based on the study of the Circular Dichiroism and Infrared spectroscopy, it indicated that the PMLG applied in this experiment possess the helix structure, and the PMLG molecules lie on the substrate. Combined with the electronic performance of the memory device, it reveals that the dipole moment in the main chain is the main source of the memory property.

Acknowledgment

This study was funded by the China postdoctal science foundation, the Beijing Natural Science Foundation (2164061), the general project of science and technology of Beijing Municipal Education Commission (SQKM201610015006), the scientific research project of Beijing college student, and the initial funding for the Doctoral program of BIGC (NO. 27170115005/017).

References

[1] B. A. Mattis , B. Subramanian, 2006, *Tech. Dig. Int. Electron Devices Meet*, 311.
[2] H. T. Lin, Z. Pei, J. R. Chen, 2007, *Int. Electron Devices Meet*, 233.
[3] Kawai, H, 1969, *Jpn. J. Appl. Phys.* **8**, 975-976.
[4] T. Yukimoto, S. Uemura, T. Kamata, K. Nakamura, N. Kobayashi, 2011, *J. Mater. Chem.* **21**, 15575–15579.
[5] L.J. Liang, K. Nakamura, S. Uemura, T. Kamata, N. Kobayashi, 2014, *Sci. Adv. Mater.* **6**, 1-4.
[6] L.J. Liang, Y. Mitsumura, T. Yukimoto, S. Uemura, T. Kamata, K. Nakamura, N. Kobayashi, 2012, *J. Korean Soc. Image Sci. Tech.* **18**, 16–24.
[7] A. J. Rogers, Z. Bao, V. R. Raju, 1998, *App. Phys. Lett*, **72**, 2716.
[8] H. Sakai, H. Isoda, Y. Furukawa, 2012, *Jpn. J. Appl. Phys*, **51**, 040210.
[9] Bamford, C. H., Brown, L., Elliott, A., Hanby, W. E., Trotter, I. F. 1952, *Nature*, **169**,357-358.
[10] Watanabe, J., Sasaki, S., Uematsu, I., 1977, *Polymer Journal*, **9**, 451-458.
[11] Nakajima, A., Fujiwara, T., Hayashi, T., Kaji, K., 1973, *Biopolymer*, **12**, 2681-2690.
[12] Tachibana, T., Oda, F., Bull., 1973, *Chem. Soc. Jpn.* **46**, 2583-2584.
[13] Watanabe, J., Sasaki, S., Uematsu, I., 1977, *Polymer Journal*, **9**, 337-340.
[14] Ishikawa, S. J., 1965, *Polym. Sci. (A)*, **3**, 4075-4092.
[15] J. J. B. P. Blais, P. H. Geil, 1968, *J. Ultrastructure Res*, **22**, 303.
[16] Martin Egginger, Mihai Irimia-Vladu, Anddreas Tanda, Irene Frischauf, Siegfried Bauer, Niyazi Serdar Sariciftci, 2008, Adv. Mater. 20, 1018-1022.
[17] Martin Egginger, Siegfried Bauer, Reinhard Schwödiauer, Helmut Neugebauer, Niyazi Serdar Sariciftci, 2012, *Monatshefte fuer Chemie/Chemical Monthly*, **140**, 735-750.
[18] C. Kim, A. Facchetti, T. J. Marks, 2007, *Science*, **318**, 76-80.
[19] H. Yan, Z. Chen, Y. Zheng, C. Newman, J. R. Quinn, M. Kastler, F. Dotz, A. Facchetti, 2009, *Nature*, **457**, 679-686.

[20] D. J. Gundlach, J. E. Royer, S. K. Park, S. Subramnian, O. D. Jurchescu, B. H. Hamadani, A. J. Moad, R. J. Kline, L. C. Teague, O. Kirillov, C. A. Richter, S. R. Parkin, T. N. Jackson, J. E. Anthony, 2008, *Nature Materials*, **7**, 216-221.

Anti-corrosion and wear properties of plasma electrolytic oxidation coating formed on high Si content Al alloy by sectionalized oxidation mode

Libin Dai, Wenfang Li[1], Guoge Zhang, Nianqing Fu, and Qi Duan

School of Materials Science and Engineering, South China University of Technology, Guangzhou 510641, China

E-mail: mewfli@163.com

Abstract. In this study, a uniform and less defective ceramic coating was prepared on high Si content aluminium alloys by a sectionalized plasma electrolytic oxidation (PEO) mode. The PEO process of Al-9 wt. % Si binary alloy was performed under constant current mode followed by constant voltage mode. The surface micrographs and chemical compositions of different samples were analysed by scanning electron microscopy (SEM) and X-ray diffraction (XRD), respectively. Micro-hardness and reciprocal-sliding testers were used to measure the coatings hardness and tribological performance. It was found that the sectionalized PEO mode could produce hard and anti-friction passive oxide layers with smaller holes and fewer cracks on the Al-Si alloy, comparing with the single constant current mode. In addition, the results of polarization curves and electrochemical impedance spectroscopy (EIS) tests conducted in 3.5 wt. % NaCl solution revealed that the coatings obtained by sectionalized PEO mode had a higher corrosion resistance and provided better corrosion protection for Al-Si alloy.

1. Introduction

The increase of silicon content in Al-Si alloy can effectively improve the wear resistance, flowability, and crack resistance meanwhile reduce its thermal expansion coefficient and specific gravity. Therefore, the Al-Si alloy with high silicon content has been widely used in the meter housing, cylinder and piston parts manufacture [1]. However, many Al-Si alloy products need surface treatment before industrial applications due to the inferior hardness and anti-corrosion property of the bare aluminium alloy [2]. Chemical conversion and anodic oxidation are conventional surface treatment techniques used for Al alloys. However, both of them are not appropriate for the high-silicon content aluminium alloy because the Si element is more difficult to be oxidized than Al in the substrate. It will hinder the formation of continuous and uniform films, further resulting in a decline of the whole film layer performance [3, 4].

Plasma electrolytic oxidation has been considered as one of the most effective and eco-friendly methods to promote anti-corrosion and tribological performance of aluminium and its alloys by forming thick and hard ceramic coatings [5, 6, 7, 8, 9]. Mistry, *et al.* [10] had prepared ceramic coatings on eutectic Al-Si alloy as a cylinder wall surface by PEO technology, which had a greater durability in the longer term than the Ni–SiC coating. Wang, *et al.* [11] also successfully fabricated ceramic coatings on ZAl12Si3Cu2NiMg by PEO and systematically studied the coating mechanism at the initial stages. The formation mechanism of PEO coating is very complicated due to the involvement of electro-, theromo- and plasma-chemical reaction in the electrolyte and metal-electrolyte interface [7, 12-14]. It was found that the coating formation of PEO was influenced by current density, additives, the size of silicon, and the defects of local uneven electrical conductivity [15-18]. However, there are few researches involving the effect of sectionalized oxidation mode (SOM) on anti-corrosion and tribological performance of PEO coatings on high Si content Al alloy. Jaspard-Mécuson and Melhem [19, 20] reported that under constant current mode, with the increase of processing time, the sparks turned to micro-arcs, and then to arcs which was detrimental to the coatings. Wang, *et al.* [21] revealed that under constant voltage mode, when the voltage reached set value and kept constant, the intensity of light emitted by arcs was reduced with the proceeding of PEO. Thus, the processing time of constant current mode and constant voltage mode has a significant effect on the microstructures and properties of the coatings on aluminium alloy. It's likely to acquire a good PEO coating on high Si content Al alloy by using both constant current and constant voltage mode and changing their processing time during PEO process.

In this work, two kinds of different PEO modes, namely, single constant current mode and constant current followed by constant voltage mode, were proposed to treat Al-9 wt. % Si alloy. The single constant current mode was set as the control test. The surface morphologies and compositions of coatings were characterized by SEM and XRD, respectively. The anti-corrosion performance and tribological property of the coating were investigated by electrochemical workstation and friction-abrasion testing machine, respectively.

2. Materials and Methods

The casting Al-9 wt. % Si binary alloy (20 mm × 20 mm × 4 mm) was selected as the substrate in this study. The samples were immersed in the electrolyte. A stainless-steel sheet was used as the counter electrode. The SiC abrasive papers of 240, 400, 600, 800 and 1200 grades were used to polish the samples in sequence. The polished samples were then cleaned with acetone for 5 min in an ultrasonic cleaner. After that the samples were thoroughly washed with distilled water and dried in hot air. The PEO process was conducted using a WHD60 bipolar pulse power with a data acquisition system to record the instantaneous current and voltage. The electrolyte was composed of 10 g/L Na_2SiO_3, 2 g/L NaOH and 4 g/L KF. The temperature was controlled below 40 ℃ during the PEO process using a circulating water cooling unit. The positive and negative current densities of the constant current mode were set as 10 A/dm^2 and 2 A/dm^2, respectively. The positive and negative voltage of the constant voltage modes were 400 V and 50 V, respectively.

The PEO experimental apparatus is shown in Figure 1.

The microstructure and surface morphology of the coating was observed by scanning electron microscopy (SEM, Quanta 200) equipped with energy dispersive X-ray spectroscopy (EDS).

The electrochemical behaviour of the coated sample was investigated by potentiodynamic polarization and electrochemical impedance spectroscopy with CHI660D electrochemical workstation in 3.5 wt. % NaCl solution.

The phase composition of the coating was analysed by Philips X'pert MPD X-ray diffraction in grazing incidence of 2°.

The micro-hardness of the coating was analysed by a HMV-2T micro-hardness tester at a load of 50 g with the loading duration of 10 s. The average value of five different positions was recorded as the final film micro-hardness.

The tribological properties of the PEO coatings were evaluated on a SFT-2M reciprocal-sliding test instrument. The GCr15 steel ball with a diameter of 4mm was used as the friction antithesis. The dry sliding was performed at a load of 2 N, a sliding speed of 300 r/min, and sliding radius of 5 mm. The wear rates were calculated by the weight loss of the whole coating.

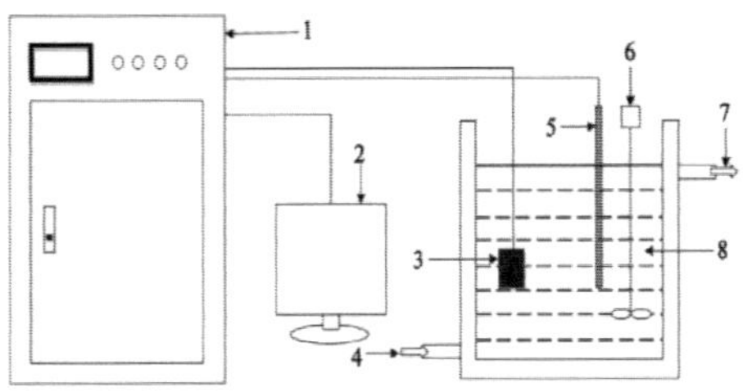

Figure 1. Schematic diagram of the PEO equipment:

1. WHD60 bipolar pulse power, 2. Data processing system, 3. Anode, 4. Circulating water cooling unit, 5. Cathode, 6. Stirring system, 7. Circulating water cooling unit, 8. Electrolyte.

3. Results and Discussion

3.1. The growth process of the coatings in SOM

Figure 2 shows the change of positive voltage over the reaction time in SOM. Table 1 shows the processing time of oxidation modes. In the early stage of the SOM, the voltage rapidly increases and a large number of fine bubbles are produced on the anode surface due to the formation of a passive film on the surface [22]. When the positive voltage reaches to about 160 V, the spark discharges start to appear on the sample surface, and gradually brighten with the increase of voltage. When the voltage is raised to about 240 V, the yellowish arc discharges begin to happen at the corners of the samples because of the high electric field. Around the discharge channel, the deposition of reaction product causes the increase of local resistance and breakdown voltage. Therefore, the sparking area is slowly extended to central part of the samples. At about 280 V, the micro-arcs evenly cover the sample surface.

At the reaction time of 9.5 min, the positive voltage is raised to about 402 V. The 1 # sample is instantly switched into constant voltage mode of 400V and the other samples remain constant current mode. Because the coating resistance is high, the current of 1# rapidly drops to 1 A or less. The average energy of the discharges is decreased, leading to generate uniformly distributed and small sized micro-arcs, as shown in Figure 3.

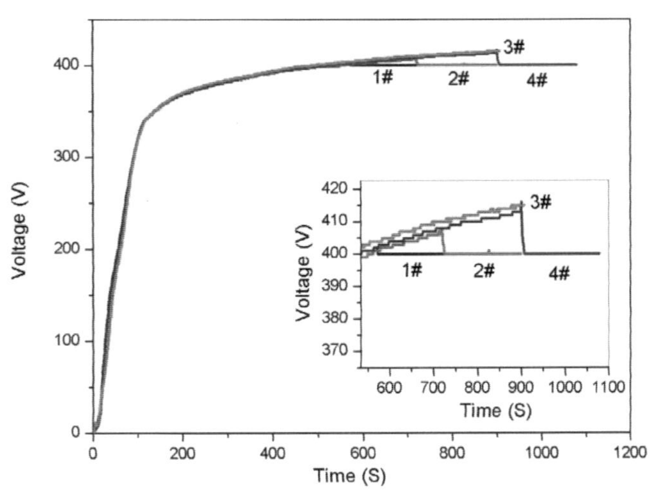

Figure 2. The variation of positive voltage with the PEO processing time for 1#, 2# &4# samples.

Table 1. The oxidation time of constant current mode and constant voltage mode of different samples.

Code of sample	Treatment time of constant current oxidation (min)	Treatment time of constant voltage Oxidation (min)
1#	9.5	5.5
2#	12	3
3#	15	0
4#	15	3

3.2. Microstructure and composition of the oxide film

The surface morphologies of the coatings obtained under different PEO processes are shown in Figure 3. It is obviously seen that the surface of 4# (shown in Figure 3(a) and (b)) is more homogeneous than that of 3# (presented in Figure 3(c) and (d)). In addition, many brightness areas are found on 3# surface. The EDS data reveal that these high brightness areas are mainly composed of silicon oxides (seen in Figure 4). Moreover, there are some cracks can be observed on the 3# surface. By comparison, no cracks are found on the 4# surface. This is attributed to the change in the number of micro-arcs before and after constant voltage oxidation. During constant voltage oxidation, the current density is low and the number of micro-arcs increases. According to the study of Lu [23], the discharges occur preferentially at relatively thin or defective locations. Namely, more micro-arcs appear on the cracked areas. When the molten material is cooled down by the surrounding solution, the residual discharge channels become small and uniformly distributed nodules. It can repair the defects left by the constant current oxidation. Consequently, the sectionalized plasma electrolytic oxidation mode is beneficial to the formation of less defective ceramic coatings on high silicon content Al alloys.

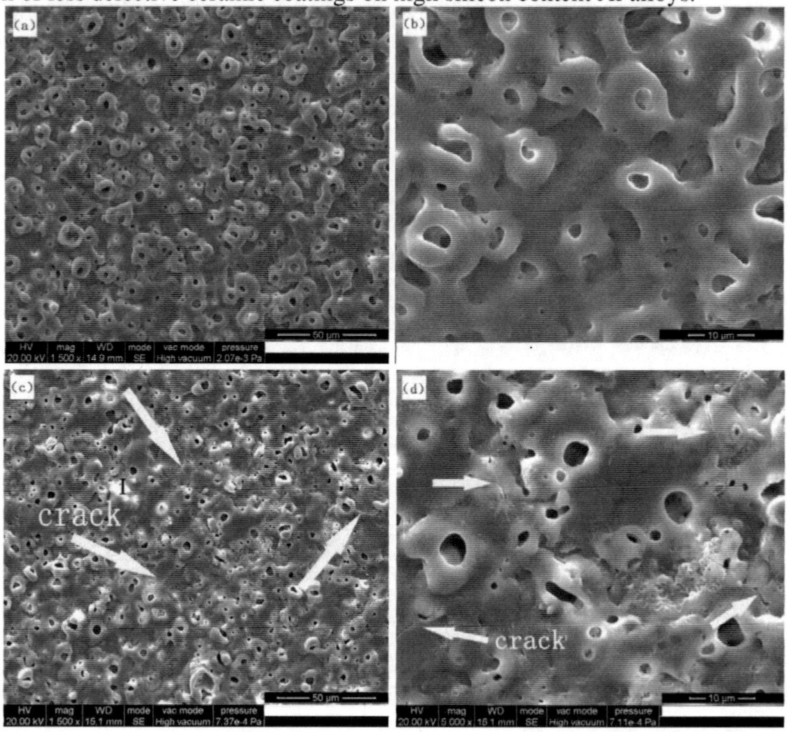

Figure 3. Morphologies of PEO coatings at different oxidation modes: (a, b) constant current oxidation for 15min followed by constant voltage oxidation of 3min. (c, d) constant current oxidation for 15min.

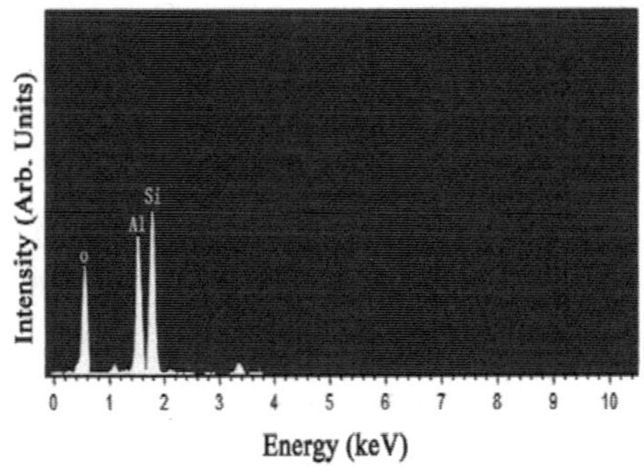

Figure 4. The corresponding EDS spectra of the area marked by I in Figure 3c.

Figure 5 shows the XRD patterns of oxide coatings made by the two different kinds of oxide modes. It can be seen that the plasma electrolytic oxidation coatings are composed of γ-Al_2O_3 and SiO_2. No α-Al_2O_3 or mullite phase is found. Thus, the introduction of sectionalized oxidation mode has little effect on the composition of PEO coating. Researchers [24] found that the content of α-Al_2O_3 and mullite in the PEO coating was increased with the extension of reaction time. The γ-Al_2O_3 does not spontaneously transform to α-Al_2O_3, since the transformation of γ-Al_2O_3 to α-Al_2O_3 needs tremendous energy provided by plasma electrolytic discharge to overcome the energy barrier. Considering that only a short time of 15min is used for the whole PEO process, the energy is not enough to complete this transformation.

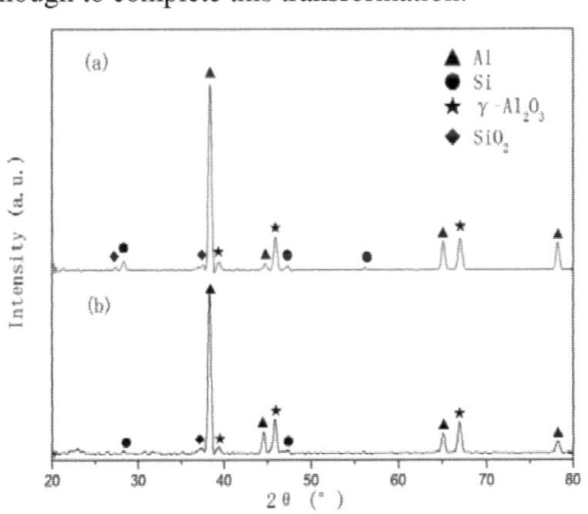

Figure 5. XRD results of the coatings: (a) treated by sectionalized oxidation mode (constant current oxidation for 12min followed by constant voltage oxidation of 3min), (b) treated by single constant current mode.

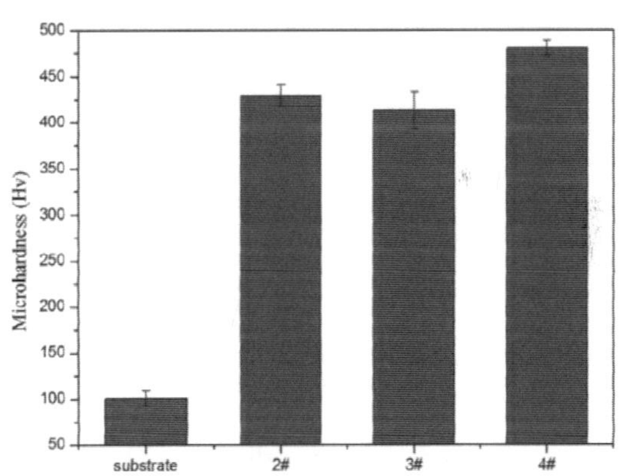

Figure 6. Micro-hardness of PEO coating of 2#, 3#, 4# samples and Al-Si alloy substrate.

3.3. The mechanical and tribological properties of bare aluminium and coatings

The micro-hardness of the films obtained by different treatment processes is shown in Figure 6. The films hardness of 2#, 3#, and 4# is 430Hv, 410Hv, 480Hv, respectively and the hardness of Al- Si substrate is 104Hv. It can be clearly seen that, after PEO treatment, the films micro-hardness is improved over four times higher than that of the Al-Si alloy substrate. This is due to the formation of hard alumina ceramic layer. Comparing 2# with 4#, it is found that longer constant current time producing PEO film with higher micro-hardness when the oxidation time of constant voltage step is constant. It is because the thickness of dense layer increases with longer galvanostatic polarization time [25]. Meanwhile, when the oxidation time of constant current is constant, the increase of the constant voltage oxidation time improves film's micro-hardness by comparing 3# with 4#. This is considered to be caused by the formation of a more compact passive coating on the specimen surface after longer constant voltage oxidation time treatment.

Figure 7 represents the variation of friction coefficients versus sliding time for the substrate and PEO coatings against Gr15 steal ball under dry friction condition. It is very obvious that there is a great difference on friction properties between the uncoated and coated substrates. The Al-Si alloy substrate records an average friction coefficient of 0.41, while other ceramic coatings under the same test conditions record much larger average friction coefficient around 0.58-0.73. Moreover, no large fluctuations are observed during the wear test. It indicates that there is no coating failure.

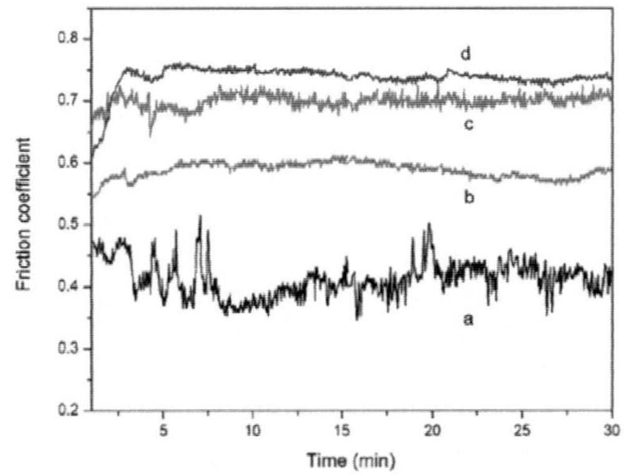

Figure 7. Friction coefficients of bare aluminium and PEO coatings formed at different oxidation modes:

(a) Al-Si alloy, (b) treated by constant current oxidation for 15min followed by constant voltage oxidation of 3min, (c) treated by constant current oxidation for 12min followed by constant voltage oxidation of 3min, (d) treated by single constant current mode.

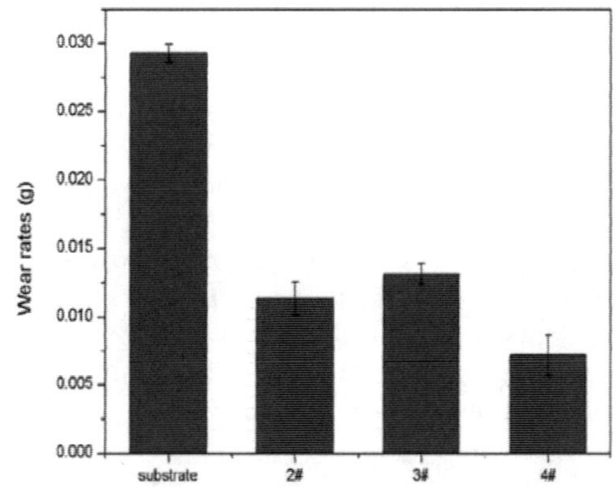

Figure 8. Wear rates of Al alloy substrate and PEO coatings.

The wear rates of bare substrate and PEO coatings are shown in Figure 8. The wear rate of the Al-Si substrate is 29.4 mg, while the wear rates of the coatings are significantly reduced, reaching 7.0-13.2 mg. This clearly indicates that PEO is an effective technology to improve the wear resistance of substrates. Furthermore, the coating treated by constant voltage reveals a smaller wear rate than that of untreated one. This is attributed to the higher micro-hardness (shown in Figure 6) and better compactness (seen in Figure 3) of the coating obtained under constant voltage mode.

3.4. Anti-corrosion study of PEO coatings

In order to study the effect of two different oxidation modes on the corrosion resistance performance of PEO coatings, electrochemical test was carried out. The polarization curves are shown in Figure 9. The corresponding self-corrosion potential and corrosion current density are summarized in Table 2. As seen from Figure 9, after PEO treatment, polarization curves move to a lower current density side. The self-corrosion current density of the PEO processing Al-Si alloy is 10-30 times smaller than that of the pristine substrate and the E_{corr} is improved over 80 mV. It shows that the anti-corrosion performance of the coatings is significantly improved after PEO treatment. In addition, the self-corrosion current densities of the coatings prepared by the single constant current oxidation mode are higher than those of the sectionalized oxidized coatings. Such results conform to the loose morphologies of the coatings obtained under single constant current oxidation mode. Without constant voltage oxidation, the coatings have more cracks on the surface. These cracks result in the degradation of anti-corrosion performance. In contrast, as mentioned before, the constant voltage oxidation can repair some defects of the coatings produced in the constant current oxidation stage. Therefore, the coating shows an enhanced blocking ability against the etching solution.

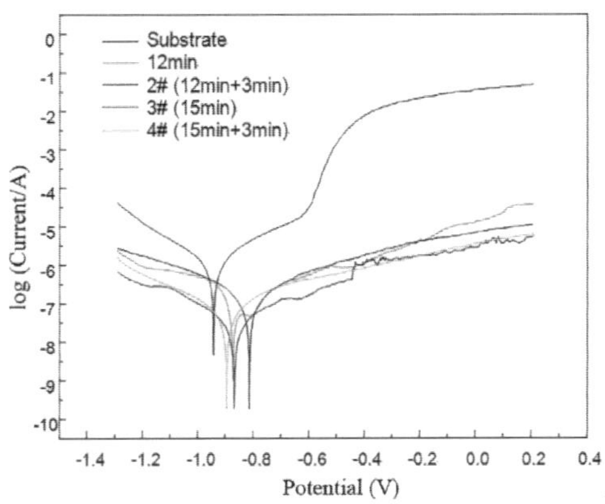

Figure 9. Polarization curves of the substrate and PEO coatings.

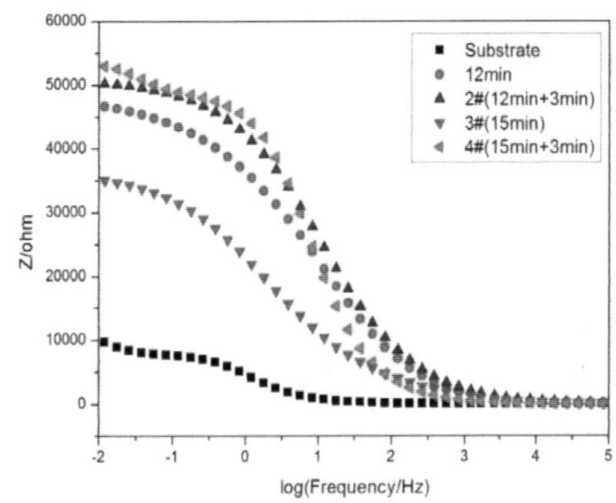

Figure 10. EIS diagrams (Bode) of Al-Si alloy and PEO coatings treated at different processing parameters.

Table 2. Potentiodynamic polarization parameters of different samples: self-corrosion current density (I_{corr}), and self-corrosion potential (E_{corr}).

Sample treatment	I_{corr} ($\mu A/cm^2$)	E_{corr} (V)
C-9% (Substrate)	1.1140	-0.9390
constant current oxidation for 12min	0.1082	-0.8398
2#	0.0335	-0.8216
3#	0.1134	-0.8309
4#	0.0572	-0.8514

The EIS behaviour of bare aluminium and coated samples was investigated after 3600s of immersion. The Bode plots are presented in Figure 10. According to V. Raj, *et al.* [26], the PEO coatings were consisted of outer porous layer and inner compact layer. The anti-corrosion performance of the coating is mainly attributed to the inner barrier layer. In the EIS, the high frequency portion of impedance curves reflects the outer layer properties and the low frequency portion represents the inner layer properties. Thus, the overall polarization resistance values of the substrate and coatings can be evaluated by the extrapolation of the impedance modulus at low frequency [27]. From Figure 10 it can be seen that the resistance of samples is greatly improved after treated by PEO, reaching three to five orders of magnitude higher than that of untreated substrates. Moreover, the introduction of constant voltage oxidation mode after constant current oxidation leads to an improvement of corrosion resistance performance of coatings, which is in good agreement with the test results of potentiodynamic polarization (in Figure 9 and Table 2).

4. Conclusions

In summary, a sectionalized PEO mode (constant current oxidation followed by constant voltage oxidation) for the treatment of Al-9 wt. % Si alloy had been successfully developed. It was found that the introduction of constant voltage oxidation could repair some defects of the coatings produced in the constant current oxidation stage. Coatings prepared by the SOM technique exhibited a fewer defects, higher micro-hardness (up to 440 Hv) and better anti-wear properties. These resulted in a significant improvement in anti-corrosion performance. The self-corrosion current density of the substrate treated under optimized condition was reduced by around 30 times (from 0.1134 μA/cm^2 to 0.0335 μA/cm^2) in comparison with the bare alloy. Moreover, the self-corrosion current density of the coating prepared under SOM mode was lower than that of the coating treated by the conventional single constant current oxidation.

References

[1] Osório W R, Goulart P R, and Garcia A 2008 *Mater. Lett.* **62** 365.
[2] Wang P, Li J, Guo Y, Wang J, Yang Z, and Liang M 2016 J. *Alloys Compd.* **657** 703.
[3] Xu F, Xia Y, and Li G 2009 Appl. *Surf. Sci.* **255** 9531.
[4] Forn A, Picas J A, Baile M T, Martin E, and García V G 2007 *Surf. Coat. Technol.* **202** 1139.
[5] Xiang N, Song R G, Zhuang J J, Song R X, and Su X P 2016 T. *Nonferr Metal Soc.* **26** 806.
[6] Yerokhin A L, Nie X, Leyland A, Matthews A, and Dowey S J 1999 *Surf. Coat. Technol.* **122** 73.
[7] Hussein R O, Northwood D O, and Nie X, 2013 *Surf. Coat. Technol.* **237** 357.
[8] Zhu M H, Cai Z B, Lin X Z, Ren P D, Tan J, and Zhou Z R 2007 *Wear* **263** 472.
[9] Hu C J, and Hsieh M H 2014 *Surf. Coat. Technol.* **258** 275.
[10] Mistry K, Priest M, and Shrestha S 2010 Proceedings of the Institution of Mechanical Engineers, Part J: *Journal of Engineering Tribology* **224** 221.
[11] Wang P, Li J P, Guo Y C, Yang Z, and Wang J L 2016 *Surf. Eng.* **32** 428.
[12] Lu X, Blawert C, Huang Y, Ovri H, and Kainer K U 2016 *Electrochim. Acta* **187** 20.
[13] Klapkiv M D 1999 *Mater. Sci.* **35** 279.
[14] Yerokhin A L, Shatrov A, Samsonov V, Shashkov P, Pilkington A, Leyland A, and Matthews A, 2005 *Surf. Coat. Technol.* **199** 150.
[15] Krishtal M M 2004 *Met. Sci. Heat Treat.* **46** 377.
[16] Pan Y K, Chen C Z, Wang D G, Yu X, and Lin Z Q 2012 *Ceram. Int.* **38** 5527.
[17] Mi T, Jiang B, Liu Z, and Fan L 2014 *Electrochim. Acta* **123** 369.
[18] Kazanski B, Kossenko A, Zinigrad M, and Lugovskoy A 2013 *Appl. Surf. Sci.* **287** 461.
[19] Jaspard-Mecuson F, Czerwiec T, Henrion G, Belmonte T, Dujardin L, Viola A, and Beauvir J, 2007 *Surf. Coat. Technol.* **201** 8677.
[20] Melhem A, Henrion G, Czerwiec T, Briançon J L, Duchanoy T, Brochard F, and Belmonte T 2011 *Surf. Coat. Technol.* **205** S133.
[21] Wang L, Chen L, Yan Z, and Fu W 2010 *Surf. Coat. Technol.* **205** 1651.
[22] Hussein R O, and Northwood D O 2014 *Developments in Corrosion Protection* **201**.
[23] Lu X, Blawert C, Kainer K U, and Zheludkevich M L 2016 *Electrochim. Acta* **196** 680.
[24] Wu Z, Xia Y, Li G, and Xu F 2007 *Appl. Surf. Sci.* **253** 8398.
[25] Xue W, Shi X, Hua M, and Li Y 2007 *Appl. Surf. Sci.* **253** 6118.
[26] Raj V, and Ali M M 2009 *J. Mater. Process. Technol.* **209** 5341.
[27] Veys-Renaux, and Rocca E 2015 *J. Solid State Electrochem.* **19** 3121.

Investigation on the formation of projections and cracks in anodic oxidation of reaction-sintered silicon carbide

Kang Peng[1], Xinmin Shen[1,3], Kazuya Yamamura[2], Xiaonan Zhang[1], Dong Wang[1]

[1]College of Field Engineering, PLA University of Science and Technology, No. 1 Haifu Street, Nanjing, Jiangsu 210007, China.
[2]Graduate School of Engineering, Osaka University, 2-1 Yamadaoka, Suita, Osaka 565-0871, Japan.

E-mail: shenxmjfjlgdx2014@163.com

Abstract. Among the present oxidation-assisted polishing (OAP) technique, anodic oxidation is a promising method to process reaction-sintered silicon carbide (RS-SiC) with the assistance of abrasive polishing. Projections and cracks are inevitably formed in the anodic oxidation of RS-SiC for the volume expansion force. Evolvements of the oxide morphologies along with different oxidation voltage and different oxidation time are investigated by scanning electron microscope (SEM). It can be found that when the oxidation time stays the same, size of the projections and cracks on the oxidized RS-SiC sample is gradually enlarged. Meanwhile, when the oxidation voltage stays the same, size of the projections and cracks on the oxidized RS-SiC sample is also enlarged. Details of projections and cracks on the oxidized RS-SiC sample are obtained by the SEM with high magnification. Based on the theoretical derivation, it can be concluded that the formations of projections and cracks depend on the inner volume expansion force and the resultant force generated from other grains. Research on formation of projections and cracks in anodic oxidation of RS-SiC can promote the process level of RS-SiC products.

1. Introduction

Reaction-sintered silicon carbide (RS-SiC) has excellent mechanical and chemical properties, which makes it a novel and promising material for optics and ceramics [1, 2]. Many technologies have been developed to process RS-SiC, such as diamond lapping [3], plasma chemical vaporization machining [4], and so on. However, few of these techniques can obtain ultrasmooth surface, because there are SiC domains and Si domains in the RS-SiC substrates generated from the fabrication process. Along with the application of RS-SiC samples in the optical and ceramic fields, the requirements of surface quality of RS-SiC products are more and more strict. Oxidation-assisted polishing (OAP) is a novel method to process RS-SiC, which includes the oxidation of RS-SiC and the removal of oxide layer [5]. Many oxidation methods have been introduced to oxidize RS-SiC, such as the plasma oxidation [6, 7], thermal oxidation [8, 9], anodic oxidation [10, 11], and so on. Among these oxidation methods, anodic oxidation is a promising method to obtain high material removal rate (MRR) and the smooth surface quality simultaneously with the assistance of abrasive polishing by the ceria slurry [12, 13], because hardness of the oxide layer is remarkably smaller than that of the RS-SiC substrate [14, 15], and the anodic oxidation rate is larger than that in the thermal oxidation or that in the plasma oxidation [16-18].

Surface morphologies of the oxides determine the obtained surface quality of the RS-SiC sample with the assistance of abrasive polishing. Similar with other oxidation processes, the anodic oxidation of RS-SiC sample is a volume expansion process, which indicates that the volume expansion force is inevitably generated [19-22]. The oxides generated from SiC/Si grains in the RS-SiC substrates extrude each other, which will result in complex oxide morphology. In this study, firstly, the oxide layer is generated by the anodic oxidation of RS-SiC in the mixture of hydrogen peroxide (H_2O_2) and hydrochloric acid (HCl). Secondly, the scanning electron microscope (SEM) is used to detect the oxidized RS-SiC sample, which can directly exhibit the oxide morphologies. Finally, a theoretical analysis on the formation of projections and cracks in anodic oxidation of RS-SiC is conducted, which is based on the oxidation processes of SiC grains and Si grains in the RS-SiC substrates.

2. Experimental apparatus and parameters

The anodic oxidation system applied in this study is shown in figure 1. The potential is supplied by the potentiostat, which is inspected by the sampling software in the controlling computer. The reference electrode (Ag/AgCl) is used to keep the oxidation potential constant by the feedback control. The electrical current is generated between the anode and cathode, and the oxide can be formed at the region of RS-SiC sample that contacts the electrolyte. The electrolyte used in this study is a mixture of the H_2O_2 and HCl, and their concentrations are 5.88wt% and 0.69wt%, respectively. In the oxidation system, H_2O_2 is treated as major oxidant, and HCl is used to increase conductance of the electrolyte.

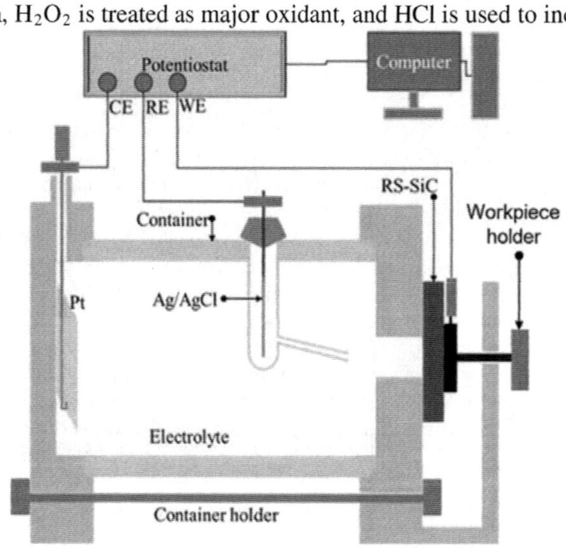

Figure 1. Schematic diagram of the used anodic oxidation system in this study.

When the oxidation voltage is 9.8V and the oxidation time is 1800s, the obtained evolvement of the oxidation current along with the increasing of oxidation time is shown in figure 2, and the sampling time interval is 0.2s. Conductance of the RS-SiC sample is larger than that of the oxide layer. It can be found that the oxidation current can reach almost 240mA at the beginning, and it decreases rapidly along with the increasing of oxidation time, which indicates that the oxide layer is quickly formed in a short time. When the oxidation time arrives at 3.6s, the oxidation current falls down to 20mA. When the oxidation time increases from 3.6s to 100s, the oxidation current falls from 19.862mA to 1.128mA, which is close to the exponential decline. Along with the further increasing of the oxidation time, the reduction of oxidation current is smaller than 2μA when the oxidation time is increasing a sampling time interval 0.2s. Finally, when the oxidation time reaches 1800s, the oxidation current is 139.772μA, which is almost 1/1717 of the initial oxidation current 240mA. From the analysis on evolvement of the oxidation current, it can be found that the anodic oxidation rate of RS-SiC is decreasing along with the increasing of oxidation time, because the formed oxidation layer prevents the further oxidation process in the thickness direction. The maximum output of the potentiostat is 9.8V. Therefore, for the purpose of analyzing the formation of projections and cracks in anodic oxidation of RS-SiC, the experiments are conducted in 12 series. In these experiments, the investigated oxidation potential is 2V to 9.8V, and the oxidation time is 10s to 3600s, as shown in Table 1.

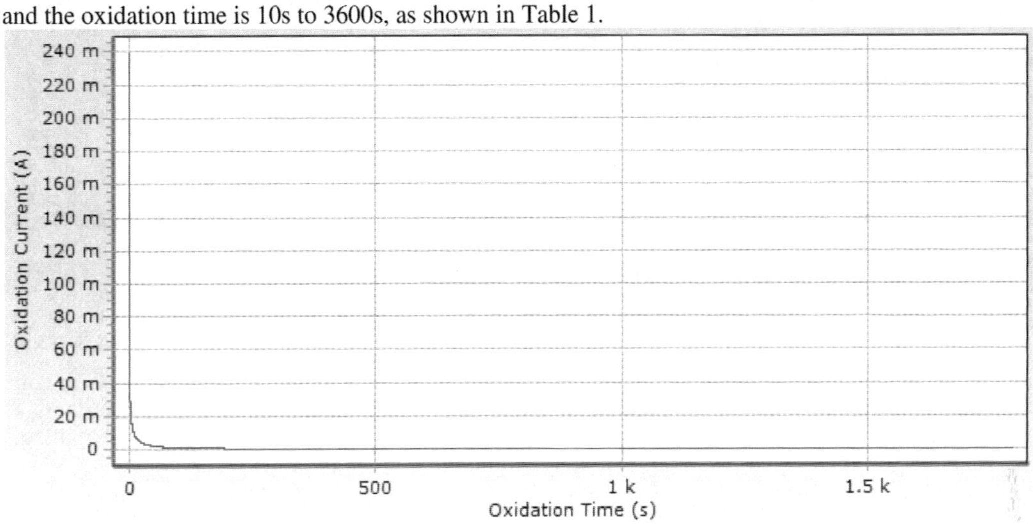

Figure 2. Evolvement of the oxidation current along with the increasing of oxidation time.

Table 1. Experimental parameters in anodic oxidation of RS-SiC

Serial number	Oxidation potential	Oxidation time	Oxidant
1	2V	30s	
2	2V	900s	
3	3V	10s	
4	3V	900s	
5	4V	10s	
6	5V	10s	H_2O_2 (5.88wt%) and HCl (0.69wt%)
7	5V	900s	
8	7V	10s	
9	7V	900s	
10	9V	10s	
11	9.8V	45s	
12	9.8V	3600s	

3. Investigation on oxide morphologies

Evolvements of the oxide morphologies corresponding to the different oxidation voltages and different oxidation times are obtained by SEM, as shown in figure 3. It can be found that when the oxidation time stays the same, as shown in figures 3(a), 3(c), 3(e), 3(f), 3(h), and 3(j), size of the projections and cracks on the oxidized RS-SiC sample is gradually enlarged. Meanwhile, when the oxidation voltage stays the same, as shown in figures 3(a) and 3(b), or 3(c) and 3(d), or 3(f) and 3(g), or 3(h) and 3(i), or 3(k) and 3(l), size of the projections and cracks on the oxidized RS-SiC sample is also enlarged.

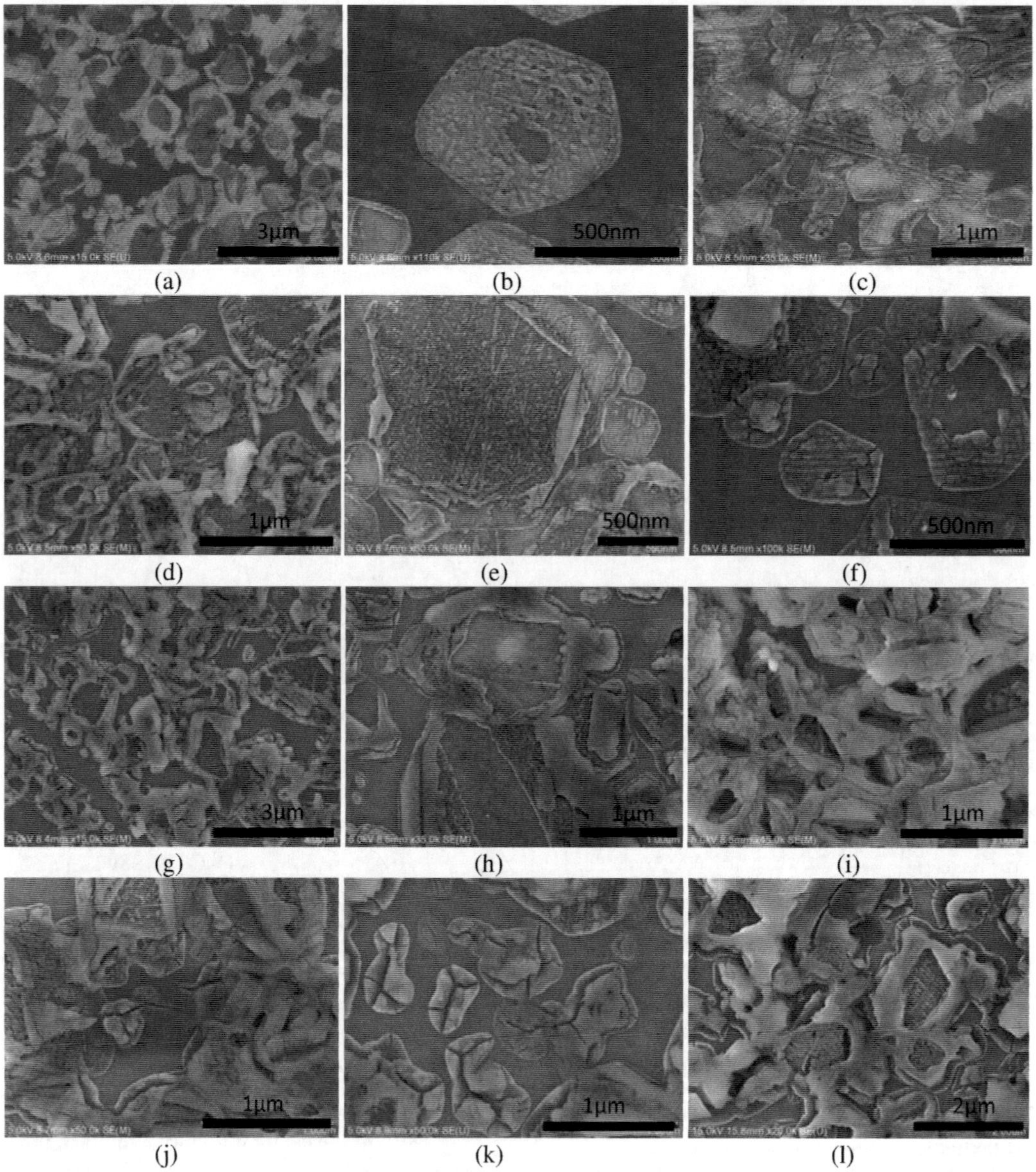

Figure 3. Evolvements of the oxide morphologies corresponding to the different oxidation voltage and different oxidation time.

(a) 2V and 30s. (b) 2V and 900s. (c) 3V and 10s. (d) 3V and 900s. (e) 4V and 10s. (f) 5V and 10s. (g) 5V and 900s. (h) 7V and 10s. (i) 7V and 900s. (j) 9V and 10s. (k) 9.8V and 45s. (l) 9.8V and 3600s.

Details of the projections and cracks on the oxidized RS-SiC sample are shown in figure 4. It can be found that most of projections are generated from initial SiC grains, and the oxides generated from Si grains have a relative smooth surface. From the figure 4(b) it can be found that there are cracks at the connecting regions among different SiC grains and Si grains, and width of the crack can be more than 50nm. Meanwhile, it can be judged from figure 4(c) that there is obvious crack on the formed projection, which is generated from the volume expansion force of the subsurface oxidized grains.

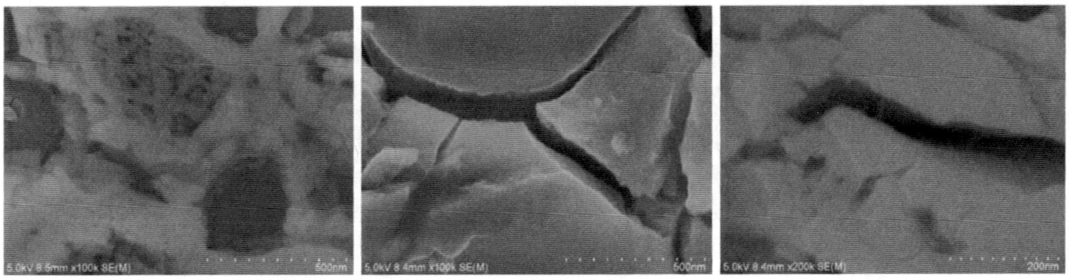

(a) (b) (c)

Figure 4. Details of projections and cracks on the oxidized RS-SiC sample obtained by SEM under high magnification.

(a) Projections. (b) Cracks. (c) Cracks on projection.

4. Study on oxidation mechanism

The major reason for formation of projection and crack is volume expansion of the oxidation process. The formulas for the anodic oxidation of the SiC grains and Si grains in the RS-SiC substrate can be summarized, as shown in formulas (1) and (2).

$$SiC + 2H_2O_2 \xrightarrow{8p^+/8e^-} SiO_2 + C + 2H_2 \uparrow + O_2 \uparrow \tag{1}$$

$$Si + 2H_2O_2 \xrightarrow{8p^+/8e^-} SiO_2 + 2H_2 \uparrow + O_2 \uparrow \tag{2}$$

The Molar mass M of SiC/Si/SiO_2 is 40g/mol, 28g/mol, and 60g/mol, respectively, and the density ρ of SiC/Si/SiO_2 is 3.22g/cm^3, 2.32g/cm^3, and 2.2g/cm^3, respectively. From formulas (1) and (2) it can be found that the oxidation of SiC/Si to SiO_2 is volume expansion process, and volume expansion coefficients of SiC and Si are 2.194 and 2.257, respectively. Supposing the volume expansion force from one grain is F , and its direction is (α,β,γ) in the 3-dimensional systems. To a certain grain, supposing its serial number is k , the inner volume expansion force generated from oxidation of itself is $F_k(\alpha_k,\beta_k,\gamma_k)$, as shown in equation (3), and the extrusion force generated from oxidation of other grains can be expressed by the equation (4).

$$F_{inner} = F_k(\alpha_k,\beta_k,\gamma_k) \tag{3}$$

$$F_{outer} = \sum_{1}^{k-1} F_i(\alpha_i,\beta_i,\gamma_i) + \sum_{k+1}^{n} F_i(\alpha_i,\beta_i,\gamma_i) \tag{4}$$

Therefore, when $F_{outer} > F_{inner}$, projections are formed, as shown in the figure 4(a). Otherwise, if $F_{outer} < F_{inner}$, cracks are produced, as shown in figure 4(b). What's more, the volume expansion force generated from oxidation of the subsurface grains can result in crack on the projection, as shown in the figure 4(c). The volume expansion force F is determined by the oxidation potential and the oxidation time. The exact value of F can be obtained in the finite element analysis (FEA) by simulating the oxidation process, which is the direction of the further research. Characters of the oxide morphologies affect the obtained MRR and surface quality in the following abrasive polishing process. On the one hand, the hardness of the oxide is obviously lower than that of the initial RS-SiC substrates, and the projections and cracks on the soft oxides increase the contact areas with ceria particles in the abrasive polishing process; thus, high MRR can be easily obtained in anodically oxidation-assisted polishing (AOAP) of RS-SiC. On the other hand, some cracks penetrate into the oxidized RS-SiC sample, which is very difficult to remove completely with the assistance of abrasive polishing; thus, it is difficult to obtain an ultrasmooth surface in the AOAP of RS-SiC. Therefore, AOAP can be treated as an efficient method for the rapid figuring of RS-SiC, which can improve the process level of the RS-SiC sample.

5. Conclusions

Projections and cracks are generated in the anodic oxidation of RS-SiC for the volume expansion force. For the oxide generated from a certain grain, if the inner volume expansion force is smaller than the resultant force generated from other grains, projections are formed. Otherwise, cracks can be produced. Since the oxidation current pass through all the RS-SiC, the volume expansion force generated from the oxidation of subsurface grains can press surface oxide to result in cracks on the projection.

Characters of the oxide morphologies affect the obtained MRR and surface quality in the following abrasive polishing process. AOAP of RS-SiC can obtain high MRR and relatively fine surface quality. Therefore, AOAP can be treated as an efficient method for the rapid figuring of RS-SiC, which can improve the process level of the RS-SiC sample and promote the application of RS-SiC products.

This work was supported by a grant from National Natural Science Foundation of China (Grant No. 51505498) and a grant from Natural Science Foundation of Jiangsu Province (Grant No. BK20150714). The authors also express their gratitude to the staffs and students of Research Center for Ultra-precision Science and Technology of Osaka University.

References

[1] Suyama S and Itoh Y 2007 NTSIC (New-Technology Silicon Carbide): evaluation of microstructure of high-strength reaction-sintered silicon carbide for optical mirror *Proc. of SPIE* **6666** 66660K

[2] Robb P, Charpentier R and California P A 1995 Three-mirror anastigmatic telescope with a 60-cm aperture diameter and mirrors made of silicon carbide *SPIE* **2543** 185-189

[3] Tam H Y, Cheng H B and Wang Y W 2007 Removal rate and surface roughness in the lapping and polishing of RB-SiC optical components *J. Mater. Process. Technol.* **192-193** 276-280

[4] Yamamura K, Yamamoto Y and Deng H 2012 Preliminary Study on Chemical Figuring and Finishing of Sintered SiC Substrate Using Atmospheric Pressure Plasma *Procedia CIRP* **3** 335-339

[5] Shen X M, Dai Y F, Deng H, Guan C L and Yamamura K 2013 Comparative analysis of oxidation methods of reaction-sintered silicon carbide for optimization of oxidation-assisted polishing *Optics Express* **22(21)** 26123-26135

[6] Deng H, Ueda M and Yamamura K 2012 Chemical and Morphological Analysis of 4H-SiC Surface Processed by Plasma assisted Polishing *Key Engineering Materials* **516** 186-191

[7] Shen X M, Tu Q Z, Deng H, Jiang G L and Yamamura K 2015 Mechanism analysis on finishing of reaction-sintered silicon carbide by combination of water vapor plasma oxidation and ceria slurry polishing *Optical Engineering* **54(5)** 055106

[8] Chen X F, Ning L N, Wang Y M, Li J, Xu X G, Hu X B and Jiang M H 2009 Thermal Oxidation of Silicon Carbide Substrates *J. Mater. Sci. Technol.* **1(25)** 115-118

[9] Shen X M, Dai Y F, Deng H, Guan C L and Yamamura K 2013 Ultrasmooth reaction-sintered silicon carbide surface resulting from combination of thermal oxidation and ceria slurry polishing *Optics Express* **12(21)** 14780-14788

[10] Woon W S, Hutagalung S D and Cheong K Y 2009 Characterization of anodic SiO2 films on P-type 4H-SiC *Thin Solid Films* **517** 2808-2812

[11] Shen X M, Tu Q Z, Deng H, Jiang G L, He X H, Liu B and Yamamura K 2016 Comparative analysis on surface property in anodic oxidation polishing of reaction-sintered silicon carbide and single-crystal 4H silicon carbide *Appl. Phys. A: Materials Science & Processing* **122** 354

[12] Ke Y, Yan F, Devaty R P and Choyke W J 2009 Surface polishing by electrochemical etching of p-type 4H SiC *J. Appl. Phys.* **106** 064901

[13] Shimozono N, Shen X M, Deng H, Endo K and Yamamura K 2015 Figuring and Finishing of Reaction-Sintered SiC by Anodic Oxidation Assisted Process *Key Engineering Materials* **625** 570-575

[14] Jacobson N S and Myers D L 2011 Active Oxidation of SiC *Oxidation of Metals* **75(1)** 1-25

[15] Tu Q Z, Shen X M, Zhou J Z, He X H and Yamamura K 2015 Efficient processing of reaction-

sintered silicon carbide by anodically oxidation-assisted polishing *Optical Engineering* **54(10)** 105113

[16] Lilov S K 2007 Anodic oxide films on silicon carbide *Cryst. Res. Technol.* **42(11)** 1054-1057

[17] Jiang G L, Shen X M, Tang J, Zhang X N, Zhang X P and Yamamura K 2016 Oxidation Characteristic and Machining Performance of Reaction-Sintered Silicon Carbide Ceramic in Anodically Oxidation-Assisted Polishing *International Journal of Electrochemical Science* **11** 1512-1529

[18] Zheleva T, Lelis A, Duscher G, Liu F D, Levin I and Das M 2008 Transition layers at the SiO_2/SiC interface *Appl. Phys. Lett.* **93** 022108

[19] French H M, Henderson M J, Hillman A R and Vieil E 2002 Temporal resolution of ion and solvent transfers at nickel hydroxide films exposed to LiOH Original Research Article *Solid State Ionics* **150(1-2)** 27-37

[20] Young J L, Molero H and Birss V I 2014 The effect of pre-oxidation treatments on the oxidation tolerance of Ni-yttria-stabilized zirconia anodes in solid oxide fuel cells *Journal of Power Sources* **271** 538-547

[21] G. W. Zhou 2009 Stress-driven formation of terraced hollow oxide nanorods during metal oxidation *Journal of Applied Physics* **105(10)** 104302

[22] Shen X M, Deng H, Zhang X N, Peng K and Yamamura K 2016 Preliminary study on atmospheric-pressure plasma-based chemical dry figuring and finishing of reaction-sintered silicon carbide *Optical Engineering* **55(10)** 105102

Preparation and catalytic application of Ag/polydopamine composite on surface of glass substrates

Jianying Yu[1],Chengyi Sun[1],Shixiang Lu[1,3],Wenguo Xu[1], Zhehan Liu[1] and Dongsheng He[2]

[1]School of Chemistry, Beijing Institute of Technology, Beijing, 100081, P.R. China.
[2]School of Mechanics and Civil. & Architecture, Northwestern Polytechnical University, Xi'an, 710072, P.R. China.

E-mail:shixianglu@bit.edu.cn

Abstract. In this work, Ag/polydopamine composite on glass substrates (Ag/PDA@slides) were formed by using polydopamine (PDA) as both reducing and stabilizing agent to reduce silver salts to silver nanoparticles (NPs) and adhesive them to slides. The morphology and chemical composition of the composite material was characterized by scanning electron microscopy (SEM) and X-ray diffraction pattern (XRD). The prepared Ag/PDA@slide was a highly active catalyst for the reduction of 4-nitrophenol (4-NP) in the presence of sodium borohydride ($NaBH_4$) aqueous solution at room temperature. The reduction rate of the optimal catalyst was as fast as 10 s and it was stable up to 6 cycles without a significant loss of its catalytic activity. By measuring the UV-Vis absorption bonds of Ag/PDA@slides, it proved that condition of the strongest surface plasmon resonance of Ag/PDA@slides is the optimal condition of catalytic reduction of 4-NP.

1. Introduction

Silver nanoparticles (Ag NPs) possess many excellent properties and are widely applied in optics, catalysis and antibacterial field [1-3]. Among these properties, catalytic activity and localized surface plasmon resonance (LSPR) have attracted considerable interests. It is a kind of effective catalyst for many practical applications, such as the reduction of 4-nitrophenol (4-NP). Nitrophenols are one of the most common organic pollutants in industrial and agricultural waste water and is highly toxic. While its reduced product 4-aminophenol (4-AP) is low toxicity as well as a chemical and pharmaceutical intermediate [4]. Therefore, reducing 4-NP to 4-AP is of significant importance. Ag NPs can catalyze the reduction of 4-NP effectively. There are many methods to synthesize Ag NPs, among which chemical reduction is a commonly used way. The reductants often used are sodium borohydride, ascorbic acid, compound with hydroxyl group or aldehyde functional group and so on [5-7]. Recently, dopamine (DA), a kind of biopolymer, was successfully used to prepare metal NPs, as which can reduce metal salts to metal NPs through successive reactions. Dopamine, a biological adhesive inspired by mussel, can form polydopamine (PDA) films by self-polymerization onto various kinds of materials in weak alkaline condition, especially pH 8.5 [8]. Dopamine has been widely studied for surface modification due to the ability to self-polymerize to form polymer films on nearly all substrates[9-11]. Using dopamine to prepare silver nanoparticles is a facile and environmentally-benign way. Ag^+ salts could be easily reduced to Ag NPs and adsorbed on the support simultaneously. For example, Li et al. used dopamine to synthesize Ag@AgCl nanophotocatalyst, which showed superior photocatalytic performance [12]. In addition, the reusability is one of important performances for catalyst. Slides, unlike common particle supports for catalysts, are easy to operate, which needn't centrifugation and filtration. Besides, slides have no effect on the catalytic reaction and are low-cost.

In this work, we prepared Ag NPs on glass substrate by using PDA as both reducing and adhesive agent. The prepared Ag/PDA@slides showed high catalytic performance for the reduction of 4-NP.

2. Experimental section

2.1. Sample preparation

2.1.1. Preparation of PDA@slides. Slides with a size of 25mm×5mm×1mm were firstly ultrasonically cleaned for 3 min in ethanol. Then it was washed with deionized water and ethanol again in the same way to remove grease and other dirty on the surfaces. 0.5 mol/L NaOH solution was injected to 5 mL 20 mg/mL dopamine solution to tune the pH to 8.5. A clean slide was vertically immersed in the dopamine solution for 1 h, and then rinsed with deionized water. The polydopamine film (PDA@slide) was obtained.

2.1.2. Preparation of Ag/PDA@slides. Determination of optimal concentration of $AgNO_3$ aqueous solution: In order to determine the optimal concentration of $AgNO_3$ aqueous solution, PDA@slides were immersed in 3 mL $AgNO_3$ aqueous solution of different concentrations for 2 h at ambient temperature. 0.01 mol/L, 0.05 mol/L, 0.1 mol/L, 0.2 mol/L and 0.4 mol/L $AgNO_3$ aqueous solutions were used to prepare Ag/PDA@slides.

Determination of optimal reaction time: PDA@slides were immersed in 3 mL 0.05 mol·L^{-1} $AgNO_3$ aqueous solution for 15 min, 30 min, 45 min, 60 min at ambient temperature respectively.

2.2. Characterization

The surface morphology was observed using a NovaNano 430 scanning electron microscope (SEM). The crystalline structure was investigated by X-ray diffractometer (XRD, D8 ADVANCE, Bruker, Germany) with Cu K_a radiation (40 kV, 40 mA, and $\lambda = 0.15418$ nm) at X'Celerator's scanning rate of 3°min^{-1}. The UV-Vis spectra were recorded with a HITACHI U-3310 UV-Vis spectrophotometer.

2.3. LSPR and catalytic study

Prepare Ag/PDA@slides obtained at different $AgNO_3$ immersion time. After drying, UV-Vis spectrophotometer was used to detect the absorption peaks to investigate the LSPR of the Ag NPs.

The catalytic reduction of 4-NP was carried out at room temperature in a quartz cuvette. Ag/PDA@slide was added in the cuvette containing 4-NP (2.5 mL, 0.1 mmol·L^{-1}) and freshly prepared aqueous solution of $NaBH_4$ (250 μL, 0.1 mol·L^{-1}). UV-Vis absorption spectra were recorded to monitor the change of the reaction mixture.

3. Results and discussion

3.1. Morphology and chemical compositions

Figure 1 shows the morphology of samples. The slide has a smooth surface as shown in figure 1(a). PDA@slide is lightly crumpled on the surface as shown in figure 1(b). Figure 1(c)shows images of Ag NPs which produced on the surface of PDA@slides that were immersed in 0.05 mol·L^{-1} $AgNO_3$ aqueous solution for 30 min. Figure 1(d) is an enlarged image of figure 1(c). There are countless nanoparticles on the surface of Ag/PDA@slides as shown in figure 1(c) and (d). The nanoparticles are spherical, and the size of the particles is from 20 to 50 nm.

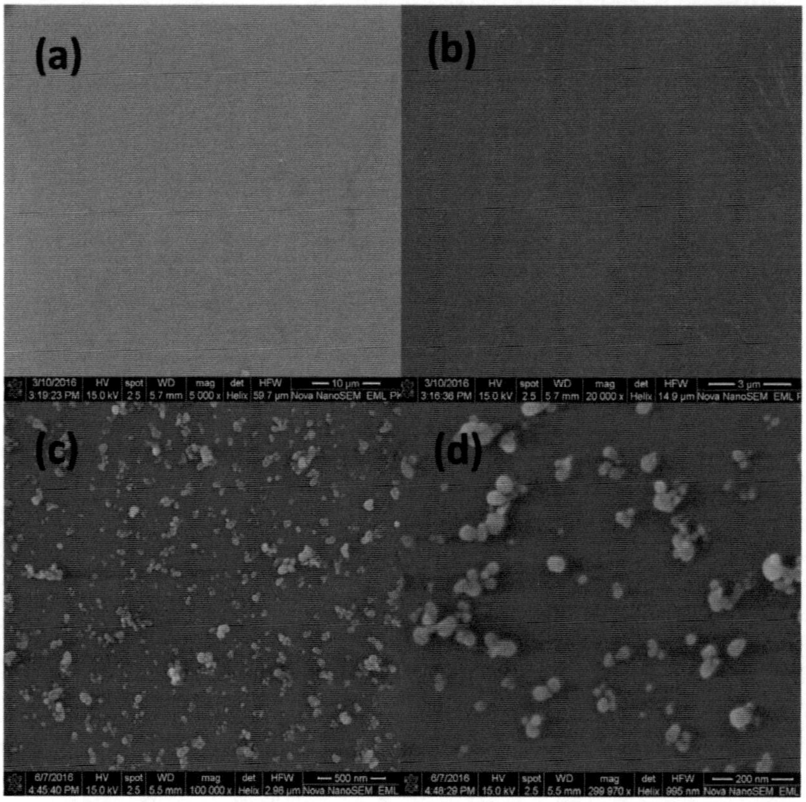

Figure 1. SEM images of slide (a), PDA@slide (b), Ag/PDA@slides immersed in 0.05 mol·L^{-1} $AgNO_3$ solution for 30 min (c,d), respectively.

Figure 2 shows XRD pattern in the 2θ region of 30-70° of Ag/PDA@slides. XRD patterns of slides, PDA@slides and Ag/PDA@slides obtained in 0.05 mol·L^{-1}$AgNO_3$ solution for 30 min are shown in figure 2(A), (B) and (C). The symbols of ∗, # and ▽represent the peaks of slides, Ag and Ag_2O respectively. The diffraction peaks of slides are shown in figure 2(A). Six peaks at 36.5°, 43.1°, 48.8°, 54.5°, 56.7°, 59.4° correspond to planes of slides. The first peak at 36.5° corresponds to hexagonal SiO_2 (110) (JCPDS 04-0836). When formed PDA film, no new diffraction peaks appear on the surface of PDA@slides compared with slides as shown in figure 2(B). For Ag/PDA@slides, new diffraction peak appeared at about 2θ = 44.2° is assigned to cubic Ag (200) (JCPDS 04-0783). The diffraction peaks at 36.5° and 38.6° correspond to hexagonal Ag_2O (JCPDS 04-0836) to (002) and (101) planes, respectively. The presence of Ag_2O is most likely due to the oxidation of Ag when it exposed to air. The peaks of Ag_2O (002) and SiO_2 (110) coincide, so the diffraction peak at 36.5° is strong.

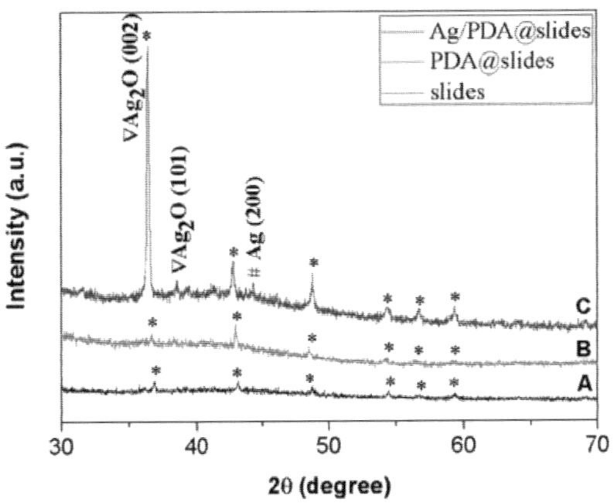

Figure 2. XRD patterns of the samples: slides (A), PDA@slides (B) and Ag/PDA@slides obtained in 0.05 mol·L^{-1} $AgNO_3$ solution for 30 min (C). The symbols of ∗, # and ∇ represent the peaks of slides, Ag, and Ag_2O respectively.

3.2. LSPR of Ag/PDA@slides

The optical properties of Ag NPs were monitored by UV-Vis spectra as presented in figure 3. Ag NPs, a kind of noble metal nanoparticles, possess compelling localized surface plasmon resonance (LSPR) properties due to the collective oscillation of conduction band electrons. LSPR of plasmonic nanostructure is strongly dependent on its size, shape, inter-particle distance, and particularly the refractive index (RI) of the surrounding environment [13]. After 15 min of $AgNO_3$ immersion time, a peak at about 400 nm appeared, corresponding to the SPR band of Ag NPs. When the reaction time prolonged to 30 min, a strong band appeared. The full width at half-maximum (FWHM) broadened when $AgNO_3$ immersion time prolonged to 60 min. As the $AgNO_3$ immersion time increased from 15 min to 60 min, the SPR band first gradually intensed and then gradually broadened. The SPR peak was strongest when the $AgNO_3$ immersion time was 30 min. It is speculated that the sharper and more intensed resonance peaks in the Ag NPs are better if applied to sensors [14]. Thus, immersing for 30 min is the best time for SPR.

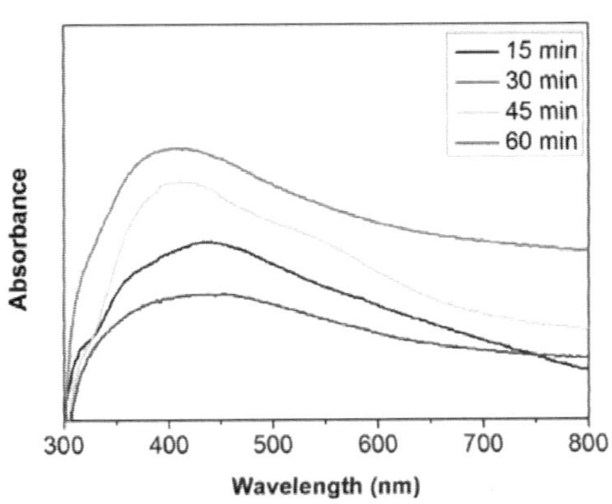

Figure 3. UV-Vis spectra of Ag/PDA@slides obtained at different $AgNO_3$ immersion time.

3.3. Catalytic reduction of 4-NP

The reduction of 4-nitrophenol (4-NP) to 4-aminophenol (4-AP) was chosen as a model reaction to evaluate the catalytic performance of the prepared Ag NPs. $NaBH_4$ was chosen as a reducing agent, and the concentration of it largely exceeded the concentration of 4-NP. Without Ag/PDA@slides, the reaction was not performed. The mixture of 4-NP and $NaBH_4$ showed a yellow color. After Ag/PDA@slide was put in, the reaction began. As the reaction proceeded, the color of the solution became fading. As monitored by UV-Vis spectra, the absorption peak of 4-NP at 400 nm gradually disappeared, and a new peak at 300 nm progressively increased, which means the reduction of 4-NP and the formation of 4-AP. The reduction of 4-NP is illustrated in figure 4.

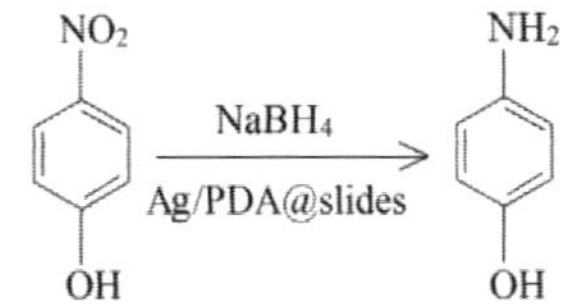

Figure 4. Illustration of the reduction of 4-NP by Ag/PDA@slides.

In order to determine the optimal concentration of $AgNO_3$, 0.01 mol/L, 0.05 mol/L and 0.1 mol/L, 0.2 mol/L and 0.4 mol/L $AgNO_3$ aqueous solutions were used to prepare Ag/PDA@slides which were next used to catalyze the reduction of 4-NP respectively. Experiments demonstrated that the fastest rate of the reduction was added in the Ag/PDA@slides which obtained in 0.05 mol/L $AgNO_3$aqueous solution, so the optimal concentration of $AgNO_3$ aqueous solution is 0.05 mol/L.

Next different $AgNO_3$ immersion time was investigated. Figure 5 shows the UV-Vis spectra of 4-NP after the addition of the as-prepared Ag/PDA@slides. According to experiments, the conversion time of 4-NP was 50 s, 10 s, 30 s and 30 s respectively when added the catalysts obtained by $AgNO_3$ immersion of 15 min, 30 min, 45 min and 60 min. According to experiments, the reduction could be completed in only 10 s when added in the Ag/PDA@slide obtained by 30 min immersion. Obviously, the immersing time in $AgNO_3$ aqueous solution plays an important role in the catalytic process, as which may impact the amount and the dispersion of Ag NPs.

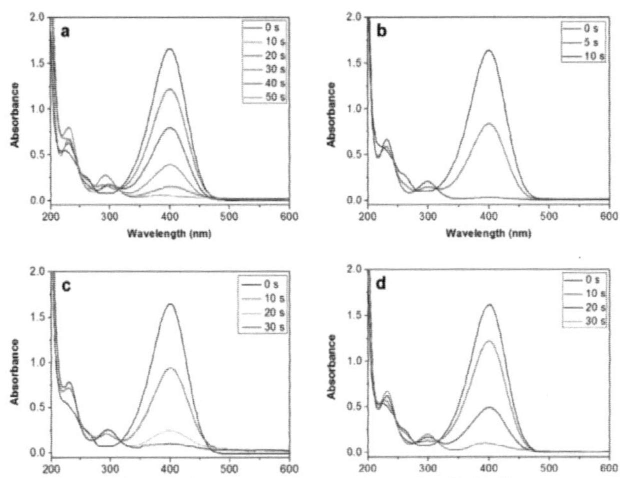

Figure 5.UV-Vis spectra of 4-NP after the addition of Ag/PDA@slides obtained by the $AgNO_3$ immersion time of 15 min (a), 30 min (b), 45 min (c) and 60 min (d).

In the experiment, A, B, C, D represent the Ag/PDA@slides which reacted with 0.05 mol/L $AgNO_3$ aqueous solution for 15 min, 30 min, 45 min, 60 min respectively. The kinetic constant *k* of the reduction of 4-NP were 6.7×10^{-2} s^{-1}, 3.9×10^{-1} s^{-1}, 9.7×10^{-2} s^{-1}, 9.3×10^{-2} s^{-1} when A, B, C, D, were used as catalysts respectively. Figure 6(a) shows the plot of ln (c_t/c_0) versus time. The kinetic equation of the reaction is as follow:

$$\ln(C_t/C_0) = \ln(A_t/A_0) = -kt \quad (1)$$

The reusability of Ag/PDA@slides was investigated by repeating the catalytic reaction with the same catalyst as shown in figure 6(b). After each reaction, the catalyst was taken out of the cuvette by a tweezer and it was used directly in the next reaction after washing with deionized water. As shown in figure 6(b), the catalyst possesses high reusability, which could be used for 6 times without obvious loss. Moreover, the prepared Ag/PDA@slides exhibited significant advancement over common particulate catalyst in the catalytic reduction of 4-NP because Ag/PDA@slides could be directly reused without filtered and centrifuged. The reuse process was much simpler.

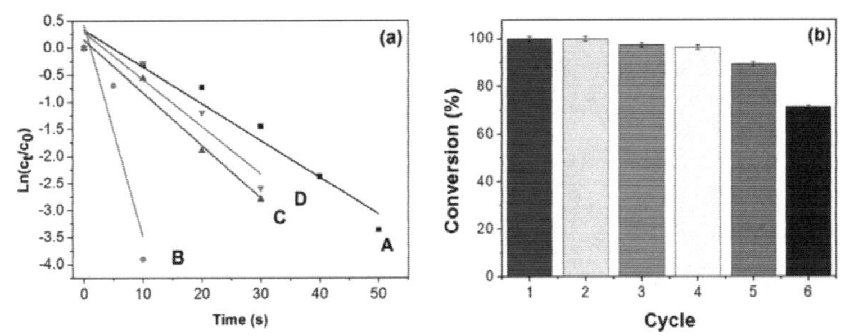

Figure 6. Relationship of $\ln(c_t/c_0)$ versus reaction time for the catalyzed reduction of 4-NP by A, B, C, D (a) and the conversion of 4-NP versus the catalyst cycles (b).

The prepared Ag/PDA@slide has higher catalytic rate for the reduction of 4-NP and simpler preparation process than other nano-Ag catalysts. To the knowledge of authors, the reduction of 4-NP used as-prepared Ag/PDA@slide immersed in 0.05 mol/L $AgNO_3$ aqueous solution for 30 min is the fastest among other nano-Ag catalysts.

4. Conclusion

In summary, an Ag/polydopamine composite has been prepared on the glass surface by self-polymerization and reduction properties of polydopamine. Ag/PDA@slide is a highly efficient catalyst which could catalyze the reduction of 4-NP in 10 s when at optimal condition. The UV-Vis absorption peak of Ag/PDA@slides proved that the condition of the strongest SPR of Ag/PDA@slides is the optimal condition of catalytic reaction. In addition, the catalyst is prepared on the surface of glass, so it is easy to be recycled.

Acknowledgements

We gratefully thank the National Natural Science Foundation of China (no. 21271027) for this work.

References

[1] Xiong S Q, Wang Y, Zhu J, Yu J R and Hu Z M2015 *Langmuir***31** 5504-5512
[2] Rostami-Vartooni A, Nasrollahzadeh M and Alizadeh M 2016*Journal of Alloys and Compounds***680**, 309-314
[3] Wan X, Zhuang L, She B, Deng Y, Chen D and Tang J 2016*Materials science & engineering. C, Materials for biological applications***65** 323-330

[4] Zelekew O A and Kuo D H 2016*Phys. Chem. Chem. Phys.***18** 4405-4414
[5] Arthur E E, Li F, Momade F W Y andKim H 2014*Energy***76** 822-829
[6] Ye W C, Chen Y, Zhou Y X, Fu J J, Wu W C, Gao D Q, Zhou F, Wang C M and Xue D S2014 *Electrochim. Acta***142** 18-24
[7] Chou K S and Lai Y S 2004*Mater. Chem. Phys.***83** 82-88
[8] Lee H, Dellatore S M, Miller W M and Messersmith P B 2007*Science* **318** 426-430
[9] Wang C, Zhu W, Lan Y, Zhang M, Tian T, Wang H and Li G T2014 *J. Phys. Chem. C***118** 10754-10763
[10] Yu X Q, Wang H, Guo L P and Wang L2014 *Chem.-Asian J.***9** 3221-3227
[11] Postma A, Yan Y, Wang Y J, Zelikin A N, Tjipto E and Caruso F2009 *Chem. Mat.* **21** 3042-3044
[12] Li W, Ma Z Y, Bai G Q, Hu J M, Guo X H, Dai B and Jia X 2015 *Applied Catalysis B-Environmental***174** 43-48
[13] Beeram S R and Zamborini F P2009 *J. Am. Chem. Soc.***131** 11689-+
[14] Lee K S and El-Sayed M A2006 *J. Phys. Chem. B***110** 19220-19225

Synthesis and Characterization of 4-(2-hydroxyphenoxy) phthalic nitrile

Yudong YANG [1,2], Zhili LAN [2,3], Liping SHENG [1], Suli XING [1] and Binrui WU [1]

[1]College of Aerospace Science and Engineering, National University of Defense Technology, Changsha 410073, China
[2]College of Chemistry and Chemical Engineering, Hunan Normal University, Changsha 410006, China

E-mail: scropotato@163.com

Abstract. 4-(2-hydroxyphenoxy) phthalic nitrile was prepared by the nucleophilic substitution reaction of catechol and 4-nitro phthalic nitrile in DMF with the existence of N_2 and K_2CO_3. The optimal reaction conditions were obtained as follows: the mole ratio of catechol, K_2CO_3 and 4-nitro phthalic nitrile is 2:1:1, with the quality of the solute concentration of 0.222g/mL and reaction temperature 80°C. Consequently, the yield of final product reached 87.64%, and the structure of the 4-(2-hydroxyphenoxy) phthalic nitrile was characterized by IR, ^{1}H NMR and DSC.

1. Introduction

With the development of aerospace industry, the high-temperature resistant resin matrix composites have attracted a lot of attentions owing to their promising applications in aerospace manufacturing. Among the wide range of high-temperature resistant resin matrix composites, nitrile resin matrix composites, which possess excellent mechanical property, high-temperature resistance, abrasion resistance, flame resistance and high char yield, are widely applied in the preparation of new-style equipment [1-2]. Therefore, many researchers [3-11] have focused on their research of phthalic nitrile. Besides, a lot of researches [12-15] showed that phenol hydroxyl could improve the self-catalyzed property of phthalic nitrile monomer. However, at present, many researches about this resin monomer are mainly at the stage of laboratory research. One important reason is the lack of basic data for extensive industrial application. In this paper, we used the catechol and 4-nitro phthalic nitrile to synthesize 4-(2-hydroxyphenoxy) phthalic nitrile, and the structure was characterized by ^{1}H-NMR and FTIR. DSC was used to confirm self-catalyzed polymerization behavior. And synthesis process has been carefully investigated, in order to provide basic data for massive industrial production.

2. Experiment

2.1. Materials

Catechol (AR) was obtained from Chinese Medicine Group Co. Ltd. Potassium carbonate (AR) by grinding was calcinated for 5 hours in the muffle furnace after 500°C. 4-Nitro phthalic nitrile (AR) was bought from Shanghai XIYA Chemical. Co. Ltd. N, N-Dimethyl formamide (DMF, AR) and dimethyl sulfoxide (DMSO, AR) were obtained from Tianjin HengXing Chemical Reagent Co. Ltd, and dried for at least a month with activated A4 molecular sieve.

2.2. *Characterization*

Solution proton nuclear magnetic resonance (^1H NMR) spectra were recorded on an Agilent DD2 400-MR spectrometer using DMSO-d6 and $CDCl_3$ as solvents with respect to tetramethylsilane as the internal standard.

For APN and HPN, Fourier transform infrared spectra (FTIR) were recorded with a PerkinElmer Spectrum TWO spectrometer in the solid state as KBr pellets. FTIR spectra of curing APN and HPN were also recorded every 2 min on the same instrument through the infrared reflectance mode with a heating furnace at a heating rate of 2°C min^{-1}.

DSC diagrams were recorded on a Mettler–Toledo DSC1 STAR System (Zurich, Switzerland) at different heating rates under nitrogen at a flow rate of 40 mL min^{-1}.

2.3. *Synthesis of 4-(2-hydroxyphenoxy) phthalic nitrile*

In the first stage, catechol (2.20g, 0.02mol), K_2CO_3 (1.38g, 0.01mol) and DMF (15mL) were added to a 50 mL two-necked flask to form a uniform solution by magnetic stirring at 80 °C for 4h with the protection of nitrogen. In the following stage, 4-nitro phthalic nitrile was added into the mixture when the solution cooled down to ambient temperature, and kept stirring at 80 °C for 5h. Next, the reaction liquid was filtered after the reaction stopped and its volume was measured with measuring cylinder. Then, 12mL filtrate was dissolved into 60mL water, and extracted 3 times with 30mL ethyl acetate each time. Finally, ethyl acetate was evaporated and 4-(2-hydroxyphenoxy) phthalic nitrile was separated from the mixture through the silica gel column, and the volume ratio of eluent (petroleum ether/ethyl acetate) is 1/0.24. The as-obtained 4-(2-hydroxyphenoxy) phthalic nitrile was dried in vacuum at 70 °C for 3 days. In the last stage, final product was weighed with analytical balance to calculate the product yield.

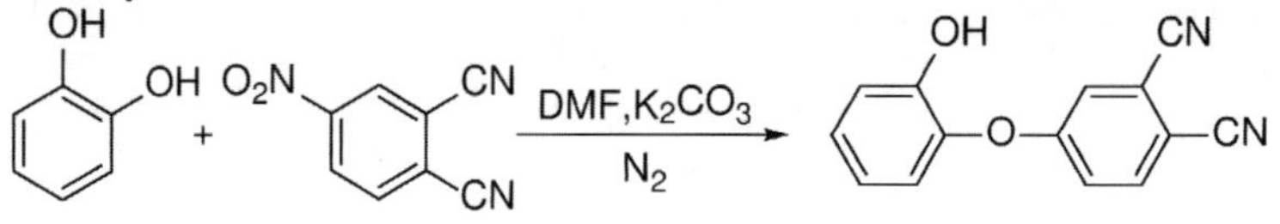

Figure 1. Synthesis of 4-(2-hydroxyphenoxy) phthalic nitrile

3. Results and discussion

3.1. *The optimal time of reaction*

Different reaction conditions, especially different temperatures, always leads to different reaction time. During this reaction, a moderate condition can result in a relatively short reaction time. To acquire maximum yield of 4-(2-hydroxyphenoxy) phthalic nitrile, the optimal reaction time should be adopted. The product and reactants of the reaction are all sorts of aromatic ramification, which own the hallmark of the obvious sample points by TLC in the UV camera analyzer, and can be beneficial to determine the optimal reaction time.

During the reaction, 0.1 mL reaction media was transformed to the centrifuge tube with pipette every 0.5 h after the 4-nitro phthalic nitrile was added. Subsequently, 200 times diluted reaction media was obtained by diluting the pristine reaction media with ethyl acetate. A certain amount of the diluent was put on the thin-layer plate by capillary glass and plate was put into the UV camera analyzer to compare the color of the product at every time point. The deepest color represent the optimal reaction time. The optimal reaction times at each temperature are shown in Table 1.

Table 1. The optimal reaction times at each temperature

Temperature/°C	40	50	80
Reaction time/h	14	12	5

An apparent decrease of the reaction time can be observed from 50 °C to 80 °C. Temperature plays a vital role in nitro nucleophilic substitution. Wang et al [16], found that the nitro nucleophilic substitution is limited when the temperature is lower than 60 °C. Thus, the effects caused by other factors can be carefully investigated at the optimal reaction time in different temperatures.

3.2. The effect of temperature on production yield

To research the effect of temperature on production yield, all the reactions shared the same reaction condition. The moles ratio of catechol, K_2CO_3 and 4-nitro phthalic nitrile is 2:1:1(the moles amount of catechol is 0.02), and solvent was DMSO. The first step took 4 hours, after adding 4-nitro phthalic nitrile, and the reaction time was determined by different temperatures (14 h at 40 °C, 12 h at 50 °C, and 5 h at 80 °C).

Table 2. The effect of temperature on yield

Temperature/°C	40	50	80
Yield/%	65.22	70.31	72.04

As is shown in table 2, the yield increase with the increase of the reaction temperature. The yields between 50 °C and 80 °C only show negligible difference, but the reaction cost more time at 50 °C. Though there are more by-products at 80 °C, they did not affect the purification of the target product in the post-treatment process. Thus, 80 °C is chosen as the optimal temperature.

3.3. The effect of the moles ratio among different ingredients on the yield

According to the above experimental results, the condition (80 °C, 4h+5h) was fixed to research the effect of the moles ratio among different ingredients on the yield. The experimental condition is as follows: The reaction temperature is 80 °C for 5 h, and the solvent is DMSO (15 ml), the moles amount of 4-nitro phthalic nitrile is 0.01. The moles ratios of catechol, K_2CO_3 and 4-nitro phthalic nitrile are shown in table 3.

Table 3. The effect of the moles ratio among ingredients on the yield

the moles ratio among ingredients	1.0:1:1	1.5:1:1	2:1:1
Yield/%	44.36	59.34	72.04

Table 3 shows the effect of the moles ratio among ingredients on the yield of final product, and the yield increases with the increase of the moles radio. The increment of catechol can increase the concentration of negative ions, which can promote the nucleophilic substitution as well as the final yield. Although the increased catechol in original reaction liquid can be redundant in final liquid, the redundant catechol can be recycled during the post-treatments. Thus, the optimal moles radio among catechol, K_2CO_3 and 4-nitro phthalic nitrile is 2:1:1.

3.4. The effect of solvent on the yield

As dipolar aprotic solvents, DMF and DMSO can promote the nucleophilic. The negative poles of polar groups of DMF and DMSO are both related to O^- (C^+-O^- of DMF and S^+-O^- of DMSO) and have no steric hindrance. However, the steric hindrance of positive poles of DMF and DMSO is bigger than that of negative poles. Consequently, the nucleophilic ability of the anions is enhanced by the preferential solvation.

The nucleophilic ability of reagent with negative charge is better than that of conjugate acid counterparts. Therefore, the K^+ of K_2CO_3 substituted the H^+ of the hydroxyl in the catechol. The stable anions were produced with the effect of the dipolar aprotic solvent. The 4-(2-hydroxyphenoxy) phthalic nitrile was obtained by nucleophilic substitution in the stable anions and 4-nitro phthalic nitrile.

According to the research results above, to study the effect of solvent on the yield, the reaction condition is optimal when the moles ratio between different ingredients is 2:1:1, with the reaction temperature 80 °C. The effects of different solvents on the yield are shown in table 4.

Table 4. The effect of solvent to the yield

solvent	DMF	DMSO
Yield/%	80.82	72.04

It can be easily found that, when utilizing DMF as solvent, the yield is 8% higher than that of DMSO. This is because the electronic ability of $-NH_2$ of DMF is better than that of $-CH_3$ of DMSO, and the increase of electronic ability impairs the solvation between the positive poles of polar groups and anions, and enhances the nucleophilic ability of anions. Thus, DMF was selected as solvent to research the effect of volume of solvent on the yield.

3.5. *The effect of solvent amount on yield*

In the final stage, the effect of solvent amount on yield was carefully investigated. Four dissimilar solvent amounts, 17.7 mL, 15.0 mL, 13.3 mL and 11.8 mL (corresponding to 0.222 g/mL, 0.262 g/mL, 0.295 g/mL, 0.333 g/mL), were utilized.

Table 5. The effect of solvent amount on yield

solvent amount/mL	17.7	15.0	13.3	11.8
Yield/%	87.64	80.82	68.93	66.32

According to table 5, more the dosage of solvent is, the higher final yield is. When the mass concentration is 0.222 g/ml, the yield reaches to its peak 87.64%. DMF and water are miscible, so, the DMF solution dissolves a bit products. In the post-processing, water was added to precipitate the products. If the DMF amount is increased, more of the products are dissolved, and the yield decreases. So, 0.222 g/ml is chosen to be the optimal mass concentration.

3.6. *Analysis of products*

The products were characterized by ^{1}H-NMR and FTIR.

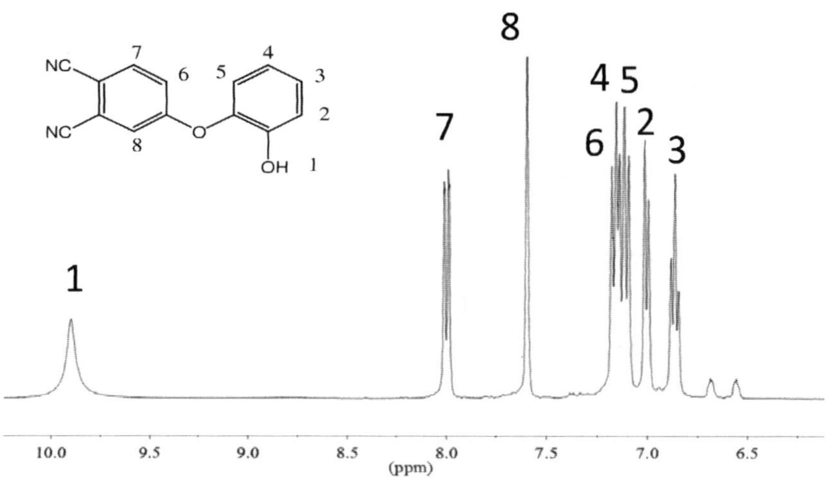

Figure 1. ^{1}H-NMR spectra of 4-(2-hydroxyphenoxy) phthalic nitrile

Figure 1 shows 1H-NMR spectra of 4-(2-hydroxyphenoxy) phthalic nitrile. According to the chemical shift values, the phenol hydroxyl protons appeared at 9.90 ppm, the aromatic protons appeared as multiple at 6.87-8.00 ppm. The results indicated the successful synthesis of 4-(2-hydroxyphenoxy) phthalic nitrile.

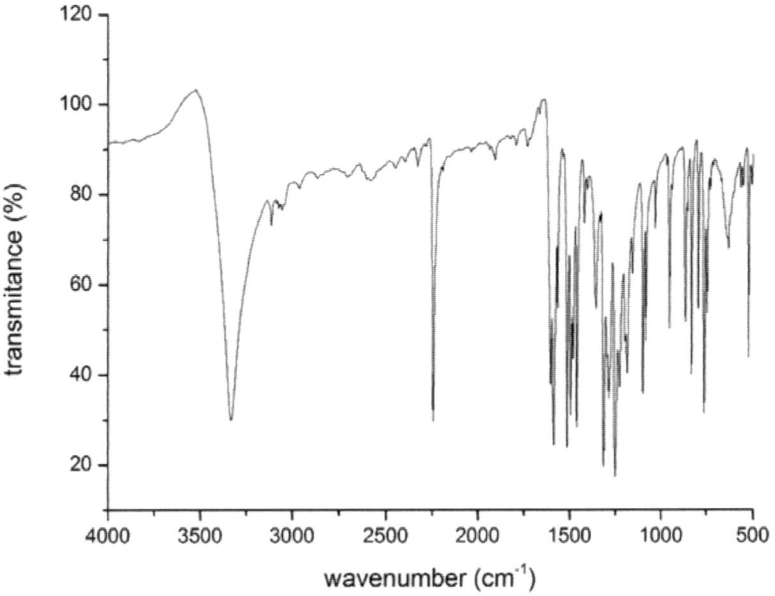

Figure 2. FTIR spectra of 4-(2-hydroxyphenoxy) phthalic nitrile

Figure 2 presents FTIR spectra of 4-(2-hydroxyphenoxy) phthalic nitrile. The peak at 3331 cm^{-1} was characteristic of the phenols, the peak at 2244 cm^{-1} was characteristic of the cyano, the peaks at 1586 cm^{-1}, 1513 cm^{-1}, 1494 cm^{-1} and 1460 cm^{-1} were characteristic of the benzene ring, the peak at 1252 cm^{-1} was characteristic of the asymmetrical vibrations of the aromatic ether, the peaks at 747 cm^{-1}, 762 cm^{-1}, 794 cm^{-1}, 831 cm^{-1}, 864 cm^{-1} and 867 cm^{-1} were characteristic of the out-of-plane C-H stretching. The characterization of FTIR proves the structure of 4-(2-hydroxyphenoxy) phthalic nitrile.

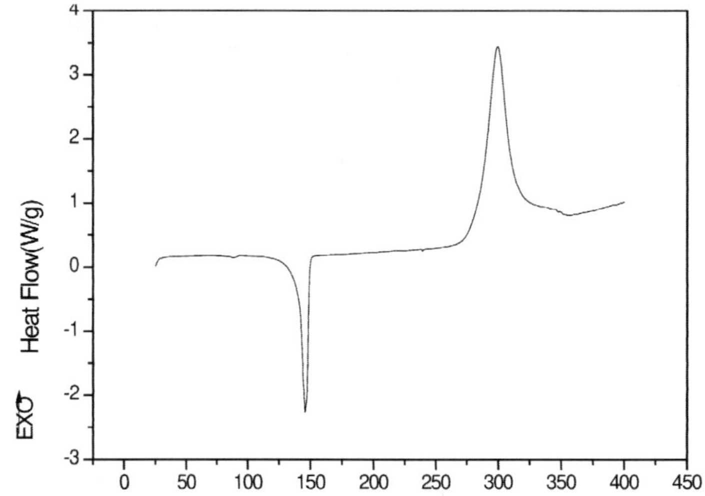

Figure 3. DSC curve of 4-(2-hydroxyphenoxy) phthalic nitrile

Figure 3 shows the nonisothermal DSC thermogram of 4-(2-hydroxyphenoxy) phthalic nitrile. There is a sharp endothermic peak at 146°C, and it is considered to be the melt peak of 4-(2-hydroxyphenoxy) phthalic nitrile. And the products are considered to have high-purity due to the narrow peak. A broad exothermic peak can be observed at 299°C. It is considered to be caused by curing behavior. There is no curing agent in 4-(2-hydroxyphenoxy) phthalic nitrile, so, the result indicates that 4-(2-hydroxyphenoxy) phthalic nitrile has excellent performance of self-catalysis.

4. Conclusion

The final product, after characterized by FTIR and ^{1}H-NMR, can be confirmed to be 4-(2-hydroxyphenoxy) phthalic nitrile. Simultaneously, it shows outstanding self-catalyzed performance according to DSC. After a rigorous study, the optimal reaction condition can be determined: the moles ratio among catechol, K_2CO_3 and 4-nitro phthalic nitrile is 2:1:1, with the quality of the DMF concentration of 0.222 g/mL and reaction temperature 80°C, and the final yield reaches to 87.64%.

References

[1] Serafini, T. T. (1987). High temperature polymer matrix composites. Elsevierence.

[2] Hergenrother, P. M., & Hergenrother, P. M. (2003). The use, design, synthesis, and properties of high performance/high temperature polymers: an overview. *High Performance Polymers*, **15**(1), 3-45.

[3] Laskoski, M., Dominguez, D. D., & Keller, T. M. (2005). Processable Phthalonitrile Resins with High-Thermal and Oxidative Stability.

[4] Laskoski, M., Dominguez, D. D., & Keller, T. M. (2005). Synthesis and properties of a bisphenol a based phthalonitrile resin. *Journal of Polymer Science Part A Polymer Chemistry*, **43**(18), 4136-4143.

[5] Dominguez, D. D., & Keller, T. M. (2008). Phthalonitrile-epoxy blends: cure behavior and copolymer properties. *Journal of Applied Polymer Science*, **110**(4), 2504-2515.

[6] Dominguez, D. D., & Keller, T. M. (2006). Lowmelting phthalonitrile oligomers: preparation, polymerization and polymer properties. *High Performance Polymers*, **18**(3), 283-304.

[7] Yang, X. L., & Liu, X. B. (2010). Study on curing reaction of 4-

aminophenoxyphthalonitrile/bisphthalonitrile. *Chinese Chemical Letters*, **21**(6), 743-747.

[8] Keller, T. M., & Dominguez, D. D. (2005). High temperature resorcinol-based phthalonitrile polymer. *Polymer*, **46**(13), 4614-4618.

[9] Du, R., Li, W., & Liu, X. (2009). Synthesis and thermal properties of bisphthalonitriles containing aromatic ether nitrile linkages. *Polymer Degradation & Stability*, **94**(12), 2178-2183.

[10] Cao, G. P., Chen, W. J., & Wei, J. J. (2007). Synthesis and characterization of a novel bisphthalonitrile containing benzoxazine.*Express Polymer Letters*, **1**(8), 512-518.

[11] Fang, Z., & Liu, X. (2010). Synthesis and curing behavior of a novel benzoxazine-based bisphthalonitrile monomer. *Journal of Applied Polymer Science*, **117**(117), 1469-1475.

[12] Zhou, H., Badashah, A., Luo, Z., Liu, F., & Zhao, T. (2011). Preparation and property comparison of ortho, meta, and para, autocatalytic phthalonitrile compounds with amino group. *Polymers for Advanced Technologies*, **22**(10), 1459–1465.

[13] Zeng, K., Zhou, K., Tang, W. R., Tang, Y., Zhou, H. F., & Liu, T., et al. (2007). Synthesis and curing of a novel amino-containing phthalonitrile derivative. *Chinese Chemical Letters*, **18**(5), 523-526.

[14] Zeng, K., Zhou, K., Zhou, S., Hong, H., Zhou, H., & Wang, Y., et al. (2009). Studies on self-promoted cure behaviors of hydroxy-containing phthalonitrile model compounds. *European Polymer Journal*, **45**(4), 1328-1335.

[15] Zhou, S., Hong, H., Zeng, K., Miao, P., Zhou, H., & Wang, Y., et al. (2009). Synthesis, characterization and self-promoted cure behaviors of a new phthalonitrile derivative 4-(4-(3, 5-diaminobenzoyl) phenoxy) phthalonitrile. *Polymer Bulletin*, **62**(5), 581-591.

[16] Wang, H. (2005). Synthesis of 4-phenoxyphthalonitrile compound. *Aerospace Materials & Technology*.

IOP Conf. Series: Materials Science and Engineering **167** (2017) 012067 doi:10.1088/1757-899X/167/1/012067

Preparation and properties of high storage stability polyester polyol dispersion for two-component waterborne polyurethane coating

H Hao, J Q Hu, F Wang, W P Tu [1]

South China University of Technology, School of Chemistry & Chemical Engineering, Guangzhou 510640, Guangdong, P R China

E-mail: cewptu@scut.edu.cn

Abstract. A new type of polyester polyol dispersion with good storage stability was prepared based on a hydrophilic monomer 5-sodium sulfodimethyl isophthalate (5-SIPM), and frequently-used monomers such as neopentyl glycol (NPG), dimethyl terephthalate (DMT), dimethyl phthalate (DMP) and trimethylolpropane (TMP) by the transpolycondensation and polycondensation method. The polyester polyol dispersion was characterized by FTIR and GPC. The proper content of these monomers were determined by the performance of polyester dispersion: the content of TMP was 15wt%, the content of NPG was 7.5wt% and the hydrophilic monomer 5-SIPM content was 5wt%. Two-component waterborne polyurethane (2K-WPU) coatings were prepared by Bayhydur® XP2487/1 and polyester polyol dispersions, which were stored before and after at 40 ℃ for 6 weeks, the prepared films have no differences in drying time, adhesion, pencil hardness, gloss and chemical resistance, the result also reveals that the polyester polyol dispersion have excellent storage stability resistance.

1. Introduction

Nowadays, as increasing demand for minimizing volatile organic compounds (VOCs) and hazardous air pollutants emissions, substantial waterborne systems have been introduced gradually to meet the requirements for environment-friendly coats, with lower toxicity and VOCs emissions compared to solvent-based products. Two-component waterborne polyurethane (2K-WPU) coatings appear better properties than one-component waterborne polyurethane coatings due to higher crosslinking density in coatings. 2K-WPU with the high performances of mechanical properties,chemical resistance and outdoor durability, has been used in industrial metal coating, plastics coating and automotive finishing [1-2]. Consequently, 2K-WPU coatings composed of polyisocyanate component and polyol dispersion component have attracted great interest from researchers. However, polyester polyol dispersion is the most widely used of polyol dispersion component. The structure of polyester polyol dispersion component has a significant effect on properties of 2K-WPU coatings, such as adhesion, hardness and impact resistance. Unfortunately, there are a lot of easily hydrolyzed ester bond in polyester polyol dispersion [3]. These ester bond hydrolysis reduces film properties and storage stability of polyester polyol dispersion. Hence, it is key to improve the storage stability of waterborne polyester in polyester polyol dispersion research. Yue [4] used trimellitic anhydride (TMA) as a hydrophilic monomer, synthesized polyester polyol dispersion by esterification process. The stability and properties of polyester polyol dispersion was decreased by the decarboxylation of TMA. Yasuhiko [5] used 2,2-dimethylol propionic acid (DMPA) as a hydrophilic monomer and avoiding decarboxylation, synthesized polyester polyol dispersion by esterification process. But the difficulty of the process is that the reaction of carboxyl group on DMPA is hard to control and reaction system is easy to be gelation, because the carboxyl group of DMPA is partially reaction with hydroxyl group of other monomer at the late stage of esterification reaction when temperature exceeding 200℃.

In this work, our group use 5-sodium sulfodimethyl isophthalate (5-SIPM) as hydrophilic monomer, trihydroxy methyl propane (TMP) and neopentylglycol (NPG) with high sterically hindered neoamyl structure to synthetize a novel type of polyester polyol dispersion through transpolycondensation and polycondensation process. The polyester polyol dispersion has superior storage stability due to high resistance ester bond. The influences of TMP, NPG and 5-SIPM content on the properties of polyester polyol dispersion were investigated. Then the polyester polyol dispersion was applied to the preparation of a new 2K-WPU coating with the Bayhydur® XP2487/1 polyisocyanate hardener. The basic properties of that 2K-WPU coating using polyester polyol dispersion before and after storage for 6 weeks at 40 ℃ were studied. The results demonstrate that the properties of 2K-WPU coatings have no obvious change in store of polyester polyol dispersion, including drying time, pencil hardness, gloss, adhesion and chemical resistance.

2. Experimental

2.1. Materials

Dimethyl phthalate (DMP), dimethyl terephthalate (DMT), 5-sodium sulfodimethyl isophthalate (5-SIPM), trihydroxy methyl propane (TMP) and neopentylglycol (NPG) were obtained from Tianjin Kermel Chemical Reagent Co., China. Zinc acetate and antimony trioxide were provided by Aladdin Industrial Co., China. Bayhydur® XP2487/1 polyisocyanate hardener (100% solid content) was kindly supplied by Bayer Co., Germany. The agent BYK-028 and BYK-346 were purchased from BYK Co., Germany. All these reagents were used as such without any further purification.

2.2. Synthesis of the polyester polyol dispersion

The polyester polyol dispersion was prepared according to transpolycondensation (Figure 1) and polycondensation process (Figure 2) with basic formulation (Table 1). The reaction was carried out in four-necked flask equipped with a variable speed stirrer, a thermometer, nitrogen gas inlet tube, and Allihn condenser. A Dean-Stark apparatus was used to remove the methanol and water of reaction.

Table 1. Basic formulation to prepare the polyester polyol dispersion

Materials	Content
DMT, DMP and 5-SIPM	1 mol (in total)
NPG and TMP	2 mol (in total)
Zinc acetate	0.04 wt%
Antimonous oxide	0.05 wt%

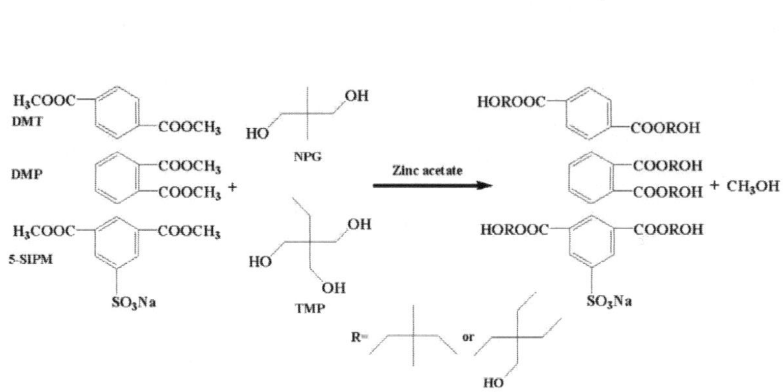

Figure 1. Transpolycondensation process of synthetizing the polyester polyol dispersion

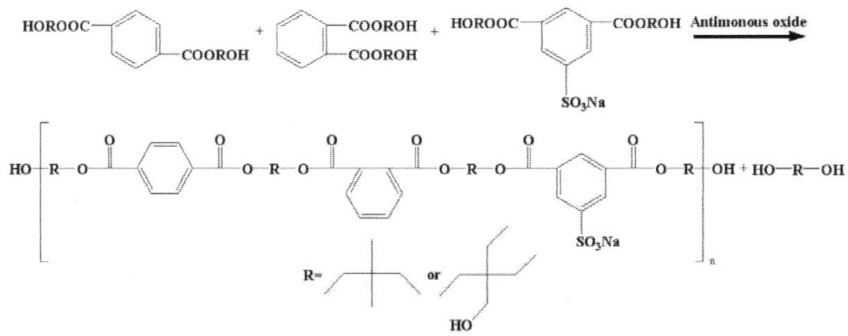

Figure 2. Polycondensation process of synthetizing the polyester polyol dispersion

Experimental polyester polyol dispersion had been synthesized according to the following steps: DMT, DMP, TMP, NPG and 5-SIPM were placed in 250 mL four-necked round bottom flask. The mixture was heated to 140 ℃ and maintained for 0.5 h to get rid of the water contained in the raw materials under N_2 atmosphere. Zinc acetate as the catalyst for transpolycondensation, was then added and slow heating flask to 195 ℃ for 2 h, methanol produced in transesterification process was distilled from flask to Dean-Stark apparatus. Continuing to heat up to 200 ℃ and adding antimony trioxide as the catalyst for polycondensation. Then the flask was evacuated to 0.1MPa and maintained until all the dihydric alcohol was distilled from flask to Dean-Stark apparatus. At last, the flask was cooled to 120 ℃ and diluted with Butyl Cellosolve to 85% solids, and then distilled water was added in flask with stirring at 2000 rpm about 1 h to form the polyester polyol dispersion dispersion with 52±2% solid content.

Bayhydur® XP2487/1 (100% solid content) was chosen as polyisocyanate hardener for the preparation of 2K-WPU coating with the polyester polyol dispersion. Table 2 illustrates the properties of 2K-WPU coating.

Table 2. Formulation to prepare 2K-WPU coating (NCO/OH=1.5)

Components	Weight/g
Polyester polyol dispersion	69.36
Byk-346	0.3
Byk-028	0.4
Water	12.61
Bayhydur® XP2487/1	17.33

2.3. Characterizations

2.3.1. Characterization of the polyester polyol dispersion
Fourier transformation infrared (FTIR) spectra of the polyester polyol dispersion was recorded on a Nicolet MAGNA-IR 560 FTIR spectrometer. Hydroxyl value of the polyester polyol dispersion was determined by chemical analysis interms of standard methods HG/T 2709-95. Molecular weights and polydispersity of the polyester polyol dispersion were measured by gel permeation chromatography (GPC, Model PerkinElmer series 2000). DMF was used as a solvent and flow rate of 1.0 mL·min-1 at 30 ℃.

2.3.2. Characterization of 2K-WPU composed of polyester polyol dispersion
The water and solvent resistance of 2K-WPU films was measured as follows [6]. The 2K-WPU films were cut into 3x3 cm pieces and determined their dry weight. These films were immersed in water or solvent after a certain period of time, and were weighed immediately when the water or solvent attached to the surface was wiped off.

Other physical properties of 2K-WPU films were evaluated according to standard test methods [7] (dry time GB/T 1728-89, gloss GB/T 1743-89, adhesion ASTM D 3359, pencil hardness ASTM D 3363).

3. Results and discussion

3.1. Characterization of polyester polyol dispersion

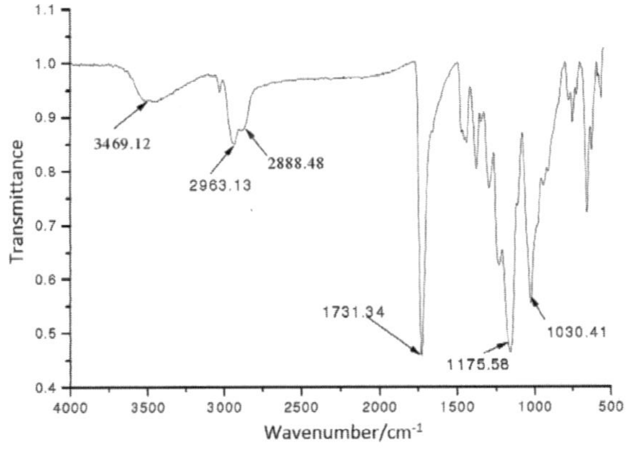

Figure 3. FTIR spectrum of the polyester polyol dispersion

In Figure 3, the following stretching vibrations are observed: The broad band at 3469 cm^{-1} is due to -OH groups. Typical absorption peaks of 2963 cm^{-1} to 2888 cm^{-1} due to $-CH_3$ and $-CH_2$- stretching vibration are observed. Obvious sharp band at 1731 cm^{-1} is due to C=O groups of ester linkage. In addition, other representative absorption peaks of 1175 and 1030 cm^{-1} are due to asymmetric and symmetric stretching vibrations of S=O groups, and it illustrate that the formation of the desired polyester polyol dispersion containing sulfonic acid sodium groups ($-SO_3^-Na^+$) through the transpolycondensation and polycondensation process [8].

Molecular weight including weight average weight (M_w), number average weight (M_n), Z-average molecular weigh (M_z) and polydispersity (PD) of samples was measured by GPC (Table 3). Table 3 shows the number average molecular weight of the polyester polyol dispersion is 2266, the polydispersity is 1.44, the molecular weight and polydispersity is desired.

Table 3. GPC results of polyester

M_n/Da	M_w/Da	M_z/Da	PD (M_w/M_n)
2266	3269	5055	1.44

3.2. Influence of TMP content

To investigate TMP content influence, a series of polyester polyol dispersions were synthesized based on different TMP content, the same ratio of other raw materials and experimental conditions. Their hydroxyl values and TMP contents are showed in Figure 4. It indicates that MPP synthesized at higher TMP content has a higher hydroxyl value. Generally, the hydroxyl value of polyester polyol dispersion is 100~150 mg·g^{-1} to ensure that there is sufficient cross-linking point to prepare 2K-WPU coating. Hence, the optimal content of TMP was determined to 15% for the synthesizing reaction.

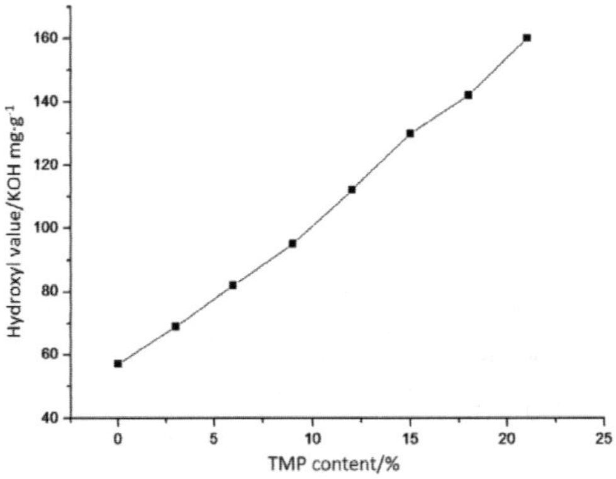

Figure 4. The relationship between the content of TMP and hydroxyl value

3.3. Influence of NPG content

The NPG content plays an important role in the synthesizing of polyester polyol dispersion, which can some what influence structure of high resistance ester bond and storage stability of the final products. Four samples of polyester polyol dispersions were synthesized at NPG contnent of 5, 7.5, 10 and 12.5wt%, respectively, and the results of influence of NPG content are exhibited in Table 4. As can be seen, when the NPG content exceeds 10wt%, the reaction system is very easy to gel. Because that carbon atoms in the side chain of NPG will also be involved in the condensation reaction at high temperatures. With the increase of NPG content, the proportion of hydrophilic monomer decreased, and the hydrophilic property was not enough to stabilize the resin in water. On the other hand, increasing of NPG content, the particle size of the polyester polyol dispersion increases and the stability decreases. Because the relatively less hydrophilic property was not enough to stabilize the polyester polyol in water with increasing of NPG content. Therefore, the optimum NPG content is found to be 7.5wt%.

Table 4. The impact of the ratio of NPG

NPG content/wt%	Appearance	Storage stability (40 ℃d)	Particle size of dispersions/nm
12.5	gel	-	-
10	white, flocculation partly	sediment	221
7.5	white, without flocculation partly	good	148
5	white, with blue light	good	91

3.4. Influence of 5-SIPM content

A series of polyester polyol dispersions were prepared at different 5-SIPM content. The influence of 5-SIPM content on particle size, stability and water resistance of polyester polyol dispersion are presented in Table 5. It can be demonstrated that 5-SIPM content increasing leads to decreased particle size and water resistance, increased the stability of polyester polyol dispersion.

The number of hydrophilic $-SO_3^-Na^+$ groups in polyester polyol dispersion increases with increasing 5-SIPM content. The $-SO_3^-Na^+$ groups attract the water molecules forming a water layer on the surface of polyester polyol dispersion particles. Then the relative size of the water layer to total particle size increases with reducing polyester polyol dispersion particle size. With reducing polyester polyol dispersion particle size, the relative size of the external water layer to total dispersion particle size increases. Meanwhile, there is an increase in the number of total dispersion particles with reducing particle size [9]. The study of stability and water resistance of polyester polyol dispersion displays that polyester polyol dispersion based on the 5-SIPM content of 5 wt% has best comprehensive performance.

Table 5. The impact of the content of 5-SIPM

5-SIPM content/wt%	Appearance	Storage stability	Water resistance (60 ℃, h)	Particle size of dispersions/nm
4.5	milky white and opaque	sediment	unchanged	202
5	milky white and translucent	without sediment	unchanged	148
5.5	milky white and translucent	without sediment	slight white	110
6	brown and translucent	without sediment	white	96
6.5	brown and transparent	without sediment	white, vesicle	78

3.5. The storage stability of the polyester polyol dispersion

Table 6 presents that the appearance, solid content, acid value and pH value of polyester polyol dispersion have no obvious change in 6 weeks at 40 ℃ store condition. It indicates that the polyester polyol dispersion has excellent storage stability.

Table 6. Properties of the polyester polyol dispersions (before and after storage at 40 ℃, 6 weeks)

Items	Before storage	After storage
Appearance	milky white, with blue light	milky white, with blue light
Solid content/%	43.7	43.7
Acid value/mg KOH/g	1.1	1.3
pH value	6.0	5.3

3.6. The properties of 2K-WPU coating

Table 7. Properties of film in different the polyester polyol dispersions

(before and after storage at 40 ℃, 6 weeks)

Properties	Before storage	After storage
Surface drying time/min	30	40
Hard drying time/h	5	6
Adhesion/grade	0	0
Pencil hardness/grade	2H	2H
Gloss(20°)/%	92.7	92.5
Gloss(60°)/%	97.3	98.0

Bayhydur® XP2487/1 (100% solid content) was chosen as polyisocyanate hardener for the preparation of 2K-WPU coating. Table 7 and 8 illustrate the some properties of 2K-WPU coating.

Drying time, adhesion, hardness, gloss and chemical resistance of 2K-WPU coatings using polyester polyol dispersions before and after storage for 6 weeks at 40 ℃ have no obvious change. It demonstrates that polyester polyol dispersion has the high content of hydroxyl groups, the high crosslinking and excellent storage stability.

Table 8. Chemical resistance of film in different the polyester polyol dispersions

(before and after storage at 40 ℃, 6 weeks)

Chemical names	Immersion time/min	Before storage	After storage
90# gasoline	10	0	0
Xylene	10	2	2
Ethanol	10	1	1
	20	2	2
Acetone	10	3	3
	20	4	4

0 represents the best effect, 5 represents the worst effect.

4. Conclusion

(1) The FTIR result reveals the polyester polyol dispersion are in line with expectations. The GPC results indicate the number average molecular weight is 2266 and polydispersity is 1.44.

(2) The proper content of TMP was 15 wt%, the proper content of NPG was 7.5 wt% and the proper content of 5-SIPM was 5wt%.

(3) Drying time, adhesion, hardness, gloss and chemical resistance of 2K-WPU coatings using Bayhydur® XP2487/1 and the polyester polyol dispersions which were stored before and after at 40 ℃ for 6 weeks, the prepared films have no obvious change. It indicates that the polyester polyol dispersion has outstanding storage stability.

References

[1] García-Pacios V, Jofre-Reche JA, Costa V, *Colera M and Martín-Martínez J M 2013*
[2] Coatings prepared from waterborne polyurethane dispersions obtained with polycarbonates of 1,6-hexanediol of different molecular weights *Prog. Org. Coat.* **76** 1484-1493
[3] Fan W, Du W, Li Z, Dan N and Huang J 2015 Abrasion resistance of waterborne polyurethane films incorporated with PU/silica hybrids *Prog. Org. Coat.* **86** 125–133
[4] Hawkins C A, sheppard A and Wood T 1997 Recent advances in aqueous two-component systems for heavy-duty metal protection *Prog. Org. Coat.* **32** 253-261
[5] Mehdipour-Ataei S and Zigheimat F 2007 Soluble polyesters with preformed ether and imide units: synthesis and properties *Des. monomers polym.* **10** 145-152
[6] Wu J R and Chen, D J 2016 Synthesis and characterization of waterborne polyurethane based on covalently bound dimethylol propionic acid to e-caprolactone based polyester polyol *Prog. Org. Coat.* **97** 203-209
[7] Huang K, Liu Y and Wu D 2014 Synthesis and characterization of polyacrylate modified by polysiloxane latexes and films *Prog. Org. Coat.* **77** 1774–1779
[8] Wu G M, Kong Z W, Chen J, Huo S P and Liu G F 2014 Preparation and properties of waterborne polyurethane/epoxy resin composite coating from anionic terpene-based polyol dispersion *Prog. Org. Coat.* **77** 315–321
[9] Lee H T, Wu S Y and Jeng R J 2006 Effects of sulfonated polyol on the properties of the resultant aqueous polyurethane dispersions *Colloids Surf., A* **276** 176–185

Main challenges in demulsifier research and application

Fusheng Zhang[1], Guoliang Liu, Junhan Ma, Jian Ouyang, Xiaoling Yi and Huimin Su

Research Institute of Petroleum Exploration and Development, PetroChina, Beijing, China

E-mail: zhfsh@petrochina.com.cn

Abstract. Main challenges in demulsifier research, such as demulsification of ASP flooding produced liquid, demulsification of heavy oil produced liquid, low temperature demulsification and fast demulsification, are summarized. Some importance technology routes to solve the challenges are proposed according to demulsification mechanisms and emulsion characteristics. The proposed routes include increasing aromaticity, molecular weight and branch degree of demulsifiers, and introducing double-function groups to demulsifiers for W/O and O/W emulsions, or groups with alkyl matching with alkyl carbon number of the crude oil into demulsifier molecule. The demulsification mechanisms of the above-mentioned research routes are described in detail.

1. Introduction

Main methods of demulsification and dehydration for the produced liquid include heat method, electrical method and heat chemical method. Demulsifiers with many structures have been developed according to different characteristics of the various produced liquids, and resolved preferably separation of oil and water. But new problems are continually arising along with the rapid development of petroleum industry, such as: (1) Water cut of the produced liquid from old fields in middle and later stages is higher than 90%. ASP flooding technology [1] has been extensively used and has made better effects on the increasing oil production and decreasing water cut, but the produced liquids are very stable and more difficult to demulsify and dehydrate; (2) With more attention being paid to the exploitation of heavy oil [2], sophisticated stimulation technologies including steam flooding, steam soaking, alkaline flooding and polymer flooding have enhanced oil recovery. However, the application of these technologies leads to the formation of more stable emulsions which makes demulsification or dehydration more difficult; (3) Heating or enhancing temperature of the produced liquid can improve the efficiency of oil-water separation [3]; but enormous energy is wasted. As saving energy is strengthening in oil fields now, many oil fields are in dire need of low temperature demulsification; (4) The production of offshore oil field arises greatly. Due to small oil-water separation tanks, oil and water must be separated in short time so as to meet the production demand. The offshore oil fields are in dire need of the demulsifiers for fast demulsification. Existing demulsifiers can hardly meet the demand of above-mentioned problems, so more severe requirements for demulsifiers are proposed and more application prospects are provided. Much research has been performed and some progresses have been made, although there are still many problems remained unresolved.

2. Demulsification of ASP flooding produced liquid

With the production of oil field being into middle and late stages, water cut in the produced liquid is increasing but production of crude oil is decreasing year by year. The application of chemical flooding technologies (including polymer flooding, surfactant flooding, binary combination flooding, ASP flooding) for enhanced oil recovery has gradually become one of the key technologies to sustainable development of oil fields. So far, polymer flooding technology has been applied in large-scale, pilot test of ASP (alkali/surfactant/polymer) flooding technologies has achieved good results on increasing-oil and decreasing-water, but the emulsion of produced liquid is very stable and difficult to demulsify.

The test of polymer flooding and ASP flooding in Daqing oilfield showed the following three problems: (1) Middle emulsion layer is thick in sedimentation tank; (2) Oil content in sewage is high (up to 3000mg/L) and cannot meet the requirement for the water reinjected; (3) The electric dehydrator is not stable and easily crushed. The above-mentioned problems will become increasingly prominent with the popularization and implementation of ASP flooding technology, and will be an important factor to restrict the application of chemical flooding technology.

2.1. Composition and emulsion state of produced liquid

There are big differences between composition and properties of produced fluid from early stage of oil field development (water flooding) and late stage(chemical flooding, such as polymer flooding and ASPA flooding): ① The produced fluids contain oil displacement agents such as alkali, polymers and surfactants; ② The content of heavy component (asphaltene and resin) from chemical flooding is higher than that from water flooding because chemical flooding can drive heavy oil (asphaltene and resin) firmly adsorbed on the rock surface; ③ Since the viscosity of produced liquid increases and the liquid carries solid particles (mechanical impurities, clay, fine sand and asphalt), content of the above-mentioned composition is higher.

The emulsion state of the produced liquid from water flooding compared with chemical flooding, has a qualitative change: emulsion state changes from W/O in early stage to O/W, W/O, and complex multiple emulsion(W/O/W,O/W/O) coexisting in the produced liquid at the same time, especially the produced fluid contains a large number of small O/W emulsion droplets which are very difficult to coalesce and separate.

2.2. Research routes

2.2.1. Study of demulsification mechanism. It is necessary to intensively study the mechanisms of complex emulsion formation and demulsification from produced liquid in ASP flooding to provide theoretical guidance for the design and synthesis of demulsifier. It is also necessary to study the influence of these changes on emulsion state, emulsion structure and emulsion stability because the composition of produced liquid between ASP flooding and water flooding is greatly changed.

2.2.2. Study of the dual-function demulsifiers. The emulsion structure of ASP flooding produced liquid in which O/W, W/O and multiple emulsion coexist, is very complex, so demulsifier of dual-function emulsion, which can meet demulsification of both O/W emulsion and W/O emulsion at the same time, could be designed and studied. There are two roles for these demulsifiers: ① Break W/O emulsion in produced liquid , resulting in water drops coalescence and water cut reduction in oil phase, ② Break the O/W emulsion in produced liquid , resulting in oil drops coalescence and oil cut reduction in water phase.

As we all know, polyether demulsifiers can meet the requirements of W/O demulsification and with coordination of electric dehydration technology, water cut in oil can meet the requirement of crude oil external transportation. Reverse demulsifiers [4] can break O/W emulsion, with coordination of filtration and sedimentation, oil cut in sewage meets the requirement of sewage discharge. The design and syntheses of dual-function demulsifiers which adapt to both W/O and O/W is an effective way to solve the problems of demulsification of the complex produced liquids.

2.2.3. Increasement of demulsifier molecule weight. In the role of oil displacement agents and other natural emulsifiers and after the pore formation shear for a long time, produced liquid forms the tiny emulsion (W/O and O/W) that is more stable and difficult to break.

According to the demulsification mechanisms and the requirements of efficient demulsifiers, demulsifiers with high molecular weight and branched structure can solve above-mentioned problems. Above-mentioned demulsifiers are unique for the following reasons: ① it can be absorbed simultaneously on the surface of many emulsion droplets, so that the emulsion droplets have more opportunities of collision and coalescence; ②demulsifiers with high molecular weight are easy to form micelles, which can solubilize emulsifiers and lead to emulsion demulsification; ③the interface film is not tight and low strength after demulsifiers with high molecular weight replace the emulsifiers on the water-oil interfaces.

3. Demulsification of Heavy Crude

Heavy oil produced liquid is very stable and difficult to demulsify because it has higher content of asphaltene and resin, higher density and viscosity. Produced liquid becomes more complex after the application of stimulation technology, such as steam soaking, polymer flooding, alkaline flooding and surfactant flooding.

World energy crisis is worsening nowadays. People are paying more attention to heavy crude oil because reserves of good quality oil explored easily are decreasing. Heavy oil production is for about 10% of total crude oil production in China [5], and the demulsification difficulty of produced liquid becomes prominent increasingly with the increase of heavy oil production, so research on special demulsifiers for heavy oil is much important.

3.1. Composition and physical properties of heavy crude oil

The nature of some paraffin crude and heavy crude are shown in following table.

Crude oil	Zhongyuan Oilfield	Daqing Oilfield	Liaohe heavy Oil	Shengli heavy Oil
Saturated hydrocarbon /%	69.5	73.2	21.4	17.7
Aromatics /%	23.6	15.1	22.4	23.9
Resin /%	6.8	10.1	51.8	54.5
Asphaltene /%	0.1	1.6	4.0	3.9
Density /g·cm^{-3}	0.8509	0.8601	0.9016	0.9926
Pour point /°C	33	28	24	23
Viscosity /mPa·s,50°C	20	30	20420	64160

Crude oil of Zhongyuan and Daqing is paraffin crude, whose saturated hydrocarbon content is greater than 50% with low density, viscosity and content of asphaltene and resin, and the ratio of asphaltene and saturated hydrocarbon is small. On the contrary, the crude oil of Liaohe and Shengli is asphaltene crude oil.

3.2. Composition and structure of asphaltene and resin

Structure and composition of asphaltene and resin have been studied deeply by various analytical methods. The basic structure of asphaltene and resin [6] is a core of fused aromatic ring which is surrounded with several cycloalkanes, aromatic hydrocarbons and heterocycles, and includes several alkane side chains on the ring and groups containing S, N, O (such as -OH, $-NH_2$, -COOH) and heavy metals complexed (such as Ni, V, Fe) in the molecules.

Because of the basic structure of asphaltene is a plane structure, the force (π-conjugated and the van der Waals force) among the molecule structure of aromatic plane is larger and hydrogen-bonds are easily formed between polar groups (-OH,-NH2,-COOH). Under the two acting forces, it is easy to form asphaltene particles (3 to 5 fragrance tablets), and then asphaltene particles form micelles by the forces of inter-molecular hydrogen bonding and inter-particle dipole, and at last micelles are interconnected and greater associations are formed [5]. The asphaltenes are in the center of micelles, and the dispersion media (resin and oil) are adsorbed into both interior and surface of micelles. Components of dispersion medium which have greater molecular weight, stronger aromaticity and stronger polarity are close to the center, while the lighter components are adsorbed around the micelles and transit to the oil phase gradually.

3.3. Influence of asphaltene and resin on the stability of heavy oil emulsion

The substances of heavy crude emulsion film are natural emulsifiers including asphaltene, resin, oil acid soap, wax crystals, carbonates and clay particles [7], among which asphaltene and resin have a greater effect on the stability of emulsion. The higher the content of asphaltene and resin, the more stable the crude oil emulsion is. Because heavy crude has a higher content of asphaltene and resin, its emulsion has higher stability. In addition, asphaltene and resin in heavy oil have strong polarity and adsorb easily impurities, the density contrast of oil and water is very smaller and multiple emulsion is easily formed, the above factors lead to very stable heavy oil emulsion.

The interfacial activity of asphaltene is not very strong with the oil-water interfacial tension of only 25 to 30 mN/m, but it has greater interfacial steric hindrance, so asphaltene has a greater effect on the stability of emulsion [8]. Interfacial film with certain viscoelasticity is formed after asphaltene is adsorbed at the oil-water interface, such interfacial films can withstand higher temperature, higher mechanical strength and higher pressure.

3.4. Research routes

3.4.1. Enhance aromaticity of demulsifiers. Process of demulsification emulsion includes demulsifier infiltrating into the interface, displacing natural emulsifiers in the interface and destroying the interfacial films, emulsified water droplets approaching and coalescing to each other, and water separation from the crude at last. Therefore, the performance of demulsifiers can be improved by enhancing the capability of diffusion in the emulsion, the capability of penetration into the interface film and the capability of replacing and solubilizing emulsifier according to demulsification mechanism.

The higher aromaticy results in demulsifiers having strong ability of diffusion and infiltration in heavy oil, main reasons are that ① heavy oil contains a large number of aromatic substances (such as resin, asphaltene), which resists diffusion of demulsifiers in heavy oil; ②asphaltene and resin are the main substances of heavy oil emulsion film, which has higher strength, and demulsifiers are not easy to penetrate. However, diffusion and infiltration of demulsifiers with higher aromaticity and similar structure with asphaltene and resin, is easier than the ones with non-aromaticy, so the capability of diffusion and infiltration can be improved by increasing the aromaticity of demulsifiers.

The higher the aromaticity degree of emulsifier, stronger it solubilizes interfacial film and reduces the film strength because the activity of film-forming substances (asphaltene and resin) depends on the its state in oil phase [9]. Asphaltene molecule has three state in crude oil: ①single asphaltene molecule is not able to stabilize emulsion; ②micelles can enhance the emulsion stability. The higher the concentration of micelle, the higher the emulsion stability;③existing in the form of condensation, asphaltene makes the greatest contribution to the emulsion stability. Therefore, the key of demulsification is to change the state of asphaltene in oil phase, and studies show that the balance of three existing states of asphaltene molecules in crude oil depends on the aromaticity of oil phase [10], oil phase with higher aromaticity has an stronger solubilizing ability to asphaltene by π-bond interactions between aromatic rings, so the film strength is reduced and the effect of demulsification can be improved.

Increasing the aromaticity of demulsifiers can be achieved by enhancing the aromaticity of the initiator of demulsifiers and modification of demulsifiers using aromatic compounds.

3.4.2. Enhance wettability of demulsifiers. The wettability of demulsifier is closely related to their molecular structure. Demulsifiers with a branch structure are not conducive to the association of molecules and the formation of micelles, but to the adsorption of molecules on solid surfaces which changes the wettability of surfaces[11,12]. Because of stronger carrying capability of heavy oil, the content of solid particles, such as asphaltene, paraffin, clay particles, metal salt and mechanical impurities, are high in the interface films of heavy oil emulsion, which increases the stability of emulsion. The enhanced wettability of demulsifiers can promote the demulsifier molecules diffusion to the emulsion droplet and penetrate through the protective layer on the solid particles and adsorb on the surface of the solid particles, because of the change of wettability of the particles surfaces and the decrease of surface energy which can drastically reduce the strength of interfacial films , the interfacial films rupture.

4. Low temperature demulsification

The treatment process of produced liquid commonly used in oilfields is demulsification and electric dehydration technology assisted with necessary heating. To improve demulsification efficiency, it is an effective way to reduce the viscosity of crude oil and enhance the temperature, but a lot of energy is wasted. A large amount of water is injected into the formation in late stage of oil field development, which results in the decline in temperature of produced liquid. To obtain good oil-water separation only by heating produced liquid to raise temperature, once the furnace is stopped, the middle emulsified layer of sedimentation tank will become thicken and the oil cut in sewage will be overproof. In order to save costs, it is generally hoped that the oil-water separation is able to meet the requirements without heating. Therefore, the study of low temperature demulsifier has practical and economic significance.

Methods of demulsifier complex and demulsifier modification are used in order to meet the requirements of low temperature demulsification [13]. For instance, the modification of demulsifiers using unsaturated acids has achieved good effect of demulsification at low temperature.

It is believed that the following technical routes are worth considering. Due to high viscosity of crude oil when temperature is very low, the diffusion rate of demulsifiers is affected. Based on carbon number distribution of crude saturated hydrocarbon and carbon chain length of substituted alkyl on the aromatic ring of asphaltene, adjusting the length of alkyl chain in the demulsifiers structure, can match the number of alkyl chain in demulsifiers with that of crude oil. Thus the hydrophilic and lipophilic keep a certain balance and the demulsifiers have faster diffusion rate in corresponding crude, improving the effect of low temperature demulsification.

5. Fast demulsification

The production of offshore oilfields increases in pace with the rapid development of offshore oil industry year by year. However, it is impossible to build large-capacity processing tanks in the small space of offshore platforms. Improving the efficiency of processing tanks requires fast demulsifiers.

A few reports about fast demulsifiers have been published, one of which is nano-demulsifier [14,15]. Nano-demulsifier is obtained by demulsifiers modification on the surfaces of nanoscale inorganic solid particle to improve its diffusion rate and permeation rate, thus the rate of demulsification is enhanced.

6. Conclusion and Suggest

- Compared with water flooding produced liquid, ASP flooding produced liquid and heavy oil produced liquid have a big difference in the emulsion structure and state and are very difficult to demulsify.
- Demulsification mechanisms of complex emulsion, which can provide fundamental knowledge for the development of new demulsifiers, needs further research.

References

[1] Cheng Jiecheng, Liao Guangzhi, Yang Zhenyu, et al. Pilot Test of ASP Flooding in Daqing Oilfield [J]. *Petroleum Geology & Oilfield Development in Daqing*, 2001,**20**(2):46-49.

[2] Li Kui, Liu Zhaobin. Preparation and Application of Demulsifier for Quickly Breaking SuperHeavyoil [J]. *Advances in Fine Petrochemicals*, **2004**, (7):34-37.

[3] Wu Zongfu, Huang Hongquan, Li Yongchang, et al. Development and Application of a New Emulsion Breaker of High Efficiency and Low Temperature [J]. *Journal of Jianghan Petroleum Institute*, 2003,**25**(1):85-88.

[4] Liu Huiying, Song Nairen, Li Weifeng, et al. Studies and Application of Demulsifier CW-01 for Breaking-Down Reverse, O/W, Crude Oil Emulsions [J]. *Design of Oilfield Construction*, 1996,**41**:39-42.

[5] Yu Liandong. Distribution of World Heavy Oil Reserves and Its Recovery Technologies and Futrue [J]. *Special Oil and Gas Reservoirs*, 2001,**8**(2):98-103.

[6] Yang Xiaoli, Lu Wanzhen. Advances in Stabilization and Destabilization of Water in Crude Oil Emulsions [J]. *Oilfield Chemistry,* 1998,**15**(1):87-96.

[7] Wang Biao. Recent Progress in Development of Demulsifiers for Crude Oils [J]. *Oilfield Chemistry*, 1994,**11**(3):266-272.

[8] Li Ming-yuan, Zhen Peng, Wu Zhao-liang, et al. Investigation of Stability of Water-in-Crude Oil Emulsions: VI. Properties of Interfacial Film and Stability of Crude Oil Emulsions [J]. *ActaPetroleiSinica (Petroleum Processing Section)*,1995,**11**(3):1-6.

[9] Yu Guo-xian, Chen Hui, Lu Shan-xiang, et al. Compound Formulation of Composite Demulsifier for Crude Oils [J]. *Speciality Petro Chemicals*, 2003,**10**(5):1-4.

[10] Joseph D.M. Effects of asphaltene solvency on stability of water-in-crude oil emulsion [J]. *J Colloid & Interface Sci.*, 1997, **189**:242-253.

[11] Gong Huijuan, Hu Gengyuan, Chen Guanxi. The Synthesis of CrosslinkedPolyethers and Their Application in Demulsifying Crude Oil [J]. *Speciality Petro Chemicals,* 2000,**7**(4):1-4.

[12] Li Shuxian, Cui Jidong, Yao Yanfeng, et al. Preparation and Use of High-Molecular Amphipathic Demulsifier BC-068 for Dehydration of Heavy Crude Oils [J]. *Oilfield Chemistry*. 2004, **21**(1):39-41.

[13] Fan Zhengzhong, Liu Qingwang, Wei Xiaoming. The Preparation and Appraisal of High Efficiency LTB Demulsifier for Heavy Crude at Low Temperature [J]. *Oil Gasfield Surface Engineering*, 2001,**20**(1):29-30.

[14] Wu Tong, Wang Xingwang, Zhang Yunan, et al. Study on the Dehydration Process of Aging Oil by Demulsifier Modified With Nanometer-Silicion[J]. *Contemporary Chemical Industry*, 2013, **42** (11):1488-1490.

[15] Sun Zhenggui. Application of Polyether Demulsifier TA1031 Modified by Nano-Al_2O_3[J]. *Journal of Petrochemical Universities*, 2008, **21**(3):9-12.

IOP Conf. Series: Materials Science and Engineering **167** (2017) 012069 doi:10.1088/1757-899X/167/1/012069

A multi-scale micromechanical investigation on thermal conductivity of cement-based composites

Jiahan Liu, Shilang Xu[1] and Qiang Zeng

Institute of Advanced Engineering Structures and Materials, Zhejiang University, Hangzhou 310058, China

E-mail: slxu@zju.edu.cn

Abstract. Cement-based composites (CBCs) are one of the most widely used materials in construction. An appealing characterization of thermal conductivity of CBCs plays an essential role to evaluate the energy consumption in buildings and to facilitate the development of novel thermal insulation materials. Based on Eshelby equivalent inclusion principle and multi-scale methodology, this paper attempted to present a generalized multi-scale micromechanical model in terms of thermal performance of the CBCs, which covers some classic models for thermal conductivity estimation. A Mori-Tanaka homogenization method was applied to investigate the thermal conductivity of the CBCs of different compounds, water-to-cement ratios and curing ages. In addition, saturation degree factor was considered. The results of this model are in good agreement with the experimental value, showing that the multi-scale model developed in this paper is able to evaluate the thermal conductivity of the CBCs in different conditions.

1. Introduction

Cement-based composites (CBCs) are one of the most extensive used construction materials in the world. Increasing demands of the CBCs and requirements of reducing energy conservation provide both scientific and engineering incentives for evaluating the thermal conductivity of the materials.

Many factors may influence the thermal conductivity of the CBCs [1, 2], including types and volume fractions of aggregates and raw binder materials, mineralogical characters of hydrates, water-to-cement ratios and saturation degrees. Although numerous previous studies have been conducted on experimental methods and calculation models of the thermal conductivity of the CBCs, these methods mostly depend on mix proportion and petrographic composition of aggregates, therefore the proposed methods are generally empirical. Alternatively, effective-medium-theory based models, with clear physical meanings, can predict the thermal conductivity of the CBCs, because the structures of the materials generally conform to the inclusion-matrix type. For example, concrete can be treated as a two-phase composite material made up of cement mortar (continues phase) and aggregates (dispersed phase). The models [3, 4] include the simple series and parallel models, Maxwell-Euken type model, Bruggeman type model, and Hasselman-Johnson model. Note that these models are remaining on a single scale, while the CBCs are typical composites structured in multi scales.

Recent and ongoing studies of the CBCs by different experimental methods allow mineralogical and structural understanding of the materials on different scales in depth. In microscale, the CBC can be treated as a porous inorganic material with calcium-silicate-hydrate (C-S-H) matrix, calcium hydrate (CH) crystals, unreacted cement clinkers and micro gel pores. In mesoscale, the CBC can be regarded as a composite consisting of a cement paste matrix and sands surrounded by interfacial transition zones (ITZs). Therefore, multiscale method might be suitable for predicting the thermal conductivity of the CSCs, as the method has been widely used in analysis of thermal and mechanical property of other composite materials [5].

Based on the principle of the effective medium theory and the multi-scale nature of the CBCs, this study attempted to evaluate the thermal properties of the materials. A multi-scale micromechanical model was developed and the modelled results were compared to the data from experimental measurements.

2. Micromechanical model of the thermal conductivity

2.1. Multi-scale nature of CBCs

The macro properties of CBCs are associated with their structural nature that can be characterized in multi-scales. By connecting the physical and mechanical properties of each scale, a systematic understanding of the nature of the materials can be established.

Generally, one can model a CBC with four elementary levels based on the microstructures of the material (see Figure 1) [6]:

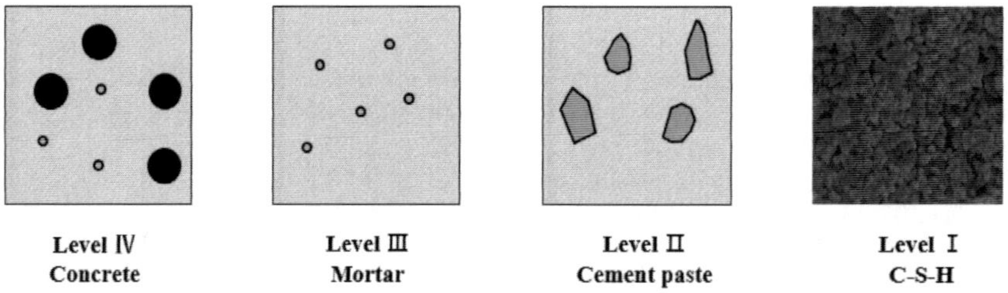

Figure 1. Four-level microstructure of CBCs [6].

- Level Ⅰ (10^{-8}-10^{-6}m): The main solid phase in this scale is the C-S-H, the main hydration product of C_2S and C_3S. It has now established that the C-S-H exists in two different forms, named the low density (LD) and high density (HD) C-S-Hs. In this case, an amorphous colloidal structure of the C-S-H (about 5 nm) is constructed by 'globules' that are composed of basic building blocks and an instra-globules porosity (about 2 nm) [7].
- Level Ⅱ (10^{-6}-10^{-4}m): The C-S-H matrix, together with the large CH crystals, cement clinker and micro capillary pores, forms the cement paste. The cement clinker may contain C_2S, C_3S, C_3A and C_4AF. In this scale, the micro and meso capillary porosity is strongly related to the w/c.
- Level Ⅲ(10^{-4}-10^{-2}m): This scale refers to mortar, which is a three phase composite material composed of cement paste, sand particle inclusions, and ITZs.
- Level Ⅳ (10^{-2}-10^{-1}m): Concrete is considered as a three phase material composed of aggregates embedded in a mortar matrix and ITZs in this scale.

According to the four elementary levels of the CBCs, admixtures can be embedded in the materials in different scales. For instance, nanoparticles can be placed in Level Ⅰ, fly ash in Level Ⅱ and the glazed hollow bead in Level Ⅲ or Ⅳ depending on their sizes. This interpretation can help to access the thermal and mechanical properties of the CBCs blended with different admixtures.

2.2. Multi-scale model of thermal conductivity of composite materials

According to Fourier's law, the amount of the heat flux through a material is proportional to the negative gradient of temperature, which gives:

$$Q = -K\frac{\mathrm{d}t}{\mathrm{d}x} = -\boldsymbol{K}\cdot\nabla T \tag{1}$$

Where Q is the heat flux, $\boldsymbol{K}$ is the thermal conductivity, and ∇T is the temperature gradient.

Based on the mean-field theory, a consistent mechano-physical framework can associate the homogenized thermal conductivity, $\boldsymbol{K}_{\mathrm{hom}}(\boldsymbol{X})$ at level i+1 to $\boldsymbol{K}_i(\boldsymbol{x})$ at level i, where $\boldsymbol{X}$, $\boldsymbol{x}$ are position vectors at level i+1 and level i, respectively. In micromechanics, continuum points can be characterized by temperature gradients, namely:

$$\text{Level i: } \boldsymbol{g} = \nabla\tau\text{; Level i+1: } \boldsymbol{G} = \nabla T \tag{2}$$

Where $\boldsymbol{G}$ and $\boldsymbol{g}$ are temperature gradients at level i+1 and level i, respectively. The temperature gradient and heat flux at level i+1 are essentially the average of these variables in level i:

$$\boldsymbol{G} = \langle\boldsymbol{g}\rangle_V\text{; } \boldsymbol{Q} = \langle\boldsymbol{q}\rangle_V \tag{3}$$

Where $\langle y\rangle_V = \frac{1}{V}\int y(x)\mathrm{d}V$ is the volume average of representative volume element (RVE), and $\boldsymbol{Q}$ and $\boldsymbol{q}$ are the heat flux at level i+1 and level i, respectively. Furthermore, temperature gradients at two levels can be linked via a linear localization condition:

$$\boldsymbol{g}(\boldsymbol{x}) = \boldsymbol{A}(\boldsymbol{x})\cdot\boldsymbol{G}(\boldsymbol{x}) \tag{4}$$

Where $\boldsymbol{A}(\boldsymbol{x})$ is a second order localization tensor that satisfies the compatibility condition:

$$\langle\boldsymbol{A}(\boldsymbol{x})\rangle = \boldsymbol{I} \tag{5}$$

For a heterogeneous material composed of homogeneous phases, it is convenient to introduce a linear phase concentration of the form:

$$\langle\boldsymbol{g}(\boldsymbol{x})\rangle_{V_s} = \langle\boldsymbol{A}(\boldsymbol{x})\rangle_{V_s}\cdot\boldsymbol{G}(\boldsymbol{x}) \tag{6}$$

$$\sum_{s=1}^{n} f_s\langle\boldsymbol{A}(\boldsymbol{x})\rangle_{V_s} = \boldsymbol{I} \tag{7}$$

Where $f_s = V_s / V$ is the volume fraction of each phase. The constitutive relation of thermal conduction with two scales can be exploited to relate $\boldsymbol{q}$ and $\boldsymbol{g}$:

$$\boldsymbol{q}(\boldsymbol{x}) = \boldsymbol{K}_i(\boldsymbol{x})\cdot\boldsymbol{g}(\boldsymbol{x}) = \boldsymbol{K}_i(\boldsymbol{x})\cdot\boldsymbol{A}(\boldsymbol{x})\cdot\boldsymbol{G}(\boldsymbol{X}) \tag{8}$$

By spatially averaging the former identity, the homogenized thermal conductivity is expressed as:

$$\boldsymbol{K}_{\mathrm{hom}}(\boldsymbol{X}) = \langle\boldsymbol{K}_i(\boldsymbol{x})\cdot\boldsymbol{A}(\boldsymbol{x})\rangle_V \tag{9}$$

This is the generalized multi-scale model of thermal conductivity tensor of composite materials. In scope of linear mechanics, the strain localization tensor is constant for an ellipsoidal inclusion embedded in an infinite matrix. Similarly, $\boldsymbol{A}_T$, for heat conduction can be given by [8]:

$$\boldsymbol{A}_T = [\boldsymbol{I} + \boldsymbol{S} : (\boldsymbol{K}_M^{-1}\boldsymbol{R}\boldsymbol{K}_T\boldsymbol{R}^T - \boldsymbol{I})]^{-1} \tag{10}$$

Where $\boldsymbol{S}$ is the Eshelby-like tensor, $\boldsymbol{I}$ is the unit tensor, and $\boldsymbol{K}_M$, $\boldsymbol{K}_T$ are the thermal conductivity of the matrix and inclusion, respectively. The rotation matrix $\boldsymbol{R}$ is used to transform the local to global coordinate systems.

The above model for a two-phase matrix-inclusion system can be extended to incorporate a multiphase system. The thermal conductivity of the volume around the inclusion phase needs to be considered. The localization tensor is written as:

$$\boldsymbol{A}_s = [\boldsymbol{I} + \boldsymbol{S}_s : (\boldsymbol{K}_0^{-1}\boldsymbol{R}\boldsymbol{K}_s\boldsymbol{R}^T - \boldsymbol{I})]^{-1}\left\langle[\boldsymbol{I} + \boldsymbol{S}_s : (\boldsymbol{K}_0^{-1}\boldsymbol{R}\boldsymbol{K}_s\boldsymbol{R}^T - \boldsymbol{I})]^{-1}\right\rangle_V^{-1} \tag{11}$$

Where $\boldsymbol{S}_s$ is the Eshelby-like tensor (s=1, 2...n, n is the number of the phase), $\boldsymbol{K}_s$ is the thermal conductivity of each phase.

According to the Mori-Tanaka scheme [9], the reference medium is chosen to be the matrix phase ($\boldsymbol{K}_0 = \boldsymbol{K}_{\mathrm{M}}$). In this case, the localization tensor can be written as

$$\boldsymbol{A}_s = [\boldsymbol{I} + \boldsymbol{S}_s : (\boldsymbol{K}_{\mathrm{M}}^{-1}\boldsymbol{R}\boldsymbol{K}_s\boldsymbol{R}^T - \boldsymbol{I})]^{-1}\left\langle[\boldsymbol{I} + \boldsymbol{S}_s : (\boldsymbol{K}_{\mathrm{M}}^{-1}\boldsymbol{R}\boldsymbol{K}_s\boldsymbol{R}^T - \boldsymbol{I})]^{-1}\right\rangle_V^{-1} \tag{12}$$

For the spherical inclusions, $S = 1/3tr\boldsymbol{I}$. With the absence of preferential orientation for inclusions, the rotation matrix is reduced to $\boldsymbol{R} = \boldsymbol{I}$. Using the above equation in (12), the localization tensor reduces to [10]:

$$\boldsymbol{A}_s = A_s\boldsymbol{I} = \frac{3K_{\mathrm{M}}}{2K_{\mathrm{M}} + K_s}\boldsymbol{I} \tag{13}$$

Combining Eqs (9) and (13), it was obtained that the Mori-Tanaka based micromechanical model of the thermal conductivity:

$$K_{\mathrm{MT}} = \frac{f_{\mathrm{M}}K_{\mathrm{M}} + \sum_{s=1}^{n} f_s K_s A_s}{f_{\mathrm{M}} + \sum_{s=1}^{n} f_s A_s} \tag{14}$$

Where f_{M} , f_s are the volume fraction of the matrix and inclusions, respectively, n is the number of the inclusion phases, K_{M}, K_s are the thermal conductivity of the matrix and inclusions, respectively.

When $A_s = 1$, the multi-scale model is in accordance with parallel model. When $A_s = K_{\mathrm{M}} / K_s$, this model is in accordance with series model. Benveniste and Miloh [11] proved that the Mori-Tanaka method with thermal boundary resistance is identical to the Hasselman-Johnson model. It indicates that this generalized multi-scale micromechanical model is able to recover some classic models for estimating thermal conductivity of composites.

3. Application to CBCs

3.1. Cement paste

Cement paste can be simplified to be a two-scale model from level Ⅰ to level Ⅱ (see Figure 1). However, it still needs homogenization two times for calculating the effective thermal conductivity. It was adopted that a classical hydration model (e.g., Ref. [12]) to estimate the volume fractions of solid phases. For simplification, in this study, the aluminate phases, such as C_3A and C_4AF, are equivalent to C_3S and C-S-H, yielding the Max. and Min. estimations of the thermal conductivity of the cement paste. The density and thermal conductivity of C_2S, C_3S, CH, C-S-H are respectively 3.55, 3.57, 2.25, 2.40 g/cm^3 and 3.45, 3.35, 1.32, 0.98 W/m·K [10]. The thermal conductivity values of water and air are 0.6 and 0.026 W/m·K under ambient condition.

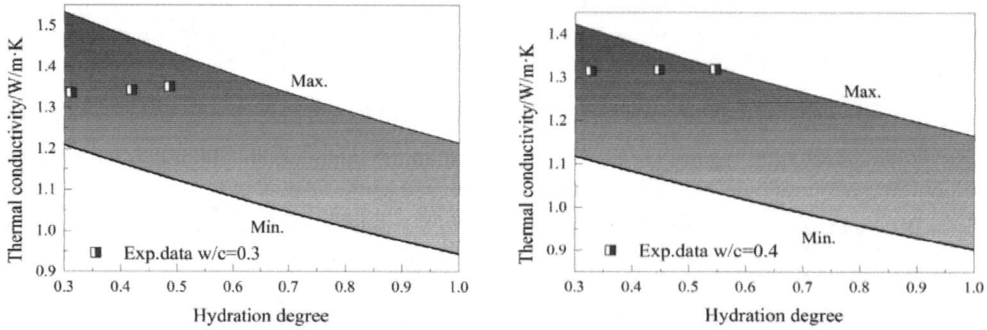

(a) (b)

Figure 2. Plots of predicted thermal conductivity of cement pastes against hydration degree

(a) with w/c ratio of 0.3; (b) with w/c ratio of 0.4. The shaded region presents the Max. and Min. estimations of the thermal conductivity of the cement paste. The experimental data were adopted from Ref. [14].

Figure 2 displays the predicted thermal conductivity of the cement pastes with different w/c ratios against hydration degree. It was found that the thermal conductivity of the cement pastes decreases with increasing the hydration degrees. This is reasonable as the thermal conductivity values of the C-S-H and CH are lower than those of the clinker phases. The result is in agreement with the experimental data of [13]. Another indication obtained in this study is that the thermal conductivity values of the cement pastes decrease with increasing the w/c ratios, which is consistent with the experimental observations in Ref. [15]. An important reason is that the thermal conductivity of water is heavily lower than those of other solid phases. Figure 2 also presents the predicted and measured thermal conductivity of the cement paste with w/c ratio of 0.3 and 0.4 in Ref. [14]. The experimental values are close to the max values in the region of the low hydration degrees. It may suggest that the thermal conductivity values of the aluminate hydrates are closed to that of C_3S. Note that, in this study, it was neglected the effect of self-desiccation that is especially severe for the cement pastes of lower w/c ratios. The differences between the measured and predicted data presented in Figure 2 are thus acceptable.

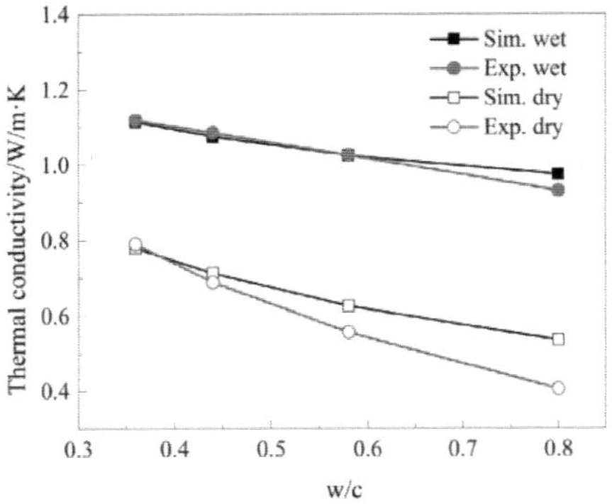

Figure 3. Predicted and measured thermal conductivity of dried and saturated cement pastes with different w/c ratio; the experimental data were adopted from Ref. [16].

Figure 3 presents the comparison between the predicted and measured thermal conductivity data of the dried and saturated cement pastes with w/c ratios of 0.36, 0.44, 0.58 and 0.8. The experimental mix proportion and material parameters were adopted from Ref. [16]. It was observed that the thermal conductivity values of the saturated cement pastes are obviously higher than the dried ones. Again, higher thermal conductivity of water than the air accounts for the observation. Another indication from Figure 3 is that the relative errors between the predicted and measured values of the dried cement pastes increase with increasing the w/c ratio. This maybe because the classical hydration model used in this study cannot fully consider the actual microstructure and mineral distribution of the cement pastes that vary with the w/c ratios, curing ages, and moisture conditions.

3.2. Concrete

Concrete has more complex structures with four-level representative scales (see Figure 1). The thermal conductivity of quartz, granite, limestone and cement pastes calculated above are 4.45, 2.50-2.65, 2.29-2.78 and 1.094 W/m·K. The density of cement pastes, sands and gravels are 3.15, 2.55 and 2.58 g/cm^3. The experimental data is selected from Ref. [17], where the mix proportion and curing conditions can be found therein. Figure 4 presents the comparison between the measured and predicted thermal conductivity data of the dried concrete with different volume fractions of coarse aggregates. Clearly, the proposed multi-scale model can capture the experimental data with relatively low errors (<5%). Note that the distribution of the coarse aggregates may be much more complex than the simple sphere particles presented in Figure 1, so it could be safe to conclude the promised accuracy of the established model for predicting the thermal conductivity of the CBCs.

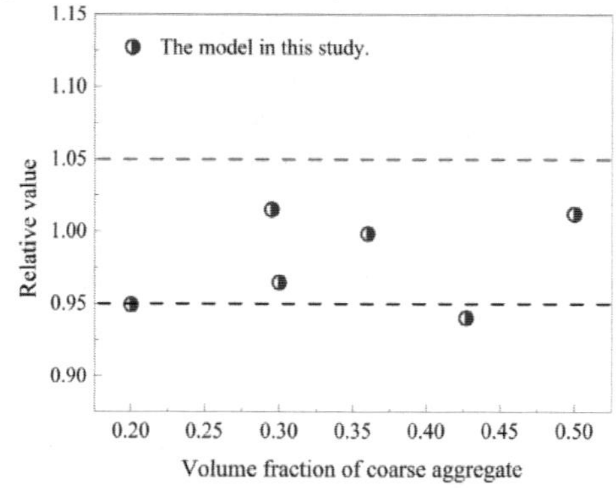

Figure 4. Relative errors between the measured and predicted thermal conductivity of the dried concrete with different volume fractions of coarse aggregate; the experimental data were adopted from Ref. [17]

For further examination of the robustness of the model, the thermal conductivity of the concretes with different types of aggregates under both dried and saturated conditions was calculated. Figure 5 shows the experimental and predicted values of the thermal conductivity of the concretes. It is found that the predicted values are in a good agreement with the experimental results for both the dried and saturated concretes. The deviations of the predicted the thermal conductivity of the dried concretes from the measured ones are in the range of (0.5%, 16.3%), while those of the saturated concretes are in the range of (5.9%, 22.2%). The relatively large errors between the predicted and measured thermal conductivity of the saturated concretes can be due to the fact that for the very high water saturation case, the spatial distribution of the pore fluids that may percolate through the material is beyond the Eshelby equivalent inclusion principle. Further investigation is deserved in the future for the water effect on the thermal conductivity of complex composites.

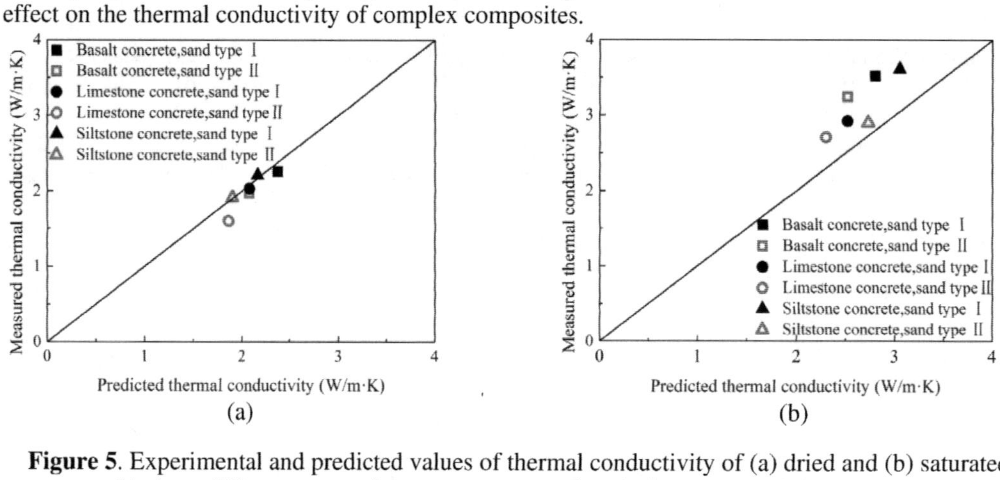

Figure 5. Experimental and predicted values of thermal conductivity of (a) dried and (b) saturated concrete with three different types of coarse aggregates (basalt, limestone and siltstone) and two types of fine aggregates (sand type Ⅰ: reddish land quarried quartz sand and sand type Ⅱ: greyish river sand); the experimental data were adopted from Ref. [18].

4. Conclusions

Based on Eshelby equivalent inclusion principle and multi-scale nature of the microstructure of the CBCs, this study attempted to evaluate the thermal conductivity of the materials by the multi-scale micromechanical approach. With a Mori-Tanaka homogenization method, a multi-scale micromechanical model was developed to capture the thermal conductivity of the CBCs.

- The multi-scale model captures the influences of the cement pastes, fine and coarse aggregates on the thermal conductivity of the concretes in different scales. Generally, the multi-scale model can predict the thermal conductivity of the pastes and concretes in relative good accuracy.
- The Eshelby equivalent inclusion principle and Mori-Tanaka homogenization method oversimplify the effects of water confined in pores that may percolate the materials spatially. This causes the relative large errors between the predicted and experimental values of the thermal conductivity of the cement pastes and concretes for the saturated conditions. Further study is needed in this aspect in the future.

References

[1] Campbell-Allen D and Thorne C P. The thermal conductivity of concrete. 1963 *Magazine of Concrete Research*. **15**(43): 39-48.
[2] Marshall A L. The thermal properties of concrete. 1972 *Building Science*. **7**(3): 167-174.
[3] Bruggeman D A G. The dielectric constants and conductivities of mixtures composed of isotropic substances. 1935 *Annals of Physics*. **24**: 639-791.
[4] Hasselman D P H and Johnson L F. Effective thermal conductivity of composites with interfacial thermal barrier resistance. 1987 *Journal of Composite Materials*. **21**(6): 508-515.
[5] Ji R, Zhang Z, Liu L and Wang X. Development of the random simulation model for estimating the effective thermal conductivity of insulation materials. 2014 *Building and Environment*. **80**: 221-227.
[6] Constantinides G and Ulm F J. The effect of two types CSH on the elasticity of cement-based materials: Results from nanoindentation and micromechanical modeling. 2004 *Cement and concrete research*. **34**(1): 67-80.
[7] Jennings H M. A model for the microstructure of calcium silicate hydrate in cement paste. 2000 *Cement and Concrete Research*. **30**(1): 101-116.
[8] Hatta H and Taya M. Effective thermal conductivity of a misoriented short fiber composite. 1985 *Journal of Applied Physics*. **58**(7): 2478-2486.
[9] Mori T and Tanaka K. Average stress in matrix and average elastic energy of materials with misfitting inclusions. *Acta metallurgica*. 1973 **21**(5): 571-574.
[10] Qomi M J A, Ulm F J and Pellenq R J M. Physical origins thermal properties of cement paste. 2015 *Physical Review Applied*. **3**(6): 064010.
[11] Benveniste Y and Miloh T. An exact solution for the effective thermal conductivity of cracked bodies with oriented elliptical cracks. 1989 *Journal of Applied Physics*. **66**(1): 176-180.
[12] Taylor H F W. 1997 *Cement chemistry*.
[13] Mounanga P, Khelidj A and Bastian G. Experimental study and modelling approaches for the thermal conductivity evolution of hydrating cement paste. 2004 *Advances in cement research*. **16**(3): 95-103.
[14] Tang S W, Chen E, Shao H Y and Li Z J. A fractal approach to determine thermal conductivity in cement pastes. 2015 *Construction and Building Materials*. **74**: 73-82.
[15] Maruyama I and Igarashi G. Cement reaction and resultant physical properties of cement paste. 2014 *Journal of Advanced Concrete Technology*. **12**(6): 200-213.
[16] Baghban M H, Hovde P J and Jacobsen S. Analytical and experimental study on thermal conductivity of hardened cement pastes. *Materials and structures*. 2013 **46**(9): 1537-1546.
[17] Zhang W, Min H, Gu X, Xi Y and Xing Y. Mesoscale model for thermal conductivity of concrete. 2015 *Construction and Building Materials*. **98**: 8-16.
[18] Khan M I. Factors affecting the thermal properties of concrete and applicability of its prediction models. 2002 *Building and Environment*. **37**(6): 607-614.

Electrochemical hydrogenation of thiophene on SPE electrodes

Haiyan Huang[1,3], Penghui Yuan[1], Ying Yu[1] and Keng H. Chung[2]

[1]State Key Laboratory of Heavy Oil Processing, College of Science, China University of Petroleum, Beijing 102249, China
[2]Well Resources Inc., Edmonton, Alberta, Canada T6R 1J8

E-mail: hhy@cup.edu.cn

Abstract. Electrochemical reduction desulfurization is a promising technology for petroleum refining which is environmental friendly, low cost and able to achieve a high degree of automation. Electrochemical hydrogenation of thiophene was performed in a three-electrode system which SPE electrode was the working electrode. The electrochemical desulfurization was studied by cyclic voltammetry and bulk electrolysis with coulometry (BEC) techniques. The results of cyclic voltammetry showed that the electrochemical hydrogenation reduction reaction occurred at -0.4V. The BEC results showed that the currents generated from thiophene hydrogenation reactions increased with temperature. According to Arrhenius equation, activation energy of thiophene electrolysis was calculated and lower activation energy value indicated it was diffusion controlled reaction. From the products of electrolytic reactions, the mechanisms of electrochemical hydrogenation of thiophene were proposed, consisting of two pathways: openingringfollowed by hydrogenation, and hydrogenation followed by ring opening.

1. Introduction

The sulfur emissions from combustion of fossil fuels are detrimental to environment and human health. In petroleum refining industry, desulfurization processes are commonly used to produce low sulfur transportation fuels, such as gasoline and diesel. Catalytic hydrodesulfurization(HDS) is a mature technology which operates under a relatively severe reaction condition; hence the process operating costs are high. As a result, tremendous efforts have been devoted to improve and optimize the conventional HDS technology, as well as to explore new desulfurization technologies for producing low sulfur gasoline and diesel.[1-6]Among the new desulfurization technologies, electrochemical desulfurization is a promising technologywhich is environmental friendly, low cost and able to achieve a high degree of automation.

There are two type of electrochemical desulfurization: electrochemical oxidation desulfurization and electrochemical reduction desulfurization. In electrochemical oxidation desulfurization, sulfur compounds are oxidized and transform to sulfoxides or sulfones. Due to their polarity changes, sulfoxides or sulfones can be removed by adsorption or extraction process. In electrochemical reduction desulfurization, sulfur compounds are electrochemically hydrogenated on the cathode under mild reaction conditions.[7-11]The reaction mechanisms of electrochemical reduction desulfurization needs further studies.

Organic sulfur compounds are common sulfur compounds in petroleum, among which thiophenic compounds are dominant. Thiophenes account for 90% and 80% of total sulfur compounds in gasoline and diesel, respectively. These compounds have relatively stable molecular structure and are refractory to desulfurization processes. In this paper, electrochemical hydrogenation reactions of thiophene on SPE electrode were investigated. Protons produced in the anode region of SPE electrode by electrolysis of water, transferred to the cathode region through a proton exchange membrane. Desulfurization of thiophene by reduction reactions occurs in the cathode region of the electrolytic tank. The hydrogen sulfide generated is further converted to elemental sulfur by oxidation of high valence cations.

2. Experimental

2.1. Materials

Polytetrafluoroethylene (PTFE) latex, 5% Nafion solution, and Nafion®117 membranes were obtained from DuPont Company. Analytical grade octane, thiophene were obtained from the Beijing Chemical Plant. Carbon cloth with 3K vertical textures was used.

2.2. SPE Electrodes Preparation

SPE electrodes were prepared by hot pressing treated carbon cloth and pretreated Nafion 117 membrane.[12] SPE electrodes were hot pressed at 0.5 MPa for 10 min at 175 °C.

2.3. Electrolysis and Performance Indicator

The SPE electrode was the working electrode and Pt was the counter electrode. The anode and cathode were separated by SPE electrode. The electrolyte acidity of anode was 0.5 mol/L H_2SO_4 and thiophene solution (octane as solvent) was in the cathode chamber. The cell has a constant temperature by circulation constant temperature water bath. The main reaction was thiophene hydrogenation on the cathode. The electrons produced during the water electrolysis from the external circuit reached the cathode where they were used by hydrogenation reactions.

A CHI650A electrochemistry test station was used to measure the cyclic voltammograms and the current of the process with the bulk electrolysis coulometry (BEC) curve. In-situ cyclic voltammetry was conducted by cycling the potential at various scanning rates (300, 200, 150, 100, 75, 50, 30,15, and 10 mV/s) between 1.2 and -1.0 V until a stable voltammogram was obtained. The potential (E) denotes the working electrode potential versus saturated calomel electrode (SCE) as a reference electrode.

The composition changes of thiophene electrolysis were analyzed by Agilent 7890 gas chromatography (GC-SCD) and Agilent 7890A 5975C gas chromatography-mass spectrometry (GC-MS).

3. Results and Discussion

Figure 1 shows the cyclic voltammetry curves of octane and thiophene-octane solution in the range of -1.0~1.2V. The results showed that there was no redox peak of octane, whereasthere were an oxidation peak and a reduction peak of thiophene-octane solution system, which were redox couples. The electrochemical reduction reactions of thiophene at the SPE electrode occurred at -0.350V and reduction reaction products oxidized at 0.106V with the change of the scanning potential.

The peak potential and peak current at various cyclic voltammetry scanning rates from 10 to 300 mV/s are shown in Table 1. The reduction peak shifted to the negative potential and the oxidation peak of anode moved to the positive potential with increased scanning rate. The electron transfer rate increased with increased scanning rate. At a constant diffusion rate,the peak current exhibited a lag phenomenon because the concentration of substance though diffusion to the electrode cannot be provided in time. Hence, these two peaks moved away from the equilibrium potential. At the same time, the $I_P/(v_{1/2})$ values of the reduction peak and oxidation peak decreased with increased scanningrate. The ratio of the oxidation peak current and reduction peak current was larger than 1, and increased with increased scanning rate. These were consistent with the mechanism of proceeding chemical reactions according the cyclic voltammetry criterion[13]

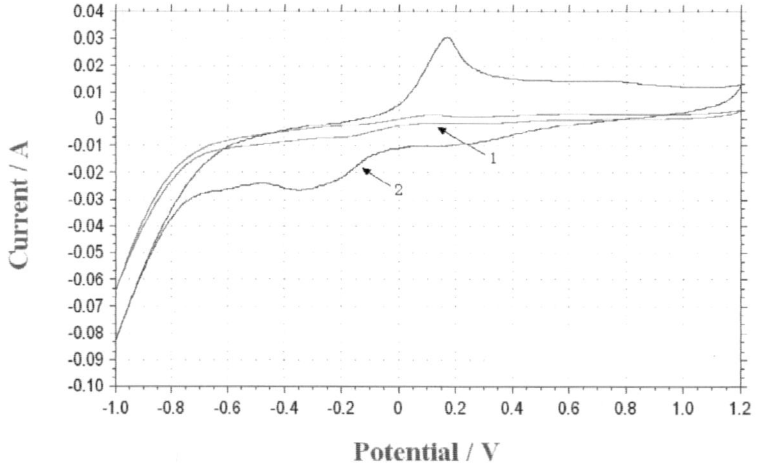

Figure 1. Cyclic voltammety curve on the SPE electrode, in octane (1) and thiophene solution (2).

Table 1. The relationgship between potential and current of peak at different scanning rate.

ScanningRate v (mV/s)	ReductionPotential (V)	Reduction I_p(mA)	Oxidation Potential (V)	OxidationI_p (mA)
10	-0.079	-6.238	0.067	7.305
15	-0.085	-6.753	0.069	8.474
30	-0.101	-8.033	0.067	11.850
50	-0.132	-7.527	0.072	14.150
75	-0.143	-9.255	0.073	17.710
100	-0.156	-10.970	0.075	22.170
150	-0.198	-13.360	0.079	26.140
200	-0.239	-14.360	0.082	34.492
250	-0.246	-16.231	0.084	39.120
300	-0.350	-16.192	0.163	30.430

Thiophene reduction peak of electrochemical hydrogenation was near -0.400V by the cyclic voltammetry. In order to improve the reaction rate and avoid the hydrogen evolution reaction under the condition of high potential, the selected potential of the bulk electrolysis with coulometry (BEC) was set to -0.800V. Table 2 shows the electrolysis current as a function of reaction temperature. The current efficiency increased with reaction temperature.

According to the Arrhenius equation, the rate constant of electrochemistry reactions can be expressed as: $\ln i=-[Ea/R]\ [1/T] + \ln A$, where i is the reaction current, Ea is the activation energy, R is the gas constant, T is the reaction temperature. The value of - $[E_a/R]$ can be obtained from the slope of $\ln i$ versus $1/T$ plot. Figure 2 shows the plot of $\ln i$ versus $1/T$ for the thiophene hydrogenation. The reaction activation energy of thiophene electrolysis hydrogenation was 14.06 kJ/mol determined by linear regression. The positive activation energy indicated that the reaction rate increases with temperature. A low activation energy value indicates the thiophene hydrogenation process is controlled by the diffusion step, not controlled by the electrochemical reaction.

Table 2. The relationship between the temperature and the electrolysis current under-0.800V.

T(℃)	*T*(K)	*i* (A)	1/*T*	ln (*i*)
25	298.15	-0.010	3.354×10^{-3}	-4.605
35	308.15	-0.012	3.245×10^{-3}	-4.423
45	318.15	-0.014	3.143×10^{-3}	-4.255
55	328.15	-0.017	3.047×10^{-3}	-4.098
65	338.15	-0.019	2.957×10^{-3}	-3.948
75	348.15	-0.023	2.872×10^{-3}	-3.781

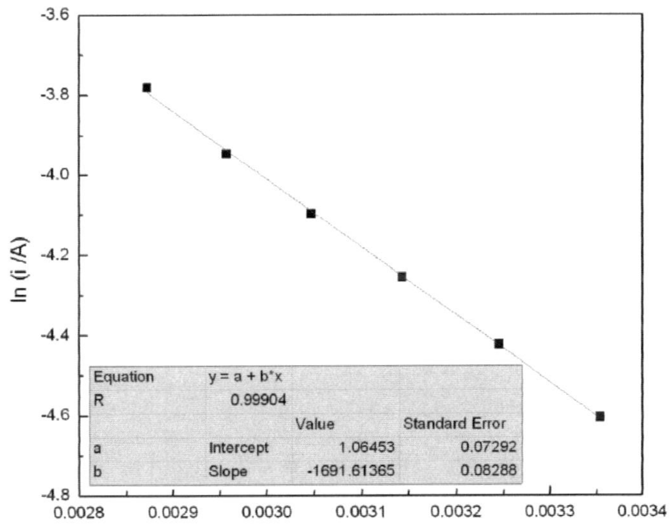

Figure 2. The relationship between ln(*i*/A)and 1/*T* at the potential of -0.800V.

No sulfur containing products were detected by gas chromatography in the thiophene solution after the reaction. An additional experiment was performed to determine the thiophene reaction products. Pure thiophene was subjected to constant potential electrolysis and the products were characterized by GC-MS. Sulfur compounds, such as 4,5-dihydrothiophene, 2-ethylhexylthiol, diethylsulfide, ethyletheynlsulfide, butenylthiol, allylmethylsulfide,2-butylthiophene, and thiophene-2-methanethiol, were found in the reaction product. In addition, some C_5-C_7 hydrocarbons, such as normal pentane, 2-methylbutane, and normal heptanes were detected. Based on the composition of reaction products, the pathway of thiophene reduction-hydrogenation reactions on combined SPE electrode was proposed, as shown in Figure 3.

The composition of the reaction products suggested that electrochemical reduction reactions of thiophene on the combined SPE electrode exhibited two reaction pathways. One pathway was started from the formation of the cyclic sulfide by hydrogenation saturation of thiophene, the cyclic sulfide was converted to various unsaturated thiols and sulfides by ring cleavage reactions, which were further converted to saturate sulfides, thiols and/or hydrocarbons. Another pathway was direct transformation of thiophene to the sulfides by ring opening,which followed by hydrogenation.In the further hydrogenation, the thiol was converted to various compounds by hydrogenation, isolation, and rearrangement reaction since it has unsaturated bond. At the same time, activated thiophene could react with unsaturated thiols and produced various thiophenic compounds with alkyl chains. These sulfur compounds could undergo further reactions, such as ring cleavage, hydrogenation, and dealkylation.

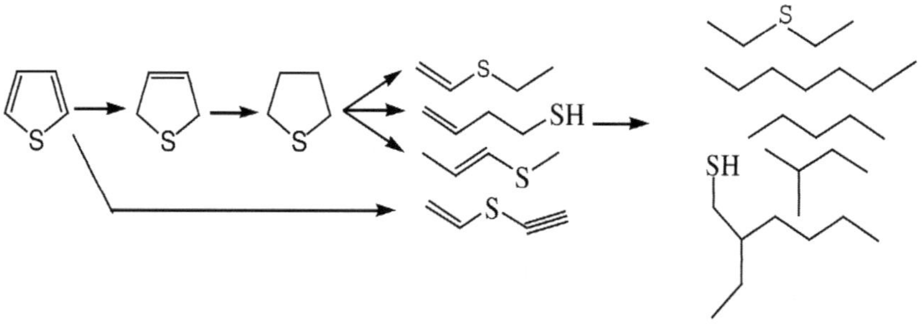

Figure 3. Reaction pathway of thiophene on the SPE electrode.

4. Conclusions

The electrochemical hydrogenation reactions of thiophene on SPE electrode were investigated using cyclic voltammetry and bulk electrolysis with coulometry (BEC) techniques. The results of cyclic voltammetry showed that the electrochemical hydrogenation reduction reaction occurred at about -0.4V. The BEC results showed that the currents generated from thiophene hydrogenation reactions increased with temperature. According to Arrhenius equation, activation potential of thiophene electrolysis was calculated and lower activation energy value indicated it was diffusion controlled reaction. From the products of electrolytic reactions, the mechanisms of electrochemical hydrogenation of thiophene on the combined SPE electrode were proposed, consisting of two pathways: opening ring followed by hydrogenation, and hydrogenation followed by ring opening.One pathway was started from the formation of the cyclic sulfide by hydrogenation saturation of thiophene, the cyclic sulfide was converted to various unsaturated thiols and sulfides by ring cleavage reactions, which were further converted to saturate sulfides, thiols and/or hydrocarbons. Another pathway was started from thiophene direct to the sulfide by ring opening, followed by hydrogenation.

References

[1] ChenY, WangL, ZhangY, LiuT, LiuX, JiangZ and LiC, A New Multi–Metallic Bulk Catalyst with High Hydrodesulfurization Activity of 4, 6–Dmdbt Prepared Using Layered Hydroxide Salts as Structural Templates 2014*Applied Catalysis A: General***474** 69-77.

[2] KimT, AliS A, AlhooshaniK, ParkJ-I, Al-YamiM, YoonS-H and MochidaI, Analysis and Deep Hydrodesulfurization Reactivity of Saudi Arabian Gas Oils 2013*Journal of Industrial and Engineering Chemistry***19(5)** 1577-1582.

[3] VillasanaY, Ruscio-VanalestiF, PfaffC, MéndezF J, Luis-LuisM and BritoJ L, Atomic Ratio Effect on Catalytic Performance of Few-Based Carbides and Nitrides on Thiophene Hydrodesulfurization 2013*Fuel***110**259-267.

[4] ZhangM, YangT, ZhaoR and LiuC, Effect of Solid-State Synthesized Alumina Properties on the Structure and Catalytic Performance of Nimo Catalyst in Hydrodesulfurization2013*Applied Catalysis A: General***468**327-333.

[5] YinC-L, ZhaiX-P, ZhaoL-Y and LiuC-G, Mechanism of Hydrodesulfurization of Dibenzothiophenes on Unsupported Nimow Catalyst 2013*Journal of Fuel Chemistry and Technology***41(8)** 991-997.

[6] ZuberC, HochenauerC and KienbergerT, Test of a Hydrodesulfurization Catalyst in a Biomass Tar Removal Process with Catalytic Steam Reforming2014 *Applied Catalysis B:*

*Environmental***156** 62-71.
[7] WangW, WangS, LiuH and WangZ, Desulfurization of Gasoline by a New Method of Electrochemical Catalytic Oxidation2007 *Fuel***86(17)** 2747-2753.
[8] ZhaoW, ZhuH, ZongZ-M, XiaJ-H and WeiX-Y, Electrochemical Reduction of Pyrite in Aqueous Nacl Solution2005 *Fuel***84(2)** 235-238.
[9] ArnanzA, MarcosM-L, DelgadoS, González-VelascoJ and MorenoC, The Effect of Thiophene Ring Substitution Position on the Properties and Electrochemical Behaviour of Alkyne–Dicobaltcarbonylthiophene Complexes 2008*Journal of Organometallic Chemistry***693(23)** 3457-3470.
[10] GreaneyM A, WangK, BielenbergJR and HissongDW, Electrodesulfurization of Heavy Oils Using a Divided Electrochemical Cell**2011**US Patents, patent number7985332.
[11] ChapuzetJ, LasiaA and LessardJ1998*Electrocatalytic Hydrogenation of Organic Compounds, Electrocatalysis* Wiley-VCH, New Yorkpp 155-96.
[12] HuangH, YuY and ChungK H, Seasonal Storage of Electricity by Hydrogen in Benzene–Water System2012 *International Journal of Hydrogen Energy***37(17)** 12798-12804.
[13] LrfajB2001 *Electrochemical Methods: Fundamentals and Applications*Wiley, New Yorkpp156-165.

IOP Conf. Series: Materials Science and Engineering **167** (2017) 012071 doi:10.1088/1757-899X/167/1/012071

Preparation of MgB_2 superconducting microbridges by focused ion beam direct milling

Xuena Zhang[1,2], Yanli Li[1,2], Zhuang Xu[3], Xiangdong Kong[1,4] and Li Han[1]

[1]Department of Institute of electrical engineering, Chinese academy of sciences, china
[2]University of Chinese academy of sciences, china
[3]LanZhou University, china

E-mail: slkongxd@mail.iee.ac.cn

Abstract. MgB_2 superconducting microbridges were prepared by focused ion beam (FIB) direct milling on MgB_2 films. The surface topography of the microbridges were observed using SEM and AFM and the superconductivity was measured in this paper. Lots of cracks and holes were found near the milled area. And the superconducting transition temperature was decreased a lot and the bridges prepared were not superconducting due to ion damage after milled with large dose. Through these works, we explored the effect regular of FIB milling and experimental parameters on the performance of microbridges.

1. Introduction

The superconductivity of MgB_2 material was found in 2001 and then much research has been done on it in a lot of countries due to its simple structure, high critical transition temperature (39 k), two weak coupling energy gap, the larger coherence length, strong and grain boundary connection and many other advantages.[1-5] MgB_2 superconductor is expected easier to be prepared compared with high temperature superconductors and the cooling cost can be greatly reduced compared with low temperature superconductors such as Nb. Meanwhile, MgB_2 has higher current density. Thus, the discovery of MgB_2 superconductivity provides a new material of great potential to make superconducting electronics based on Josephson junctions.[6]

Traditional tunnel Josephson junction has the higher capacitance relatively due to geometric feature while microbridge Josephson junction has a lower one. So the electronic devices based on microbridge Josephson junction are very suitable for application in lower noise environment.[7] Therefore, to explore a simple and repeatable preparation method to fabricate submicron size devices is quite necessary.[8] In our work, a new method is explored on this problem. MgB_2 microbridges were fabricated by focused ion beam direct milling on MgB_2 film. We observed the morphology characteristics of MgB_2 films and microbridges using SEM and AFM. The superconductivity of MgB_2 films and microbridges was measured by PPMS. We explored the influence regular rule of the new method on MgB_2 microbridge and film which has a great guiding significance for the fabrication of Josephson junctions and superconducting electronics.[9]

2. Experimental

2.1. The preparation of MgB_2 Film

The MgB_2 film used in this paper is made by HPCVD method on SiC substrate in Beijing university. The thickness of the film is about 100 nm. The superconducting transition temperature of MgB_2 film fabricated by HPCVD is up to 42 k and the zero resistance transition temperature is about 41 k. The superconducting transition width is less than 1 k. The quality of MgB_2 film prepared by HPCVD is greater than others. So it's suitable to choose MgB_2 film made by HPCVD.

2.2. The Fabrication of MgB2 microbridges

Focused ion beam milling progress is finished on the FIB/SEM dual beam 820 system. 30 KeV Ga+ ion beam was used with different beam currents from 12 pA ~ 6300 pA in the process of the fabrication of MgB_2 microbridges. First of all the ion beam with 6300pA beam current was focused on the edge of the surface, and then the screen was freezed. Next we designed and prepare the rectangle frame with $300\times800um^2$ and the trench as shown in figure 1(a). Two exactly the same microbridges were made by this method. The image of one of the microbridges is shown in figure 1(b) with its partial enlarged image shown in the insert. The size of the bridge is about $7\times10um^2$. The other one was then milled to the size of $3\times3um^2$ with 12pA. The whole process will continue for several hours. In this method we prepared a series of microbridges which can be accurate to the nanometer level.

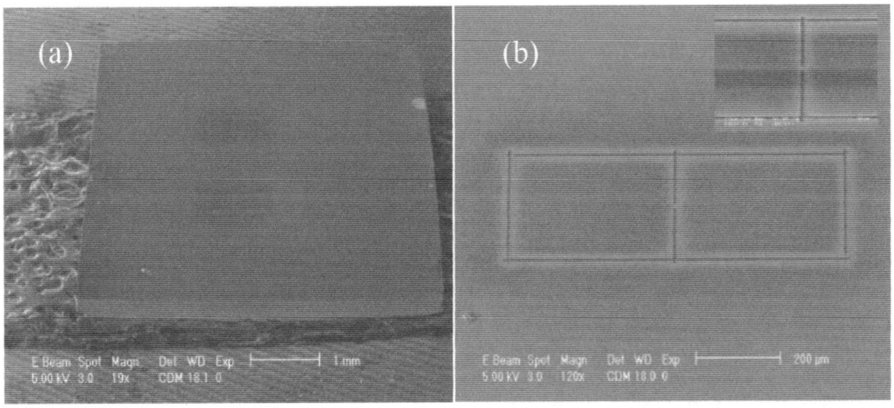

Figure 1. SEM figures of MgB_2 microbridge by focused ion beam milling without scanning.

What's more, we fabricated only the outside frame without any trench in the middle area by direct milling with the same experimental parameters. One of them was scanned quickly for several times with the current of 2700pA. We observed the surface morphology of films and micro bridges and explore the effect of focused ion beam milling directly on the superconductivity of the films and microbridges by comprehensive physical property measurement system (PPMS).

3. Results and Discussion

The superconductivity of the film and microbridges was measured by (PPMS). The R-T curves are shown in figure 2. The inset shows the detail of the superconducting transition temperature of MgB_2 film on MgO substrate. It is obvious that the microbridges are not superconducting. This maybe due to the damage of MgB_2 film in the bridge in ion beam milling process. When MgB_2 film is milled with 6300pA in three hours or longer time, thermal effect is probably produced and the bridge area is affected seriously from it and lost the superconductivity.

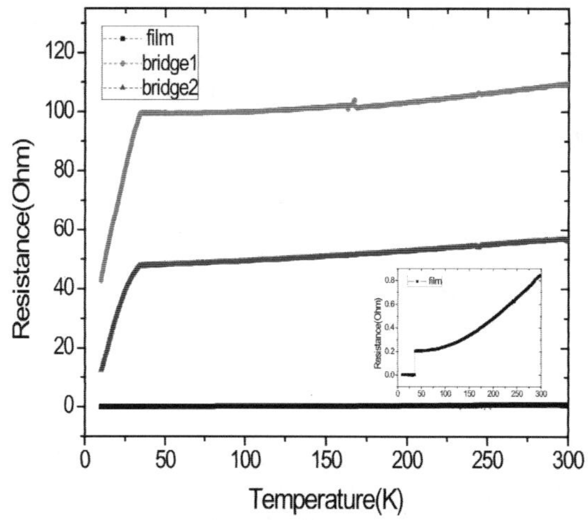

Figure 2.R-T curves of MgB_2 microbridges and film on MgO substrate. Bridge 1 is 1.5 um long and wide. Bridge 2 is 3 um long and wide.

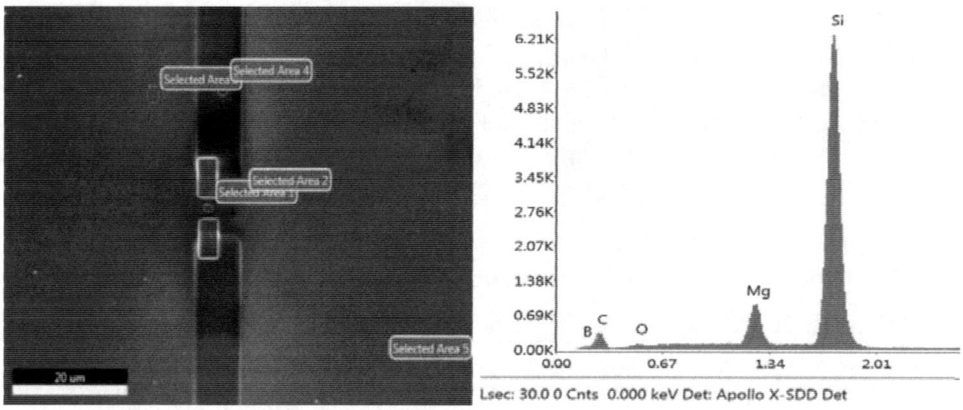

Figure 3. Spectrum image. analysis of micro bridge in scan mode

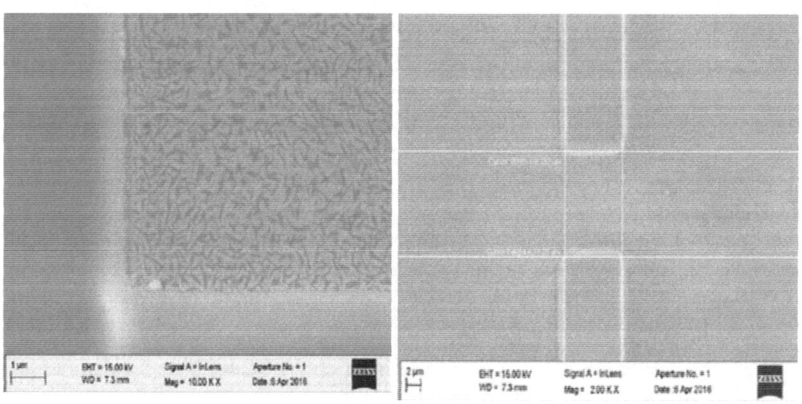

Figure 4. SEM figures of MgB_2 microbridge after measuring the superconductivity.

Figure 3 and table 1 shows the energy spectrum of the microbridge. It can be seen that content of O and Ga in the bridge is higher than other areas. This may be due to the decomposition of MgB_2 caused by thermal effect and ion beam bombarding when milling with large beam current for a long time. The SEM images of the edge of the milled frame and the area of middle bridge are shown in figure 4(a) and (b) respectively. We can see that the density of MgB_2 film becomes poorer. This illustrates that the film is damaged in milling process and thus confirms our aforementioned guess.

Table 1 The regional energy spectrum analysis of element content in scan mode

Area	1	2	3	4	5
O（Atomic %）	1.41	1.36	---	---	0.79
Mg（Atomic %）	1.98	2.64	5.26	---	2.48
Ga（Atomic %）	0.96	0.83	---	1.33	---

In order to make a compare, two rectangle frames were made only using the same experiment parameters. the width and depth of the frame were about 7um respectively and one frame was scanned with 2700pA ion beam for several times. The superconductivity was measured by PPMS as shown in figure. 5. The MgB_2 film outside the frames has a starting transition temperature of 40.6 K with the zero resistance transition temperature of 40.2 K and the superconducting transition width is about 1 K. The starting transition temperature of MgB_2 film enclosed with frames is about 34 K and the zero resistance transition temperature is 25.8 K. The superconducting transition width is 8.2 K. The superconducting transition temperature is reduced by about 15 k. But the trend of the R-T curves of the two frames is nearly the same. It indicates that ion scanning for several times has little effect on the zero resistance transition temperature of MgB_2 film. But milling with large beam and long time indeed has a big effect on the superconductivity on the film.

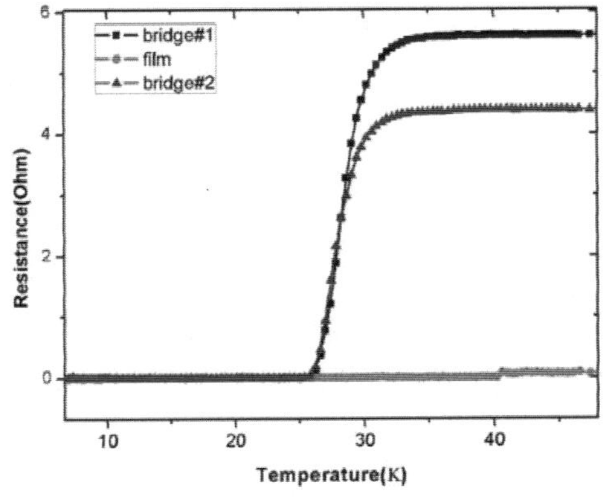

Figure 5. R-T curves of MgB_2 films outside and enclosed the frames on SiC substrate. The area of frame 2 is scanned for several times by 2700pA ion beam.

The SEM images near and far from the milled area are shown in figure 6(a) and (b). It's obvious that there are a lot of holes occurred on the film and it is more serious as near the trench. From the images, we can see the affected area is in the range of about 6 um from the milled area edge as shown in figure 6(c), and the hole's diameter is about 30 nm as shown in figure 6(d). This dedicates that the bridge area is completely damaged and is not superconducting.

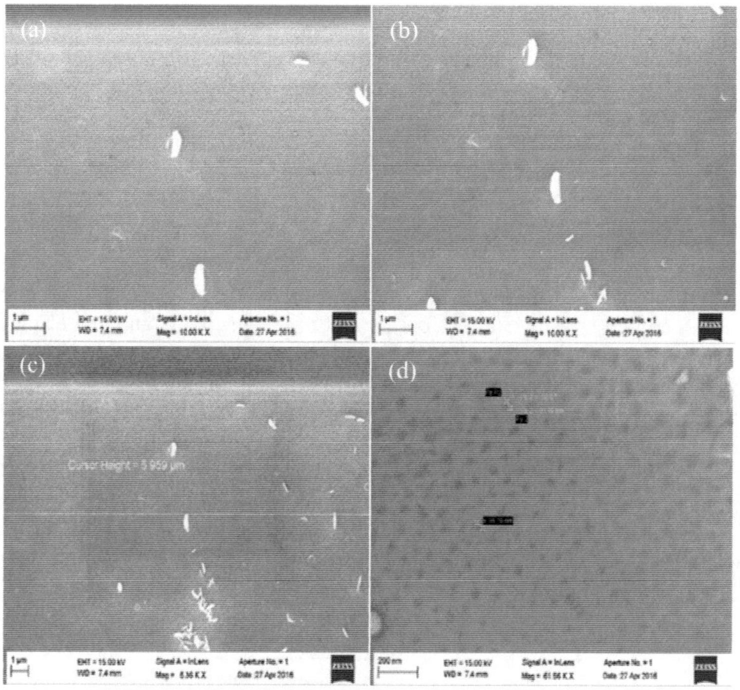

Figure 6. SEM figures of MgB_2 film.

In order to explain this phenomenon theoretically, the simulation was also carried out by using SRIM software. The penetration process and distribution of Ga^+ ion into MgB_2 film were simulated using 200、1000、5000、25000 ions respectively as shown in figure 7. We can see that the penetration depth in MgB_2 film is about 20-40nm and the ion concentration peak is appeared in the depth of 30nm with 30KeV ions. Only a few number of Ga^+ can penetrate into the depth of the 50nm. The horizontal injection radius is about 15nm. In consideration of the actual size of ion beam spot, the diffusion radius should be much larger than 15nm. According to the simulation analysis, the film especially the middle bridge area is seriously damaged and form into cracks and holes due to heat effect under the condition of large ion beam current and as long as several hours. Therefore, the affected region can be as long as several microns from the milled area edge when MgB_2 film is milled with depth and width of 7um using 6300pA beam current. It further illustrates the problem why the micro bridges prepared were not superconducting.

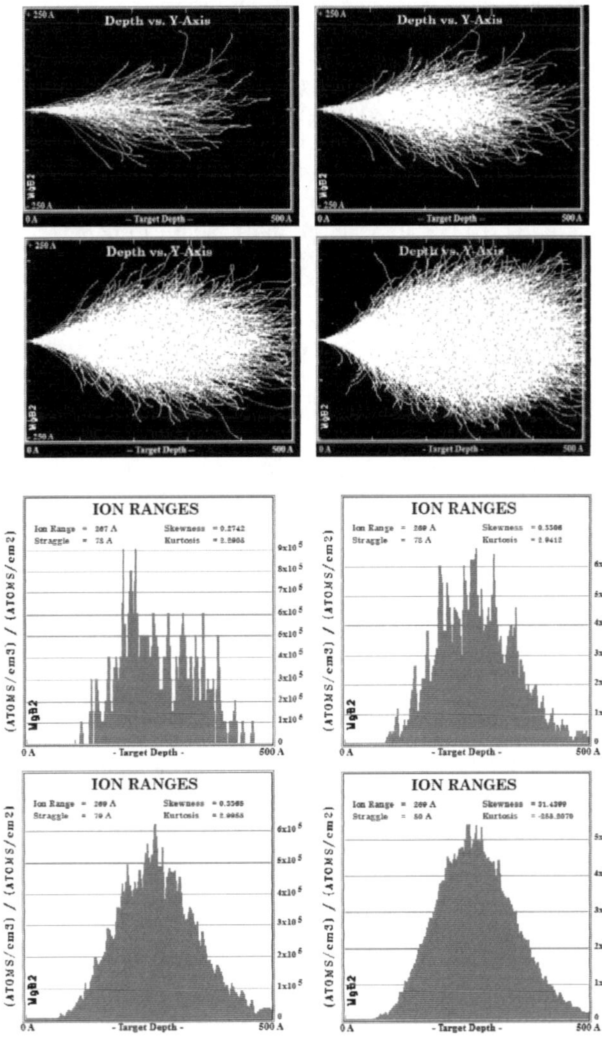

Figure 7. the dynamic process and the corresponding two-dimensional image of 200、1000、5000、25000 Ga^+ penetrating into MgB_2 film.

4. Conclusion

In all, MgB_2 microbridges were prepared by focused ion beam direct milling on MgB_2 film. It is found that the film near the milled area is seriously damaged and cracks and holes are produced after the film is milled with large beam current and long time. The superconducting transition temperature of MgB_2 film can be greatly reduced and the micro bridge is not superconducting due to the effect. Thus, although the new method to prepare micro bridges can avoid the disadvantage of traditional lithography, such as contacted with water vapor, there is still a big problem about ion damage. How to reduce the disadvantages of focused ion beam direct milling on MgB_2 film and microbridges will be our next focus. Some measures, such as changing the experiment parameters, to improve the quality of the film and microbridges, should be taken into. Our research provides an important guiding significance and innovative idea for using focused ion beam milling technology to the prepare superconducting Josephson junction and device.

Acknowledgements
The authors thank Dr. Yang Bo and Professor Feng QingRong in BeiJing University for providing the MgB_2 films used in this work. This work is supported by the Chinese National Natural Science Foundations(Foundation No. 51177160 and No.51307162).

References
[1] Kang W N, Jung C U, Kim K H P, et al. 2001 Hole carrier in MgB_2 characterized by Hall measurements. *J. Appl. Phys.* Lett. pp 982-984.
[2] Chen X H, Xue Y Y, Meng R L and Chu C W. 2001 Penetration depth and anisotropy in MgB_2. *J. Phys.* Rew. B. pp 172501-172504.
[3] Larbalestier D C, Cooley L D, Rikel M O, et al. 2001 Strongly linked current flow in poly crystalline forms of the superconductor MgB_2. *J. Nature*. pp 186-189.
[4] Finnemore D K, Ostenson J E, Budko S L, Lapertot G and Canfield P C. 2001 Thermodynamic and transport properties of superconducting MgB_2. *J. Phys. Rev.* Lett. pp 2420-2422.
[5] Karapetrov G, Iavarone M, Kwok W K, Crabtree G W and Hinks D G. 2001 Scanning tunneling spectroscopy in MgB2. *J. Phys. Rev.Lett.* pp 4374-4377.
[6] Hong S H, Lee S G Seong W K. 2010 Fabrication of MgB2 nanobridge dc SQUIDs by focused ion beam. *J. Physica C-Superconductivity and Its Applications.* Vol 470, ed Moshchalkov V V, Obradors X et al (Netherlands:Elsevier Science) pp S1036-S1037.
[7] Likharev K K. 1979 Superconducting weak links. *J. Reviews of Modern Physics.* Vol. 51, ed Gene D. Sprouse (United States:American Institute Of Physics). chapter1 pp 101-159.
[8] Malisa A, Valkeapaa M, Johansson L G and Ivanov. Z. 2004 Josephson junctions fabricated by focused ion beam from ex situ grown MgB_2 thin films. *J. Physica C.* Vol. 405, ed Moshchalkov V V, Obradors X et al (Netherlands:Elsevier Science) chapter 1 pp 84-88.
[9] Gregor M, Micunek R, Plecenik T, et al. 2008 Nano-bridges based on the superconducting MgB_2 thin films. *J. Physica C*: Superconductivity and its Applications. pp 785-788.

Study on improvement of durability for reinforced concrete by surface-painting migrating corrosion inhibitor and engineering application

Ning Song [1], Zixiao WANG [1], Zhiyong LIU [1,3], Jiyuan Zhou [2], Duo Zheng [2]

[1]School of civil engineering, Yantai University, Yantai, Shandong 264005, China
[2]Jilin Province Institute of Water Conservancy Science Research, Changchun, Jilin 130000, China

E-mail: lzy1698@163.com

Abstract. The corrosion currents of steel bar in concrete with three W/B and four chloride contents after surface-painting two migrating corrosion inhibitors (PCI-2015 and MCI-A) 14d to 150d in atmospheric condition were measured. The results showed that the corrosion current density (I_{corr}) of steel bar reduced to 0.1 μA.cm^{-2} from the initial highest 3.833 μA.cm^{-2} (W/B=0.65, NaCl-1%) after surface-painting PCI-2015 14 d, and the I_{corr} was still lower than 0.1 μA.cm^{-2} until 150d. The compressive strength and chloride migration coefficient of concrete specimens were tested. The possible reasons of the mechanisms of durability improvement for reinforced concrete by applying PCI-2015 inhibitor were PCI-2015 may be reacted with calcium hydroxide in cement concrete and lots of inhibitor particles may be adsorbed on the active sites first and then a stable protective layer may be formed. The I_{corr} of steel bars in a hydraulic aqueduct concrete structure after painting PCI-2015, MCI-A (the United States) and MCI-B (Europe) during 6 months was monitored by Gecor 8 tester. The results showed that the average values of I_{corr} of steel bars after painting the PCI-2015 150d fulfilled the specification requirements in "Design code for concrete structure strengthening (E.3) "(GB50367-2013).

1. Introduction

The corrosion of steel bars in concrete structures causing by chloride ions in marine environment and the areas using deicing salts is a main cause for early deterioration and short service life of concrete structures [1], which leads to huge economical losses. Large amounts of engineering practices showed that migration corrosion inhibitor (MCI) is an effective technique to delay and inhibit the corrosion of steel bar in concrete [2]. In recent years, some studies on MCI have been carried out, such as the anti-corrosion performance and mechanism of MCI in chloride solution [3] and the anti-corrosion performance of MCI by surface-painted to concrete [4-5,10] and the transmission model of MCI in cement concrete [6-9]. The MCIs have been used in some experimental engineering projects. With the increase of the requirements for concrete structures' durability repairing, MCI technique has gradually evolved into the principal method to improve the durability of structures.

2. Experimental

2.1. Materials and concrete mix proportion

Portland cement with the type of P.O.42.5R, river sand, stone with continuous grading between 5mm to 25mm, superplasticizer, fly ash with fineness of 20% and slag powder with the type of S95 were used in the test. Steel electrodes were made of Q235 steel with the size of ϕ12 mm × 70 mm. The surface roughness of steel electrodes was not greater than 3.2μm, The auxiliary electrodes were stainless steel sheets with the size of 10 mm×70 mm. Concrete specimens (size of 100 mm×100 mm×100mm) were prepared with three water to binder ratios (W/B of 0.45, 0.55, 0.65) and four contents of NaCl (0%, 0.5%, 0.75% and 1% of binder weight), concrete mix proportion was shown in Table 1. Two steel electrodes and one auxiliary electrode block were placed in each cubic specimen. The concrete in 1#, 5# and 9# (in Table 1) were also designed to prepare the cylinder concrete specimens (size of ϕ100 mm×250 mm). All specimens were placed in the standard curing room till the age of 28 d. After curing, the cylinder concrete specimens were cut into slice with the size of ϕ100×(50±2) mm.

Table 1. Concrete mix proportion for 1m^3 (kg)

number	W/B	water	cement	fly ash	slag	stone	sand	NaCl
1#	0.45	150	240	80	80	950	930	0.00
2#	0.45	150	240	80	80	950	930	2.00
3#	0.45	150	240	80	80	950	930	3.00
4#	0.45	150	240	80	80	950	930	4.00
5#	0.55	150	210	100	50	950	930	0.00
6#	0.55	150	210	100	50	950	930	1.80
7#	0.55	150	210	100	50	950	930	2.70
8#	0.55	150	210	100	50	950	930	3.60
9#	0.65	150	190	80	70	950	970	0.00
10#	0.65	150	190	80	70	950	970	1.70
11#	0.65	150	190	80	70	950	970	2.55
12#	0.65	150	190	80	70	950	970	3.40

2.2. Test methods

2.2.1. Electrochemical tests Linear potential resistance(LPR) of steel bars in cubic concrete specimens were tested by PARSTAT2273 electrochemical workstation. After curing, the initial LPR of the steel bars in cubic specimens were tested, and then these specimens dried in the oven at 50℃ for 24 h. After the specimens cooling to room temperature, two kinds of MCIs (PCI-2015 and MCI-A) were surface-painted to the specimens. Finally the LPR of the steel bar was tested after painting for 14 d, 28 d and 150 d. The corrosion current density (I_{corr}) of steel bar could be calculated based on the value of LPR. In order to ensure the comparability of the results, the painting amounts of PCI-2015 and MCI-A were 200 g.m^{-2}.

2.2.2. Mechanical property and durability tests Compressive strength of cubic specimens was tested based on the method in the national standard (GB/T50081-2002). Chloride migration coefficient of each specimen was tested by NTB-DAL type chloride diffusion coefficient tester and vacuum water retention system. The test method is according to the national standard (GBT50082-2009).

3. Results and discussion

3.1. Electrochemical properties

I_{corr} of steel bars in the concrete (three water to binder ratios) was changing with time after painting MCIs, as seen in Figure 1 to Figure 3. The results showed that the anti-corrosion property of PCI-2015 was better than that of MCI-A under the same condition.

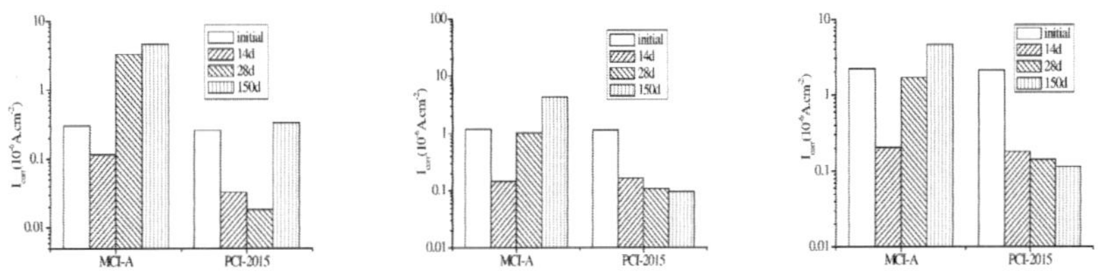

(a) 0.5%NaCl **(b)** 0.75%NaCl **(c)** 1.0%NaCl

Figure 1. Temporal variations of I_{corr} of steel bars in the concrete after painting MCIs (W/B=0.45)

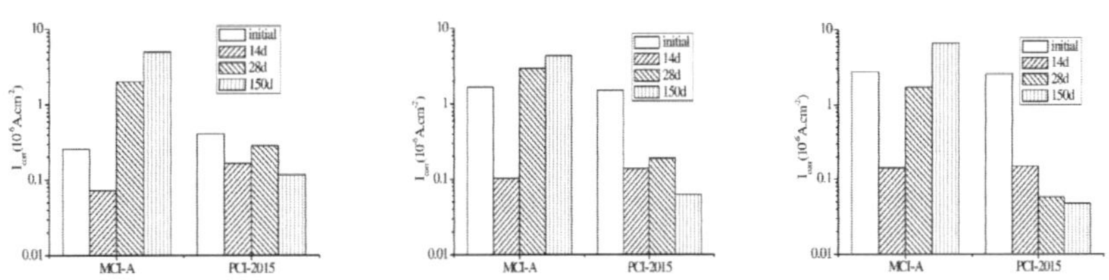

(a) 0.5%NaCl **(b)** 0.75%NaCl **(c)** 1.0%NaCl

Figure 2. Temporal variations of I_{corr} of steel bars in the concrete after painting MCIs (W/B=0.55)

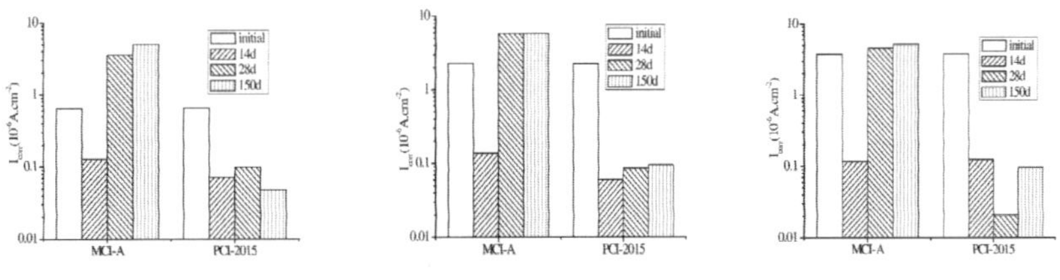

(a) 0.5%NaCl **(b)** 0.75%NaCl **(c)** 1.0%NaCl

Figure 3. Temporal variations of I_{corr} of steel bars in the concrete after painting MCIs (W/B=0.65)

3.2. Mechanical property and durability

The compressive strength of concrete was shown in Figure 4. The compressive strength of concrete decreased with the increase of W/B ratio. When concrete specimens have the same water to binder ratio, the more the content of chlorine salt content, the higher the compressive strength. This might due to the amounts of calcium aluminate hydrate were higher in concrete with lower W/B ratios than that in higher W/B ratios, causing the porosity of concrete reduced and the density of the material improved.

The chloride diffusion coefficients of concrete specimens (three water to binder ratio) painting PCI-2015 and MCI-A were shown in Figure 5. Compared with the blank group, the enhancements of concrete to resist chloride penetration capability by painting PCI-2015 was better than that of painting MCI-A group.

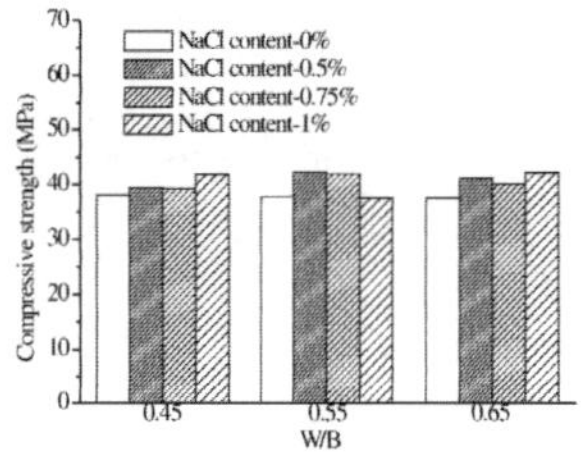

Figure 4. compressive strength of concrete at 28 day

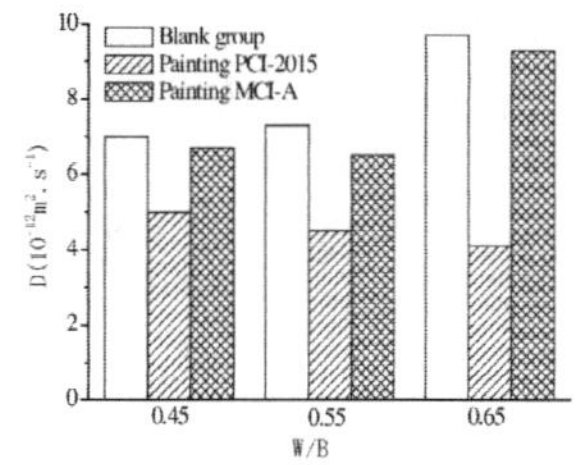

Figure 5. Chloride ion diffusion coefficients

3.3. Discussion
The above experimental results showed surface painting PCI-2015 can not only reduced the chloride diffusion coefficient of concrete but also inhibited the corrosion rate of steel in concrete. The possible reasons were as follows. First, the polycarboxylic acid in PCI-2015 may be reacted with calcium hydroxide in cement concrete[7], so plugging and hydrophobic effect was produced. So the penetration of oxygen, water and chloride ion is reduced. Second, when PCI-2015 was penetrated into the steel surface, lots of inhibitor particles may be adsorbed on the active sites first and then a stable protective layer may be formed , as a result the anodic and cathode reactions on the steel surface are hindered[11].

4. A case of engineering application
An aqueduct was built in 1970s at Hengdaohe in Jilin province. Years later, concrete of aqueduct showed a large area of aging and erosion, and the reinforcements were partially uncovered in some areas. In 2013, two framed bents were collapsed by the flood, causing a serious dislocation of the aqueduct body. In order to improve the durability and prevent further corrosion of reinforcements in the aqueduct, surface-painting MCI was selected. Three kinds of MCI products (PCI-2015, MCI-A from USA and MCI-B from Europe) were selected to the repair the durability of the aqueduct. The electrochemical performance of reinforced bars in concrete was tested online by the Gecor 8 tester. Setting 60 testing points in each section of the aqueduct body, the LPR values of the steel bar in these points were tested before and after painting the MCIs. After painting MCIs 150d, the calculated average inhibition efficiency (IE) of the three inhibitors was 86.37% (PCI-2015)、48.85% (MCI-A) and 81.09% (MCI-B) respectively under the same conditions. Obviously the IE of using PCI-2015 on concrete of the aqueduct met the requirements of the national standard[12].

Figure 6. Testing process using Gecor 8

Figure 7. Process of painting MCI

5. Conclusion
The corrosion currents of steel bar in concrete with three W/B and four chloride contents after surface-painting two organic corrosion inhibitors (PCI-2015 and MCI-A) 14d to 150d in atmospheric condition were measured. The results showed that the corrosion current density (I_{corr}) of steel bar reduced to 0.1 μA.cm^{-2} from the initial highest 3.833 μA.cm^{-2} (W/B=0.65, NaCl-1%) after surface-painting PCI-2015 14 d , and the I_{corr} was still lower than 0.1 μA.cm^{-2} until 150d. The diffusion coefficient of chloride ion of concrete surface-painted PCI-2015 decreased by 28.57%, 38.36% and

57.73%. The possible reasons of the mechanisms of durability improvement for reinforced concrete by applying PCI-2015 inhibitor were PCI-2015 may be reacted with calcium hydroxide in cement concrete and lots of inhibitor particles may be adsorbed on the active sites first and then a stable protective layer may be formed. The I_{corr} of steel bars in a hydraulic aqueduct concrete structure after painting PCI-2015, MCI-A and MCI-B during 6 months were monitored by Gecor 8 tester. The results showed that the average I_{corr} of steel bars after painting the PCI-2015 150d fulfilled the specification requirements in "Design code for concrete structure strengthening (E.3) "(GB50367-2013). In this way, the restoration project realized lower construction waste, energy saving and low carbon emission, which lead to remarkable social, economic and ecological benefits.

6. Acknowledgement

Thanks for the support of National Natural Science Foundation of China(51278443) and 948 projects of Ministry of Water Resources（201424）.

Reference

[1] LEE H S, RYU H S, PARK W J, et al 2015 Comparative study on corrosion protection of reinforcingsteel by using amino alcohol and lithium nitrite inhibitors *J Mater* **8**(1) 251-269

[2] LIU Zhiyong, MIAO Changwen, ZHOU Weiling, et al 2008 Maintenance and improvement of durability of reinforced concrete using migrating corrosion inhibitors *J Chin Ceram Soc* **36**(10) 1494–1500

[3] CHEN Cuicui, ZHOU Weiling,LIU Jiaping, et al 2011 Corrosion resistance of a new organic corrosion inhibitor to steel bar *J BUILD MATER* **14**(1) 136–139

[4] YU L, LIU Z Y, YANG W B, WANG Z X 2015 Adsorption and characterization of an organic corrosion inhibitor for inhibiting carbon steel corrosion in chloride solution *J Curr Anal Chem* **11**(4) 1-6

[5] LIU Z Y, YU L, LI Q Z 2015 Synergic mechanism of an organic corrosion inhibitor for preventing carbon steel corrosion in chloride solution *J Wuhan Univ Technol Mater Sci* **30**(2) 325-330

[6] YU Lei, LIU Zhiyong, ZUO Xiaobao, et al 2014 Transport model of migrating corrosion inhibitor in concrete *J ,B CHIN CERAM SOC* **42**(11) 1370-1376

[7] LIU Z Y, YU L, WANG Z X, YANG W B 2016 Modeling and experimental validation of MCI transport involving pore-blocking effect in cement-based materials *J ASCE J Mater Civil Eng* **28**(5) 04015187-1- 04015187-7

[8] YANG Weibin,YU Lei, LIU Zhiyong 2015 The quantitative model and application of the influence of the migration inhibitor on the corrosion rate of steel bar *J ,B CHIN CERAM SOC* **43**(6) 839-844

[9] LIU Zhiyong,WANG Zixiao, SONG Ning 2016 Optimization of migrating corrosion inhibitor with transmission and inhibition performance *J ,B CHIN CERAM SOC* **44**(3) 481-486

[10] LIU Z Y, WANG Z X, SONG N 2016 A quantitative study on MCI rehabilitating steel bars in chloride contaminated concrete based on inhibition and transport behaviors *J ASCE J Mater Civil Eng* DOI:10.1061/(ASCE)MT.1943-5533.0001695

[11] WANG Z X, YU Lei, LIU Zhiyong 2015 Effects of Organic Corrosion Inhibitor and Chloride Ion Concentration on Steel Depassivation and Repassivation in Solution *J ,B CHIN CERAM SOC*

[12] *Code for design of concrete structures.* GB50367-2013[M]. China Architecture and Building Press, 2014.

IOP Conf. Series: Materials Science and Engineering **167** (2017) 012073 doi:10.1088/1757-899X/167/1/012073

Revealing invisible photonic printing: colorful pattern shown by water

C Chen[1], Z Q Dong[1], Y Chen[1], X B Hu[1], X R Zhu[1], Z G Zhu[1,4], S H Qian[2] and W H Shih[3]

[1]School of Environmental and Materials Engineering, College of Engineering, Shanghai Polytechnic University, Shanghai 201209, P. R. China
[2]Department of Ophthalmology, Eye and ENT Hospital, Shanghai Medical College, Fudan University, Shanghai 200231, P. R. China
[3]Department of Materials Engineering, Drexel University, Philadelphia, Pennsylvania 19104, United States

E-mail: zgzhu@sspu.edu.cn

Abstract. Patterning and displaying of photonic crystals are critical for building advanced photonic materials. We developed a monolayered colloidal crystal (MCC) based photonic material through a mask-assistant photo-polymerization process. Polystyrene (PS) nanoparticles were firstly assembled on the glass slide to form highly ordered MCC, this template is invisible and could diffract visible light when infiltrated by solvent. This reversibly displaying property was successfully maintained after the MCC was surrounding by a polymer layer *via* photo-polymerization, meanwhile a pattern could also be printed on the polymer with the help of a mask. Such photonic printing could be reversibly shown and hidden, revealing the colorful pattern was mainly due to the non-uniform cross-linking density of the polymer.

1. Introduction

Photonic crystal (PC) materials have been considered to be one of the promising candidates for advanced photonic devices of the next generation due to whose prominent light trapping and waveguiding properties. As periodically ordered assembly of monodispersed colloidal particles, colloidal crystals (CC) have demonstrated to possess a region of forbidden frequencies in transmission is known as the photonic band gap (PBG)[1]. These crystalline PBG materials have potential for a wide variety of applications including chemical sensing, displaying, anti-counterfeiting, and antireflective coatings [2-5]. To achieve advanced photonic devices, many techniques have been developed for rapid, convenient and large-scale production of CCs in three-dimensional (3D) and two-dimensional (2D), such as spin-coating, interfacial assembly, and convective assembly [6-9].

For 3D CCs, extensive attentions have been focused on the fabrication of opal and inverse opal structures [10]. The 3D CCs could be embedded in a hydrogel whose volume change leads to the change of the CC lattice, and thus shift the Bragg diffraction wavelength. As the hydrogel was functionalized for recognition of the analytes such as glucose, pH, certain solvents and heavy metal cations, the hydrogel swell or shrink according to the concentration of the analytes could be transformed to the color change [11-15]. On the other hand, 2D CCs, which are monolayered architectures of colloidal particles distinguished from 3D ones, have also attracted intensive research interest due to their unique properties as PBG materials, or as templates in engineering of 2D arrays of various nanostructures [16, 17].

Patterning and displaying of photonic crystal materials are critical in building advanced photonic devices. In this paper, we report an invisible photonic printing based on monolayered colloidal crystal (MCC) that can easily shown by water. MCC was firstly self-assembled onto a glass slide *via* a classic horizon evaporation method [18, 19]. A layer of hydrogel was *in situ* photo-polymerized upon this 2D crystal to form a polymerized MCC (PMCC). The pattern of the PMCC photonic material was printed through a mask during the polymerization process. This technology enables the convenient fabrication of invisible printing with a 2D photonic crystal, allowing the fast preparation of PMCC materials.

2. Experimental

2.1. Chemicals and Materials

Styrene (St, 99%), sodium hydrogen carbonate ($NaHCO_3$, 99.5%), ammonium persulfate (KPS, 98%), acrylamide (AMD, 98.5%), N,N'-methylene-bisacrylamide (MBA, 98%), hydrogen peroxide (H_2O_2, 30%), ethanol (99.7%) and acetone (99.5%) were purchased from Shanghai Chemical Agent Co., Ltd. 2,2-Diethoxyacetophenone (DEAP, 98%) was purchased from Acros Organics. All chemicals were used as received without further purification. The glassware used in all experiments was cleaned in H_2O_2 solution for 30 min.

2.2. CC Preparation

Monodisperse polystyrene (PS) nanoparticles with diameter of 400 nm were prepared by emulsifier-free polymerization using St as monomer, $NaHCO_3$ as buffer and ammonium persulfate (98%, Shanghai Chemical Agent Co., Ltd) as initiator. Detailed reaction procedure was described elsewhere [20]. The resulting latex was dialyzed against ultrapure water (18.2 MΩ cm) for 7 days to remove the excess ions, and PS particles were diluted with ethanol after purification (V_{latex}:$V_{ethanol}$=1:3). Monolayered CC was assembled from ethanol-diluted PS particles on a glass slide by utilizing a horizon evaporation method [21]. The glass slide was wetted with water and 5 μL diluted PS was injected onto the slide, and a layer of PS CC was formed rapidly on the air-liquid interface. After the solvent was removed by evaporation, the sample was treated in an oven at 80 °C for 2 h.

2.3. PMCC Preparation and Patterning

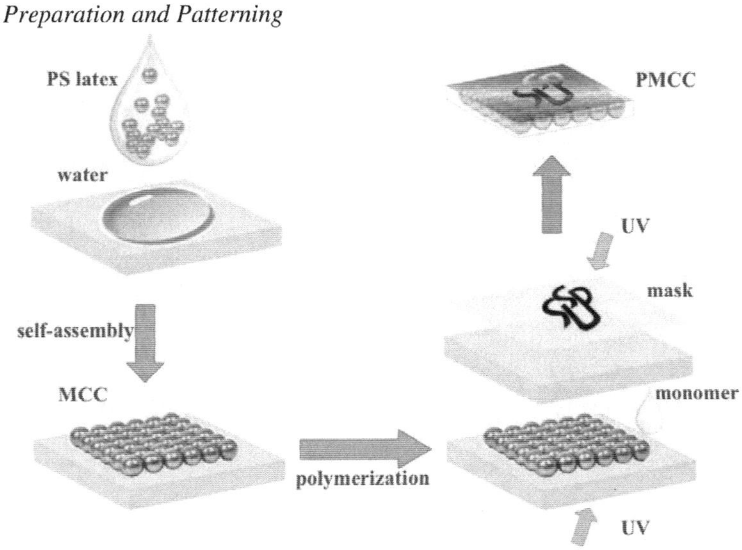

Figure 1. Schematic representation of the preparation of PMCC.

In a typical recipe, 0.702 g AMD and 0.003 g MBA (monomer and cross-linker, respectively, in a molar ratio of 50:1) were mixed and dissolved in 3 mL water in a sample vial, and 5 μL DEAP was added as photo-initiator, and the mixture was agitated and then carefully injected onto the MCC layer. Another glass slide with a mask was covered onto the mixture and MCC supporting slide, and the system was exposed to a set of UV light from Black Ray (365 nm) mercury lamps from both sides of the coupled slides for 40 min to form PMCC. The resulting photonic hydrogel film could be peeled off from the glass slide and washed with ultrapure water. The route for the fabrication of PMCC is illustrated in Figure 1.

2.4. Characterization

The optical data of photonic films were collected with a fiber optic spectrometer (USB 4000-XR1-ES, Ocean Optics). Spectra were captured between the wavelengths of 350 and 900 nm. The diffraction characteristic could be accurately predicted through Bragg's relationship:

$$m\lambda_0 = 3^{1/2} d \sin\theta \qquad (1)$$

where m is the diffraction order, λ_0 is the diffracted wavelength, d is the nearest spacing of the monolayered particle, and θ is the angle of the light relative to the normal to the MCC, during diffraction the θ is ~19°.

Structural and morphological characterizations of the crystals were performed using a field emission scanning electron microscope (FESEM, S-4800, Hitachi) operating at an acceleration voltage of 10 kV, the samples were sputter coated a thin layer of Au.

3. Results and Discussion

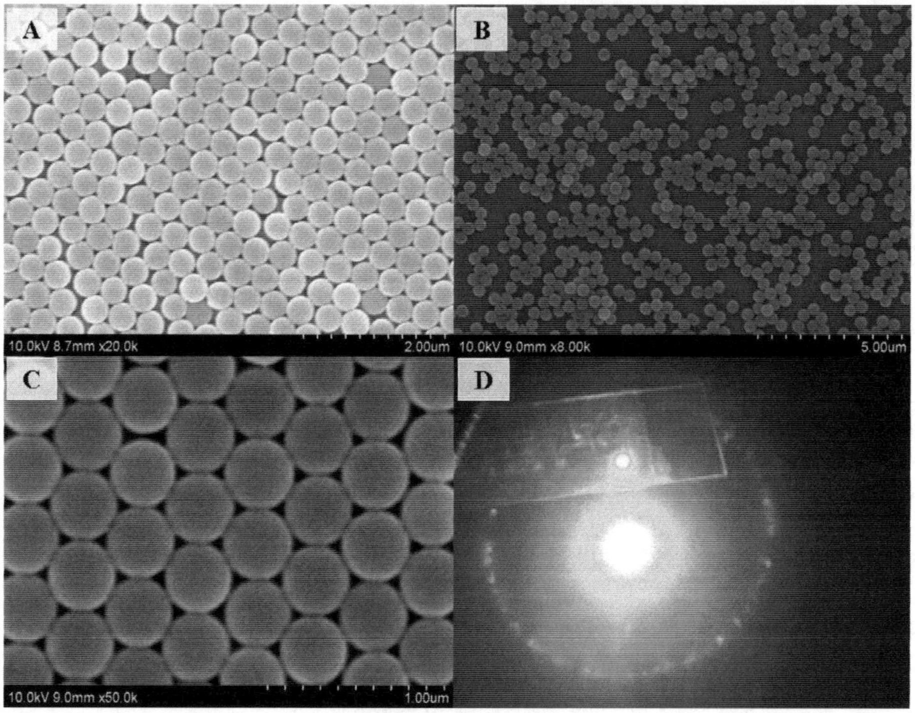

Figure 2. SEM images of the assembled PS particles at various evaporation conditions: (A) 100 °C heating, (B) nitrogen flow, and (C) ambient temperature. (D) Debye ring diffraction from the MCC irradiated by a green laser.

The work presented here describes the combination of polyacrylamide (PAMD) hydrogel and monolayered colloidal crystal (MCC) to form a polymerized monolayered colloidal crystal (PMCC) photonic crystal (PC) material. The MCC consists of an assembly of spherical polystyrene (PS) particles constructed into a generally hexagonal-close-packed (HCP) lattice such that the system minimizes its free energy. This ordered structure of could be coupled with a hydrogel matrix *via* an *in situ* photo-polymerization to form a more stable PMCC, where the MCC lattice distance, and accordingly the diffraction wavelength, reports the hydrogel volume shifting. In contrast to traditional 3D polymerized CC, 2D PMCC readout should be intrinsically more sensitive and reliable since the 2D diffracted wavelength is independent of the refractive index [22].

3.1. Self-Assembly Condition

Monodispersed colloidal spheres can be self-assembled into large-area CCs with high quality by a variety of methods. We compared evaporation conditions upon morphology of the PS assembly. As shown in Figure 2A, the sample was put on a heater at 100 °C for several seconds to remove the solvent, the PS particle readily assembled into multilayered structures since the air-liquid interface was disturbed by the ultrafast dehydration of the system, and the particles stacked repeatedly leading in a 3D ordered CC [23]. By applying gentle nitrogen flux on the top of the sample to promote the evaporation (Figure 2B), on the other hand, resulted in a 2D disordered diluted sub-monolayer. Although the assembling of PS particles on the gas-liquid interfaces was visually evidenced, the fast removal of the solvent caused the disturbance of the PS particles, and then randomly scattered across the substrate due to the capillary immersion that attracted each other, hence no continuously ordered array could be found. Figure 2C confirms a structure with long-distance order and hexagonal symmetry of the sample prepared at the natural assembly speed, long-distance ordered structure. The structure of the MCC was also studied by laser diffraction, a convenient method to confirm 2D array and it's particle spacing. If the particles form a perfect hexagonal 2D array, a diffraction of 6-spots symmetry will be shown when illuminated with monochromatic laser light. If the 2D CC exposed in laser beam is randomly oriented, since the crystallites are significantly smaller than the beam, many symmetry spots with the same diffraction diameter will form a ring, such diffraction pattern is called Debye ring [19]. We excite the MCC along its normal with a 532 nm laser pointer and the 2D crystal array diffracts the light depend on the particle spacing and the laser wavelength. As expected, a sharp circle was observed.

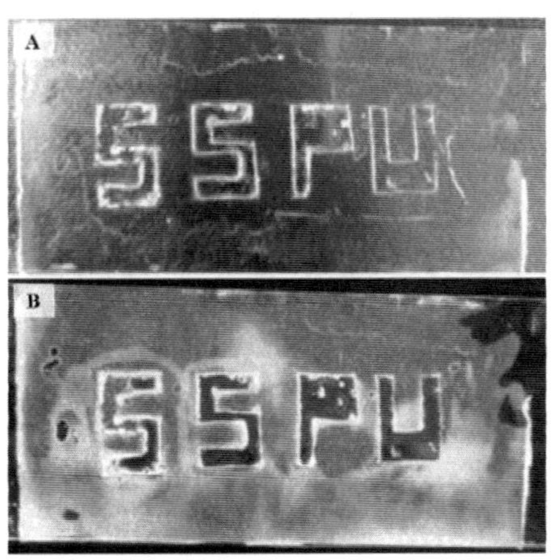

Figure 3. Photograph of (A) invisible MCC coating and (B) MCC coating with "SSPU" pattern shown by solvent.

The as-prepared MCC was shown in Figure 3A, we cut the word "SSPU" with parafilm (~125 μm thick) and covered on the glass slide, After MCC coating, the parafilms were peeled off from the slide and the covered part remain uncoated. The diffraction of the 2D array in air is in the UV region and thus cannot be visualized. The pattern could be easily shown by slightly swelling the PS particles with acetone, as shown in Figure 4b, the image of the word "SSPU" whose area excluded the MCC coating could be easily distinguished. As the diffraction of 2D array is unrelated to refractive index, the red-shift of the MCC from UV to visible region is caused by the increasing packing density due to the swelling of the PS particles.

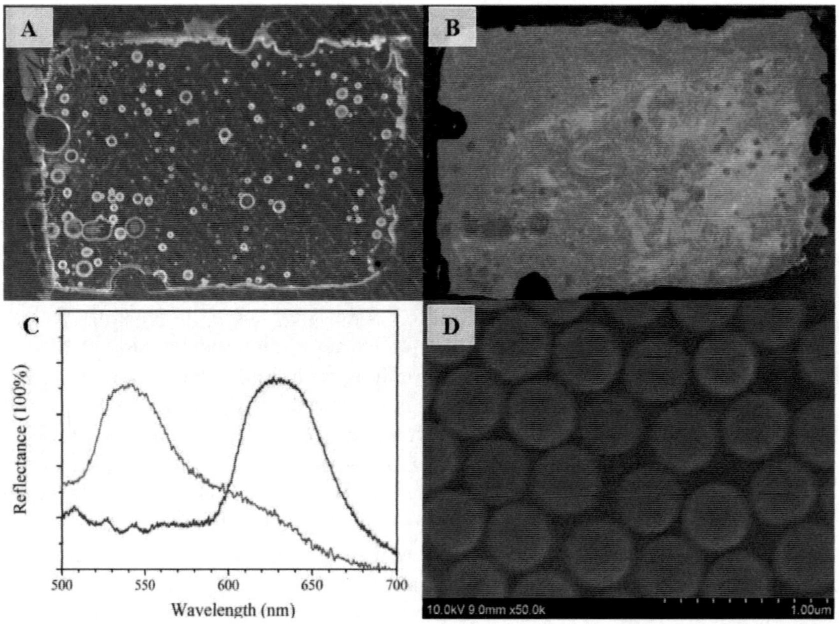

Figure 4. Photograph of PMCC at (A) dehydration state and (B) rehydration state; (C) reflection spectra of different parts of the rehydrated PMCC; and (D) SEM image of the PMCC.

It is a facile method to combine the MCC with a hydrogel, which is significantly different from the 3D CC based hygrogels. Since the ionic species could disorder the 3D array, the ionic monomers cannot be directly utilized for the formation of the hydrogel. To solve this limitation, physically gelated hydrogels were studied, which allows exclusively variation in the shape and thickness of the PC materials, however, the preparation need several hours or even longer for the subsequence modification [24-26]. The 2D CC, without the electrostatic force between layers, could be combined with nonionic monomers as well as ionic ones.

We *in situ* photo-polymerized AMD upon the MCC to form a PAMD PMCC. During the polymerization, a mask with pattern of "SSPU" logo was covered on the glass slide, and thus the pattern was printed on the PMCC due to the different degrees of polymerization. Figure 4A shows the photograph of the as-prepared PMCC, the hydrogel is transparent and the diffraction of MCC is in UV, thus the pattern is invisible. The photonic patterns in dried state could be revealed by water immersion due to the non-uniform change in photonic structure. As shown in Figure 4B, the swelled PMCC clearly displayed the pattern "SSPU" in red while other area displayed green, these two parts have very close photonic structures but different swelling capabilities, because the usage of the mask led to different crosslinkages of the polymer hydrogel, so that they could generate a large color contrast as they swell differently. Once the sample is dried, the PMCC changes back to invisible state, and such displaying is fully reversible. The sample could instantly and reversibly show the patterns by rehydration and hide by dehydration for many times. We measured the reflection spectra of the swollen PMCC (Figure 4C), the pattern reflected ~ 632 nm while the green part reflected ~ 540 nm, the different crosslinking levels made the pattern and the background both invisible at dehydrated state but a $\Delta\lambda$ (~100 nm) under rehydration, this high resolution was accomplished by the optimization of the printing time during the polymerization process based on the relationship between crosslinking level and the reflection changes during rehydration. We also examined the microstructure of the PMCC material, as can be seen in Figure 4D, we found that the 2D layer was well conjugated with the hydrogel except that the close-packed structure was converted into a slightly non-close-packed form after the polymerization. This might be the relaxation of the PAMD molecule chain after the polymerization was quenched. It took ~100 s for the revealing of the pattern, the response time is faster than those 3D CC hydrogels [15].

4. Conclusions

In conclusion, we fabricated two-dimensional photonic crystal material by first assembling the polystyrene nanoparticles into monolayered colloidal crystal at air-liquid interface, followed by polymerizing a hydrogel layer and a specific pattern could be printed under UV irradiation. The construction of PMCC maintaining highly ordered 2D array structure and the key point of the patterning technique is controlling the crosslinking density so that a non-uniform change in photonic structure could be realized under rehydration. The PMCC was initially transparent and a sharp resolution with a red pattern (~ 632 nm) and green background (~ 540 nm) could be observed by water immersion. Our progress enables the convenient fabrication of invisible printing with a 2D photonic crystal, allowing the fast preparation of PMCC materials. This printing technique is capable of creating invisible photonic prints that could be easily shown and hidden, we anticipate such PMCC would found potentially use in applications for sensing, security printing and anti-counterfeiting purpose.

Acknowledgments

This work was supported by the National Natural Science Foundation of China (61471233, 21504051), the Program for Professor of Special Appointment (Eastern Scholar) at SIHL, the Sailing Project and Basic Research Program from Science and Technology Commission of Shanghai Municipality (14YF1410600, 14JC1406402), Shuguang and ChenGuang project supported by Shanghai Municipal Education Commission and Shanghai Education Development Foundation (14SG52, 13CG62), the key subject of Shanghai Polytechnic University (Material Science and Engineering, XXKZD1601).

References

[1] Zhao Y, Shang L, Cheng Y and Gu Z 2014 *Accounts Chem. Res.* **16** 3632
[2] Ge J and Yin Y *Angew.* 2011 *Chem. Inter. Edit.* **50** 1492
[3] Ye S, Fu Q and Ge J 2014 *Adv. Funct. Mater.* **24** 6430
[4] Zhang Y, Tang B, Wu Z, Shi H, Zhang Z and Zhao G 2016 *Green Chem.* **18** 2424
[5] Heo E, Kang H, Lee J, Oh Y and Kim S 2016 *Small* **28** 3819
[6] Ye X and Qi L 2014 *Sci. China. Chem.* **1** 58
[7] Xia Y, Gates B and Li Z 2001 *Adv. Mater.* **13** 409
[8] Mastrangeli M, Abbasi S, Varel C, Hoof C, Celis J and Böhringer K 2009 *J. Micromech. Microeng.* **19** 083001
[9] Freymann G, Kitaev V, Lotsch B and Ozin G 2013 *Chem. Soc. Rev.* **42** 2528
[10] Marlow F, Muldarisnur, Sharifi P, Brinkmann R and Mendive C 2009 *Angew. Chem. Int. Ed.* **48** 6212
[11] Fenzl C, Hirsch T and Wolfbeis O 2014 *Angew. Chem. Int. Ed.* **53** 3318
[12] Cullen D, Xu Y, Reneer D, Browne K, Geddes J, Yang S and Smith D 2011 *Neuroimage* **54** S37
[13] Galisteo-López J, Ibisate M, Sapienza R, Froufe-Pérez L, Blanco Á and López C 2011 *Adv.*

Mater. **23** 30
[14] Lee J, Jin H, Desai M, Ren S, Kim S and Lee S 2015 *Nanoscale* **7** 18379
[15] Lu W, Asher S, Meng Z, Yan Z, Xue M, Qiu L and Yi D 2016 *J. Hazard. Mater.* **316** 87
[16] Jing P, Wu J and Lin L 2014 *ACS Photonics* **1** 398
[17] Ye X and Qi L 2011 *Nano Today* **6** 608
[18] Alfrey T Jr, Bradford E, Vanderhoff J and Oster G 1954 *J. Opt. Soc. Am.* **44** 603
[19] Prevo B and Velev O 2004 *Langmuir* **20** 2099
[20] Chen C, Zhu Z, Shih W, Ge Q, Liu M and Zhu X 2015 *J. Nanosci. Nanotech.* **15** 3239
[21] Denkov D, Velev O, Kralchevski P, Ivanov I, Yoshimura H and Nagayama K1992 *Langmuir* **8** 3183
[22] Kanai T, Sawada T and Kitamura K 2003 *Langmuir* **19** 1984
[23] Gallego-Gómeza F, Morales-Flórezb V, Moralesa M, Blancoa A and Lópeza C 2016 *Adv. Colloid. Interfa.* **234** 142
[24] Chen C, Zhu Y, Bao H, Zhao P, Jiang H, Peng L, Yang X and Li C 2010 *Soft Matter* **7** 915
[25] Chen C, Zhu Y, Bao H, Shen J, Jiang H, Peng L, Yang X, Li C and Chen G 2011 *Chem. Commun.* **47** 5530
[26] Chen C, Zhu Z, Zhu X, Yu W, Liu M, Ge Q and Shih W 2015 *Mater. Res. Express* **2** 046201

Influence of booster size on the total energy of RBUL-1 explosive in underwater explosion

Tongtang Cao [1], Lin Zhou [1], Xiangrong Zhang [1,3], Jiming Wang [2]

[1] State Key Laboratory of Explosion Science and Technology, Beijing Institute of Technology, Beijing 100081, China
[2] Yin Guang Chemical Industry Group Co, Ltd, Baiyin 730900, China

E-mail: zhangxr@bit.edu.cn

Abstract. To investigate the influence of the booster size on the total energy of DNAN-based insensitive melt-cast explosive RBUL-1, six groups of underwater explosion experiments with varied booster sizes were carried out to measure the total energy of RBUL-1 explosive. Experimental results show that the booster size especially the booster diameter has a great influence on the total energy of RBUL-1 explosive. An expression was proposed and calibrated to illustrate the relationship between the total energy of RBUL-1 explosive and the booster size. Besides, the booster size used in the underwater explosion experiments was optimized to achieve higher total energy results. The present investigation has practical significance for the design of the booster sequence and warhead.

1. Introduction

The recent development and performance characterization of insensitive explosives based on DNAN has led to the possible replacement of TNT and Composition B in fielded munitions. DNAN offers lower toxicity and sensitivity than TNT. DNAN-based explosives have been demonstrated to have greatly improved insensitive munition (IM) characteristics, while maintaining explosive performance requirements [1]. Thus, the use of DNAN in explosive formulations is presently undergoing a renaissance. This is exemplified by the development of a range of DNAN based explosive formulations at Picatinny Arsenal, typically referred to the melt-cast PAX explosives [2].

RBUL-1 (containing DNAN, AP, and aluminum) is a new developed insensitive DNAN-based melt-cast aluminized explosive, which owns the characteristic of larger critical diameter, wider reaction zone, and longer reaction time. The insensitiveness of RBUL-1 explosive demands that initiating ability of the booster should be enough to initiate RBUL-1 explosive reliably. Otherwise, the RBUL-1 explosive couldn't be detonated completely resulting in its energy couldn't be released fully, which will reduce its damage power.

The initiating ability of an unconstrained booster is mainly affected by the booster formulation and the booster size [3]. In this study, the size effect of booster JH-14C on its ability to initiate RBUL-1 explosive was investigated by underwater explosion experiments. The measured total energy of RBUL-1 explosive was used to characterize the initiating ability of JH-14C. An expression was proposed and calibrated to illustrate the relationship between the total energy of RBUL-1 explosive and the booster size. With the help of the expression, the booster size used in the underwater explosion experiments was optimized and the corresponding total energy of RBUL-1 explosive was maximized.

A practical use of the expression is to maximize the total energy with the limit of booster mass, which has much interest and practical significance for the design of the booster sequence and warhead.

2. Experiment

2.1. Experimental Devices

Underwater explosion experiments were conducted in a pool with a diameter of 85m and a depth of 14.5m, the main charge RBUL-1 was placed in the center of the pool and 5 m below the water surface. At this condition, the shock parameters measured by pressure sensors are free from the effect of water surface, bottom and wall of the pool [4]. The main charge was initiated by the frequently-used JH-14C booster (96.5% RDX, 3% fluororubber and 0.5% graphite by weight). The detonator is commercial plain detonator No.8. PCB138M130 pressure sensors, 482A22 charge-amplifier and Multipro MP high speed data acquisition device were applied together to record the pressure curve of the pressure sensor, which is 3m far away from the main charge. Figure 1 shows a schematic diagram of the underwater explosion experiment.

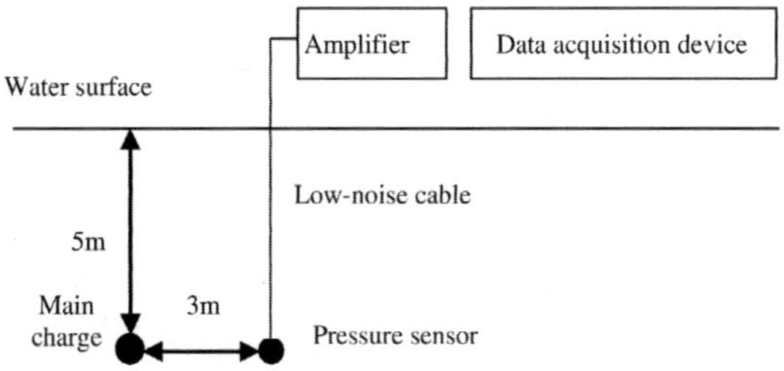

Figure 1. Schematic diagram of underwater explosion experiment.

2.2. Experimental Conditions

Table 1 listed the detail of the booster and the main charge in underwater explosion experiments. Six groups of underwater explosion experiment were conducted, and there were two parallel experiments for each group. Group 1 was used to measure the total energy of cylindrical booster JH-14C with the density of 1.65g/cm^3; groups 2~6 were used to measure the total energy of the cylindrical main charge RBUL-1 with the density of 1.84g/cm^3, which was initiated by cylindrical booster with different size (diameter and height).

Table 1. Detail of the booster and the main charge in underwater explosion experiment.

Group Number	Booster			Main charge		
	Name	Size (mm×mm)	Mass (g)	Name	Size (mm×mm)	Mass (g)
1	-	-	-	JH-14C	ϕ78×25	197
2		ϕ30×30	35		ϕ85×96	1000
3		ϕ39×39	77		ϕ85×96	1000
4	JH-14C	ϕ42×42	96	RBUL-1	ϕ114×114	2000
5		ϕ61×44	197		ϕ85×96	1000
6		ϕ78×25	212		ϕ85×96	1000

3. Results and Discussions

3.1. Evaluation of specific shock energy

The specific shock energy in the primary water shock when it passes the gauge was determined from the energy density at the gauge and assuming a spherical expansion [5].

$$E_s = \frac{4\pi R^2}{W\rho C}\int_0^{6.7\theta} P^2(t)dt \tag{1}$$

where E_s is the specific shock energy, MJ/Kg, ρ is the density of water, Kg/m^3, C is the sound speed of water, m/s, R is the distance between the main charge and pressure sensor, m, W is the mass of the main charge, Kg, $P(t)$ is the pressure at the time t, MPa, θ is the time constant, which is defined as the time it takes for the pressure to decay from its peak pressure P_m to a value P_m/e. In this study, θ was obtained by fitting the pressure curve with expression below [6].

$$P_m\left[\frac{0.368\exp(\theta/\omega)-1}{\exp(\theta/\omega)-\exp(\theta/\varphi)}\exp\left(-\frac{t-\theta}{\varphi}\right)-\frac{0.368\exp(\theta/\varphi)-1}{\exp(\theta/\omega)-\exp(\theta/\varphi)}\exp\left(-\frac{t-\theta}{\omega}\right)\right] \tag{2}$$

where φ and ω are parameters used to describe the degree of shock wave attenuation at $t > \theta$ and can be obtained from experiments.

3.2. Evaluation of Specific bubble energy

The specific bubble energy in the first bubble pulse can be defined as the work done in expanding the bubble to its first maximum against the hydrostatic pressure. The so called Willis formula [5] can be used to calculate the bubble energy from the period of oscillation t_b of the bubble in cases where we have no boundary effects. This formula can be written as:

$$t_b = 1.135\rho^{1/2}E_b^{1/3}P_h^{5/6} \tag{3}$$

where t_b is the period of the first bubble oscillation, P_h is the total hydrostatic pressure at the charge depth, ρ is the density of water. Formula (3) can be written as:

$$E_b = t_b^3/k_1 \tag{4.a}$$

where

$$k_1 = (1.135\rho^{1/2}P_h^{5/6})^3 \tag{4.b}$$

3.3. The Total Energy

The total energy is the sum of the specific shock energy and the specific bubble energy and is presented as follows:

$$E_t = E_s + E_b \tag{5}$$

The specific shock energy and bubble energy of JH-14C booster were firstly measured (group1 in table 2). Subtracting the corresponding specific shock energy and bubble energy of JH-14C, then the actual specific shock energy and bubble energy of RBUL-1 explosive can be obtained (group 2-6 in table 2).

Table 2. Results of RBUL-1 explosive in underwater explosion experiments.

Group Number	Booster Size(mm×mm)	Main Charge Size (mm×mm)	Specific Shock Energy(MJ/Kg)	Specific Bubble Energy(MJ/Kg)	Total Energy(MJ/Kg)
1		ϕ78×25	1.24	2.02	3.26
2	ϕ30×30	ϕ85×96	1.42	4.11	5.53
3	ϕ39×39	ϕ85×96	1.55	4.14	5.69
4	ϕ42×42	ϕ114×114	1.61	3.97	5.58
5	ϕ61×44	ϕ85×96	1.63	4.32	5.95
6	ϕ78×25	ϕ85×96	1.65	4.32	5.97

It can be obviously seen in table 2 that the booster size has a great influence on the total energy. Increasing booster diameter and/or height helps to increase the total energy. Therefore, given the formulations of both the booster and main charge, the initiating ability of booster increases with the increase of booster diameter and/or height. In order to quantify such ability, a mathematical formula was proposed as follows:

$$N = kx^a y^b \tag{6}$$

where N is the total energy, MJ/Kg, $x = d/d_0$, $y = h/d_0$, d and h are diameter and height of booster, respectively, d_0 is diameter of the main charge, k, a, b are constants. A MATLAB code using non-linear curve fit method has been applied to determine the constants in equation (6) based on the data in table 2, and the resultant constants are: k=6.2281 MJ/Kg, a=0.0845, b=0.0284. The correlation coefficient of the non-linear curve fit is 0.995.

From the shock to detonation transition (SDT) point of view, equation (6) tells that the shock wave developed in the interface between the booster and main charge is less affected by the lateral and rear rarefaction wave with large values of x and y compared to small values of x and y.

For the cylindrical JH-14C booster, the mass of the booster can be given by:

$$M = \rho\pi d_0^3 \frac{x^2 y}{4} \tag{7}$$

where M is the mass, ρ is the density, 1.65g/cm^3.

Usually, the maximum booster mass is not more than twenty percent of main charge mass. Therefore, for a 1000 g main charge in the study, the maximum booster mass can be taken as 200g. Combining with equation (7), the constraint of booster mass can be then written as:

$$0 < x^2 y < 0.2513 \tag{8}$$

Obviously, there is a maximum value of total energy in equation (6) with the constraint of equation (8). The maximum total energy was achieved using a MATLAB code and the result is 5.974 MJ/Kg, the booster size corresponding to maximum total energy are x=0.9176, y=0.2984, that is d=78mm, h=25.4mm.

The same method was used to optimize the booster size in table 1. The optimized booster size, the corresponding maximum total energy together with the original booster size and original total energy were listed in table 3. It is found that the total energy corresponding to the optimized booster size is higher than that of the original booster size. The difference between the optimized total energy and the original total energy seems to be slightly, this is may be because the original booster size was chosen in an appropriate way, so there is a limited room for the increment of the total energy. Future work of measuring the total energy of the main charge initiated by the booster with a bad size (bigger height but smaller diameter) and the comparison of the optimized total energy and the original total energy corresponding to the booster with the bad size is believed to be useful to confirm our assumption. Another notable result in table 3 is that the height of optimized booster decreased but the diameter of optimized booster increased, and the optimized maximum total energy is always bigger than the original one. This is because the reduction of the total energy dues to the decreased booster height is super compensated by the increased booster diameter. A bigger value of a(0.0845) in equation (6) than the b(0.0284) also confirms our assumption. Thus, we concluded that diameter and height both help to increase the initiating ability of booster and the total energy of the main charge, but the booster diameter has a greater influence on the change of the total energy than the booster height.

Table 3. Comparison of results of optimized booster and that of original booster.

Booster Mass	Optimized Booster Size		Original Booster Size		Maximum N (MJ/Kg)	Original N (MJ/Kg)
(g)	d (mm)	h (mm)	d (mm)	h (mm)		
35	32.9	25.0	30	30	5.551	5.53
77	48.8	25.0	39	39	5.739	5.69
96	54.4	25.0	42	42	5.793	5.58
212	78.0	26.9	61	44	5.984	5.95
197	78.0	25.0	78	25	5.972	5.97

4. Conclusions

Six groups of underwater explosion experiment were carried out to investigate the influence of booster size on the total energy of RBUL-1 explosive. The main conclusions are as follows:

- Booster size has a great influence on the total energy of RBUL-1 explosive. Increasing booster diameter and/or booster height both increase the total energy of RBUL-1explosive. However, the influence of booster diameter on change of the total energy is more prominent than that of booster height.
- The proposed expression can be used to appropriately describe the relationship between the total energy of RBUL-1 explosive and the JH-14C booster size. The booster size used in the underwater explosion experiments can be optimized to obtain the maximum total energy.

References

[1] Grau H, Gandzelko A, Samuels P 2014 *Propellants Explos. Pyrotech.* **39** 604–608.
[2] Davies, Phil J, and Arthur Provatas 2006 *Characterisation of 2, 4-dinitroanisole: an ingredient for use in low sensitivity melt cast formulations* (Edinburgh: Defence Science and Technology Organisation Weapons Systems Division).
[3] Beijing Institue of Technology 1974 *The physical basis of detonation* (Beijing: National Defense Industry Press)
[4] Bjarnholt G 1980 Propellants Explos. Pyrotech. **5** 67-74.
[5] R. H. Cole 1961 *Underwater explosions* (Princeton: Princeton University Press)
[6] Zhou Lin, Xie Zhong-Yuan and Wei Xue-xia 2011 *Combust. Explos. Shock Waves.* **47** 721-726.

Numerically design the injection process parameters of parts fabricated with ramie fiber reinforced green composites

L P Chen[1], L P He[1,4], D C Chen[1,2], G Lu[1], W J Li[1] and J M Yuan[1,3]

[1] State Key Laboratory of Advanced Design and Manufacturing for Vehicle Body, College of Mechanical and Vehicle Engineering, Hunan University, Changsha, Hunan 410082, P.R. China.
[2] College of Civil Engineering, Hunan University, Changsha, Hunan 410082, P.R. China
[3] College of Materials Science and Engineering, Hunan University, Changsha, Hunan 410082, P.R. China

E-mail: lphe@hnu.edu.cn

Abstract. The warpage deformation plays an important role on the performance of automobile interior components fabricated with natural fiber reinforced composites. The present work investigated the influence of process parameters on the warpage behavior of A pillar trim made of ramie fiber (RF) reinforced polypropylene (PP) composites (RF/PP) via numerical simulation with orthogonal experiment method and range analysis. The results indicated that fiber addition and packing pressure were the most important factors affecting warpage. The A pillar trim can achieved the minimum warpage value as of 2.124 mm under the optimum parameters. The optimal process parameters are: 70% percent of the default value of injection pressure for the packing pressure, 20 wt% for the fiber addition, 185 ℃ for the melt temperature, 65 ℃for the mold temperature, 7 s for the filling time and 17 s for the packing time.

1. Introduction

Along with the increasingly environmental pollution, natural fiber-reinforced polymer composites have received widely concern, especially in automotive industry. By comparison with synthetic fibers, natural fiber exhibits significant advantages, such as lower cost, high specific modulus and environmental friendliness [1-3]. Recently many researchers focus on using the natural fibers such as lignocellulosic fillers to replace glass fiber filler. Ramie fiber (RF) is a kind of natural fiber and its production is quite big in China [4]. Thus, more and more researches were carried out for the ramie fiber reinforced composites. So far, lots of work [5-8] aimed to obtain the good interfacial adhesion property of the composites through the pretreatment of ramie fiber. Limited work [9-10] tried to investigate the influence of process parameters on the performance of industry products.

Injection molding possess has been widely used to manufacture automotive parts, particularly for automotive interiors, which are made of polymer and polymer-matrix composites. The qualities of injection-molding parts were ordinarily influenced by the materials, the geometry of the parts, mold design and the process parameters. Warping deformation is one of the significant criteria for evaluating the quality of injection-molding parts [11-13]. Recently, many researchers were interested in investigating the warpage of parts made of natural fiber-reinforced composites. This is because fiber-reinforced composites are more and more widely used in various fields. Rahman et al. [14] investigated the window frame fabricated with the natural fibers by using the Moldflow Software. They reckoned that the natural fiber composite material was suitable to manufacture window frame. Azaman et al. [15, 16] evaluated the warpage of a molded thin-walled part composed of wood-filled polypropylene (PP) composites during the injection-molded process under different process conditions and fiber addition percentages. The optimum process parameters and minimum warpage were investigated, and the results showed that cooling time and packing time had less effect on warpage. Anyway, so far no simulation work addresses the parts made of ramie fiber-filled PP composites. Therefore, this work numerically eco-design the injection process parameters in order to obtain a minimum warpage deformation of A pillar trim made of ramie fiber (RF) reinforced polypropylene (PP) composites (RF/PP) with fiber aspect ratio of 20.

2. Materials and methodology

2.1. Materials

In this work, the short ramie fiber (RF) with fiber aspect ratio of 20 was used as the filler, and the polypropylene (PP) used as the matrix. The RF/PP green composites were applied to fabricate automobile A pillar trim using injection-molding process. The properties of PP were obtained from the materials database of Autodesk Moldflow as shown in Table 1. The tested properties of degummed ramie fiber were also listed, which are consistent with that reported in literature [17].

Table 1. Properties of PP and ramie fiber.

Properties	PP	Ramie fiber
(Melt) Density (g/cm^3)	0.862	1.50
Elastic modulus (MPa)	1,340	90,000
Poisson's ratio	0.4	0.3
Tensile Strength (MPa)	—	900
Recommended melt temperature (℃)	190	—
Recommended mold temperature (℃)	50	—
Fiber addition (wt%)	—	10-40

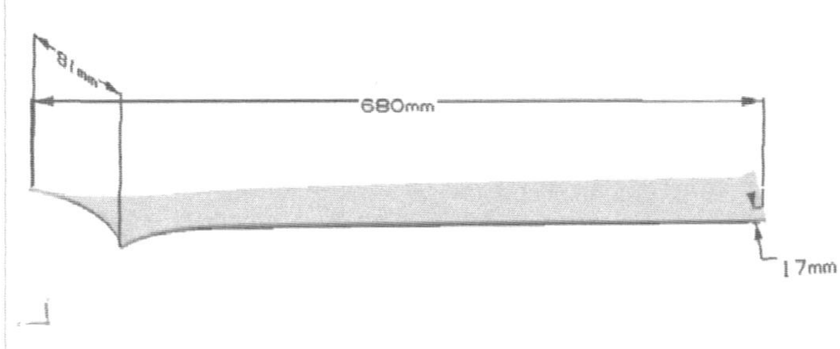

Figure 1. 3D model of an automotive A pillar trim.

2.2. Simulation model details

An automotive A pillar trim model was established using Unigraphics NX, with dimensions of 680× 81×17 mm, and a thickness of 3 mm. The 3D model is shown in Figure 1.

The warpage of A pillar trim made of RF/PP were simulated and investigated using the software Autodesk Moldflow. The simulation model was established with Moldflow software based on standard design guidelines, including a sprue, a runner and a cooling system [14], which is shown in Figure 2. The analysis model applied fusion mesh type and consisted of 8122 triangular elements.

After the establishment of the injection-molding system model, it was necessary to verify its reliability. So, a simulation was conducted through using the material parameters given in Table 1, and the mold temperature and melt temperature were set with recommended value in Table 1. The simulation results were obtained, such as the filling time, the temperature at flow front, weld line distribution and so on, which manifested that the established injection-molding system model was reasonable.

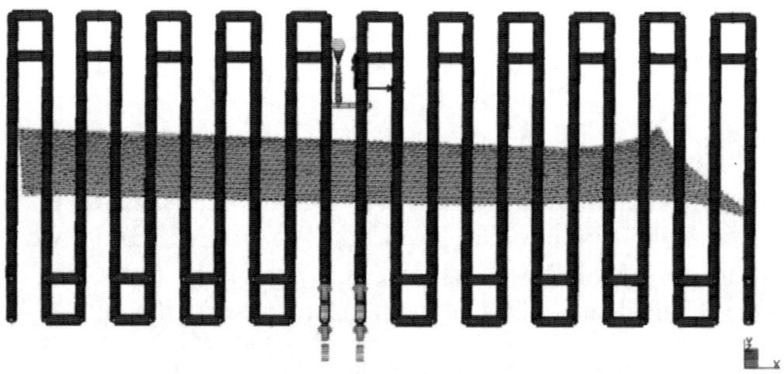

Figure 2. Simulation model of the injection-molding system.

2.3. Simulation and orthogonal experiment design

The orthogonal experiment method was applied in order to achieved the minimal warpage of the A pillar trim. The parameters used for evaluating the warpage behaviors were the packing pressure (A), melt temperature (B), mold temperature (C), filling time (D), packing time (E) and fiber addition (F). The values of these parameters were listed in Table 2, in which packing pressure was given by the percent of the default value of injection pressure. The four levels for the six process parameters were designed as an L_{32} (4^6) orthogonal array, with 32 trials.

Table 2. Process parameters and their levels.

Levels	A-Packing pressure (%)	B-Melt Temperature (℃)	C-Mold temperature (℃)	D-Filling time (s)	C-Packing time (s)	F-Fiber content (%)
1	70	180	50	5	17	10
2	80	185	55	6	19	20
3	90	190	60	7	21	30
4	100	195	65	8	23	40

3. Orthogonal experiment results and discussion

3.1. Warpage results of each group

The warpage results of the 32 trials are presented in Table 3. It can be found that the warpage of the A pillar trim varied in the range of 2.2-6.1 mm when various parameters are used. Therefore, it is necessary to optimize the parameters in order to obtain the minimal warping deformation.

Table 3. Warpage results of 32 trials.

Trials number	A	B	C	D	E	F	Warpage
1	1	1	1	1	1	1	3.945
2	1	2	2	2	2	2	2.221
3	1	3	3	3	3	3	2.461
4	1	4	4	4	4	4	2.962
5	2	1	1	2	2	3	3.276
6	2	2	2	1	1	4	3.234
7	2	3	3	4	4	1	4.701
8	2	4	4	3	3	2	2.633
9	3	1	2	3	4	2	3.124
10	3	2	1	4	3	1	5.226
11	3	3	4	1	2	4	4.312
12	3	4	3	2	1	3	3.588
13	4	1	2	4	3	4	3.796
14	4	2	1	3	4	3	3.398
15	4	3	4	2	1	2	3.365
16	4	4	3	1	2	1	6.082
17	1	1	4	1	4	3	2.421
18	1	2	3	2	3	4	2.996
19	1	3	2	3	2	1	3.847
20	1	4	1	4	1	2	2.337
21	2	1	4	2	3	1	4.861
22	2	2	3	1	4	2	2.610
23	2	3	2	4	1	3	2.852
24	2	4	1	3	2	4	3.267
25	3	1	3	3	1	4	4.026
26	3	2	4	4	2	3	3.122
27	3	3	1	1	3	2	3.514
28	3	4	2	2	4	1	5.474
29	4	1	3	4	2	2	2.804
30	4	2	4	3	1	1	5.037
31	4	3	1	2	4	4	4.269
32	4	4	2	1	3	3	4.648

3.2. Effects of process parameters on warpage

In this study, the effects of parameters on the warpage were investigated using range analysis method. The range analysis method was applied to analyze the optimal parameters [18]. In the range analysis method, $\bar{u}_{pk}$ is used to determine the optimum level of each parameter, and it can be calculated using equation (1). The minimal value among $\bar{u}_{p1}$, $\bar{u}_{p2}$, $\bar{u}_{p3}$ and $\bar{u}_{p4}$ is corresponding to the optimum level of the process parameter p.

$$\bar{u}_{pk} = \frac{1}{8}(\sum u_{pk}) \quad (1)$$

where p stands for parameters (A, B, C, D, E and F), k stands for levels (1, 2, 3 and 4), $\bar{u}_{pk}$ means the warpage when process parameter is p and level is k. Range is defined as the difference between the maximal ($\bar{u}_{pmax}$) and minimal ($\bar{u}_{pmin}$) values of $\bar{u}_{pk}$ as shown in equation (2), and it can be applied for evaluating the effects of parameters on the warpage. The larger the range is, the more prominent the effect of the parameter on the warpage will be.

$$\text{Range} = \bar{u}_{pmax} - \bar{u}_{pmin} \tag{2}$$

Taking parameter F (Fiber content) as an example, the range analysis is shown as follow:

$$\bar{u}_{F1}=\frac{1}{8}(3.945+4.701+5.226+6.082+3.847+4.861+5.474+5.037)=4.897,$$

$$\bar{u}_{F2}=\frac{1}{8}(2.221+2.633+3.124+3.365+2.337+2.610+3.514+2.804)=2.826,$$

$$\bar{u}_{F3}=\frac{1}{8}(2.461+3.276+3.588+3.398+2.421+2.852+3.122+4.648)=3.221,$$

$$\bar{u}_{F4}=\frac{1}{8}(2.962+3.234+4.312+3.796+2.996+3.267+4.026+4.269)=3.608,$$

$$\text{Range} = \bar{u}_{pmax} - \bar{u}_{pmin} = \bar{u}_{F1} - \bar{u}_{F2} = 4.897 - 2.826 = 2.071.$$

Based on the above method, the results of range analysis for all parameters are included in Table 4. Therefore, the optimal levels of each parameter can be obtained as shown in Table 5.

Table 4. Results of range analysis.

	A (Packing pressure)	B (Melt temperature)	C (Mold temperature)	D (Filling time)	E (Packing time)	F (Fiber content)
$\bar{u}_{p1}$	2.899	3.532	3.654	3.846	3.548	4.897
$\bar{u}_{p2}$	3.429	3.481	3.650	3.756	3.616	2.826
$\bar{u}_{p3}$	4.048	3.665	3.659	3.474	3.767	3.221
$\bar{u}_{p4}$	4.175	3.874	3.589	3.475	3.620	3.608
$\bar{u}_{pmax}$	4.175	3.874	3.659	3.846	3.767	4.897
$\bar{u}_{pmin}$	2.899	3.481	3.589	3.474	3.548	2.826
Range	1.276	0.393	0.069	0.372	0.219	2.071

Table 5. Optimal level of each parameter.

A (Packing pressure)	B (Melt temperature)	C (Mold temperature)	D (Filling time)	E (Packing time)	F (Fiber content)
A1 (70%)	B2 (185 ℃)	C4 (65 ℃)	D3 (7 s)	E1 (17 s)	F2 (20%)

In the previous research about the influence of process parameters on warpage, Ozcelik et al. [19] found that packing pressure was the most important process parameter affecting warpage, while Tang et al. [20] showed that melting temperature had the greatest influence. Besides, Hakimian et al. [21] reported that glass fiber gave the highest influence on the warpage and shrinkage. However, there is rare research about the influence of process parameters on the warpage behavior of component made of RF/PP. In this study, the effects of parameters on the warpage of A pillar trim fabricated with RF/PP were investigated using range analysis method. It is found from Figure 3 that fiber addition (F) and packing pressure (A) are the most important factors affecting warpage, while other process parameters have less influence.

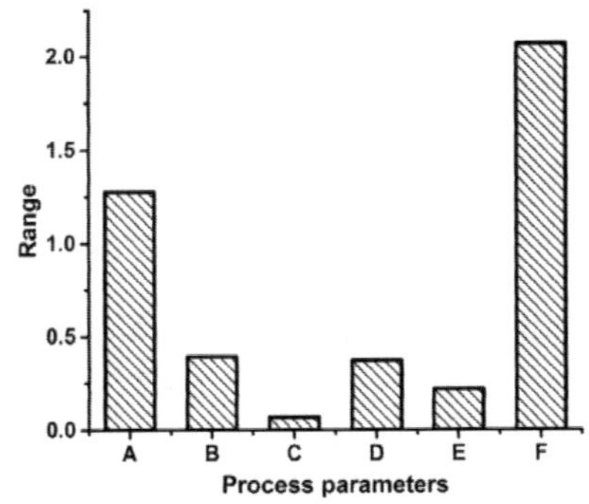

Figure 3. Range results of process parameters

The injection process was then simulated using the optimal processing parameters in Table 5 in order to achieve the minimal warpage of the A pillar trim. The minimal warpage value is of 2.124 mm, which is within the acceptable warpage as prescribed by DIN 16901 standard [22] for plastic moldings. According to the DIN 16901 standard, the maximum permissible tolerance value of the investigated A pillar trim part should be less than 7.2 mm. Thus the optimized warpage of the part meets the requirement of tolerance standard.

4. Conclusion

In conclusion, this study aimed to investigate the warpage of an injection-molded automotive interior component made of RF/PP composite and numerically design the process parameters of minimal warpage deformation using orthogonal experiment method and range analysis method. The analysis showed that the fiber percentage and packing pressure gave the most significant influence on warpage compare to melt temperature, mold temperature, filling time and packing time. The optimal packing pressure was of 70% injection pressure and 20 wt% for fiber addition. The A pillar trim part can achieved the minimum warpage value as of 2.124 mm under the optimal parameters.

Acknowledgements

The authors thank the financial support of National High Technology Research and Development Program of China (863 Program), No. 2008AA030905 from MOST, and the Natural Science Foundation of China, No. 51073051. This work is also financially supported by the State Key Laboratory of Advanced Design and Manufacturing for Vehicle Body (Hunan University), No. 71475003, and the Green Car Collaboration Innovation Center of Hunan Province.

References

[1] Guo R C, Wu N and Zhang G R 2012 *Adv. Mat. Res.* **341** 226-230
[2] Faruk O, Bledzki A K, Fink H. P and Sain M 2014 *Macromol. Mater. Eng.* **299** 9-26
[3] La Mantia F P and Morreale M 2011 *Compos. Part. A-Appl. S.* **42** 579-588
[4] Yuan J, Yu Y, Wang Q, Fan X, Chen S and Wang P 2013 *Fiber. Polym.* **14** 1254-60
[5] Li X, He L, Zhou H, Li W and Zha W 2012 *Carbohyd. Polym.* **87**(3) 2000-04
[6] Li Y, Sun J, Cheng P, Jiang Y, Zhou Z, Zhang Q and Qiu Y 2013 *J. Adhes. Sci. Technol.* **27**(22) 2387-97
[7] Wang X, Zhao Y, Li W and Wang H 2015 *Appl. Surf. Sci.* **342** 101-105
[8] Yan H, Wang H and Fang Z 2014 *Ind. Eng. Chem. Res.* **53**(51) 19961-69
[9] Gu Y, Tan X, Yang Z and Zhang Z 2014 *Mater. Design.* **56** 852-861
[10] Feng Y, Hu Y, Zhao G, Yin J and Jiang W 2011 *J. Appl. Polym. Sci.* **122**(3) 1564-71
[11] Ho M P, Wang H, Lee J H, Ho C K, Lau K T, Leng J and Hui D 2012 *Compos. Part. B-Eng.*

43(8) 3549-62
[12] Friedrich K and Almajid A A 2013 *Appl. Compos. Mater.* **20**(2) 107-128
[13] Panthapulakkal S and Sain M 2007 *J. Appl. Polym. Sci.* **103** 2432-41
[14] Rahman W AW A, Sin L T and Rahmat A R 2008 *J. Mater. Process. Tech.* **197** 22-30
[15] Azaman M D, Sapuan S M, Sulaiman S, Zainudin E S and Khalina A 2013 *Mater. Design.* **52** 1018-26
[16] Azaman M D, Sapuan S M, Sulaiman S, Zainudin E S and Khalina A 2015 *Polym. Eng. Sci.* **55**(5) 1082-95
[17] Akil H, Omar M F, Mazuki A A M, Safiee S Z A M, Ishak Z M and Bakar A A 2011 *Mater. Design.* **32** 4107-4121
[18] Shi L N, Zhang X and Chen Z L 2011 *Water. Res.* **45**(2) 886-892
[19] Ozcelik B, Erzurumlu T. Ozcelik, B and Erzurumlu T 2006 *J. Mater. Process. Tech.* **171**(3) 437-445
[20] Tang S H, Tan Y J, Sapuan S M, Sulaiman S, Ismail N and Samin R 2007 *J. Mater. Process. Tech.* **182**(1) 418-426
[21] Hakimian E and Sulong A B 2012 *Mater. Design.* **42** 62-71
[22] DIN 16901:1982. Plastics mouldings-tolerances and acceptance conditions for linear dimensions.

IOP Conf. Series: Materials Science and Engineering 167 (2017) 012076 doi:10.1088/1757-899X/167/1/012076

Rapid sintering and microstructure evolution of composite TiC cermet

L Ding[1,2], X G Liu[1,2], Y L Pan[1,2], Y W Wang[1,2] and D P Xiang[1,2,3]

[1] State Key Laboratory of Marine Resource Utilization in South China Sea, Hainan University, Haikou 570228, China

[2] Key Laboratory of Advanced Materials of Tropical Island Resources, Ministry of Education, Hainan University, Haikou 570228, China

E-mail: dpxiang@hainu.edu.cn

Abstract. Ti, Ni, activated carbon, and Mo powders were used as raw materials to prepare a composite TiC cermet in this study. The powders were mixed and prepared through high-energy ball milling and then sintered in a spark plasma sintering (SPS) system. Results revealed that ball milling time affected the raw materials. After ball milling was performed for 10 h, Ti and C particles reacted and generated TiC, meanwhile, the solid Mo solutionized in TiC and formed (Ti,Mo)C lumps. XRD results showed that the product of (Ti,Mo)C cermet with high hardness can be prepared at a low sintering temperature of 1150 °C. The microstructure of composite TiC cermet was different from the traditional core-ring structure. In particular, the developed microstructure comprises a (Ti,Mo)C-Ni dark-gray phase at the center surrounded by (Ti,Mo)C light-gray phase and dispersed Mo white phase.

1. Introduction

TiC cermet is a metal matrix composite extensively investigated as a promising material for high-temperature structural applications because of its high melting point, hardness, and corrosion resistance [1-2]. Traditional TiC cermet preparation is relatively complex, and high sintering temperature and difficult control of both product purity and batch production significantly hinder the improvement of TiC properties [3-6]. As an advanced rapid sintering technique, spark plasma sintering (SPS) requires lower temperature and shorter time than other sintering techniques do [7-8]. In our study, high-energy ball milling (HEBM) and SPS technologies were examined and used to prepare composite TiC cermet. Ti, Ni, activated carbon, and Mo powders were used as raw materials, mixed through HEBM, and sintered in a SPS system to prepare the composite TiC cermet. The effects of HEBM time on the composite powders were investigated. Meanwhile, the effects of sintering temperature on the microstructural evolution of composite TiC cermet were also determined.

2. Experimental

The raw materials used in this study were Ti, Ni, activated carbon, and Mo powders (>99.5 wt.% purity). The average particle size was approximately 3.0 um for Ti powder, 2.8 um for activated carbon, and 1 um for Ni and Mo powders. The powders were mixed, and prepared by high-energy ball milling with Ar protection. The ball-to-powder weight ratio was set to 10:1, and WC-cemented carbide balls were used as grinding media. The ball milling speed was 580 rpm for 10h of milling time. Subsequently, powder mixture was retrieved, filled in vacuum glove box mould, and eventually sintered in SPS. The sintering equipment utilized in the experiment was the SPS LABOXTM-3010K sintering system. The powders were sintered in the SPS system for 5 min at a pressure of 20 MPa, a temperature of 1150 °C to 1450 °C, and a heating rate of 100 °C/min by sintering. The samples were cooled to room temperature and removed from the furnace. Hardness was measured by the digital HRS-150 Rockwell hardness tester. All of the samples were subjected to X-ray diffraction (XRD) for phase composition analysis. Scanning electron microscopy (SEM) with a S-4800 Type II microscope was used to observe the microstructure of the sintered samples.

3. Results and discussion

3.1. Effects of mechanical ball milling time on the raw powders

Figure 1 shows the external view of the raw materials at different ball milling times. After different milling times of 5 and 10 h, the color and state of the raw materials varied. After ball milling was conducted for 5 h, the powders with a small particle size were dark gray. The mixture was very uniform, and the powder was very fine. After ball milling was performed for 10 h, the materials formed large lumps with varying sizes. The gray lumps contained apparent pores and exhibited a high hardness value. After measurement was conducted, the hardness of the raw material ball-milled for 10 h was 82.6 HRA.

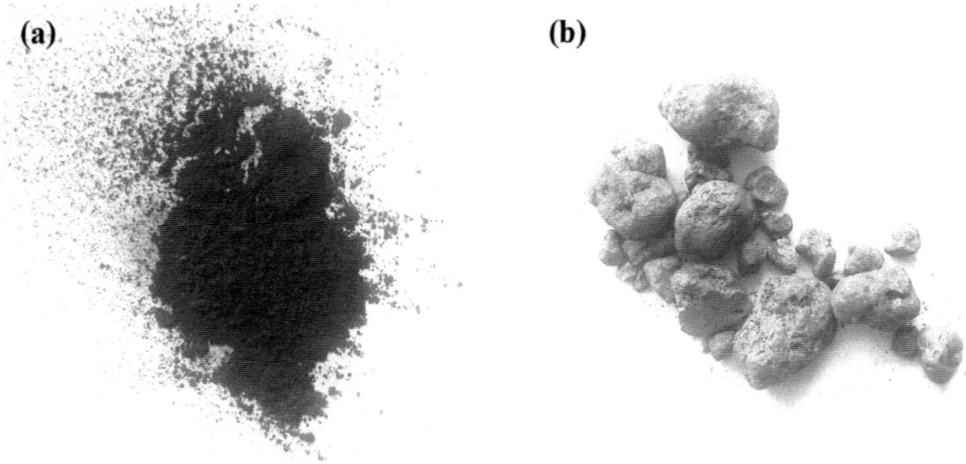

Figure 1. External view of raw materials with different ball milling durations: (a) 5 h and (b) 10 h.

The XRD image of the raw materials ball-milled for 5 and 10 h is shown in figure 2. After the raw material powders were ball-milled for 5 h, they were mainly composed of elemental Ti, Mo, and Ni. Thus, the raw materials did not undergo any reaction during the experimental procedure. After the powders were ball-milled for 10 h, the main components of the raw materials were Mo, (Ti,Mo)C, and NiTi. Therefore, the raw materials chemically reacted after the duration of ball milling was prolonged. Ball milling time affected the composition of raw materials. HEBM was conducted by using a rotating or vibrating machine to drive the grinding balls on the powders and achieve strong impact, grinding, and mixing influences; the powders were further broken down into small particles. Ball milling not only plays a significant role in mixing raw material powders but also in grinding them into small particle sizes and increasing the surface energy of particles. Particle size gradually decreases and temperature increases when milling time was extended to 10 h. As a result, Ti and C particles react and generate TiC, as expressed in the following chemical equation. In contrast to the peak of the standard PDF card, the actual peak was found at the right side. This phenomenon may be attributed to the incorporation of the Mo solution into TiC and the subsequent formation of (Ti,Mo)C. Mechanical alloying occurs between the Ni and Ti components of the raw materials and an NiTi intermetallic compound forms through continuous ball milling, as expressed in the following equation: $Ni+Ti \rightarrow NiTi$. As a result, prolonged ball milling induces the raw materials to react with one another and produce (Ti,Mo)C. Therefore, the optimum ball milling time was 10 h.

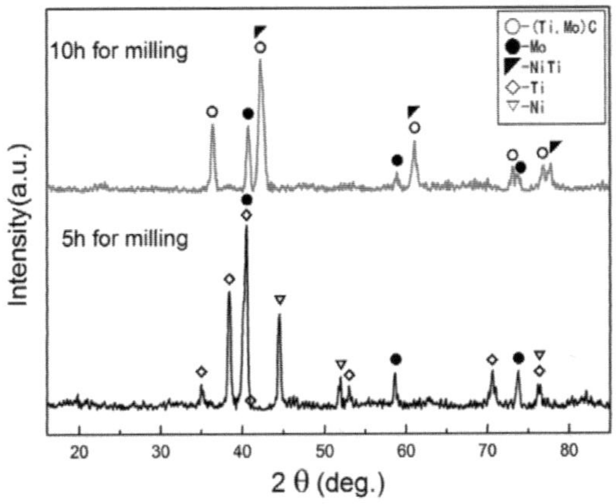

Figure 2. XRD patterns of raw materials ball-milled for 5 and 10 h.

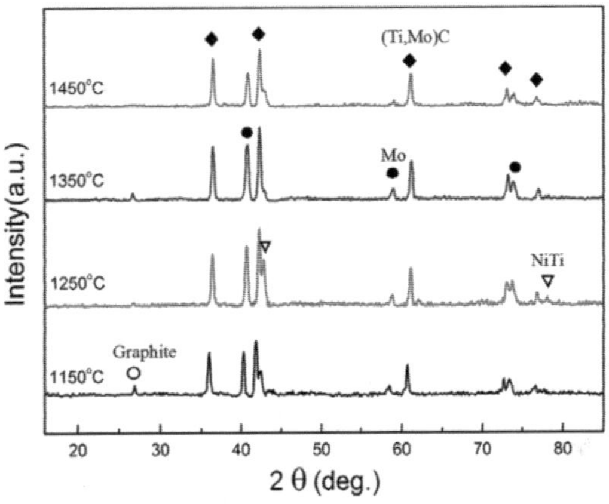

Figure 3. XRD patterns of the samples at different sintering temperatures ranging from 1150 °C–1450 °C.

3.2. XRD analysis of the samples at different sintering temperatures
The XRD patterns of the samples sintered at 1150 °C–1450 °C are shown in figure 3. At 1150 °C, the diffraction pattern reveals the diffraction peaks of (Ti,Mo)C, Mo, NiTi, and graphite. TiC was prepared. Compared with the position of the diffraction peak of the TiC standard, the position of the observed diffraction peak shifted. This phenomenon may be attributed to the incorporation of the Mo solution into the TiC and the formation of (Ti,Mo)C. As sintering temperature increases, the intensity of (Ti,Mo)C diffraction peak is enhanced to a certain extent. This result is possibly related to the growth of the grains. At high sintering temperatures, such as 1350 °C and 1450 °C, the intensity of the NiTi peak gradually weakened. This gradual weakening may be caused by a very high temperature, and some of the NiTi products were dissolved, and Ni, Ti were formed. Ti also reacts with C, and Ni is released.

3.3. Effects of sintering temperature on microstructure of composite TiC cermet
The composite TiC cermets produced by HEBM assisted SPS method possess high hardness. The sample sintered at 1150 °C has a high hardness value of 88.2 HRA. Figure 4 illustrates the SEM micrographs of the samples sintered at 1150 °C - 1450 °C. The microstructures of composite TiC cermet are composed of (Ti,Mo)C-Ni dark-gray phase, (Ti,Mo)C light-gray phase, and Mo white phase. Good wettability was observed among the different structures. The morphological characteristics of the sintered samples differed from the traditional core-ring structure; its center was the dark-gray phase surrounded by the light-gray phase and dispersed white phase. As sintering temperature increased, the grains gradually grew in size. For instance, the size of the dark-gray grains was approximately 0.5 μm (figure 4a). The average size of the particle increased to 3 μm when the sintering temperature was increased to 1450 °C (figure 4d). The grain size of two other structures also showed the same varying trend.

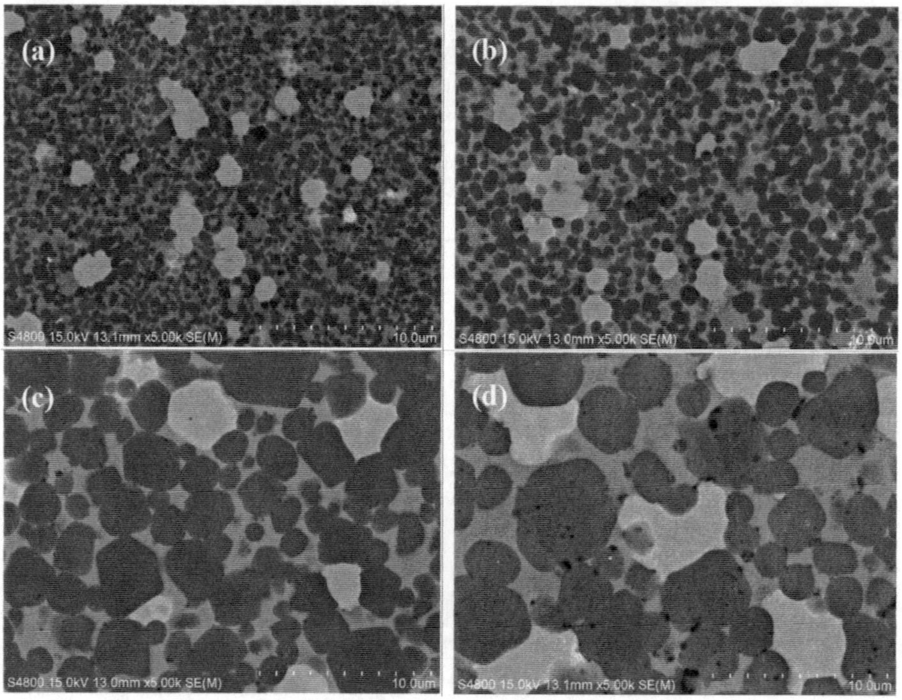

Figure 4. Microstructure of the samples at different sintering temperatures: (a) 1150 °C, (b) 1250 °C, (c) 1350 °C, and (d) 1450 °C.

4. Conclusions

HEMB time significantly affects the raw materials. The raw materials can be ground into fine powders and produce a uniform mixture when they are ball-milled for 5 h. The raw materials can react become lumpy with a certain degree of hardness, and generate the (Ti,Mo)C when they are ball-milled for 10 h. The (Ti,Mo)C cermet with a high hardness value of 88.2 HRA was prepared at 1150 °C. The morphological characteristics of the sintered samples were different from those of the traditional core-ring structure. Dark-gray (Ti,Mo)C-Ni phase was found at the center surrounded by light-gray (Ti,Mo)C phase and large white dispersed Mo particles.

Acknowledgments

This work was supported by the National Natural Science Foundation of China (Grant No. 51464010), the Natural Science Foundation of Hainan Province (Grant No. 20165207), and the Middle-West part plan subject project of China (Grant No. ZXBJH-XK009).

References

[1] R. Landfried, F. Kern and R. Gadow 2015 *Int. J. Refract. Met. Hard Mat.* **49** 334-8
[2] P. Istomin, E. Istomina, A. Nadutkin and V. Grass 2016 *Int. J. Refract. Met. Hard Mat.* **57** 12-8
[3] L. Tong and R.G. Reddy 2005 *Scripta Mater.* **52** 1253-8
[4] L. Ding, D.P. Xiang, Y.L. Pan, T.M. Zhang and Z.Y. Wu 2016 *J. Alloys Compd.* **661** 136-40
[5] D.W. Lee and B.K. Kim 2003 *Scripta Mater.* **48** 1513-8
[6] M. Razavi, M.R. Rahimipour and A.H. Rajabi-Zamani 2007 *J. Alloys Compd.* **436** 142-45
[7] Lai-Ma Luo, Xiao-Yue Tan, Hong-Yu Chen, Guang-Nan Luo, Xiao-Yong Zhu, Ji-Gui Cheng and Yu-Cheng Wu 2015 *Powder Technol.* **273** 8-12
[8] Lixia Cheng, Zhipeng Xie and Guanwei Liu 2013 *J. Eur. Ceram. Soc.* **33** 2971-77